CRC
Handbook
of
Medicinal Herbs

Author

James A. Duke, Ph.D.

Chief, Germplasm Resources Laboratory
United States Department of Agriculture
Washington, D.C.

CRC Press, Inc.
Boca Raton, Florida

Library of Congress Cataloging in Publication Data

Duke, James A., 1929-
 Handbook of medicinal herbs.

 Bibliography: p.
 Includes index.
 1. Medicinal plants. 2. Herbs. 3. Herbals.
4. Folk medicine. 5. Materia medica, Vegetable.
I. Title. [DNLM: 1. Medicine, Herbal. 2. Plants,
Medicinal. WB 925 D877h]
QK99.AlD83 1985 615′.321 84-12148
ISBN 0-8493-3630-9

Direct all inquiries to CRC Press, Inc., 2000 Corporate Blvd., N.W., Boca Raton, Florida, 33431.

© 1985 by CRC Press, Inc.
Second Printing, 1986
Third Printing, 1986

International Standard Book Number 0-8493-3630-9

Library of Congress Card Number 84-12148
Printed in the United States

INTRODUCTION

Like one old Chinese herbal, this Handbook of Medicinal Herbs treats 365 folk medicinal species. There could be many more, but I am folksy enough to believe that an herb a day, like an apple, a beet, a carrot, a citrus and/or a cole slaw a day can help keep the doctor away, maybe even cancer.

The evolution of this title has been rather interesting in itself. The first draft was entitled Borderline Herbs and it was reviewed for years under that title. It contained most of the medicinal herbs whose salubrity was being challenged, borderline often implying that the herb industry was promoting while regulatory agencies were challenging the herb. As the list of borderline herbs grew to 365, a tempting, somewhat logical title seemed to be 'An Herb a Day . . . Borderline Herbs' or Herbs of Dubious Salubrity. Both these alternative titles seemed a bit demeaning for important medicinal species, worth well over $500 million (US) a year. Now, following strenuous recommendations of the publisher, the title rests at Handbook of Medicinal Herbs. Predicting criticism will not undo the urtication of the critic's barbs. Surely I will be criticized for having omitted some important medicinal species, on both the safe and on the dangerous side of borderline. All plants contain pharmacologically active compounds and vitamins, and I clearly can't treat all of them. I have striven to include those that have been batted about in public view, with both antagonists and protagonists, with the "borderline" herbs caught in the middle of the altercation between advocates and adversaries.

As Chief of the Medicinal Plant Resources Laboratory, United States Department of Agriculture (USDA), I received many oddball questions on folk medicines or herbs, which come in by the hundreds to agencies such as the USDA. In 1981, my official connection with medicinal plants was concluded with the termination of a 25-year-old program involving the National Cancer Institute (NCI) and the USDA in the screening and collection, respectively, of plants with potential anticancer activity. But the letters on medicinal herbs keep pouring in.

One definition of the word herb is "a non-woody plant", often more strictly a "non-woody temperate plant", leaves of which are used for culinary or medicinal purposes. This stricter definition was to separate "herbs" from "spices", sometimes defined as tropical species, barks or buds or fruits of which are used for culinary purposes. I won't bother you with the many exceptions to these narrow definitions. The broadest definition of herb I have seen is *a useful plant*. Since all green plants take on CO_2 and manufacture food, liberating oxygen in the process, all such plants are useful. Thus my definition resides uncomfortably between these extremes, and I am restricting my list to include only species that have medicinal or folk medicinal uses. Assuming there are about 365,000 higher plant species (estimates range from 250,000 to 750,000), all of which manufacture sugar, I am treating in this account 0.1% of those species.

Many medicinal plants are borderline, and their safety may have been challenged or should have been challenged. I have tried to include among these herbs of dubious salubrity species that the Drug Enforcement Administration (DEA), Food and Drug Administration (FDA), Herb Trade Association (HTA), National Cancer Institute (NCI), or United States Department of Agriculture (USDA) inquired about during the years while I was actively involved in the study of medicinal plants.

During the past few years, floods of inquiries have reached me at home and in the office regarding the potential dangers of herbal ingestions. Some inquiries came from those who might be called "champions of herbs", others from "enemies of herbs", but perhaps most came from confused citizens barraged with overzealous statements, pro and con. Within one year alone, I was asked about more than 100 herbs by the FDA, trying to protect consumers from certain herbs, and/or the late HTA, trying to educate consumers about the utilities of the herbs. Often one faction would be promoting an herb or a concept the other faction was denouncing. These inquiries led me to consult some general references[1-65] and other more specific ones trying to determine which herbs were clearly safe, which clearly dangerous, and which belong in that twilight zone of "borderline" herbs. I have added to my own list of borderline medicinals those questioned by Tyler in his intelligently conservative *Honest Herbal*,[37] and Rose in her interesting yet more earthy text.[47]

Herewith are notes derived largely from the open literature[1-76] on 365 medicinal "herbs", some of which are relatively innocuous, some of which are relatively dangerous, but most of which might be found on sale somewhere in the United States. Many dangerous medicinal herbs have had official medical uses in the past, and might be reconsidered in the future. I am not promoting these herbs, nor recommending herbal self-medication. I have gathered pertinent published information, some of it obviously contradictory. Unless preceded by the short and unambiguous pronoun, I, the information is not to be attributed to me but to the reference cited. I believe that the information will help educated consumers arrive at opinions on the potential good and/or evil in some of our medicinal herbs. No herbal endorsements or recommendations are intended. He who self-medicates has a fool for a doctor. In no event am I recommending any herb, just citing other compilations. Only where followed by an exclamation point (!) can the reader assume that the medicinal suggestion is personal. If in my haste I have let the words "is recommended for" slip through, this means that somewhere in the literature "it is recommended". I have tried only to marshall the folklore and fact, side by side, so that the reader can make better-advised decisions about the relative safety or danger of a given herbal application.

Each plant species is biochemically different from each other, and embraces a wide array of intraspecific biochemical variation. All of these species contain toxic compounds,[15] and all contain vitamins.[66] I dare speculate that all contain compounds that will cause tumors in vitro and vivo and all contain compounds that will arrest tumors in vitro and in vivo (shikimic acid and beta-sitosterol, e.g., may be ubiquitous or nearly so in higher plants). I further dare speculate that all contain compounds which will raise or lower the blood pressure, the body temperature, etc. (if we analyze large quantities of all the chemicals that are present in minute quantities in plant species). Antagonists seek out the bad, protagonists seek out the good. Both will find active compounds to support their claims. Small wonder the American consumer is confused. I even believe that in some cases, the homeostatic human body can select from the herbal potpourri of compounds the few it needs.

Using many compilations in my own compilations, I have not always gone to the original sources. I feel it more productive to try to locate pertinent new material than to double-check the references, some of which are unavailable to me. My greatest difficulty has been with the very useful reference, Hager's Handbook.[33] I have not found equivalents for all of the German chemical and medical terms and instead have just transcribed them.

I did not coin the words Herbal Renaissance, but there are symptoms of such a renaissance. If there is a doubling of utilization of herbs, we can expect more bad and good experiences with herbs. I feel that an educated public can avoid some of the bad experiences with good herbs and, if they must, survive more interesting experiences with some of the bad herbs. Clearly, there has been a tremendous increase in interest in the use of herbs by groups not traditionally versed in economic botany. There are consequently cases of poisoning each year; some because of misidentification as with foxglove tea, some due to accidental or intentional contamination, some because of attempted herbal abortion, some due to overdoses of narcotics and hallucinogens. Death may result from intentional ingestion of toxic plants to experience the disturbance of the central nervous system. Many intentionally ingest insalubrious herbs, hoping to alter their consciousness. In some cases of poisoning, the herb is purchased from a health food store, as with recent cases for aloes, burdock, buckthorn, comfrey contaminated with belladonna, pennyroyal, pokeweed, lobelia, rue, senna, and others. Many of the herbs here included have no other real use here in America. Still they are sold. Some quite dangerously. Nature has many faces and some want to embrace them all.

For some, a back-to-nature movement alone is call enough to get back to the basic herbs that our forefathers knew. Recent suggestions from governmental agencies may further augment America's interest in herbs.

1. Eat less fat meat (some people even recommend vegetarianism; there is no clear demarcation between leafy vegetable and leafy herb; e.g., chicory, purslane, comfrey, are herb to some, vegetable to others). Vegetarianism could help solve our energy problems as well as many chronic disease problems. One survey shows that 97% of our legumes and 90% of our cereals go to feeding animals other than humans. A lot of biomass for gasohol production could be produced on the acres tied up in livestock production.

2. Eat more vegetables (herbs can make vegetarian meals more interesting).

3. Cut back on salt (there are herbal substitutes for salt). The American Heart Association has recommended as a salt substitute one part each of basil, black pepper, garlic powder, mace, marjoram, onion powder, parsley, sage, savory, and thyme to 1/2 part cayenne pepper.[67]

4. Increase fiber intake (herbs are often quite high in fiber).

5. Don't go short on Vitamin A and C and antioxidants, all of which have been suggested as lessening the probability of cancer, and may be frequent in some herbs.

It's more than a back to *nature* movement. Many are actively seeking alternatives to their current life styles. Three MD friends recommended to me that I take allopurinol maintenance doses (for life) when I experienced my first attacks of gout five years ago. Since then, however, there have been suspicious reports of death due to the allopurinol. I prefer dietary and herbal approaches to gout, eating cherries and strawberries while avoiding organ meats and seafoods, and only resorting to the carcinogen colchicine if and when I suffer an acute accumulation of uric acid crystals. Colchicine was derived from the fall crocus, *Colchicum autumnale*, once a folk remedy for gout.

Readers should consult a collection of articles concerning the current and future importance of medicinal plants.[68] In this collection, Farnsworth and Farley say:[68]

Yes at least 25 percent of the standard therapeutic armamentarium of the modern-day physician, as was that of his counterpart from decades ago, came into being by way of Folk Medicine. Now-A-Days, we hear very little of Folk Medicine as a part of modern day medicine. What we do hear about are increasing numbers of people who are deeply concerned over the thought of being treated with modern synthetic drugs, and their well-known side effects. These individuals are searching for a reasonable alternative for certain aspects of present day medicinal treatment.

One only need to walk through any large city health food outlet, to gain two immediate impressions. First, these outlets are well stocked with every conceivable type of herbal remedy; some 'legal' whereas others are definitely 'illegal', both types being offered for sale. The other impression that one gains is that the clientele frequenting those establishments are, more-often-than-not, middle- to upper-class income, college educated individuals, in the 20 to 40 age bracket . . . In a highly sophisticated and technological health care system, this modern breed of 'Folk Medicine' pioneer is searching for an acceptable alternative.

Sherry and Koontz[69] suggest that there are about 12,000 licensed homeopaths and naturopaths who routinely use herbal remedies in the U.S. In other parts of the world, as much as 90% of the population relies on traditional medicines, mostly herbal.

There are of course a great many herbs, clearly toxic, that are not listed, presumably because there is no question about their toxicity. I have mentioned some of the very dangerous ones that are widely used in folk medicine. Some traveler will bring it back to the U.S. full of dangerous enthusiasm.

Genera involved in fatalities are listed below from my illustrated lecture to the Second Annual Herb Symposium of the Herb Trade Association:[70] Acokanthera, Aconitum, Actaea, Aesculus, Aethusa, Agrostemma, Albizia, Alchornea, Aleurites, Andira, Aquilegia, Arisaema, Arnica, Arum, Atropa, Azalea, Blighia, Bowiea, Caladium, Calotropis, Cassia, Cicuta, Colchicum, Conium, Convallaria, Courbonia, Croton, Cryptostegia, Cucurbita, Cycas, Cytisus, Dalbergia, Daphne, Datura, Delphinium, Dieffenbachia, Digitalis, Duranta, Erythrophleum, Eupatorium, Euphorbia, Garcinia, Gaultheria, Gelsemium, Gloriosa, Gnidia, Hedera, Helleborus, Hippomane, Hoslundia, Hura, Hyoscyamus, Jatropha, Juniperus, Justicia, Kalmia, Laburnum, Lantana, Lavandula, Ligustrum, Lobelia, Lolium, Malus, Manihot, Melia, Menispermum, Mercurialis, Momordica, Narcissus, Nerium, Nicotiana, Oenanthe, Paris, Peumus, Phaseolus, Phoradendron, Physostigma, Phytolacca, Pilocarpus, Piper, Podophyllum, Poinsettia, Prunus, Ranunculus, Rheum, Rhododendron, Ricinus, Robinia, Rourea, Sambucus, Senecio, Solanum, Sophora, Spigelia, Strychnos, Tamus, Tanacetum, Taxus, Terminalia, Thevetia, Urginea, Veratrum, Zamia, and Zigadenus. Lest anyone get too hasty and try to ban all these, let me quickly add that among those *reported* to cause fatalities are apples, black pepper, elderberry, rhubarb, and tobacco. I suppose we could classify tobacco and marijuana as medicinal herbs; and alcohol, cocaine, and heroin as medicinal herbal products. These are the big killers. It's hard to find the figures on the big five. Recently a television broadcast (*Monitor*, June 18, 1983) put the American tobacco toll at 300,000 human lives a year. In one weekend alone, there can be several deaths due to heroin overdoses in Washington, D.C. Tobacco and alcohol are legal, hence rather cheap killers; cocaine, heroin and marijuana, also additive, are illegal, hence rather expensive killers, with much of the tab paid by crime to crime, with the government taking most of the losses, organized crime the profits. While I may be out of phase with the government that has sponsored me, it is my personal opinion that the government and legal citizens would be better off, and criminals worse off, if all five were legal and carefully controlled, labeled with educated warnings, not dictated by their lobbies. Not even my wife agrees with me on this point.

One new mail order catalog[71] crossed my desk as this book was going to press. It catalogued at least twenty items that "have been restricted or declared dangerous by various governmental agencies". The catalog concluded "we do not sell any of these items for food or beverage" without specifying their intended use: Arnica Flowers, Bloodroot, Broom Tops, Buckeyes, Calamus Root, Deer's Tongue Leaves, Helleborus, Horsetail, Jaborandi, Lobelia, Mandrake, Mezeron, Mistletoe, Periwinkle, Poke Root, St. John's Wort, Sassafras, Tonka Bean, Valerian Root, and Wahoo.

According to colleagues in the late Herb Trade Assocaition, the top herbs in total sales in the U.S. are Peppermint, Camomile, Hibiscus, Rose Hips, Orange Peel, Carob, Spearmint, Alfalfa, Ginseng, Parsley, Comfrey, Raspberry leaves, Chicory, Lemongrass, Fenugreek, Sage, Cayenne, Thyme, Sarsaparilla, Wild Cherry Bark, Mate, Red Clover Tops, Licorice, Garlic, Fennel Seed. Of these 25, (mostly beverage herbs) few have been challenged lately. Camomile has been reported to cause problems in some (one) hypersensitive to ragweed pollen, alfalfa has been reported to contain saponins, ginseng to cause (in overdoses) the so-called ginseng-abuse-syndrome, parsley to contain myristicin, comfrey to contain lasiocarpine, fenugreek to contain coumarins, wild cherry bark perhaps to contain cyanogenic glucosides, maté to contain caffeine, red clover to contain estrogens, licorice to raise the blood pressure, and garlic to lower it.

Too frequently, I am asked to comment on the safety or toxicity of some herb Americans are consuming. Since toxicity is a relative thing, and since all plants (even our food plants) contain numerous toxins, these are touchy questions. Still I venture to compare here, in an ever-growing list (Table 1) my personal opinions as to the safety of various herbs for oral ingestion.

Not really knowing whether most herb teas are more or less safe than booze, chocolate, coffee, cola, maté, milk, tea, or city water, I can only assign my biased opinions. I feel that two cups of coffee per day is safe enough not to be dreadfully harmful. On a safety scale of zero to three, I rate coffee "two", not real safe, not real poisonous. A score of zero means I think it is much more toxic than coffee; perhaps deadly in small doses. For a score of one, I would discourage all but the most cautious experimentation. A score of two means I am as leery of this herb as I am of coffee. Two cups a day would seem to me to be as safe (or as dangerous) as two cups of coffee. A score of three indicates that it seems to me to be safer than coffee, that I would not be afriad to drink three cups a day. Much more than three cups of anything seems immoderate. My hunches are not based on scientific toxicology, strictly on "gut reaction". Hence, they cannot possibly be considered as an endorsement or condemnation either by me, personally, or by the former sponsors of my medicinal plant research, the NCI and the USDA, both of which clearly discourage self-medication and ingestion of strange herbs. In Table 1, I have assigned scores to 365 herbs I have been asked about over the last eight years. I am not sure that I would score them exactly the same a few weeks later, as new data become available to me. For those who asked, these are my current opinions as 1983 draws to a close. For the reader's enlightenment, I also estimate the evaluation by a rather conservative author, Varro Tyler,[37] and a rather liberal author, Jeanne Rose.[47] Of course, all of these evaluations are subject to limitations. Seeds of the opium poppy, like the seedling, are relatively innocuous and would get a 2 on my scoreboard, but the latex, more likely to be abused, is clearly more dangerous. The pulp of the fruit of *Podophyllum pelatatum* is edible, but almost all other parts are poisonous. I would score the young leaves of *Phytolacca*, after boiling in two changes of water, as edible, all other parts, especially woody parts, as poisonous. My scores in Table 1 attempt to average out such details, taking into account what is most likely to be purchased by naive consumers. Recent prices for some of these herbs are also included in Table 1.

Very often, levels of chemicals in a given species may vary considerably, one specimen containing ten times as much as another. Chemical constitution varies both with the genetics and environment of the plant, and temporally and spatially within the plant. Aromatic mints will often, on a dry matter basis, generate higher levels of essential oils in dryer and/or sunnier habitats. And of course there are many species, like *Manihot esculenta*, *Melia azedarach*, and *Solanum nigrum*, reported to have poisonous "varieties" morphologically indistinguishable from nonpoisonous "varieties". Indeed one of the best-justified criticisms of herbal medicines is the difficulty in determining dosages of bioactive chemicals, due to this inherent variation. While my tabulated evaluation (Table 1) may obscure such variation, I intend to address it in the individual accounts of the species. Once an herb has been shredded and stuffed into the teabag, it is often difficult to determine the genus or species involved, much less the variety or chemovar. Talents required to make positive identification of herbs in herbal tea mixtures are few and far between.

In Tables 2 and 3, reproduced from my earlier paper in *CRC Critical Reviews in Toxicology*,[15] some of the toxins liable to find their way into the herbal tea pot are listed. In Table 2, the toxic phytochemical compounds are listed in alphabetic sequence. In Table 3, some of the more frequently cited genera containing these phytochemicals are presented alphabetically. Of course all medicines at high dosages are poisonous, and many poisons are medicinal. In Table 4, I list some of the phytochemicals and their reputedly proven bioactivities. All plants of course contain toxins, but most of those in our food chain contain negligible quantities of toxins. On the other hand, all plants contain carbohydrates, proteins, fats, and minerals and vitamins. Where I have located nutritional data, I have included it along with any toxicological data uncovered. Standing alone, these nutritional data do not mean a great deal. They are better compared to the proximate analyses of conventional foods (Table 5). In Table 5, I include proximate analyses for 75 rather conventional food items.

Averages for the various categories are presented on both an ''as-purchased basis'' (APB) and on a zero-moisture basis (ZMB) to facilitate comparisons.

The 365 species, listed alphabetically by scientific name, are numbered sequentially in the text; the serial number is followed by the scientific name and authority, the scientific name of the plant family, and one to three colloquial or common names. There are often four paragraphs, the first dealing with uses, the second with folk medicinal applications, the third with chemistry, and the fourth with toxicity. Thanks to the generosity of other authors and publishers, I have amassed figures for most of these. The figures are given the same number as the serial number of the species, but there are gaps in this enumeration, for those species for which I was unable to find suitable illustrations and/or to obtain permission from their publishers to use them. The list of credits for these figures appears after the reference list.

Let us take a look at 365 medical herbs, some of which are quite dangerous. In the sense that all plant species manufacture toxins and vitamins, perhaps all species could be considered both potentially harmful and potentially useful. The optimist in me believes that we can find safe uses for all of them, if America is not duped into turning away from the scientific investigation of medicinal herbs.

THE AUTHOR

Born in Birmingham, Alabama, in 1929, James A. Duke is a Phi Beta Kappa graduate of the University of North Carolina, where he took his AB (1952) and BS (1955) in Botany. Following a 2 1/2 year tour of military duty, which included a mycological assignment at Fort Dietrick, Maryland, he took his PhD in Botany, in 1961, at the University of North Carolina, moving on to postdoctoral activities at Washington University and the Missouri Botanical Garden, in St. Louis, Missouri, where he assumed professorial and curatorial duties, respectively. At the Missouri Botanical Garden, Duke began intensive studies of neotropical ethnobotany, which is his overriding interest to this day. From 1963 to 1965, Duke served with the USDA at Beltsville, Maryland, devoting much of his time to neotropical ecology, especially seedling ecology. In 1965, he joined Battelle Columbus Laboratories, for ecological and ethnological studies in Panama and Colombia, returning to the USDA in 1971, for crop diversification and medicinal plant studies in developing countries. As a collaborator with the Smithsonian Institution, he has lectured and taught there, with emphasis on neotropical ethnobotany and folk medicine. Considered a key figure in the "Herbal Renaissance", Duke received the Cutty Sark Science Award in 1981. Currently, he is Chief, Germplasm Resources Laboratory, USDA, Beltsville, Maryland.

Dr. Duke is an active member in the American Herb Association (Life), American Society of Pharmacognosy, Association for Tropical Biology, International Association of Plant Taxonomists (Life), International Society for Tropical Ecology, International Weed Science Society (Life), National Association for Professional Bureaucrats, Organization for Tropical Studies (Life), Society for Economic Botany (Life), Southern Applachian Botanical Club (Life), Tri-State Bluegrass Association (Life), and Weed Science Society of America.

Widely travelled in his ethnobotanical studies, Dr. Duke has more than 100 scientific publications to his credit. His books include *Isthmian Ethnobotanical Dictionary*, Harrod and Company, Baltimore, 1972 (being reissued with illustrations by Scientific Publishers, Jodphur, India); *A Handbook of Legumes of World Economic Importance*, Plenum Press, New York, 1981; *Medicinal Plants of the Bible*, Trado-Medic Books, Buffalo, New York, 1983; with E. S. Ayensu, *Medicinal Plants of China*, 2 volumes, Reference Publications, Algonac, Michigan, 1984; and *A Culinary Herbal*, Trado-Medic Books, Buffalo, New York, 1984. During 1984, Dr. Duke is also scheduled to submit copy for *A Handbook of Energy Species* to Plenum Press and *Handbook of Agricultural Energy Potential of Developing Countries* to CRC Press.

ACKNOWLEDGMENTS

My wife, Peggy, didn't have time to draw many illustrations for this book, but she was kind enough to retouch the photographs, the glossy prints required by CRC Press. She was even willing to do this painstaking work in the final month of preparation (September 1983) when the pieces of the manuscripts and materials pertinent thereto were scattered over two bedrooms, the living room, and the dining room, all at a time when record temperatures were searing our non-air-conditioned house. Thanks Peggy, I couldn't have made the deadline, or much of anything, without your help. Mr. Eric Mathis also deserves a medal of honor for culling my long list of potential illustrations scattered in over a hundred pounds of books, and preparing, mounting, and numbering more than 200 illustrations. Thanks Eric, I'm looking forward to seeing your own photographs and illustrations of your own collection from up around historic Harper's Ferry. Mrs. Ruth Nash accomplished her own miracle, in converting my extremely rough handwritten drafts into acceptable typescript. Thanks Ruth, I'm gladdened and saddened that your lovely daughter Carey keeps you close to home so you can't rejoin us back in the USDA.

TABLE OF CONTENTS

1. *ABELMOSCHUS MOSCHATUS* Medic. (MALVACEAE) — Musk Okra, Muskmallow, Ambrette

Cultivated for the seeds which have a musk-like odor and yield the exalting agent ambrette, an aromatic oil used in perfumery. Ambrette seed and its tincture are used in vermouths and bitters. The oil and its absolute are used in creams, detergents, lotions, and soaps. Arabs use the seeds for flavoring coffee. Seeds chewed to sweeten the breath.[2] Seeds burned as incense.[42] Tender leaves, shoots, and pods are eaten as vegetables. Seeds are also insecticidal, especially on woolens against moths. Stems yield a bast fiber used for sails and cloth. Plants are often grown as ornamentals.

Tagalogs of the Philippines use a decoction for cancer of the stomach.[4] The mucilaginous seeds are used for emollients and demulcents. Seeds, pulverized and made into a paste with milk, are used to treat prickly heat and itch. They are used for hysteria and other nervous disorders. For internal use, seeds are antihysteric, antispasmodic, aphrodisiac, carminative, diuretic, litholytic, stimulant, stomachic, and tonic, and are used for gonorrhea, nervous disorders, prickly heat, spermatorrhea. Seeds are inhaled for dryness of the throat. Trinidad people steep the seed in rum or water for asthma, chest congestion, cold, flu, snakebite, and worms.[42] Roots and leaves are used for poultice. Poultices are recommended for boils, cystitis, fever, headache, rheumatism, swellings, and varicose veins. The roots and leaves are decocted for gonorrhea and rheumatism.[16] Chinese use the plant for headache.

Upon distillation, oil of seeds yields furfural; the chief constituent is farnesol. Seeds contain 11.4% H_2O, 2.3% protein, 13.4% starch, 31.5% crude fiber, 14.5% fatty oil, and 0.2 to 0.6% volatile oil. The volatile oil, extracted by steam distillation, is semisolid at ordinary temperatures, with a high percentage of fatty acids, chiefly palmitic and myristic acid and some acetic acids. The characteristic musk-like odor is due mainly to the ketone ambrettolide, a lactone of ambrettolic acid, and ambrettol, Mucilage, gum and resin are also reported. Seeds also contain methionine sulfoxide and several phospholipids (alpha-cephalin, phosphatidylserine, phosphatidylserine plasmalogen, and phosphatidylcholine plasmalogen).[29] Stem bark yields a jute-like fiber with 78% cellulose.

Toxicity — Classified by the FDA[62] as a Herb of Undefined Safety: ''Seeds were formerly considered stimulant and antispasmodic but are now used only in perfumery.'' Classified by the FDA as GRAS (acronym for Generally Recognized As Safe) (§ 182.20).[29] Oil and absolute are used in baked goods, candy, frozen dairy desserts, gelatin, nonalcoholic beverages, and puddings at levels usually less than 10 ppm.[29]

2. *ABRUS PRECATORIUS* L. (FABACEAE) — Jequerity

Leaves can be used as a dangerous substitute for licorice. Bruised seeds have been used for poisonings. Boiled seeds have been eaten in famines. The red and black seeds are used in necklaces. Leaves are eaten as a potherb in East Africa. Powdered seed (ca 200 mg) used as an oral contraceptive in Central Africa, the single dose remaining effective for 13 menstrual cycles.

Seeds, seed hulls, and decorticated seed have been used for epitheliomas of the face, hand, mucosa, vagina and vulva, and for warts on the eyelids.[4] Decoction of the leaves and roots are used for colic, cold, and coughs. Leaves, high in glycyrrhizin, are chewed for hoarseness. Formerly a 5% infusion of decorticated seeds was dangerously used for granular eyelids and pannus. Infusion of the seeds, when applied to conjunctiva, may cause fatal poisoning. A suppository of the seed poultice brings on abortion. Zulu use a root decoction for pain in the chest. Luvale use vapor from crushed leaves in boiling water for inflamed eyes.[3] Fresh root chewed as an aphrodisiac. Yao use a plant decoction for gonorrhea. Ayurvedics consider the fruits aphrodisiac, tonic, and toxic, improving the complexion and taste perception, useful in biliousness, eye diseases, leucoderma, itch, skin ailments and wounds. Sharing these properties, the roots and leaves are also respected for their effects in adenopathy, asthma, caries (teeth) fever, head disorders, stomatitis, thirst, and tuberculous glands.[26] Yunani describe the fruit as ''tonic to the brain and the body; harmful to old men'', believing the roots and leaves have the same properties.[44] Asian Indians consider the root alexeteric and emetic, its aqueous extract useful in obstinate coughs. Roots are topically applied in snakebite. Leaves are steeped with mustard as a liniment for rheumatism. Since it is extremely dangerous medicine, I resort to quotation in self defense: "Taken internally by women, the seed disturbs the uterine functions and prevents conception".[26] The seed pasted with lumbago root is applied as a stimulant dressing to leprosy. According to Perry, "A decoction of the seeds has been used as a counteragent in treating conjunctivitis, or trachoma, but this is so drastic that the practice has been dropped."[16] The Indians suggest the seeds for paralysis, sciatica, skin diseases, stiffness of the shoulder, and other nervous diseases. Chinese use the seed for dropsy, fever, headache, malaria, and worms. Indochinese use the plant for conjunctivitis, diarrhea, dysentery, and malaria. Malayans chew or drink decoctions of the leaves or roots, for colic, colds, or cough. Indonesians use the leaves for hoarseness and sore throat and sprue. In Palau the root is used for arthritis and leprosy. Philippinos, in addition to the regional uses, also bathe newly-born children in the stem and leaf decoction, especially if they are weakly. Africans, like Brazilians and Indians, also use the plant in ophthalmic remedies. Is this coincidence, doctrine of signatures, or coevolution of empirical wisdom so well founded in traditional medicines? East Africans use the seed for gonorrhea and other venereal diseases. Conversely they use the root as an aphrodisiac and for snakebite. Central Africans use the seed for ophthalmia, snakebite, ulcers, and worms. West Africans use the leaves as a CNS sedative, for conjunctivitis, constipation, cough, cancer, convulsions, enteritis, hoarseness, freckles, leucoderma, stomatitis, spermatorrhea, and syphilis; the root for chest complaints, gonorrhea, hookworm, jaundice, pleurisy, rheumatism, sore throat, and snakebite. West Indians use the leaves in teas for chest cold, cough, fever, and flu, the seeds for conjunctivitis and worms.[72] Latin Americans use the plant for asthma, burn, colds, cough, fever, flu, headache, hernia, rheumatism, sore throat, trachoma, tuberculosis.[42] Peruvians add the leaves to their chew of coca leaves for the licorice flavor they impart.[72] Orientals use the leaf for colic, cold, cough; the seed for abortion, fever, ophthalmia; antimalarial in Cambodia, anthelmintic, antiperiodic, diaphoretic, expectorant, sudorific; for dropsy, fever, headache, malaria, and worms; the root for cough and cold, colic, the plant for cancer, aphtosa, and epitheliomas.[41]

Per 100 g, the leaves contain 62 calories, 79.0% water, 5.6% protein, 13.7% total carbohydrate, 3.7 g fiber, 1.7 g ash, 266 mg Ca, and 65 mg P.[21] Glycyrrhizin is reported from the root (1.25%) and leaves (10%). Abrin (N-methyltryptophane) consists of two fractions, a globulin and an albuminose. In microgram levels, subcutaneous injections of abrin are lethal. Defatted seeds contain 5.9% polygalacturonic acid, 0.2% pentosan, and 32% reducing sugar calculated as galactose. Seeds contain gallic acid and hepaphorine.[33] Abraline, abrin, abrine, abrussic acid, campestrol, 5 beta-cholanic acid, cycloartenol, precatorine, squalene, and trigonelline are also listed.[41]

Toxicity — Juice irritates the mucous membranes. Necklaces made of the pierced seeds may induce dermatitis.[6] Seeds contain the toxin abrin. Human fatalities are reported. One chewed seed may be lethal,[56] or 5 mg abrine.[42] Lethal doses may run about 0.01 mg/kg body weight.[33]

To the Physician — External cooling, fluids, electrolytes, calcium gluconate, arecoline.[1]

5

3. *ACACIA FARNESIANA* (L.) Willd. (FABACEAE) — Cassie, Huisache

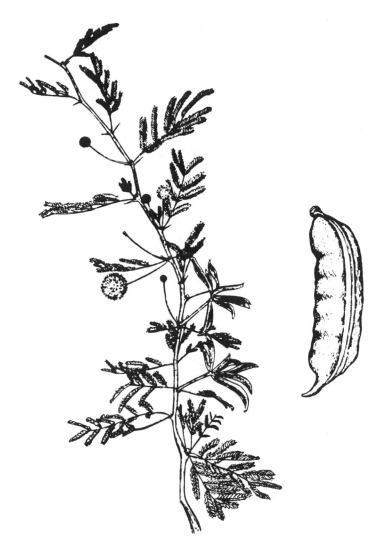

Cassie perfume, distilled from the flowers, yields cassie absolute, employed in preparation of violet bouquets, extensively used in European perfumery. Cassie pomades are manufactured in India. Pods, containing 23% tannin, are used to tan leather. Bark also used for tanning and dying leather in combination with iron ores and salts. Pods used for a black leather dye. A gummy substance obtained from pods is used as cement for broken crockery. Gum exuding from trunk considered superior to gum arabic. Trees used as ingredient in Ivory Coast for arrow poison; elsewhere they are used as fences and to check erosion. Wood, hard and durable underground, used for wooden plows and for pegs. Trees often planted as an ornamental.[40]

Bark is astringent and demulcent, and like the leaves and roots is used medicinally. Woody branches used in India as toothbrushes. The gummy roots also chewed for sore throat. Flowers, mixed with grease, are applied to the head for headache. Said to be used for alterative, antispasmodic, aphrodisiac, astringent, demulcent, diarrhea, febrifuge, rheumatism, and stimulant.[40] Vietnamese poultice the leaves onto sores and ulcers. Malayans use the flower and leaf in puerperal infusions. Philippinos use the bark decoction for leucorrhea, rectal prolapse, sores and ulcers. Burmese paste the root onto cattle hooves to prevent or kill parasites. Indonesians poultice pulp from green pods onto sores on the corners of the eye, using the leaves and juice for lumbar pain.[16]

Cassie has been reported to contain anisaldehyde, benzoic acid, benzyl alcohol, butyric acid, coumarin, cresol, cuminaldehyde, decyl aldehyde, eicosane, eugenol, farnesol, geraniol, hydroxyacetophenone, methyleugenol, methyl salicyclate, nerolidol, palmitic acid, salicylic acid, and terpineol. Dried seeds of one Acacia sp. are reported to contain per 100 g: 377 calories, 7.0 g moisture, 12.0% protein, 4.6 g fat, 72.4 g carbohydrate, 9.5 g fiber, and 3.4 g ash. Raw leaves contain per 100 g: 57 calories, 81.4 g moisture, 8.0 g protein, 0.6 g fat, 9.0 g carbohydrate, 5.7 g fiber, 1.0 g ash, 93 mg Ca, 84 mg P, 3.7 mg Fe, 12,255 μg β-carotene equivalent, 0.20 mg thiamine, 0.17 mg riboflavin, 8.5 mg niacin, and 49 mg ascorbic acid. The amino acid constitution follows: lysine, 4.7 (g/16g N); methionine, 0.9; arginine, 9.2; glycine, 3.4; histidine 2.3; isoleucine, 3.5; leucine, 7.5; phenylalanine, 3.5; tyrosine, 2.8; threonine, 2.5; valine, 3.9; alanine, 4.3; aspartic acid, 8.8; glutamic acid, 12.6; hydroxyproline, 0.0; proline, 5.1; serine, 4.1; with 76% of the total nitrogen as amino acids.[40]

Toxicity — Vietnamese think the pollen causes problems.[16]

4. *ACACIA SENEGAL* (L.) Willd. (FABACEAE) — Gum Arabic, Senegal Gum, Sudan Gum

The tree yields commercial gum arabic, used extensively in pharmaceuticals, inks, pottery pigments, water-colors, wax polishes, liquid gums; for dressing fabrics, giving luster to silk and crepe; for thickening colors and mordants in calico-printing; in confections and sweet-meats. Bark fibers yield strong cordage. Wood white, used for making tool handles. Heart-wood is black, used for making weavers' shuttles. One of the strongest of local fibers is obtained from the long flexible surface roots, used for cordage, well-ropes, fishing nets, horse-girdles, foot-ropes, etc. Young foliage makes good forage. Plants useful for afforestation of arid tracts and soil reclamation.[40]

Gum arabic is astringent, demulcent, and emollient; it is used internally in inflammation of intestinal mucosa, and externally to cover inflamed surfaces, e.g., burns, sore nipples, and nodular leprosy. Also said to be used for catarrh, colds, coughs, diarrhea, dysentery, expectorant, gonorrhea, hemorrhage, sore throat, typhoid, urinary tract.[32,40]

On a dry matter basis, the leaves contain 18.2% protein, 6.7% tar, 66.7% total carbohydrate, 11.2% fiber, 8.4% ash; the fruits contain 22.0% protein, 1.0% fat, 69.9% total carbohydrate, 39.0% fiber, 7.1 g ash; and the seeds contain 40.1% protein, 5.8% fat, 49.1% total carbohydrate, 13.7% fiber, 5.0 g ash, 0.75% Ca and 0.31% P.[21] The main component of gum arabic is arabic acid, a polysaccharide composed of L-arabinose, L-rhamnose, D-galactose, and D-glucuronic acid in different molar ratios (depending on the species the gum is produced from).

Toxicity — Gum arabic contains trypsin inhibitors.[40]

5. *ACALYPHA INDICA* L. (EUPHORBIACEAE) — Indian Acalypha, Hierba de Cancer

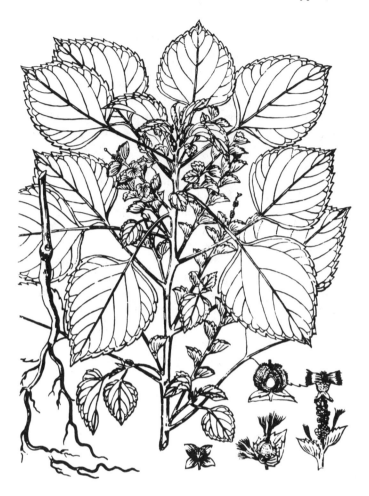

Used as a diaphoretic, expectorant, and emetic substitute for ipecac or senega. About 15 g of the decoction is said to be a safe and speedy laxative. Plant sometimes used as a substitute for ipecac or senega.[33] Said to work like catnip on cats.

The juice is applied to tumors in India.[4] Transvaal Africans used the plant for eye ailments. In Indochina the leaf is used as an anthelmintic, the root as a purgative, and the juice as an emetic. Bruised leaves introduced as an anal suppository to relieve constipated children. Perry recites a dangerous headache remedy, dribbling the acrid sap into the eye.[16] Juice of fresh leaves used against croup. The leaf, ground with salt and/or lime or lime juice, is applied externally as a parasiticide for ringworm, scabies, and maggoty wounds. The herbage is used as an expectorant in bronchitis, croup, phthisis, and pneumonia; also for eczema and pimples.[33]

Contains the alkaloid acalyphine, a resin, a tannin, a volatile oil, a cyanogenetic glucoside $(C_{14}H_{29(22)}O_{10}N_2)$ and HCN (ca 270 mg/100 mg dry weight). Triacetonamine, quebrachitol (1-inisitol monomethyl ether).

Toxicity — The plant is toxic, causing a dark chocolate coloring of the blood (due to something beside the cyanide).[33] Contact with the plant can cause dermatitis.

6. *ACHILLEA MILLEFOLIUM* L. (ASTERACEAE) — Yarrow

Swedish call it "field hop" and used it in beer manufacture, the beer thus brewed said to have been more intoxicating than that prepared with hops. Chandler, Hooper, and Harvey[73b] devoted a whole review article to the ethnobotany and phytochemistry of yarrow. "Yarrow has been employed as a popular medicine since the Trojan war (ca. 1200 B.C.) when the Greek hero, Achilles, is said to have used the leaves of a plant to check the flow of blood from the wounds of his fellow soldiers. Yarrow is used in herb tea, herbal tobaccos, and as a hair rinse and shampoo (reported to prevent baldness). Yarrow contains compounds that inhibit seed germination and have sex-pheromone qualities as well as compounds that serve as bactericides and culicides. Amchies use the plant as an insecticide on plant insects.[53] Extracts are used in some bath preparations for alleged quieting and soothing effects on the skin. Used in bitters and vermouth.[29]

The plant and/or its juices are used in folk remedies for cancers, indurations, or tumors (of breast, feet, liver, penis, spleen, uterus) condylomata, warts, and wens.[4] Chandler, Hooper and Harvey[73b] mention many uses for the plant: abortive, analgesic, anthelmintic, antiinflammatory, antispasmolytic, antiviral, contraceptive, diaphoretic, diuretic, emmenagogue, febrifuge, laxative, stimulant, and toxic, noting its use for earache, epilepsy, fever, fistula, headache, hematoma, hemorrhage, hemorrhoids, hypertension, hysteria, influenza, leucorrhea, measles, melancholy, menorrhagia, nervousness, pleurisy, pneumonia, rashes, rheumatism, smallpox, sore throat, toothache, tuberculosis, ulcers, urinary incontinence, and wounds. The plant has no significant antifertility effects in rats.[53]

Per 100 g, the forage is reported to contain on a zero moisture basis 12.8 to 14.4 g protein, 1.8 to 3.9 g fat, 71.3 to 71.8 g total carbohydrate, 20.1 to 23.6 g fiber, 10.4 to 12.5 g ash, 1,330 mg Ca, and 360 mg P. Among the alkaloids are achiceine, achilletin, betaine, betonicine, choline, homostachydrine, moschatine, stachydrine, and trigonelline. Flavonoids include[73] apigenin, apigenin glycosides, artemetin, casticin, 5-hydroxy-3,6,7,4'-tetramethoxyflavone, isorhamnetin, luteolin, luteolin glucosides, quercetin glycoside, rutin (quercetin rhamnoglucoside). The volatile oil[73b] contains allo-ocimene, allo-ocimene isomer, azulene, borneol, bornyl acetate, butyric acid, Δ-cadinene, camphene, camphor, carophyllene, chamazulene, chamazulene carboxylic acid, 1,8-cineole, copaene, cuminic aldehyde, p-cymene, eugenol, farnesene, furfural, furfuryl-alcohol, humulene, isoartemsia ketone, isobutyl acetate, isovaleric acid, limonene, menthol, myrcene, alpha-pinene, beta-pinene, sabinen, salicyclic acid, sesquiterpene lactones (8-acetoxyartabsin, acetylbalchanolide, achillicin, achillin, 8-anelooxyartabsin, austricin, balchanolide, 2,3-dihydrodeacetoxymatricin, hydroxyachillin, leucodin, millefin, millefolide, proazulenes), alpha-terpinene, terpinen-4-ol, and terpinolene. The oil contains at least linoleic-, oleic-, cerotic-, myristic- and palmitic-acids.[33] Amino acids include alanine, glutamic acid, histidine, leucine, and lysine. The analgesic salicylic acid is reported. Some of the alkaloids are said to be weakly antipyretic and hypotensive. Achilletin reduces coagulation time in canines. Fresh leaves contain 0.058% ascorbic acid, dried leaves 0.31%.

Toxicity — Though not generally considered toxic, yarrow has been indicted at least in the literature for causing the rapid death of a calf who ingested one plant.[73b] Some individuals show positive patch test reactions to yarrow and cross-sensitivity between other Asteraceae and yarrow has been demonstrated. FDA approval for use in alcoholic beverages only. Finished beverage must be thujone free. Yarrow oil normally contains little or no thujone, whereas sage oil may contain 50% (§ 172.510).[29]

7. *ACOKANTHERA SCHIMPERI* Schweinf. (APOCYNACEAE) — Arrow Poison Tree

Although the fruits are edible at the right time of the year, the rest of the plant is quite toxic. Fruits are made into jams.[17] The wood, once used in African arrow poisons, is now a major commercial source of ouabain, which has digitalis-like activity. Sometimes grown as an ornamental shade tree in East Africa.[17]

Several compounds found in this and/or related species of Acokanthera are acobioside A, acolongifloroside A, actospectoside A, acovenoside A, and B, and opposide, all active in the KB tumor system.[10]

Seed contains 0.3 to 0.74% acovenoside A, 0.3 to 0.5% ouabain, acoschimperoside N and O, and acolongifloroside G, H, and K. The wood contains 0.2% ouabain, acoschimperoxide N, P, S, T, and U, and acolongifloroside but no acovenoside A.[33] The seeds contain 1.7% acovenoside A, 0.1% ouabain, .0037% acolongifloroside G, 0.98% acolongifloroside H, and .024% acolongifloroside K acetate.[17]

Toxicity — Ingestion of dried leaves in excess of 15 g may be seriously toxic to humans. Inhalation of powdered plant material may cause fatalities. The poison is also absorbed through the skin. A small quantity of extract placed in a sheep's ear will bring death in two hours. The tree is a well-known hazard to grazing animals. Signs of cattle poisoning include rapid, shallow respiration, diarrhea, muscular spasms, grinding of teeth and salivation; death usually occurs quickly from heart failure.[17] The lethal dose of ouabain for an adult human is ca 2 mg, while African arrows may have borne overwhelming doses of 1000 to 5000 mg.[3] Africans once used the plants for homicide by smearing the poison onto prickly fruits along a path likely to be used by the intended victim. Symptoms of Acokanthera and Strophanthus poisonings come on rapidly: collapse with slow pulse and slow respiration, convulsions and death with stoppage of the heart in systole.[3] African medicine men smear their skin with ashes or juice of *Coleus* spp. to protect the skin from dermal absorption.[3]

8. *ACONITUM NAPELLUS* L. (RANUNCULACEAE) — Aconite, Monkshood, Blue Rocket

Aconite roots are relatively unimportant in the U.S., but are cultivated in Europe. Aconite was formerly used internally as a febrifuge and gastric anesthetic, and veterinarily to lower blood pressure and slow the circulation.[20] Once used for its anesthetic and irritant effects in gastralgia and neuralgia.[17] Certain species have antitumor activity in lab animals.[29]

The plant juices were folk remedies for cancer and scirrhous tumors.[4] Aconite liniment is used externally for rheumatism and neuralgia, or as a counterirritant.[1] The principal alkaloid aconitine, stimulates and then depresses the central and peripheral nervous system. It is considered anodyne, febrifuge, and sedative, and has been used as a depressant in high blood pressure of cardiac origin. Tincture of aconite is used internally as a sedative and analgesic. Dried roots are used as an analgesic. In ancient times a decoction of this toxic herb was given to criminals. A black mass is extracted from the roots of Russian aconite after boiling. A tincture of this black mass, known as ''parpi'', is used in folk medicine to cure cancer and other maladies. Anticarcinogenic properties of these plants are recognized by Russian botanists and cancer specialists according to an unpublished typescript by the USDA's late E. E. Leppik.[74] These are the cancer roots from Solzhenitsyn's ''Cancer World''. Solzhenitsyn started with low concentration, one drop from the alcoholic extract to a glass of water on the first day. On the next day, he added two drops, the next, three, and so on until he could drink 10 drops a day. After a week of use, he gradually reduced the concentration again. He repeated this procedure several times.[74]

Total alkaloidal content of the roots is 0.2 to 1.5%. The alkaloids reported include aconine, aconitine, benzaconine, ephedrine, hypoconitine, mesaconitine, napelline, napellonine, neoline, neopelline, picraconatine, pseudaconitine, and sparteine. Aconitic, caffeic, chlorogenic, citric, glyceric, itaconic, isocitric, malic, malonic, oxalic, pyrrolidoncarbonic, quinic, and succinic acids are reported, along with resins, mannite, mannitol, fructose, maltose, and melibiose.

Toxicity — *Very toxic*. Aconite is poisonous and should not be used without medical advice. Overdose should be avoided as there is no sure antidote. One mg of aconitine is said to have killed a horse, 2 mg may kill a human. The poison aconitine can be absorbed through the skin. Symptoms include intense nausea, emesis, and diarrhea, muscular spasms and weakness, weak pulse, respiratory paralysis, convulsions, and death.

To the physician — Hardin and Arena recommend gastric lavage or emesis followed by 2 mg atropine injected subcutaneously; maintain blood pressure and artificial respiration.[34]

9. *ACORUS CALAMUS* L. (ARACEAE) — Sweet Flag, Flagroot, Calamus

This long-used plant was an article of commerce in the Near East for 4,000 years. Sometimes roots are coated with sugar as candy and breath sweeteners. Leaves and roots yield oil of calamus, a yellow aromatic volatile oil, used in hair powders, perfumery, and for flavoring liqueurs, (e.g., Benedictine and Chartreuse), beer, gin, vinegars, snuff, and various other preparations. Stockton bitters, once popular in Britain, was made from a mixture of calamus root and gentian root. "Bach" of commerce, which has many medicinal uses, is prepared from the rootstock. Leaves and rootstocks make very effective insecticides for use against biting and sucking insects attacking field crops, stored grains, wool, and against household pests, like bedbugs, fleas, and flies. To make the insecticide, leaves or rootstock are finely powdered and used as a dust or in an aqueous solution as a spray. Beta-asarone, active ingredient of the root extract, sterilizes female insects by preventing ovary development. Also regarded as an insect repellent.[75] Fresh rootstock is used as a confection or as a substitute for ginger. Young leaves make a palatable salad. Menomini Indians used mature leaves in construction of wigwams. In France and other countries, calamus is a cultivated ornamental water-plant.[43] Calamus root adds a mellow, somewhat spicy odor to potpourri or sachets. Young tender inflorescences are eaten. Rhizomes used for chewing gum. Leaves sometimes used in fish sauces. Tyler concludes that "calamus has no therapeutic ability, which is not provided more effectively and more safely by other drugs. This is also true of its use as a flavoring agent, which can no longer be condoned".[37] G. R. Morgan gives an interesting account of The Ethnobotany of Sweet Flag among North American Indians.[76] Omaha gave sweet flag to their horses as a snuff, and baited their fishing nets with it, and they gave it to their watchdogs to make them fierce.[76]

The root, prepared in various manners, is a folk remedy for indurations of the liver, spleen and stomach, scirrhus of the spleen, hard swellings and tumors of the testicles, uterus, and vagina.[4] Acts as a carminative, removing the discomfort caused by flatulence and checking the growth of the bacteria which cause flatulence. Dried root chewed to relieve dyspepsia. Formerly used in ague and "low fever". Used as a mild stimulant febrifuge in typhoid. Said to be useful in all nervous complaints, colds, cough, dyspepsia, hysteria, insomnia, malaria, melancholy, neurasthenia, vertigo, headache, hypochondria, gout, rickets, rheumatism, and scrofula.[32] Orientals use the root for bronchitis and regard it as aphrodisiac.[41] Said by lowlanders to have a beneficial effect on the bladder when "a man loses control".[17] Turks carried candied calamus to prevent infection. Rhizome is emetic, stomachic in dyspepsia, colic, remittent fevers, nerve tonic, in bronchitis, dysentery of children, insectifuge and used for snakebite. Rhizomes externally applied for fever, lumbago, rheumatism.[16] Juice of the rhizome applied to buboes, carbuncles, deaf ears, and sore eyes.[16]

Valued for rheumatism in Iran.[75] Plants are anthelmintic (Brazil), aromatic, and mildly tonic, carminative and used to increase appetite and benefit digestion. Dried root is chewed, or a tea is made, to relieve dyspepsia and to clear the voice. The plant is described in Hagers Handbook as analgesic and sedative, with a hypotensive effect on anesthetized animals.[33] Said to have antiarrhythmic, anticonvulsant, antiveratrinic effects similar to quinidine.[16]

Formerly used as a strewing herb, the plant is aromatic because of essential oil containing the glucosidic bitter principle, acorin, and a phenolic ether, asarone. Roots have an agreeable aromatic odor and pungent, bitter taste which are retained after drying. Dry European roots yield 0.9 to 4.8% oil, Japanese material to 5%, powdered material down to 0.5% oil. Tetraploids may contain up to 7%. The following compounds have been identified in calamus: acolamone, acoragermacrone, acoric acid, acorine, acorone, acoroxide, asaraldehyde, asaronaldehyde, asarone, beta-asarone, azulene, calamene, calameone, calamenol, calamenone, calamol, camphene, camphor, choline, cineole, dextrin, dextrose, dimethylamine, eugenol, *n*-heptylic acid, isoacolamone, isoacorone, linalol, methylamine, methyleugenol, palmitic acid, parasarone, pinene and trimethylamine. Oil from dried rhizomes of an Indian specimen (yield 2.8% oil) contained 82% asarone, 5% calamenol, 4% calamene, 1% calameone, 1% methyleugenol, and 0.3% eugenol. Two bitter principles, acorin and acoretin, are reported. Mucilage, resins, and tannins are also reported.

Toxicity — Oil of Calamus has been shown to be carcinogenic,[17] probably due to its asarone or safrole content. Beta-asarone is the reputed mild hallucinogen in the plant.[54] Classified by the FDA as an unsafe herb. ''Oil of Calamus, Jammu variety, is a carcinogen.''[62] Oil of calamus has been shown to possess considerable toxicity in long-term feeding studies in rats. After 18 weeks the animals exhibited depressed growth, liver and heart abnormalities, and a serious effusion in the abdominal and peritoneal cavities. Malignant tumors in the duodenal region were noted initially after 59 weeks (in animals fed 500 to 5000 ppm, but not in controls).[37]

10. *ACTAEA PACHYPODA* Ell. (RANUNCULACEAE) — Baneberry

Medicinal perennial of the deciduous forest. Reported to be alterative, laxative, nervine, repellant. Is a folk remedy for cold, cough, congestion, debility, diarrhea, female ailments, gastralgia, headache, rheumatism. In my draft on Amerindian folk medicine[43] I note that Arikaras applied the pulverized roots to blood clots, and inflammation and abscess of the breast (the latter with puffballs), Blackfeet used the root decoction for cold and cough, Cherokee used the root tea as a gargle and to cure itch, Cheyenne used the decoction as a lactagogue, Chippewa used the root decoction for metrorrhagia, Cree used the fruit as purgative, Fox used the plant (root decoction) for parturition, urogenital disorders and, like the Cherokee, to revive patients near death. It was said to kill teeth of young people if not careful. Ojibwa used the root for stomach disorders and parturition. Potawatomi used the root decoction to purge in puerperium. Thompson used for rheumatism and syphilis.[43] Homeopaths use the roots for arthritis and rheumatism.[33]

Fruits and seeds contain trans-aconitic acid, "protoanemonoid" compound.

Toxicity — *Poisonous.*[11] Root stocks, sap, and berries markedly irritant if ingested. (The European A. spicata is suggested to have killed children who ingested the berries.)[14] Sometimes externally vesicant.[6] A protoanemonin-like compound inflames and blisters the skin, internally produces nausea, vomiting, gastroenteritis, dyspnea, and delirium.[33]

To the physician — Hardin and Arena suggest gastric lavage or emesis and symptomatic and supportive treatment.[34]

11. *ACTINIDIA POLYGAMA* (Sieb. and Zucc.) Planch. (ACTINIDIACEAE) Cat Powder, Silver Vine

In China the twigs and young leaves of this ornamental vine have been used for centuries to tranquilize cats in zoos. This led to adding the powder to saki to increase its euphoriogenic capacity.[54] A tea "Polygamol", a preparation made from the galled fruits, is sold in the orient as a cardiotonic and diuretic.[16] Leaves and salted fruits are eaten by the Japanese.[36] Considered tonic in Japan.[32] The dried fruit decoction is frequently used for colic and rheumatism. According to Perry only the galled fruits are used.[16]

Powdered twigs contain actinidine, which depresses the limbic system leading to sedation,[54] and metatabilacetone.[54] Fruits contain matatabic acid.[16]

Toxicity — Emboden lists this among his "narcotic" species, as a tranquilizer with mild hallucinogenic effects.[54]

12. *ADHATODA VASICA* (L.) Nees (ACANTHACEAE) — Malabar Nut, Adotodai, Pavettia

Plants grown for reclaiming waste lands. Because of its fetid scent, it is not eaten by cattle and goats. Leaves and twigs commonly used in Sri Lanka as green manure for field crops, and elsewhere in rice fields. Leaves, on boiling in water, give a durable yellow dye used for coarse cloth and skins; in combination with indigo, gives cloth a greenish-blue to dark green color. Also used to impart black color to pottery. Stems and twigs used as supports for mud walls. Wood makes good charcoal for gunpowder; used as fuel for brick kilns. Ashes used in place of crude carbonate of soda for washing clothes. In Bengal, heads are carved from wood. Leaves also used in agriculture as a weedicide, insecticide, and fungicide, as they contain the alkaloid, vasicine. As weedicide, it is used against aquatic weeds in rice-fields; as insecticide, used in same way on tobacco leaves; as fungicide, prevents growth of fungi on fruits which are covered with vasica leaves. Market gardeners place layers of leaves over fruit, like mangoes, plantains and custard-apples, which have been picked in an immature state to hasten ripening and to ensure development of natural color in these fruits without spoilage. Whole plant used in Sri Lanka for treatment of excessive phlegm, and in menorrhagia. Leaves are source of drug used as an expectorant and to relieve coughs.

Plants are used in folk remedies for glandular tumors in India.[4] Leaf used for asthma, bronchitis, consumption, cough, fever, jaundice, tuberculosis; smoked for asthma; prescribed as a mucolytic, antitussive, antispasmodic, expectorant.[41] Ayurvedics use the root for hematuria, leucorrhea, parturition, and strangury, the plant for asthma, blood impurities, bronchitis, consumption, fever, heart disease, jaundice, leucoderma, loss of memory (amnesia), stomatosis, thirst, tumors, and vomiting.[26] Yunani use the fruit for bronchitis, the flowers for jaundice, poor circulation, and strangury; the emmenagogue leaves in gonorrhea, and the diuretic root in asthma, bilious nausea, bronchitis, fever, gonorrhea, and sore eyes.[44]

Used in Indian medicine for more than 2000 years, adhatoda now has a whole book dedicated to only one of its active alkaloids.[77] In addition to antiseptic and insecticidal properties, vasicine produces a slight fall of blood pressure, followed by rise to the original level, and an increase in the amplitude of heart beats and a slowing of the rhythm. It has a slight but persistent bronchodilator effect. With a long history as an expectorant in India, vasicine has recently been modified to form the derivative bromhexine, a mucolytic inhalant agent, which increases respiratory fluid volume, diluting the mucus, and reduces its viscosity. Fluid extract of leaves liquifies sputum, relieving coughs and bronchial spasms. The plant also contains an unidentified principle agent active against the tubercular bacillus.[1,11] Adhatodine, anisotinine, betaine, vasakin, vasicine, vasicinine, vasicinol, vasicinone, vasicoline, vasicolinone, are reported.[41] Deoxyvasicine is a highly effective antifeedant followed by vasicinol and vasicine. "These plant products as antifeedants could be safely used for controlling pests on vegetable crops."[78]

Toxicity — Vasicine and vasicinol exhibit potential to reduce fertility in insects. "Vasicine is also likely to replace the abortifacient drugs in current use as its abortifacient activity is comparable to prostaglandins."[78] In large doses the leaves cause diarrhea and nausea.

13. *AESCULUS HIPPOCASTANUM* L. (HIPPOCASTANACEAE) — Horse Chestnut

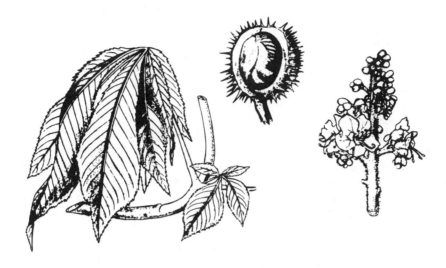

Often grown as an ornamental tree. The smooth grey-green bark has been used as a yellow dye. The soft spongy wood is sometimes used for packing cases, cutlery, soft furniture, etc. Can be made into charcoal used in gunpowder. Indians roasted the poisonous nuts among hot stones, peeled, and mashed them, and then leached the meal in lime water for several days before using the meal for making breadstuffs. According to Millspaugh, germinating the seeds made them pleasant by changing the bitter principles to sweet, like malting barley.[19] Yet he suspected that the testa of the nuts was narcotic, 10 g equalling 3 g opium.[19] Powdered seeds or bruised branches may be used as a fish intoxicant.[13]

Fruits have been used as a cataplasm for mammary indurations and other types of cancer. Bark or nut infusions used for healing ulcers.[4] The astringent bark is said to be sternutatory, tonic, narcotic, and febrifuge, hence useful in malaria. Fruits have been used for backache, neuralgia, rheumatism, and whooping cough; also for hemorrhoids and other rectal complaints. Flowers are said to be anodyne, astringent, tonic, and vulnerary. Seed tincture applied to painful piles;[8] flower tinctures for rheumatism. Seeds are said to be alterative, analgesic, febrifuge, narcotic, sternutatory, tonic, and vasoconstricting.

Seeds are said to contain 3% water, 3% ash, 11% crude protein, 5% oil, and 74% carbohydrates. A kilogram of horse chestnut meal is said to have the equivalent of ca. 1.1 kg barley, 1.25 kg oats, 1.5 kg bran, and 3.4 kg good meadow hay[2] for feeding animals like cows and sheep. Adenine, adenosine, aesculin, aesculetin, allantoin, argyrin, astragalin, carotin, choline, citric acid, epicatechin, fraxin, guanine, isoquercitrin, leucocyanidine, leucodelphinidin, phytosterol, resin, rutin, saponin, scopolin, scopoletin, tannin, and uric acid are also reported from the horse chestnut.[2,33]

Toxicity — Classified by the FDA as an unsafe herb.[62] Said to have killed children. Alkaloids, glycosides and saponins have all been held responsible for toxicity in the genus, but most problems probably result from the glycoside aesculin (6-beta-glucoside-7 hydroxycoumarin). Seeds are stated to have killed children who ingested them.[14] Symptoms of poisoning include nervous twitching of muscles, weakness, lack of coordination, dilated pupils, vomiting, diarrhea, depression, paralysis, and stupor.

To the physician — Gastric lavage or emesis and symptomatic treatment is recommended in cases of poisoning.[34] Other toxins reported from the genus Aesculus include nicotine, quercitin, quercitrin, rutin, saponin, and shikimic acid.[15]

14. *AETHUSA CYNAPIUM* L. (APIACEAE) — Fool's Parsley

Fetid annual weed, sometimes fatally confused with parsley.[2] Reported to be poison, sedative and stomachic, fool's parsley is a folk remedy for cancer, cholera, convulsions, diarrhea, enterosis, gastrosis, spasms, and tumors.[33] Used medicinally as sedative and stomachic in pediatric gastroenteritis, cholera, and diarrhea.[2] Once used in homeopathy for colic, cramps, diarrhea, dysuria, and stomachache.[33]

The active principle is the alkaloid cynopine,[2] possibly the same as cynapium. Fresh herbage contains 0.00023% coniine-like alkaloid and 0.015% essential oils, the polyine aethusin, aethusanol A and B, as well as acetylenes.[33]

Toxicity — Victims of poisoning showed symptoms of heat in the mouth and throat; post mortem exam showing redness of the lining membrane of the gullet and windpipe and slight congestion of the duodenum and stomach.[2] Millspaugh gives rather vivid details of human poisonings including one case who survived but hallucinated cats and dogs.[19]

15. *AGAVE SISALANA* Perrine (AMARYLLIDACEAE) — Mescal

Leaves provide a fiber of great commercial value, used for making binder twine and as substitute for Manila fiber for manufacturing heavier twine, ropes, and marine cordage. Sometimes fiber used as substitute for jute, and woven into rugs, mats, and fabrics for making sacks for coffee, wagon covers, floor-coverings, and for mops, brushes, paper board, and kraft paper. Sisal waste used to make cheap twines and upholstering tow as well as paper, and to manure sisal plants.[22] Waste also used as source of cortisone and other hormones. Like any of several genera in several plant families this could be used as a source of steroids. We get conflicting signals on these steroids in reading "Medicinal Plants Need Extensive Safeguarding".[79] "Most oral contraceptives — one of the biggest product groups stemming from diosgenin — are now manufactured by total synthesis . . . And after 25 years of research on steroids, manufacturers still obtain 95% of their start-point material from natural sources." Sap exuding from cut ends of flowering stalks is sweet and used to make a beer (pulque) by fermenting this sap, and a kind of brandy (mexical or mescal or mezcal). Central buds may be cooked and eaten. They can be baked with little corn oil and salt, peeling off the outer "leaves" and serving them as a side dish.[49] Sisal waste is molluscocidal[17] and may be used as a beneficial mulch on plants not requiring an acid soil.

Reported to be cicatrizant, depurative, detergent, and soporific, sisal is a folk remedy for dysentery, jaundice, leprosy, sores, sprains, and syphilis.[32]

Contains tigogenin, hecogenin, gitogenin, neo-tigogenin, sarsapogenin, sisalogenin, gloriofenin, gentrogenin, delta 9-11-hecogenin, diosgenin and yamogenin[17,33] pectin, and much vitamin C. The cuticle contains 5 to 17% of a hard wax with properties suggesting candelilla or carnauba wax.[33] The dried residue after fiber extraction contains 11% fermentable sugars from which alcohol can be prepared, perhaps not profitably.[1]

Toxicity — Raw sap highly irritating to the eyes and skin. Sisal in mattresses may cause allergic reactions in sensitive individuals.

16. *AGRIMONIA EUPATORIA* L. (ROSACEAE) — Agrimony

Stems and leaves source of a yellow dye.[8] In France, agrimony tea is drunk as much for its flavor as for its medicinal virtues.[2] Herb used in gargles by singers and speakers.[33] Sheep and goats eat the plant but cattle, horses, and swine leave it alone.[2] Said to induce sleep if placed under one's pillow.[2]

Said to alleviate condylomata, sclerosis of the spleen and liver, tumors of the internal organs, mesenteric region, scrotum, sinews, and stomach, as well as corns and warts, and cancer of the breast, face, mouth, and stomach.[4] Reported to be astringent, cardiotonic, coagulant, depurative, diuretic, emmenagogue, litholytic, sedative, tonic, vermifuge, and vulnerary; agrimony is a folk remedy for asthma, bronchitis, dermatitis, enterorrhagia, enuresis, gastrorrhagia, hematuria, hepatosis, metrorrhagia, neuralgia, neuritis, pharyngitis, rheumatism, tuberculosis, and warts.[32,33] Astringent, the herb tea is useful for internal bleeding, and/or looseness of the bowels. Has a great reputation for curing jaundice and other liver ailments. Also recommended for snakebite. Considered useful in gout, skin eruptions, pharyngitis, diseases of the blood, blotches, pimples, etc. Febrifugal and vermifugal; used for tapeworm, e.g., by the Zulu.[3] Sotho use as vermifuge with *Artemisia afra*; the leaf is used for colds. Plant is gargled for inflammation of throat and mouth. Mixed with fat, it is used as a poultice to draw out indolent ulcers.[47]

The fresh herb contains a glucoside alkaloid, a nicotinic acid amide, traces of essential oil, and organic acids, vit. B$_1$, vit. K, ascorbic acid, 1.5% triterpene (0.6% ursolic acid), and a derivative of alpha-amyrin. Fresh herb also contains agrimonolide, C$_{18}$H$_{18}$O$_5$, palmitic acid, stearic acid, ceryl alcohol, and phytosterols.[33] The seed contains 35% oil with oleic-, linoleic-, and linolenic-acids.

Toxicity — Said to induce photodermatitis in man.[11] Contains ca. 5% tannin (2.6% catechin).[3] The leaf contains catechin-tannin, quercitrin, and ellagen-tannin but woody part and roots contain ellagen-tannin. Leaf contains 0.29 to 0.12 mg/g of "nicotinic acid" complex. The tonic effect of the herbal infusion may possibly be due to the thiamine content (2.355 μg/g).

17. *AGROSTEMMA GITHAGO* L. (CARYOPHYLLACEAE) — Cockle, Corn Cockle

Rarely encouraged for the small but attractive flower, this is more often regarded as a weed. The saponin is said to have a local anesthetic effect. Flower extracts have bacteriostatic properties.

Regarded as diuretic, emmenagogue, expectorant, poison, and vermifuge, cockle has been used to treat cancer, dropsy, and jaundice. The root has been used for exanthemata and hemorrhoids.[3] Homeopathically, the seeds are used in gastritis and paralysis.[2] It is a noxious weed, growing in cereal fields, whose seeds are quite poisonous if grouped up with the cereal. Cockle, especially the seeds, has been used in Europe for cancers, hard tumors, warts, and apostemes, or hard swellings in the uterus.[4] Seeds have been placed in the conjunctival sac by patients to induce kerato-conjunctivitis; the saponin agrostemmin or githagin is irritating to the eye.[6]

Contains the saponin Sapotoxin A, with the prosapogenin githagin ($C_{35}H_{54}O_{11}$), the aglycone githagenin $C_{30}H_{46}O_4$, and agrostemmic acid $C_{35}H_{54}O_{10}$. Ripe seeds contain several aromatic amino acids including 2,4-dihydroxy-6-methyl-phenylalanine, L(+)-citrullin ($C_6H_{13}N_3O_3$), sugar, oil, fat, and starch. Seedlings, like so many other seedlings, contain allantoin and allantoic acid. Root contains in the starch up to 2.02% "lactosin".[33]

Toxicity — Historically, poisonings were frequent when the seeds of the weed contaminated wheat seed. Hog mortality has been ascribed to root ingestion. From 0.2 to 0.5% body weight of seed is lethal to young birds. Grain heavily contaminated with corn cockle has killed cows.[14] Repeated ingestion of small doses induces a chronic poisoning termed githagism, large doses causing acute poisoning with vertigo, respiratory depression, vomiting, diarrhea, salivation, and paralysis, all possibly accompanied by severe muscular pain, twitching, and coma.[3] Effects from absorption include headache, dizziness, restlessness, delirium, and finally cramps and circulatory disorder (weak pulse, strongly accelerated), and in terminal cases, respiratory paralysis.[33]

To the physician — Hardin and Arena recommend gastric lavage or emesis.[34]

18. *AJUGA REPTANS* L. (LAMIACEAE) — Bugleweed

Attractive herbal groundcover, with more historical than current use as an herbal medicine. One old saying, an earlier version of An Apple a Day, went "He that hath Bugle and Sanicle will scarce vouchsafe the chirugeon a bugle."[2]

According to Hartwell, bugle was ingested for cancer and poulticed onto indurations of the uterus.[4] Reported to be astringent, carminative, deobstruent, diuretic, hemostat, narcotic, stomachic, and vulnerary, the bugleweed is a folk remedy for cancer, fistula, fever, gangrene, hangover, hemoptysis, hepatosis, jaundice, lung ailments, quinsy, rheumatism, sores, splenosis, and ulcers.[2,33] Sixty grams herb in a liter of water was boiled for biliary disorders.[2] It has been considered good for the bad effects of excessive drinking.[2] "It has been termed one of the mildest and best narcotics in the world."[2]

Contains 15% tannin, three glycosides, cyanidin and delphinidin, as well as heterosides aucubin and asperulin.

Toxicity — Narcotic,[2] resembling digitalis in its action.

19. *ALCHORNEA FLORIBUNDA* Muell. Arg. (EUPHORBIACEAE) — Niando

Macerated root bark is both a strong intoxicant and aphrodisiac described as a powerful plant "capable of bringing great joy and the deepest sorrow."[54] Africans ate powdered niando mixed with salt or food before tribal activities or warfare. Sometimes it is steeped in banana or palm wine.[54] Reported to be aphrodisiac and a remedy for hypertension.[33]

Schultes and Farnsworth[80] questionably reported the aphrodisiac hallucinogen yohimbine, and alchorneine, isoalchorneine and alchorneinone.

Toxicity — Classed as a narcotic hallucinogen.[54] Steeped in wine, it can produce an intense excitement followed by deep depressions, known to have been fatal on several occasions.[54]

20. *ALETRIS FARINOSA* L. (HAEMODORACEAE) — Unicorn Root, Star Grass, Colic Root

An old Amerindian drug that has become world renowned, at least as a folk medicine. Reported to be carminative, cathartic, diuretic, emetic, narcotic, purgative, sedative, stomachic, and vermifuge; unicorn root is a folk remedy for ague, amenorrhea, anasarca, backache, chlorosis, colic, dysmenorrhea, dyspepsia, fever, hysteria, rheumatism, sore, stomachache, uteral prolapse, uterosis, and vaginal prolapse.[2,19,32,33] Grieve states that it proves of great use in cases of habitual miscarriage. Known as colic root because used for colic and ague root because used for rheumatism, often called ague in Colonial days; unicorn root is still taken for rheumatism, often in alcohol, in Appalachia. Amerindians valued unicorn root for narcotic, stomachic and tonic virtues. Catawbas placed leaves in water for colic, dysentery, and other stomach disorders. Cherokee used the root for colic, constipation, cough, fever, jaundice, lung ailments, rheumatism, and strangury; the root was believed to strengthen the womb and prevent abortion.[43]

Contains a saponin, with aglycone diosgenin ($C_{27}H_{42}O_3$), essential oil, a resin, alkaloids, and starch. Containing diosgenin, it is no wonder that the root shows estrogenic activity.

Toxicity — Narcotic,[2] small doses induce colic (hypogastric), stupefaction, and vertigo.

21. *ALEURITES MOLUCCANA* (L.) Willd. (EUPHORBIACEAE) — Tung, Candlenut, Candleberry, Varnish Tree

Seed yields 57 to 80% of inedible, semi-drying oil, liquid at ordinary temperatures, solidifying at −15°C, containing oleostearic acid. Said to be the most perfect drying oil known[2] the oil is quicker drying than linseed oil, and is used as a wood preservative, for varnishes and paint oil, also as an illuminant, for soap making, and for waterproofing paper; in India for rubber substitutes and insulating masses. Seeds are moderately poisonous, and press cake is used as fertilizer. Seeds are thrown into rivers to stupefy fish.[16] Kernels when roasted and cooked, are considered edible; may be strung as candlenuts. Oil is painted on bottoms of small craft to protect against marine borers. Tung oil, applied to cotton bolls, stops boll weevils from eating them. Also prevents feeding by striped cucumber beetle.

Bark used on tumors in Japan. The oil is purgative and sometimes used like castor oil. Kernels are laxative, stimulant, and sudorific. The irritant oil is rubbed on the scalp as a hair stimulant. In Sumatra the pounded seeds, burned with charcoal, are applied round the navel for costiveness. In Malaya, the pulped kernel is used in poultices for headache, fevers, ulcers, and swollen joints. Philippinos apply oiled leaves to rheumatism.[16] In Java, the bark is used for bloody diarrhea or dysentery. Bark juice with coconut milk is used for sprue. Malayans apply boiled leaves to the temples for headache, and to the pubes for gonorrhea.[5]

The oil cake, containing ca 46.2% protein, 4.4% P_2O_5 and 2.0% K_2O, is said to be poisonous. A toxalbumin and HCN have been suggested. Bark contains ca. 4 to 6% tannin. Oil also contains glycerides of linolenic (ca. 6%), oleic (ca. 54%), and linoleic acids (ca. 32%). Per 100 g, the seed is reported to contain 626 calories, 7.0 g H_2O, 19.0 g protein, 63.0 g fat, 8.0 g total carbohydrate, 3.0 g ash, 80 mg Ca, 200 mg P, 2.0 mg Fe, 0 μg beta-carotene equivalent, 0.06 mg thiamine, and 0 mg ascorbic acid.

Toxicity — According to Kingsbury, *A. fordii* is about twice as toxic as *A. trisperma*, with *A. montana* and *moluccana* intermediate.[14] Contact with the latex can cause acute dermatitis,[6] as part of the tree contains a saponin and phytotoxin. Symptoms from eating the seed are severe stomach pain, vomiting, diarrhea, debility, slowed breathing, poor reflexes, possibly death. ''A single seed can cause severe poisoning.''[34]

To the physician — Hardin and Arena recommend gastric lavage or emesis; control convulsions with parenteral short-acting barbiturates.[34]

22. *ALNUS GLUTINOSA* (L.) Gaertn. (BETULACEAE) — European Alder, Black Alder

The wood, elastic and soft, fairly light and easily worked, is used for cigarboxes, pumps, and wooden carvings, shoes and slippers. The bark, used for tanning, imparts a hard red appearance to leather. The wood is also used in making molds for glass manufacture. The tree provides habitat and food for wildlife, watershed protection, and is used in environmental forestry. Alder leaves, placed in a corner of a room, are said to repel fleas.[47] With little ornamental value, it is recommended only for wet sites.

Leaves are decocted in folk remedies for cancer of the breast, duodenum, esophagus, face, pylorus, pancreas, rectum, throat, tongue, and uterus.[4] The bark and/or roots are used for cancers and inflammatory tumors of the throat. Inner bark boiled in vinegar is used to treat pediculosis and scabies. Also used as a dentifrice.[47] Reported to be alternative, astringent, detersive, diuretic, sudorific, tonic, and vermifuge, black alder is a folk remedy for cancer, fever, foot ailments, tumors, and worms.[32] The bark decoction is taken as a gargle for angina and pharyngitis, as an enema in hematachezia.[33] Rose mentions a concoction of alder bark, with dodder, endive, fennel, smallage, and succory to cleanse and strengthen the liver and spleen.[47]

The bark contains up to 20% tannin, a flavone glycoside of the hyperoside type, a reddish dye, emodin (?), alnulin ($C_{30}H_{50}O$), protoalnulin ($C_{30}H_{48}O$), phlobaphene, taraxerol, taraxerone, lupeol, beta-sitosterol, glutinone ($C_{30}H_{48}O$), and citrullin. The leaves contain alnusfoliendiolone, 3-beta-hydroxyglutin-5-cn, delta-amyrenone, taraxerol, beta-sitosterol, wax, and sugars.[33] Gibbs reports *l*-ornithine in the roots of this species, *l*-serine in the genus.

23. *ALOE BARBADENSIS* Mill. (LILIACEAE) — Mediterranean Aloe, Barbados Aloe, Curacao Aloe

Aloe drug is the dried juice obtained from the leaves, used medicinally as an anthelmintic, cathartic (acting on large intestine), emmenagogue, stomachic and vermifuge. Fresh juice is cathartic and refrigerant, useful in fevers, treatment of abrasions, burns and skin irritations; used in gels or with lanolin for skin ailments. In Egypt, the aqueous extract of "Aloe vera" is sold as Aloderm. Produced as a creme or lotion, aloderm promotes the healing of wounds, burns, radiodermatitis, sunburn, and ulcers. It helps acne vulgaris. In the U.S., it is finding its way into all sorts of cosmetics, emollients, lotions, shampoos, etc., sometimes in combination with jojoba. Fresh latex is supposed to inhibit the activity of Corynebacterium, Salmonella, Staphylococcus and Streptococcus spp.[81] Extracts are nematicidal.[82] The cooling mucilage is poulticed onto inflammations caused by x-ray and other radiation burns. Malays value the plant for keeping their hair in good condition. Used as a flavor ingredient in most major categories of food products, including alcoholic (bitters, liqueurs, vermouths) and nonalcoholic beverages, frozen dairy desserts, candy, baked goods, and gelatins and puddings. The seed and root oils and the root extract are more commonly used; average maximum use levels are low, usually below 0.01% except for the seed extract, which is reported to be 0.2% in alcoholic beverages.[29]

A plaster of the leaf or the leaf juice is said to be a folk remedy for tumors. A decoction of the root for a bath is said to remedy stomach cancer. Used also for acrochordons, condylomata, figs, warts, and other abnormal skin growths, and for cancers or tumors of the lip, anus, breast, larynx, liver, nose, prepuce, stomach, uterus, etc. I have received more letters about this as a folk remedy for skin cancer than any other species. In our computer index the plant is cited as being considered abortive, antiseptic, cathartic, cholagogue, decoagulant, demulcent, depurative, digestive, ecbolic, emmenagogue, emollient, insecticide, larvicide, laxative, purgative, stimulant, stomachic, tonic, and vermifuge, and is reported in folk remedies for amenorrhea, asthma, boil, bruise, burn, cancer of the stomach, cold, convulsion, cough, dyspepsia, eczema, erysipelas, excrescence, hemorrhage, hepatitis, hysteria, inflammation, internal ulcer, intestine, jaundice, lash, leukemia, pile, pregnancy, radiation burn, rectitis, scald, skin sore, stomach, sunburn, tumor, ulcer, veneral, weaning, and wounds.[32] Pulp used in menstrual suppressions, and roots used for colic. Parts of plant chewed to purify the blood. The pulp, said to possess wound-healing hormonal activity and "biogenic stimulators", is used for intestinal ailments, sore throat, and ulcers. In India, it is used for treating piles and rectal fissures. It produces pelvic congestion and is used for uterine disorders, generally with iron and carminatives.[1] A decoction of the root is used for stomach cancer.[4] Slukari hunters in Africa's Congo rub their bodies with the gel to eliminate the human scent, thereby making them less likely to disturb their prey. During cold epidemics in Lesotho, natives take a public bath in an infusion of Aloe latifolia. "Aloe dissolved in spirit, is used as a hair dye to stimulate hair growth." A sweet confection prepared from the pulp of the leaves is given in piles. Pulp mixed with honey and tumeric is recommended in coughs and colds. The juice of the leaves is useful in painful inflammations and chronic ulcers."[59] Without commenting on the contradiction, Rama[59] noted both antifertility and profertility reports. Alcoholic and water extracts showed 85% reduction in fertility with experimental rats. But one compound with aloe as the main ingredient suggested was useful in cases of infertility associated with irregular menstrual cycles. The same compound improved the fertility of patients with functional sterility and improved their menstrual functions. That study, not seen by me,[83] "revealed that Aloes Compound is the drug of choice in cases of functional sterility and disturbed menstrual function."[59]

Aloe contains several anthraquinone glycosides, as aloin (barbaloin), isobarbaloin, and emodin. Aloin is a lemon-yellow to dark-yellow powder, with a slight odor of aloe and an intense bitter taste, used as a cathartic or combined with other cathartics or antispasmodic drugs. The aloin content of *Aloe barbadensis* is ca. 30%, that of *A. perryi* ca. 25 to 28%, and that of *A. ferox* ca. 10%. Aloin consists mostly of barbaloin and other pentosides, resins, saponins, and other substances. On hydrolysis, barbaloin yields aloe-emodin and D-arabinose.[17] The resins consist of resinotannols with cinnamic or p-hydroxy-cinnamic (p-coumaric) acids, etc. In the pulp (96% water), researchers have found arabinose, galactose, glucose, mannose, and xylose.[17,29,59] Guar and locust bean gums are frequently mixed with aloe (which contains a similar glucomannan) to increase its viscosity and yield.[29]

Toxicity — According to Morton, aloe is contraindicated in pregnancy and in individuals afflicted with hemorrhoids. It is apt to cause kidney irritation.[17]

24. *ALOYSIA TRIPHYLLA* Britton (VERBENACEAE) — Lemon Verbena

Fresh leaves are used as a condiment in home cookery, and as a tea, especially in Latin America. Dried leaves make a delicious hot tea, greenish and with a lemon-lime aroma and flavor. The essential oil was once employed in perfumery, especially for colognes. It is still used in flavoring liqueurs.[42] The essential oil is acaricidal[1] and bactericidal. An alcoholic leaf extract is antibiotic against *Escherichia coli*, *Mycobacterium tuberculosis*, and *Staphylococcus aureus* (antimalarial tests were negative).[3] A 2% emulsion of the essential oil kills 90% of the mite, *Tetranychus telarius*, and the aphid, *Aphis gossypii*.

Reported to be antispasmodic, carminative, digestive, expectorant, nervine, pectoral, sedative, stimulant, stomachic, and tranquilizer, the lemon verbena is a folk remedy for fever, rabies, and spasm. Latin Americans take the tea for asthma, cold, colic, diarrhea, dyspepsia, and fever.[32,33,57] North Africans also take the tea for colds, fever, and spasms.[84]

On steam distillation, the leaves yield 0.42 to 0.65%, the stems 0.34% essential oil, rich in citral (35%), borneol, cineol, dipentene, geraniol, limonene, linalool, nerol,[42] citronal, verbesone, acetic acid,[57] alpha- and beta-caryophyllene, myrcenene, pyrollic- and isovalerianic-acid.[33]

25. *ANADENANTHERA COLUBRINA* (Vell.) Brenan (MIMOSACEAE) — Vilca, Anguo Branco

Hard wood used by Amerindians for many types of articles requiring strong wood. Seeds used as medicine but also added to chicha to enhance its intoxicating effect.[52]

Reported to be astringent, diuretic, laxative, and purgative, vilca is a folk remedy for fever, hemorrhage, infertility, lung ailments, and malaria.[32,52]

The active agents are N,N-dimethyltryptamine and related tryptamines.[54] Like bufotenine, closely related to serotonin. Active tryptamines apparently reach the brain from the nasal mucosa without a general circulation through the blood stream.

Toxicity — Classed as a narcotic hallucinogen.[54]

26. *ANANDENATHERA PEREGRINA* (L.) Speg. (MIMOSACEAE) — Niopo, Cohoba Yope, Yupa

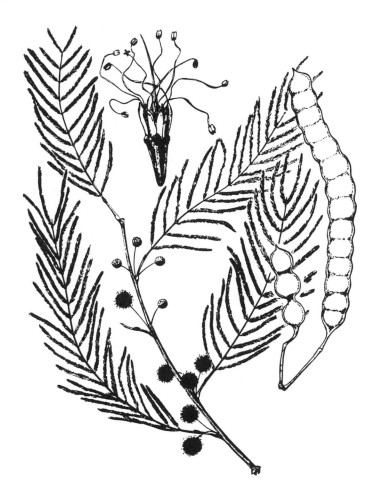

Source of a powder ingested as a snuff by Amerindians, who then became intoxicated. Sometimes the intoxicant is administered rectally. Hunters from one tribe not only took the clyster before going on the hunt, but administered it to dogs to clear the vision and render them more alert.[52] Bark used for tannin.

Reported to be hallucinogenic, intoxicant, narcotic, poison and psychedelic, niopo is a folk remedy, apparently of only ceremonial consequence.[32]

The active agents are *N,N*-dimethyltryptamine and related tryptamines[54] like bufotenine, closely related to serotonin. Polyphenolic tannins, catechol, and leucoanthocyanins, e.g., leucopelargonidol. Leaves contain homoorientine, orientine, saponaretine, and viterine.[33]

Toxicity — Deaths are reported from overdoses accompanied by wine,[52] suffocated with the snuff. Classed as a narcotic hallucinogen.[54]

27. *ANANAS COMOSUS* (L.) Merr. (BROMELIACEAE) — Pineapple

Pineapple is cultivated for fruit, used fresh, canned, frozen, or made into juices, syrups, or candied. Pineapple bran, the residue after juicing, is high in vitamin A, and is used in livestock feed. From the juice may be extracted citric acid, or on fermentation, alcohol. In Central America, the tender leaf shoots, hijo de pina, are sometimes cooked and eaten.[22] In the Philippines, a fine quality cloth is made from leaf fibers. Commercial bromelain is generally prepared from pineapple wastes. A mixture of several proteases, bromelain is used in meat tenderizers, in chill-proofing beer, manufacturing precooked cereals, in certain cosmetics, and in preparations to treat edema and inflammation. Bromelain is nematicidal.

The fruit, peel, or juice is used in folk remedies for corns, tumors, and warts.[4] Reported to be abortifacient, cholagogue, depurative, diaphoretic, digestive, discutient, diuretic, ecbolic, emmenagogue, estrogenic, hydragogue, intoxicant, laxative, parasiticide, purgative, refrigerant, styptic, and vermifuge, pineapple is a folk remedy for bladder ailments, fever, gonorrhea, hypochondria, kidney stones, parturition, scarlet fever, scurvy, sores, sprains, and venereal disease.[3,16,32,33] An antiedemic substance has been reported from the rhizome.[33] Many real or imagined pharmacological effects are attributed to bromelain: burn debridement, antiinflammatory action, smooth muscle relaxation, stimulation of muscle contractions, cancer prevention and remission, ulcer prevention, appetite inhibition, enhanced fat excretion, sinusitis relief.[29] Bromelain is given as an antiinflammatory agent following dental, gynecological, and general surgery, and to treat abscesses, contusions, hematomas, sprains, and ulcerations.[17] Pineapple juice from unripe fruits acts as a violent purgative, and is also anthelmintic and ecbolic.[16] Ripe fruit juice is diuretic, but in large doses may cause uterine contractions. Sweetened leaf decoction drunk for venereal diseases. Juice of the leaves consumed for constipation, hiccups, and worms. Juice of ripe fruit regarded as antiscorbutic, cholagogue, diaphoretic, refrigerant, and useful in jaundice. Young vegetative buds are used for respiratory ailments among Choco children.[60]

Per 100 g the fruit is reported to contain 47 to 52 calories, 85.3 to 87.0 g H_2O, 0.4 to 0.7 g protein, 0.2 to 0.3 g fat, 11.6 to 13.7 g total carbohydrate, 0.4 to 0.5 g fiber, 0.3 to 0.4 g ash, 17 to 18 mg Ca, 8 to 12 mg P, 0.5 mg Fe, 1 to 2 mg Na, 125 to 146 mg K, 32 to 42 μg beta-carotene equivalent, 0.06 to 0.08 mg thiamine, 0.03 to 0.04 mg riboflavin, 0.2 to 0.3 mg niacin, and 17 to 61 (to 96) mg ascorbic acid.[21] Cultivars may contain 1 to 5% citric acid (wild forms up to 8.6%), ca. 3.5% invert sugars, 7.5% saccharose, approaching 15% at maturity. Also reported are vanillin, methyl-*n*-propyl ketone, *n*-valerianic acid, isocapronic acid, acrylic acid, L(−)-malic acid, beta-methylthiopropionic acid methyl ester (and ethyl ester), 5-hydroxytryptamine, quinic acid-1,4-di-*p*-coumarin (my translation).[33] The aromatics from the essential oils of the fruit include methanol, ethanol, *n*-propanol, isobutanol, *n*-pentanol, ethyl acetate, ethyl-*n*-butyrat, methylisovalerianate, methyl-*n*-capronate, methyl-*n*-caprylate, *n*-amyl-*n*-capronate, ethyl lactate, methyl-beta-methylthiolpropionate, ethyl-beta-methylthiolpropionate, and diacetyl, acetone, formaldehyde, acetaldehyde, furfurol, and 5-hydroxy-2-methylfurfurol.[33] Steroid fractions of the lower leaves possess estrogenic activity.[33]

Toxicity — Workers who cut up pineapples have their fingerprints almost completely obliterated by pressure and the keratolytic effect of bromelain (calcium oxalate crystals and citric acid were excluded as the cause). The recurved hooks on the leaf margins can painfully injure one. Mitchell and Rook also restated earlier work on "pineapple estate pyosis" occurring in workers who gather the fruits, probably on acarus infestation with secondary bacterial infection. Angular stomatitis can result from eating the fruit. Ethyl acrylate, found in the fruits, produced sensitisation in 10 of 24 subjects "by a maximisation test." Ethyl acrylate is used in creams, detergents, food, lotions, perfumes, and soaps.[6] In "therapeutic doses", bromelain may cause nausea, vomiting, diarrhea, skin rash, and menorrhagia.[29] Watt and Breyer-Brandwijk[3] restate a report, unavailable to me, of unusual toxic symptoms following ingestion of the fruit, heart failure with cyanosis and ecchymoses, followed by collapse and coma and sometimes death.[3] Morton adds that unripe pineapple is poisonous, causing violent purgation.[17]

28. *ANAPHALIS MARGARITACEA* (L.) Benth. and Hook (ASTERACEAE) — Pearly
Everlasting

Sold occasionally as an herbal remedy in the U.S. Amerindians poulticed or bathed indolent
tumors with the herb.[4] Anodyne, astringent, expectorant, pectoral, and sedative, the herb
is recommended for boils, bronchitis, bruises, catarrh, cuts, diarrhea, dysentery, painful
swellings, pulmonary afflictions, and sprains.[2,33]

Essential oil, phytosterin, resin and tannin are reported.[27] Hagers Handbook reports that
tridecapentain-en, trans-dehydromatricariaester, and 5-clor-2 [octatriin-(2,4,6)-yliden]-5,6-
dihydro-2H-pyram occur in the root.[33]

Toxicity — The FDA classified this as an Herb of Undefined Safety.[62]

29. *ANDIRA ARAROBA* Aguain (FABACEAE) — Goa Powder, Araroba

Brazilian timber tree, used for bridges, construction, posts.[24] Goa powder was introduced by the English dermatologist Baimann-Squire in 1878 as a topical medicine for psoriasis.

Reported to be alterative, detergent, poison, taenifuge, and vermifuge, araroba is a folk remedy for alopecia, arthritis, dermatosis, psoriasis, rheumatism.[24,32,33] Used veterinarily for trichophytosis. As a salve the araroba is believed effective in eczema, herpes, pityriasis, psoriasis, ringworm, and other skin ailments.[33]

Hagers Handbook lists chrysophanolanthrone, emodiananthronmonomethylether, dehydroemo-dinanthranolmonomethyl ether, ararobinol, chrysaphanol, and em-dinmonomethylether, noting that chrysarobin is also the name of a resin $C_{15}H_{12}O_3$.[33] Mitchell and Rook[6] add that chrysarobin is the naturally occurring mixture derived from the wood, 30% of which is 1,8-dihydroxy-3-methyl-9 anthrone.[6]

Toxicity — Powder (or sawdust) is to be avoided as it inflames the mucous membranes. Dangerous for the eyes.[33] Internally, 200 mg can produce diarrhea, nausea, and nephritis.[33] Goa powder is irritant to the respiratory tract.[6] In industrial accidents, chrysophanic acid has produced conjunctivitis and keratitis.[6]

30. *ANDIRA INERMIS* HBK. (FABACEAE) — Cabbage Bark

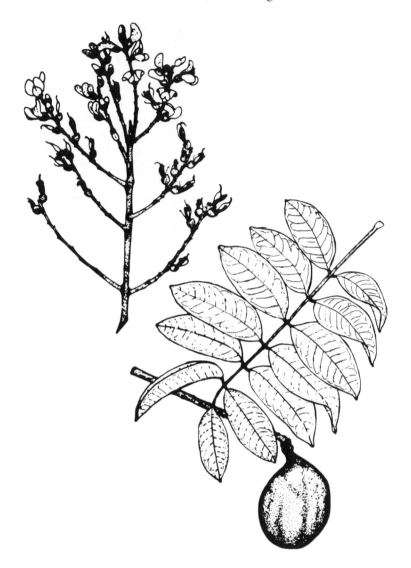

Timber tree, apparently fixing N, sometimes grown as a park tree. Wood is dark reddish-brown to black, with average density of 800 kg/m³, very resistant to attack by fungi and insects. It is used for boats, bridges, cabinets, canes, carts, cue sticks, and furniture. Bark used as a fish poison. Flowers serve as nectar source for bees, butterflies, and hummingbirds; classes as an excellent honey plant. Bats eat the fruits.

Reported to be anthelmintic, piscicidal, purgative, and vermifuge; is a folk remedy for eczema, fever, malaria, and yaws.[32,42] Cubans use the leaf decoction as an antidote to Comocladia poisoning. Trinidad natives apply the bark decoction for eczema and yaws, taking it as a vermifuge.[42]

Bark and seeds contain berberine and andirine (angelin or N-methyltyrosine). Wood contains an isoflavonoid related to dalbergiones and biochanin A.[6]

Toxicity — Narcotic;[2] large doses of bark or seed may cause violent nausea, possibly fatal. Antidotes suspected for overdoses include castor oil or lime juice.[2] Tree can cause dermatitis in wood workers.[6] Smoke from burning wood is irritating to the eyes.[42]

31. *ANEMONE PULSATILLA* L. (RANUNCULACEAE) — Pasque-Flower, Pulsatilla

Sometimes cultivated as an Easter-blooming flower. It is a popular rock garden plant in the U.S. Juice of the purple sepals gives a temporary green stain to paper, linen, and Easter eggs. Anemonin and protoanemonin are bactericidal.[11]

The leaf is used in folk remedies for warts, the plants for pterygium.[4] Reported to be alterative, antidotal, diuretic, emmenagogue, and expectorant, the pasque flower is a folk remedy for asthma, bronchitis, catarrh, diarrhea, gout, headache, hemicrania, migraine, neuralgia, ophthalmia, pertussis, rheumatism, and warts.[32] Homeopathically used for amenorrhea, bilious attacks, dysmenorrhea, dyspepsia, earache, gastritis, measles, neuralgia, otitis, rhinitis, toothache, and urticaria.[2,33]

Fresh herb contains ranunculin $C_{11}H_6O_8$ which on fermentation yields glucose and protoanemonin. Tannin, resin, and 0.19 to 0.75% saponin are also reported in the root and rhizome. Flowers contain delphinidin- and pelargonidin-glycosides. Fructose, and the acetate of beta-amyrin and beta-sitosterol, arabinose, galactose, glucose, rhamnose, and hederagenin ($C_{30}H_{48}O_4$) also occur in this or a closely related species.[33] Hartwell reports anemonin, active in the KB tumor system.[10]

Toxicity — Allergenic information has been summarized[6] indicating that patch tests (20 to 60 minutes) produced vescicular reactions in volunteers. Hyperpigmentation appeared at the test sites accompanied by conjunctivitis and nasal mucosa, due to the vapor.[6] Protoanemonin is an irritant toxic principle, potentially hazardous to livestock.[11]

32. *ANETHUM GRAVEOLENS* L. (APIACEAE) — Dill, Dill Seed, Garden Dill

Dill is used primarily as a condiment. Dried fruits (seeds) are used in pickles, soups, spiced beets, fish, and fish sauces, with eggs and in potato salads. Fresh leaves are used in salads, with cottage cheese, cream cheese, steaks, crops, avocado, cauliflower, green beans, squash, tomatoes and tomato soup, zucchini, and shrimp. Dried leaves, known as dill weed, are also used to season various foods. Oil from the seed is used chiefly as a scent in soaps and perfumes, and in the pickle industry. Weed oil, from the above-ground parts of the plant, is used in the food industry because of its characteristic dill herb smell and flavor. The essential oil showed inhibitory effects on *Bacillus anthracis, B. mycoides, B. pumilus, Escherichia coli, Pseudomonas mangiferae, Salmonella typhi, Sarcina lutea, Staphylococcus albus, S. aureus,* and *Xanthomonas campestris* but did not inhibit *Shigella* sp.

The seed, prepared in various manners, is a folk remedy for abdominal tumors, condylomata, indurations, and tumors (abdomen, anus, liver, mouth, stomach, throat). The flower, cooked in oil, is said to help tumors around the anus. A cataplasm of the leaf is said to cure grains and indolent tumors.[4] Dill oil is used for apostemes of the breast. Dill is considered balsamic, detersive, digestive, diuretic, lactagogue, laxative, narcotic, psychedelic, resolvent, seductive, stimulant, and stomachic.[32] Dillwater is used for children's ailments, as flatulence and indigestion. The plant figures also into folk remedies for bruises, colic, cough, dropsy, hemorrhoids, insomnia, jaundice, sclerosis, scurvy, sores, and stomachache.[32,33]

Dill is said to contain or yield camphene, carvone, dihydrocarvone, dillapiole, dipentene, isomyristicin, limonene, monoterpene, myristicin, and phellandrine. Embong et al.[85] compared the chemical composition of various dill seed oils grown in Alberta (Canada). They found that the oils examined contained D(+)-carvone (43.3 to 48.9%), D(+)-limonene (33.1 to 40.8%), D-alpha-phellandrene (4.3 to 9.7%), trans-anethol (0.1 to 0.7%), p. cymene (2.8 to 4.9%), D(+)-dihydrocarvone (1.4 to 2.5%), myrcene (0.7 to 1.2%), beta-phellandrene (1.4 to 2.5%), and alpha-pinene (0.2 to 0.3%). The dried residue left after oil extraction of the seed contains 16.8% fat and 15.1% protein and can be used for cattle feed. Per 100 g, dill seed is reported to contain 305 calories, 7.7 g H_2O, 16.0 g protein, 14.5 g fat, 55.2 g total carbohydrate, 21.1 g fiber, 6.7 g ash, 1,516 mg Ca, 277 mg P, 16.3 mg Fe, 256 mg Mg, 20 mg Na, 1,186 mg K, 5.2 mg Zn, 53 IU vit. A equivalent, 0.42 mg thiamine, and 0.28 mg riboflavin.[21] Of the fatty acids, there are 0.73 g saturated, 9.4 g monounsaturated, and 1.0 g polyunsaturated, plus 124 mg phytosterol but 0 cholesterol. There are 575 mg threonine, 767 mg isoleucine, 925 mg leucine, 1,038 mg lysine, 143 methionine, 670 mg phenylalanine, 1,120 mg valine, 1,263 mg arginine, and 320 mg histidine. Per 100 g, dill weed is reported to contain 253 calories, 7.3 g H_2O, 20.0 g protein, 4.4 g fat, 55.8 g total carbohydrate, 11.9 g fiber, 12.6 g ash, 1,784 mg Ca, 543 mg P, 48.8 mg Fe, 451 mg Mg, 208 mg Na, 3,308 mg K, 3.3 mg Zn, 0.42 mg thiamine, 0.28 mg riboflavin, 2.8 mg niacin, and 1.5 mg vit B_6.

Toxicity — Insects (fruit flies) exposed to parathion alone showed only 8% mortality, but those exposed to the same level of parathion and d-carvone (or 3 other dill compounds) showed 99% mortality. Can natural products react with pesticides to produce synergistic responses, harmful to humans? Dill is said to contain the alleged "psychotroph" myristicin, like several other umbels, e.g., *Carum, Levisticum, Oenanthe, Pastinaca, Petroselinum,* and *Peucedanum*. Like *Ammi, Apium, Daucus, Foeniculum, Heracleum, Pastinaca,* and *Peucedanum*, dill may be a photosensitizer and/or cause dermatitis. As with oil of fennel, in vivo amination of dill oil can result in a series of three dangerous hallucinogenic amphetamines.[54]

33. *ANGELICA ARCHANGELICA* L. (APIACEAE) — Angelica, Garden Angelica, European Angelica

Cultivated primarily as a spice, seeds are used in preparing vermouth, benedictine, chartreuse,[37] and as a flavoring in wines and perfumes. Roots are used with juniper berries in gin. Oil from the seed is used to flavor custards and bread. It is cultivated also as a fresh vegetable, or cooked and prepared like rhubarb. Leaves used to flavor fish dishes and rhubarb jams. Stems are candied and used on cakes and buns, or made into jams and jellies. Ripe fruits are used to make teas (not known to contain xanthine alkaloids). The roots and rhizomes have a pungent aromatic taste and are used commercially in medicines and confections. Roots have been used to flavor cigarette tobacco. Oil extracted from the seeds or roots (fresh roots yield 0.10 to 0.37%, dried roots, 0.35 to 1.00%), is used as an insect attractant (e.g., for Mediterranean fruit fly).

The tea made from the rhizome is a folk remedy for stomach cancer. Dietary intake is said to alleviate indolent tumors.[4] A decoction of the root is used for bronchial colds and indigestion. The plant juice is sometimes applied to dental caries. Medicinally, dried leaves and flowering tops are regarded as aromatic, carminative, diaphoretic, diuretic, and stimulant. Roots suggested to be abortifacient, carminative, diaphoretic, diuretic, emmenagogue, expectorant, spasmolytic, and stomachic. Angelica is useful for dyspepsia, enteritis, flatulence, gastritis, insomnia, meteorism, neuralgia, rheumatism, and ulcers.[33]

Roots and fruits contain several furocoumarins, e.g., angelicin, bergapten and xanthotoxin, as well as umbelliprenin and various phenols. The main constituent of the root essential oil is beta-phellandrene; others include alpha-pinene, borneol, osthenole, osthole, angelicin, methyl ethyl acetic acid, diacetyl, methanol, ethanol, and furfural. The most important aroma compound is a lactone of 15-hydroxypentadecanoic acid. The flavonoid archangelenone has recently been identified. Escher et al. identified five previously unknown phellandrene derivatives with a typical angelica smell from the root oil: 2-nitro-1,5-P-menthadiene, *cis* and *trans* 6-nitro-1(7), 2P menthadiene; *trans*-1(7)-5P menthadiene-2-yl-acetate, and 7-isopropyl-5-methyl-5-bicyclo(2.2.2.) octan-2-on.[86] The seed oil is said to contain 0.5% imperatorin, 0.1% bergaptene, 0.02% xanthotoxol, 0.04% umbelliprenin and a phenol. Phellandrene, methyl ethyl acetic acid, and hydroxymyristic acid are also reported.

Toxicity — Plant evokes photodermatitis (contains 5-methoxypsoralen and 8-methoxypsoralen). Confectioners have contracted recurrent vesicular dermatitis when gathering angelica.

34. *ANGELICA POLYMORPHA* Max. (APIACEAE) — Dong Quai

Some is now being imported to the U.S. as an herbal "remedy". It was concluded by Sung et al.[87] that there is some validity to the reported use of "Tang-Kuei" for the prevention and relief of allergic symptoms. Immunosuppressive activity was observed both by oral and i.p. administration of the root extract. To elucidate the mechanism of this suppression seems worthy of further investigation, since these results may ultimately lead to a novel approach for treating atopic diseases. Atopy, including hayfever, asthma, and atopic dermatitis, constitute those conditions in which individuals are sensitive to a variety of environmental substances.[87]

The root, considered analgesic, deobstruent, emmenagogue, sedative, is used for anemia, boils, constipation, dehydration, dysmenorrhea, injuries, metrorrhagia, rheumatism, and ulcers, in China for cancer,[4] hepatitis, sterility in the Orient. Elsewhere it is considered hematinic, sedative, tonic; used for angina, cancer, chills, diabetes, hypertension, nephritis; for anemia, amenorrhea, cancer, constipation, dysmenorrhea, headache, lumbago, menoxenia, and venous thrombosis. In China used to assure a healthy pregnancy and easy delivery.[41]

Essential oil from oriental angelica is said to contain *n*-butylphthalide, cadinene, carvacrol, *n*-dodecanol, isosafrole, linoleic acid, palmitic acid, safrole, sesquiterpene, sesquiterpenic alcohol and *n*-tetradecanol.[41]

Toxicity — Safrole is the carcinogen for which sassafras tea was banned.

35. *APIUM GRAVEOLENS* L. (APICEAE) — Celery

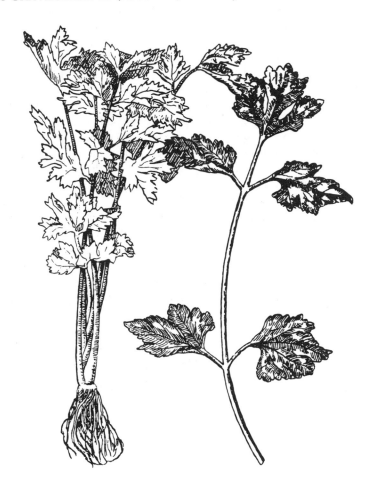

Blanched petioles of celery are eaten as a pre-dinner appetizer, as a vegetable in salads, as a relish with meats and fish, or cooked as a flavoring or vegetable in soups, stews, stuffing mixtures, and in pickles and sauces. Leaves are used as a garnish, and seeds for a condiment. Oil of celery, steam-distilled from the seeds, is used as a condiment. Oil of celery is principally used to flavor canned foods, soups, sauces, sausages, etc. It is also used in perfumery and in pharmaceuticals. An oleoresin distilled from the seeds constitutes a safer condiment. Antioxidant activity is reported from the seeds. Wild celery, considered to be poisonous, was used for centuries as a medicine to purify the blood. However, when celery is cultivated, the stalks become less bitter and more palatable. Clinically, celery juice has proven hypotensive in 14 or 16 hypertensive males taking 40 mℓ orally three times a day with honey or syrup.[29] Phthalides in celery seed oil are sedative in mice.

The leaf, prepared in various manners, is said to remedy hard tumors in the mamillae, felons, whitlows, cancerous ulcers, old tumors, corns, and inflammatory tumors.[4] The root, in syrup, is said to help indurations of the spleen and liver. An antidote prepared from the seed is said to alleviate indurations of the liver and spleen, tumors of the vulva, indurations of the uterus, and mammary tumors. The juice, prepared in various manners, is used for non-ulcerated cancer, indurations of the spleen and liver, lumps in the breast, tumors of the eye and stomach, cold tumors, and ulcerated cancer.[4] Decoction of the seed said to relieve lumbago and rheumatism. Plant has been used as emmenagogue and abortifacient. Root and leaf considered diuretic and aphrodisiac. The volatile oil produces contractions of the gravid and non-gravid uterus. The plant has been shown to possess hypoglycemic activity. Intravenous celery extracts are hypotensive in dogs and rabbits.[29] Roots in India are considered alterative, diuretic, and are given in anasarca and colic. Seeds are used for bronchitis, asthma, and for liver and spleen diseases. Oil of celery is antispasmodic, carminative, sedative, stimulant, and tonic. Seeds considered anodyne, carminative, emmenagogue, nervine, stimulant, stomachic, and tonic. Root given in anasarca and colic.

On steam distillation French seed yield 1.9 to 2.4% oil; Indian seed yield 2.14 to 2.50%. The essential oil contains the glucoside apiin. Celery seed oil contains ca. 60% limonene, 10 to 15% selenine, 1 to 3% sesquiterpene alcohols, sedanolide, sedanonic anhydride a phenol (perhaps guaiacol), and palmitic acid.[29] Seeds also contain myristicic acid, hydroxy-methoxypsoralen, and umbelliferone. Diuresis may be due to the presence of glycolic acid. Stalks may yield the photodynamically active principle bergapten. Per 100 g, the leaf is reported to contain 21 calories, 92.8 g H_2O, 1.4 g protein, 0.3 g fat, 4.2 total carbohydrate, 1.0 g fiber, 1.3 g ash, 62 mg Ca, 37 mg P, 2.5 mg Fe, 96 mg Na, 326 mg K, 1,040 μg beta-carotene equivalent, 0.06 mg thiamine, 0.07 mg riboflavin, 0.4 mg niacin, and 20 mg ascorbic acid. Garg et al.[88] report 6 coumarins in the seeds, seselin, isoimperatorin, osthenol, bergapten, isopimpinellin, and the novel apigravin. Per 100 g, celery seed is reported to contain 392 calories (1,641 kJ), 6.6 g H_2O, 18.1 g protein, 25.3 g fat, 41.4 g total carbohydrate, 11.8 g fiber, 9.3 g ash, 1,767 mg Ca, 547 mg P, 45 mg Fe, 440 mg Mg, 160 mg Na, 1,400 mg K, 6.9 mg Zn, 52 IU vitamin A, and 17.1 mg ascorbic acid. There are 60 mg phytosterol.[89]

Toxicity — Celery tops containing 3.2 to 8.7% nitrates (dry weight) have caused loss of dairy cattle in California. Celery workers may develop pruritic papulo-vesicular dermatitis. Fungus infected celery, containing 8-methoxypsoralen and 4,5′,8 trimethylpsoralen, are more liable to photosensitize, especially Caucasians. Oil GRAS (§ 182.10, 182.20).[29]

36. *APOCYNUM ANDROSAEMIFOLIUM* L. (APOCYNACEAE) — Bitterroot, Spreading Dogbane

Bark once used as a hemp substitute, for bags, fishing lines, linens, string, and twine. Flowers frequently imprison insects. The floss can be used as a cotton substitute or stuffing. Sugar, perhaps dangerous, can be obtained from the flowers. Plant has been suggested as an emergency rubber source. Dried latex makes a very flammable gum elastic.[46]

Used as an alternative in rheumatism, scrofula, and syphilis. Digitalis-like, apocynum slows the pulse and has strong action on the vaso-motor system. Powerful hydragogue, helpful in dropsy due to heart-failure. Has been called the "vegetable trocar in ascites of hepatic cirrhosis".[2] With his propensity to recommend relatively poisonous herbs, Kloss[44] describes bitterroot as good for arthritis, fever, malaria, mucositis, neuralgia, rheumatism, and typhoid, excellent for the bowels, digestion, kidneys, and liver, especially valuable in gallstones and "wonderful for diabetes". Frankly I fear the herb, considered cardiotonic, cathartic, diaphoretic, diuretic, emetic, expectorant, laxative, narcotic, poison, purgative, sudorific, and tonic. Some Indians believed that eating the boiled root would result in temporary sterility.[45] Fumes from burning dried roots over coals inhaled for headache. Boiled green fruits used for heart and kidney ailments. In his AmerIndian draft, Duke[43] notes that Cherokee used it to treat mange in their dogs. Chippewa and Ojibwa used the roots for headache, the Chippewa snuffing it, inhaling the fumes, or even applying the root to incisions in the temple. They also used the root for convulsions, dizziness, insanity, nervousness, palpitations, and twitching. Chippewa used dilute root tea for children's colds, and applied a stronger decoction topically for earache. Fox used root as a diuretic for dropsy, and the bark in a compound treatment for womb injuries and snakebite. Iroquois used for bloody fluxes. Ojibwa used the boiled root tea for headache, glossitis, and sore throat. Ojibwa also used the root decoction as a diuretic during pregnancy and applied boiled leaves for poisoning. The Potawatomi used the root or berry decoction as a cardiac and diuretic for heart and kidney ailments.[43]

Contains the glucoside apocynamarin and a bitter principle cymarin; apocynein, apocynin, volatile oils, fixed oils, and caoutchouc are also reported.

Toxicity — Dosage must be watched carefully; absorption rate may be unpredictable. Being rather irritant to the mucous membrane, it may cause catharsis and emesis. Not to be confused with or used for *A. cannabinum*. The cardioactive glycoside cymarin is responsible for toxic symptoms in livestock.[11] Sometimes held responsible for poisoning of young cattle and sheep, even when dry.[13] Certain resinoid fractions, administered orally, are said to have produced gastric disturbances, and death in a dog.[14]

37. *APOCYNUM CANNABINUM* L. (APOCYNACEAE) — Indian Hemp

This plant is an extremely poisonous cardiotonic drug. In California, the bark was used for making bags, cordage, fishing lines and nets, "linen", and twine. Indians also used the tough fibrous bark to make fishing nets, and some western Indians chewed dried bits of the latex. According to Morton[46] roasted seeds are eaten or made into meal in the southwest. Frankly I would be afraid to eat them. Honeybees gather the floral nectar which yields a light, good-quality honey. The decoction affords a permanent black or brown dye.[45] Recently, tobacco companies here in the U.S. have expressed interest in *Apocynum venenetum*, a Chinese medicinal species, which is also used as a tobacco additive in China.

The sap is used in folk remedies for condylomata and warts.[4] It has been used as a cardiac, cathartic, diaphoretic, diuretic, emetic, expectorant, febrifuge, hydragogue, laxative, masticatory, purgative, sudorific, tonic, and vermifuge. Since used by Amerindians in treating dropsy, it has been called dropsy weed. Blackfeet used root decoction as a laxative and for alopecia. Cherokee used the roots for asthma, Bright's disease, cough, dropsy, pertussis, pox, rheumatism, and uterine obstructions. Chickasaw and Choctaw swallow the juice from chewed roots for syphilis. Fox Indians used the root tea for ague, cold, dropsy, fever, headache, sore throat, etc. Potawatomi used the berry tea for heart ailments. Chippewa used boiled roots for heart palpitation. They snuffed powdered roots for head cold. Iroquois used for bloody fluxes. Menominee used the root infusion as a vermifuge. Meskwaki used the root for ague, dropsy, and womb ailments. Ojibwa used the plant in teas for pregnant women to keep the kidneys "free". Penobscot used the roots as a vermifuge. Sea Islanders used the roots for abortion.[43] Elsewhere the plant is used for alopecia, asthma, cardiac conditions, condylomata, dropsy, dysentery, dyspepsia, enteritis, fever, inflammation, malaria, pneumonia, warts, and worms.[32,33] South Carolina blacks steeped the roots in whiskey for colds, constipation, and fevers. The root decoction is administered by midwives to induce abortion. It has been prescribed as a quinine substitute in intermittent and remittent fevers.

Contains the cardioactive glycoside cymarin ($C_{30}H_{44}O_9$), apocannoside, cynocannoside, K-strophanthin, tannins, resin, 1.6% fatty oil, alpha-amyrin, lupeol, oleanolic acid, P-oxyacetophenone, saponins, androsterol, homoandrosterol, and harmalol.[33] The entire plant contains the glycosides cymarin and apocannoside, both of which have shown antitumor activity.[45]

Toxicity — Of the several resins and glycosides, some cardioactive, isolated from the plant, toxicity of some has been established in the lab. The toxic glycoside apocynamarin, markedly increases cats' blood pressure upon injection. Certain resinoid fractions administered orally produced gastric disturbances and death in a dog. Since grazing animals find dogbane distasteful, cases of poisoning are rare.[14] According to Hardin and Arena, "cases of poisoning in humans are not known".[34]

38. *AQUILEGIA VULGARIS* L. (RANUNCULACEAE) — Columbine

Attractive wildflower, more often met in cultivation today, with hay-scented flowers.

The leaf is used in folk remedies for stomach cancer, the seed for cancers of the breast or nose, and the plant for cold impostumes of the uterus.[4] Reported to be astringent, cyanogenetic, diaphoretic, diuretic, emmenagogue, narcotic, resolvent, and sudorific, columbine is a folk remedy for cancer, dysmenorrhea, hysteria, scurvy, and somnolency.[32] Spaniards ate the root, then fasted several days, to correct stones. Seed were taken with saffron in wine for hepatosis and jaundice. Leaves have been applied in sore throat and stomatosis.[2] Herbage used in homeopathy for female problems (Globus hystericus, climacteric nausea), insomnia, tremors, photosensitivity, audiosensitivity, and dysmenorrhea in young maidens. Seeds are used for buccal sores, dim vision, dysmenorrhea, eczema, fistulas, and jaundice.[33]

Herb contains delphinidin-3,5-diglucoside, traces of HCN. Seeds contain lipase, nitrylglycoside; seed oil contains the following acids: capronic-, caprylic-, caprinic- lauric-, myristic-, palmitic- (ca. 8%), palmitolic-, stearic-(ca. 1.9%), oleic-(ca. 6%), linoleic-(ca. 24%) and *trans*-5,*cis*-9,*cis*-12,octadecatrienic (ca. 60%).[33]

Toxicity — Children have "been poisoned by it when given in too large doses".[2] Plant said to be irritant.[6] Lewis and Elvin-Lewis report fatalities.[11]

39. *ARBUTUS UNEDO* L. (ERICACEAE) — Strawberry Tree, Arbutus

Beautiful evergreen shrub, whose "narcotic" fruit was once an item of diet. The young shoots were once recommended as winter fodder for goats, the rather insipid fruits eaten by blackbirds and thrushes. The wood makes good charcoal, and the bark is used for tannin. "In Spain, a sugar and spirit have been extracted from the fruit and a wine made from it in Corsica."[2] The mealy fruits are also used to make marmalades.[1] The strawberry tree is used in folk remedies for excrescences.[4] Reported to be antiseptic, astringent, intoxicant, narcotic, and tonic,[32] arbutus has proven useful in urinary infections.

Leaves contain ca. 2.7% arbutin, methylarbutin, hydroquinone, 0.2% unedoside ($C_{14}H_{20}O_9$) arbutoflavonol A ($C_{22}H_{16}O_{10}$), arbutoflavonol B ($C_{23}H_{18}O_{13}$), tannin, lupeol, myricetin, leu-codelphinidin, glycolic-, ellagic-, ursolic- and gallic-acids; nonacsanol-(15), triacontanol-(1), dotriacontanol-(1), lycopine, cryptoxanthine, violaxanthine. Bark contains up to 45% tannin.[33]

Toxicity — Narcotic.[2]

40. *ARCTIUM LAPPA* L. (ASTERACEAE) — Edible Burdock, Great Burdock, Lappa

Stalks, cut before flowering and stripped of their rinds, are eaten boiled like asparagus, or eaten like a salad with oil and vinegar. Formerly they were candied like angelica. New sprouts rising from roots in spring are eaten raw or cooked after peeling as a vegetable. Roots, containing about 45% inulin, are eaten boiled in salted water and topped with butter or sauce. Burdock is said to have antipyretic, antiseptic, antitumor, diaphoretic, diuretic, and hypoglycemia activity.[29]

Burdock, used as a folk-cancer remedy in places as far apart as Chile and China, India and Indiana, Canada and Russia, has been suggested for cancers, indurations or tumors of the breast, glands, intestine, knee, lip, liver, sinus, stomach, tongue, and uterus, as well as for corns and warts.[4] The root decoction is said to alleviate ulcerated, glandular, and white tumors.[4] Homeopaths prescribe the tincture of fresh root for acne, bunions, Dupuytren's contraction, eczema, serpeginosa, eruptions, glandular afflictions, gonorrhea, gout, impotence, leucorrhea, phosphaturia, rheumatism, ringworm, scrofula, sterility, ulcers, and uteral prolapse.[30] Tierra recommends burdock as a blood-purifier for arthritis, lumbago, rheumatism and sciatica.[28] Roots of the first year's growth are used medicinally as an alterative, antiphlegmatic, depurative, diaphoretic, diuretic, and stomachic. Kloss describes the herb as "excellent for canker sores, gonorrhea, gout, leprosy, rheumatism, sciatica, scrofula and syphilis."[44] They have been used to treat bladderstones, eczema, gallstones, gout, and skin afflictions. Tincture of seeds used for kidney diseases, acne, prurigo, and psoriasis. Orientals use the seed for abscesses, acne, constipation, dropsy, flatulence, flu, measles, scarlet fever, scrofula, smallpox, and snakebite.[16] The essential oil, said to promote hair growth, is believed to be alterative, apertif, carminative, depurative, diuretic, and sudorific. Roots said to be a certain remedy in many skin diseases, especially eczema. Tierra even calls them ''an excellent remedy for all skin diseases.''[28] Applied externally as a poultice, leaves are good for bruises, tumors, and gouty swellings. Bruised leaves used as a hysteria remedy. Orientals use the leaf decoction for rheumatism and vertigo.[16] Oily seeds said to restore the smoothness of healthy skin. Believed useful in dropsy and kidney ailments.[2] Root once used for venereal diseases. Used by Meskwaki Indians in labor. Shows experimental hypoglycemic activity.[11]

Leaves contain arctiol, dehydrofukinone, eremophilene, beta-eudesmol, fukinone, fukinanolide, and taraxasterol. Roots contain polyphenolic acids (caffeic, chlorogenic), gamma-guanidino-*n*-isovaleric, and propionic) arctic acid, and polyacetylenes (1,11-tridecadiene-3,5,7,9-tetrayne, 1,3,11-tridecatriene-5,7,9-triyne). Seeds contain 15 to 30% fixed oil, the glucoside arctiin, chlorogenic acid, two lignans (lappaol A, lappaol B), and a germacranolide. Raw roots are reported to contain per 100 g: 89 calories, 76.5% moisture, 2.5 g protein, 0.1 g fat, 20.1 g carbohydrate, 1.7 g fiber, 0.8 g ash, 50 mg calcium, 58 mg phosphorus, 1.2 mg iron, 30 mg sodium, 180 mg potassium, 0 μg beta-carotene equivalent, 0.25 mg thiamine, 0.08 mg riboflavin, 0.3 mg niacin and 2 mg ascorbic acid.[21] Plant contains inuline, and a bitter compound, lappatin; leaves and roots contain an essential oil; seeds contain a glucoside, arctin plus tannic acid. An alkaloid, arctiin, is reported as well as gum, mucilage, resin, and tannin. The root may contain up to 45% inulin, 0.07 to 0.18% essential oil, palmitic- and stearic-acids, phytosterol, 0.4 to 0.8% fat, sugar, resin, tannin, mucilage, stigmasterol, sitosterol, and caffeic acid. Of 14 polyacetylene compounds, 3 were bactericidal and fungicidal.

Toxicity — Rough hairs can irritate the skin and cornea. May cause contact dermatitis.[11] Tyler describes a reported poisoning, but the evidence he cites suggests that atropine was the culprit, not burdock.[37]

41. *ARCTOSTAPHYLOS UVA-URSI* (L.) Spreng. (ERICACEAE) — Bearberry

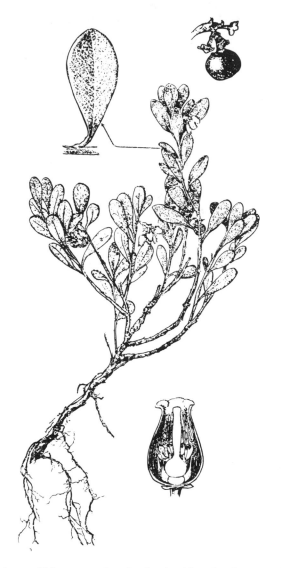

Berries mentioned as edible, roasted and mixed with animal grease.[45] Dried leaves used medicinally. Crude extracts have been suggested as effective in cancer chemotherapy.[11] Both uvaol and ursolic acid are active in the PS-125 tumor system.[10] Leaves once mixed with tobacco for smoking by Indians of Eastern North America;[45] leaves so smoked were said to intoxicate the Chippewa.[45] Leaves have been used for tanning leather in Russia and Sweden. In parts of Russia, the leaves are used to make kutai or Caucasian Tea. According to Tierra, bearberry tincture was routinely prescribed in many European hospitals following parturition to reduce hemorrhaging and help restore the womb to normal size.[28]

Once used as an astringent, to treat backache, bladder ailments, bronchitis, consumption, calculus, diabetes, diarrhea, dysentery, dysmenorrhea, dysuria, fever, hemorrhoids, hepatitis, menorrhagia, nephritis, pancreatitis, piles, polyuria, rheumatism, splenitis, ulcers, and urinary disorders.[32,45] Indians once used a tea made from the fruit for weight control. During its excretion, arbutin exercises an antiseptic effect on the urinary mucous membrane, leading to its use in cervical ulceration, cystitis, blennorrhea, enuresis, gallstones, gleet, gonorrhea, gout, incontinence, kidney stones, leucorrhea, metrorrhagia, nephritis, and urethritis.[32,48] It should be used for cystitis only when the urine is alcalic, not in acute catarrh of the bladder.[27] Users should avoid eating acid food. Homeopaths recommend bearberry in cystitis, dysuria, hematuria, incontinence, pyelitis, urethritis, and urogenital disorders.[30,33]

Contains 5 to 18% arbutin, a hydroquinone, methyl-arbutin, ericinol, ericolin, ursone (crystalline substance of resinous character) allantoin, gallic acid, malic acid, quinic acid, ursolic acid, ursone, ellagic acid, a yellow coloring principle resembling quercetin, uvaol; also hyperin, isoquercitin, myricetin, myricitrin, corilagin pyroside. Tannin content reported at 6 to 40%. Tyler suggests 15 to 20%, recommending that the leaves not be extracted with hot water, but left to steep in cold water for 12 to 24 hours to minimize the tannin content in the beverage.[37] Resins and waxes are also present.

Toxicity — Arbutin hydrolyzes to yield hydroquinone which is the active urinary disinfectant. Hydroquinone, however, is toxic causing collapse, convulsion, delirium, nausea, tinnitus, possibly even death.[29] Bearberry leaves should not be used without a doctor's advice; "excessive dosing and long term use can cause chronic impairment of the liver, especially in children."[45] "Large doses are oxytocic."[42]

42. *ARECA CATECHU* L. (ARECACEAE) — Betel-Nut Palm, Areca, Areca-Nut

Chief use of betel nut is as a breath sweetening masticatory, enjoyed for centuries by about one-tenth the human population. Slices of the nut, together with a little lime and other ingredients, are folded in a betel pepper leaf (*Piper betel*) and fastened with a clove. Sometimes nuts are ground up with other materials and carried about in a pouch similar to a tobacco pouch. Betel chewing is often an after-dinner or social affair. Chewing colors the saliva red and stains and teeth and gums black, eventually destroying the teeth. Some oriental dentists, however, believe the mastication may in fact benefit the teeth. The ground seeds are said to show up in certain European dentrifices. Used in the tanning industry. Extraction of areca nuts makes black and red dyes. Dried nuts are said to sweeten breath, strengthen the gums, and improve the appetite and taste. Husks are an important byproduct, used for insulating wool, boards, and for manufacturing furfural. Inoculated with yeast (*Saccharomyces cervisiae*), leaves used as a fermentation stimulant in industrial alcohol production. Large, tough, sheathing parts of leaf bases are used as a substitute for cardboard or strawboard for protecting packages; also used in Philippines for hats, insoles for slippers, book-covers; makes an excellent paper pulp.

The nut, in the form of ghees, powders, bolmes, or enemas, is said to help abdominal tumors.[4] Nuts are astringent, stomachic, stimulant and a powerful anthelmintic, especially in veterinary practice. An enema of the nut decoction works well in removing tapeworms. They are also considered digestive, emmenagogue, mitotic, and are recommended as cardiac, nervine, tonic, and as an astringent lotion for eyes, causing dilation of pupil; once used for glaucoma and malaria. Applied to ulcers, bleeding gums, and for urinary discharges. Burned and powdered nuts used as dentrifice in Europe. Husks used for beri-beri, dropsy, and sunstroke.[16] Once used as an antidote to abrin poisoning. An essential oil from the leaves is antiseptic and has been shown to be effective in bronchitis, diphtheria, laryngitis, and throat inflammations.[16] Mixed with sugar and coriander, the intoxicating nuts are given to induce labor in Iran.[75] Unripe fruits are cooling, laxative, and carminative.

Nuts contain the alkaloids, arecoline, arecaine, arecaidine, and arecolidine, isoguvacine, guvacine, and guvacoline; tannins (18%), (tannic- and gallic-acid), fats (14 to 18%; with glycerides of palmitic-, stearic-, myristic-, lauric-, oleic-, margaric-, nonadecanoid-, and heneicosanic-acids), choline, catechin, saccharose, mannan, galactan, other carbohydrates and proteins, and some Vitamin A.[1] Gum, mucilage, and resin are also reported. Per 100 g, the shoot is reported to contain 43 calories, 86.4 g H_2O, 3.3 g protein, 0.3 g fat, 9.0 g total carbohydrate, 1.0 g ash, 6 mg Ca, 89 mg P, and 2.0 mg Fe. Per 100 g, the mature seed is reported to contain 394 calories, 12.3 g H_2O, 6.0 g protein, 10.8 g fat, 69.4 g total carbohydrate, 15.9 g fiber, 1.5 g ash, 542 mg Ca, 63 mg P, 5.7 mg Fe, 76 mg Na, 446 mg K, 0.17 mg thiamine, 0.69 mg riboflavin, 0.6 mg niacin, and a trace of ascorbic acid.[21]

Toxicity — Classified by the FDA[62] as an Herb of Undefined Safety. Listed as a narcotic stimulant.[54] Excessive use of betel nut causes black teeth, inappetence, salivation, and general degeneration of the body. Arecaine is poisonous and affects respiration, the heart, increases peristalsis of intestines, and causes tetanic convulsions. According to Hager's Handbook, large doses (8 to 10 g seed) can be fatal; due to cardiac or respiratory failure.[33] Atropine is suggested as an antidote. On the other hand, Liu is quoted as saying 30 g is practically nontoxic.[16]

To the physician — Hardin and Arena[34] suggest 2 mg atropine subcutaneously following gastric lavage or emesis.

43. *ARGEMONE MEXICANA* L. (PAPAVERACEAE) — Prickly Poppy

The so-called chicalote opium, produced from the milky exudate of supposed hybrids between the opium and prickly poppies, has been discounted because hybrids have been discounted. Doubtless opium "heads" with a sense of taxonomy would not hesitate to smoke the latex from nonhybridized prickly poppies, just as some Iranians smoke the "bracteum" derived from Papaver bracteatum! Oil from the seeds is used as an illuminant and lubricant.[8]

Reported to be anodyne, aperient, carminative, cathartic, demulcent, depurative, diuretic, emetic, emmenagogue, expectorant, hemostat, laxative, narcotic, poison, purgative, sedative, stimulant, sudorific, and vulnerary, prickly poppy is a folk remedy for cancer, catarrh, chancre, cholecystitis, cold, colic, dysuria eruptions, excrescences, eyes, fever, headache, herpes, inflammation, itch, ophthalmia, parturition, pinkeye, puerperium, rheumatism, scabies, skin ailments, toothache, and warts.[4,16,32,42] According to Williams[22] the plant has been little used in Central America, except for the latex which is used as a purge in Costa Rica and as a cure for drunkenness in Guatemala. Curacao natives take young-leaf tea for asthma, cardosis, cough, and fever. Bahamians use it for hepatitis; Jamaicans for colds and fever; Yucatanese for hepatosis, jaundice, and pertussis; Venezuelans for cancer, epilepsy, and malaria. The latex is widely used for ophthalmia, ringworm, scabies, and warts.[42] Burmese use the seeds for dropsy. Chinese believe that the leaves can be used as contraceptive.[16]

According to Duke[91] the prickly poppy is reported to contain several alkaloids, allocryptopine, berberine, codeine (report doubtful), chelerythrine, coptisine, dihydrochelerythrine, dihydrosanguinarine, morphine (report doubtful), norargemonine, protopine (argemonine) romneine, and sanguinarine. Flowers contain isorhamnetin-3-glucoside and -3,7-diglucoside. Seeds contain 20 to 35% oil, with palmitic-, myristic-, linoleic-, physetolic- and ricinoleic acid.[33] L-glutamic acid (6% of defatted meal of oilseed cake) has been used to treat pediatric mental deficiencies.[41]

Toxicity — Narcotic, possibly hallucinogenic.[54] Seeds have caused many fatalities in poultry but are relished by quail. Where used as an accidental or intentional adulterant of other salubrious oilseeds, argemone seed oil causes edema and glaucoma.

To the physician — Hardin and Arena recommend emesis or gastric lavage and symptomatic treatment.[34]

44. *ARGYREIA NERVOSA* (Burm.) Bojer (CONVOLVULACEAE) — Wood Rose, Silver Morning Glory

For a while, Hawaii shipped the rose-like capsules to the U.S. and Europe. Since so many other Americans, in addition to poor Hawaiians, were abusing the seeds as an hallucinogen, an embargo was placed on continued shipment of the wood rose.[54] Still, they are frequently sold for use in dry flower arrangements.[51] Possibly more important to the counterculture as a herbal high,[51] induced, according to an anonymous writer, by 4 to 8 grams seed, chewed or taken in capsules. "At first you will feel weak and lethargic. If you have a sensitive stomach, you may get nauseated for about fifteen minutes. If so sip a little warm water or mint tea and allow yourself to vomit if necessary. Dramamine . . . may also help. After this has passed you will feel very relaxed and peaceful yet very aware. This state of bliss lasts for about three or four hours and is followed by a gradual descent to normality except that you will probably feel unusually relaxed and mellow for several days."[51] Sounds too good to be true! See toxicity!

Reported to be anodyne[33] in Haiti, rubefacient and vesicant in the Philippines.[6] Immature seeds contain lysergic acid amides, and, according to one anonymous author, innocuously small quantities of strychnine and other alkaloids.[51] Various species of Argyreia contain amides of lysergic acid, e.g., chanoclavine, ergine, ergonovine, and isoergine.[54]

Toxicity — Classed as a narcotic hallucinogen, packing a miserable hangover with blurred vision, constipation, nausea, physical inertia, and vertigo.[54]

45. *ARISAEMA TRIPHYLLUM* (L.) Schott. (ARACEAE) — Jack-in-the-Pulpit

Attractive woodland herb, sometimes grown as an ornamental curio. After much preparation the root was used as food by Amerindians as the so-called Indian turnip. The berries have been eaten raw but have a peppery taste.[56]

Indian turnip is a folk remedy for aphtha, asthma, bronchitis, cancer, colic, consumption, croup, cough, felons, headache, laryngitis, lungs, pertussis, polyps, rheumatism, ringworm, snakebite, sores, stomatitis, swellings, tuberculosis, and whitlows.[2,4,32] The species was widely used by the Amerindians. Algonquins mixed the root with wild cherry and snake root for cough and fever. The plant has been used as an expectorant, irritant, and diaphoretic. The Delaware used the root as a purgative for bowel compaints. The powdered root is listed as an oral contraceptive. Cherokee applied the raw root to the head or temples for headache. They also used the plant for boils, colds, consumption, cough, ringworm, scald, scrofula, sore throat and tetterworm. Pawnee sprinkled powdered root onto the affected areas. Mohegans and Penobscot boiled the root as a liniment for sore joints and muscles. Chippewa, Menominee, Ojibwa, and Iroquois used the plant for sore eyes. Mohegans gargled a dilute root tea for sore throat. Choctaw drank such tea for blood tonic. Fox used the root to reduce swelling following snakebite; Mohegans used the root as poultice for pain. Osage and Shawnee used for cough and malaria. Meskwaki used the chopped root for the swellings of snakebite.[43]

Contains a violently acrid principle (perhaps calcium oxalate), 10 to 17% starch, albumen, gum, sugar, lignin, and salts of potassium and calcium.

Toxicity — Taken internally, this plant causes violent gastro-enteritis which may end in death . . . Antidote — strong tea and stimulants.[2] According to Kingsbury, mortality in humans and livestock is not known, although mortality has been produced experimentally.[14] The needlelike crystals of calcium oxalate become imbedded in the mucous membrane if ingested, causing intense irritation and burning sensation.

46. *ARISTOLOCHIA SERPENTARIA* L. (ARISTOLOCHIACEAE) — Snakeroot, Virginia Snakeroot

Rather popular recently among herbal medicines, selling for more than $125 per kilogram in 1983.

Boiled with rum in Virginia as a cancer cure; also used for leukemia and skin cancer.[4] Root infusion gargled for malignant sore throat.[17] Stimulant, tonic roots promote perspiration. Used by Choctaw Indians to alleviate stomach pain.[8] Like other snakeroots, used for snakebites.[8] In small doses said to promote appetite, tone up digestive organs, and serve as a cardiac stimulant. Useful as an adjunct to quinine in malaria. In large doses, promotes arterial action, diaphoresis, and frequently diuresis. Useful also in amenorrhea, bilious fever, dyspepsia, erysipelas, pneumonia, smallpox, sore throat, typhoid, and typhus.[19] Steeped to make an aperient febrifuge. Used also as alterative, emmenagogue, expectorant, gastric stimulant, stomachic, sudorific and vermifuge. According to Steinmetz, snakeroot is a sedative for the genital nerves.[27] Gums, resins, and tannins are also reported.

Contains borneol, serpentarin, aristolactone, and aristolochine; also an essential oil. Mix, Guinaudeau and Shamma[92] list also aristored ($C_{19}H_{17}O_6N$) and aristolochic acid I ($C_{17}H_{11}O_7N$).

Toxicity — In too large a dose, aristolochine violently irritates the gastro-intestinal tract and kidneys, even causing coma and death from respiratory paralysis.

47. *ARNICA MONTANA* L. (ASTERACEAE) — Mountain Tobacco, Leopard's-bane

Plant yields an oil used in perfumery. Applied to the scalp, the eczema-inducing tincture is said to make the hair grow. Dried flowering heads are bactericidal (hence once used for abrasions, gunshot wounds, etc.). Mosca and Costanzo[93] report on fungicidal activity. Germany alone markets more than 100 drug preparations containing arnica extract.[37]

A plaster made from the root is said to be a folk remedy for tumors, and the tincture derived from the flower is said to help liver, stomach, and intestinal cancer in addition to tumors in general.[4] Dried flower-heads are used medicinally as anodyne, expectorant, hemostat, febrifuge, irritant, nervine, resolvent, sedative, stimulant, tonic, and vulnerary. The leaves are regarded as febrifuge, sternutatory, and vulnerary. The rootstock is said to be discutient, diuretic, febrifuge, stimulant, and vulnerary. Arnica is used externally in many home remedies for aches, acne, boils, bruises, skin rashes, sprains, and wounds; and internally as a stimulant and febrifuge. Russians use the herb to promote bile, reduce cholesterol, stimulate the CNS; and to stop bleeding, boils, and inflammation of the genitals, as well as to strengthen a weak heart.[30] Homeopathic tinctures are used for abscess, apoplexy, back pain, baldness, bed sores, black eye, boil, brain, bronchitis, bruises, carbuncles, chest ailments, chorea, corns, cramps, diabetes, diarrhea, dysentery, ecchymosis, epilepsy, excoriation, exhaustion, eye, fever, halitosis, meningitis, miscarriage, nose, paralysis, pleurodynia, purpura, rheumatism, rhinosis, sea-sickness, sore feet, sore nipples, splenalgia, sprains, stings, suppuration, taste, thirst, trauma, tumors, voice, whooping cough, and wounds.[30,94]

Arnica root contains a bitter compound, arnicin, and an essential oil, which are obtained from the rhizomes and roots. Arnisterol (arnidiol), anthoxanthine, and tannin are also reported. A yellow-green fluorescent flavonic pigment is obtained from the flowers, rhizomes, and roots, as well as an essential oil (1% in flowers, 0.5 to 1.5% in roots), and bitter principle. Flowers also contain flavones, an adrenalin-like pressor, a cardiotonic substance, choline, betaine, inulin, luteine, phytosterol, trimethylamine, and xanthophyll.

Scolimoside, cynaroside, apigenin-3-O-beta-D-gluco-pyranoside, isoquercetrin, astragalin, scopoletin, and umbelliferene are reported from the flowers. The flavone causes fall in blood pressure. Hager's Handbook adds astragalin, caffeic acid, helenien, isoquercitrin, beta-lactucerol, beta-sitosterol, taraxasterol, thymol, thymolmethylether, xanthophyllepoxide, and zeaxanthin.[33] Dihydrohelenalin, helenalin, and their esters are said to be analgesic, antibiotic, and antiinflammatory.[37] Helenalin is allergenic and may induce contact dermatitis.

Toxicity — Roots and flowerheads are irritant, producing erythema and swellings of the part to which applied, or may incite a real eczema.[6] Essential oil is a skin irritant. As Tyler puts it "studies of the effects . . . on the heart and circulatory system of small animals have verified the folly of using the drug internally for self medication. Not only did it exhibit a toxic action on the heart, but in addition, it caused very large increases in blood pressure. More recent studies have confirmed arnica's cardiac toxicity."[37] Hermann, Willuhm, and Hausen[95] in describing the new pseudoguaianolide, helenalinmethacrylate, note that like helenalinacetate, it could be considered as a sensitizer for the often-observed contact dermatitis. "The FDA classifies this as an unsafe herb, because the aqueous and alcoholic extracts of the plant contain, besides choline, two unidentified substances which can produce violent toxic gastroenteritis, nervous disturbances, change in pulse rate, intense muscular weakness, collapse and death."[62]

To the physician — Hardin and Arena[34] suggest "Gastric lavage or emesis; symptomatic." Morton notes that the herb is still used in homeopathy, though when given internally has caused fatal poisoning.[17] Steinmetz claims "the whole plant is toxic."[27]

48. *ARTEMISIA ABROTANUM* L. (ASTERACEAE) — Southernwood, Old Man

Primarily cultivated for its pleasant scented foliage which has a pleasant taste and tonic properties, resembling those of wormwood. Young leaves and shoots sometimes used for flavoring cakes and other culinary preparations. In England, an herb tea is made from leaves, devoid of xanthine alkaloids. Plant is of minor importance as a food, but more common as an ornamental, kept trimmed as hedge. Has been used to protect clothes from moths. Branches said to dye wool a deep yellow. Potted plants make good house plants and lend themselves as bonsai subjects. Ointment made with its ashes said to promote the growth of beards in young men. Boiled with barley meal, it is applied to remove pimples and wheals.[2,47]

The leaf, with olive oil, is used for indurations of the uterus. In Europe the plant is used for chilblains, indurations of the liver, spleen, stomach, and uvula, for scirrhus, tumors, wens, and whitlows.[4] Aromatic foliage said to deter drowsiness. Like wormwood, southernwood is said to ward off infections. In older herbals, bruised southernwood was said to help draw out splinters. Suggested for acne, anemia, frostbite, gout, marasmus, piles, pleuritis, scrofula, and tuberculosis.[33] In the Middle Ages, it was taken in wine as an antidote to poison.[47] Ashes said to remedy alopecia. Medicinally, considered anthelmintic, antiseptic, astringent, aromatic, bitter, choleretic, cordial, deobstruent, detergent, emmenagogue, expectorant, febrifuge, nervine, stimulant, tonic, vermifuge, and vulnerary.

Volatile oil consists chiefly of absinthol. Abrotanin, a bitter principle, adenine, adenosine, calycanthosides, choline, essential oil, guanines, isofraxidine, malates, nitrates of potassium, resin, scopoletin, scopolin, succinic acid, tannin, and umbelliferone are also reported. The herb contains ca. 0.2% essential oil (0.18% after flower, 0.19% in flower, 0.20% before flower).[33]

49. *ARTEMISIA ABSINTHIUM* L. (ABSTERACEAE) — Wormwood, Ajenjo, Absinthium

Wormwood is grown for the leaves, stems, and flower-heads, which are used as a vermouth ingredient. Formerly the principal ingredient in the liqueur absinth. Brewers once added the fruits to hops to make beer more heady.[45] Oil is a fragrance component in creams, detergents, lotions, perfumes, and soap. Crushed in vinegar and applied to the body, it repels fleas and flies.[45] Twigs serve as a moth repellent. Once used in granaries to ward off weevils, and as a strewing herb to repel fleas, wormwood does contain a repellent essential oil. Mice are said not to eat books written with ink in which wormwood was boiled.[45] The filaricidal herb has been suggested as an additive to sheep diet. Fresh branches are placed to repel mosquitos.[42] The essential oil kills house flies.[96] Stuffed into pillows for cats and dogs it repels fleas.[47] The tea is said to prevent seasickness.[45]

The herbage or its juices are used for various types of cancers and indurations of the breast, foot, larynx, liver, sinax, spleen, stomach, testicles, tongue, and uterus. The essential oil of absinth is used in medical and veterinary liniments. *Poison,* it is said to be abortifacient, antiseptic, and narcotic. Taken internally, absinthe is considered anthelmintic, deobstruent, depurative, digestive, diuretic, discutient, emmenagogue, febrifugal, lactagogue, stomachic, sudorific, tonic, and vermifuge. It expels worms in people and animals, but is considered dangerous. It has shown up in folk remedies for arthritis, bruises, diarrhea, dropsy, dysmenorrhea, dyspepsia, gout, gravel, hepatitis, inappetence, itch, jaundice, malaria, neuralgia, orchitis, rheumatism, splenosis and sprains.[45,47]

The essential oil (up to 1.7%) contains phellandrene, pinene, thujone (3 to 12%), thujyl alcohol, thujyl acetate, thujyl isovalerate, bisabolene, thujyl palmitate, camphene, cadinene, nerol, and azulene (chamazulene,3,6-dihydrochamazulene, 5,6-dihydrochamazulene). Formic and salicyclic acids occur in the saponification lyes of wormwood oil.[7] The herb also contains bitter glucosides absinthin, absinthic acid, anabsinthin, astabsin, artametin, succinic acid, together with tannin, resin, starch, malates, and nitrates of potassium and other salts.[2] Lactones include arabsin, artabin, and ketopelenolide (a germacranolide). The bitter taste is due to absinthin and anabsinthin. Since it damages the brain, absinthe liqueur was outlawed in France in 1915. Some authors propose that the mind-altering is due to thujone reacting with the same receptor sites in the brain as those which interact with THC. Thujone and THC have similar functional groups allowing them to "fit" a common receptor site without changing orientations or relative positions; they also can undergo similar oxidative reactions.[37] Inulobiose, ascorbic acid, carotene, pipecolic acid and 3,7-dioxabicyclo (3,3,0)-octane.

Toxicity — Flowers may cause a scarlatiniform eruption in sensitized people.[6] Tyler recounts an ancient medical report suggesting quite properly that 15 g of the volatile oil can cause convulsions and unconsciousness in humans.[37] Morton recounts a singular event where the "daily ingestion of Italian vermouth is suspected as a causative factor in a case of esophageal cancer."[42] Millspaugh relates that the effects of absinthe drinkers are: derangement of the digestive organs, intense thirst, restlessness, vertigo, tingling in the ears, tremblings in the arms, hands, and legs, numbness of the extremities, loss of muscular power, delirium, loss of intellect, general paralysis, and death.[19] The FDA classifies this as an unsafe herb containing "a volatile oil which is an active narcotic poison. Oil of wormwood is in absinthe, a liqueur, which can produce absinthism."[62] Habitual use or large doses cause convulsions, insomnia, nausea, nightmare, restlessness, tremors, and vertigo. Thujone-free derivatives have been approved for food use by the FDA (§ 172.510).[29]

50. *ARTEMISIA DRACUNCULUS* L. (ASTERACEAE) — Tarragon

Tender leaves used fresh or dried for flavoring, mostly in vinegars, salads, and pickles. Also used in soups, stews, cream sauces, with eggs, poultry,meats, and fish. Dried leafy tips used in perfumes, toilet waters and confectionery. Essential oil, known as estragon, disappears when herb is dried, so plants are sometimes lifted in autumn and replanted under glass to provide a continuous supply of fresh green leaves. Oil of tarragon, known also as oil of estragon, is obtained from the leaves. Commercially, the oil, which is almost exclusively available from France, is used in both flavor and fragrance preparations. As the total production is probably not more than 2 metric tonnes (MT), the amount of this oil used is rather self-limiting.[31]

Regarded as anthelmintic, apertif, carminative, diuretic, emmenagogue, hypnotic, refrigerant, stomachic, and vermifuge, tarragon has been prescribed for swellings, toothaches, and tumors.[28] Ground with oil, tarragon is used for tumors of the sinews. The herb is considered aperient, stomachic, stimulant, and febrifuge, and is used to cure bites and stings of venomous animals and for bites of mad dogs. The root was once used to treat toothache.

Herb contains about 0.3% essential oil composed of methyl chavicol (60 to 70%) and some *p*-methoxycinnamic aldehyde, a bitter substance. The following are reported in tarragon oil: alpha-pinene, beta-pinene, camphene, limonene, *cis*-ocimene, *trans*-ocimene, methyl chavicol, *p*-methoxy cinnamaldehyde, delta-4-carene, alpha-phellandrene, and linalool. The oil contains limonene, *cis*- and *trans*- ocimene, *cis*- and *trans*-alloocimene, linalool, butyric acid, methyl chavicol (major constituent), geraniol, 1,2-dimethoxy-4-allyl benzene, and eugenol. Recently the isocoumarin artemedinol has been discovered. Tarragon herbage contains rutin, which was first isolated from rue and is best known for its ability to decrease capillary permeability and fragility. It is also said to be a cancer-preventative, i.e., inhibiting tumor formation on mouse skin by the carcinogen benzopyrene, and protective against irradiation damage. Rutin is also useful to counteract edema, atherogenesis, thrombogenesis, inflammation, spasms, and hypertension. It was once official in the U.S. for arteriosclerosis, hypertension, diabetes, and allergic manifestations. It is suggested that it may be useful for stroke prevention.[29]

This has been identified as a promising source of rutin, cited[29] as an anticancer compound, and quercetin-3-glyco-gallactoside. Perhaps its greatest value is in Sauce Bernaise.[97] Per 100 g ground tarragon is reported to contain 295 calories (1,237 kJ), 7.7 g H_2O, 22.8 g protein, 7.2 g fat, 50.2 g total carbohydrate, 7.4 g fiber, 12.0 g ash, 1,139 mg Ca, 313 mg P, 32 mg Fe, 347 mg Mg, 62 mg Na, 3,020 mg K, 4 mg Zn, 4,200 IU Vitamin A, 0.25 mg thiamine, 1.34 mg riboflavin, 9.0 mg niacin, and 81 mg phytosterols, with no cholesterol.[31,89]

Toxicity — Estragole, the main constituent of the essential oil, is reported to produce tumors in mice.[29]

51. *ARTEMISIA VULGARIS* L. (ASTERACEAE) — Mugwort, Carline Thistle

Plants grown mainly for their ornamental foliage. Japanese eat boiled young shoots in spring. Used for flavoring rice cakes in the Orient, geese and pork in Europe, and dumplings elsewhere. Leaves used as tobacco substitute. Plants are somewhat poisonous to some livestock. Sheep enjoy the herbage and roots. Said to be good for poultry and turkeys. In China, the twigs are fashioned into ropes, burned to keep mosquitos away. Said to be smoked as an opium substitute in Sumatra. Experimentally hypoglycemia. For the avante-garde to experience interesting divinational dreams, Rose recommends stuffing the pillow with a half kilo and sleeping on it.[47] Parvati suggests mugwort tea with honey to induce clairvoyance and trances.[48]

The seed, soaked in distilled water, is used for old cancers.[4] Plant used for cancerous lesions, indurations, scirrhus, especially of spleen, stomach and uterus, and wens and whit-lows. Herb considered anthelmintic, antiepileptic, antiseptic, antispasmodic, choleretic, dia-phoretic, digestive, diuretic, emmenagogue, expectorant, nervine, spasmolytic, stimulant, stomachic, and tonic. Roots used as antiseptic and tonic. Infusions of leaves and flowering tops administered in nervous and spasmodic affections, in asthma and diseases of the brain. Leaves smoked to alleviate asthma.[16] Fresh herb used as an opium antidote.[47] In Java, poultices of the herb are applied to old sores, scurvy, and other skin afflictions. Used as an emmenagogue in the Philippines. Amerindians used the leaf decoction for bladder ailments, bronchitis, cold, colic, dysmenorrhea, epilepsy, fever, gout, hysteria, kidney ailments, poisoning, rheumatism, sciatica, worms, and wounds. Fresh plant juice is said to ease the itch of poison ivy.[47] Russians use the herb for abortifacient, amenorrhea, bladder-stones, cold, convulsions, cramps, diuretic, dysmenorrhea, epilepsy, fear, fever, gallstones, gastritis, inflammation, kidney stones, labor, nervousness, neurasthenia, rickets, sores, tuberculosis, and wounds. Homeopaths prescribe a tincture of the fresh root for catalepsy, chorea, convulsions, epilepsy, hydrocephalus, hysteria, somnambulism, and worms.[30] For headache, Rose recommends sticking a leaf up your nose.[47]

Plants yield 0.03 to 0.2% of volatile oil used as a larvicide and weak insecticide. Cineole is the major constituent, with quebrachitol, tauremisin, sitosterol, tetracosanol, fernenol, thujone, alpha-amyrin, stigmasterol, beta-sitosterol, and alpha- and beta-pinene.[33] The oil contains thujone. Per 100 g, the leaf is reported to contain 35 calories, 87.3 g H_2O, 5.2 g protein, 0.8 g fat, 4.5 g total carbohydrate, 3.4 g fiber, 2.2 g ash, 82 mg Ca, 40 mg P, 1.5 mg Fe, 2140 μg beta-carotene equivalent, 0.15 mg thiamine, 0.16 mg riboflavin, 3.0 mg niacin, and 72 mg ascorbic acid. On a zero-moisture basis, the tops contain 11.4% crude protein, 6.0% ether extract, 33.5% crude fiber, 6.3% ash, and 42.8% N-free extract, 0.86% Ca, 0.19% P, 2.2% K, and 0.27% Mg. Inulin, resin, tannin, and the bitter principle artemisin are also reported. Overdoses of the drug cause pain, spasms, and other disturbances.

Toxicity — In large doses, the plant is said to be toxic. Thujone can cause epileptic spasms.[16]

52. *ASCLEPIAS SYRIACA* L. (ASCLEPIADACEAE) — Common Milkweed

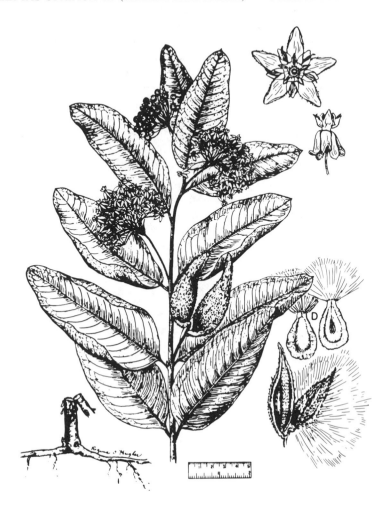

Gaertner speaks of milkweed as "the greatest underachiever among plants. Its potential appears great, yet until now it has never been continuously processed for commercial purposes."[99] The latex has been suggested as a suitable replacement for chicle in chewing gum rather than for rubber in tires. Two types of fiber are obtainable, the long, quite strong but brittle bast fiber and the seed hairs. Pulp from the fiber yields a good paper. During World War II, the seed hairs, being rather springy, light, and waterproof, were used to replace kapok in life jackets. Flowers are reported to be a source of sugar and have been used, with sugar and lemon, to make wine. Gaertner says, "Milkweed makes an attractive pot herb in many forms: first as a young shoot, then as unopened buds, and finally as pods, while these are still young, and the seeds are not yet differentiated."[99] I have tried all these, discarding my first "potlikker", and found them quite palatable. Some Indian groups dried the flower buds in summer for use in winter soups.[45] I have not tried the "boiled roots" which Kloss[44] states "taste similar to asparagus." I suspect this is a typo for shoots. But Parvati was kind, or blind, enough to carry it forward.[48]

The leaves and/or latex are used in folk remedies for cancer, tumors, and warts.[4] Reported to be alterative, anodyne, cathartic, cicatrisant, diaphoretic, diuretic, emetic, emmenagogue, expectorant, laxative, and nervine, milkweed is a folk remedy for asthma, bronchitis, cancer, catarrh, cough, dropsy, dysentery, dyspepsia, fever, gallstones, gonorrhea, moles, pleurisy, pneumonia, rheumatism, ringworm, scrofula, sores, tumors, ulcers, warts, and wounds.[42,44,45] One reported Mohawk antifertility concoction contained milkweed and jack-in-the-pulpit, both considered contraceptive. Dried and pulverized, a fistful of milkweed and three Arisaema rhizomes were infused in a pint of water for 20 minutes. The infusion was drunk, a cupful an hour, to induce temporary sterility.[45] Cherokee used the plant for backache, dropsy, gravel, mastitis, venereal diseases, and warts.[100] In homeopathy, the rhizome is used as an anti-edemic and emmenagogue, in dropsy and dysmenorrhea.[33]

The latex contains 0.1 to 1.5% caoutchouc, 16 to 17% dry matter, and 1.23% ash.[33] Also they report the digitalis-like mixture of alpha- and beta-asclepiadin, the antitumor beta-sitosterol, and alpha- and beta-amyrin and its acetate, dextrose and wax. The seed oil contains 4% palmitic-, 1% stearic-, 15% oleic-, 15% 11-octadecanoic-, 53% linoleic-, 1% linolenic-, 10% 9-hexadecenic-, and 2% 9,12-hexadecadenic-acids. Condurangin has also been reported from the seed, with at least 9 active cardenolides, among them uzarigenin, desglucouzarin, syriogenin, syriobioside; also xysmalogenin. The sprouts, eaten like asparagus (e.g., among Yugoslavs), contain asclepiadin, nicotine, beta-sitosterol, alpha- and beta-amyrin, and tannin.

53. *ATROPA BELLA-DONNA* L. (SOLANACEAE) — Belladonna, Deadly Nightshade

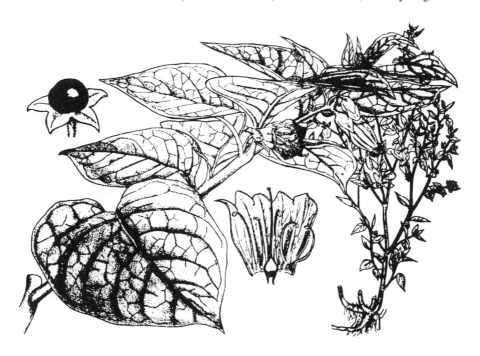

The drug obtained from roots and leaves is considered antiasthmatic, antispasmodic, anodyne, diuretic, febrifuge, mydriatic, narcotic, nervine, and sedative; and is used to check secretions. It is found in both prescription and over-the-counter drugs like antispasmodics for asthma and pertussis, cold and hayfever, collyriums, laxatives, liniments (for myalgia, neuralgia, rheumatism, sciatica), and sedatives. Used to treat Parkinson's disease.[29] Atropine and its ℓ-racemer hyoscyamine are used as preanesthetic agents to check secretion in throat and respiratory passages. Used as an antidote to depressant poisons such as chloral hydrate, muscarine, and opium. It relieves spasms and soothes muscles, for which it is given in gastritis, pancreatitis, chronic urethritis, biliary colic, and similar conditions. Hyoscine (5 to 11% of total alkaloids) is used as a truth serum. It is even said to afford protection against violent shock. Blocking central cholinergic mechanisms with scopolamine induces an amnesia in young monkeys and humans that uniquely resembles that occurring naturally in aged monkeys and humans. In 1973 atropine occurred in 1.5% of U.S. prescriptions, hyoscyamine in 0.75%, and scopolamine in 0.66%, belladonna extracts in 0.68%.[101] Compounds of atropine, ℓ-hyoscyamine and/or scopolamine occur in nearly 30% of the antispasmodic prescriptions in the U.S., 15% of the antidiarrheal, 5% of antacid prescriptions, and more than 75% of mydriatic prescriptions.[101]

The root, prepared in various manners, is used for inflammatory tumors, tumors of the rectum, other tumors, and cancer. The leaf is used for mammary and tongue cancers, nodosities of the breast, uterine cancer, carcinoma, running cancers, scirrhosity of the breast, and other cancers. The whole plant crushed is said to help hot tumors and cancer.[4] At one time, a poultice of the leaves was believed to cure cancer. Small doses allay cardiac palpitation.[2] Belladonna preparations are used for asthma, bladderstones, brachycardia, bronchitis, convulsions, epilepsy, gastric ulcers, gout, kidney stones, neuralgia, night sweats, Parkinson's disease, rheumatism, spermatorrhea, and whooping cough. The roots are applied externally for neuralgia, gout, rheumatic afflictions, and sciatica.[2] Atropine inhibits the multiplication of enveloped viruses. Kilmer[102] describes his personal encounter with mydriasis induced by belladonna. Homeopathically used for scarlet fever.

Roots and leaves contain ℓ-hyoscyamine, ($\pm$)-hyoscyamine (atropine) and ℓ-scopolamine (hyoscine) (roots also include apoatropine, belladonnine, and cuscohygrine), which in overdoses can be fatal in man.[11] Tannin is also reported. Pyridine, succinic acid, leucatropic acid, *N*-methylpyrrholine, asparagin, methylkaempferol, choline, phytosterol, beta-methylaesculetin (scopoletin), scopolin, and 7-methylquercitin have also been reported in the leaves, and hellaridine in Greek roots.

Toxicity — Sap of the plant can cause dermatitis. People handling the berries can develop vesiculo-pustular eruptions on the face with disorders of visual accommotion.[6] Belladonna can be treated with moderate doses of morphine, following evacuation via emetics.

To the physician — Hardin and Arena[34] recommend "gastric lavage (4% tannic acid solution) or emesis; pilocarpine or physostigmine for dry mouth and visual disturbance." Properly classified by the FDA as an unsafe herb.[62] Children have been fatally poisoned by eating "as many as three fruits";[17] Hager's Handbook says 3 to 4 or 10 to 20 berries.[33] Symptoms of poisoning include psychomotor unrest and excitation (sometimes sexual), euphoria, cramps, urge to be active, to dance, intense disturbance, ataxy, ranting and raving, hallucinations, shouting. Rose adds that belladonna poisoning manifests itself within 15 minutes of ingestion by dryness of the mouth, burning throat, dilated pupils, intense thirst, double-vision, giddiness, burning in the stomach, nausea, hallucinations, rambling talk, and a feeble rapid pulse. Her experience may be close to first-hand. "Two friends ate belladonna berries and described feelings of total relaxation, of floating about, of well-being and lightness. One friend hallucinated and stayed high for two hours and the other had to be hospitalized."[47]

54. *BANISTERIOPSIS CAAPI* (Spruce ex Griseb.) Morton (MALPIGHIACEAE) — Ayahuasca, Caapi

Used mostly as a mind altering drug, in healing, and divination, ayahuasca has quite an array of folk uses: to learn the location and plans of enemies, as a prewar indulgence for protective spirit, in tribal religious experiences, to tell if wives were unfaithful, to prophesy the future, for fun or aphrodisia, and to determine the cause of and cure for a disease.[103] Among Colombian Indians, where the plant is called yaje, "The drinking of yaje represents a return to the maternal womb, to the source and origin of all things. The partaker 'sees' all the tribal divinities; the creation of the universe, of the first human beings, and of the animals; and the establishment of the social order, especially regarding the law of exogamy. The Indians claim to see not only abstract designs but also the figures of people and animals, such as jaguars, alligators, snakes, and turtles, in complex mythological scenes. Having had this experience, the individual is firmly convinced of the verity of his religious beliefs. But the return to the womb is also an acceleration of time, and is equivalent to death. Upon drinking yaje, the individual 'dies', but later revives in a state of great wisdom. At the same time, the hallucinatory experience is symbolically a sexual act, essentially incestuous, in which he returns momentarily to the mythical stage of Creation."[107] The Indian, "tripping" on caapi, does not "trip" in the woods. Emboden notes that an Indian may run through the forest at night, under the influence, without stumbling. "The vision is remarkably clear and the footing sure . . . One of the usual conditions of the ayahuasca intoxication is augmentation of vision resulting in brilliant ornamentation, unusually perceptive night vision, (perhaps the military should investigate this), illusion of rapid size changes in people and objects (microscopy and macroscopy) and color perceived in shades of blues and violets."[54]

Reported to be hallucinogenic, narcotic, psychedelic, and purgative, ayahuasca is a folk remedy in psychotherapy.[32]

The bark contains ca. 2.2%, the wood ca. 1.1% of the alkaloids harmine (banisterin, telepathin, yagein) ($C_{13}H_{12}N_2O$) and harmaline $C_{13}H_{14}N_2O$, and 0.66% saponin. Harmic amide, *N,N*-dimethyl-tryptamine, *N*-methyl tryptamine, 5-methoxy-*N,N*-dimethyltryptamine, and 5-hydroxy-*N,N*-dimethyltryptamine, acetyl norhamine, ketotetrahydro norharmine are also reported. The principle alkaloids harmine and harmaline are responsible for the psychedelic effects reported. Five milligrams of harmine produces measurable sexual activity. Harmaline and harmalol have also shown sexual stimulation. Ayahuasca is as powerful as LSD.[103] Before World War II, harmine was being studied for the control of parkinsonianism.[105] Harmine is also used with paralysis (agitans) and encephalitis (lethargiac).[33] Injection of harmine in dogs induces trembling, especially of the eyes. On my first trip to the tropics, I was much impressed with the similarities between flowers on Oncidium and those of malpighiaceae. Thus I was almost amused to see Kawanishi, Uhara, and Hashimoto[106] report the orchidaceous bases shihunine and dihydro-shihunine from caapi. They also isolate six beta-carboline alkaloids harmine-*N*-oxide, harmic acid methylester, harmalinic acid, harmic amide, acetyl norharmine, and ketotetrahydronorharmine.

Toxicity — The first physical manifestations are pallor, profuse sweating, and tumors. Other manifestations are mydriasis, salivation, nausea, and defecation. Classed as a narcotic hallucinogen.[54] Garcia Barriga mentions fatalities due to the drug.[57]

55. *BAROSMA BETULINA* (Berg.) Bartl. and Wendl. f. (RUTACEAE) — Short-Leaved Mountain Buchu, Honey Buchu, Buchu

It is sometimes used as an ornamental plant, with a prolonged flowering period, aromatic serrate birch-like leaves and bushy habit. Used as a tincture in brandy, buchu Brandy appears throughout South Africa as a regular stock item upon the hotel bar shelves. It is also used for herb teas and herbal imbrocations. Buchu wine is a white sweet wine with 1% buchu extract and 1.5% of an extract of *Artemisia afra*.[3] The two extracts blend well with the wine, although retaining the titillating buchu flavor. The essential oil is antiseptic and diuretic. It is used in alcoholic and nonalcoholic beverage, frozen dairy desserts, baked goods, candies, condiments, gelatins, puddings, and relishes, with levels rarely more than 15 ppm.[29] It is also used in an artificial black currant flavor (FDA approval for food use, § 172.510).[29]

Reported to be carminative, diuretic, stimulant, and stomachic, it is a folk remedy for calculus, catarrh, cholecystitis, cholera, cystitis, hematuria, prostatitis, rheumatism, and urethritis.[29,32,33] Buchu has "been used for almost every disease which afflicts mankind".[3] Buchu brandy is highly regarded for maladies of the kidney and urinary tract, and is locally applied to bruises and rheumatic pains. The infusion has been used, chiefly in Africa, for gout and rheumatism, in chronic catarrh of the bladder, calculus, cholera, hematuria, prostatitis, and urethritis. As a liniment, buchu tea or brandy is used for rheumatism.[3] Leaves used medicinally for the buchu (boochoo) drug of commerce. Infusion of leaves used to stimulate perspiration, in rheumatism and gout, cholera, and dropsy. Most common local use in South Africa is as a stomachic tonic.

The leaf contains 1 to 3.5% essential oil of buchu, composed mainly of ℓ-pulegone, isopulegone, diosphenol, $\downarrow$-diosphenol, ℓ-isomenthone, d-menthone, d-limonene, piperitone expoxide 8-mercapto-p-menthan-3-one, 8-acetylthio-p-methan-3-one, diosmin, rutin, quercetrin-3,7-glucoside, resins, and mucilage.[29,33] On cooling, the essential oil deposits 15 to 30% of a stearoptene, barosma camphor, or diosphenol $C_{10}H_{15}O_2$. The eleaoptene, left after separation of the diosphenol, yields a hydrocarbon $C_{10}H_{18}$ or $C_{10}H_{16}$ and a ketone, probably ℓ-menthone, $C_{10}H_{18}O$. Limonene, dipentene, and thiamine are reported from the leaves.[3]

56. *BERBERIS VULGARIS* L. (BERBERIDACEAE) — European Barberry

Small red berries are used for jellies, e.g., *d'epine vinette*, a fresh jam made from a seedless cultivar. The fresh juice is sometimes used as a substitute for lemon juice. The wood and bark is a source of a yellow dye and is used for dying leather and wool. Hard wood used for mosaic, toothpicks, and turnery.

Used for cancers or tumors of the liver, neck, or stomach, and lumps of the liver.[4] Amerindians used the plant for the blood, coughs, consumption, debility, heartburn, inappetence, kidney ailments, rheumatism, sores, and ulcers. Barberry has also been recommended for arthritis, bile, bronchitis, catarrh, cholera, diarrhea, fever, flux, gout, heartburn, hepatitis, itch, jaundice, lumbago, rheumatism, ringworm, scabs, scrofula, splenitis, stomatitis, tetter, tuberculosis, typhus, and ulcers. Russians have used barberry for inflammation, metrorrhagia, hemorrhage, gall-bladder problem, hypertension, to increase the bile, and to regulate the female genitals. Mongolians use it for mucal disorders.[16] Homeopaths recommend the root-bark tincture for arthritis, biliary colic, bilious attack, bladder infections, calculus, duodenal catarrh, dysmenorrhea, dysuria, fever, fistula, gallstones, gravel, herpes, irritation, jaundice, knee pains, leucorrhea, liver, lumbago, neuralgia, ophthalmia, oxaluria, piles, polyps, renal colic, rheumatism, sacral pain, side pain, spermatic cords, spleen, tumors, and vaginismus.[30,33] Consumption of the fruits is believed to relieve itch and other skin ailments. In Iran, the berries are chewed to sweeten the breath and to promote general health. Bark of stem and root is considered alterative, antiseptic, cholagogue, diuretic, purgative, and tonic. Leaves are said to be laxative, the berries diuretic and expectorant. Berberine is a broad spectrum antibiotic for bacteria and protozoa.[11]

Contains several alkaloids in the roots: berbamine, berberine, berberubine, bervulcine. Chelidonic-, citric-, malic-, and tartaric-acids are also reported with resin and tannin, columbamine, jatrorrhizine, oxyacanthine, and palmatine. Berberine sulfate has shown activity in B1, RB, and PS[13] tumor systems. Oxyacanthine is active in the KB system.

Toxicity — Those who handle the wood may develop colic and diarrhea.[6] Symptoms of poisoning include: daze, epistaxis, nausea, diarrhea, renal stimulation, nephritis. Local anaesthesia and hyperpigmentation can follow intracutaneous injection of berberine.[6]

57. *BLIGHIA SAPIDA* Koenig (SAPINDACEAE) — Akee Apples, Seso Vegetal, Ackee

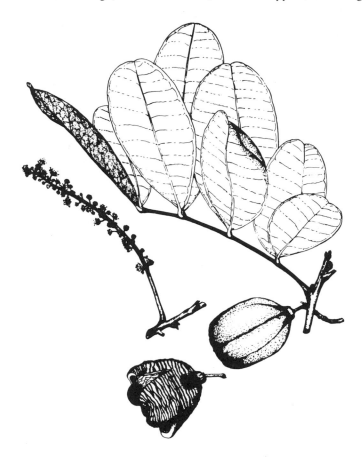

The walnut-like aril of this fruit is eaten only when exactly ripe, and may be eaten raw, roasted, or in soups. In the West Indies it is fried in butter or oil and served with salted fish, considered a delicacy known as "Codfish- or Saltfish-and-akees". Europeans like it roasted as a dessert nut. The fragrant flowers are the source of perfumed water, used as a cosmetic by natives in Africa.[3] Fruits give a good lather when rubbed in water and are used in washing clothes to fix colors. Rind used as a fish poison.[56] Dried husks and seeds are burned and their ashes used to make soap; ashes are rich in potash of a good quality. Sapwood is light-colored; and heartwood reddish-brown; timber is light in weight, hard, durable, suitable for chairs, beds, oars, boxes, and charcoal, and is used for piles and building materials; it is said to be resistant to termites. Frequently grown as an ornamental shade tree, suitable for avenues.

Colombians and Cubans use the leaf (and bark) decoctions as stomachic; Jamaicans for colds and pyorrhea. Ground arils are used in folk remedies for dysentery and fever.[42] Described as anodyne, antidote, antiemetic, poison and stimulant, akee is used in folk medicine for conjunctivitis, edema, epilepsy, migraine, ophthalmia, orchitis, smallpox, sores, tumors, ulcers, yaws, and yellow-fever.[32] Bark-pulp is used as a liniment for edemas and intercostal pains. Seasoned with Guinea-grains and ginger it is eaten to allay orchitis pains. Pulp of leafy tips is applied to cure migraine, and the juice of crushed leaves is used as eye-drops for conjunctivitis and ophthalmia.[107] A paste made of crushed leaves is applied to ulcers in yaws. Jamaicans rub boiled leaves on as an anodyne, using the tea for colds, salted as a mouthwash for pyorrhea.[72] Cataplasms of the fruits are applied to tumors of the mammary glands and testicles.[4] Arils are cooked in hot ashes to make vulnerary poultices.[42]

Per 100 g, the aril is reported to contain 196 calories, 69.2 g H_2O, 5.0 g protein, 20.0 g fat, 4.6 g total carbohydrates, 1.6 g fiber, 1.2 g ash, 40 mg Ca, 16 mg P, 2.7 mg Fe, 555 µg β-carotene equivalent, 0.13 mg thiamine, 0.14 mg riboflavin, 1.4 mg niacin, and 26 mg ascorbic acid. Per 100 g, the mature seed is reported to contain 180 calories, 71.5 g H_2O, 4.2 g protein, 17.4 g fat, 5.5 g total carbohydrate, 0.3 g fiber, 1.4 g ash, 37 mg Ca, 70 mg P, 1.7 mg Fe, 0.07 mg thiamine, 0.17 mg riboflavin, 1.6 mg niacin, and 78 mg ascorbic acid.

Toxicity — Pounded fruits are used in Africa (Sierra Leone, Nigeria, and Ghana) as a fish poison. Two poisonous substances occur in the raphe between the aril and the seed, one of which is hypoglycin A. Contains potentially useful hypoglycemic agents (Cyclopropanoid amino acids, hypoglycin A, hypoglycin B) which are too toxic for use as insulin substitute. Akee poisoning has been reported to be fatal in two hours. Hundreds of Jamaicans have lost their lives to akee. The bitter reddish raphe should be picked out as the fruit opens on the tree. Damaged, unripe, or fallen fruits should on no account be eaten. Symptoms include nausea and vomiting, then a quiescent period with drowsiness and sleep; then 3 or 4 hours later intense vomiting followed by convulsions, coma and death.

To the physician — Immediate induction of vomiting or gastric lavage, intravenous glucose for hypoglycemia, and other symptomic treatment.[56]

58. *BORAGO OFFICINALIS* L. (BORAGINACEAE) — Borage, Beebread, Bee Plant

Fresh leaves have a cucumber flavor which adds a cooling accent to salads, pickles, iced beverages, herb teas, and other vegetables. In Europe young leaves are used as pot herb, but older leaves are not palatable. Flowers may be floated in cups of wine, fruit juice, or gin and sugar, or they may be candied, or put in jams, jellies, and syrups, and for garnishing. Americans have been adding them to salads lately (!). Dried flowers used in potpourri. In America plants are grown mainly as a bee plant and as an ornamental. A refreshing tea, made from the flowers and dried stems and consumed in Europe and the Mediterranean countries, has been well known since the Middle Ages.[11] Gibbons gives recipes for borage candy, jam, jelly, and syrup.[39] The flowers in wine are considered aphrodisiac in Lebanon.[23]

The leaves and/or flowers are used in folk remedies for cancer (of breast or face), corns, scleroses, and tumors.[4] Reportedly demulcent, depurative, diaphoretic, emollient, febrifuge, lactogogue, laxative, nervine, pectoral, and sudorific, borage is a folk remedy for bladder-stone, bronchitis, corns, cough, cramps, fever, kidney stones, pain, sore throat, swellings, and urogenital ailments.[32] Lebanese use the infusion for colds, fevers, jaundice, and nephrosis, consciously adding borage to salads to "clean the blood" and as a healthy tonic.[23] They apply macerated leaves to cuts and eczema. I quote a letter from an herbal hippy friend of mine, "A weak tea of the flowers is supposed to calm the nerves, but I am hyperactive, and they are almost like amphetamines to me. It is interesting once or twice, but watch out for hepatic problems if used regularly"(!).[108] Leaves and flowers in wine are supposed to alleviate boredom, melancholy, and sadness (maybe it is the wine), often bringing on forgetfulness. In early herbals, recommended for consumption, fever, itch, jaundice, rheumatism, ringworm, snakebite, and sore throat. French use the flower infusion for feverish colds,[47] ashes boiled in mead for sore throat. Promoting kidney activity, borage is recommended for feverish catarrhs. Poultices used for inflammatory swellings. For ulcers of the mouth and throat, ashes are used in gargles. Fresh herb is made into a collyrium for conjunctivitis and a poultice for inflammation. Raw leaves sometimes used to "purify the blood." Borage is considered a cordial with exhilarating effect, pectoral, and aperient. Latin Americans take the mucilaginous juice as a sudorific and emollient in eruptive and bilious fevers and liver ailments, the flowers as demulcent, diuretic, and emollient. Argentines take the seed, powdered in white wine, as a lactagogue. Venezuelans take the flower infusion as a sudorific, the leaf and root decoction for diarrhea and fever. Brazilians take the flower decoction for bronchitis, cardosis, colds, coughs, edema, and nephrosis. Colombians use the dilute solution for conjunctivitis. Mexicans take the leaf infusion for bronchitis and eruptive fevers.[42]

Plants said to be good organic source of Ca and K. Said to emit popping sparks when burned, due to KNO_3. Contains up to 30% mucilage with glucose, galactose, and arabinose (after hydrolysis), pentoses, resin, (−)-bornesit, cyanogenic materials, acetic-, lactic-, and malic-acids, 0.003% vitamin C. Tannins may be as high as 3%, 1.5 to 2.2% silicic acid. Seedlings, like many others, contain high concentrations of allantoin.

Toxicity — The short bristly hairs on the stem may be irritating. Although I have heard yet unpublished rumors that the plant contains lasiocarpine, a carcinogen from comfrey, I would like to quote Tyler, "One need not be concerned about the relative safety of consuming borage."[37]

59. *BOWIEA VOLUBILIS* Harv. (LILIACEAE) — Climbing Onion, Zulu Potato

Poisonous plant, probably with medical potential. It has been estimated that the tuber has 30 times the intensity of activity of digitalis, the flowers 60 times. A German patent has been issued for a glucosidal preparation for human cardiac use.[3] Grown in greenhouses as curiosities with their twining aphyllous stems and large exposed bulbs.[36] Zulu males use the tuber to prepare a love attractant.[3]

Reported to be cardiotonic, poison, purgative, the climbing onion is a folk remedy for constipation, dropsy, and infertility.[33] Transvaal Africans use the tuber decoction as a collyrium. Zulu use a hot water infusion of the outer leaves of the bulbs for ascites.[3]

Contains three highly active cardiac glycosides, bovocides A, B, and C, all about as toxic as ouabain. As much as 10% sinistrin is reported. Nabigenin, isolated from the bulb, is a mixture largely composed of scilliglaucosidin; hellebrigenin is also reported. Other complex compounds and their derivatives occur.[3,33]

Toxicity — African medicine men have killed their patients with overdoses of the bulbs. Vomiting and purging were followed by death in 3 hours to 3 days. However, death may occur in a matter of minutes.[3]

60. *BRUNFELSIA UNIFLORUS* (Pohl) D. Don (SOLANACEAE) — Manaca, Manacán

Quite ornamental, the common name manacán is attributed to the most beautiful girl of the Tupi Indian tribe of Brazil and transferred to the most beautiful flower of the forest. *Brunfelsia uniflora* is the most important medicinal species, as this is the species of choice in the drug trade as manaca root. Root extracts were once used in arrow poisons. Readers are referred to Tim Plowman's interesting and detailed account.[109] It was employed by the Tupi both for magical and medicinal purposes. A perfume is extracted from the fragrant flowers with ether. R. P. Iyers' research reported in the Chemical Marketing Reporter[110] identified "hopeanine" as an active antiinflammatory agent (yield rate of only 4 mg hopeanine from one kilogram of powdered root).

Scraped bark is a strong purgative. The large root is said to stimulate the lymphatic system. Useful in syphilis, it has been called vegetable mercury. Regarded abortifacient, alterative, anesthetic, diaphoretic, diuretic, emmenagogue, hypertensive, hypothermal, laxative, narcotic, and poison; manaca is used in folk remedies for arthritis, rheumatism, scrofula, and syphilis.[32,33] Leaves are poulticed onto eczema, skin disorders, and syphilitic ulcers.

Contains the alkaloid manacine $C_{22}H_{32}N_2O_{10}$ (0.08% in the bark) and manaceine $C_{15}H_{26}N_2O_9$ aesculetin, possibly gelsemic acid, and 1.3% starch. Manacine seems to act opposite to atropine which inhibits rather than stimulates glandular secretions. Mandragorine has also been reported. The items previously mentioned, gelseminic acid and aesuletin, were probably scopoletin (6-methoxy-7-hydroxycoumarin) which occurs in all parts of the plant. Extracts show marked antiinflammatory activity, probably paralleling Dr. Iyer's hopeanine.[110]

Toxicity — Excessive doses are poisonous, causing excessive salivation, vertigo, general anesthesia, partial paralysis of the face, swollen tongue, and turbid vision. Even in small doses manacine induces strong muscular tremors and epileptiform cramps, lowered temperature, and death due to respiratory paralysis in experimental animals.

61. *BUXUS SEMPERVIRENS* L. (BUXACEAE) — Boxwood

Ornamental hedge, very popular, since Caesar's time, for topiary. Wood hard, heavy, used for musical instruments, pipes, wearing better than many metals. Decoction of leaves and bark once promoted as hair growth stimulant; boiled with lye and used to tint the hair auburn. Boxwood is said to have been used as a substitute for hops in France. Leaves make a good green manure for hops.

Oil from the wood used as a cancer treatment in Belgium.[4] Contains the alkaloid cyclo-protobuxine, which shows antitumor activity in PS-145 and WA tumor systems.[10] Leaves sometimes used to adulterate uva-ursi.[8] Wood considered cholagogue, depurative, diaphoretic, diuretic, febrifuge, laxative, narcotic, sedative, and vermifuge, given in decoction as an alterative for gout, malaria, rheumatism, and secondary syphilis. Used as a substitute for guaiacum in treating venereal diseases where sudorifics are advisable. Narcotic and sedative in full doses, emeto-cathartic, possibly fatally so, and convulsant in overdoses.[2] Tincture once used for leprosy and malaria. Volatile oil from the wood used for epilepsy, hemorrhoids, and toothache. Leaves alterative, cathartic, sudorific, and vermifuge, powdered leaves have been applied to botworms in horses; though the mixture is poisonous, it is said to improve the horse's coat.

Contains a butyraceous volatile oil and the alkaloids buxine (similar to berberine), cycloprotobuxine, parabuxine, and parabuxonidine (active in the PS and WA tumor systems).[10] Other alkaloids reported include bebeerine, bebuxine, buxalphine, buxamine, buxaminol, buxandrine, buxanine, buxarine, buxatine, buxazidine, buxazine, buxdeltine, buxenine, buxenone, buxeridine, buxetine, buxidene, buxomegine, buxpiine, buxpsiine, buxtauine, cyclobuxine, cyclobuxonine, cyclovirobuxine, dihydrocyclobuxine, dimethylcyclobuxine, dimethylcyclovirobuxine, isochondodendrine, methylcycloprotobuxine, norbuxamine, parabuxine, and parabusonidine.

Toxicity — Leaves said to have caused fatalities in grazing animals.[1] Toxic symptoms include collapse, cramps, diarrhea, nausea, shakes, and vertigo.[33]

To the physician — For poisonings, Hardin and Arena[34] suggest gastric lavage or emesis and symptomatic treatment. Acute facial dermatitis has followed application to the scalp to treat baldness. Juice of the plants can cause irritation and itch.[6] The alkaloid buxine can contribute to respiratory failure in humans.[11]

62. *CALADIUM BICOLOR* (Ait.) Venten (ARACEAE) — Heart of Jesus

Handsome tropical ornamental foliage plant, the foliage sometimes cooked and eaten. Corms, also, are carefully prepared and eaten. Powdered leaves used as an insecticide.[56] Ecuadorians place the leaves in the nostrils of their hunting dogs to make them better hunters of wild pigs.[111]

Reported to be antiseptic, ascaricidal, emetic, larvicidal, purgative, Heart of Jesus is a folk remedy for angina, catarrh, sores, splinters, and toothache.[32] In Brazil used in gargles for sore throat and angina and as a poultice on sores and wounds.

Toxicity — Contains irritant crystals of calcium oxalate and can cause dermatitis.[6] Hardin and Arena are my only source suggesting fatalities.[34]

63. *CALEA ZACATECHICHI* Schlecht (ASTERACEAE) — Mexican Calea, Dog's Grass, Bitter Grass

Used by the Chontal Indians of Oaxaca as a tea, believed to clarify the senses.[51] Emboden says they roll cigarettes from the leaves, lie down to smoke quietly, drinking the tea as well.[54] The feeling of well-being is said to persist for a day or more with no unpleasant side effects. Leaves show some experimental antiatherogenic and CNS depression activity.[42] Dried plants have been exported to Brazil where they were used for treating Asiatic cholera. Leaves are locally used as insecticides.[58]

Reported to be apertif, astringent, emetic, febrifuge, hallucinogenic, purgative, stomachic, is a folk remedy for cholecystosis, cholera, colic, diarrhea, eruptions, fever, inappetence, and malaria.[32] Yucatanese put the leaves in baths for skin eruptions. Though famed as a febrifuge, it did not check out in clinical trials for malaria.

The plant contains wax, a glucoside-like bitter principle, chlorophyll, yellow pigment, sugar, gum, acetylenes, germacrolides, and carbonic-, chlorhydric-, phosphoric-, salicylic-, succinic-, and sulphuric-acids.[42] The plant contains about 0.01% of a crystalline alkaloid, $C_{21}H_{26}O_8$.[33] Chemical studies have revealed only a large number of polyacetylenes, sesquiterpene lactones, chromenes, triterpenes, and flavonoids, none of which could account for its hallucinogenicity.[80]

Toxicity — Listed as a narcotic hallucinogen (mostly visual).[54]

64. *CALENDULA OFFICINALIS* L. (ASTERACEAE) — Pot-Marigold, Marigold, Calendula

Marigold powder, made from dried as well as fresh petals, is used like saffron to season seafoods, chowders, soups, stews, roast meats, and chicken. It is also used to color butter, cheese custards, and liqueurs. Commercially, the dried petals are used in medicines (calendula) and ointments. Fresh plant is used as an herb tea. Having eaten the flowers of most of our Maryland wild flowers, I am not at all surprised to see Rose refer to marigold sandwiches, a mixture of marigold petals, sesame seed, mayonnaise, cheese, and for nongout suffering carnivores, liverwurst.[49]

Pot-marigold is used to adulterate saffron and arnica flowers. Leaves have been eaten in salads. The leaf, in various forms (decoctions, poultices, etc.), is a folk remedy for warts and cancer. The flower, mixed with milk, is a cancer remedy. Poultices and decoctions of the whole plant are said to treat cancer of the breast and uterus as well as glandular indurations and cancerous ulcers.[4] The aqueous extracts show activity against Sarcoma 180 in mice.[33] Marigold flowers are rubbed onto bee stings. Floral infusions are said to be good for the eyes. Although now almost obsolete as a drug, pot-marigold is considered analgesic, anthelmintic, antispasmodic, astringent, bactericide, carminative, cholagogue, depurative, diaphoretic, diuretic, ecbolic, emmenagogue, febrifuge, laxative, stimulant, stomachic, styptic, sudorific, and tonic; also, has been useful in treating bleeding gums, bleeding piles, bruises, chronic ulcers, and varicose veins, and if taken internally it prevents suppuration. It is used as a liquid extract or as a tincture. The tincture of the flowers is used to heal amenorrhea, bruises, cholera, cramps, eruptions, fever, flu, gingorrhagia, hemorrhoids, jaundice, nephrosis, piles, scrofula, sprains, stomachache, syphilis, toothache, tuberculosis, typhus, ulcers, and wounds.[16,28,33]

On a zero-moisture basis, the seed contains 30.6 to 36.9% protein and 40.8 to 45.8% oil.[21] Flowers contain the amorphous calendulin, analogous to bassorin, traces of an essential oil, mucilage, oleanolic acid, a gum, resin, a saponin, a sterol, cholesterol, esters of lauric, myristic, palmitic, stearic, and pentadecylic acids, faradiol, and arnidiol.[1] Dry petals contain stigmasterol, sitosterol, 28-isofucosterol, campesterol, 24-methylenecholesterol, and cholesterol, as well as other more complex materials.[33] The pigment consists of beta-carotene, lycopene, rubixanthine, violaxanthin, saponins, and phytosterols. Fresh plant material contains the analgesic salicylic acid (0.34 mg/kg). Roots contain inulin.[1] According to Tyler, "calendula is apparently non toxic . . . One should not expect any particular therapeutic results from its use, other than those provided by the placebo effect."[37] Rose is even more positive: "It is a plant that causes no allergic reactions, and when mixed with Comfrey and Camomile is extremely useful as an internal or external application for anything. Anyone who is sensitive to plants, especially babies, will benefit from this mixture."[47]

65. *CALLIANDRA ANOMALA* (Kunth) Macbride (MIMOSACEAE) — Cabeza de Angel

Common ornamental shrub. Resin gathered from incisions made in the bark were mixed with wood ash to induce sleep.[54] The plant is used in folk remedies for condylomata and tumors.[4] Reported to be astringent, cabeza de angel is a folk remedy for chest ailments, cholera, cough, diarrhea, fever, inappetence, inflammation, malaria, nausea, ophthalmia, and ulcers.[32] Resin employed for fever and malaria; ailmentary ailments, dysentery, and eye diseases. It is used as an liniment for a swollen anus.[54]

Seeds of other species of *Calliandra* contain 26.7 to 38.9% protein, 2.2 to 16.4% oil, and 4.1 to 4.4% ash, or a zero moisture basis.[21]

Toxicity — Classed as a narcotic hypnotic.[54]

66. *CALOTROPIS PROCERA* (Ait.) R. Br. (ASCLEPIADACEAE) — Giant Milkweed, Sodomapple, Swallowwort

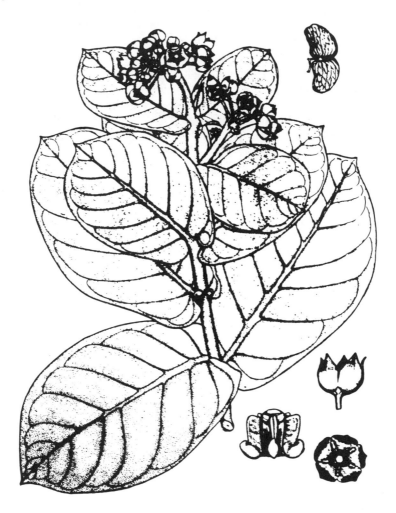

Often a weed; sometimes nearly 3 m tall, giant milkweed can be controlled by spraying with a 0.3% solution of 2,4-dichlorophenoxyacetic acid.[1] Since some of the temperate milkweeds have been considered as an energy source, it is only natural to view the tropical milkweeds as equally promising energy sources. Milkweeds are being considered as energy sources in Canada and the U.S.[112,113] The milky latex in the milkweed stem and leaves contains high concentrations of quality crude. Laboratory analyses showed a potential crop yield of 1000 gal crude oil per acre (= *circa* 60 barrels per hectare). The "milkweed crude is said to have a BTU output comparable to fossil crude." Milkweed oil is "more compatible than fossil fuel . . . Once the oil is extracted, it is refined using the same processes and equipment as in fossil fuel production. In addition to fuel and plastic production, the latex contains a protein-based substance which has high nutritional value for use in food supplements or in cosmetics."[113] Calotropis latex is used to a limited extent in the tanning industry for deodorizing, removing hair, and imparting a yellow color to hides. It is said to be an adulterant of Persian opium. The bark of the stem yields a fiber which is white, silky, strong, and durable. It is superior to cotton in tensile strength and is used for making fishing nets and lines, bow strings, and twine. The isolated fiber can be bleached to yield a white, flexible, cylinder-shaped fiber which can be spun either alone or mixed with cotton. The floss is used for stuffing matresses, pillows, etc. Being short stapled, it cannot be spun by itself; mixed with cotton it can be spun. It can also be spun after chemical treatment but cannot be bleached.[5] The floss is sometimes used to adulterate Indian kapok (*Salmalia malabarica*), but is inferior to it in resilience and water-repellent properties, even though it possesses good buoyancy. The Board of Trade, England, has stipulated that Indian kapok meant for life belts should not contain any *Calotropis* floss. The latter can be easily distinguished from kapok by its greater length (1.4 to 1.8 in.) and greater thickness (0.0018 to 0.0026 in. diameter).[1] The ash of *C. gigantea* (12.7%) is rich in potash (K_2O, 20.8%).[44] The charcoal is used in gunpowder and for fuel. In Africa, the withered (not fresh) leaves and floss are browsed by sheep and goats. Goats eat the flowers. Cattle reduced to it will eat the leaves, said to be poisonous to cattle. Fulanis of Africa use the latex to curdle milk. Stems are used for roofing rafters in dry regions of Africa, said to be immune or unattractive to termites. Stems used for charcoal and gunpowder in India. Wood has been used for firesticks, the floss for tinder. In the Sudan, corn beer is fermented with the latex. When delivering seed of *Euphorbia lathyris* as a potential energy plant for Egypt, I was amused to see *Calotropis* abundant everywhere, as a desert weed. I have seen it as a weed in the thorn forests of Honduras, Puerto Rico, and the Dominican Republic. Size alone suggests that *Calotropis* should produce more latex than *E. lathyris*, in its first and second year. If the latex is equally rich in hydrocarbons and has bonafide cardiac glycosides, *Calotropis* should be studied for total utilization, latex for medicine and hydrocarbon, leaf for protein, floss for kapok floss, stem fiber as a cotton substitute, with the residues for conversion to alcohol and soil amendments.[1]

C. procera (and/or *C. gigantea*) has served in several folk remedies for cancers, tumors, warts, and ulcerous tumors, according to Hartwell.[4] Most such reports are from India where the plants are known as "arka" and "mudar". The Arabic name is "oshar" (a word which should not be used in polite company), perhaps because the latex resembles semen. Following the doctrine of signatures, it has been suggested as a male tonic or aphrodisiac. For example, the plant is prescribed for impotency in Tanganyika. Because of its poisonous nature, such usage should be discouraged. Calotropin has shown antitumor activity in cell cultures of human epidermoid carcinoma of the nasopharynx. The plant contains active cardiac glycosides which might be extracted as a main product, with energy (hydrocarbon) as a byproduct. The glycoside (calotropin) is extremely potent and has been used for arrow poison in Africa. Minute amounts are said to cause death; hence, the plant has been used for murder and suicide. The latex can be dermatitogenic. Even honey from the flowers is said to be poisonous. The bark of the root was once used to treat leprosy. In Nigeria, it is also used for syphilis. Some Africans use the root as a chewstick, claiming that it cures toothaches. In Senegal, roots are used as a purgative, especially in the treatment of leprosy. Smoke from the roots, having been steeped in the latex, is inhaled in India by those suffering from asthma or coughs. From a pharmacodynamic point of view, the digitalis-like properties of calotropin compare with other cardiotonic substances and the ipecac-like emeto-cathartic action of the resin justify many of the therapeutic uses made of the plant. In West Africa, powdered bark is used as a carminative, lactagogue, and stomachic. Senegalese regard the plant as aphrodisiac, especially the bark. Leaves are said to cure eye ailments, headache, swollen limbs, and wounds made by rusty nails. Leaves are boiled with the fruits to make a decoction for the expulsion of guinea worms. In the Ivory Coast, the leaves are used as a vermifuge. Leaves are suggested to overcome female sterility in Ghana, and to alleviate difficult parturition. In the Sudan, the latex has been used for abortion and infanticide. Though extremely dangerous, the juice from warmed leaves is used as a nosedrop for catarrh and headache. The latex is applied for conjunctivitis in Ghana and Nigeria. Latex mixed with *Euphorbia* latex is applied in yaws. Cotton, steeped in latex, is applied to painful caries. Manna exuding from the plant is used as an asthma remedy in Gambia. The latex is used as a rubefacient for rheumatic pain in Africa. Sudanese are said to inject a swallowwart solution into the rectum thru a perforated horn for the treatment of gonorrhea. The latex is a strong irritant to the skin and mucous membrane. Injected extracts caused slowing of frog's heart and acute gastroenteritis. The latex is used in indigenous medicine in combination with *E. neriifolia* as a drastic purgative. It is also used as a local irritant. The stem, root bark, flowers, and leaves also are used in medicine. A tincture of the leaves is used in the treatment of intermittent fevers. Powdered flowers, in small doses, are useful in the treatment of colds, coughs, asthma, and indigestion. Powdered root bark gives relief in dysentery. The root bark is said to be similar to ipecacuanha in its action. In small doses (0.2 to 0.6 g) it is diaphoretic and expectorant, and in large doses (2 to 4 g) it is emetic.[1,3,32,33] The root bark, in the form of a paste, is applied in elephantiasis. Skin irritation can be allayed by the application of cold water followed by soothing preparations like glycerine belladonna. Colombians used the latex as antiodontalgic, antisyphilitic, diaphoretic, emetic, and vermifugal. West Indians may bind the leaves to the head for headache, or elsewhere for cold and rheumatisms or swellings.[42]

Analysis of the seeds (of *C. gigantea*) gave: moisture, 7.4; protein, 27; ether extract, 26.8; crude fiber and nitrogen-free extract, 32.4; and ash, 6.55%. On a zero-moisture basis, the seeds of *C. procera* contain 27.5 to 30.1% protein, 34.8 to 36.6% fat.[21] The oil extracted from the seeds is an olive-green liquid, the acid fraction of which contains: palmitic, 15; oleic, 52; linoleic, 32; and linolenic acid, 0.9%. The unsaponifiable fraction (31%) of the seed wax gave: phytosterol; stigmasterol; melissyl alcohol; and a hydrocarbon, laurane $C_{20}H_{42}$(0.6%). The seeds contain 0.437% coroglaucigenin, 0.0224 to 0.231% frugoside, 0.00665% corotoxigenin, and 0.0094% calotropine.[3] Analysis of the floss gave: moisture, 7.2; soluble matter, 4.7 to 9.7; lignin, 15.5; wax, 6.4; saccharose, 0.4; and ash, 3.64%. It contains a yellowish-brown coloring matter, chlorophyll, a resin, and a crystalline unsaturated substance. A bitter toxic substance has been reported.[1] The latex contains: water and water solubles, 88.4 to 93.0%, and caoutchouc, 0.8 to 2.5%. The coagulum contains: resins, 52.8 to 85.0, and caoutchouc, 11.4 to 22.9%. The latex contains: trypsin, an active labenzyme, and a heart poison. From the latex of the African plant, alpha-lactuceryl isovalerate and alpha-lactuceryl acetate have been isolated. On hydrolysis both yield alpha-lactucerol, $C_{30}H_{50}O$, which can be converted into isolactucerol. From the combined latex of *C. procera* and *C. gigantea*, uscharin. $C_{31}H_{41}O_8NS$, calotoxin, $C_{20}H_{40}O_{10}$, and calactin have been isolated. On hydrolysis, uscharin gives uscharidin, which with borax yields iso-anhydro calotropagenin, identical with that obtained from calotropin. Calotoxin yields pseudo-calotropagenin on treatment with sodium hydroxide. The leaves and stalks of *C. procera* contain calotropin, $C_{29}H_{40}O_9$, and calotropagenin, $C_{23}H_{32}O_6$. Other active principals include calactin ($C_{29}H_{38}O_9$), calotoxin ($C_{32}H_{42}O_8NS$), and voruscharin ($C_{33}H_{47}O_9NS$).[1,3,33]

Toxicity — Clearly a dangerous plant. Its use in traditional medicine in India may cause severe bullous dermatitis, leading occasionally to hypertrophic scars. Perkins and Payne note diarrhea, vomiting, slowed but stronger heartbeat, labored respiration, increased blood pressure, convulsions, and death.[56]

To the physician — Suggested are demulcent and mucilaginous drinks, milk, rice-guel, etc. and morphine and atropine administration to allay pain.[1]

67. *CAMELLIA SINENSIS* (L.) Kuntze (THEACEAE) — Tea

Dried and cured leaves widely used for a beverage, which has a stimulant effect due to caffeine. Used for this purpose in China for nearly 3000 years. Green tea is made from leaves steamed and dried, while black tea leaves are withered, rolled, fermented, and dried. Steam distillation of black tea yields an essential oil. Tea extract is used as a flavor in alcoholic beverages, frozen dairy desserts, candy, baked goods, gelatins, and puddings.[29] Air-dry tea seed yields a clear golden-yellow oil resembling sasanqua oil, but the seed cake, containing saponin, is not suitable for fodder. Refined teaseed oil, made by removing the free fatty acids with caustic soda, then bleaching the oil with Fuller's earth and a sprinkling of bone black, makes an oil suitable for use in manufacture of sanctuary or signal oil for burning purposes; in all respects it is considered a favorable substitute for lard, rapeseed, or olive oils. The oil is different from cottonseed, corn, or sesame oils in that it is nondrying and not subject to oxidation changes, thus, making it very suitable oil for use in the textile industry; it remains liquid below − 18°C. Tea is a potential source of food colors (black, green, orange, yellow, etc.).

The infusion, once recommended in China as a cancer cure, contains some tannin, suspected of being carcinogenic. Chinese regard tea as antitoxic, diuretic, expectorant, stimulant, and stomachic.[29] Tea, considered astringent and stimulant, acts as a nervine or nerve sedative, frequently relieving headaches. It may also cause unpleasant nerve and digestive disturbances. The infusion is, however, recommended for neuralgic headaches. Tea is reportedly effective in clinical treatment of amebic dysentery, bacterial dysentery, gastroenteritis, and hepatitis. It has also been reported to have antiatherosclerotic effects and vitamin P activity.[29] Tea bags have been poulticed onto baggy or tired eyes, compressed onto headache, or used to bathe sunburn. The plant has a folk reputation as analgesic, antidotal, astringent, cardiotonic, carminative, CNS-stimulant, demulcent, deobstruent, digestive, diuretic, expectorant, lactagogue, narcotic, nervine, refrigerant, stimulant, and stomachic; used for bruises, burns, cancer, cold, dogbite, dropsy, dysentery, epilepsy, eruptions, fever, headache, hemoptysis, hemorrhage, malaria, ophthalmia, smallpox, sores, toxemia, tumors, and wounds.[32]

Fresh leaves from Assam contain 22.2% polyphenols, 17.2% protein, 4.3% caffeine, 27.0% crude fiber, 0.5% starch, 3.5% reducing sugars, 6.5% pectins, 2.0% ether extract, and 5.6% ash. Per 100 g, the leaf is reported to contain 293 calories, 8.0 g H_2O, 24.5 g protein, 2.8 g fat, 58.8 g total carbohydrate, 8.7 g fiber, 5.9 g ash, 32.7 mg Ca, 313 mg P, 24.3 mg Fe, 50 mg Na, 2700 μg β-carotene equivalent, 0.07 mg thiamine, 0.8 mg riboflavin, 7.6 mg niacin, and 9 mg ascorbic acid. Another report tallies 300 calories, 8.0 g H_2O, 28.3 g protein, 4.8 g fat, 53.6 g total carbohydrate, 9.6 g fiber, 5.6 g ash, 245 mg Ca, 415 mg P, 18.9 mg Fe, 60 mg Na, 8400 μg β-carotene equivalent, 0.38 mg thiamine, 1.24 mg riboflavin, 4.6 mg niacin, and 230 mg ascorbic acid. Yet another give 299 calories, 8.1 H_2O, 24.1 g protein, 3.5 g fat, 59.0 g total carbohydrate, 9.7 g fiber, 5.3 g ash, 320 mg Ca, 185 mg P, 31.6 mg Fe, 8400 μg β-carotene equivalent, 0.07 mg thiamine, 0.79 mg riboflavin, 7.3 mg niacin, and 85 mg ascorbic acid.[21] Leaves also contain carotene, riboflavin, nicotinic acid, pantothenic acid, and ascorbic acid. Caffeine and tannin are among the more active constituents.[1] Ascorbic acid, present in the fresh leaf, is destroyed in making black tea. Malic and oxalic acids occur, along with kaempferol, quercitrin, theophylline, theobromine, xanthine, hypoxanthine, adenine, gums, dextrins, and inositol. Chief components of the volatile oil (0.007 to 0.014% fresh weight of leaves) is hexenal, hexenol, and lower aldehydes, butyraldehyde, isobutyraldehyde, isovaleraldehyde, as well as *n*-hexyl, benzyl and phenylethyl alcohols, phenols, cresol, hexoic acid, *n*-octyl alcohol, geraniol, linalool, acetophenone, benzyl alcohol, and citral. Does this mean that the leaves contain more dangerous substances than herb tea? More properly it only indicates that *Camellia* has been more intensively studied than most herb teas. Certain constituents, especially catechin, epigallocatechin, and epigallocatechin gallate, are said to have antitoxidative properties.[29] October 1, 1979, caffeine was trading at *circa* $9 per kilo, theobromine at about $10, and theophylline at about $12.[115] Seeds contain 8.5% albuminoids, 32.5% starch, 19.9% other carbohydrates, 22.9% fatty oil, 9.1% saponin, 3.8% crude fiber, and 3.3% mineral substances. The fatty acids of the oil are 7.6% palmitic, 0.8% stearic, 83.3% oleic, 7.4% linoleic, 0.3% myristic, and 0.6% arachidic acid.[33]

Toxicity — There is evidence that the condensed catechin tannin of tea is linked to high rates of esophageal cancer in some areas where tea is heavily consumed.[37] This effect apparently may be overcome by adding milk, which binds the tannin, preventing its deleterious effects (GRAS § 182.20).[29]

68. *CANANGA ODORATA* (Lam.) Hook. f. & Thoms. (ANNONACEAE) — Cananga, Ylang-Ylang

Siamese apply the flower infusion to the body after bathing. Flowers yield essential oils from which the perfume ylang-ylang is obtained on first distillation; second distillation yields an inferior oil, Canaga oil or Macassar oil. Ylang-ylang is considered the "Queen of Perfumes" and is extensively used in fixation of floral perfumes, making oriental attars, and in manufacture of face powders along with vetivert oils and the like, and as a modifier in manufacture or artificial perfumes. Cananga oil has a harsher odor than ylang-ylang, but is more lasting and is used for making hair oils, cheap perfumes, and for scenting soaps. Macassar oil is commonly found in bazaars and is usually a mixture of Cananga oil and coconut oil. With coconut oil and other ingredients, Macassar oil was quite familiar to the well-groomed Victorian and Edwardian male.[6] An essential oil is also prepared from the leaves. Wood is used in Sri Lanka for packing chests for produce, boxes, wooden shoes, and for turnery. It has been suggested for match sticks. In the Celebes, the bark is beaten into rope. Buccellato[116] gives a detailed account of the uses and composition of ylang-ylang.

Medicinally, the oil is used as application for gout, cephalagia, and ophthalmia. Reported to be carminative, emmenagogue, ylang-ylang is a folk remedy for asthma, boils, diarrhea, headache, malaria, ophthalmia, rheumatism, and stomach ailments.[32] Dried flowers are used for malaria in Java.[5] Seeds have been used externally in the treatment of malaria as well. In Malaya they are prescribed for asthma. In Upper Perak the leaves are rubbed onto itch. Bark is applied to scurf in West Java. The oil has been suggested as a possible substitute for quinine in malaria.[5]

Constituents of ylang-ylang oil include geraniol and linalool esters of acetic and benzoic acids, *p*-cresol methyl ester, cadinene, a sesquiterpene, and a phenol.[5] Pinen, benzyl alcohol, creosol, eugenol, isoeugenol, *p*-cresol, safrole (carcinogenic), isosafrole, benzyl acetate, methyl benzoate, methyl salicylate, benzylbenzoate, methyl anthranolate, formic acid, acetic acid, valeric acid, benzoic acid, and salicylic acid are also reported.[7]

Toxicity — Colognes, perfumes, and toilet water containing ylang-ylang essential oils can produce dermatitis in sensitized individuals. Ylang-ylang oil has been accepted as an allergen and removed from certain cosmetics.[6]

69. *CANNABIS SATIVA* L. (CANNABINACEAE) — Hemp, Marijuana

Long important as a fiber plant, oilseed, and legitimate medicinal plant, *Cannabis* is a multimillion dollar illegitimate business in the U.S. today, as a mild psychedelic fumitory. Marijuana is used to treat glaucoma and shows antibiotic activity (against Gram-plus bacteria). Cannabidiol, cannabigerol, cannabidiolic acid, and cannabigerolic acid are antibiotics.[33] Seeds have been located in North European funerary censers as early as 5 centuries B.C. Emboden describes *Cannabis* rituals from many cultures.[54]

Reported to be alterative, analgesic, anesthetic, anodyne, antidotal, apertif, aphrodisiac, cholagogue, CNS-stimulant, demulcent, diuretic, emmenagogue, emollient, hallucinogenic, hypnotic, intoxicant, laxative, narcotic, poison, psychotropic, sedative, stimulant, tonic, and vermifuge, hemp is a folk remedy for alopecia, anorexia, cancer, cold, colic, constipation, corns, cough, depression, dyspepsia, epilepsy, eruption, favus, flux, glossosis, gonorrhea, gravel, inflammation, migraine, nausea, neurosis, paralysis, polyuria, prolapsus, puerperium, rheumatism, senility, sore, spasm, tetanus, tumors, uterosis, worms, and wounds.[32,33,41] Perry mentions the seed as alterative, anodyne, anthelmintic, demulcent, diuretic, laxative, narcotic, and tonic.[16] According to Perry, ''the plant is a true sedative of the stomach, used to treat dyspepsia, cancer, ulcers as well as migraine, neuralgia and rheumatism, hemorrhage, nausea, favus, eruptions, parturition, and wounds.''[16] In Africa, some Europeans use the oil from a *Cannabis* pipe externally to treat cancers. A betaresercyclic acid has antibiotic and sedative properties, as well as antivirus effects. It also inhibits the growth of a mouse mammary tumor in egg embryo.[3] Mfengu use the leaves for snakebite. Xhosa treat horse bots with it. Zimbabwe blacks use the plant for anthrax, blackwater fever, blood poisoning, dysentery, and malaria. In Lesotho the women smoke cannabis to facilitate parturition, then give the children ground up seeds with bread or meal during weaning.[3] In the Old World, *Cannabis* was used for centuries, in moderate doses considered as a powerful aphrodisiac and exhilarent, in excessive doses or habituation leading to impotence and melancholia.[11] Who wants that?

Per 100 g, the seed is reported to contain 421 calories, 8.8 to 13.6 g H_2O, 21.5 to 27.1 g protein, 25.6 to 30.4 g fat, 27.6 to 34.7 g total carbohydrate, 18.8 to 20.3 g fiber, 4.6 to 6.1 g ash, 120 mg Ca, 970 mg P, 12 mg Fe, 5 μg β-carotene equivalent, 0.32 mg thiamine, 0.17 mg riboflavin, 2.1 mg niacin, and 0 mg ascorbic acid. Delta-tetrahydrocannabinol (Δ-THC) is the principle hallucinogen. The effective dose on smoking is 200 to 250 μg/kg, on ingestion 300 to 480 μg/kg.[11] For a summary of details on cannabinoids, see *Hager's Handbook*).[33] Seeds contain choline, inositol, phytosterols, trigonelline, and xylose.[16] The cystolith hairs so characteristic of the leaves and perianth contain calcium carbonate, which bubbles with addition of dilute hydrochloric acid.[3]

Toxicity — The pollen is classified as an aeroallergen causing allergic rhinitis, bronchial asthma, and/or hypersensitivity pneumonitis.[11] Lewis and Elvin-Lewis suggest that cannabis may disrupt DNA synthesis and cause chromosomal damage, impotence, temporary sterility and female-like mammaries in men, potential irreversible brain damage, bronchitis, and personality changes.[11] Kingsbury recounts the loss of several Greek horses and mules to *Cannabis*. The animals, with a military party, were allowed to forage for themselves. They found and ate some illicit *Cannabis*, shortly thereafter exhibiting excitation, dyspnea, and muscular trembling. Although temperature was subnormal, the animals sweated and slobbered profusely. Death followed in 15 to 30 min. Postmortem analysis showed congestion and ecchymotic hemorrhages of various organs, especially digestive.[14]

70. *CAPSICUM ANNUUM* L. (SOLANACEAE) — Chili, Sweet Peppers, Paprika

Fruit eaten as vegetable when green or red, raw or cooked, alone or in combination with other vegetables. Varieties of chili provide hot spice paprika and the cool vegetable bell pepper. Sweet peppers used for salads, pickles, and meat dishes. Paprika is higher in vitamin C than citrus. Oleoresin of paprika, extracted by solvents from pods, is a natural vegetable coloring agent used in Morocco, Europe, and California. Hungarian paprika is hotter, like a sweet cayenne pepper. Paprika also refers to a ground product and is used in manufacture of meat and sausage products and salad dressing, where a bright, natural red color is desired. Curry powder is made by grinding roasted dry chili with other condiments such as coriander, cumin, and tumeric. Plant antibiotic and bactericidal.

Regarded as aphrodisiac, carminative, CNS-stimulant, depurative, diaphoretic, digestive, rubefacient, stimulant, stomachic, styptic, and tonic, chili shows up in folk remedies for asthma, ciguatera, dyspepsia, lumbago, neuralgia, pharyngitis, pneumonia, rheumatism, scrofula, and sores. The plant is used for cancers and tumors.[4] Capsicum preparations are used as counterirritants in lumbago, neuralgia, and rheumatic disorders. Sometimes added to tannic or rose gargles for pharyngitis and sore throat. The root is used in Indonesia for gonorrhea. Leaves applied to bubo. Used in Central Africa as a calming medicine, the fruits contain capsaicin, which has been shown to have analgesic properties. Used in Hawaii for backache, rheumatism, and swollen feet. According to some authors regular ingestion of hot pepper (not green) is good for anorexia, hemorrhoids, liver congestion, varicose veins, and vascular conditions. High in vitamins A and C, pepper should be appreciated by those who suspect these vitamins help prevent cancer.

Per 100 g, the fruit is reported to contain 93 calories, 74.3 g H_2O, 3.7 g protein, 2.3 g fat, 18.1 g total carbohydrate, 9.0 g fiber, 1.6 g ash, 29 mg Ca, 78 mg P, 1.2 mg Fe, 374 mg K, 12,960 µg β-carotene equivalent, 0.22 mg thiamine, 0.36 mg riboflavin, 4.4 mg niacin, and 369 mg ascorbic acid. Per 100 g, the fruit is reported to contain 45 calories, 86.9 g H_2O, 2.0 g protein, 0.8 g fat, 9.5 g total carbohydrate, 1.7 g fiber, 0 to 8 g ash, 11 mg Ca, 47 mg P, 0.9 mg Fe, 374 mg K, 4770 µg β-carotene equivalent, 0.09 mg thiamine, 0.12 mg riboflavin, 0.4 mg niacin, and 86 mg ascorbic acid.[21] Traces of Al, Ba, Cu, Fe, Li, Mn, Si, and Ti are also reported. The coloring matter of the ripe fruit includes antheraxanthin, capsanthin, capsorubin, cryptocaprin, cryptoxanthin, lutein, neoxanthin, violaxanthin, zeaxanthin, alpha and beta-carotenes.[64] Per 100 g, the seed is reported to contain 309 calories, 7.4 g H_2O, 16.1 g protein, 1.8 g fat, 71.3 g total carbohydrate, 35.0 g fiber, 3.4 g ash, 57 mg Ca, 466 mg P, 7.0 mg Fe, 300 µg β-carotene equivalent, 0.64 mg thiamine, 0.29 mg riboflavin, 11.8 mg niacin, and 29 mg ascorbic acid. The fruit contains the very irritant capsaicin at 0.14%. This hot principle is still noticeable at 1:11,000,000. Of the capsaicinoids, *circa* 48.6% is capsaicin, 36% dihydrocapsaicin, 7.4% nordihydrocapsaicin, 2% homodihydrocapsaicin, 2% homocapsaicin, 1.5% decanoic acid vanillylamide, and 1% nonanoic acid vanillylamide. Purseglove et al. give many more details.[64] Plant contains solanidine, solanine, solasodine, and scopoletin. Chlorogenic acid is present in the stem.

Toxicity — Red pepper, marketed as a nostrum for rheumatism, can produce dermatitis. Dried fruit, even the smoke from fruits, is irritant to the mucous membranes. Where there is a high incidence of chili intake in India, submucous fibrosis of the palate and fauces has been observed. Capsaicin, the major pungent principle, stimulates salivation and sweating. In a review paper, Buchanan[117] reported that feeding rats a protein-deficient diet containing 10% chili peppers produced a 54% incidence of hepatomas, suggesting that capsaicin may contribute to the etiology of human liver cancer, particularly in those regions where dietary protein is minimal. Authors have reported that red peppers are carcinogenic or co-carcinogenic. Still, the low incidence of gastric cancers in Latin America suggests to others that hot pepper might be anticarcinogenic.

71. *CARICA PAPAYA* L. (CARICACEAE) — Papaya

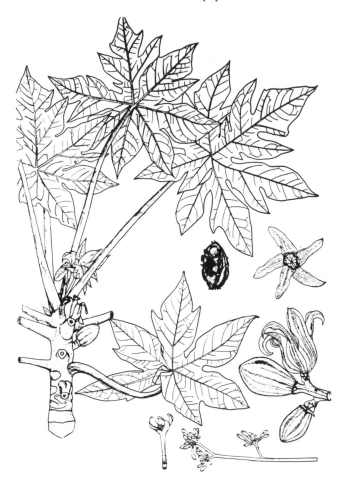

Papaya is cultivated for its ripe fruits, favored by tropical people as breakfast fruit and as an ingredient in jellies, preserves, or cooked in various ways; juice makes a popular beverage; young leaves, shoots, and fruits cooked as a vegetable. Latex used to remove freckles. Bark used for making rope. Leaves, used as a soap substitute, are supposed to remove stains. Flowers eaten in Java. Papain, the proteolytic enzyme, has a wealth of industrial uses. It has milk-clotting (rennet) and protein-digesting properties. Active over a wide pH range, papain is useful in medicine, combatting dyspepsia and other digestive disorders. In liquid preparations it has been used for reducing enlarged tonsils. Nearly 80% of American beer is treated with papain, which digests the precipitable protein fragments and then the beer remains clear on cooling.[1] Papain is also used for degumming natural silk. But most of the papain imported in the U.S. is used for meat tenderizers and chewing gums. Also, used to extract the oil from tuna liver. Cosmetically, it is used in some dentrifices, shampoos, and face-lifting preparations. Used to clean silks and wools before dying, and to remove hair from hides during tanning. It is also used in the manufacture of rubber from Hevea.[17] According to Morton,[50] dried shredded leaves are rather widely exploited for beverage purposes, though in their tropical homeland they are used mostly medicinally. Morton mentions one combination known as ''Papsady'', a combination of sassafras roots and papaya leaves. In 1982 chymopapain was approved[118] for intradiscal injection in patients with documented herniated lumbar intervertebra discs and who had not responded to ''conservative'' therapy. (However, anaphylactic shock was reported in *circa* 1% of 1400 experimental patients, with fatalities in 2.)[118]

The juice is used for warts, cancers, tumors, corns, and indurations of the skin. Sinapisms prepared from the root are also said to help tumors of the uterus.[4] Green fruit said to be ecbolic. Vermifugal seeds said to quench thirst. Leaves poulticed onto nervous pains and elephantoid growths. Roots said to cure piles and yaws.[1] In Asia, the latex is smeared on the mouth of the uterus as ecbolic. The root infusion is used for syphilis in Africa. Leaf smoked for asthma relief in various remote areas.[3] Philippinos use the root decoction for piles and yaws.[16] Javanese believe that eating papaya prevents rheumatism. Dietary papaya does reduce urine acidity in humans. Flowers have been used for jaundice.[3] Experimentally, papaya is hypoglycemic.[11] Inner bark used for sore teeth. Latex used in psoriasis, ringworm, and prescribed for the removal of cancerous growths in Cuba. Japanese use the latex for splenomegaly and stomach catarrh.[16] Externally, the latex is used for burns, corns, eczema, freckles, skin blemishes, and warts.[16] Reported to be abortifacient, amebicidal, anodyne, antibiotic, antiphlogistic, bactericide, cardiotonic, carminative cholagogue, decoagulant, digestive, discutient, diuretic, ecbolic, emmenagogue, expectorant, fungicide, insecticide, laxative, pectoral, pediculicide, proteolytic, stomachic, suppurative, tonic, and vermifuge, papaya is a folk remedy for asthma, cancer, catarrh, constipation, corns, diarrhea, diptheria, dysentery, dyspepsia, elephantiasis, enteritis, epithelioma, fever, flu, freckles, headache, hemoptysis, hypertension, itch, leucoderma, fever, madness, oliguria, piles, psoriasis, rheumatism, ringworm, splenomegaly, splenosis, toothache, tuberculosis, tumors, ulcers, venereal disease, warts, worms, wounds, and yaws.[16,32,33]

Per 100 g, the green fruit is reported to contain 26 calories, 92.1 g H_2O, 1.0 g protein, 0.1 g fat, 6.2 g total carbohydrate, 0.9 g fiber, 0.6 g ash, 38 mg Ca, 20 mg P, 0.3 mg Fe, 7 mg Na, 215 mg K, 15 μg β-carotene equivalent, 0.02 mg thiamine, 0.03 mg riboflavin, 0.3 mg niacin, and 40 mg ascorbic acid. Per 100 g, the ripe fruit is reported to contain 45 calories, 87.1 g H_2O, 0.5 g protein, 0.1 g fat, 11.8 g total carbohydrate, 0.5 g fiber, 0.5 g ash, 24 mg Ca, 22 mg P, 0.7 mg Fe, 4 mg Na, 221 mg K, 710 μg β-carotene equivalent, 0.03 mg thiamine, 0.05 mg riboflavin, 0.4 mg niacin, and 73 mg ascorbic acid. Per 100 g, the leaves are reported to contain 74 calories, 77.5 g H_2O, 7.0 g protein, 2.0 g fat, 11.3 g total carbohydrate, 1.8 g fiber, 2.2 g ash, 344 mg Ca, 142 mg P, 0.8 mg Fe, 16 mg Na, 652 mg K, 11.565 μg β-carotene equivalent, 0.09 mg thiamine, 0.48 mg riboflavin, 2.1 mg niacin, and 140 mg ascorbic acid.[21] Vitamin E is reported at 36 mg/100 g. Fresh leaf latex contains 75% water, 4.5% caoutchouc-like substances, 7% pectinous matter and salts, 0.44% malic acid, 5.3% papain, 2.4% fat, and 2.9% resin.[1] Per 100 g, the seeds are reported to contain 24.3 g protein, 25.3 g fatty oil, 15.5 g total carbohydrate, 17.0 g crude fiber, and 8.8 g ash. The seeds yield 660 to 760 mg BITC (bactericidal aglycone of glucotropaeolin benzyl isothiocyanate), a glycoside sinigrin and enzyme myrosin, and carpasemine.[119]

Toxicity — Externally, the latex is irritant, dermatitogenic, and vesicant. Internally, it causes severe gastritis. Some people are allergic to the pollen, the fruit, and the latex. Papain can induce asthma and rhinitis. Carpaine can cause paralysis, numbing of the nerve centers, and cardiac depression[16] (GRAS § 182.1585).[29] Analysis of dried papaya leaves at the University of Florida showed only 0.5 to 0.6% tannin and no saponin.[50] Fresh leaves contain the proteolytic enzyme papain which tenderizes meat (meat so tenderized is laxative). Fresh leaves contain 0.286% ascorbic acid, 0.036% vitamin E, and up to 0.4% the bitter glycoside carpaine ($C_{28}H_{50}N_2O_4$), a CNS-depressant and heart depressant.[50] Supporting my not-too-exciting hypothesis that if one studies any plant species enough, he will find something to raise the blood pressure, something to lower it, etc., etc., is an item recounted by Perry.[16] Acetone-dried powdered latex contains one factor which accelerates blood clotting, another which prevents it.[16] Can the homeostatic human body selectively take the one it needs?

72. *CASSIA ANGUSTIFOLIA* Vahl (CAESALPINIACEAE) — Indian Senna, Tinneyelly Senna

Cultivated for the medicinal senna, sold and used as a laxative in the U.S. The plant is used as a green manure crop. The dianthrone-glucosides, sennoside A and sennoside B, are of medicinal interest because of their strong laxative properties. Despite the availability of a number of synthetic laxatives, sennosides or formulas containing them are among the widest used today, and their importance is increasing. The use of laxatives is widespread among occidentals and, as a consequence, sennosides rank among the more important pharmaceuticals of plant origin.[120]

Juice or powdered leaves used for cancer and tumors.[4] Aloe-emodin is active in PS-127 and WA tumor systems, beta-sitosterol in CA, LL, and WA.[10] Senna is used for habitual costiveness, increasing the peristaltic colon movement. A paste made from powdered leaves in vinegar is applied to skin ailments and to remove pimples. A mixture of powdered senna seed and gurmala (*Cassia fistula*) made with curds is a useful ointment for the cure of ringworm.

Leaves render myricyl alcohol, a flavonol portion containing isorhamnetin and kaempferol, and an anthraquinone portion (1.5 to 3.0%) containing rhein and a little aloe emodin. Rhein is antibiotic against *Staphylococcus aureus.*[16]

Sennoside A and B are believed to be laxative principles; sennoside C and D may be present as well.[37] Sennosides act chiefly on the lower bowel and are, therefore, specially suitable for habitual constipation. The glycosides are absorbed from the intestinal tract and the active anthraquinones are liberated in the course of their breakdown. These are excreted into the colon, when these stimulate and increase peristaltic movements of the colon by its local action upon the intestinal wall. This also results in decreased absorption of water and, thereby, a bulkier and softer fecal mass.[59] Mannitol, sodium potassium tartarate, salicylic acid, chrysophanic acid, saponins, and an ethereal oil and resin are also reported. Like *C. senna,* this species is reported to contain rhein, aloe-emodin, kaempferol, isorhamnetin, chrysophanic acid, sennacrol, sennapicrin, and cathartomannite. According to Anton and Haag-Berrurier,[121] many of the therapeutic uses for natural anthraquinones touted in folk archives may have a scientific basis. Hydroxyanthraquinones form complexes with RNA and have demonstrated the ability to inhibit RNA synthesis and incorporation of nucleic acid precursors in Ehrlich ascites cells. The plant also contains 0.33% beta-sitosterol, an anticancer compound.[10]

Toxicity — Nursing mothers who take sena pass the laxative agent to the infant. "Leaves may cause severe and painful dermatitis on contact."[17] May cause griping unless combined with carminatives like coriander, cloves, or ginger. Prolonged use may lead to colon problems.

73. *CASSIA SENNA* L. (CAESALPINIACEAE) — Alexandrian Senna

Source of commercial senna, an important laxative drug. Leaves and juice used as cancer nostrums.[4] Aloe-emodin has shown activity in the PS-127 and WA tumor systems.[10] Leaves and pods are the source of Alexandrian senna of commerce, a drug generally preferred over East Indian senna, as it is milder, but has the same action. It is used as a laxative and cathartic, generally combined with aromatics and stimulants to modify its griping effects. Used also for ascites and dyspepsia.[32] In Tehran, leaves are used as a purgative, mixed with rose leaves and tamarind. Dried, pulverized leaves are applied to wounds and burns. Entire plant used as a febrifuge or as a purge to allay fever.

Does not differ significantly from *C. angustifolia*, containing 2.5 to 4.5% sennosides in the pods. The purgative qualities are due largely to anthraquinone derivatives. The plant is reported to contain, also, aloe-emodin, anthraquinones, cathartic acid, cathartin, cathar-kaempferol, chrysophanic acid, emodin, isorhamnetin, kaempferin, kaempferol, mucilage, phaeoretin, rhein, sennacrol, and sennapicrin. Senna leaves contain free anthraquinones (aloe-emodin, chrysophanol, rhein, etc.) and their O- and C-glycosides, free sugars (fructose, glucose, pinitol, sucrose). Seeds do not contain anthraquinones.[59] Shortly after germination chrysophanol, then aloe-emodin, and finally rhein are formed in the young plant. Sennosides are reported to decline, with the onset of flowering. A decrease in the rainy season has been attributed to leaching. Newly sprouted leaves after the rains are high in sennosides which decline as the leaves mature.[59] The seeds are reported to contain trypsin inhibitors. The valve of the pods are reported to contain trypsin and chymotrypsin inhibitors. Dried seeds of a species of *Cassia* called "senna" contain per 100 g 346 calories, 10.3% water, 18.0% protein, 1.3% fat, 66.9% total carbohydrate, 6.5% fiber, 3.5% ash, 205 mg Ca, 184 mg P, and 1.04 mg thiamine.[21]

Toxicity — Similar to *C. angustifolia*.

74. *CATHA EDULIS* Vahl (CELASTRACEAE) — Khat

Source of a stimulant which predates coffee in Arabia by more than a century. Like cocoa, it allows the user to go without food for extended periods. Yemen and Aden tribes retire every evening for khat breaks, sometimes rather extended. Daily export of khat to Aden lies behind the founding of the Ethiopian Airlines.[54] Tradition says its use as a social stimulant originated in the Harar region of Ethiopia.[122] The wood is pale yellow to dark brown in color, moderately hard and strong, and suitable for cabinetry. It is also useful for making high-class blotting paper.[1]

Reported to be anorexiac, aphrodisiac, astringent, CNS-stimulant, narcotic, poison, and stimulant, khat is a folk remedy for asthma, chest ailments, cough, debility, diabetes, flu, lethargy, and stomach ailments.[32] Cathine is said to open the bronchial passages, curb the appetite, and raise the blood pressure.

Per 100 g, the leaves are reported to contain 5.2 g protein, 2.7 g fiber, 1.6 g ash, 290 mg Ca, 18.5 mg Fe, 1800 μg β-carotene equivalent, 0.05 mg thiamine, 0.05 mg riboflavin, 14.8 mg niacin, and 161 mg ascorbic acid.[122] Leaves contain three alkaloids, d-nor-isoephedrine, formerly called cathine (0.27%), cathinine (0.15%), and cathidine (0.32%), besides reducing sugars, tannin, and a volatile oil.[1] Myricitin resins, mannitol, dulcitol, caoutchouc, up to 14% catechins, vitamins B and C are also reported from the plant. Seeds contain 50% oil.[33] Krikorian and Getacun gives a historical tabulation of the compounds in khat.[123] Emboden states that scopolamine is the active principle.[54] Elsewhere he lists the following "euphoriants": dexedrine (I thought that was synthetic), ephedrine, d-norpseu-doephedrine, and pervitin.[54]

Toxicity — Emboden classifies the plant as a narcotic stimulant leading to hallucinations, ending in somnolence,[54] stupidity, laziness, mindlessness, even insanity.[33] Addition of 0.1% khat extract to a rooster's diet decreased semen output and sperm concentration. Semen production stopped completely after 63 days treatment, but the testis regained normal function after withdrawal of khat extracts.[127] The *Bulletin on Narcotics*[125] devoted an entire issue of 99 pages to khat with emphasis on chemistry and pharmacology. Khat in humans induces mydriasis, tachycardia, extrasystoles, elevated blood pressure, transient facial and conjunctival congestion, headache, hyperthermia, increased respiration (through central stimulation, bronchodilation, and counterregulation of hyperthermia), inhibition of micturition, yet increased diuresis (from intake of large quantities of fluids together with khat). Reinforcing effects include euphoria, logorrhea, improvement of association, excitement, and insomnia. "Toxic psychosis occurs very rarely, if at all."[125] Would that the obese Jim Duke could chew khat and coca, dare I call it "coca cata" instead of lunch!

75. *CATHARANTHUS LANCEUS* Pichon (Boj.) (APOCYNACEAE) — Lanceleaf Periwinkle

Studied as a potential medicinal plant. In South Africa, the plant is considered antidysenteric, astringent, bitter, emetic, and lactagogue.

A single plant may (always I suspect) contain antagonistic compounds. Catharanthine is diuretic while ajmalcine is antidiuretic. According to Emboden this plant contains up to 5% yohimbine. *Hager's Handbook* lists ajmalinine, horhammericine, yohimbine, lanceine, cathalanceine, catharanthine, ammocalline, pericalline, perimivine, vinosidine, vincoline in the root, pericyclivine, periformyline, leurosine, perivine, lochnerinine, tetrahydroalstonine, vindoline, vindolinine, chatkavine, 3,4-dimethoxyphenylacetamide from the leaves. In animal studies, the plant extract had hypotensive and hypoglycemic activities.[33] Leurosine is antineoplastic, and 3,4-dimethoxyphenylacetamide cytotoxic against the 9-KB-carcinoma.[33]

Toxicity — "Unfortunately, the 'high' obtained from smoking Catharanthus has severe debilitating side effects. Ataxia, loss of hair, tingling skin sensations, a sensation of burning, and muscle deterioration followed extended use of this material. Misused, it presents the possibility of untold horrors."[54]

76. CATHARANTHUS ROSEUS (L.) G. Don (APOCYANACEAE) — Periwinkle, Madagascar or Cape Periwinkle, Old Maid

Attractive ornamental, widely cultivated and naturalized in the tropics. In Florida, juveniles very dangerously smoked dried periwinkle leaves as a marijuana substitute.[17] Extracts of entire dried plant contain many alkaloids of medicinal use. As the most important plant source of anticancer drugs, leaves are consumed in the U.S. at an estimated 1000 MT, with another 1000 for England, Italy, Netherlands, and West Germany.[59] The principal alkaloid is vincaleukoblastine (vinblastine sulfate), sold as Velban. Vinleurisine, vinrosidine, and vincristine possess demonstrable oncolytic activity. Extracts of the plant show beneficial growth inhibition effects in certain human tumors. Vinblastine sulfate is used experimentally for treatments of neoplasms, and is recommended for generalized Hodgkin's disease and resistant choricarcinoma. Vincristine sulfate (Oncovin) is used in treatment of leukemia in children. Using vincristine and vinblastine in combination chemotherapy has resulted in 80% remission in Hodgkins disease, 99% remission in acute lymphocytic leukemia, 80% remission in Wilm's tumor, 70% remission in gestational choriocarcinoma, and 50% remission in Burkitt's lymphoma.[11] In some cases, people cured of Hodgkin's disease entirely lost interest in sex. Serpentine is hypotensive, sedative, and tranquilizing. Emboden may have triggered some counterculture interest by recounting that side effects of this therapy were euphoria and hallucination. When this information became generally known, there was an outbreak of periwinkle smoking in Miami.[54]

Interestingly, though this is *the* most important plant to date in the war against cancer, it is not among the folk cancer remedies so ably tallied by Hartwell of the NCI,[4] nor would the extracts have indicated anticancer activity in the screens.[11] The NCI screened extracts and found no activity because the screening program did not include the P-1534 leukemia, and the periwinkle alkaloids were not active against the tumors in the screen at that time.[11] Decoction of roots is astringent, diaphoretic, emmenagogue, stomachic, and tonic, and is said to cause abortion.[16] Leaves are used for diabetes and as a purgative in chronic constipation, dyspepsia, and indigestion. Roots are used for dysentery. Drug, also, has a slight digitalis reaction and is said to act as a heart poison. Root decoction said to be febrifugal. Leaf used for menorrhagia and rheumatism in Africa.[3] Surinamese boil ten leaves and ten flowers together for diabetes.[42] In Cuba flowers are used in a collyrium for the eyes of infants. Bahamians take the flower decoction for asthma and flatulence, the plant for tuberculosis. In Colombia, the infusion is gargled for chest complaints, laryngitis, and sore throats.[42] In Mauritius, the leaf infusion is given for dyspepsia and indigestion. In Vietnam it is taken for diabetes and malaria. In many places the plant is used to regulate menstruation, the root considered ecbolic in the Philippines.[16] Curacao and Bermuda natives take the plant for high blood pressure.[42] Indochinese use the stalks and leaves for dysmenorrhea.[16]

Quite a few alkaloids have been isolated from the plant, a number of them interesting pharmacologically: ajmalicine, akuammicine, akuammine, alstonine, ammocaline, carosidine, carosine, catharanthine, catharicine, catharine, catharosine, cavincidine, cavincine, cleavamine, deacetylvincaleucoblastine, deacetylvindoline, dihydrositsirikine, isoleurodine, isositsirikine, leurocristine, leurosidine (vinrosidine), leurosidine, leurosine, leurosivine, lochnericine, lochnerine, lochnerinine, lochnerivine, lochrovidine, lochrovine, maandrosine, mitraphylline, neoleurocristine, neoleurosidine, pericalline, permividine, perimivine, perivine, perosine, pleurosine, reserpine, rovidine, rovidine sulfate, serpentine, sitsirikine, tetrahydroalstonine, tetrahydroserpentine, vinaphamine, vinaspine, vinblastine, vincaleucoblastine, vincamicine, vincamine, vincarodine, vincathicine, sulfate, vinceine, vincolidine, vincoline, vincristine, vindolicine, vindolidine, vindoline, vindolinine, vindorosine, vinosidine, vinsedecine, vinsesine, virosine, yohimbine. The root bark contains 2.0% of a phenolic resin and 0.03% d-camphor. The leaves yield an oleoresin and a small amount of volatile oil containing aldehydes, sesquiterpenes, and sulfur compounds, furfural, lochnerol and lochnerallol, two glycosides (adenosine, roseoside), deoxyloganin, loganin, tannin, carotene, sterols, ursolic acid, and a flavone derivative. Some of the alkaloids have both antitumor and tumorigenic attributes, suggesting what I have come to believe, that a given species will contain many pairs of compounds which are antagonistic. Catharanthine is diuretic, ajmalicine is antidiuretic. Will the body of the patient who needs the diuretic select catharanthine or reject ajmalicine? Ajmalicine (raubasine) has a broad application in circulatory diseases, especially in the relief of obstruction of normal cerebral blood flow. In combination with rauwolfia alkaloids it has been used to lower high blood pressure.[59]

Toxicity — Same as *Catharanthus lanceus*.

77. *CAULOPHYLLUM THALICTROIDES* (L.) Michx. (BERBERIDACEAE) — Blue Cohosh, "Yellow Ginseng" Squaw Root, Papoose Root

Pea-sized seeds roasted to make a coffee-like beverage.[2] Powdered rhizomes sold in limited quantities in U.S. herb stores, for use in uterine disorders. Extracts have exhibited antiinflammatory activity in rats. Sometimes suggested as a substitute for goldenseal, the plant is noted for its estrogenic and antispasmodic properties.[33]

Component of an old cancer nostrum.[4] Said to be anthelmintic, antispasmodic, diaphoretic, diuretic, emmenagogue, estrogenic, expectorant, laxative, parturifacient, vermifuge. Has been used for bronchitis, cholera morbus, colic, convulsions, cramps, dropsy, dysmenorrhea, dyspepsia, epilepsy, hiccups, hysteria, leucorrhea, metrorrhagia, neuralgia, nervousness, parturition, rheumatism, sore throat, stomatitis, uteritis, uterine cramps and infections, and vaginitis.[2,8,30,37,44,45] Has been used to expedite child delivery when the delay is due to debility, fatigue, or lack of uterine nervous energy.[2] North American Indians used the root to promote menstruation, using it also for lingering parturition.[18] Chippewa used a strong decoction for contraception.[48] According to Dharmanada,[126] blue cohosh is better for uterine pains, black cohosh (cimicifuga) for lower back pains, in menstruation. Homeopaths recommend the root tincture for after-pains, amenorrhea, barrenness, cholasma, cholera, dysmenorrhea, false conception, feet, gonorrhea, gout, hands, intramammary pain, labor, leucorrhea, metrorrhagia, neuralgia, ovaries, pityriasis, pregnancy, rheumatism, threatened abortion, uterine spasm, and uterine atony.[30,33] Parvati reports good results with hundreds of women in helping them regularize their cycles, steeping one spoonful blue cohosh and one spoonful pennyroyal per cup for 30 min, drinking one cup in later afternoon, another in the evening for the five consecutive days preceding the period, concluding "it is not abortive to my knowledge, in the dosage recommended."[48]

Leaves and seeds contain glycosides and the alkaloids anagyrine, baptifoline, magniflorine, and methylcytisine (caulophylline). Methylcytisine stimulates respiration and intestinal motility and raises the blood pressure.[37] Leontin, citrullol, saponin (caulosaponin, which is oxytocic), gum, starch, phosphoric acid, phytosterol, and resin are also reported.

Toxicity — Children are said to have been poisoned by eating the blue seeds.[11] Powder of dried blue cohosh must be handled carefully as it is strongly irritating, especially to the mucous membranes. Should be avoided by pregnant women. Caulosaponin provokes strong uterine contractions recalling those provoked by ergot.[45]

To the physician — For treatment, Hardin and Arena recommend gastric lavage or emesis along with treatment of symptoms. Caulosaponin constricts the coronary blood vessels, causing intestinal spasms and mycardial toxicity in animals.[37] Anything that constricts blood vessels should perhaps be contraindicated in high blood pressure. Still, Kloss suggests it is good for high blood pressure.[44] *Caveat emptor!*

78. *CENTELLA ASIATICA* (L.) Urb. (APIACEAE) — Gotu Kola, Fo Ti Tieng

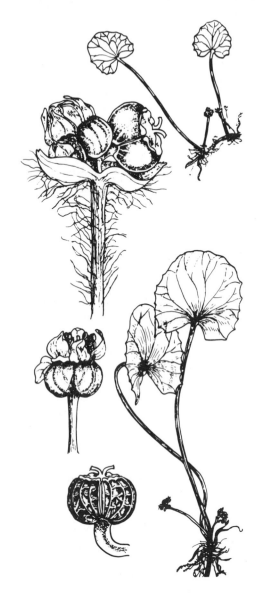

    One of the reported miracle "elixirs of life", Fo Ti Tieng (with gotu kola, ginseng, and goldenseal) seems to arouse more curiosity among American consumers than many other herbs. Since some scientists call it *Centella*, and others call it *Hydrocotyle*, many herbalists assume that two genera or species are involved. Modern research does not support such a contention. It is used as a cover crop in tea and rubber, but some farmers consider it more weed than cover crop. Although eaten, raw or cooked, in salads and curries, the leaves are also said to be insecticidal and to be used as fish poison. In the Philippines, leaves are eaten raw but more usually made into teas.[127] Chai recounts one story of the notorious herb tea "fo-ti-tieng", a trademarked name for a tea made up of gotu kola, cola, and meadowsweet, which was marketed out of England for some 40 years.[128] Mowrey[129] reports that another famous mixture of capsicum, ginseng, and gotu kola helped mice overcome the effects of fatigue-stress situations. He concluded that gotu kola and ginseng in combination seem to increase vitality from day to day, perhaps by decreasing the effects of stress on the endocrine system of the body.[129]

Described as representing "a whole apothecary shop", *Centella* is used as a restorative in both salads and teas. Gotu kola has been regarded as a treatment for leprosy. The carcinogen asiaticoside is "active in the treatment of leprosy," probably dissolving the waxy coating of the bacterium. Brazilians take the plant for uterine cancer.[42] It is also said to be useful for pruritis, sores, tuberculosis, and wounds.[33] In large doses the drug is said to be stupefying, narcotic, producing giddiness and sometimes coma.[1] In folk medicines it is used for abscesses, asthma, boils, bronchitis, cancer, cataracts, catarrh, convulsions, dropsy, dysentery, earache, elephantiasis, eczema, fever, gonorrhea, headache, hypertension, insanity, jaundice, kidney problems, leprosy, leucoderma, leucorrhea, liver, lungs, pleuritis, rheumatism, ribache, scrofula, skin diseases, spasms, syphilis, tuberculosis, tumors, ulcers, and urethritis.[1,22,37,40,42] Seeds are used for dysentery, headache, and fever. Perry notes that if a small quantity is eaten, it stimulates the appetite, aids digestion, and alleviates bowel trouble in children.[16] It is valued in cleaning and healing sores and ulcers and other skin ailments. In the Konkan of India, one or two leaves are given every morning to cure stuttering, and the juice is applied to skin eruptions. Ayurvedic medicine suggests the plant for anemia, asthma, biliousness, blood disorders, bronchitis, fever, inflammations, insanity, leucoderma, splenomegaly, thirst, and urinary discharges.[26] In Unani, it is used for asthma, bronchitis, dysuria (scalding), headache, hiccough, and inappetence. Decoctions of young roots are administered for hemorrhoids. Used as a poultice on sores and on joints pained with rheumatism. Hot juice of the roots is used to treat infected cuts. It is considered alterative, digestive, diuretic, refrigerant, restorative, and tonic. The leaf juice is employed in the Philippines for sclerotic wounds. The old report of Chinese LiChing Yun, who lived many years and had many wives and drank Fo Ti Tieng,[128] is responsible for many inquiries from people seeking the Fountain of Youth. Maybe LiChing Yun would have lived much longer and had more wives had he not drank the tea. Tyler concludes[37] "there is currently no evidence to support the use of gotu kola as a longevity promoter or to substantiate any of the other extravagant claims for revitalizing and healing . . . Substantive data on its safety and efficacy are simply nonexistent."

Nutritional analyses of the leaves reveal per 100 g: 34 calories, 89.3 g water, 1.6 g protein, 0.6 g fat, 6.9 g carbohydrate, 2.0 g fiber, 1.6 g ash, 170 mg Ca, 30 mg P, 3.1 mg Fe, 414 mg K, 6580 μg β-carotene, 0.15 mg thiamine, 0.14 mg riboflavin, 1.2 mg niacin, and 4 mg ascorbic acid.[21] Quisumbing quotes an earlier analysis of air-dried leaves:[127] 7.2% ether extract, 17.2% protein, 9.1% ash (1% P_2O_5; 1.5% Ca, 0.1% Fe_2O_3, etc.), 9.1% crude fiber, 4.6% reducing sugars, 2.0% nonreducing sugars, and 50.9% undetermined. He reported absence of cyanophoric glucosides, alkaloids, and saponins, with some resin, tannin, and volatile oil. It was uncommonly rich in vitamin B.[127] Sitosterol is found in the alcoholic extract of the herb along with a fatty oil consisting of the glycerides of oleic, linoleic, linolenic, lignoceric, palmitic, and stearic acids. Centoic acid ($C_{30}H_{48}O_6$) and centellic acid ($C_{30}H_4O_6$) have also been identified.[1] The glycoside madecassoid is antiinflammatory.[33] A bitter principle, vellarine, is present in the leaves and roots, along with pectic acids and resins.

Toxicity — The glycosides, asiaticoside and centelloside, have also been reported, and an alkaloid, hydrocotylin ($C_{22}H_{33}NO_8$). Asiaticoside is carcinogenic. Yet it is said to stimulate wound healing; it is also active against the tubercular bacillus. Some species are said to be poisonous to sheep. Under the name of *C. coriacea*, it has been reported to contain the poison hydrocyanic acid.[3] Tyler[37] notes two saponin glycosides with sedative activity, brahmoside and brahminoside. Morton[42] mentions, also, thankuniside, reporting that asiaticoside and methyl-5-hydroxy-3-6-diketo-23(or 24)-nor-urs-12-3nZ-28-oate cause consistent reduction of fertility in female mice.

79. *CHAMAEMELUM NOBILE* (L.) All. (ASTERACEAE) — Roman Camomile, English Camomile, Camomile

Formerly cultivated as the source of Roman camomile oil, this herb is used in the U.S. as a tea and home remedy. It ranks among the top selling five herbs in the U.S. according to the HTA. Tyler suggests it is the European ginseng.[37] Dried flower heads serve chiefly for herb teas. Oil occasionally used in cosmetics, liqueurs, and perfumes. Extracts used as flavoring in bitters, benedictine, vermouth, nonalcoholic beverages, baked goods, candy, frozen dairy desserts, gelatins, and puddings. Extracts used in bath preparations, cosmetics, hair dye, formulas, mouthwashes, shampoo, and sunscreens. The essential oil is used in creams, detergents, lotions, perfumes, and soaps. Pharmaceutically used in antiseptic unguents and lotions to treat cracked nipples, inflammations, sore gums, and wounds.[29]

The oil of the plant, when mixed with flour is a folk remedy for indurations of the liver, stomach, and spleen. Used in a poultice with rose oil, it is said to help indurated tumors of the parotid glands. A cataplasm prepared from the root is said to aid indurations. The flowers used in poultices or cataplasms are also said to help cancer.[4] Three of the sesquiterpene lactones exhibit anticancer activity in vitro against human tumor cells. Considered anodyne, antiseptic, antispasmodic, carminative, diaphoretic, digestive, diuretic, nervine, stimulant, stomachic, sudorific, and tonic; said to relieve colic, debility, dysmenorrhea, dyspepsia, headache, hysteria, nervousness, neuralgia, pertussis, spasms, and toothaches. The flowers are sometimes applied externally to indolent ulcers and for irritation or inflammation of the abdominal viscera. Compresses applied for gout, inflammation, lumbago, and skin problems. Tea brewed from the herb is used for indigestion. Lotion of camomile applied to earache, neuralgia, and toothache. The oil is considered antispasmodic, carminative, cordial, and sudorific.

May contain up to 1.75% volatile oil, *circa* 0.6% of germacranolides (nobilin, 3-epinobiline, 1,10,epoxynobilin, 3-dehydronobilin), flavonoids (apigenin, apigetrin, apiin, luteolin-7-glucoside, quercitrin, scopoletin-7-beta-glucoside). Tyler notes that apigenin, luteolin, patuletin, and quercitin are antispasmodic.[37] The volatile oil contains *circa* 85% esters of angelic and tiglic acids (amyl, butyl, hexyl, and isoamyl angelates or tiglates); oleic-, linoleic-, cerotinic-, palmitic-, and stearic-acids are also recorded, along with chamazulene, choline, 1,8-cineole, coumarin (scopoletin-7-beta-glucoside), dihydrocinnamic acid, farnesol, fatty acids, inositol, nerolidol, alpha-pinene, ℓ-trans-pinocarveol, ℓ-trans-pinocarvone.[29,33]

Toxicity — Tea may cause anaphalaxis (this statement has been challenged) and an enema may cause asthma and urticaria.[6] The plant yields nobilin, a potentially allergenic sesquiterpene lactone. Contact with the plant has caused dermatitis in farmers. Camomile ointments can cause dermatitis which can be exacerbated by drinking camomile tea.[6] "Generally nontoxic when applied externally"[62] (GRAS § 182.10 and 182.20).[29] According to Tyler[37] "tea . . . may cause contact dermatitis, anaphylaxis, or other hypersensitivity reactions in allergic individuals . . . Persons known to be allergic to ragweeds . . . should be cautious about drinking tea prepared from the camomiles or yarrow."[37]

80. *CHAMAESYCE HYPERICIFOLIA* Millsp. (EUPHORBIACEAE) — Spurge

Guadelupe physicians use this spurge as a substitute for ipecac.[42] Reported to be astringent, narcotic, purgative, this spurge is a folk remedy for abrasions, cancer, colic, coma, diarrhea, dysentery, leucorrhea, menorrhagia, ophthalmia, sclerosis, stomatosis, tumors, uteralgia, and warts.[4,32,33] Colombians use the plant as emmenagogue, febrifuge, and sudorific; Cubans as a diuretic and hemostat. Jamaicans give this to malnourished children. Venezuelans use the plant for buccal ulcers, stomatosis, and toothache. Guadelupans use it for female, intestinal, respiratory, and stomach ailments. West Indians use it for leucorrhea.[42] Colombians use the plant as a cataplasm on gangrenous ulcers.[57] The latex is applied to calluses, ringworm, and warts.[42]

Latex contains alkaloids, caoutchouc, gallic acid, glycosides, phorbol esters, resin, and tannin.[2,42]

Toxicity — Narcotic.[2] Juice may cause temporary blindness if introduced into the eyes.

81. *CHELIDONIUM MAJUS* L. (PAPAVERACEAE) — Celandine, Great Celandine, Nipplewort

Attractive, shade-loving, ornamental, long popular in folk medicine. Seeds contain 50 to 66% fatty oil which is not currently utilized. Medieval alchemists tried to use the golden latex to convert inferior metals to gold.

A popular medicinal herb, even in the Middle Ages, celandine was most used for treating warts. Chelidonic acid derived from celandine is an acrid irritant substance which has been used to treat warts. Repeated subcutaneous injections to treat an eyelid epithelioma caused pain and reaction following each injection; the lids sloughed up and scarred up to the orbital margin. It is very popular in Russia, "when it is said to have proved effective in cases of cancer."[2] Few plants have greater notoriety as folk remedies for cancer.[4] Celandine is used in folk treatments for cancer of the breast, colon, jaw, kidney, lip, liver, mouth, neck, nose, ovary, parotids, penis, rectum, skin, spleen, stomach, testicles, tongue, urethra, uterus, in addition to related afflictions such as egilops, calluses, condylomata, corns, eschars, indurations, polyps, tumors, and wens.[4] Four antitumor alkaloids have been isolated from celandine, i.e., berberine, chelidimerine, coptisine, and sanguidimerine (active against the KB culture, i.e., human epidermoid carcinoma of the nasopharynx). The juice has been used to treat eczema, gout, jaundice, ringworm, scurvy, scrofula, and other diseases. Experimentally, the plant is said to show hypoglycemic activity. A decoction of the plant is gargled for toothache. Powder of the dried roots was once applied to teeth to effect extraction. One European natural healer used celandine, along with cow parsnip, savory, peppermint, and plantain, in a cream massage for massaging the base of the spine, along with foot baths and vaginal douches, said to work miracles for frigid females.[11] In spite of the reputed allergenicity, the fresh yellow latex is applied to correct summer freckles. The powdered drug is considered alterative, diaphoretic, diuretic, expectorant, purgative, and sedative. My counterparts in Manchuria seemed cognizant of its anticancer potential and added that it was used there as an analgesic and for skin ailments. In the 14th century, it was supposed to be used as an antiseptic in wounds and as a collyrium to improve opacities. In milk it was used to clarify the cornea. With anise seed in wine, it was said to remove obstructions from the liver and gall. With sulfur, it was applied to itch. The roots, leaves, and flowers, extracted in boiling lard, have been used to make an unguent for piles. The plant is an important ingredient in oriental remedies for peptic ulcers, as well as gastric cancers.[16]

Numerous alkaloids have been reported from the celandine: allocryptopine, berberine, cheladimine, chelamine, chelerythrine, chelidamine, chelidonine, chelilutine, chelirubine, choline, coptisine, corysamine, dihydrosanguinarine, homochelidonine, hydroxychelidonine, hydroxysanguinanine, methoxychelidonine, oxychelidonine, oxysanguinarine, protopine, sanguinarine, sparteine, stylopine, and tetrahydrocoptisine.[91] Since chelerythrine is described as "narcotic" by Grieve (meaning poisonous), there has been quite an interest in celandine in the counterculture. Sanguinarine stimulates intestinal paralysis and salivary secretions.[16]

Toxicity — Occasional poisonings of humans (rarely fatal) have been reported.[14] The orange-yellow sap is extremely acrid and causes extreme stomatitis and gastroenteritis if ingested. Externally, the juice is escharotic, irritant, and vesicant, and the plant itself can cause contact dermatitis. The alkaloid protopine produces brachycardia. It may also cause an increase in blood pressure followed by decrease, with possible heart damage and injury to the vasomotor center. But chelerythrine is the main toxin, possibly causing paralysis and muscle spasms. Chelidonine and homochelidonine have a morphine-like action.[16]

82. *CHENOPODIUM AMBROSIOIDES* L. (CHENOPODIACEAE) — Wormseed

The major modern use[29] (perhaps explaining an anonymous call to me looking for several tons) is as a fragrance component in creams, detergents, lotions, perfumes, and soaps, with maximum use level of 0.4% reported for perfumes.[29] Regardless of its rank odor, the plant is used to flavor soups, a little bit imparting a good flavor. Shoots are even used as a potherb.[22] Brazilians feed the plant to fasting pigs to rid them of parasites.[42] Powdered seed used as an anthelmintic and insecticide. Oil of *Chenopodium* is a world-renowned vermifuge, especially effective in ancylostomiasis. It is also effective against dermatopathogenic fungi, indicating the wisdom of its folk usage for athletes foot. A culicidal incense is made from the oil.[16] Indochinese farmers mix the infructescence with fertilizer to deter development of insect larvae detrimental to their truck crops. Leaves are used in hair care.[16]

According to Hartwell, the leaves and roots are used in folk remedies for tumors.[4] Reported to be abortifacient, amebicide, analgesic, anthelmintic, antispasmodic, ascaricide, carminative, diaphoretic, diuretic, emmenagogue, fungicide, lactogogue, narcotic, nervine, stimulant, stomachic, sudorific, tonic, wormseed is a folk remedy for amebiasis, amenorrhea, anemia, arthritis, asthma, bugbite, colic, dyspepsia, dysentery, dysmenorrhea, dyspnea, fatigue, fever, hookworm, neurosis, palpitations, puerperium, rheumatism, roundworm, sores, stomachache, and tumors.[3,32] According to *WHO Chronicle*, a decoction of 20 g rapidly expels parasites without apparent side effects.[130] Zulu use the infusion as a enema for intestinal ulceration. Xhosa use the green leaf tincture as an antitussive. Sotho take the infusion for colds and stomachache. South Africans use the plant as a poultice to remove sandworm.[3] Japanese drink the leaf tea as a vermifuge. Philippinos use the crushed leaves in a back rub for malaria.[16] Interestingly, some South American Indians bathe with the decoction to reduce fever. Venezuelans boil the plant with Lepidium for burns and as a stomach tonic. Yucatan natives use it for asthma, catarrh, chorea, and other nervous afflictions. New Mexicans take salted leaves for abortion and puerpereal pain. Suppositories for appendicitis are made from the leaves with spearmint and salt. Mexicans chew the root for toothache, and use the root decoction for proctorrhagia. Curacao natives take the plant, sometimes rectally, for fever, measles, and worms. Bolivians drink the tea after meals as a stomachic.[42] Trinidad natives take the tea for dyspnea, dyspepsia, dysentery, palpitations, and asthma.[42] Chinese use the fresh root for articular rheumatism.[41]

Per 100 g, the leaves are reported to contain 42 calories, 85.5 g H_2O, 3.8 g protein, 0.7 g fat, 7.6 g total carbohydrate, 1.3 g fiber, 2.4 g ash, 304 mg Ca, 52 mg P, 5.2 mg Fe, 1 mg Na, 4840 μg β-carotene equivalent, 0.06 mg thiamine, 0.28 mg riboflavin, 0.6 mg niacin, and 11 mg ascorbic acid.[21] The essential oil may contain up to 90% ascaridole an unsaturated terpene peroxide. Austrian material yielded 0.3% oil containing 65 to 73% ascaridole; Russian material 76 to 86% ascaridole. Up to 1% Mg PO_4 is reported in Chilean material. All organs contain saponins, the root *circa* 2.5%. Geraniol, 1-limonene, myrcene, *p*-cymene, and *d*-camphor are also reported, with butyric acid, spinasterol, terpinene, triacontal alcohol, methyl salicylate, urease.[29,33]

Toxicity — The oil is quite poisonous, causing fatalities in overdoses, preceded by cardiac disturbance, convulsions, respiratory disturbances, sleepiness, vomiting, and weakness. Severe poisoning is characterized by coma, extreme dizziness, headache, prostration, somnolence, and stupor.[3] Lewis and Elvin-Lewis add that the therapeutic dose is close to minimum toxic levels and death may result from such undesirable respiratory abnormalities.[11]

83. *CHIONANTHUS VIRGINICA* L. (OLEACEAE) — Fringe Tree, Old Man's Beard

Ornamental shrub, with several folk medicinal applications. Reported to be alterative, aperient, cholagogue, diuretic, narcotic, and tonic, fringe tree is a folk remedy for bilious fever, cancer, cirrhosis, diarrhea, fever, hepatosis, icterus, inflammations, malaria, skin ailments, sore, sorethroat, stomatitis, typhoid, and wounds.[4,32] Used in homeopathy for hepatitis, hepatosis, hepatogenic headache, and icterus.[33]

The plant contains phillyrin (chionanthine, phyllyroside, forsythin); the aglucone phillygenin $C_{21}H_{24}O_6$. Said to contain a glucoside and saponin.[2]

Toxicity — Regarded as slightly narcotic.[2] Overdoses cause severe frontal headache, sore eyes, nausea, flatulence, severe vomiting, black stools, slow pulse, cold sweats, and weakness.[56]

84. *CHRYSANTHEMUM CINERARIIFOLIUM* (Trevir.) Vis. (ASTERACEAE) — Pyrethrum, Dalmatian Insect Flower

Pyrethrum is cultivated for the dried inflorescences which furnish on insecticide, used in manufacture of insecticides and parasiticides, in live stock sprays, meat processing, ointments for scabies, and aerosol sprays against fleas, flies, lice, and mosquitos. Pyrethrum use evolved in Iran, and was introduced into Europe and the U.S. in the 19th century. Kerosene extracts are usually used, but mixing pyrethrum with sesame-oil synergists reduces the cost while enhancing toxicity of pyrethrins, and requires less pyrethrum per unit kill. Toxicity to higher animals is minimal, so that pyrethrum is an excellent insect deterrent for use around foods. Smoke of burning flowers is as effective as the powder in keeping down insects.[2] The flower has shown antibiotic activity against mycobacterium tuberculosis.[3] Rarely cultivated as border plant; common in botanical gardens. Recent interest in organic or "natural" pesticides has stimulated interest in, e.g., combinations of pyrethrum, rotenone, and ryania.

Used as an external application for pediculosis and scabies.[32] Used veterinarily as an ascaricide, insecticide, and vermifuge. *Hager's Handbook* also mentions D,L- and L-stachydrin ($C_7H_{18}NO_2$) and a pyrrolidine alkaloid (mixture of stachydrine and choline = chrysanthemin). Flowers also yield (+) sesamin and beta-cyclopyrethrosin.[33]

Shoots contain on a zero-moisture basis — 13.0 g protein, 0.5 g fat, 79.4 g total carbohydrate, 23.6 g fiber, 7.1 g ash, 530 mg Ca, and 240 mg P per 100 g.[21] Flowers contain pyrethrin (0.7 to 2.0%, palmitic, pyrethrotoxic, and linoleic acids, and some volatile oil. Alcohol extracts of pyrethrum flowers yield several sesquiterpene lactones, among them beta-cyclopyrethrosin, chrysanolide, and chrysanin and beta-amyrin. Many sesquiterpenes are allergenic.

Toxicity — Urticaria, asthma, and rhinitis have been reportedly induced by pyrethrin and by working with pyrethrin preparations.[3] In humans, overdoses cause headache, tinnitus, facial pallor, epigastralgia, nausea, syncope, asphyxia. Overdoses of the powder will cause unconsciousness, albuminuria, pallor, collapse, slow and heavy heartbeat, respiratory difficulties, and nausea.

Antidote — Emetic.[33]

85. *CHRYSANTHEMUM PARTHENIUM* Pers. (ASTERACEAE) — Feverfew, Altamisa

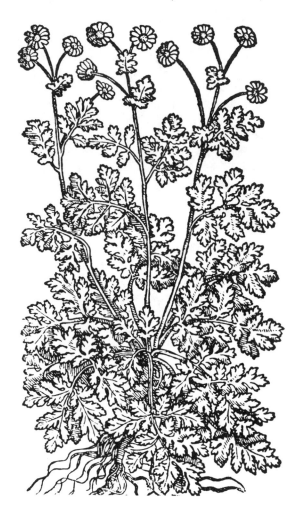

Introduced to the Americas as an ornamental medicinal herb. Used as an insecticide in Mexico.[42] Said to be antiseptic.[33] Once used as an antidote for overindulgence in opium.[42]

Reported to be abortifacient, antiseptic, aperient, carminative, depurative, emmenagogue, insecticidal, laxative, resolvent, stomachic, tonic, and vermifuge, feverfew is a folk remedy for anemia, bilious conditions, bruises, cancer, cold, colic, diarrhea, dysmenorrhea, dyspepsia, earache, fever, spasms, swellings, and worms.[33] Europeans use the infusion as aperient, carminative, sedative, and vermifuge, using it for hysteria and parturition. Costa Ricans consider the decoction cardioactive, digestive, and emmenagogue. Plant used as an enema for worms. Guatemalans boil the plant with *Salvia microphylla* for diarrhea. New Mexicans used the decoction in a sitz bath to halt menstruation. Old Mexicans regard it as antispasmodic, emmenagogue, and tonic. Venezuelans use the decoction for earache.[42] Homeopaths may use this like true camomile.[33]

Flowering herbage contains 0.02 to 0.07% yellow to greenish essential oil with L-camphor, L-borneol, some terpenes, and esters. Flowers contain parthenolide $C_{15}H_{20}O_3$, cosmosiin (apigenin-7-glucoside), and santamarin $C_{15}H_{20}O_3$.[33] Dry seeds contain 22.2% protein and 31.2% fat.[21]

Toxicity — Contact with the plant may cause skin irritations.[42]

86. *CICUTA MACULATA* L. (APIACEAE) — Water Hemlock

Sometimes transplanted to the wildflower or bog garden.[36] Too dangerous for any medicinal use at anything higher than homeopathic doses. Reported to be contraceptive, is a folk remedy for headache, neuralgia, and tumors.[4,32,33] Cherokee Indians chewed the roots for 4 days as a contraceptive, to become forever sterile. Iroquois used the plant for reducing inflammations and sprains.[43]

Said to contain cicutine and another alkaloid in the seed identical with conine.[2] Cicutine is said to be a C_3-epimere anthricine (hernandione, silicoline, desoxypodophyllotoxin) $C_{22}H_{22}O_7$.[33] Dry seeds contain 23.8% protein and 15.9% fat.[21]

Toxicity — Children have been fatally poisoned by eating the roots. Symptoms include frothing at the mouth, diarrhea, nausea, abdominal pain, dilated pupils, delirium, tremors, and periodic violent convulsions alternating with brief relaxations. Convulsions may be so violent that vomiting is prevented, and the tongue may be chewed up. Death is due to cardiac or respiratory failure.

To the physician — Convulsions must be managed to permit gastric lavage or emesis. Anesthesiologists may be necessary. Treat symptomatically, watching for oxygen deficiency. Recovery should be complete in 24 hr.[56] Administration of barbituates may prove helpful.[1]

87. *CIMICIFUGA RACEMOSA* (L.) Nutt. (RANUNCULACEAE) — Black Cohosh, Black Snakeroot

   Used only as medicine, a small amount is apparently sold in U.S. and Canadian herbal stores. It was one of the main components of Lydia Pinkham's Vegetable Compound. Leaves laid above a room are said to drive bugs away.[47]

   Amerindians used the rhizome for general malaise, gynecopathy, kidney ailments, malaria, rheumatism, and sore throat. American colonists used it for amenorrhea, bronchitis, chorea, dropsy, fever, hysteria, itch, lumbago, malaria, nervous disorders, snakebite, uterine disorders, and yellow fever.[18] Small doses are said to be adequate for menstrual cramps. It is said to promote quick delivery and to ease pain during childbirth. Occasionally, it is used to treat the symptoms of menopause, but prolonged use may irritate the uterus. The *Merck Index* (8th ed.) describes the root as an astringent bitter.[20] Experimentally antiinflammatory and hypoglycemic. In rats and mice, the drug has a strengthening effect on the female reproductive system and acts as a sedative. A chloroform-soluble resinous fraction is reported to have hypotensive activity in animals and to have peripheral vasodilatory effects on man. Perhaps such data lead to Tierra's statements: "It is an excellent remedy for high blood pressure"[28]? closely paralleling Kloss' statement: "wonderful remedy for high blood pressure."[144] Root is used in a cataplasm for scirrhous tumors (N.Y. city), or in a tincture or decoction by southern Amerindians.[4] The root is variously considered alterative, antispasmodic, antidotal, antitussive, aphrodisiac, astringent, diaphoretic, diuretic, emmenagogue, expectorant, narcotic, nervine, sedative, stomachic, and tonic. A tea from the root has been recommended for sore throat and rheumatism. Root is supposed to be specific against rattlesnake bite and St. Vitus' dance (chorea). Given as a syrup in various infantile disorders,[2] e.g., diarrhea. Said to be effective also for whooping cough, ringing ears, and to ease the paroxyms of chronic coughs.[2] Homeopathically recommended for stimulating the female system, e.g., in amenorrhea, dysmenorrhea, menorrhagia, and menopausal difficulties like arthrosis and rheumatism, and in parturition.[33]

Dry seeds contain 16.2% protein and 32.6% fat.[21] Said to contain the alkaloid *N*-methylcytisine and other unnamed alkaloids. The root contains 15 to 20% of an amorphous resinous substance called cimicifugin (macrotin) and a bitter principle, racemosin. Two triterpene glycosides, actein and cimigoside, are reported. Actein lowers blood pressure in cats and rabbits but not dogs.[37] Drug also contains two resins, fat, wax, starch, gum, sugar, and an astringent principle.[2] Acetic, butyric, formic, oleic, tannic, gallic, salicylic, palmitic, and isoferulic acids are also reported.

Toxicity — Overdoses produce nausea, vomiting, and dizziness.[28] May reduce the pulse and induce perspiration. Not listed as poisonous.[6,14] Overdose during pregnancy can cause premature birth.[126] Classified by the FDA as an herb of undefined safety: ''No pharmacologic evidence of any therapeutic value.''[62]

88. *CINCHONA* Sp. (RUBIACEAE) — Quinine

Quinine wine has long been regarded as a powerful tonic. Cinchona does have astringent, bitter tonic, analgesic, and anesthetic properties along with the antimalarial, antiarrhythmic, and febrifugal characters. Quinine is used in collyria as an anesthetic, astringent bactericide. Quinine sulfate is used to treat colds and leg cramps.[29] Quinine inhibits meat decay and yeast fermentation.[17] Morton adds that a mixture of quinine and urea hydrochloride (59% quinine) is injected as a sclerosing agent for internal hemorrhoids, hydrocele, varicose veins, and pleural vacities after thoracoplasty. Quinidine is also used to treat hiccups.[17] In 1973 0.18% (2,758,000) of all prescriptions in the U.S. (1.532 billion) contained quinidine.[98] As a matter of fact, more than 25% of all prescriptions contained one or more active constituents obtained from seed plants. Powdered bark used as a dentifrice, cinchona, extracts in hair tonics, promotes, and stimulates hair growth. Quinic and cinchona extracts are used in tonic waters, bitters, and liqueurs (up to 278 ppm red cinchona extract in alcoholic beverages); also, in baked goods, candies, condiments, frozen dairy desserts, and relishes.[29] Cinchona bark is still used for tanning after extraction of the alkaloids.[17]

The bark and/or its extracts are used in folk remedies for cancer (e.g., breast, glands liver, mesentery, spleen) carcinomata and tumors.[4] Reported to be anesthetic, antiperiodic, antiseptic, astringent, contraceptive, febrifuge, insecticide, schizonticide, stomachic, tonic, and uterotonic, quinine is a folk remedy for adenopathy, amebiasis, cancer, carditis, cold, diarrhea, dysentery, dyspepsia, felons, fever, flu, hangover, lumbago, malaria, neuralgia, neuritis, pertussis, piles, pinworms, pneumonia, sciatica, septicemia, sore throat, stomatitis, tumors, typhoid, and varicosities.[1,32,33] Bark is used as a bitter and stomachic; in small doses it is a mild irritant and stimulant of the gastric mucosa.[11]

The bark contains up to 16% (mostly 6 to 10%) total quinoline alkaloids (quinine, quinidine, cinchonine, cinchonidine). Other alkaloids include epiquinine, epiquinamine, hydroquinidine, hydroquinine, quinamine, etc. Tannins, quinovin, quinic acid, starch, resin, wax, and other items are also reported. According to *Hager's Handbook,* cuscamine, cuscanoidine, homocinchonine, javanine, dicinchonine, dicinchonine, and pericine are dubious names from the old literature. The *Handbook* devotes more than 20 fine-print pages to just the alkaloids of *Cinchona.*[33] Dry seed contains *circa* 18% protein, 16% fat, and 6% ash.[21] Serendipitously, malaria patients treated with Cinchona bark were found to be free of arrhythmia. Quinine and, moreso, quinidine regulate atrial fibrillation and flutter. Quinidine will suppress abnormal rhythms in any heart chamber. Quinine has been used to treat hemorrhoids and varicose veins.[11] Quinine and quinidine are also oxytoxic, "but the high incidence of fetal distress and intrauterine death associated with their use indicates that they should not be administered to induce labor unless the fetus has died in utero."[11] Quinine is supposed also to be prophylactic for flu.

Toxicity — Chronic use will lead to cinchonism (abdominal pain, disturbed vision, headache, nausea, skin rashes, tinnitus).[11] Quinic wine may cause gastric intestinal irritation.[16] Ground cinchona can cause contact dermatitis, urticaria, and other hypersensitive reactions in humans.[29] Eight grams of quinine can kill an adult in one dose. Acute hemolytic anemia and fatality from uremia have followed taking quinine as an abortifacient.[17] Cinchona barks approved for use in beverages only, not to exceed 83 ppm in the finished beverage (§ 172.510 and 172.575).[29]

89. *CINERARIA ASPERA* Thunb. (ASTERACEAE) — Mohodu-Wa-Pela

Attractive ornamental. Leaves are smoked as an intoxicant. Reported to be a fumitory and hallucinogen, the plant is a folk remedy for asthma and tuberculosis.[32] Sotho people inhale smoke from a pot of smoldering leaves to relieve congestion in acute cases of asthma and tuberculosis, probably affording temporary relief as a vasodilator.[54]

Toxicity — Emboden lists it as a narcotic hallucinogen of questionable status.[54]

90. *CINNAMOMUM CAMPHORA* (L.) J. S. Presl (LAURACEAE) — Camphor, Hon-Sho

Camphor is cultivated for the essential oil obtained from the wood by distillation. Camphor oil of commerce is oil from which the camphor has been removed. The oil, containing camphor and safrole, is used in the preparation of expensive perfumes. Several oils may be obtained by different processes: *white oil* — boiling below the camphor fraction, contains mainly cineol, limonene, and pinene; *brown oil* — obtained after camphor fraction distills over, contains mainly safrole; a terpineol fraction; and *blue oil* — obtained by direct fire distillation of the residue, made up mainly of sesquiterpene alcohols. Residue after distillation of the leaves may be mixed with guano, ammonium sulfate, lime, and pen manure as fertilizer. Camphor is used extensively (*circa* 65% of world production) as a plasticizer to make celluloid, and in the preparation of explosives, disinfectants, and other chemicals. Wood, highly esteemed for cabinet making, is easily worked and polished, and resistant to insect attack. It is widely used to make Japanese carved chests. It is also used as a fumigant and quite dangerously as a fumitory.[51] According to Purseglove et al., much of the world camphor is now produced synthetically from pinene, a turpentine derivative, or from coal tar. Safrole, produced from the residual oil after camphor extraction, is used to manufacture soap and perfume.[64]

The camphor, in a solution of wine, and used as a liniment, is said to be a folk remedy for tumors.[4] Camphor is extensively used in external substances as a counterirritant in treatment of muscular strains, gout, rheumatic conditions, and inflammations, in relieving itching skin, and as a weak antiseptic. In small doses respiration is stimulated, especially when respiration is depressed by ether, morphine, or other drugs. Hence, it is employed for asthma, bronchitis, emphysema, lung congestion, and rhinitis. It is used as an analeptic in cardiac depressions, myocarditis, and paralysis, has a calmative influence in cholera, convulsions, epilepsy, hysteria, and nervousness, and is used in treatment of serious diarrhea, favus, and toothache.[16] Leaves, steeped in alcohol, rubbed onto rheumatic pains. Camphor in olive oil is a popular Mexican remedy for bruises and neuralgia. The gum from the tree is considered anodyne, anthelmintic, antirheumatic, antispasmodic, carminative, diaphoretic, sedative, stimulant, and vulnerary. Homeopathically used for colic, collapse, cramps, flu, and rhinitis.[33] Parvati cites camphor as anaphrodisiac ("Arabians use it to lessen sexual desire") and cinnamon as aphrodisiac ("upon inhalation, the oil and incense is reported to act as a sexual stimulant to the female").[48]

Indian camphor leaves yield *circa* 22% camphor (2-bornanone, a crystalline ketone containing 97% or more $C_{10}H_{16}O$) and camphor oil containing caryophyllene, cineol, dipentene, pinene, and terpineol. Formosan leaves yield 44% camphor and oil with 28% cineol, 0.4% aldehyde (acetaldehyde, caprinaldehyde, isovaleraldehyde, myristinaldehyde, and stearinaldehyde), camphene, dipentene, limonene, phellandrene, and pinene. Examining the sesquiterpene alcohols, Takaoka et al.[131] found alpha- and beta-bisabolol, cadineol, t-cadinol, epicubenol, suvenol, t-muurolol, and nerolidol. *Hager's Handbook* adds *p*-cymol, orthodene, geraniol, ethylguaiacol, cuminalcohol, piperonal, piperonylacrolein, caffeic acid, quercetin, kaempferol, leucocyanadin, as well as caprylic-, lauric-, myristic-, citronellic-, and piperonlylic-acids.[33] Laurolitsin is reported from the roots. Camphor seeds yield 42% of an aromatic yellowish-white fat, consisting mainly of laurin.

Toxicity — The carcinogen safrole occurs in the oil, more in the roots than in other parts. Toxic doses of camphor taken internally result in convulsions, accompanied by vertigo and mental confusion, and may lead to delirium, and even to coma and death. As little as 700 mg can cause narcosis, 2000 mg inducing dulled senses, hallucinations, cramps, and unconsciousness. There may be intense burning in the stomach accompanied by illness and nausea. Paralysis may follow, finally affecting the respiratory and vasomotor centers.[33] It is interesting to compare *Hager's Handbook's* words above with those from "Herbal Highs", neither specifying whether they mean a gram of camphor chemical or a gram of camphor bark. Dangerously, "Herbal High" recites "one gram produces a pleasant, warm, tickling sensation on the skin, ecstatic mental excitation and an impulse to move about. Two grams brings on thought floods, ego loss, vomiting, amnesia, delirium and convulsions."[51] Death is not mentioned in the "Herbal High's" description.

To the physician — Overdoses should be followed by gastric lavage, then 15 to 20 g magnesium- or sodium-sulfate, with 500 mℓ water. The patient should be kept warm. Cramps can be controlled with barbiturates.[33]

91. *CINNAMOMUM VERUM* J. S. Presl (LAURACEAE) — Ceylon Cinnamon

The bark as the condiment, cinnamon, is used in food, dentrifices, incenses, and perfumes. Cinnamon bark oil, distilled from chips and bark of inferior quality, used in foods, perfumes, soaps, cordials, and in drug and dental preparations. Cinnamon leaf oil, distilled from dried green leaves, is a powerful germicide, and is also used in perfumes, spices, and in the synthesis of vanillin.[61] The essential oils are antiseptic, and ether-soluble fraction antioxidant.

Ceylon cinnamon considered aromatic, astringent, carminative, and stimulant. A fragrant cordial, useful for weakness of stomach and diarrhea, checking nausea and vomiting, and used in other medicinal mixtures.[61] Is a folk remedy for amenorrhea, arthritis, asthma, bronchitis, cancer, cholera, coronary problems, cough, diarrhea, dysentery, dyspepsia, fever, fistula, lumbago, lungs, menorrhagia, nephritis, phthisis, prolapse, proctosis, psoriasis, spasms, tumors, vaginitis, warts, and wens.[3,32,33] Cinnamaldehyde is antipyretic, hypothermic, and sedative. Cinnamon oil is antifungal, antiviral, bactericidal, and larvicidal. Eugenol is antiseptic. At a 0.1% concentration, a liquid CO_2 extraction of cinnamon bark completely suppressed growth of candida, escherichia, and staphylococcus.[29]

Cinnamon bark contains up to 4% volatile oil (*circa* 1%); tannins, consisting of polymeric 5,7,3',4'-tetrahydroxyflavan-3,4-diol units; resins; mucilage; gum; sugars; calcium oxalate; two insecticidal compounds (cinnzelanin and cinnzelanol); and coumarin.[29,33] The bark oil contains *circa* 60 to 75% cinnamaldehyde; also, eugenol, eugenol acetate, cinnamyl acetate, cinnamyl alcohol, methyl eugenol, benzaldehyde, cuminaldehyde, benzyl benzoate, linalool, caryophyllene, safrole, etc.[29,33] Of monoterpenes the root bark oil contains camphene, delta-3-carene, limonene, alpha-phellandrine, alpha and beta-pinene, sabiene, alpha-terpinene; of oxygenated monoterpenes borneol, camphor, 1:8-cineole, geraniol, linalol, piperitone, alpha-terpineol, and terpinen-1-ol; of aromatic monoterpenes, para-cymene, cinnamaldehyde, cinnamyl acetate, ethyl cinnamate, eugenol, eugenol acetate, safrole, benzyl benzoate, beta-caryophyllene, alpha gumulene, and gamma-ylangene.[64] The leaf oil contains eugenol, cinnamaldehyde, cinnamyl alcohol, cinnamyl acetate, ethyl cinnamate, eugenol acetate, benzaldehyde, etc., monoterpene hydrocarbons (e.g., camphene, carene, pinene, phellandrene, cymene, etc.), terpinene, humulene, isocaryophyllene, alphalylangene, coinferaldehyde, methyl cinnamate, and ethyl cinnamate.[29] Of the various types of cinnamon bark oils, that of *C. zeylanicum* has the largest amount of eugenol. Eugenol is reportedly absent in cassia bark oil.[29,64] Per 100 g, commercial ground cinnamon (mixture of cassia and cinnamon) contains 261 calories (1094 kJ), 9.5 g H_2O, 3.9 g protein, 3.2 g fat, 79.8 g total carbohydrate, 24.4 g fiber, 3.6 g ash, 1.228 mg Ca, 61 mg P, 38 mg Fe, 56 mg Mg, 26 mg Na, 500 mg K, 2 mg Zn, 260 IU vitamin A, 0.02 mg thiamine, 0.14 mg riboflavin, 1.3 mg niacin, and 28 mg ascorbic acid. There are 26 mg phytosterols.

Toxicity — Cinnamic aldehyde in perfumes can cause dermatitis, in toothpaste can cause sensitivity. Following ingestion of cinnamon, contact dermatitis may flare up as "pompholyx".[6] Eugenol has been reported to be irritant and a weak tumor promoter.[29]

92. *CITRULLUS COLOCYNTHIS* (L.) Schrad. (CUCURBITACEAE) — Colocynth, Bitter Apple, Wild Gourd

This is the wild gourd of two kings, mistaken for melons by one of the prophets during the death of Gilgal, and shred into a pot of pottage making it impotable.[35] Pulp of ripe fruits dry to form a powder used as a bitter and drastic purgative. This powder is inflammable enough to be collected for kindling. Arabs smear a bitter black extract of the rind of the fruit onto waterbags to repel camels.[59] The fruit is used to repel moths from wool. Seed, often removed from the poisonous pulp and eaten in Central Sahara regions, contains a fixed oil, which can be used as an illuminant. It is also used to darken gray hair.[59] The bitter fruits are eaten by grazing animals, the seeds gathered by desert rodents. Bedouins are said to be able to survive nearly 2 weeks on the seed after soaking in water, though probably with diarrhea. Goats and wild game eat the stem and leaves.[59]

Considered cathartic, ecbolic, emmenagogue, febrifuge, hydragogue, purgative, and vermifugal, colocynth is used in folk remedies for amenorrhea, ascites, bilious disorders, cancer, dropsy, fever, jaundice, leukemia, rheumatism, snakebite, tumors (especially of the abdomen), and urogenital disorders. The plant figures into remedies for cancer, carcinoma, endothelioma, leukemia, corns, tumors of the liver and spleen, even the eye.[4] This folk cancer "remedy" contains three antitumor ingredients: cucurbitacin B (active against PS and KB tumor systems), cucurbitacin E (active against LL and KB systems), and the D-glucoside of beta-sitosterol (active against CA, LL, and WA tumor systems).[10,132] For rheumatism the Bedoins tie a slice of fresh gourd onto the heel before retiring (I was told that Sinai Bedouins could taste the bitter gourd the following morning as a result). Milk soaked in the empty gourd is said to serve as a vermifuge. A tar-like exudate of the seeds is used to treat camels. Arabs use the fruit as an abortifacient. A pinch of powdered colocynth in cider was said to have arrested dropsy in Tripoli. Lebanese used the powdered pulp in resinous gum for piles. In Beirut, the pulp was used for open varicose veins. Lebanese also allude to its use for cancer, gangrene, and wounds. In Algeria, colocynth is used as a gargle and a mouthwash, a counterirritant in chest cold plasters. The rind with salt is poulticed onto frostbite. Ayurvedics use the root for arthritic pain, breast inflammation, ophthalmia, and uterine pain, the fruit for adenopathy, anemia, ascites, asthma, bronchitis, constipation, dyspepsia, elephantiasis, fetal atrophy, jaundice, leucoderma, splenomegaly, throat diseases, tubercular glands, tumors, ulcers, and urinary discharges.[26] Unani, considering the fruit abortifacient, carminative, and purgative, use it for brain disorders, epilepsy, hemicrania, inflammation, leprosy, ophthalmia, and weakness of the limbs.[38]

Active drug contains an ether-chloroform soluble resin, a phytosterol glycoside (citrullol), other glucosides (elaterin, elatericin B, and dihydroelatericin B), pectins, and albuminoids. Bitter substance is colocynthin and colocynthetin. Roots contain alpha-elaterin, hentriacontane, and saponins. Per 100 g, the seed is reported to contain 556 calories, 6.7 g H_2O, 23.6 g protein, 47.2 g fat, 19.5 g total carbohydrates, 1.5 g fiber, 3.0 g ash, 46 mg Ca, 580 mg P, and 0.11 mg thiamine.[21] The oil contains oleic, linoleic, myristic, palmitic, and stearic acids. Seeds contain phytosteroline, two phytosterols, two hydrocarbons, a saponin, an alkaloid, a polysaccharide or glycoside, and tannin. The two phytosterols may be the beta-sitosterol and lanosterol detected in all parts of the plant by Nag and Harsh.[133] *Hager's Handbook* lists also quercetin, and salts of acetic-, malic-, citronic-, and tartaric-acids, as well as caffeic-, chlorogenic-, ferulic-, and *m*-coumaric acids and alpha-spinasterol.

Toxicity — The purgative action is so drastic as to have caused fatalities.[3] One woman who took 120 gr to induce abortion died in 50 hr.[3] In case of poisoning, stomach evacuation is recommended, followed by oral or rectal administration of tincture of opium, followed by stimulating and mucilaginous beverages.[33]

93. *CLEMATIS VITALBA* L. (RANUNCULACEAE) — Traveler's Joy

Sometimes cultivated as a woody trailing or climbing vine with aromatic greenish-white flowers. Wyman describes it as an excellent vine for growing over fence or garden pergola where dense foliage is desired.[35] According to Rose,[47] small boys smoke the plants (perhaps the fuzzy seed mass). According to *Hortus III*,[36] young sprouts of the Eurasian variety *taurica* are sometimes eaten (I presume cooked with at least one change of water).

According to Hartwell,[4] the plant and/or leaf is used in folk remedies for cancers, indurations, and tumors. Reported to be diuretic, poison, purgative, rubefacient, sudorific, and vesicant, traveler's joy is a folk remedy for blood disorders, cancer, fever, itch, nephrosis, renosis, scrofula, ulcers.[2,32,33] Grieve describes the boiling of bruised roots and stems in water, one steeping in sweet oil, as a preparation for itch.[2] Homeopathically it is used for disorders of the penis, adenopathy, and indolent ulcers.[33] Rose says it is used as a Bach Flower Remedy for "the demused, far-away feeling often preceeding a faint or loss of consciousness."[47]

Dry seeds contain *circa* 15% protein and 14% fat.[21] Said to contain the active ingredients anemonin and protoanemonin, caulosaponin glycoside ($C_{54}H_{86}O_{16}$) which yields caulosapogenin ($C_{42}H_{66}O_6$) and glucose on hydrolysis, stigmasterol glycoside, ceryl alcohol, myricylalcohol, glucose, beta-sitosterol, trimethylamine, behenic-, caffeic-, chlorogenic-, and melissic-acid, *n*-triacontane, *n*-nonacosane, ginnone ($C_{29}H_{58}O$), ginnol ($C_{29}H_{60}O$), and campesterol.[33]

Toxicity — According to Lewis and Elvin-Lewis,[11] *Clematis* species have substances resembling protoanemonin, and may be fatal (the juice taken internally acts as a violent purgative). As noted above, this species does contain protoanemonin.

94. *COFFEA ARABICA* L. (RUBIACEAE) — Arabica Coffee, Arabian Coffee, Abyssinian Coffee

Dried seeds or "beans" are roasted, ground, and brewed to make one of the two important beverages in the western world. In its native Ethiopia, used as a masticatory since ancient times, it is also cooked in butter to make rich flat cakes. In Arabia a fermented drink from the pulp is consumed. Coffee is widely used as a flavoring, as in ice cream, pastries, candies, and liqueurs. Source of caffeine, dried ripe seeds are used as a stimulant, nervine, and diuretic, acting on central nervous system, kidneys, heart, and muscles. Indonesians and Malaysians prepare an infusion from dried leaves. Coffee pulp and parchment, used as manures and mulches, are occasionally fed to cattle in India. Coffeelite, a type of plastic, made from coffee beans. Wood is hard, dense, durable, takes a good polish, and is suitable for tables, chairs, and turnery. Coffee with iodine is used as a deodorant.[33] Caffeine has been described as a natural herbicide, selectively inhibiting germination of seeds of *Amaranthus spinosus*.[134] Caffeine is a widespread additive in over-the-counter diet pills, pain killers, and stimulants.[61] In 1982, radio news reports announced that an enzyme from green coffee beans can convert Type O to Type B blood. Experiments with animals and three human volunteers confirmed the long-sought possibility (WTOP News, January 14, 1982).

Reported to be analgesic, anaphrodisiac, anorexic, antidotal, cardiotonic, CNS-stimulant, counterirritant, diuretic, hypnotic, lactagogue, nervine, and stimulant, coffee is a folk remedy for asthma, atropine-poisoning, fever, flu, headache, jaundice, malaria, migraine, narcosis, nephrosis, opium poisoning, sores, and vertigo.[32,33] Coffee enemas have been suggested in asthma and cancer.[49]

Hager's Handbook[33] devotes pages of fine print to the chemicals reported from coffee, some of the more hazardous ones, perhaps, being acetaldehyde, adenine, caffeine, chlorogenic acid, guaiacol, tannic acid, theobromine, and trigonelline. Tyler[37] produces a chart comparing various caffeine sources to which I have added rounded figures from Palotti.[135]

Source	Caffeine (mg)
Cup (6 oz.) espresso coffee	310
Cup (6 oz) boiled coffee	100
Cup (6 oz) instant coffee	65
Cup (6 oz) tea	10—50
Cup (6 oz) cocoa	13
Can (6 oz) cola	25
Can (6 oz) Coca Cola®	20
Cup (6 oz) maté	25—50
Can (6 oz) Pepsi Cola®	10
Tablet caffeine	100—200
Tablet Zoom® (800 mg) (Paullinia cupana)	60

In humans, caffeine, 1,3,7-trimethylxanthine, is demethylated into three primary metabolites: theophylline, theobromine, and paraxanthine. Since the early part of the 20th century, theophylline has been used in therapeutics for bronchodilation, for acute ventricular failure, and for long-term control of bronchial asthma. According to Tiscornia et al.,[136] the sterol fraction of coffee seed oil contains 45.4 to 56.6% sitosterol, 19.6 to 24.5% stigmasterol, 14.8 to 18.7% campesterol, 1.9 to 14.6% 5-avenasterol, 0.6 to 6.6% 7-stigmasterol, and traces of cholesterol and 7-avenasterol. Coffee pulp is a valuable cattle feed, unpalatable to cattle at first. The pulp is comparable to corn in total protein, and superior to it in calcium and phosphorus content. In India, cattle feed on the pulp with no apparent ill effects. The ash of the "cherry" husk is rich in potash and, therefore, forms a valuable manure. Air dry coffee pulp contains 1.34% N, 0.11% phosphoric acid (P_2O_5), and 1.5% potash (K_2O). After compositing these values change to 0.91% N, 0.31% P_2O_5, 0.71% K_2O.[1] Leaves and reject seed may also be used as compost.[1] Leaves are reported to contain, per 100 g, 300 calories, 6.4% water, 9.3% protein, 5.5 g fat, 66.6 g total carbohydrate, 17.5 g fiber, 12.2 g ash, 1910 mg Ca, 170 mg P, 96.6 mg Fe, 2360 mg carotene equivalent, 0.00 mg thiamine, 0.21 mg riboflavin, and 5.2 mg niacin. Seeds contain, per 100 g, 203 calories, 6.3% water, 11.7 g protein, 10.8 g fat, 68.2 g total carbohydrate, 22.9 g fiber, 3.0 g ash, 120 mg Ca, 178 mg P, 2.9 mg Fe, 20 mg β-carotene equivalent 0.22 mg thiamine, 0.6 mg riboflavin, and 1.3 mg niacin.[21] Raw coffee contains *circa* 10% oil and wax extractable with petroleum ether. The fatty acids consist chiefly of linoleic, oleic, and palmitic acids, together with smaller amounts of myristic, stearic, and arachidic acids. From the unsaponifiable matter, a phytosterol, sitosterol, cafesterol, caffeol, and tocopherol have been isolated. Among the identified components of the volatile oil present in roasted coffee are acetaldehyde, furan, furfuraldehyde, furfuryl alcohol, pyridine, hydrogen sulphide, diacetyl, methyl mercaptan, furfuryl mercaptan, dimethyl sulfide, acetylpropionyl, acetic acid, guaiacol, vinyl guaiacol, pyrazine, *n*-methylpyrrole, and methyl carbinol. All these substances do not preexist in the unroasted coffee beans; some are, undoubtedly, the products of the roasting process and others are produced by the decomposition of the more complex precursors.[1]

Toxicity — Classed as a narcotic stimulant.[54] As a long-term drinker of 5 to 10 cups of coffee a day, I do not think I do myself any favors by drinking coffee. Tyler[37] cites "some evidence linking coffee and cancer of the pancreas . . . Caffeine . . . in large amounts produces many undesirable side effects — from nervousness and insomnia to rapid and irregular heartbeats, elevated blood sugar and cholesterol levels, excess stomach acid, and heartburn. It is definitely a teratogen in rats."[37] At 100 mg/kg theophylline is fetotoxic to rats, but no teratogenic abnormalities were noted. In therapeutics, theobromine has been used as a diuretic, as a cardiac stimulant, and for dilation of arteries. But at 100 mg, theobromine is fetotoxic and teratogenic.[137] Leung reports a fatal dose in man at 10,000 mg, with 1000 mg or more capable of inducing headache, nausea, insomnia, restlessness, excitement, mild delirium, muscle tremor, tachycardia, and extrasystoles. Leung also adds "caffeine has been reported to have many other activities including mutagenic, teratogenic, and carcinogenic activities; . . . to cause temporary increase in intraocular pressure, to have calming effects on hyperkinetic children . . . to cause chronic recurring headache . . . Coffee drinking has also been linked to myocardial infarction . . . cancer of the lower urinary tract (e.g., bladder), ovaries, prostate, and others."

Most of these reports have been challenged.[29] Jacobson cites numerous studies on pregnant animals and humans in which the equivalent of three to four daily cups of coffee caused birth defects such as cleft palate and missing bones.[138] Colonic irrigation with coffee can be quite hazardous. In some individuals, caffeine causes nervousness, restlessness, excitement, and insomnia. Patients with peptic ulcers, hypertension, and other cardiovascular and nervous disorders are usually advised by their physicians to refrain from drinking coffee. After noting that her husband had indigestion for 2 years, until he stopped his 8-cup-a-day habit, Rose quotes, "Coffee squeezes the adrenals dry and the next day when you need the energy of the adrenals they have nothing to give — resistance lowers and illness strikes with difficulties like circles under the eyes."[49] *Science News*[139] reports evidence that caffeine acts as a chemical "antagonist" blocking the normal action of adenosine, resulting in a hyperactive adenosine system, which causes abnormal sedation (and a craving for caffeine), whenever the effects wear off. Other scientists report an, as yet, unidentified compound in both regular and decaffeinated coffee (but not in tea or cocoa) that inhibits the binding of the opiate antagonist naloxone to opiate binding sites in the brain. The substance may also act as an opiate antagonist, blocking the opiates which work as both painkillers and mood elevators.[139] Chlorogenic acid may induce rhinitis and dermatitis in workers engaged in roasting, sorting, or grinding coffee.[17] The role of chlorogenic acid in the respiratory symptoms was discounted.[6] Still, workers develop asthma, dermatitis, rhinitis, and urticaria.[6] Inhalation of coffee bean dust can produce coffee worker's lung, a type of allergic alveolitis. Lest this review seems to bias against my caffeine-containing friends: cocoa, cola, and coffee, I should cite Ponte's recent paper "All About Caffeine".[140] He first discusses the evidence that caffeine is addictive, that more than 2 1/2 cups of daily coffee can cause nervousness, anxiety, and shortness of breath; that more than 10 cups can cause ringing in the ears, mild delirium, flashes of light, rapid irregular heartbeat, rapid breathing, muscle tension, and trembling; that under extreme conditions caffeine may poison pregnant rats and induce birth defects; that caffeine does affect the heart; that it may be linked to cancer of the pancreas. Ponte then notes the conclusion of one of the world caffeine experts, B. A. Kihlman, University of Upsala, Sweden: "When all the evidence is weighed," he has written, "only pregnant women, and those with some special health problems such as arrhythmia, seem even remotely at risk from heavy consumption of regular coffee."[140] Coffee extracts are GRAS (§ 182.20), but the GRAS status of caffeine (§ 182.1180) is being reassessed.[29]

95. *COLA ACUMINATA* (Beauv.) Schott & Endl. (STERCULIACEAE) — Kola Nuts, Cola, Guru

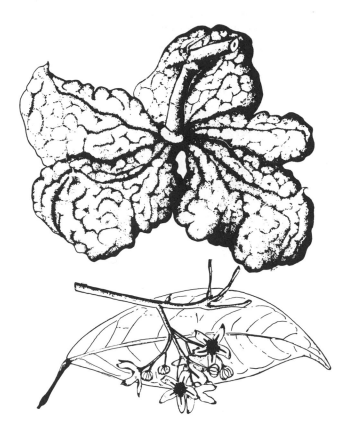

Widely used as a flavor ingredient in cola beverages, but also used in baked goods, candy, frozen dairy desserts, gelatins, and puddings,[29] kola plays important role in social and religious life of Africans. Beverage made by boiling powdered seeds in water, equal in flavor and nutriment to cocoa. Seeds also used as a condiment. Dye utilized from red juice. Wood valuable, light in color, porous, and used in shipbuilding and general carpentry. Tree often planted as ornamental. Cola is said to render putrid water palatable.[61]

Nuts used as diuretic, heart tonic, and masticatory to resist fatigue, hunger, and thirst. The powdered bark is used for malignant tumors and cancer. The tea made from the root is said to remedy cancer.[4] A small piece of nut is chewed by Africans before mealtime to improve digestion. On the other hand, it is chewed as a stimulant and appetite depressant, e.g., during religious fasts. Jamaicans take grated seed for diarrhea.[42] Powdered cola is applied to cuts and wounds.[2] Formerly used as a CNS-stimulant and for diarrhea, migraine, and neuralgia.[29] Regarded as aphrodisiac, cardiotonic, CNS-stimulant, digestive, diuretic, nervine, stimulant, and tonic.[32] The fresh drug is used, especially in its native country, as a stimulant, social drug, being mildly euphoric.[33]

Kola nut important for its caffeine content and flavor; caffeine content 2.4 to 2.6%. Nuts also contain theobromine (<0.1%) and other alkaloids, and narcotic properties. Seeds also contain betaine, starch, tannic acid, catechin, epicatechin, fatty matter, sugar, and a fat-decomposing enzyme. From a bromatological point of view, cola fruits contain, per 100 g, 148 calories, 62.9% water, 2.2% protein, 0.4% fat, 33.7% carbohydrates, 1.4% fiber, 0.8% ash, 58 mg Ca, 86 mg P, 2.0 mg Fe, 25 μg carotene, 0.03 mg thiamine, 0.03 mg niacin, 0.54 mg riboflavin, and 54 mg ascorbic acid.[21] *Hager's Handbook* suggests 1.5 to 2%

caffeine, up to 0.1% theobromine, 0.3 to 0.4% D-catechin, 0.25% betaine, 6.7% protein, 2.9% sugar, 34% starch, 3% gum, 0.5% fat, 29% cellulose, and 12% water.[33]

Toxicity — Caffeine in large doses is reported to be carcinogenic, mutagenic, and teratogenic.[29] Caffeine is also viricidal, suppressing the growth of polio, influenza, herpes simplex, and vaccinia viruses, but not Japanese encephalitis virus, Newcastle disease, virus, and type 2 adenovirus. In 1978, an FDA advisory panel concluded that caffeine, as it is added to cola soft drinks, should be subject to a more restrictive regulatory approach. Removal of caffeine from the GRAS list "was urged".[141]

96. *COLA NITIDA* (Vent.) Schott & Endl. (STERCULIACEAE) — Gbanja Kola

Containing more caffeine, this species is more valued than *C. acuminata*. Nuts are used in West Africa to sustain people during long journeys or long hours of work. Kola, cola, or kolanut is the dried cotyledon of *C. nitida*, or of some other species of *Cola*. In the U.S. the kolanut is used in the manufacturing of nonalcoholic beverages. The tree is valued for its wood which is whitish, sometimes slightly pinkish when fresh; the heartwood is dull yellowish-brown to reddish-tinged. Wood is suitable for carpentry and some construction work as house building, furniture, and boat building. Wooden platters, domestic utensils, and images are often carved from the wood. Sometimes trees are planted for ornamental purposes.[61]

In the tropics where it grows, the fresh nut is chewed as a stimulant, similar to the betelnut. According to *Herbal Highs,* Africans believe kola acts as an aphrodisiac on males, while promoting conception in women.[51] Said to be astringent, nervine, poison, restorative, sedative yet stimulant, stomachic, and tonic, this cola is used in folk remedies for dysentery, dyspepsia, exhaustion, malaria, nausea, and toothache.[32]

Yields no less than 1% anhydrous caffeine, usually about 3.5%, and nearly 1% theobromine.

97. *COLCHICUM AUTUMNALE* L. (LILIACEAE) — Autumn Crocus, Meadow Saffron, Colchicum

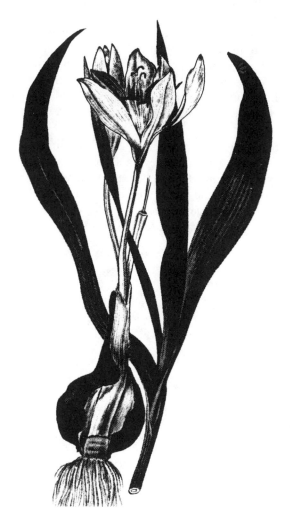

Colchicum has long been used medicinally for its cathartic and emetic properties as well as a remedy for gout, but not for arthritis. Colchicine is one of several compounds said to both cause and "cure" experimental cancers. 3-Desmethylcolchicine (isolated from *Colchicum speciosum*) has certain advantages over other antitumor colchicine derivatives, one of which (demecolcine or *N*-deacetyl-*N*-methylcolchicine) has been recognized, temporarily at least, as a clinically active antitumor agent.[10] In bioassay, 3-desmethylcolchicine, has shown significant activity against the L-1210 lymphoid leukemia in mice and cytotoxicity against cells derived from human carcinoma of the nasopharynx.[10] Powdered seed have been introduced into alcoholic beverages for homicidal purposes. It has also been used for rheumatism, and as an alterative, cathartic, diuretic, emetic, and sedative.[32] Corms are usually sold in transverse slices, notched on one side and sometimes reniform in outline, white, and starchy internally. Widely grown as an ornamental. The poisonous alkaloid, colchicine, is used in manipulating the genetic structure of plants and animals. Drug of choice for acute gout, colchicine is also suggested for Bright's disease, cholera, colic, skin ailments, and typhus.[59]

The paste of the bulb in a cataplasm is said to be a folk remedy for chronic and indurated tumors.[4] A poultice of the leaf is said to remedy corns. A decoction of the seed is said to remedy leukemia.[4] Reportedly alterative, diuretic, laxative, poison, sedative, and sudorific, colchicum has been advocated in many diseases: Bright's disease, cancer, cholera-typhoid, colic, enlarged prostate, gout, palsies, rheumatism, and skin complaints.[32]

Seeds contain up to 0.8% colchicine and corms about 0.6%. Taste of the poisonous corm is sweetish, then bitter and acrid; odor of fresh root is radish-like.

Toxicity — Colchicine occurs as pale yellow, amorphous scales or powder which gradually turns darker on exposure. Overdoses of colchicine may cause intestinal pain, diarrhea, vomiting, and may even cause death. Several other alkaloids are reported: colchamine, colchiceine, colchicerine, demecolcine, 2-demethylcolchicine, O-demethyl-N-deacetylcol-chine, N-dea-cetyl-N-methylcolchicine, N-formyl-Ndeacetylcolchicine, alpha- and beta-lum-icolchi-cine, and 3-methylcolchicine. As little as 7 mg colchicine has proved lethal, but the usual lethal dose is closer to 65 mg. The lowest human lethal dose (LDLO = 186 μg in 4 days).[15] Fresh corms are irritant to the skin and mucous membranes. Leaves are said to cause dermatitis. Demecolcine is irritant to the intact skin.

To the physician — Hardin and Arena suggest emesis or gastric lavage, shock therapy, supportive, and symptomatic treatment.[34]

98. *COLEUS BLUMEI* Benth. (LAMIACEAE) — El Nene

Common ornamental foliage plant. Mazatec Indians are said to have used the plant (chewed leaves or ingested tea) as a hallucinogen.[51]

Reported to be anodyne, el nene is a folk remedy for asthma, colic, conjunctivitis, cough, dyspepsia, and elephantiasis.[32] Listing the species under *Coleus scutellariades,* Perry[16] notes that Solomon Islanders apply the sap to cuts and sores. New Guinea natives apply this or a related species for myalgia, and as a snuff for colds. Philippinos apply heated leaves to foot infections and powdered leaves to bruises and headaches. In Indonesia, women are said to swallow the leaf sap as a contraceptive; the decoction, regarded as collyrium, depurative, and emmenagogue, is taken for hemorrhoids; the leaves warmed with oil are applied to boils. Malayans also use the leaves as a collyrium for conjunctivitis; the decoction for abdominal distension, dyspepsia, parturition, and nausea, the boiled root for distension of the stomach. Malayans poultice pounded leaves onto cardalgia or gastralgia.[16]

An uncharacterized alkane, sterol, and triterpene and the flavonoids aromadendrin and cyanidin 3,5-di-*O*-beta-D-glucosyl-*p*-coumarate are reported, but would not explain the pharmacological effects.[80]

Toxicity — Classed as a narcotic hallucinogen.[54]

99. *CONIUM MACULATUM* L. (APIACEAE) — Hemlock

The poison which Socrates took, hemlock is too dangerous for herbal administration by the uninitiated. Unripe fruits were used for capital punishment in ancient Greece. And there is penal punishment as well: "crushed and placed upon a man's genitals, it does away with the lust for women."[45] Sometimes used as an antidote to strychnine, and to the poisons of rabies and tetanus.[42]

Greek and Arab physicians used the hemlock for indolent tumors, pain, and swellings in the joints, skin affection, and it was often poulticed onto ulcers and cancers. Praised by Avicennia as a cure for breast tumors. Millspaugh[19] related that 26 of 46 cases of cancer were ameliorated by *Conium*. Few plants have more anticancer citations in Hartwell's "Plants Used Against Cancer".[4] Hartwell reports that hemlock is used for cacoethes, cancer of the breast, intestine, nose, parotids, skin, sternum, and uterus; carcinoma; indurations of the breast, glands, liver, and viscera; scirrhus of the breast, liver, pancreas, and spleen; tumors of the breast, ganglia, liver, mannae, mesentery, neck, penis, scrotum, and testicles; ulcers; and wens. In Iran the fruits are applied externally as an anodyne. Inhaled for asthma, bronchitis, whooping cough. Used as an alterative in scrofula. Dioscorides is said to have used *Conium* mixed with wine as a cataplasm in erysipelas. The drug is described as analgetic, anaphrodisiac, anodyne, antispasmodic, aphrodisiac, nervine, and sedative. Has also been recommended in carditis, chorea, delirium, epilepsy, erysipelas, glandular swellings, jaundice, larygismus stridulus, leprosy, mania, nervous diseases, neuralgia, palpitations, paralysis, pertussis, rheumatism, sclerosis, sores, spasms, syphilis, tetanus, ulcers, and wens.[32,42,45] Homeopathically suggested for adenopathy, amenorrhea, cachexia, colic, coughs, dysmenorrhea, eczema, hyperlactation, hepatitis, icterus, impotence, styes, and vertigo.[33] Conine has been used in convulsive and spasmodic diseases, like asthma, chorea, epilepsy, pertussis, and tetanus.[45]

Dry seeds contain 20.6 to 24.9% protein and 11.6 to 17.9% fat.[21] The alkaloid coniine is the most important constituent, and may constitute up to 1.65 to 1.75% of the plant. Other alkaloids are conhydrine, conicein, ethyl piperidine, methyl-coniine, and pseudoconhydrine. *Hager's Handbook* lists, also, diosmin, caffeic acid, chlorogenic acid, lignin, uronic acid, pentosan, arabinose, xylose, arabion-4-O-methylglucuronoxylan, 4-cumarine, falcarinone, and falcarinol-on.[33]

Toxicity — Classed appropriately by the FDA as an unsafe herb[62] containing the poisonous alkaloid coniine, and other closely related alkaloids. Plant can also cause contact dermatitis. Ingestion may cause debility, drowsiness, nausea, labored respiration, paralysis, asphyxia, and death.[20] Following lethal doses, animals rapidly begin to show symptoms; among them: paralysis of the tongue, mydriasis, head pressure, giddiness, nausea, vomiting, diarrhea, and collapse into central paralysis, first the feet and legs, then the buttocks, arms, then paralysis of the swallowing and speech. With increasing dyspnea and cyanosis, death ensues through central respiratory paralysis. Lethal dose is about 500 to 100 mg coniine for man.[33]

To the physician — "Gastric lavage or emesis; saline cathartic; keep airway clear, oxygen and artificial respiration; anticonvulsive therapy, with short-acting barbiturates."[34] Keep patient warm. Strychnine and picrotoxin are suggested as antidotes,[33] as well as lemon juice and vinegar.[45]

100. *CONSOLIDA AMBIGUA* (L.) P. W. Ball & Heywood (RANUNCULACEAE) — Larkspur

Powdered seeds are found in some commercial lice remedies.[6]

Reported to be cathartic, emetic, insecticide, parasiticide, pediculicide, poison, larkspur is a folk remedy for scorpion sting and toothache.[32] In "Lebanese Folk Cures", mashed leaves are wadded into little rolls as suppositories for hemorrhoids.[23] Guatemalans crush the leaves in water and apply them to toothache. Brazilians prescribe the seed infusion as a tonic for debility.[42]

Seeds contain 39% fatty oil. Contains ajacine ($C_{34}H_{48}N_2O_9 \cdot 2H_2O$), ajaconine ($C_{22}H_{33}NO_3$), ajacinine ($C_{22}H_{35}NO_6$), ajacinoidine, delcosin, acetyldelconsine (alkaloid B) ($C_{26}H_{41}NO_8$), alkaloid D ($C_{48}H_{64}N_2O_{11}$), and delsolin.[33]

Toxicity — Seed and leaf can induce dematitis. Alkaloids ajacine and delphinine, occurring in seed and young plants of most *Delphinium* spp. Ingestion may cause stomach upset and nervous symptoms, death may occur if the plant is eaten in large quantities, especially by children.[11] In Asia the seeds are reported to poison cattle; still they are used as cathartic and emetic.[16]

To the physician — Emesis or gastric lavage, 2 mg subcutaneous atropine, repeating as needed, maintenance of blood pressure, and artificial respiration.[34]

101. *CONVALLARIA MAJALIS* L. (LILIACEAE) — Lily-of-the-Valley

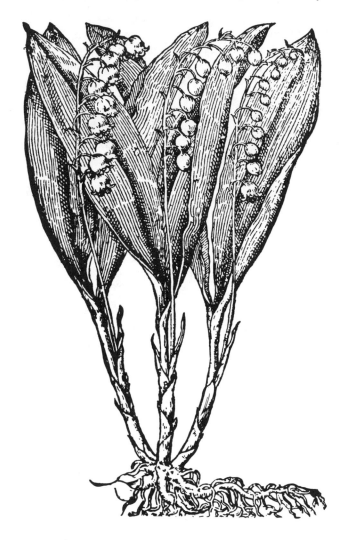

Dark green ornamental ground cover, with attractive, aromatic white flowers. Flowers source of essential oil. Dried flowers an ingredient of German snuffs. Can be used as a substitute for digitalis, when digitalis fails to act on the heart as desired. Once used as an antidote in gaseous warfare. The herb is said to accumulate silver.

Ointment of the root in lard used for burns, scalds, and ulcers to prevent scar tissue. Valued as cardiotonic, diuretic, emetic, and nervine. Flowers have been used like digitalis, and veterinarily as a cardiac tonic and diuretic.[20] As a digitalis substitute, it is strongly recommended in valvular heart disease, cardiac debility, and dropsy.[2] Used for epilepsy in Russia.[33] Flower decoction used for urinary obstructions, as an internal application for apoplexy, palsy, and weak memory. Roots suggested for apoplexy, convulsions, dropsy, epilepsy, heart ailments, palsy, and vertigo.[44] Flowers steeped in wine have been recommended as an external application to rheumatism and sprains. Decoction of the bruised roots in wine recommended for fevers. Homeopathically suggested for arrhythmia, angina pectoris, cardiac insufficiency.

Contains more than 20 cardiac glycosides, including convallamarin, convallatoxin, convallarin, and volatile oil. Asparagin, resin, and saponins are also reported. *Hager's Handbook*[33]

adds rutin, isorhamnetic kaempferolglycoside, chlorogenic-, caffeic-, ferulic-, chelidonic-, malic-, citronic-, and acetidine-2-carbonic-acids; calcium oxalate, choline, carotene, and wax.

Toxicity — Though usually listed among the poisonous plants, there are few reported cases of toxicity. Symptoms include: nausea, vomiting, diarrhea, diuresis, dazedness, giddiness, oppressiveness of the chest, feeble heart, lethal doses leading to collapse, and death. *Hager's Handbook* suggests treating convallaria poisoning like digitalis poisoning.[33]

To the physician — Hardin and Arena[34] suggest gastric lavage, or emesis, potassium, procainamide, quinidine sulfate, disodium salt of edetate have been used effectively. Although frequently handled by florists and gardeners, the plant causes few or no authenticated cases of dermatitis.

102. *CORIARIA THYMIFOLIA* Humb & Bonpl. (CORIARIACEAE) — Shanshi

Although the fruit is considered toxic almost everywhere, some Ecuadorians eat the berries (give a sensation of flight).[54] Colombians are reported to use doses of 1 mg as a stimulant in case of collapse.[57] A red dye, used as the indelible ink "chani", is produced from the red juice of the berries.[22,38]

Reported to be astringent, canicidal, poison, and rodenticidal, shanshi is a folk remedy for collapse and diarrhea.[32,57]

Active principles are the sesquiterpenes coriamyrtine, coriatine, pseudotatine, and tutine.[54] Hallucinogenic activity may be due to an unidentified glucoside.[58] Ellagic-, gallic-, acetic-, coriaric-, and succinic acid are reported, along with quercetin. Seeds contain 22.6% oil.[33]

Toxicity — Classified as a narcotic hallucinogen (giving flight sensations).[54] The LD_{50} for the leaves is 3.75 mg/kg, mature fruits 1.55, and green fruits 0.45 mg/kg. Frequent symptoms of intoxication include stupor, vertigo, convulsion. Death may result from asphyxia, respiratory paralysis, and heart failure.[22,57]

To the physician — Garcia Barriga suggests nembutal.[57]

103. *CORNUS FLORIDA* L. (CORNACEAE) — Dogwood, American Boxwood

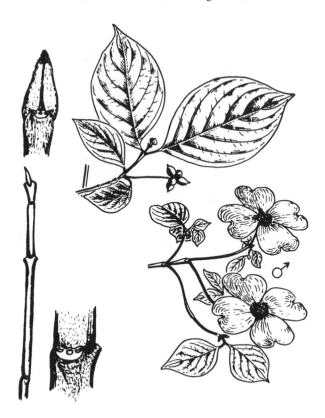

White- and pink-flowered forms planted as ornamentals. Hard wood used for wheels, turnery, barrel-hoops, bearings, etc. Root bark once used as a substitute for quinine. Inner bark once used to make black ink, mixed with iron and gum arabic in rainwater. Scarlet dye obtained from the root bark.[45] Twigs used as a chew stick. Cited by the *Merck Index* as an astringent bitter.[20] Berries, boiled and pressed, yield a limpid oil. Grieves *Herbal* makes the improbable statement that the oil is used in soups in Italy.[2] Leaves said to make good fodder. Flower infusion drunk as a substitute for camomile tea. The hard compact wood is used for farm tools and instruments.

Bark, especially of the root, smoked for breast cancer (Ohio) or taken as a tea for various cancers.[4] Amerindians used the bark like cinchona, as an aperient, astringent, antimalarial, stimulant, and tonic. As an antiseptic it is poulticed onto anthrax, erysipelas, and indolent ulcers. Ripe fruit steeped in brandy, are used as a stomachic. Tincture of the berries said to restore tone to the alcoholic stomach.[45] In lowland South Carolina, bark soaked with black cherry bark in whiskey is used as a remedy for chill and fever.[46] A dogwood bark infusion is used for nerves. Steeped in whiskey, the bark has been recommended for dysentery and typhoid. Homeopathic tinctures recommended in dyspepsia, intermittent fever, and pneumonia.[30] Recommended for ague, cancer, colic, dyspepsia, fever, hepatitis, hysteria, inappetence, jaundice, malaria, myalgia, stomatosis, and typhus.[32,33,45]

Bark contains gallic and tannic acid, resin, gum, oil, wax, lignin, lime, the glucoside cornin, kaempferol, quercetin, scyllitol, inositol, *n*-nonacosan, and ursolic acid.[2,33] The root contains tannin (3%), gallic acid, betulinic acid, resin, and the glycoside verbenalin.[33] Seeds contain *circa* 35% protein, 49.4% fat, 73.3% total carbohydrate, 16% fiber, 3.6% ash, 2,590 mg Ca, and 150 mg P.[21]

104. *CRATAEGUS OXYACANTHA* L. (ROSACEAE) — Hawthorn

Ornamental tree with edible fruits, often made into marmalade. A vitamin C-rich beverage like a tea is made from the rind and pip. Berries have been made into excellent brandies and liqueurs. Leaves have been used to adulterate tea.[2] The wood is yellowish-gray, close and even-grained, smooth, durable, hard, and heavy (weight 46 to 56 lb/ft³). It is used for axe handles and walking sticks. It is suitable for engraving and is as good for this purpose as boxwood. According to Grieve's *Herbal* it makes the hottest fire of any fuelwood known.[2] Its charcoal is said to melt pigiron without the aid of a blast.

Reported to be astringent, cardiotonic, cyanogenetic, depurative, diuretic, fumitory, stomachic, tonic, hawthorn is a folk remedy for arteriosclerosis, dropsy, dyspnea, hypertension, nephrosis, sore throat.[2,32,37] *The Wealth of India*[1] describes the fruit extract as a cardiac tonic, used for organic and functional heart diseases such as dyspnea, hypertrophy, valvular insufficiency, and heart oppression. (After stopping the drug, the blood pressure returns to its previous level.) More recently, Tyler summarizes: "it acts on the body in two ways: First it dilates the blood vessels, especially the coronary vessels, reducing peripheral resistance and thus lowering the blood pressure. It is thought to reduce the tendency to angina attack. Second, it apparently has a direct, favorable effect on the heart itself which is especially noticeable in cases of heart damage. It therefore seems to be a relatively harmless, mild heart tonic which apparently yields good results in many conditions where this kind of therapy is required."[37]

Fresh fruit contains citric-, crataegus-, and tartaric-acids, pectins, and fatty oil (0.76%). Sugars, predominantly fructose and glucose, constitute *circa* 35% (ash *circa* 32%) of DM. Cyanogenetic glucosides are present in the young shoot, amygdalin and emulsin in the seed. Leaves contain calcium oxalate. Bark fat contains palmitic-, stearic-, myristic-, and resinic-acid, ceryl alcohol, tanning matter, invert sugar, polysaccharides, salts of oxalic and tartaric acids, esculin (crataegin), and a lupeol-like compound. Wood contains 25% pentosans. Flowers contain 0.157% essential oil, quercitin, quercitrin, and trimethylamine. Hawthorn marmalade has up to 0.15% vitamin C,[1] oligomeric procyanadins (dehydrocatechin) are probably foremost among the active ingredients. *Hager's Handbook* adds many other compounds, some of them pharmacologically active: ethylamine, dimethylamine, trimethylamine, isobutylamine, isoamylamine, ethanolamine, beta-phenylethylamine, choline, and acetylcholine; also the purine derivatives adenosine, adenine, guanine, resinic-acid, chlorogenic acid, and caffeic acid and the flavone quercetin-3-galactoside and 0.51 to 0.62% beta-sitosterol (0.65 to 0.78% in the flower), sorbitol, 0.14 to 0.25% vitamin C, 0.15% amygdalin, catechin, tannins, pectin, chlorogenic and caffeic acid, etc.[33] Fruit contains fatty acids with oleic-, linoleic-, stearic-, palmitic-, myristic-, and arachidic acids and the universal phytosterols, as well as wax, tannin, mucilage, essential oil, nonacosanol-10-ol, paraffin sorbitol, nicotinic acid. Fruits share with the flowers the triterpene acids, ursolic-, oleanolic-, and crataegolic acid, as well as "remarkable" quantities of Al, Na, K, Ca, and P salt, tartaric, and citric acid.[33]

Toxicity — Roots, leaves, and flowers all contain cardioactive compounds. One paper cited in Mitchell and Rook[6] notes that corneal scratches with the thorns led to blindness in 88 of 132 Irish accidents. With cyanogenic and cardioactive compounds, not to mention tannin, scattered throughout the plant, it should not be taken lightly.

105. *CROCUS SATIVUS* L. (IRIDACEAE) — Saffron, Saffron Crocus

Saffron is cultivated for the coloring dye obtained from the stigmas of the flowers; about 100,000 flowers yield 1 kg saffron. Dye used chiefly as a coloring agent and spice in cookery (especially Spanish), soups, stews, especially chicken dishes, and in confectionery to give color, flavor, and aroma. Used in cosmetics for eyebrows and nail polishes, and as incense. Dioscorides mentions its use as a perfume. Dissolved in water, it is used as an ink and is applied to foreheads on religious and ceremonial occasions. The gold of cookery is now almost as expensive as the gold of jewelry.

Saffron is an extremely often cited folk remedy for various types of cancer, e.g., tumors of the abdomen, bladder, ear, eye, kidney, liver, neck, spleen, stomach, and tonsils, as well as cancers of the breast, mouth, stomach, and uterus, venereal condylomata, warts, etc.[4] Saffron is used to promote eruption of measles, and in small doses is considered anodyne, antihysteric, antiseptic, antispasmodic, aphrodisiac, balsamic, cardiotonic, carminative, diaphoretic, ecbolic, emmenagogue, expectorant, nervine, sedative, stimulant, and stomachic, but overdoses are narcotic, and saffron corms are toxic to young animals. Apoplexy and extravagant gaiety are possible aftereffects. Saffron is not included in American and British pharmacopeias, but some Indian medical formulas still include it. It is sometimes used to promote menstruation. In India, saffron is regarded for bladder, kidney, and liver ailments, also for cholera. Mixed with ghee, it is used for diabetes. Saffron oil is applied externally to uterine sores. Saffron enters folk remedies for blood disorders, catarrh, cold, depression, fear, fever, hepatosis, hysteria, measles, melancholy, menoxenia, neurosis, puerperium, sclerosis, shock, skin ailments, spasms, and splenomegaly.[32]

Crocus sativus has been reported to contain colchicine and quercitin but the colchicine is more properly belongs to *Colchicum*. Spanish saffron with 15.6% water contains 12.6% protein, 4.7% fixed oil, 0.8% volatile oil, 57.3% N-free extract, 12.0% starch equivalent, 4.9% fiber, 4.0% ash. Another sample with 11.9% water showed 310 calories per 100 g, 5.8 g fat, 65.4 g total carbohydrate, 3.9% fiber, 5.4% ash, 111 mg Ca, 252 mg P, 11.1 mg Fe, 148 mg Na, and 1724 mg K.[21] Saffron contains a mixture of yellow glycosides, crocin being a yellow-red pigment. On hydrolysis, crocin yields gentiobiose and crocetin. *Hager's Handbook* mentions, also, 8 to 13% fatty oil (with glycerin esters of palmitic-, stearic-, lauric-, and oleic-acids, 56 to 138 γ/g vitamin B_2, 0.7 to 4.0 γ/g vitamin B_1, 13.3% starch, alpha-, beta-, and gamma-carotene, hentriacontane, xanthophyll, lycopine, and zeaxanthin.[33] Pfander and Schurtenberger[142] isolated phytoene, phytofluene, tetrahydrocopene, beta-carotene, zeaxanthin, and crocetin. They interpreted the absence of C_{20} hydrocarbon precursors of crocetin to indicate a degradation pathway for the biosynthesis of crocetin. Aroma is due to a volatile oil, safranol, present in minute quantities in the stigmas. The essential oil (0.4 to 1.3%) contains pinene, safranal, and cineole. The bitter-tasting principle is picrocrocin which hydrolyzes to glucose and safranal. Saffron odor is produced when picrocrocin separates at time of drying of stigmas.

Toxicity — Fatalities have resulted from use of saffron as an abortifacient.

106. *CROTON ELEUTHERIA* Sw. (EUPHORBIACEAE) — Cascarilla

Bark extracts used in beverages and bitters not to exceed *circa* 800 and 100 ppm. Essential oil used, up to *circa* 75 ppm used in alcoholic and nonalcoholic beverages, baked goods, candy, condiments, desserts, frozen dairy products, and relishes.[29] Prominent in some oriental and masculine perfumes, up to 0.4%, creams, detergents, lotions, perfumes, and soaps. Also, used as an aromatic bitter and tonic, sometimes added to tobacco.[29] Sometimes added to cinchona to arrest vomiting caused by the cinchona. Morton[42] states that it was used instead of quinine when the quinine caused vomiting. Bark yields a good black dye.[2] Essential oil is antiseptic.[29]

Reported to be balsamic, digestive, hypotensive, narcotic, stomachic, tonic, is a folk remedy for bronchitis, debility, diarrhea, dysentery, dyspepsia, fever, leprosy, malaria, nocturnal emissions.[2,32,57] Colombians use the bark for dysentery. Bahamans take the leaf and bark tea for colds, fever, and flu.[42]

Contains 1.5 to 3.0% essential oil containing primarily *p*-cymene, dipentene, *d*-limonene, beta-caryophyllene, gamma-terpineol, and eugenol; also, cascarillic acid, cascarillidone, carcarillone, cineole, cuparophenol, methythymol. Bark also contains *circa* 15% resin, cascarillin A (a bitter substance), vanillin, betaine and tannin, gum, wax, pectic acid, lignin, potassium chloride.[2,29,33,42]

Toxicity — Possibly narcotic.[2] As an aromatic additive to tobacco, it is said to have caused intoxication and vertigo[42] (GRAS § 182.20).[29]

107. *CRYPTOSTEGIA GRANDIFOLIA* R. Br. (ASCLEPIADACEAE) — Rubber Vine

An attractive tropical vine, planted as an ornamental in the U.S., formerly considered as a source of rubber. Plants also yield useful resins and fibers, and might be as promising for whole plant utilization as its relatives Asclepias and Calotropis. The latex makes an indelible stain on cloth.[42] It is also suggested as the source of cardioactive drugs,[33] formulas of some digitalis-like compounds being listed in *Hager's Handbook*.[33]

Although listed in *Indian Medicinal Plants,*[26] there are no indications of uses. In Curacao the latex is applied daily almost like a hand lotion, to remove calluses. Latex is also used for athlete's foot and eczema. Leaves are bound to the temples to relieve headache. A weak leaf decoction is also taken for stomachache and given to children, perhaps as a vermifuge.[42]

The coagulum (*circa* 5%) of the latex (*circa* 95% water) consists mostly of rubber, resin, and protein, usually *circa* 80% rubber, 10% resin. The serum of the latex contains proteins, amino acids, malic and citronic acids, and phenolics.[33] In addition to cryptograndosides A and B, the plant contains the carcinostatic steroids gitoxigenin, 16-anhydrogitoxigenin, 16-propionylgitoxigenin, oleandrigenin, and 3-rhamnoside.[42] Leaves contain a wax which yields ursolic acid on saponification. The seed oil contains 38.4% oleic-, 27.6% linoleic-, 8.6% palmitic-, 9.5% stearic-, and 3.7% arachidic-acid. In addition to oil, the seeds contain 4.7% rubber and 17.5% resin.

Toxicity — All parts may cause severe stomach and intestinal upset, and cases of death have been reported from India.[34] The leaf is toxic.[26] Dry vine emits an eye-irritating dust which may induce coughing and swelling.[56]

To the physician — Hardin and Arena recommend gastric lavage or emesis and symptomatic treatment.[34]

108. *CYCAS REVOLUTA* Thunb. (CYCADACEAE) — Sago Cycas

The seed meal is washed to remove formaldehyde and other toxins and then eaten in India.[1] The leaves, after silvering, are made into funereal decorations. Both roots and stems have been viewed as potential, but dangerous, starch sources. Chinese believe the sago is rejuvenating and inducive to longevity.[41]

Reported to be emmenagogue, expectorant, and tonic, sago cycas is a folk remedy for hepatoma and lung tumors.[32,33] Chinese regard the seed as antirheumatic, emmenagogue, expectorant, and tonic. They use the terminal shoots as an astringent diuretic.[41] Although the plant contains carcinogens, products extracted from the seeds are used to inhibit growth of malignant tumors,[16] reflecting observations elsewhere that most plants that contain tumor-arresting compounds also contain tumorgenic compounds or that tumor-arresting compounds are themselves tumorgenic.[10]

Mucilage from the stem contains galactose, glucose, and xylose. Seed contains 0.2 to 0.3% cycasin ($C_8H_{16}N_2O_7$), neocycasin A, neocycasin B, and macrozamin ($C_{13}H_{24}N_2O_{11}$). The red pigment in the sarcotesta is zeaxanthin ($C_{40}H_{56}O_2$) along with β-carotene and cryptoxanthin. The sarcotesta oil is dominated by palmitic-, stearic-, oleic-, and behenic-acids. The sarcotesta contains 61.8% water, 3.9% oil, and 3% sugar, the kernel 72.1%,

0.12% lipids, 6.8% starch, 3.9% starch, and 1.15% protein.[33] Seeds contains 14% crude protein, 18% soluble nonnitrogenous substances, and 0.11 to 0.22% combined formaldehyde.[1] The root contains *circa* 18% starch.[1]

Toxicity — Cycasin is carcinogenic if orally administered to rats and pigs. With cattle neurotoxic effects are obvious. Other toxic symptoms include anemia, depression, diarrhea, jaundice, gastroenteritis, hemorrhage, nausea, coma, partial paralysis, and possibly death.[33,56] Apparently it is mutagenic to onion root tip cells.[33] Frequent use of the starch is suspected to cause cancer and hepatosis.[56]

109. *CYPRIPEDIUM* Sp. (ORCHIDACEAE) — Yellow Ladyslipper

Popular wild ornamental, becoming quite rare in some of its old haunts. Formerly a sedative.[20] In Appalachia, the tea made from the root is taken for headache and nervousness.[18] Powdered root (in sugar water) regarded as an anodyne, antispasmodic, aphrodisiac, nervine, sedative, stimulant, soporific, and tonic.[18] Once ladyslippers were thought to be very effective nerve sedatives, widely used for hysteria and neuralgia.[11] Cherokee used the root decoction for worms in children.[45] Acting like valerian, though less powerful, as an antispasmodic, the ladyslipper was called American valerian in Europe. Used also in epilepsy, fever, hemicrania, and tremors.[19] Hutchens[30] describes ladyslipper as of special value in reflex functional disorders or chorea, hysteria, headache, insomnia, low fevers, nervous unrest, hypochondria, and nervous depression accompanying stomach disorders. It relieves pain and restores calm and tranquility so rapidly that some suspect it contains narcotics.[30] Homeopaths use the tincture of autumn-gathered roots of *Cypripedium* for brain affections, chorea, convulsions, debility, delirium tremens, ecstasy, epilepsy, erythemia, insomnia, mental despondency, nervous debility, neuralgia, odontitis, sleeplessness, spermatorrhea, and styes.[30,33] According to *Hager's Handbook,* used for nervous hyperesthesia, particularly in morbid sensitivity of the eyes. Also, used to calm coffee and tea jitters and nervous stomach.[33]

Cypripedin is a complex resinoid substance obtained from the rhizome.[2] Tannic and gallic acids are also reported, along with volatile oils, glycosides, and resins.

Toxicity — According to Rose,[47] the oil is thought to paralyze the brain. The roots can cause psychedelic reactions and, in large doses, giddiness, restlessness, headache, mental excitement, and visual hallucinations.[47]

110. *CYTISUS SCOPARIUS* (L.) Link (FABACEAE) — Scotch Broom

The alkaloid sparteine was used as a cardiac depressant, cathartic, diuretic, and as an oxytocic to induce labor,[17] stimulating uterine contractions.[37] Sparteine is occasionally used as a quinidine substitute in stubborn cases of atrial fibrillation. Derivatives like anagyrine, lupanine, and oxysparteine have similar actions. Scoparoside, present largely in the flowers, is diuretic.[37] Ornamental, used to hold steep, barren banks, even in the salt spray zone. Wood has been used in veneer for cabinetry; also for thatch. Bark has been used in tanning leather. Before the advent of hops, the tender green tops were used to impart bitterness to beer and make it more intoxicating. Leaves and young tops give a green dye. Seeds have served as a dangerous coffee substitute. Flowers used in hair rinses, supposed to brighten and lighten the hair.[29] Moldy blossoms, aged about 10 days in a sealed jar, are pulverized and smoked like marijuana, for relaxation and hallucination or at least color enhancement.[37]

Flower, seed, herbage, and root used for tumors and indurations of the spleen.[4] Applied externally to abscesses, sore muscles, and swellings.[29] An unguent made from the flowers was used for gout; Henry VIII used to drink water made with the blossoms for gout. Juice of the fresh herb is diuretic.[33] Fluid extracts used for coronary insufficiency and other heart ailments, including postinfectuous myocarditis.[33] Decoction also recommended for ague, dropsy, gout, gravel, jaundice, kidney, liver, renitis, rheumatism, sciatica, splenomegaly, and toothache.[44] According to Steinmetz, the herbage is cardiosedative (cardiotonic),[16] diuretic, poison, purgative, and vasoconstricting; the flowers diuretic and cardiosedative, like the root and seeds.[27] In homeopathy, the essence is used for ailments of the heart muscles (acceleration, arrhythmia), hypertony, and congestion of the head, throat, and chest, occasionally in diptheria.[33]

Contains up to 1.5% sparteine, the flavone, scoparin; genisteine, sarothamnine, tannin, wax, and volatile oil.[17] Tyramine, hydroxytyramine and epinene, flavoxanthin and taraxanthin are also reported.[29] From the flowers, dopa, oxytryamine, isosparteine, sarothamnin, scoparine, tyramine, and tyrosine are reported.[33] Dry seeds contain *circa* 32.5% protein, 6.0% fat, and 3.8% ash.[21]

Toxicity — The FDA classes this an unsafe herb, because of the alkaloids hydroxytyramine, isosparteine, and sparteine.[62] Ths plant is narcotic, stupefying sheep who eat it. The tops are cathartic and emetic.[29] Toxicity symptoms suggest nicotine poisoning: circulatory collapse with tachycardia, paralysis of the ileus, nausea, diarrhea, vertigo, and headache.[33] Even *Herbal Highs*[51] says what is true of any beverage, food, medicine, or herb: "too much of it is definitely injurious".

111. *DAEMONOROPS DRACO* Bl. (and other species) (ARECACEAE) — Dragon's Blood

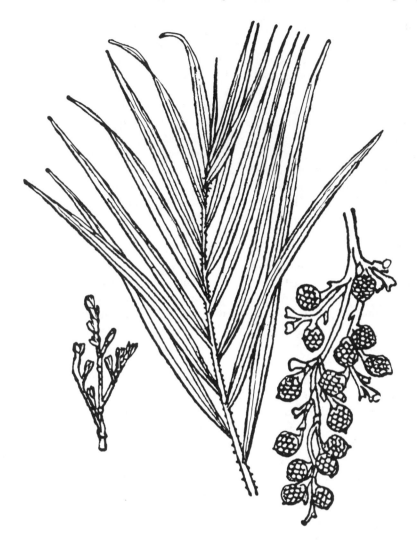

Resins derived from the scales of the fruit, used to tint plasters, tinctures, toothpaste, tortoiseshell (imitation), and varnishes, even staining marble a deep red. In the orient bamboo furniture is stained with the resin. The astringent resin was once used in collyriums, dentrifices, and mouth washes.[8] Some of the dragon's blood species are used in dart poisons.[5] The bright crimson powder processed from the fruits is easily ignited.

Latex once used for cancer or tumors, especially of the uterus, in China.[4] Once used as an astringent in diarrhea, dysentery, and enterorrhagia, and as a component of a complex for severe syphilis and as a hemostat. Used in Malaysia for indigestion.[5] Recommended in wine as a hemostatic in China,[12] where it is considered both sedative and tonic.[16] Malayans use for dyspepsia, hematuria, sprue, and stomachache.[16]

Dragon's blood consists of 56% resin alcohol dracoresinotannol — associated with benzoic and benzoylacetic acids; also, benzoyl acetic ester, dracorsene, dracoalban, and cinnamic acid.[12,16] Abietic acid has been isolated from the resin acids. The principal pigment is dracocarmin, $C_{31}H_{26}O_5$, an anthocyanadin, with another dracorubin, $C_{28}H_{24}O_7$, also reported.[1]

112. *DAPHNE MEZEREUM* L. (THYMELAEACEAE) — Mezereon

Widely cultivated, but dangerous, ornamental shrub, hardy in England.[2] Quills of bark have found their way into European medicine, the root bark being most active.

Reported to be abortifacient, alterative, carcinogenic, diuretic, poison, purgative, sialogogue, stimulant, and sudorific and vesicant, mezereon is a folk remedy for dropsy, neuralgia, rheumatism, sclerosis, scrofula, snakebite, syphilis, toothache.[2,32] Siberian veterinarians apply mezereon to horses' hooves. Japanese use the bark as a stimulant to treat rheumatism.[16]

Reports probably confusing seeds and fruits, suggest that on a dry basis, they range from 7.5 to 24.4% protein and 6.0 to 65.4% fat.[2] Contains the glycoside daphnin (7-glucoside-7,8-dioxycumarin)($C_{15}H_{16}O_9$) which yields daphnetin ($C_9H_6O_4$) and glucose, also umbelliferone, a yellow dye, malic acid, oil, wax, gum, and mezerein resin. Plant also contains hesperidin, beta-sitosterol, saccharose, a trihydroxyflavone, $C_{15}, H_{17}O_5$, and daphnopside. Seed contain 0.02% daphnetoxin.[33] 12-Hydroxdaphnetoxin and mezeriin have shown activity in the PS tumor system.[10]

Toxicity — "Thirty berries are used as a purgative by Russian peasants, though French writers regard fifteen as a fatal dose."[2] The berries have proved fatal to children.[2] Symptoms following oral ingestion of the berries include pain, inflammation and irritation of the lips, mouth, and tongue, stomatitis, salivation, diarrhea, stomachache, nausea, colic, watery hematochezia, etc.[33] The glucoside causes burning or ulceration of the throat and stomach, vomiting, internal bleeding with bloody diarrhea, weakness, coma, and death.[34]

To the physician — Hardin and Arena suggest demulcents and gastric lavage or emesis, with symptomatic and supportive treatment.[34] Mezerein is carcinogenic in animals.[11]

113. *DATURA CANDIDA* (Pers.) Saff. (SOLANACEAE) — Borrachero, Floripondio

Mixed with tobacco to form a strong hallucinogen, used in the treatment of diseases believed to have been induced by witchcraft.[52] Suggested as an industrial source of scopolamine.[33] According to Bristol,[143] Colombian Indians plant the fruits of the cultivar they call "Buyes" with corn to prevent grubs and other pests from eating the germinating seeds. They mix powdered leaves and flowers with their hunting dogs' food to get them high for the hunt.

Reported to be intoxicant, narcotic, and poison, borrachero is a folk remedy for arthritis, asthma, chest ailments, fractures, insomnia, piles, and tumors.[32] Leaves of the cultivar "Buyes" are powdered and applied topically with other drugs for rheumatic pain. The Inga Indians use the "Munchira" cultivar as a carminative, emetic, psychotrophic, and vermifuge, using it for erysipelas and rheumatism.

The isolation of scopolamine as the major alkaloid, with atropine, meteloidine, noratropine, norscopolamine, and oscine, from the aerial parts and 3-alpha, 6-beta-ditigloyloxy-tropane-7-beta-ol, scopolamine, norscopolamine, 3-alpha-trigolyloxytropane, meteloidine, atropine, oxcine, noratropine, from the roots, indicates that this differs little from related species. Total alkaloids run 0.3 to 0.55% of DM, with scopolamine at 30 to 60%.[58]

Toxicity — Said to induce insensibility, hallucinations, and madness.[52]

To the physician — Hardin and Arena suggest emesis or gastric lavage, pilocarpine or physostigmine for dry mouth and visual disturbance, sedation and temperature reduction, if necessary, for *Datura* intoxication.[34] See *D. stramonium*.

114. *DATURA INNOXIA* Mill. (SOLANACEAE) — Thorn Apple

Used as a commercial source of scopolamine, used in India as a preanesthetic in childbirth surgery, in ophthamology, and in prevention of motion sickness. It is a cerebral depressant useful in agitated or maniacal conditions. It is used to produce amnesia and partial analgesia in labor (0.5 to 1 mg orally or subcutaneously as preanesthetic or sedative). A 0.1 to 0.3% solution is used in ophthalmology.[1]

Reported to be analgesic, anesthetic, demulcent, expectorant, narcotic, sedative, spasmolytic, is a folk remedy for asthma, boils, bronchitis, cough, dandruff, earache, motion sickness, ophthalmia, parkinsonianism, phthisis, piles, pyorrhea, seasickness, sores, spasms, and tumors.[32]

Roots contain 24 to 60% scopolamine, 22 to 48% hyoscyamine, 2 to 26% ditigloyteloidin, 1 to 10% meteloidin, 5 to 30% cuscohygrine, and 1.5% of unidentified alkaloids; also, apoatropine, atropine, pseudotropine, tigloidine, tropine, 7-hydroxy-3,6-ditigloyloxytropane, (−)-3 gamma, 6-beta-ditigloyloxytropane. Seeds contain 0.3% total alkaloids with 0.09% scopolamine and 0.21% hyoscyamine, as well as 22% fatty oils with linoleic-, oleic-, and palmitic-acids.[33] The dark residue after alkaloid separation contains reducing sugars, oxalates, and nitrates, but no tannin. Psychoactive compounds include atropine, hyoscyamine, and scopolamine.[54] Alkaloidal components differ only slightly than those from *D. metel*.[58] Dry seeds contain *circa* 20% protein, 12% fat.[21]

Toxicity — Narcotic hallucinogen and hypnotic.[54] See *D. stramonium*.

115. *DATURA METEL* L. (SOLANACEAE) — Unmatal, Metel, Hindu Datura

Dried leaves and flowering tops official in the B.P.C. under the name *Daturae Folium.*
''*Datura metel* was found useful in luring virgins into prostitution; subsequently, they would
employ extracts of the herb as 'knockout' drops to take advantage of their clients.''[54]

Reported to be analgesic, anesthetic, anodyne, bronchodilator, demulcent, expectorant,
hypnotic, intoxicant, narcotic, poison, sedative, unmatal is a folk remedy for anasarca,
asthma, boils, congestion, convulsions, cough, cramps, dandruff, delirium, dermatophytosis,
dropsy, earache, epididymitis, epilepsy, headache, hemiplegia, hydrocele, madness, myalgia, myopathy, numbness, orchitis, otitis, parkinsonianism, parotitis, phthisis, piles, pimples, proctosis, prolapse, rabies, rheumatism, ringworm, snakebite, sores, spasms, syphilis,
tumors, venereal disease, and wheezes.[32]

Alkaloidal details are reported in *Hager's Handbook.*[33] Total alkaloids = 0.12% of the
fruit, 0.2 to 0.5% of the leaves, 0.1 to 0.2% of the roots, 0.2 to 0.5% of the seed. Scopolamine
was the main alkaloid, with meteloidine, hyoscyamine, norhyoscyamine, norscopolamine,
cuscohygrine, and nicotine.[58] Dry seeds contain 12.5 to 13.9% protein, 17.2 to 18.8%, and
4.5 to 5.6% ash.[21]

Toxicity — See *D. stramonium.*

116. *DATURA STRAMONIUM* L. (SOLANACEAE) — Jimson Weed, Thorn Apple, Stramonium

Mostly considered a noxious weed, stramonium has few legitimate uses outside the drug trade. The seeds are so toxic as to have been used in suicides and homicides; the victim experiences dry throat, giddiness, hallucination, staggering; his voice becomes unrecogniz-able and the vision is affected; he lapses into coma which may be terminal. Still, gamblers are said to have nibbled on the seeds for clairvoyance. Sexual functions are said to be excited, more especially in women, in whom it may cause nymphomania.[19] Perhaps this led Rose to publish her recipe for a stramonium douche adding "it could be dangerous."[47] Atropine, a strategic drug in the U.S., can be obtained from jimsonweed. It is mydriatic, antisialogogue, and depresses the muscles of the bladder, thereby controlling urination. It is also used (in China) for flatulence, hyperacidity, and nightsweats of tuberculosis.[16]

Has quite a history in cancer folk medicine, the leaves being poulticed onto various types of cancers, indurations, and tumors, especially of the breasts.[4] Considered anodyne, anti-spasmodic, and narcotic, stramonium has the vices and virtues of hyoscyamine and can be used like belladonna. Stramonium is recommended in asthma cigarettes. The atropine is said to paralyze the endings of the pulmonary branches; it is antispasmodic, relieving the bronchial spasm. The leaves may be smoked in a pipe, alone, or with various admixtures like belladonna, cubebs, sage, and tobacco. Seeds macerated in alcohol are used as an ointment for abscesses, fistula, hemorrhoids, neuralgia, and rheumatism. Mexican Indians take the leaf decoction for the pains of parturition.[42] Costa Ricans apply crushed leaves to tumors and ulcers and gargle the infusion for sore throat. Arabs of central Africa are said to smoke the dried herb to treat both asthma and influenza. Juice of the fruit is used to prevent baldness and juice of the flowers for earache. Stramonium has been used for epilepsy, mania, and psychosis. In *Merck's 8th Index,* it is considered anticholinergic, antiparkinsonian, and antiasthmatic.[20] Russians use the herb for asthma, bronchitis, epilepsy, pain, radiculitis, and stenocardia (angina pectoris). They even administer the seeds in vodka for children "scared stiff" (paralyzed by fright). Orientals use *D. stramonium* and *D. metel* interchangeably, the leaves anodyne, possibly maturative, mitigant, resolvent.[16] Chinese gather the petals after the dew has evaporated in the morning, drying them in the sun, using the decoction for skin problems, the powder as a fumitory for asthma, cough, and shortness of breath. Chinese are said to use a half and half mixture of *Datura* and *Cannabis* in wine as an anesthetic permitting painless cauterizations and operations. Flowers and seeds are used to wash rectal prolapse and swollen feet.[16] Homeopaths prescribe the tincture for anasarca, aphasia, apoplexy, burns, catalepsy, chorea, delirium tremens, diaphragmitis, ecstacy, enuresis, epilepsy, erotomania, esophageal spasm, eye ailments, headache, hiccup, hydrophobia, hysteria, lochia, locomotor ataxia, mania, meningitis, nymphomania, scarlatina, stammering, strabismus, sunstroke, tetanus, thirst, tremors, trismus, and typhus.[28,30]

Hyoscine is usually the most frequent alkaloid; leaves contain 0.055 to 0.25%, seeds 0.12 to 0.5%. Small amounts of hyoscyamine and atropine also present.[17] Other sources indicate that hyoscyamine is the chief alkaloid (0.3 to 0.5% alkaloids). Perry says there is more hyoscine in *D. metel*, more atropine in *D. stramonium*.[16] Leaves, stems, flowers, and ovular integument contain chlorogenic acid. Datugen and datugenin have also been isolated. Dry seeds contain 14.0 to 19.4% protein, 18.4 to 28.5% fat, and 2.7% ash.[21] The fatty oil seeds contains 87.7% fatty acids (including oleic, linoleic, palmitic, stearic, and lignoceric) and 2.6% unsaponifiable matter containing sitosterol. Tannin is also reported.

Toxicity — Usual consequences of poison are dimness of vision, dilation of the pupils, giddiness, and delirium, sometimes amounting to mania. Many fatal instances are recorded. Someone handling the foliage may dilate his pupils accidentally by rubbing his eyes. Thieves often used the seeds to knock out their victims in India. New settlers in Jamestown, Va. experimentally ate the leaves and experienced such trauma that they called it "Devil's apple" (also, Jamestown alias Jimson weed). Iranians sometimes call the seed "Kachola", their name for Strychnos, because both are used for killing dogs. Even sheep and goats can be poisoned by as little as 10 kg/day foliage and fruits.[144] Datura seed each weighs *circa* 5.7 mg (5700 μg) of which 0.2 to 0.5% is total alkaloid, or a conservative 25 μg total alkaloid. If we assume all this is atropine, the lowest toxic dose, i.e., one introducing psychotropic responses in humans, is 100 μg atropine per kilogram body weight. In a heavy individual (100 kg), that would indicate the ingestion of 400 seeds, assuming each seed contained 25 μg atropine and assuming all that atropine was metabolized into the body. Alkaloids probably vary anyway, as will people's and animals' reactions to them. In 1981, six New Jersey teenagers consumed jimsonweed seed (1/2 to 2 "pods of seeds") and alcohol (1 to 2 oz whiskey).[145] Illness was characterized by hallucination (all six teenagers), dry mouth, (six) thirst, (five) blurred vision, (five) flushed skin, (four) inability to urinate, (four) and slurred speech (four cases). Illness began *circa* 30 to 105 min after seed ingestion and lasted 18 to 216 hr. Blurred vision was the longest-lasting symptom. Rose describes the trip: "A person who uses this drug becomes unconscious and must be watched and kept from hurting himself. He may think he's flying. The first symptom is a dry throat, then gaiety, laughter, black objects appear green. The experience is usually entirely forgotten. Overdose is usually fatal".[47] Those who indulge in jimsonweed seed may be taking off on a one-way trip.

To the physician — For poisoning, Hardin and Arena[34] recommend "Gastric lavage or emesis; pilocarpine or physostigmine for dry mouth and visual disturbance; sedation and reduction of temperature with cool water sponging if necessary". Perry suggests licorice as an antidote.[16]

117. *DAUCUS CAROTA* subsp. *SATIVUS* (Hoffm.) Arcang. (APIACEAE) — Cultivated Carrot, Queen Anne's Lace (Wild)

Cultivated for the enlarged fleshy taproot, eaten as a raw vegetable or cooked in many dishes. Eaten sliced, diced, cut up, or shoe-stringed, carrots are used in many mixed vegetable combinations They are sold in bunches, or canned, frozen, or dehydrated. They may be baked, sauteed, pickled, and glazed, or served in combination with meats, in stews, roasts, soups, meat loaf, or curries. Roasted carrots have served as coffee substitutes. Carrot juice is beneficial. Britishers once brewed a good wine from carrot. Humans are said to eat the leaves in Java.[1] Essential oil is used to flavor liqueurs and perfumes. Carrotseed oil, blended with cedarwood oil, is a good imitation of orris. Roots and tops may be fed to livestock.[61]

Seeds are aromatic, carminative, diuretic, emmenagogue, and stimulant, and are used for dropsy, chronic dysentery, kidney ailments, and worms.[32] Also, as an aphrodisiac, a nervine tonic, and for uterine pain. Roots are refrigerant and are used in infusion for threadworm. Diuretic and eliminating uric acid, carrots belong in the diet of gout-prone people. Local stimulant for indolent ulcers; other ingredients of carrot lower blood sugar; hence, carrot might be increased to good advantage in the prevention of cancer, diabetes, dyspepsia, gout, and possibly heart disease. Elsewhere, the root, prepared in various manners, is used for tumors, cancerous ulcers, cancerous wounds, tumors of the testicles, mammary carcinoma, and skin cancer. The juice of the root is applied to carcinomatous ulcers of the neck and uterus, cancer of the bowels, and stomach cancer.[4] Scraped roots are used to stimulate indolent ulcers. Cancer fearers may be reinforced by the knowledge that carrots are relatively high in fiber, retinoid-like substances, and the seeds also contain the rather ubiquitous beta-sitosterol, which has shown activity in Ca, LL, and WA tumor systems. Having heard from three different sources that wild carrot seeds were used as a morning-after contraceptive in Pennsylvania, I was particularly interested to read that, "At doses of 80 and 120 mg/mouse, the seed extract, if given orally from day 4 to 6 postcoitum, effectively inhibits implantation". Experimentally hypoglycemic, a tea made from Queen Anne's Lace was believed to help maintain low blood levels in humans, but it had no effect on diabetes artifically induced in animals. Wild carrot has been recommended for bladder and kidney ailment, dropsy, gout, gravel; seeds are recommended for calculus, obstructions of the viscera, dropsy, jaundice, scurvy. Carrots of one form or another were once served at every meal for liver derangements; now we learn that they may upset the liver.[41]

Per 100 g, the carrot is reported to contain 86.0 g H_2O, 0.9 g protein, 0.1 g fat, 10.7 g carbohydrate, 1.2 g fiber, 1.1 g ash, 80 mg Ca, 30 mg P, 1.5 mg Fe, 2000 to 4300 IU vitamin A, 60 IU vitamin B_1, 3 mg niacin, and 3 mg ascorbic acid.[21] The *Wealth of India* reports thiamine (56 to 101 μg/100 g), riboflavin (50 to 90 μg/100 g), and nicotinic acid (0.56 to 11 mg/100 g) among the B vitamins. Vitamin C is in a protein-ascorbic acid complex. Vitamin D, a substance with the characteristics of vitamin E and a phospholipoid of vitamin reactions corresponding to A and D and containing calcium, phosphorus, and nitrogen in organic linkage, is also present. Carrots contain *circa* 5.27% ZMB of phytin. Sixteen percent of the phosphorus is present as phytic acid phosphorus. The lipids extracted from raw carrots are characterized by a low nitrogen content (0.33 to 0.72%) and by the absence or low content (0.52%) of choline, while those extracted from steamed roots are rich in nitrogen (1.1 to 1.3%) and choline (4.2 to 4.4%). Pectin isolated from carrots (yield, 16.82 to 18.75% on dry weight) has no gelling property. Ash of carrots gave (on fresh weight basis): total ash, 0.92; K_2O, 0.51; Na_2O, 0.06; CaO, 0.07; MgO, 0.02; and P_2O_5, 0.09%. Trace elements reported to be present include: Fe, Al, Mn, Cu, Zn, As, Cr, I, Br, Cl, U, and Li (C.S.I.R., 1948 to 1976).

Toxicity — *Daucus* has been reported to contain acetone, asarone, choline, ethanol, formic acid, HCN, isobutyric acid, limonene, malic acid, maltose, oxalic acid, palmitic acid, pyrrolidine, and quinic acid. Reviewing research on myristicin, which occurs in nutmeg, mace, black pepper, carrot seed, celery seed, and parsley, Buchanan[117] noted that the psychoactive and hallucinogenic properties of mace, nutmeg, and purified myristicin have been studied. It has been hypothesized that myristicin and elemicin can be readily modified in the body to amphetamines. Handling carrot foliage, especially wet foliage, can cause irritation and vesication. Sensitized or sensitive persons may get an exact reproduction of the leaf on the skin by placing the leaf on the skin for a while, followed by exposure to sunshine.

118. *DIEFFENBACHIA SEGUINE* (Jacq.) Schott (ARACEAE) — Dumbcane

Widely planted tropical foliage ornamental. According to Walter and Khanna,[146] Latin Americans have used the plant to induce temporary sterility in themselves or permanent sterility in their enemies, concepts which led to experimentation by the Third Reich in Germany (expressed juice led to sterility in male rats for 40 to 90 days, in females for 30 to 50 days). Farnsworth[147] relates a story from a botanist who served during World War II in Germany. The Germans established well-populated concubinary recreation camps for army officers in areas far removed from the war front. Officers admitted to such camps were ordered to chew a leaf of dumbcane once daily, which was believed to induce sterility for 24 hr. Another botanist in South America noted his guide chewing a leaf of dumbcane, to make him sterile for 1 day. Two other members of the dumbcane family are said to exhibit this oral contraceptive attribute. West Indians have used both as a female aphrodisiac (placing the juice on the genitals) and a contraceptive.[148] In Cuba the dumbcane is used to poison cockroaches and rats. Amerindians of many tribes used dumbcane in arrow poisons.

Reported to be aphrodisiac, caustic, contraceptive, insecticidal, poison, and sterilant, dumbcane is a folk remedy for angina, cancer, coma, dropsy, dysmenorrhea, edema, frigidity, impotence, internal ulcers, prurigo, swellings, varicosities, warts, and yaws.[32] In Cuba, where called maté de cancer, dumbcane juice is applied to corns and warts.[4] Morton,[42] however, drawing on a Cuban classic, says "The plant has no medical use in Cuba". The homeopathic tincture is used for frigidity and sexual impotence. Yet natives of the West Indies are said to chew the roots to bring on temporary male sterility. Guatemalans poultice mashed stems and leaves to animal bites, gout, and rheumatism. Salvadorans apply the seed oil to burns, inflammations, and wounds. Costa Ricans apply the sap, like the Cubans, to tumors and warts. For angina, Brazilians gargle the leaf decoction.[42]

Said to be cyanogenetic.[32] Contains L-asparagine and a proteolytic enzyme termed dumbcain (suggestive of the mucunain from Mucuna).

Toxicity — Acicular crystals of calcium oxalate, accompanied by a protein (enzyme) or asparagine, may cause severe burning in the mouth and throat. Swelling of the mouth may be severe enough to cause fatal choking.[34,56] The sap, under some conditions, is irritant and vesicant.[56] Contact with bruised plants seemed to be necessary for irritation.[6] Panama natives to this day blame an indolent ulcer I suffered on the ankle on my walking through a patch of dumbcane that had recently been cut. On several occasions I applied fresh latex to my wrist with no problem.

To the physician — Hardin and Arena recommend "Gastric lavage or emesis; symptomatic, give demulcents; cold pack to lips and mouth, antihistamines or epinephrine".[34]

119. *DIGITALIS PURPUREA* L. (SCROPHULARIACEAE) — Common Foxglove, Digitalis

Often planted as an ornamental. Dried leaves are used medicinally because of the effect on the central nervous system, the heart, and blood vessels. It contains glucosides which act as heart tonic, stimulant, and diuretic. Its most important use is in congestive heart failure because it increases the force of systolic contractions. It also provides more rest between those contractions. Also, lowers the venous pressure in hypertensive heart disease. Elevates blood pressure due to impaired heart function and reduces the size of dilated hearts. Diuresis and reduction of edema follow as a result of improved circulation. According to Farnsworth and Morris, digoxin occurred in *circa* 11 million (0.73% of all prescriptions), digitoxin in *circa* 5 million (0.33%), and digitalis extracts in 2.5 million (0.16%).[98] The drug digitoxin is 1000 times stronger than the powdered leaves. Used as an antidote for aconite. Nineteenth century artist, van Gogh, is speculated to have taken digitalis for epilepsy. Digitalis causes yellow vision, visions of yellow rings, as well as other visual and psychic disturbances, some of which show up in his art.

The seed is described as cordial and diuretic. The ointment made from the leaf is a remedy for indurations. The leaf, with other plants, is applied as an ointment to remedy hard breasts and indolent tumors.[4] Digitoxin shows antitumor activity in the KB tumor system.[10] Homeopathically the plant is used for gastritis, hydropsy, and icterus.[33] The leaf is said to be cardiotonic, diuretic, sedative, yet vascular stimulant. Leaf infusion once used for dropsy, epilepsy, scrofula, and sore throat. In Colombia, for example, a tea with three or four flowers of foxglove is administered to epileptic children over a period of 9 days. Plant also used for asthma, dropsy, edema, fevers, insanity, nephrosis, neuralgia, palpitations.[32]

Contains several glycosides: e.g., digitalein, digitaline, digitin, digitonin, and digitoxin. Seeds contain, also, gitonin and tigonin. Dry seeds contain 14.8 to 16.2% protein, 38.9 to 40.4% tar, and 3.6% ash.[21] For a detailed tabulation of the important compounds found in various *Digitalis* species, the reader is referred to *Hager's Handbook*.[33] In addition to the complex of glycosides, we find reference to luteolin, luteolin-7-D-glucoside, L-7-glucuronide and L-7-glucosyl-glucuronide, and others: choline, acetylcholine, P-cumaric acid, caffeic acid, ferulic acid, chlorogenic acid, mucilage and inositol, glucose, galactose, xylose, oxydase, peroxidase, catalase, invertase, and diastase.[33]

Toxicity — Many fatalities have resulted from foxglove ingestion. In 1977, a couple died who mistakenly drank foxglove rather than comfrey tea. The action of *Digitalis* and its cumulative potential must be carefully watched. Symptoms of digitalis poisoning[34] include nausea, diarrhea, stomachache, severe headache, irregular heartbeat and pulse, tremors, convulsions, and death.

To the physician — "Gastric lavage or emesis; supportive; atropine, potassium, procainamide, quinidine sulfate, disodium salt of edetate (Ca_2EDTA) have all been used effectively."[34]

120. *DIOSCOREA COMPOSITA* Hemsl. (DIOSCOREACEAE) — Barbasco

Tubers of some species are used as coffee substitutes,[11] possibly after scorching. Barbasco has been cultivated experimentally as a source of sapogenins and contains 0 to 13% diosgenin and sometimes some yamogenin. Sometimes used as a fish poison. Diosgenin is converted by chemical means and elaborate processing to provide the final products which relieve arthritis, asthma, and eczema, regulate metabolism, and control fertility. "Few realize the great contribution made by yams . . . in stabilizing or decreasing world population, but perhaps no postwar development has been so relevant in changing the life style of those at reproductive age."[11] Yams and other diosgenin producers provide the steroid building blocks for developing human sex hormones, now cheaply available for oral contraception, treating menopause, dysmenorrhea, premenstrual tension, and testicular deficiency. Cortisones and hydrocortisones are also used for Addison's disease, allergies, bursitis, contact dermatitis, psoriasis, rheumatic fever, rheumatoid arthritis, sciatica, and skin diseases. Androgens are given for impotency, prostatic hypertrophy, psychosexual disturbances in males, and to offset excess estrogen production in females.[17] Nearly 15% of prescriptions in the U.S. in 1973 contained steroids (95% from diosgenin; more than 200 million prescriptions).[149] Farnsworth[150] tells the interesting story of how the herb Dioscorea and the eccentric chemist, R. E. Marker, teamed up to launch what was later incorporated as Syntex S. A. "Mainly because of Marker, by the year 1945 the retail price of sex hormones had been cut in half."

Although this species has no folk history in our computer file,[32] it probably shares with the many other species of *Dioscorea* many folk medicinal virtues. Few taxonomists, if any, even fewer natives, can distinguish *all* the species of *Dioscorea*.

High-sapogenin yams are not consumed as human food; once sapogenins are removed, the residues might possibly be edible. Edible yams run, per 100 g, 70 to 135 calories, 65 to 86 g H_2O, 1.5 to 3.0 g protein, 0.1 to 0.2 g fat, 15 to 30 g total carbohydrate, 0.5 to 1.5 g fiber, 0.5 to 1.5 g ash, 10 to 70 mg Ca, 15 to 50 mg P, 0.7 to 5.2 mg Fe, *circa* 8 mg Na, 300 to 365 mg K, 0.10 μg β-carotene equivalent, 0.05 to 0.15 mg thiamine, 0.01 to 0.04 mg riboflavin, 0.4 to 0.8 mg niacin, and 3 to 18 mg ascorbic acid.[21] Mexican yams are rich in the glycoside saponin from which are derived by partial synthesis; steroidal saponins, primarily diosgenin, *D. composita* averages 4 to 6% total sapogenins, *D. floribunda*, 6 to 8%.

Toxicity — Tubers are bitter and toxic. Steroidal drugs may produce serious side effects in the long run.

121. *DIPTERYX ODORATA* (Aubl.) Willd. (FABACEAE) — Tonka Bean, Tonga, Cumaru

Tonka beans are cultivated for the seed which yield coumarin, used to give a pleasant fragrance to tobacco, a delicate scent to toilet soaps, and a piquant taste to liqueurs. The extract is also used in foodstuffs, e.g., cakes, candies, preserves, and as a vanilla substitute, as a fixing agent in manufacturing coloring materials, in snuffs, deodorants, and in perfume industry. The most important use of coumarin in the U.S. is for flavoring tobacco.[40] The timber is said to resist the marine borer, perhaps, because it contains 0.01% SiO_2. Williams describes the ''maypole'' festivities of the Black Caribs of Nicaragua, celebrated when the Central American species *D. oleifera* is ripe. The Caribs make a paste of the seeds, mix it with coconut water or milk, and make a rich nut-flavored beverage, more satisfying than a malted milk.[22]

Said to be used for cachexia, cramps, fumigation, narcotic, nausea, spasms, whooping cough, tonic. Fluid extract has been recommended in whooping cough, regarded as a spasmolytic. Seed extracts are used rectally for schistosomiasis in China.[40] Guyanese use the astringent gum for sore throat. Brazilians apply the seed oil for buccal ulcers and earache. Black Caribs are said to use the fruits as an aphrodisiac.[22]

The active constituent of the seed may be coumarin. ''Dietary feeding of coumarin to rats and dogs causes extensive liver damage, growth retardation, and testicular atrophy.''[62] Coumarin may constitute 1 to 3% (rarely 10%) of the seed. Five isoflavones have been isolated from the heartwood: retusin, retusin 8-methyl ether, 3'-hydroxyretusin 8-methyl ether, odoratin, and dipteryxin. A sample of Malaysian tonka bean oil showed 6.1% palmitic-, 5.7% stearic-, 59.6% oleic-, 15.4% linoleic-, and 13.2% C26 to C24 acids. Sitosterol and stigmasterol are also reported.[33] Nineteen samples of beans from Venezuela and Trinidad (uncured) contained concentrations of coumarin ranging from 2.1 to 3.5% of the dry weight. Moisture content varied from 6.9 to 8.4%. One sample of Brazilian beans contained 10.2% moisture and 2.7% coumarin. Variation is reported: Surinam beans contained 2.4% of coumarin and 29.2 to 33.7% of alcoholic extractive; 2.6% coumarin in Imburana beans, and 1.3% of coumarin and 49.5% of alcohol extractive in Trinidad beans. From the cotyledons *o*-coumaric acid and a beta-glucoside are reported. The resin exuding from the bark contains lupeol, betulin, and a mixture of 75% methyloleate, 3% methyllinolenate, and 10% conjugated methyllinoleate.[33,40] Sullivan[151] reports umbelliferone (7-hydroxycoumarin) in the methylene chloride extract of the seed.

Toxicity — Narcotic, the fluid extract can paralyze the heart if infused in large doses.[2] The reddish sawdust once caused a workman's hair, when wetted, to turn bright green.[6]

122. *DRYOPTERIS FILIX-MAS* (L.) Schott (POLYPODIACEAE) — Male Fern

Hardy ornamental fern, sometimes sold as an herbal drug. Said to serve as a substitute for hops.[27] Rhizomes and stipes contain an oleoresin which paralyzes the intestinal voluntary muscles and the analogous contractile tissue of the tapeworm, which is then easily dislodged by a purgative. Filicin is an active vermifuge, especially effective with tapeworm, to all forms of which it is poisonous. It is administered in capsules, in oil medium, or in pills of 12-g doses for adult. Tenia are expelled a few hours later. It is combined with calomel to insure both purgative and vermifugal action. The drug is also used in veterinary practice.[1] Rhizome used as insecticide in Vietnam.[16]

Considered an old folk remedy for cancerous tumors.[4] Known as a vermifuge, especially good for tenia, back to the days of Theophrastus. Said to be aperient, astringent, cyanogenetic, pectoral, poison, tenifuge, as well as vermifuge.[27,32] Used in China for epistaxis, menorrhagia, puerperium, and wounds.[16]

Contains 6.5% oleoresin, with albaspidin, filicic acid, filicin (mostly margaspidin), filix red, filmarone (tenifuge), flavaspidic acids, paraaspidin, deaspidin, resin,[8] aspidinol, phloraspin, and tannin.[27,29] Attributing most of the medicinal activity to the phloroglucides, *Hager's Handbook* lists and shows the structure for more than two dozen.[33]

Toxicity — "In too large doses, however, it is an irritant poison, causing muscular weakness and coma, and has been proved particularly injurious to the eyesight, even causing blindness."[2] Other symptoms include nausea, diarrhea, vertigo, delirium, tremors, convulsion, and cardiac or respiratory failure.[29] Rose states: "It has caused allergy reactions in some and can be fatally poisonous if misused. It should be used only by prescription from a doctor."[47]

123. *DUBOISIA MYOPOROIDES* R. Br. (SOLANACEAE) — Corkwood Tree, Pituri

Leaves of the pituri, roasted, moistened, and rolled into a quid, ward off hunger, pain, and fatigue.[54] Leaves also smoked as a stimulant.[2] Aborigines of Australia capture game by poisoning the waterholes with pituri.[54] The alkaloids are sometimes used as a substitute for atropine.[2] Until recently this was a major source of Australia's scopolamine, but recently the hybrid with *D. leichhardtii* is a preferred source.[152]

Reported to be hypnotic, intoxicant, mydriatic, and sedative, the corkwood tree is a folk remedy used in parturition and psychiatry.[32] Homeopathy employs the tincture for eye afflictions (especially when a red spot interferes with vision).[2]

The alkaloids are mostly hyoscyamine and hyoscine.[2] According to *Hager's Handbook*, the stem contains valtropine, scopolamine, atropine, noratropine, tropine, valeroidine, nor-hyoscine, apohyoscine, hyoscyamine, and butropine; the root bark atropine, hyoscyamine, tropine, scopolamine, tetramethylputrescine ($C_8H_{20}N_2$), valtropine, apohyoscine; the root wood hyoscyamine, scopolamine, tropine, apohyoscine, tetramethylputrescine, valtropine, and valeroidine.[33] The leaves contain more than 2% alkaloids, scopolamine dominating. In addition to those, poroidine, isoporoidine, the CNS-sedative tigloidine, L-anabasine ($C_{10}H_{14}N_2$), DL-isopelletierine ($C_8H_{15}NO$), nicotine, nornicotine. Younger leaves may be high in valtropine ($C_{13}H_{23}NO_2$), scopolamine, tigloyltropine, valeroidine, and hyoscyamine, while older leaves also contain butropine ($C_{12}H_{21}NO_2$), poroidine, isoporoidine, acetyltropine, nor-hyoscyamine, tropine, and apohyoscine. In animal experiments tigloidine hydrobromide is active against parkinsonianism.[33] The major psychoactive compound is scopolamine, an alkaloid which, even in therapeutic doses, causes excitement and hallucinations. In larger doses it may become fatal.[54] Recently the *Chemical Marketing Reporter* (CMR) discussed its use in treating seasickness.

Toxicity — Narcotic stimulant, secondarily a hallucinogen.[54] "Although the poisoned dreams of pituri represent a flirtation with death, they are preferred to the sting of harsh reality."[54]

To the physician — Antidote: coffee and lemon juice.[2]

124. *DURANTA REPENS* L. (VERBENACEAE) — Golden Dewdrop

Grown as an ornamental shrub or hedge in tropical and subtropical areas of the world for its handsome blue flowers and orange-yellow berries. Birds are said to eat the poisonous fruits. The fruit juice is said to be toxic to mosquito larvae.[42] The juice can be used as a larvicide in ponds and swamps, being lethal to larvae at dilutions of 1:100 parts water.[1] The hard, heavy, to pale brown wood is easily worked and suitable for turnery.

Reported to be detergent, insecticidal, larvicidal, and stimulant, golden dewdrop is a folk remedy for fever.[3,32] Mexicans and Guatemalans take the flower decoction as a stimulant, the fruit decoction as a febrifuge.[42] Chinese use the fruits for malaria, Indochinese consider the plant diuretic.[16] Closely related *D. mutisii* is used to make emollients for varicose ulcers in Colombia.[57]

Perkins and Payne say the fruit (and leaves) contain "a saponic glycoside".[56] Ripe fruits contain an alkaloid similar to 5,6-dihydro-7H-2-pyridine derivatives.[42] Dry seed contain 13.1% protein and 1.4% fat.[21] HCN has been reported in the leaf and fruit.[3]

Toxicity — Berries contain a saponin causing drowsiness, fever, nausea, vomiting, and convulsions. Deaths of children from eating the berries are recorded.[34,56] Symptoms include sleepiness, hyperthermia, dilated pupils, rapid pulse, swellings of lips and eyelids, and convulsions.

To the physician — Hardin and Arena recommend emesis or gastric lavage and symptomatic treatment.[34]

125. *ELAEOPHORBIA DRUPIFERA* (Thonn.) Stapf (EUPHORBIACEAE) — Dodo, Kankan, Toro

Fruits eaten by African antelopes. Crushed fruits and leaves used as a fish poison.

Reported to be antidotal, caustic, filaricidal, piscicidal, poison, purgative, dodo is a folk remedy for blindness, bugbite, guineaworm, ringworm, snakebite, sores, and warts.[4,32] Equatorial Africans use latex to treat ringworms, stings, and warts. The latex is sometimes mixed with eggs as a purgative.[54]

Toxicity — Classed as a narcotic hallucinogen(?).[54] The caustic latex can cause permanent blindness if introduced into the eyes.[54]

126. *ELEUTHEROCOCCUS SENTICOSUS* (Rupr. & Maxim.) Maxim. (ARALIACEAE) — Wujia Ginseng, Siberian Ginseng, Spiny Ginseng

Sold in the U.S. as Siberian ginseng and Wujia ginseng, as well as Eleuthero ginseng.

A folk remedy for bronchitis, heart ailments, hemiplegia, hypertension, insomnia, and rheumatism.[24] Regular use said to restore vigor, memory, good appetite, and increase longevity. (Harbin, !) According to representatives from the China National Native Produce Corporation, this species caused 90% improvement in neurasthenia, 60 to 95% in hypertension, 75% improvement in hypercholesterolemia, 65% in impotence, 93% in hypoxia, while increasing the life expectancy of stomach cancer patients by 1 to 4 years (China, !). Not knowing whether it belongs in the realm of folk medicine, I repeat Baranov's[154] recitation that "one very important property of eleuthero is its carcinostatic effects. A decoction of the roots can be used to weaken the vitality of cancer cells and slows the spread of metastases into other tissues." Research between 1972 and 1974 showed that the roots were effective against bronchitis and heart disease, and had no side effects. Tierra cited Chinese usage for arthritis, bronchitis, chronic lung ailments, hypertension, hypercholesterolemia, impotence, low blood oxygen, and stress.[28] Chinese believe that regular use will increase longevity, improve general health, improve the appetite, and restore memory. They also regard it as a preventive medicine and tonic (!). Following the 3rd Wujia Symposium in Harbin (1981), I was presented with a scroll with a Chinese calligraphic inscription of an ode to Wujia as follows (!):

ODE TO WUJIA

By **Ye Zhishen** (Qing Dynasty)

From earth and heavens the quintessence Five folioles clustering your leaves,
And pretty little thorns wrapped whole your shoots;
Oh, what a jackal's gaunt leg looks much alike.
How wonderful is Winzhang-grass, the Eleuthero-ginseng
Dispensing in liquor for drinking,
And decocting with Burnet for daily using,
It will keep your virgin face younger
And prolong your life for ever and ever;
Even if a cartload of gold and jewels,
That can not estimate your price of nature.

I am not sure, however, that these Chinese of the Qing Dynasty attributed as much to *E. senticosus* as to other quinquefoliolate herbs. It is rarely, if ever, singled out in the Chinese herbals I have studied. It is not even mentioned in Perry's fine compilation.[16] Still, Tierra says, "Eleuthero is considered by the Chinese to be the best medicine for treating insomnia."[28] Russian scientists seem to be promoting the adaptogenic attributes of *Eleutherococcus* more than its specific attributes; e.g., Vereshchagin et al. (1982) conclude that adaptogens like *Eleutherococcus* increase the efficacy of antibiotic therapy in children with dysentery.[153]

Baranov[154] reports that Soviet phytochemists have found coumarins in the stems and roots, eleutherosides A (daurosterin), B, C, D, E, F, and G in the bark and wood of stems and root. Sesamin occurs in the root. Roots contain 0.8% essential oil, fruits 0.5%, leaves 0.31%, and stems 0.26%. Roots contain isofraxin, wax, carotenoids, pectins, and resins. Saponins from this plant have shown "reported" pharmacological activities, including effects on arousal and performance, energy metabolism, and cardiovascular system, and macromolecular synthesis in the liver, testes, and bone marrow. Constituents have been found to increase adrenal capacity in stressed animals. Soviet scientists reported a doubling of survival time with mice treated with *Eleutherococcus* or *Panax ginseng* during chronic irradiation with a total of up to 7000 rads, hence, suggesting its usage in actinotoxemia. "*Eleutherococcus* is of value during therapeutic irradiation." Hsu reports antineoplastic activity in China,[51] but Duke reported at the 2nd Wuija Symposium that the NCI detected no significant anticancer activity in the leaves and woody specimens he collected with Lewis, Dharamanda, and Zaricor in Manchuria.[41]

Toxicity — Wherever I went in China, I was told that eleuthero was even more innocuous than Panax (!).

127. *EPHEDRA GERARDIANA* Wall. ex Stapf (EPHEDRACEAE) — Pakistani Ephedra

Fruits said to be edible.[1] This high-yielding source of ephedrine nearly gives Pakistan a monopoly for this naturally produced drug. Ephedrine is extracted from the green branches and possesses the same properties as ephedrine from *E. sinica*, with a higher total alkaloid content, from 1.0 to 2.5%. Ephedrine excites the sympathetic nervous system, depresses smooth and cardiac muscles, produces a lasting rise in blood pressure, and diminishes hyperemia. Ephedrine acts like epinephrine or adrenalin but can be given orally as well as by injection. Pseudoephedrine and ephedrine, both now synthesized, appeared, respectively, in nearly 14 million (0.90% of all prescriptions) and nearly 12 million (0.77%) prescriptions in the U.S. in 1973.[98] Prescribed in the U.S. for asthma, emphysema, hay fever, and rhinitis. Orally ephedrine has helped in enuresis, epilepsy, myasthenia gravis, and urticaria accompanying angioneurotic edema.[17] D-Pseudoephedrine is cheaper and less toxic than ephedrine and has been used in asthma with good results. The rhizomes may have football-sized knobs used as fuel in Tibet.[1]

Chinese, like the Pakistanis and others, have long treasured ephedra for allergy, asthma, cold, cough, diarrhea, eruptions, fevers, hayfever, headache, malaria, pertussis, and rheumatism.[33] Infusion used to prevent low blood pressure in flu, pneumonia, etc. Said to be alterative, diuretic, mydriatic, stomachic, sudorific, and tonic.[27] Homeopathically suggested for exophthalmia, headache, and struma.

In *E. gerardiana*, total alkaloids vary from 0.8 to 1.4%, about half ephedrine with D-pseudoephedrine, L-*N*-methylephedrine, L-norephedrine, D-*N*-methylpseudoephedrine, and D-norpseudoephedrine. Tannins (chiefly gallic and ellagic acid), saponin, catechins, and an essential oil containing a terpineol.[33]

Toxicity — Classified by the FDA[62] as an Herb of Undefined Safety: "Used as an antisyphilitic. Also used as an astringent. A Chinese species, *Ephedra sinica*, called 'ma-huang' in China, contains the alkaloid ephedrine, a powerful decongestant." In large doses, ephedrine causes headache, indigestion, nervousness, flushing, numbing of the extremities, nausea, tingling, palpitations, and vertigo.[1] Contact dermatitis may occur at onset of application or as much as 2 years later.[17]

128. *EPHEDRA NEVADENSIS* S. Wats. (EPHEDRACEAE) — Mormon Tea, Nevada Jointfir

Source of Mormon tea, Popotillo, Squaw tea, Teamster tea, or Whorehouse tea. According to the USDA's Plant Handbook,[155] *E. nevadensis* and *E. viridis* are the most important forage ephedras in the U.S. They are described as moderately palatable to all classes of domestic livestock and deer, sometimes important in the winter diet. Amerindians ground and roast the root to make bread.[47]

Reported to be astringent, depurative, fumitory, Mormon tea is a folk remedy for cold, enterorrhagia, gonorrhea, headache, nephritis, and syphilis and other urogenital ailments. The fluid extract and/or infusion is diuretic in humans.[32,37] Mexicans mix the "leaves" with tobacco and smoke for headache. Amerindians drank infusions of this and related species for venereal disease. Nevada Indians poulticed powdered leaves onto sores and took the tea as a diuretic.[156]

Per 100 g, the dry leaf is reported to contain 5 g protein, 5810 mg Ca, and 500 mg P.[21] While some authors attribute the activity of the herb to ephedrine, a drug which constricts the blood vessels, dilates the bronchioles, and stimulates the CNS, others attribute it to (+) neopseudoephredine, an even more potent CNS-stimulant. Tyler unhesitatingly concludes that the plant is "alkaloid-free".[37] Siegel lists ephedrine under *E. nevadensis*.[55] Our alkaloid compilations do not. Mormon tea contains resins, much tannin, and a volatile oil.

Toxicity — According to Morton,[50] an infusion of what was recorded as either *E. nevadensis* or *E. viridis* (reported to contain ephedrine and pseudoephedrine) produced a "prompt and extensive contraction of uterine muscle" when applied to smooth muscle strips of virgin guinea pig uteri. In experimental dogs and rabbits, the infusion caused a marked rise in blood pressure. Rose suggests that frequent use may result in nervousness and restlessness. It should only be used on the advice of a physician, particularly if you suffer from high blood pressure, heart disease, diabetes, or thyroid trouble.[47]

129. *EQUISETUM ARVENSE* L. (EQUISETACEAE) — Field Horsetail

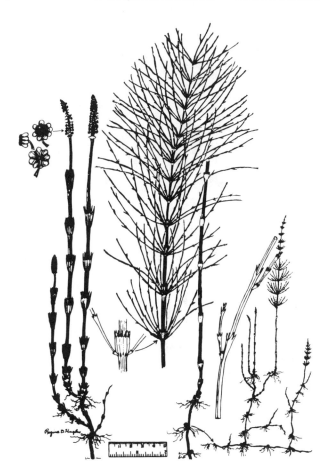

Strobili are boiled and eaten in Japan; also boiled, salted, and kept in vinegar and mixed with soy. Meskwaki Indians fed wild geese the plant, said to fatten them in a week.[45] It was also said to improve the gloss of their ponies' hair. Roots used as food by Indians in New Mexico. Plants generally collected and used for medicinal purposes. Herb is used either fresh or dried, being most effective when fresh. Because of its high silica content, the herb is used for scouring and polishing.[27] Not only is horsetail rich in silica, it is also reported to accumulate gold (0.03 to 0.075 ppm) and silver (0.23 ppm) compared to norms for woody angiosperms of <0.00045 and 0.06, respectively. I recommend experiments with horsetails to mine such metals from sewage sludge. It has been suggested that horsetail may accumulate more gold than other plants (*circa* 125 g per fresh ton).[157] Tyler concludes, "The plant is a weak diuretic and little else."[37]

Decoctions, poultices, and teas are used for various forms of cancer including polyps, abdominal, and oral cancers.[4] In Guatemala, used for bone cancer. Also, used for cancer of breast, intestines, kidney, lip, liver, stomach, and tongue.[4] Decoction of plant is astringent and is used as a styptic and to reduce the swelling of eyelids; is diuretic and is used in bladder and kidney affections, and is also used for cystic ulceration, cystitis, dropsy, dysuria, fever, gonorrhea, gout, hematopoietic ailments, hemoptysis, gravel, rheumatism, and tuberculosis. Ashes of plant used for acidity of stomach and dyspepsia. Cooling and astringent, it is used for cystic ulceration, hemorrhages, and ulcers in urinary passages. Externally, the decoction will stop bleeding and heal wounds, and reduce the swelling of eyelids.[2] Fractured bones are said to heal faster when horsetail is taken.[28] The high silica content is said to render it especially effective in pulmonary consumption,[27] but Tyler concludes no evidence supports the hypothesis that the silica and silicic acid derivatives in the drug promote the healing of bleeding tubercular lesions in the lung.[37]

Fresh forage, per 100 g, contains 20 calories, 93.7 g H_2O, 1.0 g protein, 0.2 g fat, 4.4 g total carbohydrates, 7.1 g fiber, 0.7 g ash, 58 mg Ca, 93 mg P, 4.4 mg Fe, 300 μg β-carotene, 0.0 mg thiamine, 0.07 mg riboflavin, 5.6 mg niacin, and 50 mg ascorbic acid.[21] Said to contain 5 to 8% silicic acid, 5% equisetonin (a saponin), nicotine, isoquercitrin, galuteolin, equisetine, equisetrine, nicotine, resin, silica, starch, tannin, and wax;[27,33,37] beta-sitosterol, aconitic-, oxalic-, malic-, and tannic-acids.

Toxicity — Poisonous to livestock, especially to horses and cattle, causing equisetosis. Thiamase is the major toxin in horsetail. Hay composed of 20% or more *E. arvense*, produces symptoms in horses in 2 to 5 weeks. Unthriftiness is followed by weakness, especially in the hind quarters, ataxis, and difficulty in turning. Appetite, however, remains normal. In fatal cases, death is preceded by a lapse into quiescence and coma.[16]

130. *EQUISETUM HYEMALE* L. (EQUISETACEAE) — Shavegrass, Great Scouring Rush

Meskwaki fed the herb to their ponies, said to fatten them in a week.[45] Young shoots said to be edible like asparagus.[13] Due to the siliceous nature of the plant, it is used for scouring. Used for polishing furniture and wooden floors in Europe. In Canada used to clean copper vases and to polish metal.[45] Plant once recommended for tanning leather. Indians sometimes burned the plant as a disinfectant.[45]

Tea consumed or bathed onto cancers and ulcers in Austria and Germany.[4] Said to be astringent, depurative, diaphoretic, diuretic, and hemostat.[27] The high silica content is believed to make it particularly effective in treating pulmonary consumption.[27] Used as a cataplasm for fistulas, hemorrhoids, menstrual embolisms, dysentery, and eye ailments in China.[16] Used homeopathically in cystitis, enuresis, cholecystosis, dysuria, incontinence, stones, and urethritis.[83]

Silicon oxide is present in large quantities, from 7.5 to 41.2%. In addition to the "carcinogen" tannin it contains aconitic-, caffeic-, ferulic-, and silicic-acids and the alkaloids equisetine, nicotine, and palustrine.[2,33] Equisetine is a nerve poison, and the perils of nicotine are well known. Dimethylsulfone is also reported.[33]

Toxicity — Classified by the FDA as an Herb of Undefined Safety: "Infusion of whole plants used sometimes in dropsical and renal diseases but the diuretic action is very feeble."[62] Cattle overdosed for diuresis have voided blood. Said to have caused poisoning in California.[14] Horsetails are said to develop a powerful nerve poison, aconitic acid. Sheep and cattle are poisoned by grazing the fresh plant; horses, usually by eating the dried plant in hay. It produces, especially when dried, sudden symptoms of weakness and loss of appetite followed, after a few weeks, by loss of muscular control, excitement, and falling, and in acute cases, labored respiration, rapid, weak pulse, diarrhea, convulsions, coma, and death.[50]

131. *ERYTHRINA FUSCA* Lour. (FABACEAE) — Coral Bean, Gallito, Bois Immortelle

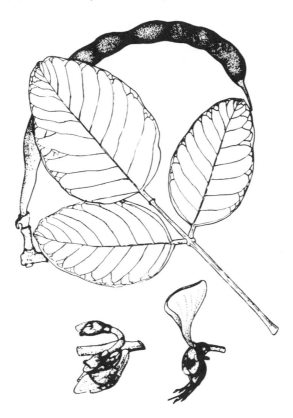

Occasionally planted as a hedge and as a support for the betel vine in Assam and Bengal. Young leaves are eaten, raw or boiled, as a vegetable in Java. In Puerto Rico, trees have been planted in pastures and along fences and roads as ornamental shade trees. Elsewhere, they are used for cacao and coffee shade and living fence posts. Heartwood is light yellow to yellowish-brown and moderately soft. The lightweight wood is weak, not durable, and scarcely suitable for lumber.[63]

Seeds are used in folk remedies for cancer in Annam.[4] Reported to have the same medicinal attributes as *E. indica,* whose bark is used for fever, hepatosis, malaria, rheumatism, toothache; also, for boils and fractures. Perry cites many more uses for *E. indica.*[16] The bark is used for poulticing fresh wounds in Malasia. Boiled roots are taken internally or externally for beri-beri. Grated wood used for hematuria.[16] The root is used for rheumatism. Bark and leaves serve as a vermifuge.[33]

Per 100 g, the leaves are reported to contain 60 calories, 81.5 g H_2O, 4.6 g protein, 0.8 g fat, 11.7 g total carbohydrate, 4.1 g fiber, 1.4 g ash, 57 mg Ca, 40 mg P, 1.8 mg Fe, 2300 µg β-carotene equivalent, 0.24 mg thiamine, 0.17 mg riboflavin, 4.7 mg niacin, and 3 mg ascorbic acid.[21] On a zero-moisture basis, the leaves contain 325 calories, 24.9 g protein,4.3 g fat, 63.3 g total carbohydrate, 22.2 g fiber, 7.6 g ash, 308 mg Ca, 222 mg P, 5.2 mg Fe, 0.91 mg thiamine, 0.52 mg riboflavin, 6.54 mg niacin, and 78 mg ascorbic acid.[21]

Toxicity — Seeds contain the alkaloid erythralin. Erysodine, erysonine, erysopine, erysothiopine, erysothiovine, erysovine, erythraline are also reported. Those species containing indoles and isoquinolines are classed as narcotics, capable of inducing hallucinogenic stupor.[54]

132. *ERYTHROPHLEUM SUAVEOLENS* (Guill. & Perrot.) Brenan (CAESALPINIACEAE)
— Sassy Bark, Ordeal Tree

The bark, an article of commerce, though highly poisonous, is used as an antidote to Strophanthus poison.[3,17] It has long been used by Africans as an arrow poison, a poison for trial by ordeal, and as a tanning material. The alkaloids in the bark act like digitalis. Dilutions of erythrophleine as low as 1:800,000,000 produce vasoconstriction. Antelopes are said to eat the pods. The hard heavy timber seasons well, seems relatively resistant to fungi and termites, and has been used for cabinetry, construction, flooring, implements. Fresh wood sinks in water.

Reported to be anesthetic, astringent, cardiotonic, emetic, purgative, sassybark is a folk remedy for colic, diarrhea, dysentery, dyspnea, ophthalmia, snakebite.[32,33] Snuff made from the bark was once sold in Africa as a sternutatory for cold relief. The bark is used for worms in Tanganyika, the leaves for snakebite. Zulu and others use the powdered bark for headache. Sierra Leone natives use it for rheumatism; West Africans for skin disease.

Alkaloids present include cassaidine ($C_{24}H_{41}O_4N$), nor-cassaidine ($C_{23}H_{41}O_5N$), cassaine ($C_{24}H_{39}O_4N$), cassamine ($C_{25}H_{35}O_5N$), erythrophalamine ($C_{25}H_{39}O_6N$), erythrophleine ($C_{24}H_{39}O_5N$), homophleine ($C_{56}H_{90}O_9N_2$). *Hager's Handbook* adds norcassamine ($C_{24}H_{37}NO_5$), norerythrosaumine ($C_{24}H_{37}NO_6$), and dehydro-norerythrosaumine, luteolin, tannin, sugar, palmitic-, stearic-, oleic-, linoleic-, and cerotinic-acid.[33]

Toxicity — Highly toxic, in humans leading to a primary slowing of the heart followed by acceleration, dyspnea with labored respiration, and death due to respiratory arrest.[3] Workers may develop a dermatitis from handling the wood — tall, fair men being more susceptible than short, dark ones. The erythrophleum alkaloids are said to be powerfully analgesic to the mucosae, and generally more potent than cocaine. With a marked stimulant effect on isolated rabbit uterus, erythrophleine has been suggested as a devitalizing agent in dentistry. Cassaine has convulsant activity. Erythrophleine and homophleine have some degree of hemolytic activity. All the alkaloids are said to be strongly antiseptic, especially cassaidine.[17]

133. *ERYTHROXYLUM COCA* Lam. (ERYTHROXYLACEAE) — Coca

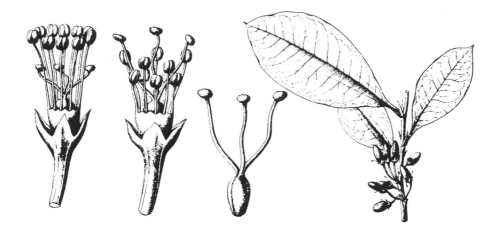

Cocaine, the widely known anesthetic, is derived from the shrub. Inca Indians regarded coca as a divinity. Coca is the only source of cocaine, which rapidly stimulated the higher levels of the brain, giving one a sense of boundless energy and freedom from fatigue. "It was one of the ingredients of Coca Cola, together with Cola nitida, until 1904 when the U.S. courts ruled against its use."[54] Cocaine ointment has been used for eczema, pruritus, urticaria, the pain of hemorrhoids, and facial neuralgia. Cocaine exerts a powerful bactericidal action on Gram-negative and coccus organisms but not on sporogenic organisms.[1]

Reported to be anesthetic, anodyne, aperient, aphrodisiac, astringent, bactericidal, carminative, deobstruent, depurative, digestive, diuretic, hallucinogenic, mydriatic, narcotic, nervine, psychedelic, stimulant, coca is a folk remedy for asthma, cancer, conjunctivitis, dyspepsia, edema, epistaxis, fractures, gout, headache, hoarseness, hypochondria, itch, melancholy, nausea, nervousness, neuralgia, neurasthenia, neuritis, piles, rheumatism, sideache, sores, soroche, splenosis, stomachache, stomatitis, swellings, syncope, throat, and wounds.[32,33] Wine fortified with coca was once sold as Vin Mariani as analgesic and anesthetic; claimed, also, to alleviate gastrosis, gingivitis, and stomatitis.

Per 100 g, the leaves are reported to contain 305 calories, 6.5 g H_2O, 18.9 g protein, 5.0 g fat, 60.6 g total carbohydrate, 14.4 g fiber, 9.0 g ash, 1540 mg Ca, 911 mg P, 45.8 mg Fe, 41 mg Na, 2020 mg K, 11,000 μg β-carotene equivalent, 0.35 g thiamine, 1.91 mg riboflavin, 1.29 mg niacin, and 1 mg ascorbic acid.[21] Leaves contain 0.4 to 2.5% alkaloids, with cocaine constituting half the total alkaloids. Also present are benzoylecgonine, cinnamylcocaine, ecgonine, hygrine, tropacocaine, truxilline, nicotine, methylecgonine, hygroline. Young leaves may contain 0.13% methyl salicylate, also carotene, palmityl-beta-amyrin, and various acids,[33] as well as isoquercitrin, quercitrin, and rutin.

Toxicity — Classed as a narcotic euphoriant and hallucinogen.[54] Small doses may produce some pleasant excitement. Large doses cause restlessness, tremors, and hallucinations and sometimes death due to paralysis of the respiratory center. In April 1884, Sigmund Freud ordered 1 g of cocaine from Merck and took 1/20 g internally. He also administered it to his colleague, Dr. Fleischl, to rid him of morphine addiction. Testing the influence of cocaine on muscular strength, Freud and Koller, a collaborator, noticed its numbing effect on the mouth and lips. Koller pursued the observation that cocaine could be used to desensitize the eye tissue which made it a viable anesthetic for surgeons, especially ophthalmologic surgeons. In 1885, a lifelong friend of Freud, Konigsteih, operated on Freud's father to relieve his glaucoma, while Koller and Freud administered the anesthetic.[158]

134. *ESCHSCHOLZIA CALIFORNICA* Cham. (PAPAVERACEAE) — California Poppy

This handsome ornamental annual or biennial is the state flower of California. West coasters may smoke the leaves and petals when marijuana is scarce.[51] It is said by an anonymous writer[51] to offer a high lasting about 30 min. Amerindians are said to have used the plant as a fish poison.[3]

Reported to be analgesic, anodyne, diaphoretic, diuretic, soporific, and spasmolytic, the alkaloids present in the roots are said to have feeble narcotic and respiratory effects, but the plant is said to have no therapeutic importance.[1] Mixed with black pepper, the plant has been used for ague, jaundice, and skin ailments.[32,33]

Dry seeds contain 25% protein, 46.8% fat.[21] Flowers contain *circa* 5% rutin[59] and a purple-red pigment eschscholtz-xanthin.[1] According to Watt and Breyer-Brandwijk[3] and Duke,[91] the California poppy contains several alkaloids, among them allocryptopine, berberine, bisnoragemonine, californidine, chelerythrine, chelidonine, chelilutine, chelirubine, codeine (report doubtful), coptisine, cryptocavine, cryptopine, escholine, eschscholtzidine, eschscholtzine, fumarine, glaucine, ionidine, lauroscholtzine, morphine (report doubtful), protopine, and sanguinarine.[3] A glycoside and succinic acid, sugars, coloring matter, resinous substances, and cyanide occur in the roots, the whole plant containing HCN;[1] KNO_3, fumaric acid, and rutin are also reported.[3] Cryptopine is uterotonic in guinea pigs at dilutions of 1 ppm.[33]

135. *EUCALYPTUS GLOBULUS* Labill. (MYRTACEAE) — Eucalypt, Tasmanian Bluegum

All species of *Eucalyptus* contain essential oils used for internal and external medicines, insectidides, and insect repellents.[56] A type of kino is extracted from the tree in Argentina. Grown for firewood in India.[1] The timber is used for platforms, sheds, stations, fences, and carpentry. Essential oil, widely used in cough drops is antiseptic, rubefacient, and stimulant.[42] The leaves are antibiotic. Africans use finely powdered bark as an insect dust. Mexicans chew the leaves to strengthen the gums. Said to be a good honey plant.

Bluegum is a folk remedy for arthritis, asthma, boils, bronchitis, burns, cancer, catarrh, cold, cough, cystitis, diabetes, dysentery, dyspepsia, fever, flu, grippe, inflammation, laryngitis, leprosy, malaria, miasma, phthisis, rhinosis, sore, sore throat, spasms, tuberculosis, vaginitis, wounds, and worms.[3,32,33,42] Venezuelans take leaf decoction for chest ailments or colds, Guatemalans for coughs and grippe, Jamaicans for cold. Cubans use the essential oil for bronchitis, bladder and liver infections, lung ailments, malaria, stomach trouble.[42] Homeopaths use the plant for bronchitis, colds, flu, laryngitis, and rheumatism. In Asia, the leaf oil, clearly poisonous in large quantities, is regarded as anesthetic, antibiotic, antiperiodic, expectorant, febrifuge, and vermifuge, and it is used for asthma, bronchitis, influenza, and tuberculosis.[16]

Leaves contain 70 to 80% eucalyptol (cineol). Also, includes terpineol, sesquiterpene alcohols, aliphatic aldehydes, isoamyl alcohol, ethanol, and terpenes.[42] Tannin is not so copious in the leaves as of many other *Eucalyptus* species. The kino, containing 28.7% kino-tannin and 47.9% catechin, contains the antibiotic citriodorol.[3]

Toxicity — In large doses, oil of eucalyptus, like so many essential oils, has caused fatalities from intestinal irritation.[42] Death is reported from ingestion of 4 to 24 m𝓁 of essential oil, but recoveries are also reported for the same amount. Symptoms include gastroenteric burning and irritation, nausea, vomiting, diarrhea, oxygen deficiency, weakness, dizziness, stupor, difficult respiration, delirium, paralysis, convulsions, and death, usually due to respiratory failure.[56] Sensitive persons may develop urticaria from handling the foliage and other parts of the plant.[3]

136. *EUONYMUS ATROPURPUREA* Jacq. (CELASTRACEAE) — Wahoo, Burning Bush

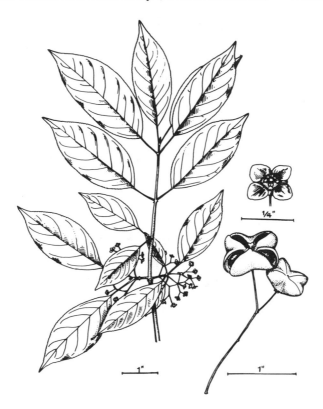

Wahoo Root Bark of Commerce is derived from this rather ornamental species. As the "Burning Bush" the shrub is planted for its handsome autumnal coloration in the U.S.

Formerly used as a cathartic and diuretic,[20] considered alterative, anodyne, antibilious, cathartic, cholagogue, collyrium, diuretic, emetic, expectorant, laxative, pediculicide, and tonic.[2,32] Used for dropsy and dyspepsia,[30,32] fever, hepatosis, secondary syphilis, and uterosis. Euonymin is said to stimulate appetite and gastric juices. It is used as a purgative in cases of constipation with liver disorder. Especially valuable for liver ailment following fever and dyspepsia. Said to promote the biliary functions and intestinal secretions, increasing capillary circulation generally. Used for the expulsion of parasites, including, apparently, malarial parasites. Homeopaths prescribe the tincture of fresh bark and root for albuminuria, bilious fever, biliousness, cholera morbus, gallstones, levitation, liver afflictions, and vertigo.[28,30]

Contains the alcohol euonymol ($C_{21}H_{30}O_4$), euonysterol [($C_{31}H_{5k}O(OH)$], atropurpurol [$C_{27}H_{44}(OH)_2$], atropurpurin, asparagine, homoeuonysterol ($C_{40}H_{70}O_2$), phytosterol, galactitol, triacetin, ·citrullol [$C_{32}H_{36}O_2(OH)_2$], dulcite, and cerotic-, citric-, euonic-, linoleic-, malic-, oleic-, palmitic-, and tartaric-acids. The resin, called euonymin, is used medicinally. The "carcinogen" tannin is also reported.[27,33]

Toxicity — This species is probably toxic. In large doses, euonymin is irritant to the intestine and cathartic. Ingestion of fruits may cause vomiting, diarrhea, weakness, chills, and convulsions followed in about 12 hr by unconsciousness. No deaths are reported in the U.S.

To the physician — Hardin and Arena recommend gastric lavage or emesis and symptomatic treatment.[34]

137. *EUPATORIUM PERFOLIATUM* L. (ASTERACEAE) — Boneset, Ague Weed

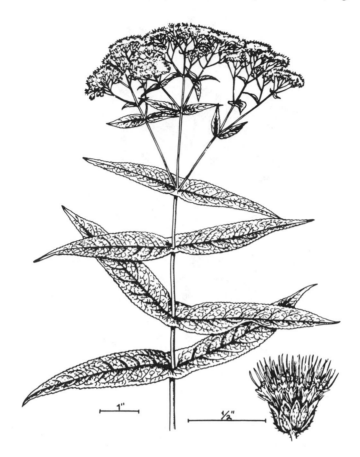

Sold to a very limited extent in the American herb trade, herbalists drink the tea to break a cold or flu, and curb the aches and pains.

Used sometimes as a fomentation with hops, for tumors.[4] Said to be aperient, diaphoretic, emetic, expectorant, febrifuge, hemostat, laxative, nervine, purgative, stimulant, sudorific, and tonic. Used for ague, backache, catarrh, chills, colds, cramps, debility, dengue, dropsy, dyspepsia, fever, gastritis, indigestion, influenza, intemperance, malaria, pneumonia, rheumatism, snakebite, typhoid, typhus, and yellow fever.[32,45] Used for the expulsion of tapeworm. Said to have worked better than quinine on persistent cases of malaria.[19] Might be tried as a quinine substitute, should we lose our supplies of that drug. Homeopaths prescribe the tincture for anal herpes, back pain, bilious fever, bladder ailments, boneache, cough, dengue, diarrhea, fever, flu, fracture, gastritis, gout, hiccup, hoarseness, indigestion, infection, influenza, intermittent fever, jaundice, liver pain, measles, mouth, ophthalmia, relapsing fever, remittent fever, rheumatism, ringworm, spotted fever, syphilitic pains, thirst, and wounds.[28,30,33]

Contains eupatorin, tannic acid, gallic acid, a bitter glucoside, resin, gum, sugar, vitamin B_1, and a volatile oil. Various flavonoids, sterols, and triterpenes have been reported but none are therapeutically exciting.[37] Toxic concentrations of nitrates are reported for the plant.[14]

Toxicity — Classified by the FDA (Health Foods Business, June 1978) as an Herb of Undefined Safety: "Has diaphoretic effect. Emetic and aperient in large doses. Household remedy never prescribed by the medical profession." Eupatorin has cytotoxic properties.[33]

138. *EUPHORBIA LATHYRIS* L. (EUPHORBIACEAE) — Mole Plant, Petroleum Plant, Caper Spurge

Sold, for example, by Maryland nurseries as the Mole Plant, this attractive poisonous evergreen perennial belongs to a family, many species of which are believed to repel moles. Castor bean is another Euphorbiaceous plant believed to repel moles and other subterranean pests. Recently,[159] Nobel Laureate Melvin Calvin has suggested that the mole plant could be the "petroleum plant", producing a hydrocarbon substance very much like gasoline. Hydrocarbon produced by the plant could probably be used directly in existing refineries, once separated from the water. His estimates of 10 to 50 barrels of oil per acre per year seem optimistic as do his cost estimates of $3 to $10 per barrel. *If* Calvin's optimistic estimates are proven, the petroleum plant should have virtues enough to outweigh its noxious attributes. Guayule rubber has been estimated to give 6000 kg/ha latex equivalent to only about ten barrels oil per hectare (four barrels per acre). It is difficult to believe that *E. lathyris* would produce ten times more latex than guayule. However, the latex in Maryland-grown material is extremely copious. Calvin suggests that the plants could be simply cut down and run through a crushing mill, new plants growing from the stumps. My Maryland material always dies like a biennial. Yet Calvin suggests that replanting would be necessary every 20 years or so. At 40 barrels an acre, he estimates it would take an area the size of Arizona to meet our current gasoline requirements. Seeds are said to yield a fine clear oil known as oil of *Euphorbia*, obtained by expression or by ether or alcohol extraction. The latex is said to be depilatory. Seeds, said to be used as a coffee substitute, are *poisonous*! The oil has been used in soap manufacture.

French country folk are said to take 12 to 15 seeds as a purgative, perhaps, similar in action and toxicity to castor oil. The root is equally purgative and emetic. The leaves, like the latex, are vesicant, and have been used by beggars to incite pity-producing blisters. Latex is used in folk remedies for cancers and warts.[4] Reported to be antiseptic, cathartic, diuretic, emetic, POISON, and purgative, mole plant is a folk remedy for cancer, corns, diarrhea, gangrene, melanoma, skin ailments, sores, and sore throats.[32] Homeopathically, the seeds are used for erysipelas, paralysis, and rheumatism. The seed oil is applied to burns.

The L isomer of dopa [3-(3,4-dihydroxyphenyl) alanine] is said to occur at 1.7% of the fresh weight of the latex. L-Dopa is used for symptomatic relief of Parkinson's disease, and is said to have produced some astounding rejuvenating effects, even priapism, on some senile males who took it. Leaves contain quercetin, quercetin-3-beta-D-glucuronide, kaempferol, kaempferol 3-glucuronide, beta-sitosterol, *p*-coumaric acid, and ferulic acid. Stalks contain hentriacontane, taraxerone, taraxerol, beta-sitosterol, and betulin. The energy-promising latex contains 0.5% 3,4-dioxyphenylalanine. Sachs et al.[160] got 6.2% rosin content (hydrocarbons) in nonirrigated, compared to 4.4% on irrigated plots. The benzene extract of the leaves is said to contain 0.1% (% of plant dry weight) rubber, and 0.2% wax; the acetone extract 13.7% glycerides, 2.2% isoprenoids, and 8.3% other terpinoids. The acetone extract of the seeds contains 40% glyceride. The seed is reported to contain 15% protein and 40 to 47.5% fat. Seed contain beta-sitosterol, 7-hentriacontane, and daphnetin.[33] I believe that if all plants are studied in detail, they will be found to contain both carcinogens and antitumor or cytotoxic compounds. This one contains the antitumor compound beta-sitosterol and the cocarcinogen ingenol-3-hexadecanic acid ester ($C_{36}H_{58}O_6$).

Toxicity — Human overdoses result in burning mouth, nausea, diarrhea, dilated pupils, and collapse with pallor, rigidity, frigidity, cold sweats, arrhythmic pulse, vertigo, delirium, alternating hot and cold flashes, cramps, etc.[33] The seeds are said to contain aesuletin, and an unnamed alkaloid and the seed oil is violent *POISON*. The plant has caused poisoning in children. Five California women are said to have pickled the fruits, believing them to be capers; severe but not fatal poisoning ensued.

139. *EUPHORBIA PULCHERRIMA* Wild. ex Klotsch. (EUPHORBIACEAE) — Poinsettia

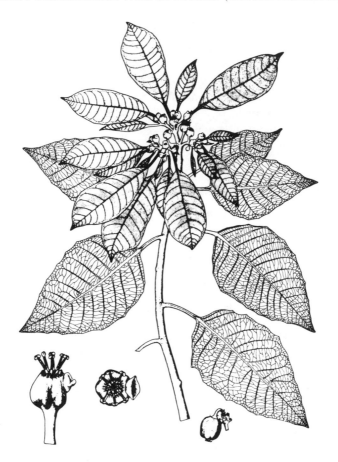

Widely grown as a subtropical ornamental, especially important at Christmas time. Some tropical people eat the colorful bracts, dangerous as they are claimed to be, as a vegetable. Latex sometimes used as a depilatory. Chinese use the plant as a fish poison. Plant yields a red dye.

Reported to be anodyne, bactericidal, depilatory, emetic, lactagogue, and poison, poinsettia is a folk remedy for erysipelas, skin ailments, toothache, and warts.[4,32] Puerto Ricans use the plant as a febrifuge;[3] Guatemalans use the latex as antiodontalgic, depilatory, emetic. Mexicans use the leaf decoction as a lactagogue.

The latex contains 5 to 15% caoutchouc and some resin; latex-containing pulcherol ($C_{24}H_{40}O$) and a mono-*N*-acetyl derivative of alpha-gamma-diaminobutyric acid. Leaves contain 0.43% rubber, stems 0.08%, twigs 0.6%, and gum 7.0%; caffeic acid, germanicol, alpha-amyrin, pseudotaraxasterol, octaeicosanol, and beta-sitosterol.[1,33]

Toxicity — Proponents of poinsettia say it is harmless, others say it's a killer. Having eaten the bracts raw myself, I lean toward the proponents point of view. Morton says,[42] "The fresh latex may irritate or blister sensitive skin. It causes great inflammation in the eyes and sometimes temporary blindness. Internally it inflames the mouth and causes anal and rectal itching. In large quantities, it is a severe gastric and intestinal irritant, producing vomiting, diarrhea and delirium preceding death." While doubting Morton and the authorities on whom she draws, I would rather err on the safe side. Just because I am foolish enough to enjoy eating poinsettia bracts does not mean that I endorse such tomfoolery!

140. *EUPHORBIA TIRUCALLI* L. (EUPHORBIACEAE) — Aveloz, Petroleum Plant, Milkbush

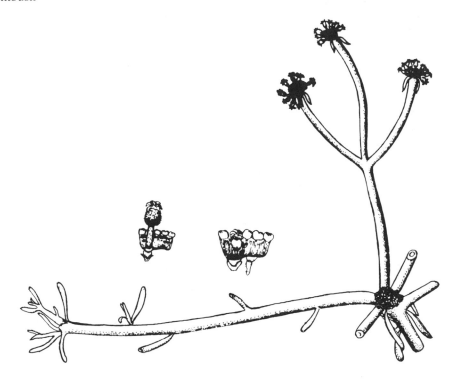

Familiar as a subtropical and tropical ornamental, aveloz has recently made headlines as a potential ''cancer cure'' and as a ''petroleum plant'', both optimistic headlines. The species makes a good living fence post, grown as a hedge in Brazil. According to Calvin,[159] a Nobel-laureate chemist, these plants grow well in dry regions or land that is not suitable for growing food, and should yield as much petroleum as *E. lathyris*. The latex is toxic to fish and rats. Africans regard the tree as a mosquito repellent. Rice boiled with the latex is used to kill birds. Aqueous wood extracts are antibiotic against *Staphylococcus aureus*. The wood, weighing 34 lb/ft³, is used for rafters, toys, and veneer. The charcoal derived therefrom can be used in gun powder. The whole plant might be harvested (if the tree coppices well) to produce rubber, petroleum, alcohol, and resins, which may find use in the linoleum, oil skin, and leather industries. In Brazil, *E. gymnoclada*, very similar to *tirucalli* (both are called aveloz), is used for firewood.

Recently, de Montmorency[161] kindled long-sleeping interests in aveloz (*Euphorbia* spp. including *tirucalli*) inferring that it ''seems to literally tear cancer tissue apart.'' Several Brazilian *Euphorbias*, *E. anomala*, *E. gymnoclada*, *E. heterodoxa*, *E. insulana*, *E. tirucalli*, known as aveloz, have local notoriety as cancer ''cures'', and often find their way into the U.S. press as cancer cures. I fear they are more liable to cause than cure cancer. Still, some Brazilians believe aveloz alleviates cancer, cancroids, epitheliomas, sarcomas, tumors, and warts. Hartwell[4] mentions *E. tirucalli* as a ''folk remedy'' for cancers, excrescences, tumors, and warts in such diverse places as Brazil, India, Indonesia, Malabar, and Malaya. The rubefacient, vesicant latex is used as an application for asthma, cough, earache, neuralgia, rheumatism, toothache, and warts in India. Bahamians spread the latex on warts, and later scrub them off.[42] In small doses it is purgative, but in large doses it is an acrid irritant, and emetic. A tender branch or root decoction is used for colic and gastralgia. Ashes are applied to open abscesses. Tanganyikans use the latex for sexual impotence (but users should recall "the latex produces so intense a reaction . . . as to produce temporary blindness lasting for several days." In Zimbabwe, one African male is said to have died of hemorrhagic gastroenteritis after swallowing the latex to cure sterility).[3] The latex is used as an emetic and antisyphilitic. Malayans apply boiled stems to painful places. Pounded stems are applied to scurf and swelling, and to extract thorns. The root infusion is used for aching bones. The root is used for snakebite, a poultice of the root or leaves for nose ulcers and hemorrhoids. The wood decoction is used for leprosy and for paralysis of the hands and feet following childbirth. Javanese use the latex for skin ailments and rub latex over bone fractures. Philippinos use root scrapings with coconut oil for stomach pain.[16]

The latex contains 53.8 to 79.9% water and water solubles and 2.8 to 3.8% caoutchouc. Fresh latex contains a terpenic alcohol, isoeuphoral ($C_{30}H_{50}O$), identical with euphol from *E. resinifera*. Dried latex contains no isoeuphorol but a ketone euphorone ($C_{30}H_{48}O$). Taraxasterol ($C_{30}H_{50}O \cdot CH_3OH$) and tirucallol ($C_{30}H_{50}O$) have also been isolated. Resin, however, is the principle constituent (75.8 to 82.1%) of the dried latex. The stem contains hentriacontene, hentriacontanol, the antitumor steroid beta-sitosterol, taraxerin, 3,3'-Di-*O*-methylellagic acid, ellagic acid, and a glycoside fraction which hydrolyzes to give kampferol and glucose, and *circa* 0.1% sapogenin acetates. The whole plant contains 7.4% citric acids with some malonic and some bernstein (= succinic?) acids.[33]

Toxicity — Exudes a caustic latex proven to produce uveitis and kerato-conjunctivitis in dog eyes and to exhibit tumor-promoting activity on mouse skin. Latex collected in Colombia contained 12-O-2Z-4E-octadienoyl-4-deoxyphorbol-13-acetate which exhibits an irritant potency equivalent to that of the standard irritant, phorbol-12-tetradecanoate-13-acetate.[162] Fatalities are reported following ingestion of the latex for folk medicinal purposes.[3,42]

To the physician — Hardin and Arena[34] suggest emesis or gastric lavage, and mineral oil.

141. *EUPHRASIA OFFICINALIS* L. (SCROPHULARIACEAE) — Eyebright

In the 14th century, eyebright was supposed to cure "all evils of the eye". In Queen Elizabeth's time, there was even an eyebright ale. The dried herb is one component of a British Herbal Tobacco (said to be "smoked most usefully for chronic bronchial colds"). Overenthusiastic supporters suggest that it might restore the sight of people 70 and 80 years old. Would that it restored my waning eyesight!

Reported to be astringent, collyrium, laxative, ophthalmic, and tonic, eyebright is a folk remedy for allergy, cancer, cough, conjunctivitis, earache, epilepsy, headache, hoarseness, inflammation, jaundice, ophthalmia, rhinitis, skin ailments, and sore throat.[2,28,32,33] Homeopathically used for catarrh, cold, conjunctivitis, hay fever, keratitis, mucositis, and scrofula. Lebanese pound the plant in a mortar, squeeze out the juice, and apply it to herpes, runny eyes, scalp scabs, and weeping eczema. They use the dry herb to smoke wounds, asthma, and coughs.[23] Tierra boldly suggests "the tea be taken liberally and daily for all eye problems."[28]

The dried herb is reported to contain 9.5% moisture, 27.7% water extract, 15.6% alcohol (abs.) extract, 20.2% diluted alcohol extract, 5.9% inorganic substances soluble in water, 10.07% volatile oil, 2.2% substances reducing Fehlings solution, 12.4% tannins, 8.6% ash, and 0.4% acid-insoluble ash. The tannins give the reactions of gallotannin. Vitamin C, β-carotene, a bitter acrid principle, an aromatic resinous substance, and a glycoside (0.05% on fresh weight) are present. The glycoside is decomposed by emulsin and is probably similar to thinanthin or aucubin.[1] Grieve[2] reports mannite and glucose. *Hager's Handbook*[33] adds gallotannins, and, from the closely related *E. rostkoviana,* aucubin, aucubigenin; also, nonacosane, ceryl alcohol, beta-sitosterol, oleic-, linoleic-, and linolenic-, palmitic-, and stearic-acids, fumaric acid, isoquercitrin, quercetin, and rutin, and a glycoside of the pseudoindican type $C_{27}H_{36}O_{15}$. Tyler adds caffeic and feurlic acids.[37]

Toxicity — German experimentation[19] suggested that 10 to 60 drops of the tincture could induce confusion of the mind and cephalalgia; violent pressure in the eyes with lacrymation, itch, redness, and swellings of the margins of the lids, dim vision, photophobia, weakness, sneezing, coryza, nausea, toothache, constipation, hoarseness, cough, expectoration, dyspnea, yawning, insomnia, polyuria, and diaphoresis. Tyler,[37] with good reason, discourages ophthalmic application of eyebright, especially of nonsterile homemade lotions containing miscellaneous poorly known compounds.

142. *FERULA ASSA-FOETIDA* L. (APIACEAE) — Asafetida

With a taste stronger than onion or even garlic, asafetida is still used as a spice in the Middle East. Iranians rub asafetida on warmed plates on which meat is to be served.[7] Alcoholic tinctures of the gum-resin, or the oil and/or fluid extract are reportedly used, at very low levels, in baked goods, beverages, candies, frozen desserts gelatins, meat and meat products, relishes, sauces (e.g., Worchestershire sauce), and spices.[29] Its main use, however, is as a fixative or fragrance component in perfumery. Appears to have hypotensive activity (like the similar smelling garlic) and to increase blood clotting time.[29] Guenther, who should know, adds that the volatile oil has not attained commercial importance because the flavoring and pharmaceutical industries utilize, instead, the tincture of asafetida.[7] Used in veterinary practice to repel cats and dogs.[17]

Asafetida was used for tumors of the abdomen, indurations of the liver and spleen, corns, calluses, as well as felons and whitlows of the fingers, and carcinomata of the gums.[4] Reported to be anodyne, antispasmodic, aperient, aphrodisiac, carminative, diuretic, emmenagogue, expectorant, laxative, nervine, sedative, stimulant, stomachic, and vermifuge, asafetida is a folk remedy for amenorrhea, asthma, bronchitis, cholera, colic, convulsions, cramps, croup, epilepsy, gas, hemiplegia, hysteria, insanity, lungs, neurasthenia, pertussis, polyps, rinderpest, sarcomas, sluggish intestines, spasms, and splenitis.[17,32,33] "Herbal Highs"[51] recommends 1/2 teaspoon in warm water as a tranquilizer. Malays take asafetida for abdominal trouble, broken bones, and rheumatism, Javanese for stomachache and worms.[41] Asafetida is used as an enema for intestinal flatulence. Homeopathically used for gas, osteopathy, and stomach cramps.[33]

Contains 40 to 64% resinous material composed of ferulic acid, umbelliferone, asaresinotannols, farnesiferols A, B, and C, etc., about 25% gum composed of glucose, galactose, 1-arabinose, rhamnose, and glucuronic acid; volatile oil (3 to 17%) consisting of disulfides as its major components, notably, 2-butyl propenyl disulfide (E- and Z-isomers), with monoterpenes (alpha- and beta-pinene, etc.), free ferulic acid, valeric acid, and traces of vanillin.[29] The chief constituent of the oil is secondary butylpropenyl disulphide ($C_7H_{14}S_2$, 40 to 45%); other disulphides ($C_8H_{16}S_2$, $C_{10}H_{18}S_2$, $C_{11}H_{20}S_2$), a trisulphide ($C_6H_{10}S_3$), pinene, another terpene, and an unidentified compound ($C_{10}H_{16}O)_n$. The disagreeable odor of the oil is reported to be due mainly to the disulphide $C_{11}H_{20}S_2$.[1] Analysis of bazaar samples from Mysore gave the following values: ash, 4.4 to 44.3% and alcohol soluble matter, 20.8 to 28.0%; samples obtained from Teheran contained: ash, 6.3 to 8.9% and alcohol soluble matter, 28.3 to 40.9%.[1]

Toxicity — The gum may induce contact dermatitis.[11] Generally not regarded as toxic. Ingestion of 15 g produced no untoward effects,[29] but see *F. sumbul*, which produced narcosis at 15 g. Approved by FDA for use in food (§ 182.20).

143. *FERULA SUMBUL* Hook. (APIACEAE) — Sumbul, Musk Root

The root is used as a substitute for musk in the perfume industry and as a fixative in the liqueur industry.[33] Iranians use as incense and perfume during religious ceremonies. Back in the days of Grieve's *Herbal*, a tincture of 10% sumbul with two volumes of alcohol, one volume water was used as an antispasmodic nervine.[2]

Reported to be aperitif, nervine, sedative, spasmolytic, stimulant, and tonic, sumbul is a folk remedy for asthma, bronchitis, diarrhea, dysentery, dysmenorrhea, hypertony, hysteria, nausea, neurosis, overarousal, and pneumonia.[2,32,33] Said to have specific activity on the pelvic region and to stimulate the mucous membrane. Lebanese, considering it nearly a panacea, use it mainly for female problems, for hysteria, nervousness, and as a uterine tonic.[23] They know the seeds are vermifugal but use the cheaper seed of pumpkin and wormseed.[23] ''Herbal Highs''[51] suggests simmering 2 to 4 teaspoons root in a pint of water for about 5 min as a tranquilizer. Also, used homeopathically.[33]

Contains 17% resin, sumbul acid, free umbelliferone, acetic-, angelic-, valerianic-, butyric-, methylcrotonic-, tiglic-, and vanillic-acids, alkaloids, fats, sugar, and phytosterol.[2,33] Roots contain only *circa* 9% gum or resin, and 0.2 to 1.37% volatile oil.[7] Betaine has been detected in the resin,[2] cerotic-, linoleic-, oleic-, palmitic-, stearic-, and tiglic-acids in the nonvolatile oil.

Toxicity — *Circa* 15 g tincture produces narcotic symptoms, confusing the head and causing tingling sensation and a tendency to snore when awake.[2]

144. *FILIPENDULA ULMARIA* (L.) Maxim. (ROSACEAE) — Meadowsweet, Queen-of-the-Meadow

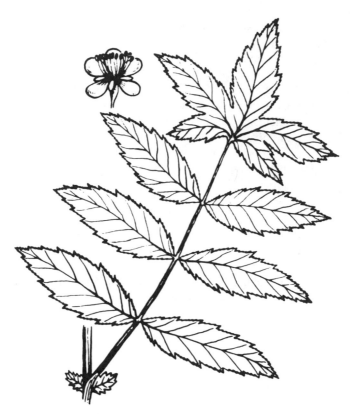

Aromatic, ornamental wildflower, once used as a strewing herb. Double-flowered form used as an ornamental. Formerly called meadwort, the flowers were often put into wine and beer, especially herbal beers. In Scandanavia, the flowers are added to beer and mead for aroma.[33] Leaves were sometimes added as a spice to claret wines. A tea of 60 g leaf to the liter of water, sweetened with honey, is said to make a pleasant diet drink, both for the healthy and invalid. Source of volatile oil known as Oil of Meadowsweet. Bees visit the flowers for pollen.

According to old herbals, the plant was used in folk remedies for cancer and tumors.[4] Once said to relieve ague and melancholy, the herb was regarded as alterative, antacid, antispasmodic, astringent, diaphoretic, diuretic, febrifuge, sedative, stomachic, styptic, sudorific, and tonic.[8,27] It has been used for dropsy, fever, and strangury. Flowers and stems taken to purify the blood. Roots supposed to be good for kidney ailments and lung ailments, shortness of breath, hoarseness of the throat, cough, phlegm, and wheezing. Root and herbage are used for epilepsy, goiter, blenorrhea, bladder and kidney stones, dropsy, tapeworm, and respiratory ailments.[32] Flowers are said to be used for rheumatism, cramps, cystitis, pyelitis, and nephritis, while the herbage is used as a diaphoretic and diuretic in ascites, edema, dropsy, renitis, gall bladder ailment, heart ailment, irregular pulse, and rheumatism.[33] The roots are homeopathically used for gout, acne, rheumatism, headache, congestion, vertigo, rheumatic heart, gastritis, gall bladder afflictions, and renitis. Decoction of flowers was at one time used as a diuretic. The roots are astringent and have been used in treatment of diarrhea. The herb is said to soothe the sympathetic nerves, restoring elasticity to the muscles.

Salicin, the popular analgesic derived from poplars and willows, probably decomposes, after ingestion, to salicylic acid, a compound first isolated from meadowsweet flowerbuds in 1839. Salicylic acid is an active disinfectant used to treat various ailments, especially skin diseases like eczema. Antirheumatic, salicylic acid is said to help fever, headache, myalgia, sciatica, neuralgia.[11] The word aspirin owes its origin to spiraea, having been coined from *a* for acetyl and *spirin* from spiraea. The herb contains salicylic aldehyde, methyl salicylate, salicylic acid, heliotropin, and vanillin. The essential oil contains salicin, gaultherin, spiraein, spiraeoside, isosalicin, anthocyanin, anthocyanadin, quercitin, heliotropin, vanillin, benzol, ethyl benzoate, phenyethyl, phenylacetate, methylsalicylate, salicylaldehyde (up to 70%), salicylic acid, citric acid, pyrogallol, isobutylamine, and isoamylamine.[33] The leaves contain avicularin, hyperoside, filalbin, as well as the compounds listed for the flowers (except spiraeoside).

Toxicity — Classified by the FDA as an Herb of Undefined Safety.[62] Like salicylates, in general, methyl salicylate has analgesic, antiinflammatory, and antipyretic activity, but is much more toxic. Methyl salicylate can be absorbed through the skin, resulting in fatalities. Children's fatalities can be induced by as little as 4700 mg methyl salicylate.[29]

145. *FOENICULUM VULGARE* Mill. (APIACEAE) — Fennel, Finocchio

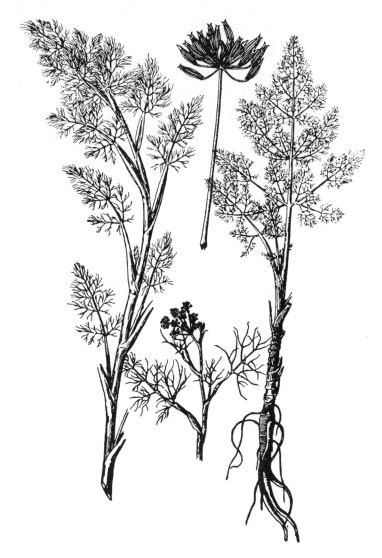

Seeds (fruits) are used to flavor liqueurs, vinegars, breads, pastries, candies, and pickles. Leaves and stems occasionally serve as vegetable, salad, or potherb. The herb has long been an important adjunct for cooking with strong fish. Oil is used in culinary articles, cordials, and toilet articles. Finocchio or vegetable sweet anise is prepared by quartering the bulbs and eating them raw or with salt, or made into salads, braised, boiled, steamed, or served with cream sauce. Powdered fennel said to drive fleas away from kennels and stables. The oil can be used to protect stored fruits and vegetables against infection by pathogenic fungi.[61]

Reported to be abortifacient, anodyne, aphrodisiac, balsamic, cardiotonic, carminative, diaphoretic, digestive, diuretic, emmenagogue, expectorant, lactagogue, pectoral, restorative, stimulant, stomachic, tonic, and vermicide, fennel is a folk remedy for aerophagia, amenorrhea, backache, cholera, colic, dyspepsia, enteritis, enuresis, flux, gas, gastritis, gonorrhea, hernia, nausea, nephrosis, parturition, snakebite, sore, spasm, splenosis, stomachache, strangury, tenesmus, toothache, and virility.[33] A plaster of the juice is a folk remedy for indurations of the mammary glands, carcinomata, and cancerous wounds. Seeds said to help indurations of the liver and spleen, tumors of the uvula, spleen, and fauces, and condylomata.[4] Latin Americans boil them in milk as a lactagogue.[42] Seeds are toasted and powdered to make a collyrium.[42] The root, in syrup, is said to alleviate indurations of the spleen and liver.[4] In Africa, the tincture is used for cramp, diarrhea, and stomachache.[3] Jamaicans take the plant for colds.[42] Seed, leaves, and roots once recommended in teas and broths for obese people. Fennel juice was once a popular cough remedy. The oil is recommended for hookworm.[1] The seeds are regarded as aromatic, carminative, emmenagogue, stimulant, and stomachic almost anywhere the spice is encountered. Leaves are said to be diuretic, roots purgative, and oil carminative and vermicidal.

Residues remaining after oil extraction contain 14 to 22% protein and 12 to 18.5% fat. The fatty oil contains 4% palmitic, 22% oleic, 14% linoleic, and 60% petroselenic acid. Fruits contain 1.5 to 3.0% essential oil and 9 to 21% fatty oil, high in petroselenic acid (67 to 69%), and 6.5% unsaponifiables containing 6-oxychroman derivatives. Tocopherols totaled 50 to 60 mg/100 g, oil, with 75% gamma-tocotrienol (rather high for vegetable oils studied), 7.9% alpha-tocopherol, 9 to 0.0% alpha-tocotrienol, 1.5 to 2% beta-tocopherol, O beta-tocotrienol, and 5 to 6% gamma-tocopherol.[163] Per 100 g, the leaves contain 31 calories, 80.2 g H_2O, 2.9 g protein, 0.5 g fat, 5.6 g total carbohydrate, 0 5 g fiber, 1.8 g ash, 114 mg Ca, 54 mg P, 2.9 mg Fe, 338 mg K, 2610 μg β-carotene equivalent, 0.12 mg thiamine, 0.15 mg riboflavin, 0.7 mg niacin, and 34 mg ascorbic acid. Seeds yield 1.0 to 6.0% oil, averaging 3.5%, the oil containing anethole, anisaldehyde, anisic acid, camphene, dipentene, fenchone, fenchyl alcohol, limonene, methyl chavicol, phellandrene, pinene. Fruits also contain 0 to 13% fixed oil, pentosan and pectin, trigonelline, and choline. Fruits also contain 28 μg iodine, 139 IU vitamin A per 100 g, and traces of Al, Ba, Cu, Li, Mn, Si, and Ti.

Toxicity — Narcotic causing "epileptiform convulsions and hallucinations" (an unidentified oil distillate).[54] Fennel oil, in quantities as small as 1 to 5 mℓ, has caused pulmonary edema, respiratory problems, and seizures. "For this reason, self-medication with fennel should be restricted to moderate use of the fruits (seeds); the volatile oil should not be used."[37] When the distilled oil of fennel was used to treat a variety of ills in Morocco, therapeutic doses occasionally induced epileptiform madness and hallucinations. "Dill, anise, and parsley all have similar oils, and it has been demonstrated that in vivo amination of these ring-substituted oils can result in a series of three hallucinogenic amphetamines."[54]

146. *FRANGULA ALNUS* Mill. (RHAMNACEAE) — Buckthorn

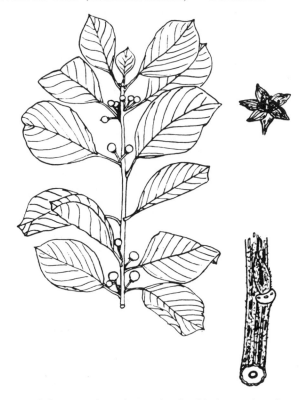

Bark and leaves used for a yellow dye; mixed with iron salts, for a black dye. Unripe berries give a good green color; ripe, blue or gray. Leaves eaten avidly by goats; flowers visited by bees. Wood makes a relatively inflammable charcoal used by old gunpowder makers. Used also for shoe-lasts, wooden nails, and veneer. Bark, used medicinally, should be aged at least 1 year before use. Extracts are said to be useful in sunscreen preparations.[29] According to Denee and Hulzing,[164] anthracene-derivative-containing drugs like *Cassia* spp., *F. alnus, Rhamnus purschianus,* and *Rheum palmatum* are still widely used, as crude extracts, for their purgative effect. The anthraquinone glycosides from these extracts travel to the large intestine where they are hydrolyzed: the aglycones which are liberated act as stimulators of the colon.

The plant has been used for internal cancers and indurations of the liver and spleen.[4] Aloe-emodin shows activity in lymphocytic leukemia and Walker carcinosarcoma tumor systems.[10] Bark cathartic, depurative, emmenagogue, laxative, poison, purgative, and tonic, used in yellow ale for jaundice. Long a home remedy for constipation, liver disorders, piles, and for clearing the blood. Berry juice considered aperient. Kloss recommends the bark or fruit, perhaps, dangerously for appendicitis; also for dropsy, gout, itch, rheumatism, skin ailments, warts, and worms.[44]

Activity is due to the 3 to 7% anthraquinone glycosides (frangulin A and B, glucofrangulin A and B, emodin-1-glucoside, emodin-8-glucoside, emodin-8-0-beta-gentiobioside, etc.). An alkaloid armepavine is present in fresh but not in dried bark. In the fruits, one encounters glucofrangulin, frangulin, emodinanthranol, jesterin; chrysophanol, aloeemodin, aloeemodindianthrone, palmidin B, and saponin. Dry seeds contain 12.9 to 15.7% protein and 10.7 to 15.8% fat.[21] The seed contains amygdalin, robinin, kaempferol, helicin, and tannin.[33] In addition to steroids and carotenoids there are 20 mg tocopherol per 100 g. Tannic acid is also present.

Toxicity — Emodin can cause dermatitis.[6]

147. *GALIUM ODORATUM* (L.) Scop. (RUBIACEAE) — Woodruff, Sweet Woodruff, Waldmeister

Dried leaves and flowers are used in candies, sachets, snuffs, wines and liqueurs, herb "teas", and fruit beverages. Used as fragrance component in European perfumes. Fresh sprigs are also used as garnish in certain summer beverages and wines. Woodruff is a component of bitters, maywines, and vermouth.[29] Powdered leaves are put in potpourri. Dried herb is also placed among linens to protect against insects. The plant contains coumarin, which has the fragrance of new mown hay, and is used as a fixative for other odors. The herb has proven bactericidal activity.[33] Asperuloside has antiphlogistic activity.

Plant used for inflammatory tumors.[4] Entire plant used as a sedative for old and young alike.[8] Medicinally, bruised leaves are applied to cuts and wounds.[47] A strong decoction is said to be used as a stomachic and to remove biliary obstructions of the liver. The tea is used for bladderstones, hepatosis, insomnia, migraine, neuralgia, restlessness, and stomachache.[29] If enough drunk, it is said to incite venery.[47] Employed homeopathically for colpitis and metritis, woodruff is said to be antiphlogistic, antispasmodic, apertif, aromatic, cordial, depurative, diaphoretic, diuretic, hypnotic, sedative, and stomachic.[27,29,32]

In addition to the toxic coumarin, woodruff contains citric-, malic-, rubichloric-, and tannic-acid; asperulisin, tannin, asperuloside (0.05%); monotropein, anthracene and napthalene derivatives, emulsion, and traces of nicotinic acid.[33] The root contains a red dyestuff of the alizarine type.

Toxicity — Dietary feeding of coumarin to experimental animals causes extensive liver damage, growth retardation, and testicular atrophy. FDA approved for use in alcoholic beverages only (§ 172.510).[29]

148. *GARCINIA HANBURYI* Hook. F. (CLUSIACEAE) — Gamboge

Formerly widely used in medicine, it is now used primarily as a coloring matter. The gum resin, known as gamboge, is rarely administered anymore unless mixed with aloes or calomel.[16] Said to have bactericidal and protisticidal activities (against *Aerobacter, Enterovirus, Micrococcus, Mycobacterium*). Long used in lacquer, metal finishes, and watercolors in China since the 13th century.[33] It was used to make the golden yellow ink of Thailand. Gamboge paint is an emulsion in water.[5]

Reported to be diuretic, emetic, poison, purgative, vermifuge, gamboge is a folk remedy for edema.[2]

Resin contains gambogic acid and three garcinolic acids.[16] With the resin is found 15 to 25% of a gum, similar to gum acacia.

Toxicity — The resinous bark exudates are drastic purgatives, causing, also, griping, nausea, and vomiting.[11] Too large doses can be fatal.[16] "It certainly purges, and if the dose is increased causes vomiting also; four grammes are recorded as proving fatal by the production of gastro-enteritis."[5]

149. *GAULTHERIA PROCUMBENS* L. (ERICACEAE) — Wintergreen, Teaberry, Boxberry

The oil of wintergreen is used as a flavoring agent in candies (and I would guess teaberry chewing gum), soft drinks, and dental preparations, especially combined with menthol and eucalyptus. One root beer remedy called for 4 drachms wintergreen oil, 2 sassafras oil, 1 clove oil, and *circa* 120 g alcohol. Methyl salicylate, the main ingredient, has been employed in baths, liniments, and ointments, for pain relief in gout, lumbago, rheumatism, and sciatica.[17] Amerindians ate the berries, even in the snow. Berries used to make pies. Leaves used to make an herbal tea (mountain tea), as a condiment, and a nibble. If children chew the roots for 6 weeks each spring, it will help prevent tooth decay.[11] Amerindians smoked and chewed the leaves, prepared by passing them through the top of the fire, more leisurely dried over the fire, without allowing them to burn. Algonquin guides chew wintergreen leaves to improve their breathing during hunting.

Oil used as an anodyne, antiseptic, counterirritant in rheumatism; also, useful in lumbago and sciatica. Medicinally, the whole plant is used as a antiseptic, carminative, diuretic, emmenagogue, galactagogue, nervine, rubefacient, stimulant, and antirheumatic, and as a flavoring in medicine. Kloss recommends the tea as a gargle for sore throat and stomatosis, as a douche for leucorrhea, as a collyrium for conjunctivitis.[44] Small doses stimulate the stomach, large doses cause vomiting. It may be used for diarrhea, and as an infant's carminative. Quebec Indians rolled the leaves around aching teeth.[45] Leaves are used in the treatment of asthma. Homeopaths prescribe the leaf tincture for gastritis, neuralgia, pleurodynia, rheumatism, and sciatica;[30] also, suggested for epididymitis, orchitis, diaphagmitis, arthritis, and dysmenorrhea.[33] Kloss[44] mentioned the use of the leaves for catarrh, diabetes, dropsy, fever, gonorrhea, rheumatism, sciatica, scrofula, and skin ailments. With leaves shaped like South America's coca, these wintergreen leaves and/or fruits were used by North American Indians to keep their breath when portaging heavy loads.[45]

The active ingredient is methyl salicylate, now made synthetically. (Commercial oil of wintergreen or oil of checkerberry is obtained from distillation of the twigs of black birch.) The volatile oil contains 98 to 99% methyl salicylate. Arbutin, ericolin, gallic acid, gaultherine, gaultherilene, gaultheric acid, mucilage, tannin, wax, an ester, triacontane, and a secondary alcohol are also reported.[17] Other acids reported include *O*-pyrocatechusic-, gentisinic-, salicylic-, *p*-hydroxybenzoic-, protocatechusic-, vanillic-, syringic-, *p*-cumaric-, caffeic-, and ferulic-acids.[33]

Toxicity — Death from stomach inflammations have resulted from frequent and large doses of the oil. The highest average maximum use level is *circa* 0.04%, in candy. Not listed under FDA § 172.510, 182.10, or 182.20.[29] Wintergreen has lectinic, including mitogenic properties. Salicylism usually marked by tinnitus, nausea, and vomiting; may result from excessive dosage of salicylic acid and/or its salts.

150. *GELSEMIUM SEMPERVIRENS* (L.) Ait. (LOGANIACEAE) — Yellow Jessamine

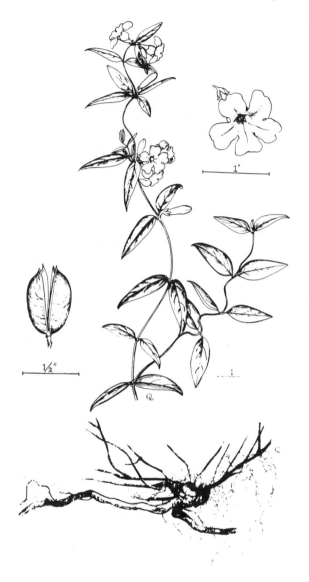

State flower of South Carolina, the yellow jessamine has a very attractive benodorous but dangerous flower. It was once exported from the U.S. by the bale as a medicine.

The plant is used in folk remedies for hard lumps, sarcomas, and wens.[4] Reported to be analgesic, anodyne, antispasmodic, CNS-depressant, diaphoretic, febrifuge, mydriatic, nervine, sedative, and tonic, yellow jessamine is a folk remedy for asthma, bilious fever, cephalalgia, chorea, convulsion, cough, croup, dysmennorrhea, epilepsy, fever, gonorrhea, hangover, headache, hyperemia, hypertension, hysteria, insomnia, malaria, migraine, neuralgia, oliguria, parturition, pertussis, pleurisy, pneumonia, poliomyelitis, rheumatism, shortbreath, spasms, stomachache, typhoid, and wens.[2,16,32] Homeopathically used for dysmennorrhea, grip, myocarditis, neuralgia, and rheumatism.[33]

Contains the indole alkaloid gelsemine ($C_{20}H_{22}N_2O_2$), gelsemicine ($C_{20}H_{26}N_2O_4$), gelsidine ($C_{19}H_{24}N_2O_3$), gelsevirine ($C_{21}H_{24-26}N_2O_3$), sempervirine ($C_{19}H_{16}N_2$). Root contains beta-methylaesculetine, $C_{10}H_3O_4$, the rhizome *n*-pentatriacontane, $C_{35}H_{72}$, a phytosterin, $C_{27}H_{46}O$, emodinmonomethylether, ipuranol ($C_{23}H_{38}O_2(OH)_2$), a fatty oil with palmitic-, stearic-, oleic-, and linoleic-acid, 0.5% essential oil, and tannin.[33]

Toxicity — ''Honey from the nectar is deadly.'' All plant parts contain toxic alkaloids, some of which depress and paralyze the motor nerve endings. Symptoms include headache, vertigo, dilated pupils, double vision, dry mouth, difficulty in swallowing and talking, muscular weakness, nausea, sweating, dyspnea, weak pulse, convulsions, oxygen deficiency, and death due to respiratory collapse.

To the Physician — Gastric lavage or emesis; control respiration and convulsions; morphine has been helpful.[56] Grieve suggests prompt evacuation of the stomach by an emetic, if the patients condition permits; and, secondly and equally important, artificial respiration, aided by the early administration, subcutaneously, of ammonia, atropine, digitalis, or strychnine.[2] The margin between therapeutic and toxic doses is narrow and extreme caution is needed with the use of the drug.[1]

151. *GENISTA TINCTORIA* L. (FABACEAE) — Dyer's Broom

According to Grieve's *Herbal*, the buds are prepared and served as seasoning.[2] Romans employed broom for dyeing, all parts, especially the flowers, yielding a good yellow dye. Seeds used as a coffee substitute. Iberians make cloth from the fibers. If cows eat the plant, it may make the milk (and butter and cheese) bitter. Still, the plant is said to enrich poor soil,[2] probably via nitrogen fixation.

Reported to be diuretic, emetic, laxative, purgative, and sudorific, dyer's broom is a folk remedy for cancer, dropsy, gout, hepatosis, nephrolithiasis, nephrosis, rheumatism, splenosis, and wens.[4,32,33]

Contains 0.3% alkaloids, anagyrine ($C_{15}H_{20}N_2O$), cytisine ($C_{11}H_{14}N_2O$), *n*-methylcytisine ($C_{12}H_{16}N_2O$), lupanine ($C_{15}H_{24}N_2O$), isosparteine ($C_{15}H_{26}N_6$), and tinctorin, luteolin, genistein, genistin, luteolin-5-glucoside 7-beta-*d*-glucopyranoside, and alpha-5,7-diglucoside.[33] Dry seeds contain *circa* 41.2% protein, 9.7% fat.[21] Contains 0.03% essential oil, sugar, wax, tannin, and mucilage.[2]

Toxicity — Some Genista alkaloids are hallucinogenic.

152. *GENTIANA LUTEA* L. (GENTIANACEAE) — Yellow Gentian

   Grown as an ornamental in flower gardens, yellow gentian is also cultivated for the drug extracted from dried rhizomes and roots. The extract is used to improve the appetite and stimulate gastric secretion, especially in Eurasia. Gentian extracts exhibit choleretic activities in animals. Used in gentian bitters, liqueurs, and "angostura bitters".[29] Gentian extract is used at levels *circa* 0.02% in nonalcoholic beverages, frozen dairy desserts, candy, baked goods, gelatins, and puddings. Gentian extracts are also used in some antismoking compounds and cosmetics.[29] One patent[165] covers an artificial sweetener containing saccharin with enough gentian extract to reduce saccharin's bitter aftertaste. Gentiopicrin, said to be lethal to mosquito larvae, has been used as an antimalarial.

   Dried rhizomes are used medicinally as bitter tonic and a stomachic. Gentian is a folk remedy for cancers, carcinomas, tumors, and indurations of the liver and spleen.[4] Gentian is used to stimulate gastric secretion, improve appetite and digestion, and alleviate debility. Reported to be anthelmintic, antiseptic, apertif, emmenagogue, febrifuge, stimulant, stomachic, and tonic, gentian shows up in folk remedies for: anorexia, blood disorders, cancer, cold, convulsions, debility, diarrhea, dogbite, dysmenorrhea, dyspepsia, fever, gastritis, gout, heartburn, inappetence, jaundice, malaria, oliguria, sideache, snakebite, splenitis, stitch, stomachache, syncope, and wounds.[2,29,32,33,44] Parvati suggests that herpes is alleviated by application of gentian violet flowers.[48]

   Dry seeds contain 31% protein, 26% fat.[21] Rhizomes contain the bitter glucoside, about 2% gentiopicrin $C_{16}H_{20}O_9$), an alkaloid, gentianine (0.6 to 0.8%), yellow xanthone pigments (gentisin, isogenitisin, and its glycoside gentioside), tannins, and sugars (including trisaccharide gentiodise), but no starch; also, gentiamarin ($C_{16}H_{22}O_{10}$ or $C_{16}H_{20}O_{10}$) and gentiin ($C_{25}H_{23}O_{14}$); pectin, gentianose, and sucrose. Roots also contain the alkaloid gentianine.[29,33] Gentianine exhibits strong antiinflammatory activity in lab animals.

Toxicity — Gentian is "unlikely to produce undesirable side effects" in normal individuals, but the drug may not be tolerated well by those with very high blood pressure or by expectant mothers. "Actually these people should be very cautious about using any medication, herbal or otherwise."[37] Approved by the FDA for food use (§ 172.510).[29]

153. *GERANIUM MACULATUM* L. (GERANIACEAE) — Cranesbill

Attractive wild flower with highly astringent roots and leaves with folk medicinal applications. One herbal catalog lists cranesbill at *circa* $20.00/kg in 1983.[71]

Reported to be astringent, diuretic, styptic, and tonic, cranesbill is a folk remedy for cancer, cheilitis, cholera, diarrhea, dysentery, enterorrhagia, gingivitis, hemorrhoids, leucorrhea, plague, renal bleeding, sores, sore throat, stomatosis, swellings, ulcers, and wounds.[2,4,32,33]

Root contains 10 to 28% tannin which hydrolyzes to gallic acid and geranium red; gum resin, starch, sugar, and calcium oxalate.[2,33]

Roots and rhizomes, containing much tannin, are astringent, antiseptic, styptic, and diuretic, and are used to treat diarrhea. In Appalachia, the tea is used for dysentery and sore throat. Folk remedy for cancers and tumors. Cherokee used the root decoction with wild grape for thrush; also, they used it for canker sores and wounds. Chippewa powdered the root to apply to sore mouths in children. Fox used the root infusion for burns, diarrhea, gingivitis, hemorrhoids, neuralgia, pyorrhea, and toothache. They poulticed bruised roots on the anus, causing protruding piles to receed. Menominee ate the root for diarrhea and stomatitis; Pillagers used it for flux and stomatitis. Following childbirth, by a month, Nevada Indians filled a trench with warm ashes, in which the mother lay, relaxing, and drinking wild geranium tea to keep her safe from pregnancy until the baby's first birthday.[43] "Drunk as a tea, the 'cranesbill' was probably the most widely used medium for birth control, the user being said to be safe from pregnancy for a year."[166]

Toxicity — The high tannin content would dictate against chronic use.

154. *GLECHOMA HEDERACEA* L. (LAMIACEAE) — Alehoof, Ground Ivy

Although a serious weed at Herbal Vineyard, choking out more valuable mint species, alehoof is sometimes used for herbal teas. It was used by Saxons to clarify their beer before hops had been "invented". The refrigerant gill tea is made from 60 g ground ivy infused in a liter of boiling water, and sweetened with honey or licorice. Sometimes used in making herbal wreaths, remaining green long into the winter.

Plant or its juice or leaves used for various types of cancer[4] including, also, corns and epithelial cancer. Combined with yarrow or camomile, it is poulticed onto tumors. Alehoof juice, honey, and calendula, boiled together, are said to clean fistulas, ulcers, and control the spreading or eating away of cancers and ulcers.[30] Once famous for purifying the blood, gill tea is still considered stimulant and tonic, useful in digestive, kidney, and pulmonary ailments, as well as stones; said to be excellent for asthma, arthritis, backache, bruises, cold, consumption, contusions, corns, diabetes, epithelioma, fever, headache, inflammation, marasmus, nephritis, osteosis, pneumonia, rheumatism, rickets, scurvy, stones, swellings, trauma, and urogenital ailments.[32] Alexiteric, alterative, anodyne, antiscorbutic, astringent, cardiotonic, diaphoretic, diuretic, febrifuge, pectoral, refrigerant, sudorific, and tonic. Used as a gargle for sore throat. Decoction used for earache and toothache in China.[16] Juice or dried leaves snuffed for headache, where all other remedies failed.[2] Expressed juice used for black eyes and other bruises. American house painters once used an infusion of the herb as preventitive and cure for "lead colic". Others believe it prevents premature aging.[30] The infusion is also used for sore and/or weak eyes, both internally and as a collyrium. Drunk with wine, it was an old and probably inefficacious remedy for gout, sciatica, and rheumatism.[2] Said to be used in China to dissolve bone tissue.[9] Used homeopathically for piles.[33]

Contains 0.03 to 0.06% essential oil with (−)-pinocamphone, (−)-menthone, (−)-pulegone, alpha pinene, beta pinene, limonene, *p*-cymene, isomenthone, isopinocamphone, linalool, menthol, alpha-terpineol, ursolic acid, beta-sitosterol, palmitic-acid, succinic acid, choline, proline, valine, tyrosine, asparagic- and glutamic-acid, marrubiin, glechomin, gum, saponin, rutin, tannin, wax.[27,33]

Toxicity — Has been considered poisonous to horses,[33] at least in Europe,[14] causing death in 5 to 8 days.

155. *GLEDITSIA TRIACANTHOS* L. (CAESALPINIACEAE) — Honey Locust

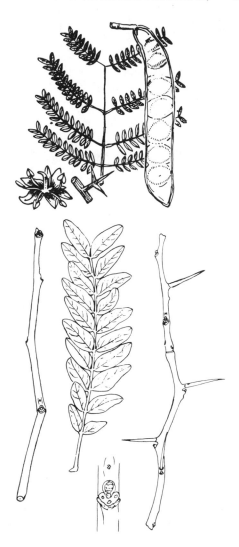

Widely introduced as a fast-growing tree for fuel, fodder, fence posts, ornament, shade, and soil reclamation. The wood is said to be coarse-grained, durable, hard, and resistant to soil decay. Hence, it is used for fence posts, railroad ties, and hubs for wheels. South Africans sometimes plant orchards of the tree for fodder. The gum from the seeds has been suggested as an emulsifying substitute for acacia and tragacanth. Flowers very attractive to bees. The pulp has always attracted the sweet tooth of animal and man alike, when better sweets were not available. A potable or energy alcohol can be made by fermenting the pulp. Seeds have been roasted and used as a coffee substitute.[63]

Sokoloff et al.[167] note that recently, Soviet investigators have been studying the biological factors present in the fruit and leaves of *G. triacanthos*. The alcoholic extract of the fruits of the Kirgis honey locust, after elimination of tannin, considerably retard the growth, up to 63%, of Ehrlich mouse carcinoma. However, the cytotoxicity of the extract was quite high and the animals, besides losing weight, showed dystrophic changes in their liver and spleen. The alcoholic extract of the fruit exerted moderate oncostatic activity against sarcoma 180 and Ehrlich carcinoma at the total dose 350 mg/kg per body weight per mouse. Weight loss was considerable. Epicatechol-3-D-glucoside dihydride, isolated from the flowers, exhibited no oncostatic or cytotoxic activity. The pigment tentatively identified as dihydroxy-4-methoxyisoflavone, isolated from the fruit, exerted considerable oncostatic activity (and cytotoxicity). Triacanthine from the leaves was highly toxic (LD_{50} *circa* 35 mg/kg) and of questionable oncostatic activity.[167] Fruit pulp is used for catarrh of the lung.[33] Powdered seed used as a snuff for head cold. Some people, probably having seen the erroneous report of cocaine in the leaves, state that "ingestion of a suitable preparation of the leaf increases the capacity for muscular work and delays the onset of fatigue." Reported to be anodyne, mydriatic, narcotic, and experimentally oxytocic,[32] honey locust pods are a folk remedy for dyspepsia and measles among the Cherokee. The bark tea is used for whooping cough. Delaware Indians used the bark for blood disorders and coughs, the Fox for colds, fevers, measles, and smallpox. Chinese probed tumors and abscesses with the thorns of *G. sinensis*, considering them counterirritant.[63]

Per 100 g, the fruit is reported to contain on a zero moisture basis, 23.1 g protein, 4.6 g fat, 66.9 g total carbohydrate, 12.7 fiber, 5.4 g ash. The seed is said to contain 10.6 g protein, 0.8 g fat, 84.7 g total carbohydrate, 21.1 g fiber, 3.9 g ash, 280 mg Ca, and 320 mg P.[21] Fodder yields 0.81% of a mixture of two pigments, one named acrammerin $C_{16}H_{12}O_8$ and another olmelin $C_{16}H_{12}O_5$, a flavonoid glucoside. Pod contains some tannin but no alkaloid and gives negative hemolysis tests. Bark contains a trace of alkaloid and the flower spikes 0.2%. The fruit is reported to yield 3.9 to 4.44% of glucose. The gum of the fruit and seed compare. The gum exuding from wounds contains calcium and strontium salts of D-glucuronic acid bound with anhydro-D-galactose and anhydro-L-arabinose.[33] Seeds contain mannogalactan and enzymes, 3% of fat, 21% of albumin, possibly a glycyrrhizin-like substance but no alkaloid, 1.9% of 1-epicatechol 3-D-glucoside dihydrate. Albumen of the seed on hydrolysis yields 94.2% reducing sugars which contain 70% galactose and 23% mannose. Light petroleum shows an oil, and phytosterol $C_{30}H_{50}O$ to 0.5 H_{20}. The oil contains dihydroxystearic acid, sativic acid, and an isomeric tetrahydroxystearic acid. The seed coat contains acetic acid, glucose, phlobaphen, and tannin, as well as polyphenols. Pericarp may contain alkaloids 0.02 to 0.05%; glucosides and anthraglucosides 2.5%; saponins and tannins 3.11%; proteins and other substances. The seed germ yields 7% of oil containing 0.056% tocopherols, while the hull-free cotyledon yields 4.9% oil containing 0.04% tocopherols. The leaf apparently contains two active principles, hypoxysin with oxytocic properties, and a neutral principle of gum-like consistence and having a direct depressant action on the blood vessel muscle resulting in vasodilation. The leaf contain much tannin.

Toxicity — Air-dried leaf yields 0.5% of an alkaloid triacanthine $C_8H_{10}N_4$ which in intravenous doses of 0.1 mg/kg depresses the action of the cat heart, the intensity apparently depending on the intensity of effect on the vasomotor center. Heart-wood contains 4 to 4.8% tannin and, also, fustin ($C_{15}H_{12}O_6$) and fisetin ($C_{15}H_{10}O_6$).[3,33] The alkaloid gleditschine is said to produce stupor and loss of reflex activity in a frog. Stenocarpine has been used for local anesthesia. "It also contains cocaine."[2] To the best of my knowledge it does not contain cocaine! *N*-methyl-beta-phenylethylamine and tyramine are, however, reported.[33,63]

156. *GLORIOSA SUPERBA* L. (LILIACEAE) — Glory Lily

Ornamental lily, being considered as an alternative source of colchicine. Indians and Sri Lankans use the flowers in religious ceremonies. Javanese poison dogs by mixing the grated fruits with their food. Juice of ground leaves is used to destroy head lice.[26]

Reported to be abortifacient, alexiteric, alterative, anodyne, bactericide, cholagogue, laxative, mitogenic, pediculicidal, poison, stomachic, tonic, and vermifuge, the glory lily is a folk remedy for bruises, cancer, colic, erysipelas, gonorrhea, labor, leprosy, malaria, neuralgia, parturition, piles, scabies, skin ailments, snakebite, sore, splenitis, syphilis, tumors, ulcers.[3,32] Indians and Sri Lankans use the root for abortion or to expedite parturition. South Africans use juice from the root stock mixed with leaf juice from Anamirta for killing guinea worms. They use the root itself for ascites and parasites. Iranians use the rootstock for epistaxis, impotence, and nocturnal emissions, Indians for bruises, colic, gonorrhea, leprosy, malaria, piles, snakebites, and sprains. Indians also use the root for abortion, homicide, and suicide.[3]

Leaves contain 3-desmethylcolchicine $C_{21}H_{23}NO_6$, colchicine $C_{22}H_{25}NO_6$, and chelidonic acid. Flowers contain beta-lumicolchicine, $C_{22}H_{25}NO_6$, *N*-formyldesacetyl-colchicine, $C_{21}H_{23}NO_6$, and 3-desmethylcolchicine. Tubers contain 0.3% colchicine, *N*-formaldesacetylcolchicine, 2-desmethylcolchicine, beta-lumicolchicine, gloriosine $C_{33}H_{38}N_2O_9$ or $C_{15}H_{17}NO_4$, gloriosol, $C_{33}H_{56}O_6$ which may represent a mixture of sitosterol and stigmasterol, traces of choline, benzoic acid, salicylic acid, 6-methyloxysalicylic acid, fatty acids, glucose, starch, γ-resorcylic acid, monomethylether, etc.[33] In *Cultivation and Utilization of Medicinal Plants*,[59] Gupta describes a new aqueous extracting method to obtain colchicine, which has been advocated in arthritis, Bright's disease, cholera, colic, skin ailments, and typhus. Compared to the older alcoholic extraction process, the aqueous method has the following advantages: (1) cheap solvent, (2) smaller quantities of extract are obtained but five times as much colchicine, (3) isolation of colchicine is easy.[59]

Toxicity — Root stock, leaf, and stem are all poisonous, with human fatalities reported. Human symptoms include tingling, then numbness of the lips, burning epigastric pain, skin numbness, intense nausea followed by nausea and bloody diarrhea, giddiness, loss of power in the limbs, heaviness of the eyelids, photophobia, respiratory problems, a feeble quick pulse, convulsions, and loss of consciousness. The temperature drops, tetanic tremblings and convulsions, may precede death in 20 to 40 hr. Death may not occur for 1 to 3 days after first appearance of symptoms.[3]

To the physician — Hardin and Arena suggest gastric lavage or emesis, shock therapy, and symptomatic and supportive treatment.[34]

157. *GLYCYRRHIZA GLABRA* L. (FABACEAE) — Common Licorice, Licorice Root, Spanish Licorice Root

Licorice, grown primarily for its dried rhizome and roots, is a condiment and used to flavor candies and tobaccos. Roots contain glycyrrhizin, 50 times sweeter than sugar. Glycyrrhizin potentiates the flavor of cocoa, replacing 25% cocoa in manufactured products.[17] In India, it is chewed with betel. Spent licorice serves in fire-extinguishing agents, to insulate fiberboards, and as a compost for growing mushrooms; also, in feed for cattle, horses, and chickens. Most licorice production is used by the tobacco industry and for the preparation of licorice paste, licorice extract, powdered root, and mafeo syrup.[40] Most "licorice" candy contains little or no licorice, but is flavored with anise oil.[37] According to Rose,[47] quantities of licorice were stored in King Tut's tomb. Singers chew the root to strengthen the throat.

It is considered demulcent, expectorant, estrogenic, and laxative. Used in treating Addison's disease, as a tonic and blood purifier, for internal inflammations, topical dressings, thirst, colds, and sore throats. Also, said to be used for alexeritic, alterative, appendicitis, asthma, bladder ailments, bronchitis, cough, deodorant (lf), depurative, diabetes, diuretic, dropsy, emollient, eyes, febrifuge, pectoral, refrigerant, scalds (lf), stomach ulcer, sudorific, and tuberculosis. With a long history of use for indigestion and inflamed stomach, licorice provides two derivatives which reduce or cure ulcers.[40]

Glycyrrhiza contains saponin and tannic acids; the seeds contain typsin inhibitors and chymotrypsin inhibitors. *G. glabra* contains 2-beta-glucuronosyl glucuronic acid, glycyrrhizin, and isoliquiritigenin-4-glucoside. One analysis showed 20% moisture, 12 to 16% glycyrrhizin, 8% reducing sugars, 8% nonreducing sugars, 30% starch and gums, 5% ash, and 12 to 17% undetermined. Rose[47] reports that licorice contains estrogenic materials. Glycyrrhizine is used in Egypt as a cortisone substitute, without the withdrawal syndrome of cortisone. Dry seeds contain 21 to 30% protein, 1 to 11% fat.[21] Excessive licorice ingestion can lead to cardiac dysfunction and severe hypertension.[11] According to Mitscher et al.,[168] licorice also contains glabridin, glabrene, glabrol (potent against *Staphylococcus aureus, Mycobacterium smegmatis*), formononetin, phaseollinisoflavone (phytoalexin), salicylic acid, *o*-acetyl salicylic acid (0.15%; perhaps enough to be pharmaceutical), and hispaglabridin (rather potent in vitro against *Staphylococcos* and *Mycobacterium*).

Toxicity — While licorice extract may alleviate peptic ulcers, swelling of the face and limbs may be a side effect. Tyler relates several tales.[37] One man who ate two or three 36-g licorice candy bars for 6 or 7 years became so weak, he couldn't get out of bed. Another man who consumed 700 g over 9 days was hospitalized for 4 days. Without exonerating some of the toxins found in pure tobacco, Tyler also noted the case of the man who chewed 24 to 36 oz of tobacco a day (containing *circa* 8% licorice paste equivalent to *circa* 1 g glycyrrhizin well within the toxic range). He, too, was unable to sit up.

158. *GOSSYPIUM BARBADENSE* L. (MALVACEAE) — Sea Island Cotton

Sea Island cotton is primarily cultivated for its vegetable seed fiber, a raw material for fine textile products. This cotton embraces a large group of cotton fibers which produce the finest quality lint known: creamy white in color, silky, and lustrous, with a staple length of 3 cm or more, stronger and firmer than any other cotton. For these reasons, it is used for the finest textiles and yarns, of value in the lace industry. It is also useful in the pneumatic tire industry. Also, cultivated for the seed itself, which provides a semidrying, edible oil making up 50 to 55% of the concentrate for livestock. Fuzz which is not removed in ginning becomes linters for the chemical industry and miscellaneous textile uses. Hulls provide roughage for livestock and as bedding, fertilizer, and fuel. Cottonseed oil, a satisfactory substitute for sesame oil, is one commercial source of vitamin E.[61]

The leaf and the seed, used as a fumigant, are said to be folk remedies for lymphatic tumors. Gossypol has shown antitumor activity in the new Lewis lung carcinoma (in mice), Walker's carcinosarcoma 256 (in rats), and lymphocytic leukemia P380 (in mice) tumor systems.[10] Hot leaf infusion recommended for uterine colic. Seeds in the form of an emulsion are given in dysentery and are claimed to be pectoral. Oil from seeds used to clear spots and freckles from skin. Recent news from China suggests that they are using gossypol as a male contraceptive. Seed decoction used for ague and fever. Roots of cotton are considered abortifacient, emmenagogue, lactagogue, oxytocic, and parturient. Cotton flowers are recommended for hypochondriasis.[27] Venezuelans use the floral or foliar infusion for bronchial inflammations.[42] Depending on the strength, Mexicans take root decoction as abortifacient, emmenagogue, and parturient. Cubans use the seeds for bronchitis and fever. Darien blacks use the leaves for rheumatism. Surinamese take the leaf decoction for hypertension. In Curacao, the leaf decoction is taken for colds, diarrhea, dysentery, strangury, and stomachache.[42] Homeopaths recommend the root bark tincture for abortion, amenorrhea, dysmenorrhea, labial abscess, ovarian pain, nausea of pregnancy, sterility, tumors, and uterine ailments.[30]

Commercial cottonseed contains approximately 92% dry matter, 16 to 20% protein, 18 to 24% oil, 30% carbohydrates, and 22% crude fiber. After ginning, cottonseed includes unginned lint, fuzz, 16% crude oil, 45.5% cake or meal, 25.5% hulls, and 8% linters. The roots contain chromogene, olein, resin, and tannin. Gossypol is concentrated in the glands. Salicylic acid, 2,3-dihydroxybenzoic acid, betaine, bisabolol, curcurmene, triacontane, oleic acid, palmitic acid, cerylalcohol, and acetovanillin are reported from cottonroots; quercetin-3-sophoroside, quercimeritrin, isoquercitrin, rutin, nicotiflorin, trifolin, gossypitrin, hirsutrin, quercetin-3'-glucoside, and chrysanthemine are reported from *G. barbadense*.[33]

Toxicity — Cotton seed and roots may be ecbolic and poisonous. Intraperitoneally in rats the LD_{50} of gossypol is 10 to 20 mg/kg. Post-mortem analyses in acute poisonings include hemorrhagia, serum, oily droplets and plaques in the peritoneum, hemorrhagic condition of all internal viscera, congestion of the lungs, and cardiac dilation.

159. *GOSSYPIUM HIRSUTUM* L. (MALVACEAE) — American Upland Cotton

Cultivated primarily for its vegetable seed fiber, this species is considered the most important cotton-yielding plant, providing the bulk of commercial cottons. Seeds yield a semidrying and edible oil, used in shortening, margarine, salad and cooking oils, and for protective coverings. Residue, cottonseed cake, or meal is important protein concentrate for livestock. Low-grade residue serves as manure, bedding, and fuel. "Fuss", which is not removed in ginning, become "linters" in felts, upholstery, mattresses, twine, wicks, carpets, surgical cottons, and in chemical industries such as rayons, film, shatterproof glass, plastics, sausage skins, lacquers, and cellulose explosives.[61]

Cottonseed and roots have been used in nasal polyps, uterine fibroids, and other types of cancer.[4] Gossypol has shown anticancer activity in the new LL, WA, and PS-150 tumor systems. Mucilaginous tea of fresh or roasted seeds used for bronchitis, diarrhea, dysentery, and hemorrhage. Flowers diuretic and emollient, used for hypochondriasis. Leaves steeped in vinegar applied to the forehead for headache. Often used by early American slaves for abortion, apparently with no serious side effects. About 100 g root was boiled in about a liter of water until reduced by 1/2. Fifty grams of the resultant witches' brew was then drunk about every half hour. Root decoction used for asthma, diarrhea, and dysentery. Root bark, devoid of tannin, astringent, antihemorrhoidal; used as an emmenagogue, hemostat, lactagogue, oxytocic, parturient, and vasoconstrictor. Gossypol is being used in China as a male contraceptive.

Per 100 g, the seed contains 398 calories, 8.1% water, 32.9% protein, 16.1% fat, 36.7% total carbohydrate, 4.8% fiber, 6.2% ash, 149 mg Ca, 1022 mg P, 10.2 mg Fe, 0 μg β-carotene, 0.14 mg thiamine, 0.03 mg riboflavin, 3.6 mg niacin.[21] Root bark contains *circa* 3% of a reddish acidic resin, a volatile oil, a phenolic acid (probably 2,3-dihydrobenzoic acid; salicylic acid, a colorless phenol, betaine, a fatty alcohol, a phytosterol [$C_{27}H_{46}O$], a hydrocarbon [probably triacontane], ceryl alcohol, and oleic and palmitic acids).[3] *Hager's Handbook* also lists isoquercitrin, quercimeritrin, quercetin-3'-glucoside, hirsutin, isoastragalin, palmitic acid, oleic acid, linoleic acid, alpha-pinene, beta-caryophyllene, bisabolol, caryophyllenepoxide, bisabolenoxide, abscissin II, serotonin, chrysanthemine, gossypicyanin, and histamine.[33]

Toxicity — Gossypol, the toxic dihydroxyphenol, occurring in seeds and the glands of seedlings, must be removed before cottonseed can be used for feed. Hogs have died from eating raw seed.[46]

160. *GRINDELIA SQUARROSA* (Pursh) Dunal (ASTERACEAE) — Rosinweed, Gumweed

Despite moderate abundance in some regions, rosinweeds and gumweeds possess little forage value and are relatively unpalatable to all classes of livestock, even goats. Local abundance is often an indication of depletion due to severe and prolonged overgrazing. Some species are cultivated as ornamental.[155] The curlytop gumweed, *G. squarrosa*, is an Amerindian medicinal species.

Seven *Grindelia* spp. are used in folk remedies for cancers of the spleen and stomach.[4,33] Reported to be antidotal, antitussive, expectorant, sedative, spasmolytic, stimulant, rosinweed is a folk remedy for asthma, bronchitis, burns, cancer, colds, fever, gonorrhea, hepatosis, nephrosis, pertussis, pneumonia, rash, renosis, rheumatism, smallpox, sores, stomachache, syphilis, and tuberculosis.[32,33] Used homeopathically for asthma, bronchitis, emphysema, and splenomegaly.[33] Indians used the resinous buds to treat asthma and bronchitis.

Contains about 20% resin with grindelic-, oxygrindelic-, 6-oxygrindelic-, and 7-alpha-8 alpha oxodihydrogrindelic acids. In the resin-free portion, matricarianol ($C_{10}H_{10}O$) and matricarianol acetate, with *circa* 0.3% essential oil (largely borneol) tannin, saponins, the alkaloid grindeline; *p*-oxybenzoic acid in the leaves.[33] Dry seeds contain 13.8% protein, 19.6% fat, and 3.4% ash.[21]

161. *HAMAMELIS VIRGINIANA* L. (HAMAMELIDACEAE) — Witch Hazel

According to Grieve, the black nuts contain a white seed which is oily and edible.[2] Used in toilet water, face and body lotions, e.g., aftershave lotions and mouth washes.[45] Avery[169] suggests that witch hazel is our most widely known botanical (I would have said ginseng): "over one million gallons of witch hazel are sold each year."[169]

Reported to be alterative, anodyne, antiseptic, astringent, hemostat, refrigerant, sedative, tonic, witch hazel is a folk remedy for backache, bruises, burns, cancer, diarrhea, dysentery, dysmenorrhea, gleet, hemoptysis, hemorrhoids, inflammation, lameness, menorrhagia, metrorrhagia, myalgia, ophthalmia, phlebitis, phthisis, tumors, ulcers, varicose veins, and wounds.[4,32,33] Amerindians poulticed the plant onto painful swellings and tumors. In the treatment of varicose veins, witch hazel should be applied on a lint bandage, which must be kept constantly moist . . . "a pad of Witch Hazel applied to a burst varicose vein will stop the bleeding and often save life by its instant application." An enema of witch hazel tea "is excellent for inwardly bleeding piles, the relief being marvelous and the cure speedy."[2] The twigs, leaves, and bark are used to prepare witch hazel extract, used to treat bruises and sprains. Fresh leaves are very astringent. Cherokee used the tea for colds, fevers, periodic pain, sore throat, and tuberculosis, and to wash sores and wounds. They bruised the leaves and rubbed them on scratches. Amerindians used the bark for painful tumors.[43] Menominee used the twigs for bad back, the seeds in sacred ceremonies. Mohegan used the twigs and leaves for bruises, cuts, and insect bites. Osage used in sores, tumors, and ulcers; Potawatomi used in steam bath for rheumatism and sore muscles; Seneca used it as a sedative.[43] "The many varied uses of a watery infusion of Witch-hazel bark were fully known to the aborigines, whose knowledge of our medicinal flora has been strangely correct as since proven. Its use in hemorrhages, congestions, inflammations, and hemorrhoids is now generally known . . . "[19]

Leaves contain *circa* 8% hamamelitannin $C_{20}H_{20}O_{14}$, which yields 68% gallic acid, also free gallic acid and hamamelose $C_6H_{12}O_6$, 0.2% choline, saponin; quinic acid, *circa* 7% of the resinoids hamamelin and hamamelidin. There is an essential oil with *circa* 40% alcohol, *circa* 15% esters, *circa* 25% carbonyl compounds, myricetin, quercetin, kaempferol, leucodelphinidin, leucocyanidin, myricetin-3-glucoside, isoquercitrin, astragalin, myricetrin, quercitrin, afzelin, spiraeoside, *n*-hexen-2-al-(1), acetaldehyde, alpha-ionone, beta-ionone, safrole, 6-methylheptadien-3-5-on-(2), and sesquiterpene. Bark contains 1 to 3% hamemelitannin, ellagic tannin, catechin-3-gallate, hamamelose, gallic acid, phlobaphene, 0.6% fatty oil, wax, saponin, 0.5% essential oil with sesquiterpene, and compounds like eugenol and safrole and *circa* 16% hamamelin.[33]

Toxicity — Contains the carcinogen safrole (though in much smaller quantities than sassafras), eugenol, and acetaldehyde.

162. *HARPAGOPHYTUM PROCUMBENS* DC (PEDALIACEAE) — Devil's Claw, Grapple Plant

Sold as an herbal tea ingredient in Europe and, with FDA misgivings, in the U.S. The herbage "is not poisonous and the young shoot and leaf may be browsed in time of necessity."[3] German clinical studies suggest that the plant does have antiinflammatory properties comparable to antiarthritic phenylbutazine.[37] Reduction of hypercholesteremia, high uric acid levels, and pain were also reported.[37]

The tubers are used in an ointment for cancerous growths in southern and eastern Africa.[4] One South African woman with skin cancer is the subject of anecdotes detailing skin cancer cures from devil's claw. An ointment made from the tuber is applied to the abdomen of pregnant women who expect a difficult delivery. Dried tubers (*circa* 250 mg) given twice daily to relieve pain in pregnant women, and in smaller doses, post partem. Prized among Africa's Europeans as a remedy, especially for indigestion. The tuber, taken orally, is purgative, but an infusion is taken for relief of fevers of all sorts (including malaria) and for blood diseases. South Africa farmers dig up the root, cut it into rings, dry and pulverize it, and apply the powder to open wounds on themselves or their livestock. The purgative plant is used for boils, fevers, sores, and ulcers.[8] In Europe, the tea is suggested for metabolic diseases, allergies, arteriosclerosis, arthritis, cholecystosis, diabetes, dysmenorrhea, gastroenteritis, heartburn, hepatitis, lumbago, neuralgia, renitis, cystitis, rheumatism, and senescence.[37]

Dried roots contain the bitter harpagoside ($C_{24}H_{30}O_{11}$), the sweet harpagide ($C_{15}H_{24}O_{10}$), and procumbide ($C_{15}H_{24}O_{11}$), 0.3% raffinose, 0.6 to 0.8% stachyose, saccharose, glucose, and fructose. The plant extract and harpagide show positive effects on arthritic rats; harpagoside and harpagide are antiinflammatory, harpagoside exhibits analgesic activity.[33]

Toxicity — The tough stem may form masses in horses' gut, causing gastrointestinal obstruction. Fruits with their hooks are even more troublesome.

163. *HEDEOMA PULEGIOIDES* (L.) Pers. (LAMIACEAE) — American Pennyroyal

American pennyroyal is a gently stimulant aromatic, its oil being used for scenting soap, perfumes, and liqueurs, and as one of many sources of synthetic menthol. Used to repel chiggers and alleviate the itching of their bites. Also, said to be insecticidal.[33]

American Indians believed that pennyroyal would cure pain in any limb to which it was applied. Onondaga Indians use pennyroyal tea for headache, a practice still recommended by Tierra.[28] Dried leaves used as a decoction for flatulence, and as a diaphoretic in catarrh, colds, and rheumatism.[17] Also, applied externally for rheumatism.[30] Catawba Indians boiled the herb as a cold remedy.[11] Mixed with linseed oil, pennyroyal oil makes a good unguent for burns.[30] Tea also used for bowel disorders, colic, gastritis,[8] itching eyes, and pneumonia.[18] Used to reduce cramps and their accompanying pain.[43] Regarded as abortifacient, antitussive, antispasmodic, carminative, corrective, diaphoretic, diuretic, emmenagogue, nervine, and stimulant.[27,28,30,32] Homeopaths prescribe for amenorrhea, dysmenorrhea, and leucorrhea.[30]

Contains up to 2% essential oil, the primary ingredient of which is pulegone (85 to 92%), with pinene, limonene, dipentene, 1-methyl-3-cyclohexanone, menthone, isomenthone, diosmin, and formic-, acetic-, butyric-, isoheptylic-, octoic-, decylic-, and salicylic-acids.[7,33]

Toxicity — Pennyroyal oil, American or European, can cause dermatitis and, in large doses, abortion and death (see *Mentha pulegium*). Though used for regulating the menstrual flow and relieving cramps, pennyroyal should not be used by women with a tendency toward excessive menstruation.[28]

164. *HEDERA HELIX* L. (ARALIACEAE) — Ivy

Once upon a time the leaves of this ornamental evergreen vine formed the poet's crown, as well as the wreath of Bacchus, to whom the plant was dedicated. Ivy was bound around the brow to prevent intoxication. A garland of ivy hung outside ale houses indicated that wine was sold inside. Greek priests presented an ivy wreath to newly married persons to symbolize fidelity. Sheep and deer are said to eat the leaves in winter, though cows often will not.[38] Containing *circa* 10% saponin, the leaves have been used for washing wool. Leaves boiled with soda are used to wash clothes. Young twigs yield a yellow and brown dye. Hardwood can be used in engraving as a boxwood substitute. Recently, an ivy extract has found its way into French massage creams and soaps.[38]

Regarded as antiseptic, astringent, cathartic, contraceptive, diaphoretic, emetic, emmenagogue, laxative, pediculicide, purgative, stimulant, sudorific, vasoconstrictor, vasodilator, and vermifuge, ivy is used for rheumatism, sclerosis, scrofula, toothache, etc. The leaf is used for cacoethes, calluses, cancer, cancromas, chironies, corns, warts, and wens; the juice for cancer or polyps of the nose. South African whites apply the vinegar-steeped leaves to cancerous growths and corns.[3] Chinese use the leafy shoot decoction for cough and headache.[16] The plant is also used for various indurations and cancers (lymph, mammaries, uterus).[4] Ivy leaves were once bruised, gently boiled in wine, and drunk to alleviate intoxication by wine. Flowers, decocted in wine, were used for dysentery. Plant said to have been used as an emetic and narcotic in at least three continents. Yellow berries used for jaundice and hemoptysis. Infusion of the fruits is used for rheumatism. In the Mediterranean, ingestion of 1 g powdered fruit is said to result in sterility.[3] Still, the resin has been believed to be aphrodisiac. Tender ivy twigs, boiled in butter, were used to treat sunburn. The resin from old stems is placed on toothache and is believed to be aphrodisiac, emmenagogue, and stimulant. Leaf has been applied to destroy vermin, e.g., head lice. Slight antimalarial activity is reported.[38]

Leaves contain the saponin alpha-hederin $C_{41}H_{66}O_{12}$ and beta-hederin, the bisdesmoside hederacoside B and C, germacrene B, beta-elemene, and elixen. Twigs contain rutin, isoquercitrin, kaempferol-3-rhamnoglucoside, chlorogenic- and isochlorogenic-acids, derivatives of ferulic- and *p*-coumaric acids, and scopoline. Wood and leaves contain hederacoside A ($C_{41}H_{66}O_{13}$) which hydrolyzes to hederagenin, arabinose, and glucose. Leaves and leafstalks contain vitamin E and provitamin A. The emetic alkaloid emetine has been reported. Stalks and roots contain falcarinone $C_{17}H_{22}O$.[33] The pericarp of immature fruits is rich in lecithin (1.7%). The sapo-glycoside hederin isolated from the leaf is said to be intensely hemolytic. Alcoholic extract of the fruit lowers the blood pressure. Dry seeds contain 16.2% protein and 35.1% fat.[21] The seeds contain a semidrying oil with 5.1% palmitic, 62% petroselenic, 20% oleic, and 13% linoleic acids. Perry adds hedera-tannic acid.[16]

Toxicity — An extract of the leaves, used as a corn cure, caused dermatitis, which recurred when the patient touched the leaves. A scarlatiniform eruption followed ingestion of the leaves. Handling may cause dermatitis, even blistering and inflammation.[3] Berry said to cause poisoning in children, leaves in cattle.[3] Large quantities taken internally may cause diarrhea, excitement, nervousness, labored respiration, convulsions, coma, and possibly death.[56]

To the physician — Hardin and Arena suggest gastric lavage or emesis, symptomatic and supportive treatment, 2 to 10 cc paraldehyde I.M., oxygen, and artificial respiration as necessary.[34]

165. *HEIMIA SALICIFOLIA* (HBK) Link and Otto (LYTHRACEAE) — Sinicuichi

Mexican highlanders use the herb to take a "trip to the past" macerating the leaves in water, squeezing out the juices, and letting the concoction to ferment in the sun. Partakers then go on the "yellow trip" in which yellow suffuses the vision. The initial giddy state gives way to one with darkened surroundings and a state of deafness. Pleasant drowsiness results in almost complete withdrawal from reality without hangover or offensive side effects. "Sinicuichi is to Mexico what Rauvolfia ('sarpaganda') has been to the people of India, only more profoundly so."[54]

Reported to be antidotal, astringent, depurative, diaphoretic, diuretic, emetic, hallucinogenic, hemostat, intoxicant, laxative, psychotomimetic, purgative, sudorific, tonic, and vulnerary, sinicuichi is a folk remedy for bronchitis, dysentery, dyspepsia, fever, poisoning, sores, and syphilis.[32,33]

Dry seeds contain 21.6% protein, 28.7% fat.[21] Leaves contain the potent tranquilizer sinicuichine which relaxes muscles, relieves experimental anxiety, and stabilizes blood pressure.[54] Several alkaloids are reported: abresoline, anelisine, cryogenine, dehydrodecadine, heimine, lyofoline, lythridine, lythrine, nesodine, sinine, and vertine. Cryogenine is anticholinergic, antispasmodic, and vasodilatory, tranquilizing and relaxing skeletal muscle.[33,58]

Toxicity — Classed as a narcotic hallucinogen; long-term use may impair the memory. *Hager's Handbook* lists it as a "Giftdroge".[33]

166. *HELIOTROPIUM EUROPAEUM* L. (BORAGINACEAE) — Heliotrope, Turnsole

Bedding annuals forming dense tufts with white to pale blue flowers. Mere inhalation of the herb is said to soothe the nerves.[47]

Leaves and juice used for cancer, polyps, ulcers, warts,[4] the plant is even called "herbe du Cancer" in Europe. Regarded as cholagogue, emmenagogue, febrifuge, and poison.[32] Seeds used against fever. Herb used for gravel, snakebite, and scorpion stings.[33]

Another species (*H. indicum*) is the source of the alkaloid indicine-*N*-oxide which shows activity in the Bl, LE, PS, and WA tumor systems.[10] Contains cynoglossin and an essential oil. Dry seeds contain 20.9% protein, 23.8% fat.[21]

Toxicity — Classed as unsafe poisonous herb, containing alkaloids which produce liver damage.[62] The pyrrolizidine alkaloid affects livestock through liver toxicity and atrophy, often resulting in death. *Hager's Handbook* lists echinatine, europine, heleurine, heliosupine, heliotridine, heliotrine, lasiocarpine, retronecine, supinidine, supinine, under the discussion of *H. europaeum*.[33]

167. *HELLEBORUS NIGER* L. (RANUNCULACEAE) — Christmas Rose, Black Hellebore

Ornamental, the pure white flowers sometimes blooming at Christmas. Once supposed to protect cattle from evil spirits, powdered hellebore was also supposed to have been scattered to the breeze by sorcerers to make themselves invisible.

Root, rootbark, or seed used for cancer or indurations, especially of the spleen, tumors, ulcers, and warts.[4] Boiled with red wine and mixed with honey, applied to cancers of fleshy parts. Said to have been used as a purgative in mania (1400 B.C.). Dried powdered rhizome causes violent sneezing. Said to be narcotic. Used as a drastic hydragogue cathartic, cardiotonic, diuretic, emmenagogue, mydriatic, narcotic, nervine, poison, rodenticide, and vermifuge.[32] In homeopathy, it is suggested for meningitis, encephalitis, nephritis, epilepsy, hydrocephaly, psychosis, melancholy, collapse, cardiac insufficiency, dementia praecox.[2,33]

Contains two poisonous glucosides, helleborin and helleborein.[2] Resin, fat, saponin, and starch are also reported. *Hager's Handbook*[33] lists hellebrin (hellebrigenin-glucorhamnoside) from the herb, kaempferol-3,7-diglucoside, kaempferol-3-sophoroside, and kaempferol-3-sambubioside-7-glucoside from the flowers. Dry seeds contain 21.2% protein, 35.2% fat.

Toxicity — Internally violently narcotic; applied locally, the fresh root is ''violently irritant''.[2] According to Steinmetz,[27] the herb is an acro-narcotic poison in large doses. There are many cases of human poisoning from confusion with other herbs, through consuming the seeds or through medicinal overdoses. Symptoms include dry or scratchy throat and mouth, salivation, nausea, stomachache, vomiting, colic, diarrhea, irregular slow pulse, weak heartbeak, dyspnea, vertigo, ringing ears, mydriasis, disturbed vision, excitement in fatal doses, coronary arrest, and collapse.

To the physician — *Hager's Handbook* suggests emesis or gastric lavage, warm towels, spasmolytics for the colic, otherwise as with *Digitalis purpurea*.[33]

168. *HIBISCUS SABDARIFFA* L. (MALVACEAE) — Roselle

Source of a red beverage known as jamaica in Mexico (said to contain citric acid and salts, serving as a diuretic). Calyx called ''karkade'' in Switzerland, a name not too different from the Arabic. Karkade is used in jams, jellies, sauces, and wines. In the tropics the fleshy calyces are used fresh for making roselle wine, jelly, syrup, gelatin, refreshing beverages, pudding, and cakes; dried roselle is used for tea, jelly, marmalade, ices, ice cream, sherbets, butter, pies, sauces, tarts, and other desserts. Calyces are used in the West Indies to color and flavor rum. Tender leaves and stalks are eaten as salad and as a pot-herb and are used for seasoning curries. Seeds have been used as an aphrodisiac coffee substitute. Fruits are edible.[3] Perry[16] cites one study showing roselle's usefulness in arteriosclerosis and as an intestinal antiseptic. Roselle is cultivated primarily for the bast fiber obtained from the stems. The fiber strands, up to 1.5 m long, are used for cordage and as a substitute for jute in the manufacture of burlap.[3,61]

Reported to be antiseptic, aphrodisiac, astringent, cholagogue, demulcent, digestive, diuretic, emollient, purgative, refrigerant, resolvent, sedative, stomachic, and tonic, roselle is a folk remedy for abscesses, bilious conditions, cancer, cough, debility, dyspepsia, dysuria, fever, hangover, heart ailments, hypertension, neurosis, scurvy, and strangury. The drink made by placing the calyx in water, is said to be a folk remedy for cancer. Medicinally, leaves are emollient, and are much used in Guinea as a diuretic, refrigerant, and sedative; fruits are antiscorbutic; leaves, seeds, and ripe calyces are diuretic and antiscorbutic; and the succulent calyx, boiled in water, is used as a drink in bilious attacks; flowers contain gossypetin, anthocyanin, and the glucoside hibiscin, which may have diuretic and choleretic effects, decreasing the viscosity of the blood, reducing blood pressure, and stimulating intestinal peristalsis. In Burma, the seed are used for debility, the leaves as emollient. Taiwanese regard the seed as diuretic, laxative, and tonic. Philippines use the bitter root as an aperitive and tonic.[16] Angolans use the mucilaginous leaves as an emollient and as a soothing cough remedy. Central Africans poultice the leaves on abscesses. Alcoholics might consider one item recounted by Watt and Breyer-Brandwijk: simulated ingestion of the plant extract decreased the rate of absorption of alcohol, lessening the intensity of alcohol effects in chickens.[3]

Per 100 g, the fruit contains 49 calories, 84.5% H_2O, 1.9 g protein, 0.1 g fat, 12.3 g total carbohydrate, 2.3 g fiber, 1.2 g ash, 172 mg Ca, 57 mg P, 2.9 mg Fe, 300 μg β-carotene equivalent, and 14 mg ascorbic acid. Per 100 g, the leaf is reported to contain 43 calories, 85.6% H_2O, 3.3 g protein, 0.3 g fat, 9.2 g total carbohydrate, 1.6 g fiber, 1.6 g ash, 213 mg Ca, 93 mg P, 4.8 mg Fe, 4135 μg β-carotene equivalent, 0.17 mg thiamine, 0.45 mg riboflavin, 1.2 mg niacin, and 54 mg ascorbic acid. The inflorescence, per 100 g, is reported to contain 44 calories, 86.2% H_2O, 1.6 g protein, 0.1 g fat, 11.1 g total carbohydrate, 2.5 g fiber, 1.0 g ash, 160 mg Ca, 60 mg P, 3.8 mg Fe, 285 μg β-carotene equivalent, 0.04 mg thiamine, 0.06 mg riboflavin, 0.5 mg niacin, and 14 mg ascorbic acid.[21] Seeds contain 7.6% moisture, 24.0% crude protein, 22.3% fat, 15.3% fiber, 23.8% N-free extract, 7.0% ash, 0.3% Ca, 0.6% P, and 0.4% S. Seed extracted with ether contained 0.7% fat, 29.0% protein, and 32.9% N-free extract.[170] Component acids of the seed lipids were 2.1% myristic-, 35.2% palmitic-, 2.0% palmitoleic-, 3.4% stearic-, 34.0% oleic-, 14.4% linoleic-, and three unusual HBr-reacting fatty acids (cis-12,13-epoxy-cis-9-octade-cenoic [12,13-epoxoleic] 4.5%; sterculic, 2.9%, and malvalic, 1.3%).[171] Salama and Ibrahim[172] report on the sterols in the seed oil, 61.3% beta-sitosterol, 16.5% campesterol, 5.1% cholesterol, and 3.2% ergosterol (said to be rare in vegetable oil but the most common mycosterol in most fungi, including yeast). Seed has properties similar to those of cotton seed oil, and is used as a substitute for crude castor oil. Karkade (dried-flowers minus-ovary) contains 13% of a mixture of citric and malic acid, two anthocyanins gossipetin (hydroxyflavone), and hibiscin, and 0.004 to 0.005% ascorbic acid. Petals yield the flavonal glucoside hibiscritin, which yields a crystalline aglycone — hibiscetin ($C_{15}H_{10}O_9$). Flowers contain phytosterols. The dried flower contains *circa* 15.3% hibiscic acid ($C_6H_6O_7$). Root contains saponins and tartaric acid. Calyces contain 6.7% proteins by fresh weight and 7.9% by dry weight. Aspartic acid is the most common amino acid. Dried fruits also contain vitamin C and Ca oxalate.[3,33]

169. *HIPPOMANE MANCINELLA* L. (EUPHORBIACEAE) — Manchineel

Latex once used for arrow poison;[2] also, used to poison fish. It is difficult to find much use for this poisonous tree, but, stretching the point, I have seen it stabilizing eroding sandy shore lines in tropical America! The wood has been employed for cabinetry, construction, furniture, and interior finish. According to USDA Handbook #249, this is classed as a honey plant, the honey reported to be nontoxic.[173]

Reported to be cathartic, diaphoretic, diuretic, emetic, piscicide, poison, vermifuge, vesicant, manchineel is a folk remedy for cancer, dropsy, infections, paralysis, rash, scabies, sores, syphilis, tetanus, venereal disorders, and warts.[32,33]

According to Perkins and Payne, all parts contain an alkaloid similar or identical to physostigmine and a sapogenin.[56] Bark contains essential oils and fats. Fresh leaves contain alpha-carotene and *circa* 0.09% phloracetophenon-2-4-dimethyl ether, $C_{10}H_{12}O_4$. Fresh fruits contain 0.02% physostigmine $C_{15}H_{21}O_2N_3$. Toxicity may be due to hippomanin A and B $(C_{27}H_{22}O_{18})$.[33] According to Kinghorn,[174] mancinellin, one of a series of toxins in manchineel, was reported as the 13-hexadeca-2,4,6-trienoate of 12-deoxy-5-hydroxyphorbol-6γ,7α-oxide. Like factor M2, a daphane ester, this may be cocarcinogenic.

Toxicity — Sap causes severe dermatitis, burning, swelling and vesication, temporary blindness if introduced into the eye. Smoke from burning plants may injure the respiratory system, skin, and eyes. The fruits may be confused with crabapples by novices (even Jim Duke). Ingestion may cause, in 1 to 2 hr, vomiting, bloody diarrhea, abdominal pain, intense gastrointestinal inflammation and irritation, shock, and sometimes death.[56]

To the physician — Hardin and Arena suggest gastric lavage or emesis, mineral oil.[34] For external irritation, wash the affected spot within 1/2 hr with salt water or soapy water.[8]

170. *HOMALOMENA* SP. (ARACEAE) — Homalomena

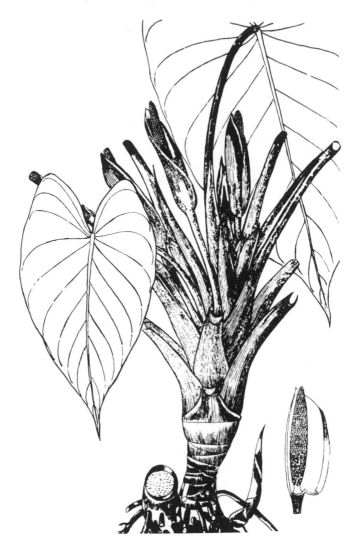

The plant seems to possess antiseptic properties. Leaves mixed with Himantandra bark and leaves in Papua. The resultant concoction, if drunk, produces a violent intoxication accompanied by spectacular vision and dream-like states that end in deep somnolence.[54] Some species are used to scent tobacco.

Reported to be expectorant, stimulant, various species of *Homalomena* show up in folk medicine for colic, distended stomach, enterosis, fever, hoarseness, lumbago, parturition, sore feet, sores, yaws.[5,32] Species of *Homalomena* have been used to treat skin ailments, miscarriage, syphilis, sores, tired feet, and wounds in Indonesia. Chinese use *Homalomena* species for numbness and rheumatism; Indochinese to treat diarrhea; Malayans for fever, hoarseness, leg sores, lumbago, parturition, sore feet, and stomach distention; Philippinos for cough, fever, pain, rheumatism, and snakebite.[16]

Toxicity — Perhaps narcotic.[54]

171. *HOSLUNDIA OPPOSITA* Vahl (LAMIACEAE) — Kamyuye

The red fruit is said to be edible. Tanganyikans are reported to use the plant homicidally.

Reported to be cholagogue, diuretic, laxative, kamyuye is a folk remedy for abdominal ailments, blennorrhea, chestpain, cold, conjunctivitis, cough, cystitis, enteralgia, epilepsy, fever, gastrosis, gonorrhea, hepatosis, herpes, hookworm, jaundice, mental disorders, shingles, snakebite, sore, sore throat, swellings, vertigo, venereal diseases, wounds, and yellow fever.[3,32] Boiled leaves are said to cure constipation.

Kamyuye oil consists mostly of sesquiterpenes and sesquiterpene alcohols.

Toxicity — "Bark has proved fatal when used to prepare a stomach medicine. Sheep are also killed by this aromatic shrub."[11]

172. *HUMULUS LUPULUS* L. (CANNABINACEAE) — Common Hops

In 1976, *circa* 100 million kg of hops were grown, solely for the brewing industry.[175] Bitter substance obtained from glandular hairs of strobilus used by brewers for giving aroma and flavor to beer. Originally used for their preservative value, the hops were only later noted to impart a flavor to beer. There is one German patent for adding hops to sausages as a "natural" preservative. Substance prevents Gram-negative bacteria from growing in the beer or wort. Amount of essential oil varies from 0.2 to 0.5%. Oil of hops also used in perfumes, cereal beverages, mineral waters, and tobacco. Stems are source of fiber which falls into class with soybean stalks, cotton stalks, flax shives, and similar agricultural residues and has been suggested for pulp or biomass.

Fiber has relatively high lignin and low pentosan content, with a cellulose content lower than any of them. Sometimes used for filler material in corrugated paper or board products, but is unsuitable for corrugating paper because of low pulp yield and high chemical requirement or for production of high-grade pulp for speciality paper. Young bleached tops used as a vegetable, especially in Belgium. Romans ate the young shoots like asparagus. Chopped very fine and dressed with butter or cream "the young shoots are excellent."[13] Alcoholic extracts of hops in various dosage forms have been used clinically in treating numerous forms of leprosy, pulmonary tuberculosis, and acute bacterial dysentery, with varying degrees of success in China.[41] Hops extracts are said to have various biological activities (antimicrobial activities due to the bitter acids, especially lupulone and humulone), strong spasmolytic effects on isolated smooth muscle preparations; hypnotic and sedative effects (disputed by one report); estrogenic properties were not observed in a more recent study; and allergenic activity on humans, causing contact dermatitis due to the pollen. Extracts are used in skin creams and lotions, in Europe, for alleged skin-softening properties. Extracts and oil are used as flavoring in nonalcoholic beverages, frozen dairy desserts, candy, baked goods, gelatins, and puddings, with the highest average maximum use level of 0.072% reported for an extract used in baked goods.[29] Hops steeped in sherry makes an excellent stomachic cordial.[2] Recently, counterculture entrepreneurs have apparently succeeded in grafting hops tops on marijuana bottoms and getting a "heady hop". Conversely, they might have succeeded in getting a perennial marijuana by grafting the annual herb onto the perennial hop.

Dried strobili used medicinally as a bitter tonic, sedative, hypnotic. The decoction from the flower is said to remedy swellings and hardness of the uterus. A cataplasm of the leaf is said to remedy cold tumors. The dried fruit, used for poultices and fomentations, is said to remedy painful tumors. The pomade, made from the lupulin, is said to remedy cancerous ulcerations.[4] Reported to be anaphrodisiac, anodyne, antiseptic, diuretic, hypnotic, nervine, sedative, soporific, stomachic, sudorific, tonic, and vermifuge, hops is a folk remedy for boils, bruises, calculus, cancer, cramps, cough, cystitis, debility, delirium, diarrhea, dyspepsia, fever, fits, hysteria, inflammation, insomnia, jaundice, nerves, neuralgia, rheumatism, and worms.[32] Moerman[176] gives interesting insight on Amerindian uses of a plant alien to them originally. Delaware Indians heated a small bag of leaves to apply to earache or toothache. More interesting was the Delaware use of hops as a sedative, drinking hop tea several times a day to alleviate nervousness. Cherokee, Mohegan, and Fox also used the plant as a sedative. George III is said to have slept on a pillow stuffed with hops to alleviate some symptoms of his porphyria. I would personally not hesitate to drink a camomile-hop-valerian tea as a sedative or herbal sleeping potion, but I would never recommend it to anyone else. The antibiotic principle lupulone is tuberculostatic.[60]

According to the *Wealth of India,* hops contain 6 to 12% moisture, 11 to 21% resins (*no* tetrahydrocannabinols), 0.2 to 0.5% volatile oils, 2 to 4% tannins, 13 to 24% protein, 3 to 4% fructose and glucose, 12 to 14% pectins, and 7 to 10% ash.[1] Hops contains 0.3 to 1% volatile oil; 3 to 12% resinous bitter principles composed of alpha-bitter acids (humulone, cohumulone, adhumulone, prehumulone, posthumulone, etc.), and beta-bitter acids (lupulone, colupulone, adlupulone, etc., in decreasing concentration); other resins, some of which are oxidation products of the alpha- and beta-acids; xanthohumol (a chalcone); flavonoid glycosides (astragalin, quercitrin, isoquercitrin, rutin, kaempferol-3-rutinoside, etc.); phenolic acids; tannins; lipids; amino acids; estrogenic substances; and many others.[29] The volatile oil is made up mostly of humulene (alpha-caryophyllene), myrcene, beta-caryophyllene, and farnesene, which together may account for over 90% of the oil. Other compounds present number over 100, including germacratriene, alpha- and beta-selinenes, selina-3,7(11)-diene, selina-4(14),7(11)-diene, alpha-copaene, alpha- and beta-pinenes, limonene, *p*-cymene, linalool, nerol, geraniol, nerolidol, citral, methylnonyl ketone, other oxygenated compounds, 2,3,4-trithiapentane (present only in oil of unsulfured hops in *circa* 0.01%), S-methylthio-2-methylbutanoate, S-methylthio-4-methylpentanoate, and 4,5-epithiocaryophyllene.[29] Countering claims that the "wonder cure" GLA (gamma-linoleic acid) is found only in mother's milk and evening primrose, I consulted the USDA lab at Peoria and learned that GLA was also in hops and borage, to mention just a few of the other vegetable sources.

Toxicity — Listed as a sedating, soporific narcotic.[54] Hops dermatitis has long been recognized. Not only hands and face, but legs have suffered purpuric eruptions due to hop picking. Although only 1 to 3000 workers is estimated to be treated, one in 30 is believed to suffer dermatitis.[6]

173. *HURA CREPITANS* L. (EUPHORBIACEAE) — Sandbox Tree

Sap and/or seeds are used to poison fish and undesirable animals. Smoke from the burning wood repels insects.[42] With its thorny trunk and a wide folk reputation as a tree that "injects" poison, the sandbox tree is, nonetheless, a commonly cultivated shade tree.[42] The soft wood is used in boats, crates, fence posts, firewood, furniture, plywood, and veneer. Strengthening Morton's criticisms[42] of books that don't cite their sources, I have forgotten where I read that the latex was once used in the U.S. to prepare tear gas, an item I reported without reference in my *Isthmian Ethnobotanical Dictionary*.[60] (Free!) *Caveat emptor*!

Reported to be anodyne, antidotal, aperient, canicidal, carcinogenic, dentifuge, emetic, insecticidal, piscicidal, poison, purgative, and vermifuge, sandbox tree is a folk remedy for elephantiasis, leprosy, rheumatism, skin ailments, and sores.[32,33] Cubans put the leaves in herbal baths. Guatemalans drink the leaf decoction as a purgative. Brazilians poultice the leaves onto rheumatism, applying the fresh male flowers to boils. Colombians apply the leaves to rheumatism and skin ailments. Venezuelans apply the leaves topically for myalgia and neuralgia.[42] In Darien it is said that the latex will make rotten teeth fall out.[60]

Latex contains hurine (similar to cardol), inositol, huratoxin ($C_{34}H_{48}O_8$), a proteinase called hurain, 24-methylencycloartenol, cycloartenol, and butyrospermol. Leaves contain kaempferol, *p*-coumaric acid, and ferulic acid. Seeds contain oil and a toxalbumin, crepitin, similar to ricin.[33] Extracts are mitogenic and useful in lymphocytic stimulation. Hura seeds contain several very active proteases and the agglutinating ability of these acting with red blood cells is enhanced by pretreatment with trypsin. Huratoxin is a cocarcinogenic daphnane ester.[174]

Toxicity — The sap may cause blistering, irritation, and swellings (trials on my wrists were negative). Introduced in the eyes, the sap may cause temporary blindness. Smoke from burning wood may irritate the eye. Ingestion of seed may cause bloody diarrhea, abdominal pain, vomiting, dim vision, rapid pulse, delirium, convulsions, collapse, and death. Two to three seeds induced moderate poisoning (but not fatality?).[56]

To the physician — Hardin and Arena recommend emesis or gastric lavage and symptomatic treatment.[34]

174. *HYDRANGEA ARBORESCENS* L. (SAXIFRAGACEAE) — Seven Barks, Smooth Hydrangea

Ornamental shrub, the pithy stalks sometimes used for pipe stems. Tyler[37] notes a renewed interest in *Hydrangea*, unfortunately, due to the trend for some people to smoke some hydrangea leaves as a euphoric of dubious efficacy.

The fluid extract is reported to have "cured" cancer of the tongue and metastasis in axillary glands following breast amputation.[4] Rutin is said to inhibit tumor formation[29] and protect against x-ray irradiation.[29] Roots said to be cathartic, diuretic, laxative, and tonic, removing calculus and gravel. Use of seven bark has resulted in expulsion of 120 calculi from one person alone.[2] Fluid extract used for alkaline urine, chronic gleet, and mucous irritations of the bladder in aged persons. Leaves said to be alterative, bactericide, cathartic, diaphoretic, diuretic, laxative, narcotic, nephritic, purgative, sialogogue, stomachic, and tonic.[32] Amerindians gave the root decoction to women experiencing unusual dreams during their menstrual period. Said to relieve backache caused by kidney, rheumatism, scurvy, and dropsy.[30] Cherokee introduced it to the early settlers as a remedy for kidney stones.[37] American settlers used the roots to treat dyspepsia. Homeopaths prescribe the young-shoot tincture for bladderstone, catarrh, diabetes, gravel, incontinence, and prostatic ailments.[30] Tyler concludes, perhaps with a bit of hyperbole, "As a drug of use or abuse, the roots or leaves of any Hydrangea species have no merit."[37]

Devoid of tannin, the root contains two resins, gum, kaempferol, quercitin, rutin, saponin, sugar, starch, albumen, soda, lime, potassium, magnesia, sulphuric and phosphoric acids, a protosalt of iron, and the cyanogenic glycoside hydrangein (glycoside $C_{34}H_{25}O_{11}$), plus fixed and volatile oils.[2,33]

Toxicity — Classified by the FDA as an Herb of Undefined Safety: "Contains hydrangin. Overdose is said to cause vertigo and feeling of oppression of the chest."[62] Hydrangin may cause gastroenteritis.[11]

To the physician — Hardin and Arena[34] recommend gastric lavage or emesis and treatment for cyanide poisoning in poisoning cases. In Florida a family was poisoned when children tossed hydrangea buds in a tossed salad.[14] Wood said to be injurious to woodworkers.[6]

175. *HYDRANGEA PANICULATA* Seib. (SAXIFRAGACEAE) — Peegee

Hardy ornamental shrub, native of China and Japan. Reported to be diuretic, peegee is a folk remedy for cough and malaria.[32]

Dry seeds contain 25.0% protein and 31.3% fat.[21] Flowers contain *circa* 4.1% rutin.[59] Extraction of dried flowers with benzene gave 2.5% of a phenolic $C_9H_6O_3$. A glycoside pseudohydrangin or parahydrangin occurs in the roots, neohydrangin and mucilage in the bark. Hydrangin yields umbelliferone and *d*-glucose on hydrolysis. The mucilage contains a polyuronide containing *d*-galacturonic acid, galactose, and rhamnose in the molecular ratio of 10:7:3.

Toxicity — Even the anonymous author of "Herbal Highs"[51] adds that smoking the leaves could be dangerous: "Hydrangea leaves contain a chemical that belongs to the cyanide family. The high derived from this is an example of subtoxic inebriation, in which there is a fairly narrow margin between pleasurable and toxic doses."[51]

176. *HYDRASTIS CANADENSIS* L. (RANUNCULACEAE) — Goldenseal

Plant used as source of yellow dye. Indians used it for their clothing and implements of warfare. It imparts a brilliant durable yellow to linens. With indigo, goldenseal is said to impart a fine green to wool, silk, and cotton. Once used for external cleansing (e.g., for acne and eczema). Cherokee Indians are said to have used goldenseal mixed with bear's grease as an insect repellant. The alkaloids show anticonvulsive activity on mice intestines and uteri.[29] Hydrastis extracts and hydrastine hydrochloride have been used in stopping uterine hemorrhage and[20] alleviating menstrual pain. Still used as a component in collyriums.

Tyler describes it as an interesting but valueless drug.[37] Kloss[44] and Tierra[28] describe it as one of the most wonderful remedies, a real cure-all. Used variously for cancers, especially of the ovary, stomach, and uterus.[4] I personally have heard three testimonials for goldenseal for remission in uterine cancer, but, doubtless, there have been more remissions without goldenseal. Alkaloids in *Hydrastis* show some anticancer activity. Berberine, or its sulfate, has shown activity in the Bl, KB, and PS tumor systems.[10] Canadine is often known as tetrahydroberberine. "Two tetrahydroprotoberberine alkaloids possessing a transquinolizidine conformation displayed in vitro cytotoxicity against KB cells derived from a human epidermoid carcinoma."[177] Goldenseal, long used by the Cherokee for eye and skin ailments, is regarded as alterative, antiperiodic, antiseptic, aperient, astringent, detergent, diuretic, and hemostatic. It has served as bitter tonic, as an alterative to the mucous membranes, and has been used as a douche for uterine hemorrhage and for hemorrhoids. Used in some eye lotions. Mixed with red pepper, goldenseal is used as a remedy for chronic alcoholism. Some Indians employed the herb as diuretic, escharotic, and stimulant, using the powder for blistering and the infusion for dropsy. Cures for boils, hemorrhoids, psoriasis, ringworm, and ulcers have been reported. Said to prevent the pitting caused by small pox. Homeopaths prescribe the fresh root for alcoholism, asthma, cancer, catarrh, chancroids, constipation, corns, deafness, dyspepsia, ear, eczema, faintness, fistula, gastric catarrh, glossitis, hemorrhoids, impetigo, jaundice, leucorrhea, lip cancer, liver ailments, lumbago, lupus, menorrhagia, metrorrhagia, mouth sore, nail affections, ozaena, placental adherence, post-nasal catarrh, rectal ailments, sciatica, seborrhea, sore mouth, sore nipple, sore throat, stomach ailments, stomatitis, taste disorders, throat ailments, tongue ailments, typhus, ulcers, and uteral afflictions.[30] Parvati suggests adding goldenseal and sage to cider vinegar and acidophilous as a douching solution.[48] For a brief period, goldenseal was "pushed" as an herb to prevent the detection of morphine in urinalysis, but there is no scientific evidence for such claims.[37] But for the other side of the coin, witness the following newspaper clipping.[178] Dateline San Diego (WNS): newspapers reported that Ethan Nebelkopf was using goldenseal (with comfrey, mullein, orange peel, and spearmint) as a "detox" tea to help addicts kick their cocaine, heroin, and methadone habits. Nebelkopf said, "People who are into herb-based drugs like heroin and cocaine seem to take naturally to herbal tea." I wonder, however, how many physicians or nurses would inject goldenseal tea into your bladder with a rubber catheter two or three times a day for bladder ailments, as Kloss recommends.[44]

Alkaloids isolated from hydrastis include hydrastine (1.5 to 4%), berberastine (2 to 3%), berberine (0.5 to 6%), canadine, candaline, hydrastinine, 5-hydroxytetrahydroberberine, meconine, reticuline, and xanthopucine. Volatile oils, chlorogenic acid, phytosterins, and resins are also recorded. Leung states that goldenseal is used for inflammation of mucous membranes (uteral and vaginal), hemorrhoids, nasal congestion, sore eyes, sore gums, acne, dandruff, etc.[29]

Toxicity — An overdose (even externally) can cause severe ulceration of any surface it may touch. Such notes should discourage those who recommend goldenseal as a douche. According to Tierra, women who have a tendency to miscarry should avoid the use of goldenseal.[28] Excessive use over a prolonged period can diminish vitamin B absorption.[28] People with a history of high blood pressure should avoid goldenseal.[28]

177. *HYOSCYAMUS NIGER* L. (SOLANACEAE) — Henbane, Henblain, Jusquaime

Dried leaves of henbane contain the alkaloid hyoscyamine, used in the treatment of alcohol and morphine habits, as a mydriatic, sedative, pain killer, and in motion sickness. In Europe, henbane is administered for earache and rheumatism. Leaves are said to repel mice. Pliny hinted at henbanes use as an intoxicant, more than four leaves in a drink putting the drinker beside himself. According to Hocking, henbane was used as an anesthetic in the Middle Ages (with aconite, datura, hemlock, mandrake, opium, etc.), either as a "soporific soponge" or "sleeping apple" pomander, acting through inhalation. It was also employed illegally as a sleeping potion or "Mickey Finn". Arabian familiarity with the herb was indicated in "The Thousand and One Nights" where guards were drugged by one who cast henbane leaves onto the fire. A nearly inconceivable contraceptive was formed by mashing the seed into paste with mare's milk and tying the paste in a piece of wild bull's skin.[179]

The juice, prepared in various manners, is used for indurations of the mammary glands, carcinomata, parotid tumors, condylomata, and cancerous wounds. A cataplasm of the leaf is used for tumors of the scrotum, cancerous ulcers, painful swellings, glandular tumors, and cancer. The tea made from the plant is said to remedy tumors.[4] Leaves are considered anodyne, antidiabetic, antispasmodic, carminative, hypnotic, mildly laxative, mydriatic, narcotic, and sedative. The herb is used for asthma, bronchitis, cough, hydrophobia, neuralgia, and rheumatism.[30] In India, leaves are sometimes smoked with ganja; powdered seeds and smoke from burning seeds are applied for toothache, and seeds mixed in wine are applied to gout swellings.[1] Russians smoke a mixture of henbane, sage, and jimsonweed for asthma.

A seed suppository is applied for pains of the uterus. After narcotic principles are expressed from seeds, an edible oil, gums, and resins are also obtained. Seeds are also anodyne and narcotic, and have been used for centuries for such purposes. The narcotic is so strong that washing ones feet in it is said to induce sleep.[30] Chinese apply the plant externally for neuralgia and odontalgia.[16] Homeopaths prescribe the tincture for amaurosis, angina pectoris, asthma, bladder paralysis, bronchitis, chorea, coma, cough, delirium tremens, diarrhea, dysmenorrhea, enteritis, epilepsy, epistaxis, erotomania, eye affections, fever, hemoptysis, hemorrhage, hiccup, hydrophobia, hypochondria, lochial suppression, mania, meningitis, mental afflictions, neuralgia, nightblindness, nymphomania, paralysis, paralysis agitans, parotitis, pneumonia, psychoses, rage, schizophrenia, sleep disorders, and visual defects.[30,33] Hocking[179] notes that henbane has been used as a sedative in delirium, epilepsy, hypochondria, hysteria, insomnia, melancholia, nervous fever, paralysis, and priapism. It is used as an anodyne in angina, arthritis, ataxis, cephalgia, colic, gastralgia, gout, lead poisoning, orchitis, swellings, and teething. Further, it was used as an antispasmodic in asthma, chorea, constipation, cough, croup, orchitis, pertussis, and tetanus.

Young plants contain more hyoscine and less hyoscyamine; in mature plants, hyoscyamine is the main active alkaloid (especially in the petiole). Leaves contain 0.04 to 0.08% total alkaloids, roots 0.16%, seeds 0.06 to 0.1%.[58] Choline, mucilage, resin, and tannin are also reported. Potassium nitrate is presumed to cause the characteristic sparking effect when the leaf is ignited. Seeds contain about 25% fatty oil and a little volatile oil. One report suggests that dry seeds contain 21.2% protein and 38.9% fat.[21] *Hager's Handbook* adds for the leaves apoatropine, atropine, cuscohygrine, choline, tetramethylputrescine, methyl pyrroline, methyl pyrrolidine, pyridine, tropine, scopine, and scopoline. In addition to the usual amino acids, there is a bitter glycoside hyoscypicrin, a butyric acid ester, coumarin, wax. Old literature adds bi- and trimethylamine and scopetol.[33]

Toxicity — Classified by the FDA as an unsafe poisonous herb,[62] containing the alkaloids atropine, hyoscine (scopolamine), and hyoscamine. People poisoned by ingesting the root, confusing them with parsnip or chicory, reported seeing red for 2 or 3 days. Convulsions and mania have resulted from smoking the leaves. Symptoms of poisonings include watering of the mouth, headache, nausea, rapid pulse, convulsions, coma, and death.[34]

To the physician — Hardin and Arena recommend gastric lavage (4% tannic acid solution) or emesis, symptomatic and supportive.[34] Hocking[179] says caffeine or morphine may be used an an antidote, with care.

178. *HYPERICUM PERFORATUM* L. (HYPERICACEAE) — St. John's Wort, Klamath Weed

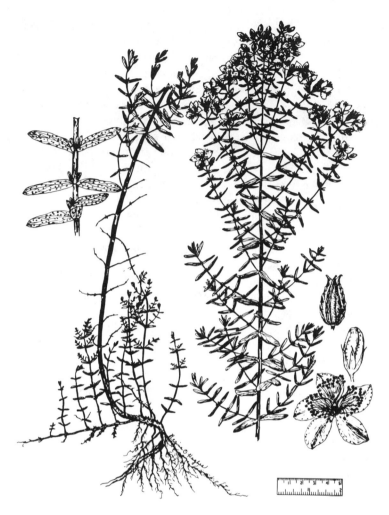

Formerly used as a source of the antidepressant, hypericin, also known as hypericum red. Hypericum red dyes wool and silk deep violet red. Boiled with alum, however, it is said to produce a yellow dye.[180] Very small quantities appear to have a tonic and tranquilizing action on the human organism.[20] Success has been reported in treating vitiligo by oral and topical administration application of hypericum extracts.[6] Blossom, capsule, leaf, and stalk are said to be antibiotic (patented as a possible food preservative). The Greek name Hypericum means ''over an apparition'' in the belief that the herb was so obnoxious to evil spirits that only a whiff would cause them to fly away. Aqueous extracts inhibit the growth of *Mycobacterium tuberculosis*. Of late the herb tea and the olive-oil floral extract have gained new popularity in Europe for anuria, anxiety, depression, gastritis, and unrest, the oil extract also used externally to promote healing and reduce hemorrhoids and inflammation.[37]

In the Hartwell file, there are 14 references to this species as a folk remedy for cancer, for hard breast, carcinoma, indolent ulcers, uterine and stomach cancers, ovary carcinoma, lymph tumors including five cases ''cured'').[4] Tests by the National Cancer Institute indicate little promise against cancer. The plant is described as anodyne, antiseptic, aromatic, astringent, cholagogue, digestive, diuretic, expectorant, nervine, resolvent, sedative, stimulant, vermifuge, and vulnerary.[32] Said to be used in bladder troubles, bronchitis, catarrh, consumption, diarrhea, dysentery, dysmenorrhea, enuresis, hemoptysis, hemorrhage, hysteria, insomnia, jaundice, nervous depression, neurasthenia, neuralgia, oliguria, phthisis, piles, rabies, sciatica, and worms. Used for chronic catarrh of the bowels, lungs, or urinary passages. Russians recommend it for bronchial asthma. An aqueous herb extract has been applied as a hair restorer. For bedwetting, children were once given the tea at bedtime. Oils in which the plant has been boiled are applied externally for gout and rheumatism. Oleum hyperici, an infusion of the fresh flowers in olive oil, is applied externally to sores, ulcers, and wounds. Though internally it could cause photosensitization, it is said to be valued externally as a sunburn oil. Oleum hyperici has been recommended as a cosmetic skin tightener. Homeopaths prescribe the tincture for adenitis, asthma, bites, breast ailments, brain concussion, bruise, bunion, compound fracture, corn, coxalgia, diarrhea, gunshot wounds, headache, hemorrhoids, hydrophobia, hypersensitivity, impotence, labor, mental ailments, neuralgia, panaritium, paralysis, rheumatism, scars, sciatica, spastic paralysis, spinal concussion, spinal irritation, stiff neck, tetanus ulcerations, whooping cough, and wounds.[30]

Hypericin yields are 0.0095 to 0.466%. Flowers may contain as much as 2400 mg/kg. Stems contain 3.8% tannin, leaves 12.4%, flowers 16.2%, and whole herb 8.2 to 9.3%. Dry seeds contain 18.1 to 20.7% protein, 32.8 to 33.2% fat.[21] Seeds also contain 39.5 mg vitamin C, 16.5 mg carotene, 12.1 g tannin, 0.33 g volatile oil per 100 g, as well as little saponin and the glucoside hyperin. The herb itself is said to contain choline, rutin, glucosides, tannin, phobaphene, pectin, beta-sitosterol, alkaloids, a fixed oil in addition to the volatile oil, 0.13% vitamin C, up to 13 mg/100 g vitamin A, and various pigments. The fixed oil contains glycerides of stearic, palmitic, and myristic acids, ceryl alcohol, phytosterol, and two hydrocarbons. The volatile oil contains pinene, cineol, myrcene, cadinene, gurjunene, hypericin, and esters of isovalerianic acid. Flowers contain the carotenoids lutein, violaxanthin, luteoxanthin, cistrollixanthin, and trollichrome.[33]

Toxicity — In sheep, may cause shedding of wool, swelling of the face, generalized skin irritation, loss of appetite, and sometimes loss of eyesight. Cattle may be more susceptible than sheep to the toxicity. Mice show severe, even fatal, photodynamic effects following doses of 0.2 to 0.5 mg. Illustrations of animals poisoned by photosensitization[3] are quite dramatic. Perhaps hypericin should be investigated in the treatment of psoriasis, like some of the psoralens.

179. *ILEX OPACA* Ait. (AQUIFOLIACEAE) — Holly

Planted as an impenetrable hedge and ornamental evergreen, perhaps tracing its days back to the Druids who planted evergreens like *I. aquifolium,* the European holly, in order to decorate their huts with evergreens during winter as an abode for sylvan spirits. A Christian legend has it that holly first sprang up under the footsteps of Christ, when he trod the earth. Its thorny leaves and scarlet berries, like drops of blood seem to symbolize the Savior's sufferings. If planted near a house or farm, it was supposed to repel lightning, poison, and witchcraft (Pliny).[2] Deer, sheep, some cattle eat the leaves. In winter the bruised branches are given to cattle as food. A holly twig, placed in a rabbit hutch, is supposed to whet the rabbits appetite. Wood, valued for ornamental ware and inlay, is very hard and susceptible to high polish. Scorched leaves have been used as a coffee or tea substitute (most common tea substitute during Civil War). Inner bark from young shoots fermented to make a bird lime.

Root bark, leaf, and berries used as a cataplasm for cancer (North Carolina).[4] Fresh fruit and bark were once gathered before the first frost, mixed with two parts by weight alcohol, put into a stoppered bottle, and allowed to stand in a dark place for 8 days, then filtered. This laxative decoction was used to treat constipation, coughs, fever, gout, pleurisy, rheumatism, tumors, and worms.[18] Once used as diaphoretic, holly leaf infusions were given in catarrh, pleurisy, rheumatism, and small pox. The alkaloid ilicin is said to act like quinine for malaria and rheumatism. Juice of the fresh leaves applied in jaundice. South Carolinian lowlanders often boil the leafy twigs, add lemon, with or without pinetops, as a favored remedy for colds and coughs.[46] The bark of holly root has been described as an excellent demulcent. Berries used for dropsy.[1] Rootbark decoction used to treat colds, coughs, and tuberculosis.

The alkaloids ilicin and caffeine have been reported from closely related species. *I. aquifolius* contains pelargonidin-3-glucoside, pelargonium-3-xyloxylglucoside, and cyanidine-3-xyloxyloglucoside.

Toxicity — Berries are said to be toxic. According to *Hager's Handbook,* children poisoned by eating the berries of *I. aquifolius* may suffer gastroenteritis, even fatalities.[33]

180. *ILEX PARAGUARIENESIS* St. Hil. (AQUIFOLIACEAE) — Yerba Mate, Paraguay Tea, South American Holly

Yerba maté is primarily cultivated for the leaves, esteemed for an aromatic, stimulating beverage, with somewhat bitter taste. The tea, made from finely ground leaves steeped in hot water, is served hot or cold, plain or sweetened, with lemon or cream added. Maté, as a stimulant, affects the muscles by reducing fatigue; substitutes for alcoholic drinks; and is recommended for persons suffering from debility and neurasthenia. With beef, maté takes the place of bread and vegetables. It is said to dispel hunger and invigorate the body. Maté-flavored products are produced, e.g., a whole wheat bread and a carbonated soft drink. Maté means gourd (Spanish), alluding to the receptacle from which it is drunk by means of a tube, called a ''bombilla'', made of carved wood, silver, or other metal.

Reported to be aperient, astringent, diuretic, *poison*, purgative, stimulant, and sudorific; maté is a folk remedy for rheumatism and scurvy.[32] As Paraguay Tea, it is promoted for such things as heart, nerve, and stomach ailments and diabetes.[33]

Porter[181] reported an analysis of the leaves showing 5.5% moisture, 4.8% resin, 17.68% crude fiber, 0.30% volatile oil, 4.25% total ash, 3% N (*circa* 19% protein), 12.4% tannin, 1.2% caffeine, 0.05% theophylline, and 1.9% chlorophyll. Per 100 g, the leaves (13 cases) averaged 1.23 mg carotene, 2095 IU vitamin A, 222.7 IU vitamin B, 404.3 IU riboflavin, 11.9 mg ascorbic acid, and 6.9 mg nicotinic acid. According to *Hager's Handbook*, the leaves contain 0.34 to 1.72% caffeine, much of which is bound to tannins, to be freed in the fermentation process. Leaves also contain 0.096 to 0.19% theobromine, some theophylline, 4 to 16% tannin, chlorogenic and nonchlorogenic, the antitumor compounds ursolic acid, and beta-amyrin, rutin, traces of vanillin, resinous substances, ascorbic acid, riboflavin, pyridoxine, nicotinic acid, panthotenic acid or its derivatives, inosital, traces of trigonellin, and choline.[33] Among the fats, there is butyric-, isobutyric-, isovalerianic-, isocapronic-, and 4-oxolauric-acids, with 2,5-xylenol, together with three monophenols and three polyphenols.

Toxicity — Lewis and Lewis classify maté with poisonous plants (like coffee and tea).[11] One kilogram prepared leaves contains only 2.5 g caffeine, cf. 4.6 g in black tea, 2.6 g in ground coffee. Gastric disturbances may also be caused by essential oils in coffee; maté contains only 10 mg/kg essential oil, cf. 410 for coffee and 6000 to 7900 for tea. Emboden lists maté as a stimulant with narcotic effects.[54]

181. *ILLICIUM VERUM* Hook. f. (MAGNOLIACEAE) — Star-Anise, Chinese Anise

Dried fruit has a pleasant, aromatic, anise-like aroma and taste. Used whole, not ground, as a flavoring agent in confections, candy, chewing gum, and tobacco. Orientals chew the seeds after meals to promote digestion and sweeten the breath. Oil is used in animal feeds, in scenting soaps, toothpaste, creams, detergents, perfumes, etc., to improve the flavor of some medicines. Japanese use the ground bark as incense. Used in cough medicines and cough drops. Highest use levels are *circa* 570 ppm in alcoholic beverages and 680 ppm in candies (numbers derived from Pimpinella but according to Leung,[29] anise oil and star anise oil are used interchangeably in the U.S., both being officially recognized as anise oil).

Reported to be anodyne, antiseptic, carminative, diuretic, expectorant, lactagogue, pediculicide, piscicide, stimulant, and stomachic, star anise is a folk remedy for cholecystitis, colic, constipation, dysentery, dyspepsia, extrophy, favus, halitosis, hernia, insomnia, lumbago, otalgia, rheumatism, scabies, spasms, and toothache.[32,41] A medicinal tea is made from the leaves in China.

Star anise contains *circa* 5% volatile oil (*circa* 2.5% in seed, 10% in follicle) with *trans*-anethole as its major ingredient. Other constituents include estragole, 1,4-cineole, beta-bisabolene, beta-farnesene, alpha-copaene, *cis*- and *trans*-alpha-bergamotene, caryophyllene, nerolidol, methylanisoate, *trans*-methylisoeugenol, cadinene, foeniculin, 3-carene, *d*-alpha-pinene, phellandrene, alpha-terpineol, hydroquinone, traces of *cis*-anethole, and safrole. Presence of safrole disputed.[29] *Hager's Handbook* adds *n*-decylic-, palmitic-, tiglic-, benzoic-, salicylic-, and anisic-acids, a reddish brown resin, tannin, protocatechuic acid, shikimic acid (carcinogenic), and quinic acid. The seed contains 20% fatty oils with 45% oleic-, 24% linoleic-, 23% palmitic-, and 2.5% stearic-acid. Leaves contain 1% essential oil with an ethole and anisaldehyde.

Toxicity — Chinese star anise should not be confused with Japanese star anise *I. lanceolatum* A. C. Smith, which is highly poisonous. A 10 to 15% aqueous extract is used as an agricultural insecticide in China.[29] Toxicities of anethole, isosafrole, and safrole were discussed by Buchanan[117] (GRAS § 182.10 and 182.20).[29]

182. *INDIGOFERA TINCTORIA* L. (FABACEAE) — Common or Indian Indigo

Indigo is cultivated for the well-known blue dye. In southern India, indigo is grown as a cover or green-manure crop in coffee plantation and rice fields.[40] Used as a cover crop, it is said to be an effective nematicide.[65] Once this was a primary source of Indian indigo, but it was largely replaced by *I. arrecta*. Leaves are rich in potash, and the plant is palatable to cattle.[40] Root infusion used to kill vermin and a seed tincture to kill lice.[3]

Leaf or juice used for cancer or tumors, especially of the ovaries or stomach.[4] Medicinally, juice of leaves is used as a prophylactic against hydrophobia, and a decoction for blennorrhagia; also, an extract of the plant is given for epilepsy, nervous disorders, bronchitis, and as an ointment to heal sores, old ulcers, and hemorrhoids.[33] Leaves applied with turmeric to bruises, inflammations, and scabies in Indochina.[16] Roots are used for asthma, dysentery, fevers, gravel, hepatitis, ichthyosis, kidney stones, scorpion bites, urinary complaints, and worms. Plant has been used to produce nausea and vomiting. In Tanganyika, the root is used for syphilis. Juice of the leaf, mixed with honey, used for enlarged liver and spleen, and for toothache.[3] Poulticed onto boils, fevers, and yaws. The plant is said to be antiseptic, antispasmodic, astringent, deobstruent, emetic, febrifuge, larvicidal, nervine, pediculicidal, purgative, stimulant, and vermifuge.[32,40]

The leaves contain (dry basis): N, 5.11%; P205, 0.78%; K20, 1.65%; and CaO, 5.35%. It is a rich source of potash, and ash (4.4%) containing as much as 9.5% of soluble potassium salts. The indigo refuse, obtained after dye extraction, gave on analysis: N, 1.8%; P_2O_5, 0.4%; and K_2O, 0.3%.[1] The plant contains indigotin and a glucoside, indican ($C_{14}H_{17}NO_6$).[33,40]

Toxicity — Indigo appears to be, at most, mildly irritant to the eye. Dermatitis is common among indigo dyers but there is no direct evidence that indigo or its derivatives is responsible. Women who shell indigo develop an exceptionally long right thumb.[6] The teratogen indospicine has been reported from *I. spicata*.

183. *IPOMOEA PURGA* (Wender) Hayne (CONVOLVULACEAE) — Jalap

Handsome ornamental, more grown as purgative than ornamental. The dry jalap is a resin extracted from the root.[2]

Root strongly cathartic, diuretic, and purgative, used for constipation, colic, colitis, dysentery, gastritis, inflammations, pain in the bowels, sluggishness, and sores.[32,42] Given in sugar or jelly, it forms a "safe purge" for children.[2] Said to be an excellent purge in rheumatism. Given with wormwood or calomel as a vermifuge for children.[2]

Contains 7 to 18% resin; gum, sugar. Source of convolvulin (rhodeoretin), an amorphous white to yellow powder which causes sneezing.[20] Alkaline hydrolysis yields, tiglic-, acetic-, propionic-, 2-methylbutanoic-, isobutyric-, isovaleric-, and valeric-acids, and a complex oligosaccharidic acid consisting of two parts: (1) ipurolic acid (3,11-dihydroxymyristic acid), convolvulinolic acid (a hydroxypentadecanoic acid), glucose, fucose, and rhamnose; and (2) purginic acid (a mixture of ipurolic acid, a hydroxylauric acid, and rhamnose).

184. *IPOMOEA VIOLACEA* L. (CONVOLVULACEAE) — Tlitliltzen, Ololiuqui

Seed taken for divination or for ritual curing,[52] washing them and steeping in water for *circa* 8 hr, straining through a cloth, and drinking. Zapotec Indians, calling it badoh negro for the black seeds, use it in religious rites as did, apparently, the Aztecs.[54] Reported to be psychedelic and psychomimetic,[32] tlitliltzen is used by Mexican Indians for magicomedicinal purposes and as a hallucinogen.[33]

Seeds of certain varieties contain substances similar to LSD. According to one anonymous writer: ''Flying Saucers'', ''Heavenly Blue'', ''Pearly Gates'', and ''Wedding Bells'' are the most hallucinogenic varieties, hence, the most toxic.[51] Emboden adds ''Blue-Star'' and ''Summer Skies''. All contain amides of lysergic acid, principally ergine (*d*-lysergic acid diethylamide) and isoergine (*d*-isolysergic acid diethylamide). Hofmann[182] compared the ''LSD'' chemistry of seeds of *I. violacea* with *Rivea corymbosa*.

	Rivea (%)	Ipomoea (%)
Ergine (*d*-lysergic acid amide)	0.0069	0.035
Isoergine (*d*-isolysergic acid amide)	0.0020	0.005
Chanoclavine	0.0005	0.005
Elymoclavine	0.0005	0.005
Lysergol	0.0005	—
Ergometrine	—	0.005

The alkaloids are concentrated in the embryo of the seed and are absent from the shells which are occasionally attacked by fungi. Hofmann experienced tiredness, apathy, a feeling of mental emptiness and of the unreality, and complete meaninglessness of the outside world, following ingestion of 2 mg isoergine. Ergometrine is used as a uterotonic and hemostat in obstetrics.

Toxicity — Classed as a narcotic hallucinogen.[54] According to Hardin and Arena, *circa* 300 seeds give the effect of 200 to 300 µg of LSD-25. Nausea and acute chronic psychotic reaction follow ingestion.[34]

185. *IRIS VERSICOLOR* L. (IRIDACEAE) — Blue Flag

Ornamental. The blue flowers yield a blue infusion which can substitute for litmus paper. Source of iridin or irisin of commerce, a bitter aperient, diuretic. Iridin acts on the liver and is milder on the bowel than podophyllum.[2]

Roots have been used in teas (dangerous), poultices, and ointments for cancer, felons, and tumors, especially of the breast and kidney.[4] Root alterative, cathartic, depurative, diuretic, emetic, laxative, purgative, resolvent, sialogogue, and stimulant. Has been recommended for blood impurities, cancer, constipation, dropsy, gastritis, liver ailments, rheumatism, scrofula, skin diseases, and syphilis.[32] Poulticed onto felons to relieve the pain, even when suppuration is far advanced. This herb, which is said to increase the rate of fat catabolism, was used for obesity by Indias Caraka Samhits as early as 3000 B.C.[11] Homeopaths prescribe the root tincture for anal fissures, bilious attack, constipation, crusta lactea, diabetes, diarrhea, dysentery, dysmenorrhea, dyspepsia, eczema, fistula, gastrodynia, headache, hemicrania, hepatosis, impetigo, liver, migraine, morning sickness, nausea, neuralgia, nocturnal emission, pancreas, parotitis, pregnancy, psoriasis, rectal burning, rheumatism, salivation, sciatica, whitlow, and zoster.[30,33]

Rhizome contains essential oil (with furfurol), gum, resin, starch, tannin, alpha-phytosterol, myricylalcohol, heptacosan, ipuranol, isophthalic acid, salicylic acid, lauric, palmitic, and stearic acids.[33]

Toxicity — Root contains the poisonous iridin. The glycoside can be poisonous to livestock which ingest it. Sometimes confused with sweet flag, with disastrous results. Full doses of iridin may cause nausea and severe prostration.

186. *JATROPHA CURCAS* L. (EUPHORBIACEAE) — Physic Nut, Purging Nut

According to Ochse, "the young leaves may be safely eaten, steamed or stewed."[183] They are favored for cooking with goat meat, said to counteract the peculiar smell. Though purgative, the nuts are sometimes roasted and dangerously eaten. In India, pounded leaves are applied near horses' eyes to repel flies. The oil has been used for illumination, soap, candles, adulteration of olive oil, and making Turkey red oil. Nuts can be strung on grass and burned like candlenuts.[3] Gaydou et al. discuss the possibilities of the species as an energy source.[184] Mexicans grow the shrub as a host for the lac insect. Ashes of the burned root are used as a salt substitute.[42] It has been used for homicide, molluscicide, piscicide, and raticide.[32] The latex was strongly inhibitory to watermelon mosaic virus.[185] Bark used as a fish poison.[3] In South Sudan, the seed as well as the fruit is used as a contraceptive.[33] Sap stains linen and can be used for marking.[6]

Extracts are used in folk remedies for cancer.[4] Reported to be abortifacient, anodyne, antiseptic, cicatrizant, depurative, diuretic, emetic, hemostat, lactagogue, narcotic, purgative, rubefacient, styptic, vermifuge, and vulnerary, physic nut is a folk remedy for alopecia, anasarca, ascites, burns, carbuncles, convulsions, cough, dermatitis, diarrhea, dropsy, dysentery, dyspepsia, eczema, erysipelas, fever, gonorrhea, hernia, incontinence, inflammation, jaundice, neuralgia, paralysis, parturition, pleurisy, pneumonia, rash, rheumatism, scabies, sciatica, sores, stomachache, syphilis, tetanus, thrush, tumors, ulcers, uterosis, whitlows, yaws, and yellow fever.[32,33] Latex applied topically to bee and wasp stings.[3] Mauritians massage ascitic limbs with the oil. Cameroon natives apply the leaf docoction in arthritis.[3]

Colombians drink the leaf decoction for venereal disease.[42] Bahamians drink the decoction for heartburn. Costa Ricans poultice leaves onto erysipelas and splenosis. Guatemalans place heated leaves on the breast as a lactagogue. Cubans apply the latex to toothache. Colombians and Costa Ricans apply the latex to burns, hemorrhoids, ringworm, and ulcers. Barbadians use the leaf tea for marasmus, Panamanians for jaundice. Venezuelans take the root decoction for dysentery.[42] Seeds are used also for dropsy, gout, paralysis, and skin ailments.[3] Leaves are regarded as antiparasitic, applied to scabies; rubefacient for paralysis, rheumatism; also, applied to hard tumors.[4] Latex used to dress sores and ulcers and inflamed tongues.[16] Seed is viewed as aperient; the seed oil emetic, laxative, purgative, for skin ailments. Root is used in decoction as a mouthwash for bleeding gums and toothache. Otherwise used for eczema, ringworm, and scabies.[16,41] I received a letter from the Medical Research Center of the University of the West Indies shortly after the death of Jamaican singer Robert Marley, ''I just want you to know that this is not because of Bob Marley's illness, why I revealing this . . . my dream was: this old lady came to see me in my sleep with a dish in her hands; she handed the dish to me filled with some nuts. I said to her, 'What were those?' She did not answer. I said to her, 'PHYSIC NUTS.' She said to me, 'This is the cure for cancer.''' This Jamaican dream is rather interesting. Four antitumor compounds, including jatropham and jatrophone, are reported from other species of *Jatropha*.[41] Homeopathically used for cold sweats, colic, collapse, cramps, cyanosis, diarrhea, leg cramps.[33]

Per 100 g, the seed is reported to contain 6.6 g H_2O, 18.2 g protein, 38.0 g fat, 33.5 total carbohydrate, 15.5 g fiber, and 4.5 g ash.[21] Leaves, which show antileukemic activity, contain alpha-amyrin, beta-sitosterol, stigmasterol, and campesterol, 7-keto-beta-sitosterol, stigmast-5-ene-3 beta, 7-alpha-diol, and stigmast-5-ene-3 beta, 7-beta-diol.[42] Leaves contain isovitexin and vitexin. From the drug (nut?) saccharose, raffinose, stachyose, glucose, fructose, galactose, protein, and an oil, largely of oleic- and linoleic-acids;[33] Curcasin, arachidic-, linoleic-, myristic-, oleic-, palmitic-, and stearic-acids.[16]

Toxicity — Poisonous seeds can cause death due to the toxalbumin, curcin. The poisoning is irritant, with acute abdominal pain and nausea about 1/2 hr following ingestion. Diarrhea and nausea continue but are not usually serious. Depression and collapse may occur, especially in children. Two seeds are strong purgative. Four to five seed are said to have caused death, but the roasted seed is said to be nearly innocuous. One frequent and deserved criticism of herbal medication is that one has no way to gauge dosage. It has been concluded in Florida that there are two strains of this species, one with toxic seeds and the other with seeds that are not poisonous. The two strains cannot be distinguished by sight.[14] Bark, fruit, leaf, root, and wood are all reported to contain HCN.[3]

187. *JATROPHA GOSSYPIIFOLIA* L. (EUPHORBIACEAE) — Tua-Tua

Native to Brazil, but widely naturalized as a weed and medicinal plant in the tropics. Cultivated in gardens for ornament, easily raised from seed, flowering and fruiting during the rainy season.[1] Seeds eaten by doves and other birds. The oil from the seeds, though purgative, can be used in lamps. Barbadans use the leaf juice in mending iron pottery.[42] Ether extract of the shoot is active against *Staphylococcus aureus* and *Escherichia coli*. Aqueous extract is insecticidal. Leaves used in the Philippines as a fish poison.

Extracts have long been used as a folk remedy for cancer.[4] The seeds are used like the purging nut, *J. curcas* L. In Colombia, there is a saying that if the red leaf is infused with the top up, it will be emetic; with the top down, purgative. Leaf decoctions used for venereal disease. Latex applied to piles and burns. Used in Curacao for asthma, cold, constipation, diabetes, diarrhea, disinfectant, gall, gargle, hematochezia, laxative, leucorrhea, sores, sore throat, stomachache, ulcers, venereal disease, and wounds. Used in Trinidad for depurative, dyspepsia, piles, rashes, rectitis, sores, and venereal ailment. In Haiti, the plant is used for a tea for dropsy and the leaves are applied to eczema and sores. The roots are used for leprosy in Venezuela, the leaf decoction for prickly heat. Mundas of India use the plant for urogenital complaints. Leaves, or the juice therefrom, are applied to carbuncles, swollen nipples, sores on the tongues of babies, etc. The plant is a popular folk rememdy for cancer, hepatitis, and venereal diseases in Costa Rica. Filipinos apply a cataplasm of fresh leaves to swollen mammae. Gold Coast natives use the seeds and leaves as a purgative. Elsewhere it is used for fever, especially malaria. Mexicans apply the root to dermatitis induced by *Hippomane*.[42] West Indians use the plant as ecbolic and emetic.[32,33,42,127]

The bark contains the alkaloid, jatrophine, and the seeds contain an emetic, purgative oil. The diterpene jatrophone ($C_{20}H_{24}O_3$) is active in lymphocytic leukemia P388 in mice and in cell cultures of human epidermoid carcinoma of the nasopharynx. Jatrophine ($C_{14}H_{20}O_6N$ — yield 0.4%) is similar to quinine. The bark also contains resins, isophytosterol (0.35%), and tannin. The latex (total solid 13.38%) is poisonous and contains 2.5% alcohol-soluble matter. Tender leaves contain a pentose glycoside of cyanidin. Leaves contain apigenin, histamine, isovitexin, tannin, vitexin.[33,42]

Toxicity — Contact can induce severe histamic reactions.[33] Contact with sap can also induce dermatitis.[6] Curcin, jatrophine, and jatrophone are all toxic. Seeds are toxic but less so than those of *J. curcas*.

188. *JUNIPERUS COMMUNIS* L. (CUPRESSACEAE) — Common Juniper

Berries of *J. communis* contain 0.2 to 3.42% volatile oil, the principle flavoring agent in gin. Those who prefer the taste of gin (but the price of vodka) might upgrade cheap vodka by steeping a berry or two. The berries are also used in alcoholic bitters. Extracts and oils are used in most food categories, including alcoholic and nonalcoholic beverages, frozen dairy desserts, candy, baked goods, gelatins, puddings, meat, and meat products. Highest average maximum use level reported for the oils is 0.006% in alcoholic beverages and 0.01% for the extract in alcoholic and nonalcoholic beverages.[29] The oil is a fragrance component in soaps, detergents, creams, lotions, and perfumes (maximum use level 0.8%). Swedes make a "wholesome" beer from cedar. In hot climates, the incised tree yields a gum or varnish.[2] With some hyperbole, Kloss states that juniper berries are excellent as a spray or fumigation in a room in which there has been a patient with an infectious disease, "as it thoroughly destroys all fungi."[44] Deer and moose graze the plant. Sheep readily eat the fruit.[2] Used as a spice in pickled fish, kraut, and gravies.

Berries, wood, and oil are used in folk remedies for cancer, indurations, polyps, swellings, tumors, and warts.[4] Reported to be carminative, cephalic, deobstruent, depurative, diaphoretic, digestive, diuretic, emmenagogue, stimulant, and sudorific, juniper is a folk remedy for arteriosclerosis, arthritis, blenorrhea, bronchitis, cancer, cholecystosis, colic, dropsy, dysentery, dyspepsia, dyspnea, gastroenterosis, gleet, gonorrhea, gout, gravel, hysteria, leucorrhea, lumbago, lungs, nephrosis, pyelitis, renal calculus, rheumatism, rhinitis, scrofula, skin, snakebite, tenesmus, tuberculosis, tumors, worms, and urogenital and venereal diseases.[2,28,32] If juniper really does prevent uric acid build-up, and red wine is really bad for gout, perhaps gout sufferers might try juniper tea (or maybe even a martini) in lieu of red wine! Remember, though, that all alcohol is bad for gout.

Berries contain 0.2 to 3.42% (usually 1 to 2%) volatile oil, depending on the geographic location, altitude, degree of ripeness, and other factors; sugars (glucose and fructose); glucuronic acid; L-ascorbic acid; resin (circa 10%); gallotannins; geijerone (a C_{12} terpenoid); 1,4-dimethyl-3-cyclohexen-1-yl methyl ketone; diterpene acids (myrceocommunic, *cis-* and *trans*-communic, sandaracopimaric, isopimaric, torulosic acids, etc.); beta-elemen-7 alphaol; and others. The volatile oil is composed mainly of monoterpenes (*circa* 58%) which include alpha-pinene, myrcene, and sabinene as the major components, with limonene, *p*-cymene, gamma-terpinene, beta-pinene, alpha-thujene, camphene, and others also present in minor amounts; small amounts of sesquiterpenes; 1,4-cineole; terpinen-4-ol; esters; and others.[29] Dry seeds contain 30.9% protein and 53.9% fat.[21]

Toxicity — The diuretic principle is 4-terpineol or terpinen-4-ol, excessive doses of which may produce kidney irritation. Juniper and extracts should not be used by expectant mothers because they increase intestinal movements and uteral contractions. "This drug is no longer recommended for various kidney disorders by the medical profession. Since much safer and more effective diuretic and carminative drugs exist, the use of juniper in folk medicine should also be abandoned."[37] Symptoms of external poisoning caused by the essential oil on the skin include burning, redness, inflammation with blisters, and swelling. Internally, symptoms are from overdose, pain in or near the kidney, strong diuresis, albuminuria, hematuria, purplish urine, accelerated heartbeat and blood pressure, and, rarely, convulsive apparitions, metrorrhagia, and, more rarely, abortion.[33]

189. *JUNIPERUS SABINA* L. (CUPRESSACEAE) — Sabine, Savin

Ornamental evergreen shrub, source of an essential oil.

Reported to be abortifacient, anthelmintic, diuretic, emmenagogue, hemostat, uterotonic, and vermifuge, sabine is a folk remedy for afterbirth, alopecia, cancer, condylomata, gout, metrorrhagia, neuralgia, paralysis, pediculosis, polyps, rheumatism, and warts.[32,33] Homeopaths use it for arthritis, colic, dysmenorrhea, gonorrhea, gout, leucorrhea, menorrhagia, periosteosis, strangury.

Contains 3 to 5% essential oil with 20% (+)-sabinene and 40% (+)-sabinylacetate, and traces of (+)-alpha-pinene, myrcene, (+)-limonene, γ-terpiene, *p*-cymol, (+)-iso-thujone, (+)-terpineol-(4), (+)-sabinol, (+)-citronellol, (+)-cadinene, and (−)-elemol, even lower levels of α-thujene, camphene, α-terpinen, 1-8-cineole, terpinolene, thujone, geraniol, δ-cadinol, carvacrol, perillalcohol, caprinaldehyde, and diacetyl; Russian samples show beta-pinene, Δ 3-carene, sabinylacetate, and cedrol. Tannin, wax with sabine-acid, juniperin-acid, and thaspia acid, sugar, calcium salts, pinipicrin, gallic acid, sabinene ($C_{20}H_{16}O_6$), and podophyllotoxin.[33]

Toxicity — It should never be used in pregnancy as it causes abortion . . . "It is an energetic poison leading to gastroenteritis, collapse, and death."[2] Only six drops of the oil can be toxic to man. Rubdowns can cause intoxication, details of which are spelled out in *Hager's Handbook*.[33]

To the physician — *Hager's Handbook* suggests prudent doses of narcotics (at the onset of cramps), cardiac stimulants and analeptics at the onset of circulatory collapse, keeping the body warm, and styptics and transfusions for strong bleeding.[33]

190. *JUSTICIA PECTORALIS* Jacq. (ACANTHACEAE) — Bolek-Hena, "Leaves of the Angel of Death"

An attractive red-flowered herb. Emboden lists the variety *stenophylla* among his narcotic plants,[54] usually ingested as a snuff by Indians of the upper Orinoco. Sometimes used to adulterate or synergize other snuffs.

Reported to be aphrodisiac, hallucinogenic, hemostat, nervine, pectoral, resolvent, sedative, bolek-hena is a folk remedy for cachexia, chest ailments, chest colds, cough, fever, flu, nausea, pneumonia, stomachache, and wounds.[32]

Per 100 g, the leaves are reported to contain 44 calories, 85.0 g H_2O, 3.9 g protein, 0.6 g fat, 8.2 g total carbohydrate, 2.8 g fiber, 2.3 g ash, 663 mg Ca, 35 mg P, 7.4 mg Fe, 2670 µg β-carotene equivalent, 0.04 mg thiamine, 0.20 mg riboflavin, 2.5 mg niacin, and 28 mg ascorbic acid.[21] Dried and powdered leaves may contain tryptamines. If *N,N*-dimethyltryptamine is assured for *Justicia,* it will be the first time that tryptamines or any hallucinogenic principle have been reported for the Acanthaceae.[58]

Toxicity — Hallucinogenic, possibly due to tryptamines.[54] Orinoco witch doctors are said to have died from overdoses of the narcotic snuff.

191. *KAEMPFERIA GALANGA* L. (ZINGIBERACEAE) — Maraba

Attractive spice plant used in various culinary applications, more rarely as an hallucinogen. The aromatic essential oils of the roots are used widely in perfumery, as a condiment, and as a folk medicine. Asians employ the rhizomes and leaves as a perfume in cosmetics, hair washes, and powders. They are used, also, to protect the clothing against insects. They are chewed with the betelnuts.[1]

Reported to be carminative, diuretic, expectorant, pectoral, pediculicidal, stimulant, stomachic, tonic, maraba is a folk remedy for chills, cough, dandruff, dyspepsia, fever, headache, inflammation, malaria, odontalgia, ophthalmia, rheumatism, scabies, sore throat, swelling, tumors.[32] The rhizome called "gisol" has been used to treat sore throat. Philippinos use the plant for headache and parturition. Philippinos mix the rhizome with oils as a cicatrizant, applying it to boils and furuncles.[58]

The essential oil contains *n*-pentadecane, ethyl-*p*-methoxycinnamate, ethyl cinnamate, carene, camphene, borneol, and *p*-methoxystyrene.[1]

Toxicity — Narcotic hallucinogen.

192. *KALMIA LATIFOLIA* L. (ERICACEAE) — Mountain Laurel

Beautiful evergreen ornamental, namesake of Laurel, Md. The hard wood has had many uses.

Reported to be alterative, astringent, cardiac, narcotic, poison, sedative, mountain laurel is a folk remedy for diarrhea, dysentery, eruptions, fever, flux, hemorrhage, infection, jaundice, neuralgia, ophthalmia, rash, skin ailments, syphilis, and warts.[2,32,33] Ointments for skin ailments were prepared by stewing the leaves in lard; salves made from the laurel's juices are applied to rheumatism.[2] Homeopaths use the laurel in heart preparations.

Per 100 g, the toxic forage is reported to contain, on a dry basis, 8.7 g protein, 9.5 g fat, 78.2 g total carbohydrate, 14.7 g fiber, 3.6 g ash, 960 mg Ca, 110 mg P, and 5.0 mg Fe.[21] Poisonous principles may be andromedotoxin (acetyl andromedol), arbutin, tannic acid.[2] Leaves contain 2.7% of the dihydrochalkone phloridzine, phlorizine ($C_{21}H_{24}O_{10}$), asebotin ($C_{22}H_{26}O_{10}$), catechin-tannin, rhodotoxin, grayanotoxin-I (also in the pollen), pectin, saccharose, invertin, emulsin.[33]

Toxicity — Narcotic[2] once used by Amerindians to commit suicide. "Whiskey is the best antidote." Experimental administration (injection) to lower animals produces salivation, lacrymation, emesis, convulsions, and, later, paralysis of the extremities and labored respiration.[2] Leaves, shoots, and fruits are dangerous to cattle and birds which feed on the plant, as honey made from the flowers may be poisonous to man.[2]

To the physician — Hardin and Arena suggest gastric lavage or emesis, activated charcoal, atropine, and hypotensive drugs.[34]

193. *LABURNUM ANAGYROIDES* Medic. (FABACEAE) — Golden Chain Tree

Attractive ornamental shrub or tree. Reported to be diuretic and emetic, golden chain tree is a folk remedy for asthma and nausea.[33] Used in homeopathy as a nerve tonic, to correct depression, vertigo, and cramps.[33]

Flowers and seed contain cytisine.

Toxicity — After yew (Maryland's number-one poisonous plant), it is considered the "most poisonous tree in Britain". Loss of human life is reported. Cytisine is excreted in the milk.[14] Cytisine causes excitement, stomach and intestinal irritation with nausea, severe vomiting, and diarrhea, irregular pulse, convulsions, coma, and death in overdoses.

To the physician — Hardin and Arena suggest emesis or gastric lavage, activated charcoal, artificial respiration, oxygen, and symptomatic treatment.[34] *Hager's Handbook* suggests treating as for nicotine poisoning, oral or rectal uzara (no opium), etc.[33]

194. *LACHNANTHES TINCTORIA* Ell. (HAEMADORACEAE) — Redroot

Formerly source of a red dye. Apparently, the plant was used as a hallucinogen by the Seminoles, to cause brilliancy and fluency of speech.[2] Millspaugh[19] said, "The root was esteemed as an invigorating tonic by the Aborigines, especially the Seminoles, in whom it is said to cause brilliancy and fearless expression of the eye and countenance, a boldness of fluency of speech, and other symptoms of heroic being, with, of course, the natural opposite after-effects."

Reported to be cardiotonic, hypnotic, poison, and stimulant, redroot is a folk remedy for cerebrosis, cough, fever, laryngitis, neck ailments, pneumonia, rheumatism, typhoid, typhus.[32,33] The root contains 4-hydroxy-3-methoxy-5-phenyl-1,8-naphthaline-anhydride ($C_{18}H_{12}O_5$) as well as di-, tri-, and tetra-methoxyphenylnapthalides, lachnanthoside, lachnanthofluoren, and lachnanthocarpone ($C_{19}H_{12}O_3$), which also occurs in the fruit, with alpha-carotene, tristearin, and chelidonic acid.[33]

Toxicity — Grieve reiterates the interesting story that white pigs die but not black pigs, from eating the redroot, the bones turning pink and the hoofs falling off.[21] Narcotic,[2] the unidentified compound produces a peculiar form of cerebral stimulation or narcosis. A few drugs of the tincture may cause mental exhilaration, followed by ill-humor, vertigo, and headache (as Duke suggests, the usual hangover that follows whatever kind of "intoxication"[186]).

195. *LACTUCA VIROSA* L. (ASTERACEAE) — Bitter Lettuce, Wild Lettuce

Once used to falsify opium, there is some evidence that the latex possesses some opiate qualities, at least as a sedative. The seed oil has been used as a substitute for wheatgerm oil. Egyptians use the seed oil in cooking.[2] Externally, the latex is used as a collyrium.[33] Latex is the source of the drug Lactucarium, used medicinally, being a mild hypnotic, mild sedative, expectorant, anodyne, and diuretic. Drug is used as an anodyne and hypnotic when opium cannot be given. Hippies apply the appellation ''head lettuce'' because it is smoked by ''heads''. Drug is obtained by cutting the stem in sections and collecting the latex. Drug is usually in angular fragments or quarters, curved on one side, indicating removal from a cup or saucer in which it was collected and dried. Externally, the pieces are dull reddish or grayish-brown, but internally, light brown, opaque, waxy, and somewhat porous with a bitter taste and a distinct opium-like odor.

Reported to be anodyne, antitussive, diaphoretic, diuretic, expectorant, hypnotic, lactogogue, laxative, narcotic, POISON, sedative, and soporific, bitter lettuce is a folk remedy for asthma, cancer, colic, cough, dropsy, gout, and jaundice. The seed oil has been used for arteriosclerosis.[33] The latex dissolved in wine is said to be a good anodyne. Homeopaths use the herb for coughs, laryngitis, enlarged liver, pertussis, strangury, and tracheitis.[33] Used in home remedies for asthma, cancer, coughs, dropsy, gout, and jaundice; frequently used in syrups to allay irritable coughs.

The seed oil contains 27% oleic-, 58% linoleic-, 12 to 15% palmitic-, and stearic-acids, beta-sitosterol, alpha-tocopherol, and squalene; also, contains caoutchouc and a mydriatic alkaloid similar to hyoscyamine.[1] Lactucerin is considered to be a mixture of acetates of alpha- and beta-lactucerol. Alpha-lactucerol is identical with taraxasterol.[1] *Hager's Handbook* adds beta-amyrin, germanicol (isolupeol), and, without questionmark, hyoscyamine, mannitol, up to 1% oxalic-, citric-, and malic-acids, sugar, resins, and proteins in the latex.[33] Dried milky juice contains two bitter principles: lactucopicrin and lactucin, along with the very bitter lactucic acid and lactuceral.

196. *LAGOCHILUS INEBRIANS* Bunge (LAMIACEAE) — Intoxicating Mint

Branches, hung to dry during the Russian winter, become increasingly more fragrant. Dried leaves are boiled with sugar and honey to make a tea.[54]

Reported to be antispasmodic, hemostat, hypotensive, and sedative, the intoxicating mint is a folk remedy for allergy, glaucoma, nervousness, skin disorders.[33,58] The tincture is official in Russia, and used for hypertensive states and nervous disorders,[54] said to reduce permeability of blood vessels and to aid blood coagulation.[58]

Contains 1 to 3% lagochilin, a polyhydroxyalcohol $C_{24}H_{44}O_6$ (sedative in humans at doses of *circa* 30 mg),[54] 1 to 3% lagochilin tetraacetate, 0.2% stachydrine, 0.03 to 0.2% essential oil, carotene, vitamin C and K, 2 to 4% tannin, 3 to 9% resin, 0.7 to 3% organic acids, 1 to 2.7% sugar, 2 to 2.7% calcium salts, and 0.13 to 0.4% iron salts, but no alkaloids.[33]

Toxicity — Classed as a narcotic tranquilizer and/or hallucinogen (?).[54]

197. *LANTANA CAMARA* L. (VERBENACEAE) — Lantana

Natives of tropical countries are said to eat the ripe blue-black berries, but human fatalities have been attributed to ingestion of the green berries.[56] Though considered ornamental by some, it is mainly a weed of plantation crops and pasture, infesting nearly 4 million hectare in Australia and 160,000 in Hawaii.[157]

The plants are used in folk remedies for cancer and tumors.[4] Reported to be alexiteric, antibiotic, carminative, depurative, diaphoretic, digestive, emmenagogue, expectorant, hemostat, nervine, pectoral, piscicide, sedative, stimulant, stomachic, sudorific, tonic, and vulnerary, lantana is a folk remedy for anemia, asthma, catarrh, chicken pox, colds, consumption, diarrhea, dysentery, dysmenorrhea, dyspepsia, eczema, fever, fistula, flu, hemorrhage, hypertension, inflammation, itch, jaundice, leprosy, malaria, measles, mumps, neurodermatitis, parotiditis, parturition, rheumatism, snakebite, sore, spasm, stomachache, stomatitis, tetanus, toothache, tumor, wounds, and yellow fever.[32,33,42] Venezuelans poultice the leaves on sores. Bahamians apply the leaves to chicken pox and measles. Surinamese use the plant for fevers; Panamanians for colds; Mexicans for rheumatism; Costa Ricans for asthma, high blood pressure; Venezuelans for diarrhea and dysmenorrhea; Jamaicans for colds and fevers; and Salvadorans for venereal diseases.[42] South Chinese use the plant decoction externally for leprosy and scabies. Indonesians consider the herb emmenagogue, stimulant, and vermifuge. Malayans poultice the antiseptic leaves to cuts, rheumatism, and ulcers. Leaves are considered antispasmodic, diaphoretic, and stomachic. Wilted leaves are said to relieve asthmatic breathing, fresh leaves to prevent fever.[16]

Leaves contain 0.2 to 0.7% lanthanine $C_{35}H_{52}O_5$ (lantadene A) and 0.2% lantadene B ($C_{35}H_{52}O_5$), icterogenin, 0.05 to 0.2% essential oil with citral and other sesquiterpenes, up to 80% caryophyllene, and 10 to 12% phellandrene, dipentene, terpineol, geraniol, linalool, cineole, eugenol, furfural, and phellandrene; tannins, resins, dyestuffs, reduced sugars, and 1.7% lantadene, methyl-3-oxo-ursolate, lantonol acid ($C_{30}H_{46}O_4$), and lantanic acid. Flowers contain anthocyanin, carotene, 0.07% essential oil, the stembark resin, the rootbark tannin.[33] Dry seeds contain 35.1% protein and 48.0% fat.[21] Lantanine depresses the circulation and lowers the temperature.[16]

Toxicity — Human poisonings, including a death in Tampa, Fla., are attributed to ingestion of green fruits. Symptoms may appear in 2.5 to 5 hr; vomiting, lethargy, dilated pupils, then weakness, labored, slow respiration, occasional diarrhea, circulatory disturbance, circulatory collapse, and death. Late stages are suggestive of atropine poisoning.

To the physician — Gastric lavage.[56]

198. *LARREA TRIDENTATA* (DC.) Cov. (ZYGOPHYLLACEAE) — Creosotebush, Chaparral

Important as the dominant vegetation on large areas, serving as an efficient soil protector and stabilizer. Unfortunately, it is nearly worthless for forage. Material soaked in 0.1 normal solution of NaOH for 2 hr, then washed to remove the resin, was quite palatable, as is or ensiled, with crude protein running around 18%. The reddish-brown resin deposited by scale insects was used by the Pima Indians for mending pottery, cementing arrowheads, and coating baskets. Bark contains a reddish dyestuff. Wood favored for smoking game. Contains the antibiotic nordihydroguaiaretic acid (NDGA) active against skin bacteria. Tierra[28] classes chaparral as one of the best herbal antibiotics, being used against bacteria, viruses, and other parasites — internally and externally. NDCA is a powerful antioxidant used to preserve fats and oils, once important in the baking industry, approved as a food additive both nationally and internationally.[63]

The leaf and stem tea is used in folk remedies for leukemia and cancers of the kidney, liver, lung, and stomach.[4] Tierra[28] notes that Amerindians used it for cancer. Reported to be antiseptic, diuretic, and emetic, creosotebush is a folk remedy for arthritis, bruises, cancer, chafing, cold, cramps, dandruff, diarrhea, dysmenorrhea, dyspepsia, dysuria, eczema, gastroenteritis, hematochezia, inflammation, influenza, itch, nephrosis, rheumatism, scabies, snakebite, sores, tuberculosis, urethritis, venereal diseases, and wounds.[28,32,37,47] Dry powdered leaves are applied to sores. Mentioned as an unproved method of cancer treatment is the so-called chaparral tea, obtained from steeping leaves and stems in hot water.[11] Tierra[28] calls NDGA an antitumor agent. Tyler[37] quotes data to suggest that Larrea might be used to purge the system of hallucinogens and prevent recurrences.

Per 100 g, the forage is reported to contain 47.9 g H_2O, 6.8 g protein, 1.5 g fat, 38.7 g total carbohydrate, 9.5 g fiber, 5.2 g ash, 1310 mg Ca, 50 mg Mg, 140 mg P, 21 mg Fe, 720 mg K, 1.8 mg β-carotene equivalent, 310 mg arginine, 570 mg aspartic acid, 100 mg cystine, 1360 mg glutamic acid, 420 mg glycine, 470 mg isoleucine, 780 mg leucine, 420 mg phenylalanine, 100 mg tryptophane, 310 mg tyrosine, and 420 mg valine. Fresh material, at an early bloom stage (58.7% DN), contained, per 100 g, 790 mg Ca, 90 mg Cl, 58 mg Fe, 60 mg Mg, 80 mg P, 510 mg K, 40 mg Na, and 160 mg S.[187] Dry seeds contain 30% protein, 11.7% fat, and 5.8% ash. *Hager's Handbook* reports a resin content of 20%.[33]

Toxicity — In the "Range Plant Handbook",[155] it is reported that sheep are poisoned by this plant when compelled to eat it during years of forage scarcity, the greatest mortality among pregnant ewes. May cause contact dermatitis, NDGA causing allergic contact dermatitis.[6] NDGA induced lesions in the mesenteric lymph nodes and kidneys, so the compound was removed from the FDA's GRAS list in 1970.[37]

199. *LATUA PUBIFLORA* (Gris.) Phil. (SOLANACEAE) — Latua

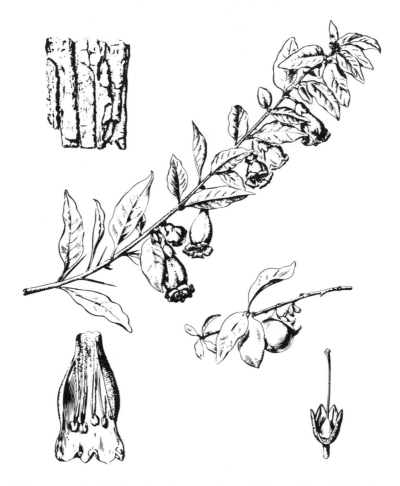

Beautiful but narcotic shrub endemic to southern Chile, used mostly by shamans in their curing rituals in the evening, taking an infusion of the green leaves and bark. Becomes a weed along roadsides and in open places, spreading easily by adventitious branches from underground parts, thereby thwarting efforts to eliminate it by cutting.[188] Once used as a fish poison.

Said to be aphrodisiac and employed as an ingredient in love potions.[188]

The psychoactive principles are atropine and scopolamine.[54] Leaves contain 0.18% hyoscyamine and 0.08% scopolamine.[58] Leaves may contain 0.07 to 0.185% total alkaloids, atropine constituting 70 to 86% of the total, scopolamine the remainder. Seeds contain 0.08% total alkaloids, 86% of which is atropine, 14% scopolamine; stems contain 0.24 to 0.50% alkaloids, with 87 to 92% atropine.[188]

Toxicity — Narcotic hallucinogen,[54] causing delirium. Aborigines believe fresh fruits are highly toxic.[58] Called a virulently toxic agent capable of producing delirium, hallucinations, and, on occasion, permanent insanity.[58]

To the physician — *Solanum nigrum* and/or *oxalis* might be suggested as an antidote.[188]

200. *LAURUS NOBILIS* L. (LAURACEAE) — Bay, Grecian Laurel, Green Bay

In biblical times, the bay was symbolic of wealth and wickedness.[38] In the ancient Olympic games the victorious contestant was awarded a chaplet of bay leaves, placed on his brow. The Roman gold coin of 342 B.C. has a laurel wreath modeled on its surface. The evergreen leaves, when broken, emit a sweet scent and furnish an extract used by the Orientals in making perfumed oil. Dried bay leaves are used to flavor meats, fish, poultry, vegetables, soups, and stews; also, as an ingredient in pickling spices and vinegars. They are particularly popular in French dishes. Leaves have served as a tea substitute. An essential oil, distilled from the leaves, is used in perfumery and for flavoring food products, such as baked goods, confectionary, meats, sausages, and canned soups. The oil can be measured more precisely and provides more uniform results. The fat from the fruits has served in soapmaking and veterinary medicine. The wood, resembling walnut, can be used for cabinetry.[38]

Regarded as apertif, carminative, diuretic, emetic, emmenagogue, narcotic, nervine, stimulant, stomachic, and sudorific, bay has found its way into folk remedies for amenorrhea, colic, condylomata, hysteria, impostumes, polyps, scleroses, spasms, and wens.[32,38] Methyl eugenol, which constitutes 4% of bay oil, is narcotic and sedative in mice, producing sedation at low doses and reversible narcosis at higher doses. The essential oil has bactericidal and fungicidal properties. An ointment of unguent derived from the plant is said to remedy sclerosis of the spleen and liver and tumors of the uterus, spleen, parotid, testicles, liver, and stomach. The fruit, prepared in various manners, is said to help uterine fibroids, tuberosities of the face, scirrhus and scleroma of the uterus, scirrhus of the liver, indurations of the joints, spleen, and liver, internal tumors, wens, and tumors of the eye. Leaves and fruits, said to possess aromatic, stimulant, and narcotic properties, were once employed for amenorrhea, flatulent colic, cough, and hysteria. In small doses, leaves are diaphoretic; in large doses, emetic. Bay oil sometimes used as a liniment or anodyne for earache. In Lebanon the leaves and berries are extracted to a carminative liver and stomach tonic, tightly corked, and steeped in brandy in the sun for several days. The residue, after subsequent distillation, is used as a liniment for rheumatism and sprains, the distillate as an emmenagogue. Lebanese mountaineers are said to use raw berries to induce abortion. Berries macerated in flour were poulticed onto dislocations.[38] Pech and Bruneton[189] reported on the alkaloidal constituents, the leaves containing mostly reticuline, with some boldine, *N*-methylactinodaphnine, (+)-isodomesticine, (+) neolitsine, actinodaphnine, nor-isodomesticine, launobine, nandigerine, and cryptodorine; actinodaphnine constituted about half of the stem alkaloids, with some reticuline and launobine; the inflorescence contained mostly actinodaphnine, with some reticuline, launobine, and nandigerine. 3,4-Dimethoxyallylbenzene produces sedation in mice at low doses; a reversible narcosis at higher doses. It prevented the death of mice treated with lethal convulsant doses of strychnine. It may have relatively specific control nervous or myoneural effects, perhaps, suggesting a clinical potential.

Per 100 g, the leaves are reported to contain 188 calories, 45.2 g H_2O, 4.2 g protein, 1.2 g fat, 47.1 g total carbohydrate, 4.6 g fiber, 2.3 g ash, 187 mg Ca, 70 mg P, 5.3 mg Fe, 1050 µg β-carotene equivalent, 0.04 mg thiamine, 0.21 mg riboflavin, 1.7 mg niacin, and 54 mg ascorbic acid. Laurel leaf oil (yield *circa* 0.5%) has been reported to contain 12% terpenes, 45% cineole, 18% free alcohols, 13% esters (mainly acetates), 0.53% eugenol, 1.1% eugenol acetate, 3% methyl eugenol, and 3 to 4% sesquiterpenes.[7] Pinene, phellandrene, cineole, linalool, terpineol, geraniol are also reported.[7] El-Feraly and Benigni,[190] studying the sesquiterpene lactones of the leaves, identified the major one as costunolide, with artemorin, regnosin, santamarine, and verlotorin present in smaller quantities. The berry (30% pericarp, 70% seed) yields 20 to 34% of an aromatic fat, the fatty acids of which are 30 to 35% lauric, 10 to 11% palmitic, 33 to 40% oleic, and 18 to 32% linoleic.[1]

Toxicity — Bay leaves used as a culinary herb have caused stomatitis and cheilitis. Laurel oil in toothpastes and hat bands has caused dermatitis. *Science News* reports that the most active cockroach-repelling compound in bay leaves is cineole, a two-ringed structure also known as eucalyptol in certain cough drop formulations. Reporting on the work of C. E. Meloan, the *Science News* article[191] goes on to state that a repellant based on two compounds that we eat (cineole and *trans*-2-noneal from cucumber skins) has obvious advantages. It avoids the problem of spraying synthetics around food. It may also prove a more long-lasting solution than pesticides, which invoke resistant cockroach strains.[191] (I don't see why natural pesticides should not also invoke the evolution of strains resistant to the natural pesticide as well.)

201. *LAVANDULA ANGUSTIFOLIA* Mill. (LAMIACEAE) — Lavender, True or Common Lavender

Lavender is primarily grown for the oil made from the flowering stalks, that from the flowers being the best quality; used in perfumes, soaps, aromatic vinegars, herbal tobaccos, and scented sachets. Under the provocative title "Light on the Shroud?," Gorkin[192] mentions the use of oil of lavender to make a reinforced plastic, durable for thousands of years. Soak linen in oil of lavender that contains asphalt; let dry in the sun. To the artisans of ancient Egypt it was one trick to wrapping mummies. Dried flowers are also used in sachets. Fresh flowers and stalks used as flavorings in beverages and jellies. The English use bouquets of the fresh flowers in their homes. The oil is also used to give fruity flavors to beverage, baked goods, confectionery, dairy desserts, gelatins, and puddings, usually at levels below 45 ppm.

Leung states that lavender, a folk remedy for acne, colic, flatulence, giddiness, migraine, nausea, neuralgia, nervous headache, pimples, rheumatism, sores, spasms, sprains, toothache, and vomiting, is used as antispasmodic, carminative, diuretic, sedative, stimulant, stomachic, and tonic.[29,32] Essential oil is carminative and stimulant, and used as an insect repellent. Flowers have been used for indurations or tumors of the breast, liver, sinews, and spleen and other types of cancers.[4] Flowering tips, considered diuretic, have been used for colic and flatulence. Seeds once recommended for worms. The inhalation of lavender was once recommended for colic, faintness, giddiness, nervous palpitations, poor appetite, and spasms. A drop of lavender vinegar in a hot footbath was once recommended for fatigue. Outwardly applied, oil of lavender has been suggested for neuralgia, rheumatism, sprains, and toothache. A few drops rubbed on the forehead is supposed to cure headache as is a tea of the tops.

Lavender contains 0.5 to 1.5% volatile oil, tannins, coumarins (coumarin, umbelliferone, and herniarin), flavonoids (e.g., luteolin), ursolic acid.[29] The principal ingredients of lavender oil are linalyl acetate (30 to 60% in French, 8 to 18% in English, 25% in Kashmir), linalool, geraniol and its esters, lavandulol, nerol, cineole, caryophyllene, coumarin, limonene, beta-ocimene, furfural, ethyl amyl ketone, thujone, and pinocamphone. Leaves contain 0.7% ursolic acid. Dry seeds of *Lavandula* species contain 23.3 to 29.4% protein, 20.0 to 21.8% fat.[21] Several articles on lavender chemistry are reviewed by Lawrence.[193]

Toxicity — In large doses lavender oil is a narcotic poison that can cause death by convulsion.[2] The oil can cause dermatitis.

202. *LAWSONIA INERMIS* L. (LYTHRACEAE) — Henna, Egyptian Privet, Mignonette

Leaves provide an important cosmetic dye. In India and Pakistan henna is widely used by both men and women for coloring nails, fingers, hands, and hair. Hair is dyed a brownish-chestnut shade which turns black on conjunction with indigo. An infusion of dried leaves with a little lime juice is used to dye the hair. Henna leaves dye fingers, nails, hands, and feet a dull orange. A deep red color may be obtained when henna is mixed with catechu. Infusion of leaves also used for dyeing cotton fabrics a light reddish-brown. Wool and silk may also be dyed by henna. Some Egyptian mummies were wrapped in henna-dyed cloth.[8] Leaves also used in manufacture of perfumed oils and as a tanning agent. Dried leaves are the source of a green powder used in cosmetics.[8] Flowers give an essential oil (mehndi oil) long used in Indian perfumery. Plants grown as hedge plants throughout India. Wood used to make tool handles, tent pegs, and other small articles. Gallic acid from the leaves inhibits *Streptococcus aureus* slightly. Shihata et al.[194] "proved that Lawsonia inermis extracts possessed cardio-inhibitory, hypotensive, intestinal antispasmodic, and uterine sedative effects. Moreover the extracts significantly increased the body weight gain of growing mice and ruminal gas production in sheep."

Fruit oil is a folk remedy for indurations of the liver and diaphragm. The ointment made from the oil is said to remedy tumors of the mouth. Poultices of the leaf are said to remedy various types of tumors.[4] Lawsone shows anticancer activity against Sarcoma 180 in mice and Walker 256 carcinosarcoma.[29] In Malaysia, the leaf decoction is used after childbirth, and for beri-beri, rheumatism, skin disorders, stomach disorders, and venereal disease. In Indonesia, leaves are used for jaundice, leprosy, and scurfy affections.[16] Mixed with the poisonous *Plumbago*, it is said to be abortifacient. Henna powder sprinkled onto rheumatic pains. Used for herpes in Java. Elsewhere, the plant is used for amebiasis and headache.[29] Cambodians drink a root decoction as a diuretic. Bark is given in jaundice and for enlargement of spleen, and an an alterative in skin diseases and leprosy. In Arabic medicine, the bark decoction is used for jaundice and nervous symptoms.[8] Henna leaves sued as prophylactic against skin disease. The tea made from the leaves is said to prevent obesity.[16] Powdered seed said to be a cerebral stimulant. Decoction of leaves used as astringent gargle in sore throat. Oil and essence rubbed over body to keep body cool. Flowers are refrigerant and soporific. Fruit said to be emmenagogue. Leaves said to act as an antiperspirant and sedative. Leaves are poulticed onto gatherings under the nails, resulting from the loss of the nail.[27]

Seeds contain 10.6% water, 5.0% protein, 10 to 11% fatty oil, 33.6% carbohydrate, 33.6% fiber, and 4.8% ash. The oil contains 1.7% behenic, 9.6% arachidic, 15.8% stearic, 9.1% palmitic, 34.7% oleic, and 29.3% linoleic acids. Mehndi oil is high in alpha- and beta-ionones. Air-dried leaf powder contains 9.0% water, 14.8% ash, and 10.2% tannin. Gallic acid has been isolated from the leaves. The principle coloring matter is lawsone (2-hydroxy 1,4-napthaquinone), which has been used as a topical sunscreen. Also, reported are 1,4-napthoquinone, tannin, laxanthone-I and laxanthone-II, and lacoumarin. Mahmoud et al.[195] isolated from the leaves luteolin, luteolin-7-*O*-glucoside, acacetin-7-*O*-glucoside, glucoside of beta-sitosterol, laxanthone-I and -II, and lawsone. Chakrabartty et al.[196] isolated dihydroxy-lupene and lupane from the bark.

Toxicity — Henna dust can irritate the skin and cause contact dermatitis. When used as a dye for eyebrows and eyelashes, there is a risk of injury to the eyes, perhaps due to the astringent tannin.[6] The LD_{50} of alcoholic extract of henna 1P in mice is 665 mg/kg.[194] Henna leaves exhibit antifertility activity in female rats.[29] Approved for use as a color additive exempt from certification, to be used in cosmetics (hair) only (§ 73.2190).[29]

203. *LEDUM PALUSTRE* L. (ERICACEAE) — Marsh Tea, Marsh Cistus, Wild Rosemary

Germans once added the leaves to their beer to make it more intoxicating. Strong decoctions of some *Ledum* species are said to kill lice.

Reported to be used like Labrador tea, *L. latifolium*. Considered abortifacient, diaphoretic, diuretic, emetic, expectorant, lactagogue, narcotic, pectoral, and vulnerary, marsh tea is a folk remedy for arthrosis, bronchitis, bugbites, chest ailments, cold, cough, dysentery, dyspepsia, eruptions, fever, gout, inflammations, itch, leprosy, malignancies, rheumatism, sore throat, and whitlows.[32,33] Koreans use the leaves to treat female disorders.[16] Homeopaths use for acne, gout, intercostal neuralgia, rheumatism, skin ailments.

Leaves contain 0.3 to 2.5% volatile oil, including *Ledum* camphor (ledol), palustrol, a stearopten, with valeric and volatile acids, ericolin, and ericinol. The tannin is called led-itannic acid,[2] but *Hager's Handbook* lists catechin tannin, with quercetin, hyperoside, a flavonal, arbutin (the ericolin of Reference 2), tartaric, malic-, and citric-acid.

Toxicity — Leaves said to have narcotic properties.[2] Overdoses may cause violent headache and symptoms of intoxication.

204. *LEONOTIS LEONURUS* R. Br. (LAMIACEAE) — Dagga, Lion's Tail

Grown as an ornamental. Formerly used as a drug. Powdered leaf has been chewed for its intoxicating effects. Zulu sprinkle a leaf decoction around as a snake repellent.[3] Leaves yield a dark green residue smoked as a euphoriant by the Hottentots of Africa.[54] Eating or drinking the root is also said to make them giddy and/or sleepy, but not to madden them.[53] Europeans in Africa sometimes take the herb in the belief it is slimming.

Reported to be emmenagogue, purgative, and vermifuge, dagga is a folk remedy for asthma, bilious ailment, bronchitis, chest ailments, cough, epilepsy, flu, headache, herpes, indigestion, jaundice, ophthalmia, piles, scabies, snakebite, sores, syphilis, tuberculosis.[3,32] Decoction used for asthma, cardiac conditions, cold, cough, epilepsy, leprosy, snakebite, and tapeworm.[32,54] In Mauritius the leaf decoction is used as emmenagogue and purgative. Early South African colonists used the decoction to treat skin diseases, perhaps even leprosy. Twigs are added to baths for those suffering itch, cramps, and skin and muscle ailments. The tincture is also used for hemorrhoids. The infusion is also used for cholecystosis and hepatosis. Fresh juice is applied to sores, including those of syphilis. Europeans suffering partial paralysis smoke the herb.

Leaves contain two phenolic substances, $C_9H_{10}O_5$ and $C_8H_{10}O_5$, a resin, and a reddish oil, marrubin (0.4%), and two other diterpenes, $C_{20}H_{28}O_5$ and $C_{20}H_{28}O_3$.

Toxicity — Listed as a mild narcotic hallucinogen.[54] Toxicity and hallucinogenic properties seem rather insignificant.[3] Although reputed to be narcotic, the plant is quite repulsive, giving off a nauseous, volatile vapor when smoked. One scientist took a 10-g dose with no ill effects.

277

205. *LEONURUS CARDIACA* L. (LAMIACEAE) — Motherwort

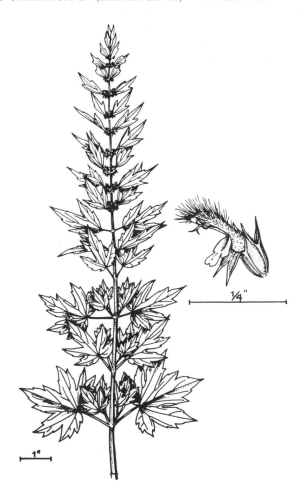

Medicinal plant. Source of a dark green dye, used along the Danube River.

Regarded as analgesic, antispasmodic, astringent, diaphoretic, diuretic, emmenagogue, expectorant, febrifuge, hypotensive, laxative, nervine, sedative, tonic, and vermifugal, motherwort is especially regarded for female illnesses, e.g., amenorrhea, dysmenorrhea, epilepsy, hysteria, neurasthenia, palpitations, urinary cramps, and weakness. It is said to brace up the uterine membrane (leonurine is uterotonic). Old herbalists recommended motherwort for gladdening and strengthening the heart. Useful in asthma, cholecystitis, convulsion, debility, delirium, insomnia, neuritis, neuralgia, palpitations, rabies, rheumatism, sciatica, spasms, spinal diseases, stomachache, and recovery from fevers where other tonics are inadmissable.[30,32] In Russia, the plant is believed effective in cardiosclerosis, epilepsy, goiter, hypertension, rabies, and tuberculosis.[30] Taken as an infusion for asthma and heart palpitations.[18] Homeopathically suggested for amenorrhea, anemia, angina pectoris, dysentery, dysmenorrhea, hemorrhage, menopause, meteorism.[30,33]

Dry plant contains 0.35% alkaloids, 2.14% tannins (pyrogallol and catechins), 0.05% essential oils, 2.8% carbohydrates, saponins, and traces of vitamins A and C (10.042%).[1] Stachydrine is identified among the alkaloids. Leonurin ($C_{14}H_{21}N_3O_5$) resins, saponins, and waxes are also reported. According to *Hager's Handbook*, the herb also contains choline, malic-, citric-, ursolic-, and oleic-acids, and in the seeds palmitic-, oleic-, stearic-, linoleic-, and linolenic-acids, as well as vitamin E.[33] Stachyose occurs in the root. Rutin has been reported in var. *villosus*. Leaves and flowers contain caffeic acid 4-rutinoside.[197] Marrubin, genkwanin, and the C_{15}-iridoid glucoside leonurid are also reported.

Toxicity — Classified by the FDA as an Herb of Undefined Safety: leaves can cause contact dermatitis. The fragrant lemon-scented oil from the plant can cause photosensitization. Many who were pushing herb teas were talking as though they were tannin free. Many mints do contain tannins. Three bufenolides l ($C_{36}H_{56}O_{12}$, $C_{25}H_{51}O_{12}$, and $C_{36}H_{54}O_{12}$) are also reported.[33]

206. *LIGUSTRUM VULGARE* L. (OLEACEAE) — Privet

This is the common deciduous privet, so widely planted in the U.S. as a hedge. Several cultivars have been developed. The fruits, though poisonous, are used to tint wines.[33]

The plant is used in folk remedies for tumors of the uvula.[4] Reported to be astringent, detersive, laxative, and vulnerary, privet is a folk remedy for pharyngitis, stomatitis, and tumors.[32,33] The leaf and flower have been used in treating mouth and throat infections.[3]

Fruits contain ligustrin, ligustrone, syringin, syringopicrin.[56] Leaves contain syringin $C_{17}H_{24}O_9 \cdot H_2O$, malvidin-3-glucoside, 6 to 10% tannin. Bark contains ligustrone, syringopicrin, syringic acid, behenic acid, *circa* 7% tannin, resin, mannite.

Toxicity — Horses are said to have died after eating the foliage. Flowering shrubs are respiratory irritants. Fruits and leaves are gastroenteric irritants, inducing vomiting, diarrhea, pain, drowsiness, loss of coordination, weak pulse, low temperature, convulsions, twitching, even death (in European children).

To the physician — Gastric lavage or emesis, correct for dehydration, symptomatic and supportive treatment.[56]

207. *LOBELIA INFLATA* L. (CAMPANULACEAE) — Indian Tobacco, Asthma Weed

One Avante Garde publication "Herbal Highs"[51] notes that it has been substituted for tobacco by people trying to kick the nicotine habit. It has a mildly euphoric marijuana-like quality while conferring to the mind a great sense of clarity. Taken as a tea, its effect is even more pronounced, acting simultaneously as both a relaxant and a stimulant. Three spoons of the stems are simmered in a liter of water.[31] Feelings of mental clarity, happiness, and well-being are supposed to result from drinking the tea or taking lobelia capsule.[37] If a bit of the leaf is chewed, there is a delayed stinging reaction like that from chewing tobacco leaves.[1] It has been used to treat angina pectoris, asthma, bronchitis, convulsions, diptheria, epilepsy, erysipelas, hysteria, narcosis, ophthalmia, pneumonia, spasm, tetanus, tonsilitis, and whooping cough. It is convulsant, diaphoretic, expectorant, nauseant, and sedative.[32] Lobeline is used for respiratory failures resulting from anesthesia or narcotic overdose. American Indians used the powdered leaves to treat dysentery. The minute seeds, which impart a greasy stain to paper, contain a higher percentage of the alkaloid lobeline than the rest of the plant. This alkaloid occurs in certain commercial preparations used to stop the tobacco smoking habit. Once again, there is scientific reason for our ancestors prescribing

Indian tobacco to those who wish to give up true tobacco. The FDA has allowed the sale of pills containing lobeline as a smoking deterrent ("Washington Post", January 11, 1982).

Poultices and teas of the whole plant are folk remedies for breast cancer.[4] The embrocation prepared from the plant is a folk remedy for cancer, while the powdered plant, with elm bark, is said to cure felons.[4] Seeds are used as remedy for asthma. Seeds in vinegar are emetic; used for such skin ailments as poison ivy, erysipelas, salt-rheum, and in spasmodic afflictions such as chorea, convulsions, cramps, epilepsy, hysteria, spasms, and tetanus. A few drops of the tincture is said to relieve earache.[28] Oil of lobelia said to be valuable in tetanus. Useful as an expectorant in bronchitis. Has been recommended in asthma, bronchitis, croup, hepatitis, hernia, hydrophobia, meningitis, nephritis, neuralgia, periostitis, peritonitis, phrenitis, pleurisy, pneumonia, spasms, tetanus, and whooping cough. The herb is considered antiasthmatic, antispasmodic, emetic, expectorant, nervine, sialogogue, and sudorific.[27] Homeopaths recommend the tincture for alcoholism, alopecia, amenorrhea, angina pectoris, asthma, cardialgia, cough, croup, deafness, debility, diarrhea, dysmenorrhea, dyspepsia, emphysema, faintness, gallstones, gastralgia, hangover, hayfever, heart ailments, hemorrhoidal discharge, hysteria, meningeal headache, morning sickness, morphinism, nausea, palpitations, pleurisy, pregnancy, psoriasis, seborrhea, shoulder pain, urethral stricture, vaginitis, wens, and whooping cough.[30]

Contains several dangerous alkaloids, from 0.36 to 2.25% lobeline, 8-10-diethyl lobelidiol, 8-ethyl norlobelol-I, isolobinanidine, isolobinine, lelobanidines I, II, lobelanidine, lobelanine, lobeline, lobinalidine, lobinaline, lobinanidine, lobinine, 8-methyl-10-ethyl-lobelidiol, 8-methyl-10-phenyl-lobelidiol, norlelobanidine, norlobelanidine, and norlobelanine. HCN tests were negative. Caoutchouc, resin, fat, inflatin, and wax have also been identified.

Toxicity — "Toxic, in large doses may cause medullary paralysis; can be fatal."[17] In excess, may produce great depression, nausea, cold sweats, and possibly death.[2] Hardin and Arena[34] list the symptoms of poisoning, nausea, progressive vomiting, exhaustion and weakness, prostration, stupor, tremors, convulsions, coma, and death. There's an old story of Sam Thompson who fatally poisoned one of his patients, Ezra Lovett, by overadministration of lobelia. He was tried for murder but released because no one should have been dumb enough to take him seriously.[19] Few things mentioned by Jethro Kloss in his "Back to Eden" seem more dangerous to me personally (I have chewed fresh lobelia on many occasions) than some of his lobelia recommendations" . . . very beneficial if given in connection with other measures, as an enema of catnip infusion morning and evening. The enema should be given even if the patient is delirious. It will relieve the brain."[44] Kloss devotes more than 25 pages to lobelia, describing it as "the most powerful relaxant known among herbs that have no harmful effects." Both Tierra[28] and Kloss[44] recommend some rather dangerous herbs as enemas, a practice I cannot recommend, especially with poisonous herbs or herbs of dubious salubrity. Remember the warning: *Death has resulted from improper use of this drug as a home remedy.*

To the physician — They recommend "gastric lavage or emesis; artificial respiration; atropine 2 mg IM as needed."

208. *LOBELIA TUPA* L. (CAMPANULACEAE) — Tupa, Devil's Tobacco

Mapuche Indians smoke the leaves for a narcotic effect.[54]

Reported to be narcotic. Expressed juice used to treat toothache.[54]

According to Emboden, the active narcotic principles are lobeline and its keto- and dihydroxy derivatives.[54] While there is uncertainty as to its hallucinogenic properties, Emboden speculates that lobeline, lobelamidine, and neolobelamidine may be the psychoactive compounds.[54]

Toxicity — Classed as a narcotic inducing hallucinogenic stupors.[54]

209. *LOLIUM TEMULENTUM* L. (POACEAE) — Darnel, Tares

Affords a nutritive feed for livestock, but animals should not be allowed to graze after the seeds set. Human deaths are attributed to eating the infected seed, ground up with wheat. Its use as chicken and pigeon feed is discouraged.[38] In Lebanon, where commonly believed to be the tares of the Bible, and where the danger of ergotism is well known, Lebanese hint that there is a mountain mystic cult which infuses the grass or soaks the seeds to extract the ergot. It is then used to induce religious ecstasy.

Regarded as anodyne, narcotic, POISON, the tare is a folk remedy for gangrene, headache, meningitis, neuralgia, rheumatism, and sciatica.[32] It has quite a reputation for "cancer remedies". Moroccans use the decoction for hemorrhage and urinary incontinence.[3] Used for cancer, eczema, indurations, kernels, "knots", putrid flesh, scirrhus (liver and spleen), tumors, and wens.[4] According to homeopaths, it is used for St. Vitus' dance and idiocy, having been dropped by the allopaths. Lebanese women made a tea of the whole grass for children with colic. Adults used the ground seed for blood poisoning, leprosy, migraine, rheumatism, and toothache.[30] Seed meal has been used as a poultice in skin diseases and taken internally during menopause.[3] The poultice is said to bring out splinters and broken bones.[2] Homeopathically used for pains of neuralgia and rheumatic pains, colic, epistaxis, and vertigo.[33]

Dry seeds contain 10.9% protein, 2.6% fat, 56.2% carbohydrate, 39.0% fiber, and 11.9% ash.[21] Grains contain loliine, perloline, (0.06%) temuline, $C_7H_{12}N_2O$ temulentine and temulentic acid, starch, fructose, a saccharose galactoside, fatty oil with free fatty acids, and tannin. Herbage contains perloline, $C_{20}H_{18}N_2O_3$, *circa* 2.5% ash, of which 30 to 50% is acetic acid (33.56). Sap contains gelatinase and labenzyme.[33]

Toxicity — Grains have caused giddiness, weakness, dizziness, dilated pupils, headache, confusion, trembling, vomiting, delirium, and death due to respiratory collapse (in animals and humans).[56]

To the physician — *Hager's Handbook* suggests emetic or laxative as well as cardiac stimulants and analeptics.

210. *LOPHOPHORA WILLIAMSII* (Lemaire) Coult. (CACTACEAE) — Peyote

For centuries the peyote has been a religious object affording ''religious'' trips to certain Amerindians who consider that it makes them braver, stronger, and more immune to the pangs of hunger and thirst. One of the sources quoted by Schleiffer says that peyote was sometimes used in infanticide, and to kill off the sickly.[52] On a more pleasant chord, Emboden relates how some North American Indians beat the Supreme Court to establish the native Church of North America with a legal peyote-consuming ritual. Observers have been impressed by the contemplative nature of the participants and the efficacy of the cactus in permitting temporary escape from life's disenchantments, partially induced by the white man who wished to deprive red man of his peyote.[54] Experimentally hypoglycemic.[11]

Reported to be anodyne, antibiotic, cardiotonic, hallucinogenic, intoxicant, lactagogue, narcotic, poison, psychedelic, sedative, tonic, peyote is a folk remedy for alcoholism, angina, arthritis, backache, burns, corns, fever, headache, rheumatism, snakebite, sunstroke, tickling throat.[32,33,50]

The active principle is the alkaloid mescaline. At least eight other alkaloids have been isolated and may play minor roles in hallucinations.[54] *Hager's Handbook* lists mescaline ($C_{11}H_{17}NO_3$), N-methylmescaline ($C_{12}H_{19}NO_3$), N-formylmescaline ($C_{12}H_{17}NO_4$), N-acetyl-mescaline ($C_{13}H_{19}NO_4$), 3-demethylmescaline ($C_{10}H_{15}NO_3$), N-formyl-3-demethylmescaline ($C_{11}H_{15}NO_4$), N-acetyl-3-demethylmescaline ($C_{12}H_{17}NO_4$), 3,4-dimethoxyphenethylamine ($C_{10}H_{15}NO_2$), tyramine ($C_8H_{11}NO$), N-methyltyramine ($C_9H_{13}NO$), hordenine ($C_{10}H_{15}NO$), candicine ($C_{11}H_{19}NO_2$), anhalamine ($C_{11}H_{15}NO_3$), N-formylanhalamine ($C_{12}H_{15}NO_4$), N-acetylanhalamine ($C_{13}H_{17}NO_4$), anhalanine ($C_{12}H_{17}NO_3$), etc.[33]

Toxicity — There are unconfirmed assumptions by nonpeyote-worshippers that peyote can cause such things as addiction, disease, intoxication, immorality, laziness, mental disease, and teratogenesis; even death. Intoxication may last 2 or 3 days.[50] Listed as a narcotic hallucinogen.[54] Hardin and Arena state: ''Although peyote may not cause a physiological addiction, the psychotic reactions and long-range effects are dangerous, and it can be psychologically habit forming.''

To the physician — Harden and Arena recommend early gastric lavage or emesis if nature doesn't provide.[34]

211. *LYCOPERSICON ESCULENTUM* Mill. (SOLANACEAE) — Tomato

Fruits widely eaten in salads, sauces, soups, and stews, both raw and cooked, green (sometimes regarded as dangerous), and ripe. Seeds yield *circa* 25% oil, used for margarines, salad oils, and soaps. The residues are used as feed and mulch.[61] A process for converting tomatidine into allopregnenolone has been worked out; the latter can be transformed into progesterone or testosterone. Tomatine is used as a precipitating agent for cholesterol.[1] Tomato juice inhibits the germination of some weedy crucifers. The leaf infusion has been effectively sprayed in cabbage caterpillars.[16] Tomatine is fungicidal.[42]

Reported to be antiseptic, aperient, depurative, digestive, diuretic, pectoral, poison, tomato is a folk remedy for asthma, boils, cancers, chilblain, corn, cough, flu, gonorrhea, gravel, ophthalmia, otitis,' palpitations, phthisis, piles, tumors, typhoid, ulcers, vitiligo, yellow fever.[16,32,42,60] Tomato juice is one home ''remedy'' for oral cancers. Chinese take a decoction of old woody parts for toothache. Indonesians, apparently not sensitized to the plants, rub the leaves on their faces for sunburn. Green fruits are applied to ringworm, ripe fruits to burns.

The pulp constitutes 85.4% (average) of the whole fruit and contains 6 to 7% total solids. Analyses of the edible portions of green and ripe fruits gave the following values: green fruit — consisting of 92.8% moisture, 1.9% protein, 0.1% fat, 4.5% carbohydrate, and 0.7% mineral matter; 20 mg calcium, 40 mg phosphorus, 2.4 mg iron, 320 IU carotene (as vitamin A), 69 μg thiamine, 0.4 mg nicotinic acid, 60 μg riboflavin, and 31 mg/100 g ascorbic acid. The ripe fruits contain 94.5% moisture, 1.0% protein, 0.1% fat, 3.9% carbohydrate, and 0.5% mineral matter; 10 mg calcium, 20 mg phosphorus, 0.1 mg iron, 320 IU carotene (as vitamin A), 120 μg thiamine, 0.4 mg nicotinic acid, 60 μg riboflavin, 32 mg/ 100 g ascorbic acid. Tomatoes contain folic acid, pantothenic acid, biotin, vitamin K, and inhibitols, related to vitamin E. Ripe tomatoes contain glucose and fructose as principal sugars, with some sucrose and ketoheptose, raffinose occurring in the ripe fruit. Polysaccharides include cellulose, pectin, an araban-galactan mixture, and a xylan-rich fraction. Ripe and unripe fruits contain all essential amino acids, except tryptophan; cystine, tyrosine, aspartic acid, glutamic acid, serine, glycine, γ-aminobutyric acid, and pipecolic acid have also been identified. A globulin is also present. Carotenoids, beta-carotene, and lycopene constitute the chief coloring matters of tomato. Lycopene is absent in green tomato and appears when the fruit begins to turn red. Several other coloring principles, e.g., neolycopene A, neolycopene B, prolycopene, alpha-, gamma-, and χ-carotenes, neo-beta-carotene B, neo-beta-carotene U, lycoxanthine, phytofluene, and neurosporene have been reported. Minerals present in fresh ripe tomatoes are 2.8 mg sodium, 288 potassium, 13.3 calcium, 11.0 magnesium, 0.43 iron, 0.10 copper, 21.3 phosphorus, 10.7% sulfur, and 51.0 mg/100 g chlorine; traces of aluminum, manganese, cobalt, zinc, boron, arsenic, and iodine are reported. Seeds contain 8.95% moisture, 27.62% protein, 24.40% fat, 0.56% lecithin, 21.41% N-free extracts, 13.60% fiber, and 4.02% ash. Seed oil contains 14.7 to 18.0% saturated acids (C_{16-22}) and 76.1 to 80.6% unsaturated acids (mainly oleic and linoleic). Tomato seeds contain 2 globulins; albumin and glutelin are absent. Arginine is the principal amino acid present in the globulins (alpha-globulin, 13.97%, and beta-globulin, 10.65%); other amino acids present in alpha- and beta-globulins are 1.28 and 1.14 cystine; 1.15 and 1.45 tryptophan; 1.16 and 3.80 lysine; and 4.89 and 6.35% histidine.

Toxicity — Handling wet plants may cause dermatitis in sensitive individuals. Fruit processors may also develop dermatitis.[6] The raw plant has caused fatalities in livestock. Hippies have been attracted by the reports of narcotine, 5-hydroxytryptamine, and tryptamine. Tomatoes contain a gluco-alkaloid, tomatine ($C_{50}H_{83(81)}O_{21}N$), and traces of solanine ($C_{45}H_{73}O_{15}N$); narcotine is in unripe fruits.

212. *MAHONIA AQUIFOLIA* (Lindl.) Don (BERBERIDACEAE) — Oregon Grape

Planted as an ornamental. According to *Hager's Handbook*, the fruits are edible and used for wines and brandies.[33] Berberine has several uses, having been used internally as a carminative, febrifuge, and antimalarial, and externally as a dressing for indolent ulcers. Berberine or its salts stimulate bile secretion in humans, behaves as a sedative and hypotensive in experimental animals; also, acts as an anticonvulsant and uterotonic; amebicidal, bactericidal, trypanocidal.[29]

California Indians used the berry decoction to stimulate the appetite.[11] Said to be alterative, anaphrodisiac, antibilious, antiseptic, cholagogue, depurative, diuretic, expectorant, febrifuge, laxative, purgative, and tonic.[27] The root is used for cholecystosis, diarrhea, dyspepsia, dysuria, fever, gravel, leucorrhea, and uterosis.[32,48] The tincture is used for acne, arthritis, bronchitis, congestion, eczema, hepatitis, herpes, psoriasis, rheumatism, syphilis, and vaginitis.[2,28,48] The root is used for pain or burning of the gall and urinary passages, especially with tendencies to gallstone. Homeopathically, the bark is used for psoriasis and other skin ailments, and for the diathesis of uric acid.

Contains the alkaloids berbamine, berberine, canadine, corypalmine, hydrastine, isocorydine, mahonine, and oxyacanthine. Resin and tannin are also reported.

Toxicity — Berberine can cause intense pain when infiltrated into the eye to treat trachoma or into the skin for leishmaniasis. Injected as a local anesthetic, berberine occasionally produces hyperpigmentation. Ingestion of large quantities of berberine can cause fatal poisoning.

213. *MALUS SYLVESTRIS* Mill. (ROSACEAE) — Apple

Apples are most valued as a fresh dessert fruit, and may be made into jams, jellies, wines, ciders, vinegars, fresh juice, applesauces, apple butter, brandies, pies, and cakes. They may also be baked, fried, stewed, spiced, candied, or used in mincemeat or chutney. The hard wood is used for turnery, canes, and pipes. It is said to make one of the best of firewoods. Apples are a good detergent food for cleaning teeth. The oil from the seeds is used for cooking and illumination.

Regarded as apertif, bactericide, carminative, cyanogenetic, depurative, digestive, diuretic, emollient, hypnotic, laxative, POISON, refrigerant, sedative, and tonic. Apple is said to be a folk remedy for bilious ailments, cacethes, cancer, catarrh, diabetes, dysentery, fever, flux, heart, malaria, pertussis, scurvy, spasm, thirst, and warts.[32] In Europe scraped apple has been used extensively to treat infant intestinal disorders, such as diarrhea, dysentery, and dyspepsia.[33] Root and bark are considered anthelmintic, hypnotic, and refrigerant, and a bark infusion is given Indians suffering from bilious ailments, intermittent and remittent fevers. Apple leaves contain an antibacterial substance called phloretin which is active in doses as low as 30 ppm. Fruit eaten to obviate constipation.

Per 100 g, the fruit is reported to contain 58 calories, 84.4 g H_2O, 0.2 g protein, 0.6 g fat, 14.5 g total carbohydrate, 1.0 g fiber, 0.3 g ash, 7 mg Ca, 10 mg P, 0.3 mg Fe, 1 mg Na, 1.0 mg K, 54 mg β-carotene equivalent, 0.03 mg thiamine, 0.04 mg riboflavin, 0.02 mg niacin, and 10 mg ascorbic acid. From seeds can be extracted HCN and a bright-yellow semidrying oil with odor of bitter almonds; they also contain amygdalin and the glucoside phlorizin (up to 8%). May contain up to 17% pectin, pectic acid, tannins, wax, traces of essential oil, quercetin, isoquercitrin, ursolic-, oleanolic-, and pomolic-acid $C_{30}H_{48}O_4$, pomonic acid, alpha-farnesene, shikimic acid, leucocyanadin, cyanidin-3-galactoside, epicatechin, catechin, chlorogenic acid, quercetin 3-glucoside, quercetin 3-rhamnoglucoside, *p*-coumaric acid.[33]

Toxicity — One who apparently relished apple seeds ingested a cupful and died of cyanide poisoning.[56] Because of the unfortunate notoriety for cyanide poisoning in the U.S. in 1983, I repeat here the symptoms of cyanide poisoning. Lethal amounts may cause spasms and death due to respiratory failure within an hour. Smaller amounts cause stimulated respiration, changing to weak and irregular, gasping, excitement, then depression, weakness, staggering, pupil dilation with glassy prominent eyes, convulsions, spasms, twitching, and coma.

To the physician — Glucose, potassium permanganate, hydrogen peroxide, sodium thiosulfate, sodium nitrite, nerve stimulation, and respiratory support, properly administered, have been beneficial.[56]

> An apple a day, keeps the doctor away,
> Or at least that's what some people say.
> But one man we read, ate a cupful of seed,
> And this man he died, overdosed cyanide.
> His doctor's away, since that day, so they say.

214. *MALVA ROTUNDIFOLIA* L. (MALVACEAE) — Dwarf Mallow

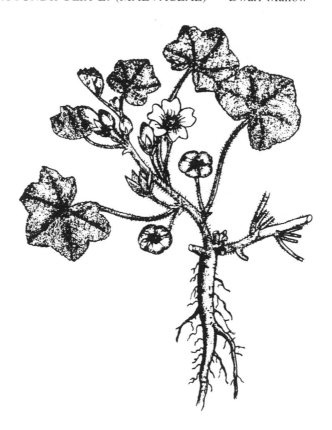

Leaves edible as a mucilaginous potherb. Tender shoots are eaten as salad. Root extracts inhibit the growth of *Mycobacterium tuberculosis*.[3]

Reported to be demulcent, emmenagogue, emollient, dwarf mallow is a folk remedy for abscesses, bronchitis, cholecystosis, cough, glycosuria, ophthalmia, piles, sore throat, stomach, and tumors.[3,4,32] Europeans in the Transvaal apply the leaf poultice to mammary inflammations.

Air-dried plant contains 0.41 to 0.42% N, 17.25 to 17.45% ash, 40.0% water-soluble extractives. Leaves contain 117.5 mg/100 g of ascorbic acid. A fatty oil contains oleic-, palmitic-, and stearic-acids, a wax comprised mainly of octacosane, a phytosterol, arabinose, KNO_3, KCl, $CaSO_4$. Seeds contain 7 to 8% fatty oil. Flowers contain tannin.[1]

215. *MANDRAGORA OFFICINARUM* L. (SOLANACEAE) — Mandrake, Loveapple

As indicated in the biblical story of Leah and Rachel, mandrake was once throught to induce fertility. Arabs called it "devil's apples" because of its supposed powers to excite voluptuousness. The yellow plum-like fruits, ripe in Israel during wheat harvest, are sickeningly sweet, otherwise, rather insipid. Eaten in quantity, fruits produce dizziness and may even stimulate to insanity. Thought to stimulate conception, they have long been used in love potions and incantations. As late as 1630 in Hamburg, Germany, three women were executed for possession of mandrake root, supposed "evidence" that they were practicing witchcraft. From early times, the large man-shaped root has been an object of superstition. Jews considered the mandrake a charm against evil spirits. Others believed that mischief-making elves would find it unbearable. Once esteemed for its medicinal and narcotic properties, mandrake still may have orgiastic and magical applications among cults.[11,38]

Mandrake is regarded as anesthetic, aphrodisiac, cathartic, cholagogue, emetic, hypnotic, mydriatic, narcotic, nervine, POISON, purgative, refrigerant, sedative, and stimulant. Fresh roots were once used for procuring sleep in continued pain, convulsions, rheumatic pain, and scrofulous tumors. Crushed leaves and boiled roots used to treat tumors.[4] Boiled in milk, the roots were poulticed onto indolent ulcers. Mixed with brandy, the root is used for chronic rheumatism. Used for asthma, arthritis, colic, coughs, hayfever, hepatitis, schizophrenia, and sclerosis.

Reported to contain several alkaloids, among them apoatropine, atropine, belladonine, cuscohygrine, hyoscyamine, mandragorine, norhyoscyamine, scopine, and scopolamine. Total alkaloids run around 0.4% with hyoscyamine, scopolamine, and atropine in a 36:5:2 ratio.[58] Fatty oils contain oleic and linoleic acid, scopoline, scopoletine, hyoscine, 3-alpha-tigloyloxytropane, 3,6-ditigloyloxytropane, as well as "chrysatroposaure" and "tropassaure".[33] Dry seeds contain 22.1% protein and 22.6% fat.[21]

Toxicity — "The fruit has been accounted poisonous, but without cause . . . " According to most authors, the plant is toxic, as the list of alkaloids would suggest. Atropine is anticholinergic, both central and peripheral. It tends to reduce secretions (gastric, intestinal, nasal, saliva, sweat, tears), decrease gastric and intestinal motility, and increase heart rate. It also causes pupil dilation, increase in intraocular pressure, and photophobia; 1-hyoscyamine and 1-scopolamine have essentially the same activities except that scopolamine is a powerful hypnotic and usually slows the heart rate.[29] Scopolamine-containing plants have been used as anesthetics for centuries in traditional Chinese medicine.

216. *MANIHOT ESCULENTA* Crantz (EUPHORBIACEAE) — Cassava, Tapioca, Manioc

Cassava is grown primarily for the tubers which are used as foodstuff. Tubers may be eaten raw, boiled, fried, or in baked goods. Since there may be HCN in the skin even of sweet varieties, they must be peeled before eating. In bitter varieties the HCN is throughout the root, and these must be cooked before using. From the manioc tuber are obtained starch, farina, a whole flour, grated manioc, and tapioca. Tapioca is used as a thickener in puddings and soups; an industrial starch is used in baked goods, laundry, and paper industries, and for sizing cotton fabrics and other textiles; from the starch a glue is prepared useable on postage stamps. Young leaves are high in vitamin B and are a good remedy for beri-beri, but they have the highest HCN content; however, they are sometimes used as a green vegetable and for hog feed. In the Philippines tubers are reduced to a pulp, wrapped with shredded coconut meats and sugar in banana leaves, and boiled; then served as a dessert (suman). Roots may be used as a fodder for livestock. Cassava is also the source of alcoholic beverages and/or power alcohol. Plant used as a fish poison.

Medicinally, the poisonous juice is boiled down to a syrup and given as an aperient. Philippinos used the bark decoction for rheumatism.[16] Indonesians use the root for body pain.[16] Fresh rhizome poulticed to sores. The flour cooked in grease, the leaf stewed and pulped, and the root decocted as a wash are said to be folk remedies for tumors. Cassava is used in folk remedies for cancerous affections, condylomata, excrescences of the eye, tumors, and whitlows.[4] Reported to be antiseptic, cyanogenetic, demulcent, diuretic, and POISON, cassava is a folk remedy for abscesses, angina, boils, conjunctivitis, diarrhea, dysentery, eczema, erysipelas, flu, hernia, hepatitis, inflammation, marasmus, neuralgia, prostatitis, snakebite, sore, spasm, swellings, testicles, toothache.[32,42] Brazilians use the juice from the tuber for ascariasis, ascites, eczema, scabies, and sycosis.[3]

Milky juice contains an essential oil (0.13%), saponin (1.14%), glucosides, and dyes, the essential oil containing sulfur in organic combination. Per 100 g, the leaves are reported to contain 60 calories, 81.0 g H_2O, 6.9 g protein, 1.3 g fat, 9.2 g total carbohydrate, 2.1 g fiber, 1.6 g ash, 144 mg Ca, 68 mg P, 2.8 mg Fe, 4 mg Na, 409 mg K, 8280 μg β-carotene equivalent, 0.16 mg thiamine, 0.32 mg riboflavin, 1.80 mg niacin, and 82 mg ascorbic acid. Per 100 g, the root is reported to contain 135 calories, 65.5 g H_2O, 1.0 g protein, 0.2 g fat, 32.4 g total carbohydrate, 1.0 g fiber, 0.9 g ash, 26 mg Ca, 32 mg P, 0.9 mg Fe, 2 mg Na, 394 mg K, 0.05 mg thiamine, 0.04 mg riboflavin, 0.6 mg niacin, and 34 mg ascorbic acid.

Toxicity — Manihot is reported to contain the following toxins: acetone (oral LC_{50} 5300 mg/kg), hydrocyanic acid (oral LD_{50} 3.7 mg), oxalic acid (700 mg; human oral LDLo), saponin (mouse oral LDLo, 3000 mg), and tryptophane (oral rat TDLo 1100 mg/kg). Bitter cassava caused death, due to cyanide poisoning. For symptoms and antidotes, see apple (*Malus*).

217. *MARANTA ARUNDINACEA* L. (MARANTACEAE) — Bermuda Arrowroot, West Indian Arrowroot, Arairut

Arrowroot starch, derived from the rootstock, is an important ingredient in the preparation of baby foods, biscuits, cakes, puddings, and pastries. Rhizomes may also be eaten boiled or roasted. The starch prevents the coarse curdling of milk. The starch grains average about 30 to 40 μm in diameter, making them more digestible for infants and convalescents. Arrowroot starch used as a base for face powder and glues. Sometimes used as a suspension agent.

Medicinally, arrowroot is demulcent, rubefacient but nonirrigating, acrid, and is so named because the Indians used the tuberous root as an application to wounds inflicted with poison arrows. Rhizomes rubefacient, used as a vulnerary. Pounded rhizomes poulticed onto ulcers and wounds.[1] Regarded as alexiteric, demulcent, depurative, refrigerant, rubefacient, and vulnerary, arrowroot is used for convalescence, dysentery, erysipelas, hoarseness, sores, sore throat, and sunburn.[32] The rhizomes are eaten in the Yucatan for ailments of the bladder and urethra. In Trinidad the starch is applied to erysipelas and sunburn. Antillean Indians apply the starch to dermatitis (caused by *Hippomane*), gangrene, and wasp stings.[42]

Per 100 g, arrowroot is reported to contain 157 calories, 57.2 g H_2O, 2.4 g protein, 0.1 g fat, 39.0 g total carbohydrate, 1.9 g fiber, 1.3 g ash, 20 mg Ca, 24 mg P, 3.2 mg Fe, 0.08 mg thiamine, 0.03 mg riboflavin, 0.7 mg niacin, and 9 mg ascorbic acid.[21] The refuse left after starch extraction contains 12.5% moisture, 2.2% ash, 0.3% fat, 14.0% fiber, 3.7% protein, and 64.0% starch (still!).

Toxicity — The starch of the root is said to produce respiratory allergy.[6] Classified by the FDA as an Herb of Undefined Safety: "Mild; easily digested. Suited for bowel complaints because of its demulcent properties."[62]

218. *MARSDENIA REICHENBACHII* Triana (ASCLEPIADACEAE) — Condurango

Pounded bark sun-dried and sold in quills as a medicine. According to Millspaugh, "Cundurango (*Gonolobus cundurango*), the Spanish Mataperro (the plant that — being announced and lauded as a cure for cancer — caused such a furor in medicinal and general circles in 1871; now considered worthless in cancer or any other disorder by those who were foremost in its advancement and use)."[19] Yet, Hayashi et al.[199] report antitumor activity in the Sarcoma 180 and Ehrlich carcinoma systems for two condurango glucosides labeled AO and CO.

Has been considered a potential cancer remedy, especially in early stages, but ineffective in later stages.[2] Recommended, however, by some for cancers of the breast, epithelium, esophagus, face, lips, neck, pylorus, skin, stomach, and tongue, as well as lymphadenomas.[4] Bark used in Latin America for chronic syphilis (an early Indian remedy for veneral disease). Considered alterative, antiseptic, analgesic, diuretic, hemostatic, nervine, stomachic, tonic.[32,33] Used in Cundurango wine as a stomachic. Also, used for beri-beri, gastritis, inappetence, rheumatism, and snakebite. Homeopathically suggested for gastroenteritis, loss of appetite, and ventricular ulcers.[33]

Tannin occurs with a glucoside and a strychnine-like alkaloid. Steinmetz reports caoutchouc (6% of the latex), conduragin, condurit (0.5%), essential oil, phytosterin, and resin as well.[27,33] *Hager's Handbook* reports that the bark contains condurangoglycoside, the agylcone of which bears as sugar moieties D-cymarose, D-thevetose, and D-glucose. Sitosterol is also present.[33]

Toxicity — Overdoses may produce convulsions, ending in paralysis, vertigo, and disturbed vision.[2] Two condurango glycosides (AO and CO) from the bark exhibited LD50s of 75 and 375 mg/kg, respectively.

219. *MATRICARIA CHAMOMILLA* L. (ASTERACEAE) — Hungarian Camomile, German Camomile, Manzanilla

Flowerheads contain an aromatic bitter principle (anthemic acid), regarded as a mild tonic, but may be emetic in large doses. In Latin America, "tea" made from *Matricaria* is taken as an after dinner beverage. Oil of camomile is used in cordials, perfumes, and shampoos. Both negative and positive bactericidal tests are reported on *Mycobacterium tuberculosis, Salmonella typhimurium*, and *Staphylococcus aureus*.[3] Azulene has an antiphlogistic action.[3] Chamillin is said to be the spasmolytic agent, perhaps the same as apigenine.[3] The oil has candicidal properties as well.[29] Chamazulen is said to possess anodyne, antiinflammatory, antiseptic, antispasmodic, and vulnerary properties. Alpha-bisabolol has antiinflammatory, antimicrobial, and antipeptic activities. Some of the cylic ethers in chamomile are antianaphylactic, antiinflammatory, antimicrobial, and antispasmodic. Umbelliferone is fungistatic. Camomile tea itself is hypnotic.[29,33] Extracts are used in bath preparations, hair dyes, mouth washes, shampoos, and sunburn preparations. Oils are used to impart fragrance to creams, detergents, lotions, perfumes, and soaps, and as a flavoring in beverages, baked goods, candies, frozen dairy desserts, gelatins, liqueurs (benedictine, bitters, vermouths), and pudding, with average food use usually below 0.002%.[29]

Internally, it serves as an anthelmintic, antispasmodic, carminative, diuretic, expectorant, sedative, stimulant, and tonic. Externally employed as a counterirritant liniment for bruises, hemorrhoids,   inflammations, and sores. Drug is regarded as an antiseptic and antiphlogistic. In China, the leaves are used as a depurative. The oil is said to be helpful for apostemes of the breast. A plaster made from the flower is used for indurations. The hot aqueous extract of the whole plant is said to cure tumors of the digestive tract.[4] In South Africa alone, camomile is used for colic, convulsions, croup, diarrhea, diptheria, gout, hyperacidity, hysteria, insomnia, lumbago, ophthalmia, rheumatism, sciatica, sore throat. It is mixed with buchu for pains in the bladder and colic. Costa Ricans take manzanilla to alleviate dysmenorrhea and insomnia. Dieting Costa Ricans take manzanilla tea to curb the appetite.[42] The juice is said to have healed lesions produced by the war gas dichlorodiethylsulfide (no better or worse than chloramine).[3]

It contains azulene, bisabolol, cadinene, choline, farnesene, furfural, paraffin hydrocarbons, sesquiterpenes, sesquiterpene alcohols, triacontane, methyl ether ($C_{10}H_8O_3$), and umbelliferone. German camomile oil contains apigenin ($C_{15}H_{10}O_5$), apigetrin (apigenin-7-D-glucoside), apiin (apigenin-7-apiosylglucoside), rutin (quercetin-3-rutinoside), luteolin, and quercimeritrin (quercetin-7-D-glucoside); coumarins including umbelliferone (7-hydroxy-coumarin) and its methyl ether (herniarin); proazulenes (matricin, matricarin, etc.); plant acids and fatty acids; a polysaccharide containing D-galacturonic acid; choline; amino acids; etc. The volatile oil contains chamazulene, farnesene, alpha-bisabolol oxide A, alpha-bisabolol oxide B, alpha-bisabolone oxide A, and en-yn-dicycloether.[29]

220. *MEDICAGO SATIVA* L. (FABACEAE) — Alfalfa

Highly valued legume forage, alfalfa has been heralded as having the highest feeding value of all commonly grown hay crops. In parts of China and Russia tender alfalfa leaves serve as a vegetable. Sprouts are consumed in many countries. Alfalfa is grown as a cover crop to reduce soil erosion, often increasing yields of succeeding crops, as potatoes, rice, cucumber, lettuce, tomatoes (increased by 10 MT/ha), corn, apples, and oranges. Valued as a honey plant. Extracts produce antibacterial activity against Gram-positive bacteria. Powdered alfalfa is used as a diluent to adjust strength of digitalis powder, and the root has been used as an adulterant of belladonna root. Seeds yield 8.5 to 11% of a drying oil suitable for making paints and varnish. Seeds contain a yellow dye. Alfalfa fiber has been used in manufacturing paper.[40] Alfalfa extracts are used in baked goods, beverages, candy, frozen desserts, gelatins, liqueurs, meat, produce, and puddings. Alfalfa is one source of commercial chlorophyll[42] and carotene.[33]

Seeds contain the alkaloids stachydrine and 1-homostachydrine, and are considered emmenagogue and lactigenic. They are used as a cooling poultice for boils in India. For weight gain, a cupful of 1:16 alfalfa/water infusion has been recommended. Alfalfa contains saponins but is sold in many ''health food'' stores. It is considered antiscorbutic, aperient, bactericide, cardiotonic, cyanogenetic, deobstruent, depurative, diuretic, ecbolic, emetic, emmenagogue, estrogenic, stimulant, stomachic, and tonic, and said to aid peptic ulcers, as well as urinary and bowel problems.[40] In Colombia, the mucilaginous fruits are used for cough. Elsewhere it is used for arthritis, boils, cancer, dysuria, fever, gravel, and scurvy.[32] Chinese consider their alfalfa deobstruent, depurative, diuretic, and stomachic, using it to treat the five viscera and, especially, intestinal and renal disorders.[16]

Tender shoots of alfalfa are reported to contain, per 100 g, 52 calories, 82.7% moisture, 6 g protein, 0.4 g fat, 9.5 g total carbohydrate, 3.1 g fiber, 1.4 g ash, 12 mg Ca, 51 mg P, 5.4 mg Fe, 3410 IU vitamin A, 0.13 mg thiamine, 0.14 mg riboflavin, 0.5 mg niacin, and 162 mg ascorbic acid. Green forage of *M. sativa* is reported to contain, per 100 g 80.0% moisture, 5.2 g protein, 0.9 g fat, 3.5 g fiber, and 2.4 g ash. Silages contain, per 100 g, 69.5% moisture, 5.7 g protein, 1.0 g fat, 8.8 g fiber, and 2.4 ash. Alfalfa whole meal and leaf meal are reported to contain, per 100 g, 66 and 77 calories, 7.5 and 8.0% moisture, 16.0 and 20.4 g protein, 2.5 and 2.6 g fat, 27.3 and 17.1 g fiber, 9.1 and 11.5 g ash, respectively. Many other details will be found in the references.[29,33,40] The "betaine fraction" of alfalfa contains: 0.785% stachydrine, 0.063% choline, 0.0069 trimethylamine, and 0.0052% betaine. The following purines have been identified: adenine, guanine, xanthine, and hypoxanthine; the primidine, isocytosine; and the ribosides adenosine, guanosine, inosine, and cytidine. These factors stimulate the growth of *Bacillus subtilis*. The three most abundant compounds in alfalfa juice were adenine (0.17%), adenosine (0.25%), and guanosine (0.36%). One study cites 5.1 g arginine/16 g N; 3.1 g cystine and methionine, 1.5 g histidine, 4.6 g isoleucine, 7.2 g leucine, 5.6 g lysine, 4.6 g phenylalanine, 4.1 g threonine, 1.5 g tryptophane, and 4.6 g valine for alfalfa hay. Alfalfa is a valuable source of vitamins A and E; it contains beta-carotene 6.24, thiamine 0.15, riboflavin 0.46, niacin 1.81, and alpha-tocopherol, 15.23 mg/100 g; pantothenic acid, biotin, folic acid, choline, inositol, pyridoxine, vitamin B_{12}, and vitamin K are also present. One report gives total crude lipids as 5.2% of total dry weight, with 11% fatty acids, 10% digalactolipids, 16% monogalactolipids, 8% phospholipids, 44% neutral lipids, and 12% others. In the chloroplasts, linoleic, linolenic, and palmitic acids are the predominant acids, whereas stearic and oleic are low. The triglyceride fraction of alfalfa meal contains 16.9% linoleic acid 32.2% linolenic acid, 31.0% oleic acid, and 19.9% saturated fatty acids. The phospholipid fraction (0.24% dry alfalfa meal) contained 35.2% linolenic acid, 36.8% oleic acid, 14.7% linoleic acid and 13.3% saturated acids. Good alfalfa hay contains 138 to 198 mg choline per 100 g hay. Alfalfa meal contains 21 mg A-spinasterol/100 g. Five xanthophylls comprise 99% of the xanthophyll fraction of fresh alfalfa, viz., 40% lutein, 34% violaxanthin, 19% neoxanthin, 4% cryptoxanthin, and 2% zeaxanthin. Alfalfa volatiles include acetone, butanone, propanal, pentanal, 2-methyl-propanal, and 3-methylbutanal.[40] Alfalfa honey gave the following average values: water 16.56, invert sugar 76.90, sucrose 4.42, dextrin 0.34, protein 0.11, acid (as formic) 0.08, and ash 0.07%. Alfalfa seeds contained: moisture 11.7, protein 33.2, fat 10.6, N-free extract 32.0, fiber 8.1, and mineral matter 4.4%.

Toxicity — Nitrate concentrations greater than 0.2% can harm livestock. Heavy N fertilization may increase nitrates to 1.0%. Tannin concentrations between 2.7 and 2.8% have been reported. Several coumestans occur in alfalfa, 4'-*O*-methylcoumestrol, 3'-methoxycoumestrol, lucernol, medicagol, sativol, trifoliol, and 11,12-dimethoxy-7-hydroxycoumestan. Four isoflavones are reported in alfalfa (daidzein, formononetin, genistein, and biochanin A) and they, like coumestrol, produce an estrogenlike response, possibly contributing to reproductive disturbances of cattle. Seeds are reported to contain trypsin inhibitors. Some people are allergic to the dust generated when alfalfa is milled. Medicagol has fungicidal properties. Some alfalfa saponins exhibit hypocholesterolemic activity.[42] Tricin(e) inhibits the smooth muscles and has antioxidant and estrogenic activity[33] (GRAS § 182.10 and 182.20).[29]

221. *MELALEUCA LEUCADENDRON* L. (MYRTACEAE) — Cajeput

Evergreen ornamental sometimes becoming a weed. Source of oil of cajeput, used as a mosquito repellent, effective, also, against lice and fleas.[1] The wood is strong and durable in contact with damp ground and sea water. It is used for piles, posts, railway sleepers, and ships. Bark serves in lieu of cork as an insulating material, also, used in floats, life belts, and stuffing cushions, mattresses, and pillows.[1]

Reported to be antiseptic, astringent, carminative, emollient, rubefacient, sedative, stimulant, sudorific, and vermifuge, cajeput is a folk remedy for acne, bronchitis, bruise, cholera, cold, colic, cough, diarrhea, earache, eczema, gout, headache, hiccup, inflammation, laryngitis, malaria, neuralgia, paralysis, pharyngitis, pityriasis, pleuritis, pneumonia, psoriasis, rheumatism, rhinitis, scabies, scurvy, skin ailments, sore throat, spasms, sprains, toothache, and tumors. Burmese mix cajeput oil with camphor for gout. Indochinese use the oil for arthritis and rheumatism, inhaling the oil for colds and rhinitis. Cambodians use the leaves for dropsy. Indonesians apply the oil externally for burns, cramps, colic, earache, headache, pain, skin disease, and toothache. Softened bark is applied to boils as a suppurative. New Guinea natives rub the oil on the body for malaria. Philippinos use the leaves for asthma.

Indonesians use the fruit for stomach disorders. Malayans use the oil as a pain killer and stomachic, dropping a bit onto sugar lumps for cholera and colic.[16]

Leaves contain *circa* 1.3% essential oil with 14 to 27% cineole (or eucalyptol), 1-pinene and terpineol, and aldehydes.[16,33] Besides these two, oil contains 1-limonene, dipentene, sesquiterpenes, azulene, sesquiterpene alcohols, valeraldehyde, and benzaldehyde.[1] The bark contains betulinic acid (melaleucin).

222. *MELIA AZEDARACH* L. (MELIACEAE) — Chinaberry

Fruit oils have been used in candles and paints, seed for hair tonic, soaps, and insecticides. Mexicans use the bark as a fish poison.³ Aqueous leaf extract is said to protect fruit orchards, kitchen gardens, and date oases from insect damage. Acridians do not eat leaves sprayed with *Melia* extract. Leaves are sometimes mixed in with woolen goods to discourage moths. The pits of the fruits have been strung as rosaries and "worry beads" now prohibited in Greece.⁴² The bark and foliage is used to intoxicate fish.²² Wood used for carpentry and fuel.³ ·

The plant is used in folk remedies for tumors in Colombia and India.⁴ Reported to be abortifacient, anodyne, antiseptic, ascaricidal, astringent, deobstruent, depurative, diuretic, emetic, emmenagogue, insecticide, laxative, narcotic, pediculicide, piscicide, pulicide, purgative, resolvent, sedative, stimulant, stomachic, tonic, and vermifuge, chinaberry is a folk remedy for asthma, cold, cough, eczema, eruptions, fever, headache, heatrash, hernia, hysteria, infection, leprosy, marasmus, rash, rheumatism, ringworm, scrofula, splenosis, stones, swellings, tumors, ulcers.³²,³³ Costa Ricans use the leaf-flower infusion as an emmenagogue; Puerto Ricans as a febrifuge and sedative in hysteria. Cubans use the leaf decoction in baths for rheumatism. Brazilians apply the leaves to tumors and inflamed glands. Bermudans took powdered bark decoction for diarrhea and dysentery. Curacao women apply the leaf decoction to skin rashes.[42] Flowers are used to treat head lice and scalp ailments.[3] Koreans use the bark decoction for intestinal worms and skin ailments. Chinese use it for atrophy, intestinal disorders, and stomachache; they use the fruits for cystitis, delirium, fever, hernia, pollution, uterosis, and worms.[16]

Per 100 g, the seed is reported to contain, on a zero-moisture basis, 28.7 g protein, 44.0 g fat, 58.6 g total carbohydrate, 11.6 g fiber, 3.4 g ash, 2310 mg Ca, and 220 mg P.[21] Leaves contain paraisine, fruits azadirine and resin, bark margosine and tannin.[56] The fruit oil consists chiefly of glycerides of palmitic-, oleic-, linoleic-, and stearic-, but no lino-lenic-, chaulmoogric-, or hydnocarpic-acids.[3] The fruit also is said to contain azadirachtin which inhibits feeding of the desert locust.[42] Vanillic acid, an ascaricidal and anthelmintic[16] compound, is found in the cortex with dl-catechol. The pericarp contains bakayanin and bakayanic acid.[16]

Toxicity — Ripe fruits are more poisonous than green ones, sometimes causing human fatalities. Fruit poisonings cause nausea, abdominal pain, vomiting, bloody diarrhea, thirst, cold sweats, poor coordination, weak pulse, difficult irregular breathing, paralysis, some-times accompanied by convulsions and death due to respiratory failure in 12 to 15 hr. A lethal dose for a 25-kg hog is said to be 150 g fruit or 6000 mg/kg body weight.[22] Leaf decoctions may cause burning of the mouth, oliguria, hematemesis, lethargy, and, occa-sionally, death.

To the physician — Gastric lavage or emesis, egg whites and milk to reduce the effects; prevent shock, symptomatic treatment.[56]

223. *MELILOTUS OFFICINALIS* Lam. (FABACEAE) — Yellow Sweetclover

Cultivated for forage, hay, pasture, and soil improvement, plants are excellent for soil improvement and erosion control, roots developing copious nitrogen-fixation nodules. Herb used to flavor cheeses and put in tobacco snuff. Seeds used as substitute for Tonka beans. Roots used as food by Kalmuks. Plants useful as honeybee pastures (flowers good source of honey). Used as a moth repellent.[40]

Plants considered alterative, anodyne, anticoagulant, aromatic, astringent, antispasmodic, carminative, diuretic, emollient, fumitory, laxative, nervine, poison, stimulant, styptic, tonic, and vulnerary. The plant is used in asthma cigarettes; as a decoction in lotions and enemas. Seeds have been suggested as a remedy for the common cold. Said to be used for arthritis, asthma, boils, brachialgia, bronchitis, carminative, colic, cramps, digestive, diuretic, expectorant, eyewash, headache, hemorrhoids, phlebitis, poison, rheumatism, swelling, thrombosis, tumors, ulcers, varicose veins, and wounds.[32,33,40]

The hay contains 5.4% H_2O, 20.3% protein, 4.6% fat, 38.3% N-free extract, 22.0% fiber, and 9.4% mineral matter.[1] Dry seeds 12.8 to 41.6% protein, 1.9 to 7.6% fat, 4.0 to 14.3% ash. Canavanine and trigonelline, seedlings contain cordianine, uric acid, and allantoic acid. Quercetin, kaempferol, mucilage, choline, resin, and tannin are also reported.[33]

Toxicity — Seeds are said to poison horses.[1] Cattle are poisoned by eating moldy hay. Toxicity is attributed to a fungus and bishydroxycoumarin. Hemorrhages occur both internally and externally, and sudden death may result (such hemorrhage may be controlled by quick administration of vitamin K). The plant contains 0.9 to 2% coumarin, 3,4-dihydrocoumarin (melilotol, melilotine, $C_9H_8O_2$), melilotic acid (*o*-hydrocoumaric acid $C_9H_{10}O_3$), coumaric acid ($C_9H_8O_3$), hydrocoumarin, melilotoside ($C_{15}H_{18}O_8$), and resin. With humans, 4000 mg (coumarin?) induces headache, nausea, vomiting, and weakness.[33]

224. *MENISPERMUM CANADENSE* L. (MENISPERMACEAE) — Moonseed

Birds eat the fruits readily, but "contrary to popular belief what a bird eats is not necessarily safe for humans."[34]

Reported to be alterative, cyanogenetic, diuretic, laxative, nervine, poison, stomachic, sudorific, tonic, moonseed is a folk remedy for cancer, debility, gout, scrofula, skin ailments, syphilis, venereal diseases.[32,33]

Contains the alkaloid dauricine $C_{38}H_{44}N_2O_6$, presumably, tetrandrine $C_{38}H_{42}N_2O_6$, and *circa* 1.7% viburnitol. Above-ground parts contain acutumine $C_{19}H_{24}NO_6$, the rhizome acutomidine, daurinoline, *N'*-desmethyldauricine, magnoflorine, an aporphine *N*-methyllindcarpinmethiodide, the protoberberine dehydrocheilanthifoline. Seeds contain *circa* 16% oil with a high percentage of unsaturated fatty acids and 13.1% protein, 2.6% ash.

Toxicity — The fruits are dangerous if eaten in quantity.[34] "Giftdroge".[33] The fruits, resembling grapes, have been eaten by children; Kingsbury reported fatalities in Ohio and Pennsylvania.[14]

To the physician — Hardin and Arena suggest gastric lavage or emesis and symptomatic treatment.[34]

225. *MENTHA PULEGIUM* L. (LAMIACEAE) — Pennyroyal

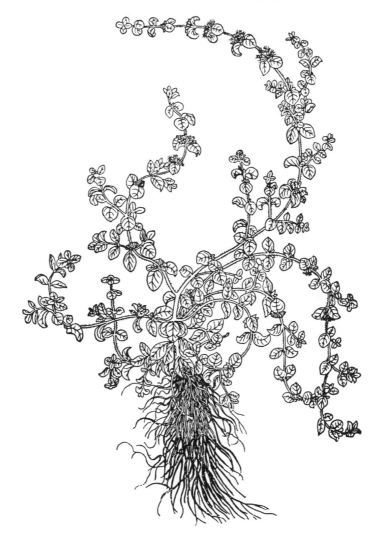

Leaves used for flavoring, for a tea, and for a spice. Once used to flavor pork puddings, especially mixed with pepper and honey. Plant has a strong peculiar odor resembling that of spearmint, but less agreeable. Taste is, at first, warm, aromatic, and bitterish, later with a cooling sensation. Pennyroyal oil has a bitter taste followed by a cooling sensation. Pennyroyal oil has scent of citronella and is used for scenting soaps, in perfumery of detergents, as a repellant and pesticide for fleas, mosquitoes, gnats, and ants, and as a strewing herb. Oil used in production of menthol. The herb was once added to water as a "water purification pill". Yet, Tyler concludes "the herb has nothing to recommend it."

The whole plant decoction used for uterine tumors, uterine fibroids, cacoethes of the uterus, and indurations of the uterus. The root, ground with vinegar, is a tumor remedy. A poultice of the leaf is said to alleviate corns.[4] Infusions of leaves used for spasms, cramps, and colds. It is used in fainting, flatulence, gall ailments, gout, hepatitis, and nervous disorders. In older days it was hung in the rooms of convalescents; said to induce health more rapidly than roses. Worn around the head, garlands of pennyroyal were said to alleviate dizziness and headache. Pennyroyal tea was once used for coughs, colds, and menstrual disorders. However, it must be used with discretion as it is toxic and dangerous if used in quantity. In the U.S., leaf and oil cannot be sold legally; in Europe, it is considered antiseptic, carminative, cholagogue, diaphoretic, digestive, emmenagogue, pectoral, stimulant, and sudorific.[27]

Plant yields 1 to 2% of a yellow or greenish volatile oil. Pulegone content ranges from 9% in Brazilian varieties, to 16 to 30% in the U.S. varieties, to 80 to 94% in European varieties.[3] Fujita and Fujita report the following among the autoxidation products and neutral compounds of the essential oil: alpha-pinene, beta pinene, limonene, 3-octanone, *p*-cymene, 3-octylacetate, 3-octanol, 1-octen-3-ol, 3-methylcyclohexanone, menthone, isomenthone, isopulegone, pulegone, piperitone, *cis*- and *trans*-pulegone oxide, piperitenone, dehydrox-ymethofuran-oxide, menthofuran oxide, caryophyllene, beta-humulene, and paraffins. Among the acidic compounds and autoxidation products were lauric acid, myristic acid, palmitic acid, beta-methyl-adipic acid, beta-methyl-delta-isobutyryl-valeric acid, phenol, *o*-cresol, *p*-cresol, salicylaldehyde, and eugeonal. Diosmin and hesperidin are also reported.[33] Dry seeds contain 24.6% protein, 26.6% fat, 11.1% ash, 0.9% Ca, 0.5% Na, and 1.5% K.

Toxicity — In 1978, a young lady in Colorado died as a result of ingestion of pennyroyal oil in an attempt to induce abortion. Like various other volatile oils, pennyroyal oil is reported to be ecbolic but abortion is impossible by a dose which is safe. An amount of this oil, capable of inducing abortion, is likely to produce irreversible renal damage. A teaspoon of the oil produces serious effects: delirium, unconsciousness, tetanic spasms, opisthotonus, shock. "Less than one ounce is likely to kill the patient . . . "[3] "While pennyroyal oil may indeed induce abortion, it does so only in lethal or near-lethal doses. Such amounts would ordinarily not be obtained from drinking a tea prepared from the herb . . . "[37] American pennyroyal contains up to 2% of a volatile oil, and European pennyroyal up to 1% of an even more disagreeably smelling volatile oil. Both oils consist of 85 to 92% pulegone and are, therefore, quite toxic, causing severe liver damage, even in relatively small amounts. Two tablespoonfuls of pennyroyal oil caused the death of an 18-year-old expectant mother in spite of intensive hospital treatment initiated just 2 hr after she took it. As little as 1/2 teaspoonful of the oil has produced convulsions and coma in one individual.[37]

226. *MERCURIALIS ANNUA* L. (EUPHORBIACEAE) — Annual Mercury, French Mercury, Garden Mercury

Common garden weed, difficult to eradicate; the seeds "taste like those of hemp."[2] The leaves were "boiled and eaten as spinach, and it is still eaten in this way in parts of Germany, the acrid qualities being dissipated, it is believed, by boiling."[2] French feed the plant to hogs.[2]

The plant is used in folk remedies for tumors.[4] Reported to be cholagogue, cyanogenetic, diuretic, emetic, emmenagogue, hydragogue, laxative, poison, purgative, and vermifuge, annual mercury is a folk remedy for cancer, dropsy, dysmenorrhea, eczema, inappetence, mucous congestion, and rheumatism.[32,33] Brazilians use the tuber juice for ascariasis, ascites, eczema, scabies, and sycosis. Italians use the latex to stop lactation and as a laxative.[3] French once made purgative syrup from the fresh herbage. Homeopathically used for rheumatism.[33]

Dry seeds contain 18.9% protein, 37.1% fat, and 11% ash.[21] Leaves and stems contains urease. The juice from peeled tubers contains 0.08% volatile oil, 0.12% fixed oil, 2.8% gummy substances, 2.25% saponins, 1.35% cyanogenetic glucosides, 3.35% tannins, and traces of alkaloids and dyestuff. Herbage and seeds yield methylamine,[3] also, ethyl-, propyl-, isobutyl-, and isoamylamine.[33] Ashes contain 15% SiO_2. Glucose, isorhamnetid, rhamnose, and xylon are also reported.[33]

Toxicity — Somewhat poisonous, diuretic, and cathartic. Roots and rhizomes are strongly laxative.[33] Said to be cyanogenetic.[32] Symptoms of livestock poisonings include severe gastric irritation with diarrhea, and hematoma with profound anemia developing.[14]

227. *METHYSTICHODENDRON AMESIANUM* R. E. Schultes (SOLANACEAE) — Culebra

Used by the Inga and Kamsa Indians of Colombia as a medicine.[4] According to Sibundoy Indians, this narcotic must be taken only during the waning of the moon. The intoxication is resorted to only for very difficult or important cases of divination, profecy, or therapy.[201]

The leaves and flowers are plastered onto tumors and swellings, especially of the joints.[4] Reported to be intoxicant and narcotic.[32] Also, used for persistent chills and fever, perhaps, advanced tuberculosis. Bristol found the Sibundoy Indians use it for colds, cramps, erysipelas, infections, myalgia, rheumatism, and swellings.[202]

Active principles are solanaceous alkaloids of the tropane series. Leaves and stems yield *circa* 0.3% total alkaloids (80% scopolamine, some atropine, and two other alkaloids). The high proportion of scopolamine may account for the high psychotrophic potency. Scopolamine may produce excitement, hallucinations, and delirium, even at therapeutic doses, while doses of atropine border on toxic before hallucinations and central excitations occurs.[58]

Toxicity — "Dangerously active narcotic tree." Death of one witch doctor has been attributed to an overdose. Intoxication usually lasts two full days and may persist for four — with a long period of complete lack of consciousness.[201]

228. *MIMOSA HOSTILIS* Benth. (MIMOSACEAE) — Jurema, Yurema

Roots are cleaned, macerated, steeped in water, and squeezed to give a strong hallucinogenic beverage.[54]

Reported to be astringent. Country folk in Brazil once used the bark to cure fatigue. Others believed it fortified the uterus.[203]

Contains *N,N*-dimethyltryptamine $C_{12}H_{16}N_2$, earlier called nigerine. The plant material contains 16.11% crude protein, 3.08% ether extract, 11.83% crude fiber, 65.46% *N*-free extract, and 3.44% ash, with an in vitro DM digestibility of only 21.0%.[204]

Toxicity — Narcotic hallucinogen,[54] said to evoke fantastic and agreeable dreams. The plant, if ingested by cattle, inhibits digestion of grass, e.g., buffel grass and jaragua.[204]

229. *MIRABILIS MULTIFLORA* (Torr. Gray (NYCTAGINACEAE) — Soksi

Hopi medicine men chew on the roots to enable them to render visionary diagnoses of their patients' diseases.[54]

Reported to be hallucinogenic, soksi is a folk remedy for stomachache.[32] Hopi Indians use the roots for stomach ailments.[54]

Toxicity — Narcotic hallucinogen.[54]

230. *MITCHELLA REPENS* L. (RUBIACEAE) — Partridgeberry, Squawvine

Evergreen ground cover, quite attractive all winter with its red, barely edible fruits.

Sometimes combined with *Caulophyllum thalictroides* and *Eupatorium aromaticum,* in folk nostrums for female problems, e.g., hysteria, chronic uteritis, or parturition. Amerindians took the herb before confinement in order to render parturition safe and easy. Tierra mentions its use as a wash for sore eyes.[28] The herb has also been suggested for amenorrhea, diarrhea, dropsy, dysentery, dysmenorrhea, dysuria, gonorrhea, gravel, hysteria, leucorrhea, menorrhagia, polyuria, uterosis, and vaginitis.[28,32,43] It has been called astringent, diuretic, tonic, especially to the uterus. Parvati mentions this as her favorite pregnancy herb, combined with a raspberry and chilled.[48] Tierra suggests a combination of partridgeberry, lobelia, raspberry leaves, and wild yam to prevent miscarriage;[28] Kloss, the berry tea with olive oil or cream to apply to sore nipples.[44] Parvati reports giving to hundreds of women coming off the pill or a pregnancy (either abortion, miscarriage, or nursing) a tea consisting of equal parts of squawvine, black haw, holy thistle, licorice, and sarsaparilla — with positive results. Steep 1 tablespoon of the mixture in a cup, taking 4 to 5 cups the first week, 2 to 3 cups the second week, 1 to 3 cups the third week, tapering off the mere sips in the fourth week. "This may also increase your fertility."[48]

Contains dextrin, mucilage, resin, and wax.[2] Not listed in most of the poisonous plant books.

231. *MITRAGYNA SPECIOSA* Korth. (RUBIACEAE) — Katum, Kratum, Kutum

Mambog is a potent viscous distillate of the tea of kratum leaves, used as an opium substitute. Leaves are chewed fresh, smoked when dried, or distilled into mambog.

Reported to be analgesic, fumitory, masticatory, katum is a folk remedy for opium addiction.[32,33]

Speciociliatine is analgesic, like codeine, and also strongly antitussive, while mitragynine works like cocaine on the CNS. The psychoactive compound is mitragynine and eight similar compounds, ajmalicine, corynantheidine, isomitraphylline, mitrophylline, paynantheine, speciophylline, speciofoline, and speciogynine.[54] Mitraspecine, rhynochophylline, speciociliatine, mitrafoline, isomitrafoline, isospeciofoline, stipulatine, mitraversine are also reported.

Toxicity — Narcotic hallucinogen.[54] Extended use leads to abdominal distention, darkened lips, dry skin, emaciation, numbness in peripheral areas, pallor twitching, and cardiac irregularities. Overdoses of mambog results in vertigo and vomiting.[54]

232. *MOMORDICA CHARANTIA* L. (CUCURBITACEAE) — Balsam Pear

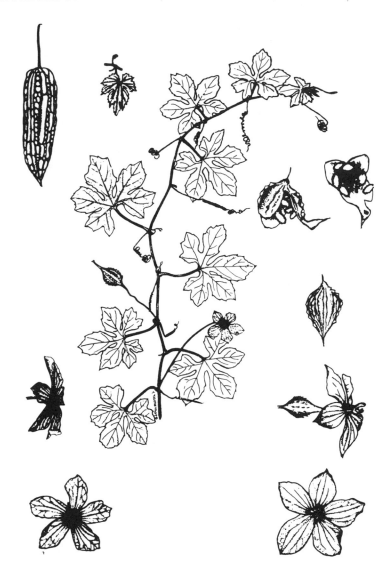

Unripe fruits are cooked and eaten in the orient, and cooked leaves after a change of water are said to be edible. The red arils around the seeds are dangerously eaten by children, and not so dangerously fed to caged birds.[42] Fruits, making lather, are used as a soap substitute. Roots and leaf extracts have shown some antibiotic activity.[3,42] Plant is used in Haiti as an insecticide.[3]

The fruits are used in folk remedies for tumors in Brazil; the plant for malignant ulcers in Guam.[4] Reported to be abortifacient, antibiotic, aperitive, aphrodisiac, astringent, carminative, depurative, digestive, hypotensive, insecticidal, lactagogue, laxative, poison, purgative, refrigerant, stomachic, styptic, and vermifuge, balsam pear is a folk remedy for asthma, boils, burns, calculus, cancer, catarrh, chilblain, cholera, cold, colic, colitis, dermatosis, diabetes, dysentery, dysmenorrhea, dyspepsia, earache, eczema, eruption, fever, gonorrhea, gout, halitosis, headache, hepatitis, hyperglycemia, hypertension, itch, jaundice, leprosy, malaria, melancholy, piles, psoriasis, rheumatism, roundworm, scabies, splenitis, thrush, ulcers, urethritis, etc.[3,4,32,33] Arubans take the decoction for hypertension, Puerto

Ricans for diabetes, Cubans for colitis, fever, and hepatosis, Hondurans as depurative, Peruvians for colic and worms, Venezuelans as a tonic for colds and fever, Jamaicans for colds, constipation, dysmenorrhea, fever, and stomachache, Bahamians for cold, flu, and fever. Yucatan natives regard the root as aphrodisiac, Cubans as litholytic.[42] Congolese use the leaf for colic, the seed for roundworm. Japanese use the plant for constipation, headache, and skin ailments. On the Indian Peninsula, the plant is used for scabies, psoriasis, and other skin diseases. Natives of the Pearl Islands of Panama take the bitter juice as a malaria preventive, while some Colombians regard it as highly as quinine for malaria treatment.[60] Poulticed onto burns, scalds, skin diseases; leaf juice gargled for sprue;[16] cooling, strengthening aqueous extracts of the fruit show lipolytic and anticholesterolemic properties. Used for halitosis, considered a male aphrodisiac.[41] Recently, balsam pear has attained a favor in China as a monoherbal medicine (dried fruit, powdered, and made into pills, 18 g/day) for diabetes mellitus, effective with mild and moderately chronic cases, reducing glucose in the blood and urine, and the frequency of urination. The herb does not promote the secretion of insulin, but increases carbohydrate utilization.[41] Clinical trials with 160 diabetics of fresh fruit juice controlled, but did not heal, diabetes.[41]

Per 100 g, the leaves are reported to contain 44 calories, 84.6 g H_2O, 5.0 g protein, 0.6 g fat, 7.0 g total carbohydrate, 1.6 g fiber, 22 g ash, 288 mg Ca, 54 mg P, 5.0 mg Fe, 19 mg Na, 510 mg K, 5085 μg β-carotene equivalent, 0.13 mg thiamine, 0.46 mg riboflavin, 1.56 mg niacin, and 170 mg ascorbic acid.[21] Per 100 g, the fruit is reported to contain 19 calories, 94 g H_2O, 0.8 g protein, 0.1 g fat, 4.5 g total carbohydrate, 1.6 g fiber, 22 g ash, 288 mg Ca, 54 mg P, 2.3 mg Fe, 2 mg Na, 270 mg K, 110 μg β-carotene equivalent, 0.06 mg thiamine, 0.07 mg riboflavin, 0.3 mg niacin, and 57 mg ascorbic acid.[21] Fruits contain the hypoglycemic principle charantin.[42] Unripe fruits contain luteolin as the bitter alkaloid momordicin.[42] Seeds yield about 32% oil (purgative),[3] which contains 17% stearic acid, and linoleic and oleic acids. Dried roots contain 12.8% ash, dried fruits 11.7%, the ashes quite naturally containing Ca, Fe, and P.

Toxicity — Juice given to an Indian child caused vomiting, diarrhea, and death.[56]

To the physician — Hardin and Arena suggest emesis or gastric lavage and symptomatic treatment.[34]

233. *MYRICA CERIFERA* L. and other sp. (MYRICAEAE) — Bayberry, Wax Myrtle

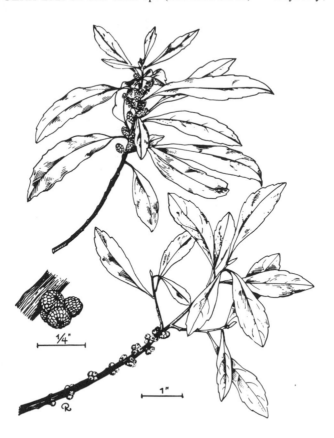

American settlers made wax and candles from the berries. (One bushel of fruit yields 4 lb wax.) French settlers in Canada added the leaves to soups as a condiment. ''The leaves and berries form an attractive and agreeable substitute for the tropical bay-leaves for use in savoring soups, etc. They are really good!''[13] Leaves are also steeped to make a tea. Leaves of the European *M. gale* were used for hop in producing beer, and as moth repellent and insecticide. Recommended as a good foliage plant, evergreen or nearly so, and susceptible to pruning and shaping into hedges. The principle use of the wax in 1958[205] was for Christmas candles, but it was also in soap, ointments, leather-polishing formulations, etching, and medicinals. It seems as though the uses of *M. cerifera*, *M. gale*, and *M. pensylvanica* are all interspersed interchangeably with each other in the literature and probably in nature as well. To the north, *M. pensylvanica* replaces *M. cerifera*. Middle-aged Americans will be delighted to learn, if what Tierra says is true, that a fermentation of bayberry tea ''can be applied externally at night to relieve, cure, and prevent varicose veins.''[28]

The bark was used in folk remedies for nasal polyps, powdered roots, and wax for cancerous ailments.[4] The powdered root, with or without blood root, has been applied to indolent ulcers and scrofulous tumors. Reported to be alterative, astringent, cordial, cardiostimulant, deobstruent, diaphoretic, emetic, laxative, narcotic, sternutatory, stimulant, stomachic, sudorific, tonic, and vermifuge. The bayberry is a folk remedy for boils, cankers, carbuncles, catarrh, cholera, diarrhea, dysentery, fever, goiter, headache, hemorrhage, hematochezia, hematoptysis, hepatosis, jaundice, leucorrhea, metrorrhagia, pharyngitis, pyorrhea, scarlet fever, scrofula, sores, sore throat, stomatitis, typhoid, ulcers, and uterosis.[32] Morton says the astringent, slightly narcotic wax has been ''successfully employed in epidemic typhoid dysentery," also, in diarrhea and jaundice. South Carolina blacks drink the leaf tea for backache, colds, diarrhea, fever, and nephrosis, the root decoction for headache.[46] Kloss[44] recommends snuffing powdered bayberry for adenoids.

Per 100 g, the wax myrtle leaves are reported to contain, on a zero moisture basis, 10.8 g protein, 2.2 g fat, 74.3 g total carbohydrate, 29.8 g fiber, and 12.7 g ash.[45,205] Leaves contain an essential oil with alpha-pinene, myrcine, limonene, gamma-terpinene, *p*-cymene, linalool, caryophyllene, and mimulene. The fat outside the fruit contains glycerides of stearic-, palmitic and myristic acids. The bark contains myricine, starch, gum, albumen, a red dye, tannin, and an astringent resin. Bayberries probably contain 5 to 10% wax, but up to 25% has been reported (32% in Erichsen-Brown). Dry seeds contain 4.9% protein, 13.3% fat, 74.3% total carbohydrates, 29.8% fiber, and 12.7% ash.[21] The wax melts at 47 to 49°C.

Toxicity — Bayberry wax is said to be irritant and sensitizing.[6]

234. MYRISTICA FRAGRANS Houtt. (MYRISTICACEAE) — Mace, Nutmeg

The well-known spices nutmeg and mace yield an essential oil used for flavoring foods and liqueurs. Alcoholic extracts of nutmeg are bactericidal, aqueous extracts kill cockroaches, while the volatile oils from the leaf are herbicidal. Myristicin enhances the toxicity of pyrethrum to house flies. In Ayurvedic documents nutmeg was called madashaunda, a term meaning "narcotic fruits". Betel chewers in India often add nutmeg, and it is also added to chewing tobaccos and snuffs. The pericarp of the fruit is used for jellies and pickles. Juice from the pericarp is an efficacious mordant for fixing dyes. Both nutmeg and mace are used as spices in many American and exotic dishes.

Nutmeg — On the Island of Banda, a pap is made of the bark preserved with sugar and tasting like sour apples. The bark is pickled in brine in Java but tends to induce sleep. Dried seed grated to flavor milk dishes, cakes, punches, possets, and vegetables, such as cabbage, cauliflower, and spinach. Recovered pericarp is used for sweet dishes. Oil of nutmeg of distilled for external medicine and perfumes. Nutmeg butter, derived from broken seeds and poor-grade mace, is used in medicinal ointments, suppositories, and perfumery. Four or five grams of myristicin, a narcotic, produce toxic symptoms in man. Nutmeg, taken as a psychotropic, often causes reactions similar to those of other hallucinogenic drugs quite unlike the classic account of myristica poisoning. Myristicin alone does not give the reaction, but eating the whole seed does. "stirred into a glass of cold water, a penny matchbox full of nutmeg had the kick of three or four reefers" (Malcolm X, as quoted by Schleiffer).[63]

Mace — Dried ground aril is used to flavor savory dishes, pickles, and ketchups. Fruits are sometimes gathered before maturing to make jelly. Peel of the fruit, in the tender stage, provides an aromatic pickling material. Oil, distilled from the leaves, has a spicy, aromatic, pleasant flavor; used in making toilet and medicinal products, as a flavoring essence, and in chewing gum.

Chinese use powdered seeds for pediatric and geriatric fluxes, for cardiosis, cold, cramps, and chronic rheumatism. Indonesians boil the powdered seed for anorexia, colic, diarrhea, dyspepsia, dysentery, malaria. Seed oil is rubbed on the temples for headache or dropped in tea for dyspepsia and nausea. Indonesians use the leaf tea for flatulence. Malayans use the nutmeg for madness, malaria, puerperium. rheumatism, and sciatica.[16] Arabians, as early as the seventh century, recommended nutmeg for digestive disorders, kidney troubles, lymphatic ailments, etc. Even earlier, Indians used it for asthma, fever, heart disease, and tuberculosis.[58] Nutmeg is considered aphrodisiac, astringent, carminative, narcotic, and stimulant. The fruit is said to be a folk remedy for indurations and tumors.[4] Mace has been used for putrid and intermittent fevers and mild indigestion. The expressed oil of nutmeg is used externally as a stimulant. They are used to allay both flatulence and nausea. Mixed with lard, grated nutmeg is applied to piles. Roasted nutmeg is used internally for leucorrhea. In India nutmeg is prescribed for dysentery, flatulence, malaria, leprosy, rheumatism, sciatica, and stomachache.[1] The essential oil is recommended for inflammation of the bladder and urinary tract. Arabs still use nutmeg as an aphrodisiac in love potions. Yemenites recommend its use for the liver and spleen, for colds, fevers, and respiratory ailments. Many women, in hopes of inducing abortion, have failed, yet suffered the intoxication due to myristicin. Ingestion of 1 to 2 oz ground nutmeg produces a prolonged delirium, disorientation, and drunkeness. Nutmeg may alleviate some symptoms of certain types of cancer, a case study presented in the *New England Journal of Medicine* suggests.[206] Dr. Ira Shafran and Daniel MaCrone of Ohio State University say, "further study may substantiate the speculation that inhibition of prostaglandin E_2 by nutmeg affords symptomatic improvement of hypercalcemia in medullary carcinoma of the thyroid, and other prostaglandin-secreting neoplasms." They prescribed 4 to 6 tablespoons of nutmeg per day to their patient, because nutmeg is known to improve diarrhea associated with medullary carcinoma of the thyroid. The patient also suffered from hypercalcemia which did not respond to standard calcium-reducing treatment. After 12 days of nutmeg therapy the calcium levels were reduced by almost one third. The medical team says that medullary carcinoma of the thyroid is known to produce "copious amounts of prostaglandin E_2. . . . (and) inhibition of prostaglandin E_2 may be nutmeg's antidiarrheal mechanism of action."[206]

One hundred grams nutmeg contains *circa* 200 mg safrole, 90 mg methylleugenol, 25 mg eugenol, 55 mg methylisoeugenol, 30 mg isoeugenol, 1050 mg myristicin, 350 mg elemicin, 15 mg isoelemicin, and 40 mg methoxyeugenol. Nutmeg contains about 14.3% water, 7.5% protein, 36.4% fat, 28.5% carbohydrates, 11.6% fiber, 0.2% calcium, 0.24% phosphorus, and 4.5 mg iron per 100 g. Mace contains 15.9% water, 6.5% protein, 24.4% fat, 47.8% carbohydrates, 3.8% fiber, 0.18% calcium, 0.10% phosphorus, and 12.6 mg iron per 100 g. The fat from Indian nutmegs contain 0.4% lauric acid, 71.8% myristic acid, 1.2% stearic acid, 4.8% hexadecenoic acid, 5.2% oleic, and 1.5% linoleic acid. Amylodextrins, pectins, resins, and reducing sugars are also present. The volatile oils contain numerous ingredients, some of them toxic, camphene, cymene, dipentene, eugenol, geraniol, isoeugenol, linalool, myristicin, pinene, safrole, and terpineol. See Purseglove et al.[64] and Lawrence[193] for details. Per 100 g, ground mace contains 475 calories (1989 kJ), 8.2 g H_2O, 6.7 g protein, 32.4 g fat, 50.5 g total carbohydrate, 4.8 g fiber, 2.2 g ash, 252 mg Ca, 110 mg P, 14 mg Fe, 163 mg Mg, 80 mg Na, 463 mg K,2 mg Zn, 800 IU vitamin A, 0.31 mg thiamine, 0.45 mg riboflavin, 1.35 mg niacin, 0 mg cholesterol, and 73 mg phytosterols. On the other hand, ground nutmeg contains 525 calories (2196 kJ), 6.2 g H_2O, 5.8 g protein, 36.3 g fat, 49.3 g total carbohydrate, 4.0 g fiber, 2.3 g ash, 184 mg Ca, 213 mg P, 3 mg Fe, 183 mg Mg, 16 mg Na, 350 mg K, 2 mg Zn, 102 IU vitamin A, 0.35 mg thiamine, 0.06 mg riboflavin, 1.30 mg niacin, 0 mg cholesterol, and 62 mg phytosterols.

Toxicity — Doses exceeding 1 teaspoonful take effect within 2 to 5 hr, producing time-space distortions, feelings of unreality, and sometimes visual hallucinations accompanied by dizziness, headache, illness, and rapid heartbeat.[51] The "carcinogen" safrole, responsible for the banning of sassafras, also occurs in nutmeg. Reviewing the work on safrole, Buchanan concluded that it is the most thoroughly investigated methylenedioxybenzene derivative. The oral LD_{50} for safrole in rats is 1950 mg/kg body weight, with major symptoms including ataxia, depression, and diarrhea, death occurring in 4 to 5 days. With rats, dietary safrole at levels of 0.25, 0.5, and 1% produced growth retardation, stomach and testicular atrophy, liver necrosis, biliary proliferation, and primary hepatomas. Reviewing research on myristicin, which occurs also in black pepper, carrot seed, celery seed, and parsley, Buchanan notes that the psychoactive and hallucinogenic properties of mace, nutmeg, and purified myristicin have been studied. It has been hypothesized that myrisiticin and elemicin can be readily modified in the body to amphetamines. The oral LD_{50} for nutmeg oil in rats, mice, and hamsters is 2600, 4620, and 6000 mg/kg, respectively. Buchanan et al.[207] found no significant evidence of mutagencity of myristicin.

235. MYROXYLON BALSAMUM VAR. PEREIRAE (Royle) Harms (FABACEAE) — Balsam of Peru

Peru balsam, produced by injured trees, exudes from trunk and limbs or is extracted from bark. Peru balsam, not produced in Peru, received its name because it was originally assembled and shipped to Spain from the Port of Callao, Peru. Oil (cinnamein) is used in perfume, cosmetic, and soap industries. Balsam and its essential oil have been used as flavorings for baked goods, candy, chewing gum, gelatin, ice cream, pudding, soft drinks, and syrups. Balsam is an excellent fixative, blending very well into perfumes of the oriental type and floral perfumes of the heavier type. Balsam wood is close-grained, handsomely grained, nearly mahogany in color, but redder, with a pleasant odor, retained for a long time, it takes a good polish, and is highly esteemed for cabinet work. Sometimes cultivated as coffee shade or as an ornamental. At one time balsam was so popular as an incense that papal edicts forbade the destruction of the trees. Seeds are used to flavor aguardiente, a popular Latin American alcoholic beverage.[40] Balsam fern is used in dental preparation to treat dry socket (postextraction alveolitis). Also, used in feminine hygeine sprays and as a fixative in creams, detergents, lotions, perfumes, and soap.[29] Choco Indians use the powdered bark as an underarm deodorant.

Peru balsam used extensively as a local protectant, rubefacient, parasiticide in certain skin diseases, antiseptic, and vulnerary, and applied externally as an ointment, or in alcoholic solutions; internally, rarely used as an expectorant. Dried fruits are sold in Guatemala for itch. An alcohol infusion is rubbed on in Cuba to alleviate headache and rheumatism. The resin is used for asthma, catarrh, rheumatism, gonorrhea, and to heal cuts and wounds. The balsam is used to treat venereal sores. Blended with castor oil or prepared as a tincture, it is used for chilblains, pediculosis, ringworm, and scabies. In suppositories, it is used for hemorrhoids and anal pruritis. Once used internally for amenorrhea, bronchitis, diarrhea, dysentery, dysmenorrhea, laryngitis, and leucorrhea. It is no longer used internally, at least in the U.S. Alcoholic extracts inhibit *Mycobacterium tuberculosis*.[40] Yucatan natives take an alcoholic extract of pulverized fruits for amenorrhea and dysmenorrhea using the resin for osteomyelitis, wounds, and ulcers.[42] Mexicans use the balsam for asthma, catarrh, and rheumatism. Choco apply the resin on the cheek for toothache. Choco Indians use it for the umbilical cord and uterine hemorrhage.[42]

Peruvian balsam contains about 60% cinnamein, a volatile oil, consisting mainly of benzyl cinnamate with some benzyl benzoate, resin esters (30 to 38%), vanillin, free cinnamic acid, and peruviol. Once the balsam is removed, the spent wood yields an essential oil containing 68 to 70% nerolidol. The seeds yield a "balsam" composed of 67.7% resin, 14.8% wax, 11.9% acid resin, 0.4% coumarin, 0.4% tannin, and 4.6% water.[40] A few years ago Wahlberg et al. examined a hydrocarbon fraction containing: alpha-pinene, styrene, *cis*-ocimene, *p*-cymene, alpha-bourbonene, alpha-copaene, beta-bourbonene, beta-elemene, carophyllene, alpha-curcumene, gamma-muurolene, alpha-muurolene, beta-selinene, δ-cadinene, alpha-cadinene, calamenene, and alpha-calacorene; a minor polar fraction containing cadalene, and 1,2-diaphenylethane and an oxygenated fraction containing benzylaldehyde, benzyl alcohol, ethyl benzoate, cinnamaldehyde, cinnamyl alcohol, methyl cinnamate, ethyl cinnamate, benzyl benzoate, benzyl cinnamate, cinnamyl benzoate, and triterpenoid constituents. Analysis of a weak acid fraction revealed eugenol, vanillin, benzyl ferulate, and a number of triterpene acids. The only two compounds present in a strong acid fraction were benzoic and cinnamic acids. The latter compound, which was found to be the predominant compound in tolu balsam, was considered to be an important flavor constituent.[31,208]

Toxicity — Early writers described systemic toxicity from absorption of the balsam, which has been applied to the nipples of a nursing mother and for scabies. Of 103 cases, 27 reacted to patch tests with cinnamic acid, 19 with benzyl cinnamate, 15 with vanillin, 12 with benzyl benzoate, 10 with benzoic acid. Fifty-four of the 103 reacted to one or more, while 49 were negative to all. Coumarin has been cited as a carcinogen.

236. *NARCISSUS TAZETTA* L. (AMARYLLIDACEAE) — Daffodil, Chinese Sacred Lily, Polyanthus Narcissus

Bulbous ornamental, grown in water and gravel, can be forced for Christmas. We received just such a narcissus, bulbs on top of gravel, for Christmas, 1982. Of the eight bulbs, flowers extended from January to March, aromatically in the kitchen window. The essential oil is used in high-grade French perfumes. Twig extracts are viristatic,[33] probably due to narciclasine.

Flowers thought to benefit the hair; used in female fevers in China.[16] Bulbs absorbent, diuretic, emetic, purgative.[1] Chinese chop and apply the bulbs externally as antiphlogistic and analgesic to abscesses, boils, itch, mastitis, bulbs poulticed to sores, swellings, ulcers, shaped into demulcent bolus "to carry bones out of the oesophagus."[16,33] Juice of bulb used for eye ailments. A female anticancer remedy,[41] considered abortifacient.[33]

Reported to contain lycorine ($C_{16}H_{17}O_4N$), narcitine, pseudolycorine, suisemine ($C_{17}H_{19}O_5N$), tazettine ($C_{18}H_{21}O_5N$);[1,16] also, galanthamine, narzettine, nartazine, haemanthamine, homolycorine, fiancine, hippeastrine, pluviine, galanthine, panceratine, narcissidine. Flowers contain narcissin (isorhamnetin-3-rutinoside, $C_{28}H_{32}O_{16}$), isorhamnetin, tazettine. Leaves contain 0.23% alkaloids. Flowers of var. *chinensis* (Chinese sacred lily) contain rutin, isorhamnetin, and carotenoid pigments. The essential oil contains 25% linalool, 24.8% benzylacetate, 8% benzyl alcohol, 4.3% alpha-terpineol, 3.9% cineol, 3.3% alpha-phenylpropylalcohol, 1.7% phenylpropylacetate, 1.5% indol, 0.4% phenylethylacetate, 0.3% heptylalcohol, 0.1% benzol(?), 0.08% phenylethylalcohol, 0.02% nonanal, traces of undecalactone.[33] The essential oil of the typical variety contains eugenol, benzyl alcohol, cinnamyl alcohol, benzaldehyde, and free and esterified benzoic acid. Bulbs contain 65.85 mg narciclasine per kilogram.[209]

Toxicity — POISON, in animals, induces gastralgia, gastroenteritis, accelerated pulse, and pyrexia; larger doses produce convulsions, paralysis, and death.[41] Lycorine is cytotoxic in the KB tumor system.[10] The volatile oil has such a strong odor as to cause headache. Bulbs of various narcissus species will cause nausea, vomiting, diarrhea, trembling, convulsions, and may be fatal.[34]

237. *NEPETA CATARIA* L. (LAMIACEAE) — Catnip, Catnep, Catmint

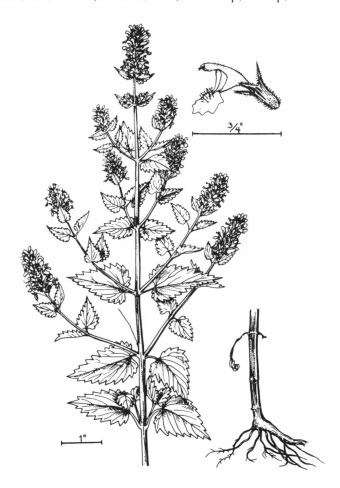

Dried leaves and flowering tops are used as a spice to prepare a sedative "tea" and as a tidbit for domestic cats. Leaves and shoots are used for flavoring sauces, soups, and stews. Catnip tea was frequent in England before the advent of modern Chinese tea. Of the thousands of tea combinations I have tried, one of my favorite is one part catnip, two parts beebalm, and four parts lemon balm. Whole fresh plants contain neptalactone, which is irresistible to the cat family (on the other hand, asafetida and mustard oil are supposed to repel cats). An essential oil (about 0.3%) is used as bait for wild cats. Some young Americans are reported to smoke the dried leaves, reported to have effects in man similar to those observed with cannabis. Showing that catnip increased the frequency of catnaps in chicks, Sherry and Koontz[69] concluded: "it is clear that 'catnip tea' possesses significant and potentially psychotropic activity." Rats are said to be repelled by catnip; hence, it is suggested as a rat-preventive border crop. Catnip is also a good insect repellent; of 27 insects tested, 20 were repelled, 7 unaffected by catnip.

A catnip salve is said to be a folk remedy for cancer, the tea a "cure" for internal cancer. A poultice of the whole plant is said to help corns.[4] Catnip is said to be antispasmodic, carminative, diaphoretic, digestive, emmenagogue, nervine, pectoral, refrigerant, sedative, soporific, stimulant, stomachic, sudorific, and tonic. The leaves may be chewed to alleviate toothache. In Appalachia, the tea is used for colds, hives, nerves, and stomach.[210] The leaves are smoked for respiratory ailments. Catnip tea has been recommended for amenorrhea, anemia, bronchitis, catarrh, colds, colic, debility, diarrhea, dysmenorrhea, fever, flatulence, headache, hysteria, indigestion, insanity, insomnia, nervousness, neuralgia, neurasthenia, nightmare, scurvy, stress, and tuberculosis. Leaves were sometimes chewed to alleviate toothache. Little sacks of catnip are tied around babys' necks, the aroma said to settle the stomach. Equal parts of catnip tea and saffron were once recommended for scarlet fever and smallpox. A tablespoon of the leaf juice, two or three times a day, is claimed to restore the menstrual flow when all else fails. A catnip poultice is said to reduce swelling. The herb should be infused only; boiling is said to spoil it.[44] The tea should be covered to slow the escape of volatile ingredients. According to Kloss, ''a high enema of catnip will relieve hysterical headaches. Such enemas, not condoned by the medical profession, are also suggested for colic, convulsions, dysmenorrhea, dyspepsia, dysuria, fever, fits and worms.''[44] One wireless flash[211] states ''catnip tea, for example, has a pick-me-up effect, and has been used as a folk remedy for years to treat colds, stomach ailments or jittery nerves. The tea is fortified with vitamins and is high in calcium and iron.''

Freshly harvested flowering tops yield 0.3 to 1.0% volatile oil by steam distillation, the major constituent nepetalactone (70 to 99%). Dry seeds contain 18.4% protein, 21.2% fat, and 3.2% ash. The seeds contain 57% linolenic, 18% linoleic, 12% oleic acid, and 6% saturated fatty acids.[33] The variety *citriodora* contains acetic acid, butyric acid, citral, citronellol, dipentene, geraniol, limonene, nerol, tiglic acid, and valeric acid.

Toxicity — Classified by the FDA (*Health Foods Business*, June 1978) as an Herb of Undefined Safety. Tyler, however, says, ''There just may be some basis in fact for the cup of hot catnip tea taken at bedtime to insure a good night's sleep (nepetalactone is somewhat similar in its chemical structure to the valepotriates, the sedative principle of valerian). Besides its relatively inexpensive, it tastes good, and no harmful effects from using it have been reported.''[37] Even Tierra, who advocates catnip enemas, says that smoking catnip ''may cause headaches.''[28] The LD_{50} intraperitoneally injected for catnip oil is 1300 mg/kg, the minimal lethal dose is 1000; for nepetalactone the LD_{50} is 1550, the minimum lethal dose is 1500; for nepetalic acid the LD_{50} is 1050 mg, the minimal lethal dose is 500.[212] Citronellal, geraniol (oral LD_{50} in rats 3600 mg), and citral (oral LD_{50} in rats 4960 mg) are also reported. List and Horhammer[33] report, also, carvacrol (oral LD_{50} 810 mg), pulegone (ipr LD_{50} 150 mg), thymol (oral LD_{50} in), caryophyllene, epinepetalactone, isodihydronepetalactone, ''methyl nepetonat''. In addition to the toxins carvacrol, saponin, tannin, and thymol, *N. cataria* contains camphor, beta-caryophyllene, dihydronepetalactone, epinepetalactone, humulene, nepetalactone, nepetalic acid, nepetalic anhydride, nepetol, and pulegone.

238. *NERIUM OLEANDER* L. (APOCYNACEAE) — Oleander, Rose

According to careful biblical students, the oleander is the "rose of the waterbrooks" — the "rhododendron" or "rose tree" of the Greeks. To the Spanish it is known as "laurel", and is their favorite shrub for parks and gardens. An evergreen summer favorite, "tough and attractive", it does well in almost any soil. In Greece, India, and Italy, it is a funeral plant. It is used to decorate Hindu temples. Palestineians secure from it a very active cardiac glucoside used in pharmacy. It is used as a rat poison in Europe. It is widely planted as an ornamental in tropical and subtropical countries. It may be the "willow of the brook" of Leviticus used for constructing booths for the Feast of Tabernacles. Also, regarded as the Jericho rose, since on the eastern side of Jordan the oleander becomes a tree 25 ft tall.[38]

A very POISONOUS plant, regarded as cardiac, cardiotonic, cyanogenetic, diuretic, emetic, emmenagogue, insecticidal, insectrepellant, parasiticide, purgative, sternutatory, stimulant, and tonic, oleander finds its way into folk remedies for apostemes, asthma, atheroma, carcinomata, corns, dysmenorrhea, eczema, epilepsy, epitheliomas, eruptions, herpes, malaria, psoriasis, ringworm, scabies, skin, snakebite, sore, tumors, and warts.[4,41,49] With such a fabulous folk repertoire of anticancer activity, oleander will probably be found to contain more proven anticancer agents than just the rutin and ursolic acid[10] reported from *N. indicum*. Leaves, flowers, and stembark possess cardiotonic properties, especially oleandrin, which is diuretic and stimulates the heart. In Ethiopia, Guatemala, and Curacao, the leaves are used for dressing skin diseases. The flavonol glycosides influence vascular permeability and possess diuretic properties. Cornerine has proved effective against cardiac ailments in clinical trials, particularly improving the heart muscle functions.[38] Venezuelans boil the leaves, inhaling the steam to alleviate sinus problems.[42] Leaves are dangerously applied to cutaneous eruptions; the decoction is used to destroy maggots in wounds. In Lebanon, as, perhaps, elsewhere, informants contradict, consider it calming yet irritating, a cause yet a cure for sore eyes, a medicine yet a poison.

Dry seed contain 30% protein, 29.4% fat.[21] Morton notes that all parts contain the cardiac glycosides oleandrin and neriin; also, folinerin, rosagenin, cornerin, pseudocuramine, rutin, cortenerin, and oleandomycin.[42] Toxicity and activity of many of these are detailed by Watt and Breyer-Brandwijk.[3] Additionally, it contains HCN, 0.049% caoutchouc, 0.014% sterol, 4.3% ursolic acid, quercetrin-3-rhamnoglucoside, kaempferol-3-rhamnoglucoside. Traces of vitamin A and K, 0.288% vitamin C are also reported.[3] Seeds contain 17.43% fat, plus a sitosterol. Tittel and Wagner,[213] analyzing leaves for cardenolides, reported that gentiobiosyloleandrin, odoroside A, and oleandrin were the chief glycosides present.

Toxicity — According to Morton,[42] children have died from eating a handful of flowers, adults from using the twigs as meat skewers. The nectar makes a toxic honey. A horse can be killed by 15 to 20 g fresh leaves, a cow by 10 to 12 g, sheep by 1 to 5 g.[3,42]

To the physician — Hardin and Arena suggest emesis or gastric lavage and symptomatic and supportive treatment, adding that potassium, procainamide, quinidine sulfate, disodium salt of edetate (Na_2 EDTA, dipotassium EDTA is preferable) have all been used effectively.[34]

239. *NICOTIANA GLAUCA* R. Grah. (SOLANACEAE) — Tree Tobacco

Introduced to South Africa as an ornamental. Wild and cultivated forms can serve as a source of anabasine, an insecticide. Useful in the breeding of disease-resistant tobaccos.

Reported to be anodyne, hirudicidal, insecticidal, and POISON, tree tobacco is a folk remedy for boils, headache, piles, sores, and wounds. Philips tells an interesting Lebanese tale of entopsychiatry[23] reflecting a Lebanese sheik who used the wild tobacco much as did the Amerindians. He put his patients into a trance with the infusion (and smoke) of the green leaf, then prayed to God asking for healing help, then answered God and the Angels as though they were conversing, "making use of hypnotic suggestion, with an indubitable mixture of sincerity and charlatanry."[23]

Anabasine and nicotine are the dominant alkaloids. Fruits and leaves are richest in anabasine (1.2 and 1.1%, respectively), followed by roots, flowers, and stems.[214] Leaves contain 1.2 to 2.1% rutin.[59]

Toxicity — Poisonous to cattle, horses, ostriches, and sheep, a few leaves have proven fatal to an ox.[215] Symptoms of poisoning in pregnant cows and pigs fed up to 700 mg/kg body weight were irregular gait, recumbency, excessive salivation (in cows), and tremors (in sows). Abnormalities in offspring were arthrogryposis of forelimbs and spinal curvature in calves, and arthrogryposis of fore and hind limbs in piglets.[216] Anabasine has an LD_{100} s.c. in guinea pigs of 22 mg/kg. Toxic symptoms of anabasine in humans include increased salivation, vertigo, confusion, disturbed vision and hearing, photophobia, cold extremities, nausea, vomiting, diarrhea, syncope, and clonic spasms.[20]

240. *NICOTIANA RUSTICA* L. (SOLANACEAE) — Aztec Tobacco, ''Turkish Tobacco''[54]
Indian Tobacco

Long before the conquest, leaves were dried and powdered and rubbed over the body in ceremonial ablution. The leaves were also chewed as a euphoriant.[54] According to the *Wealth of India*, this was the first tobacco grown in Virginia for export to Europe.[1] More recently it is cultivated in Australia, the Balkans, Burma, India, New Zealand, Pakistan, and the U.S.S.R. Rustica types are used for hookah, chewing, and snuff, being unsuitable for biris, cigarettes, and cigars.[1]

According to Hartwell, the leaves are used in folk remedies for cancerous wounds, indurations, and cancerous ulcers of the spleen, and scirrhous tumors in the arm and hip.[4] In Lebanon the wild Turkish tobacco, used more for snuff than smoking, is used as an emetic, expectorant, and externally as a poultice.[23]

Rustica types are generally high in nicotine.[1]

Toxicity — Classed by Emboden as a narcotic protoplasmic poison and retardant to neural transmission.[54]

241. *NICOTIANA TABACUM* L. (SOLANACEAE) — Tobacco, Virginia Tobacco, Tabac

Tobacco is primarily grown for the leaves. When cured, leaves are used for smoking, as cigars, cigarettes, or in pipes, or chewed, or used as snuff along with other ingredients. Cured tobacco is classified as flue-cured, fire-cured, air-cured, cigar filler, cigar binder, and cigar wrapper, and each of these categories can be broken into finer groups. Tobacco dust is widely used on vegetable crops as an insecticide, or made into a liquid form, commonly known as black leaf 40. Tincture of tobacco is used in Latin America to remove ticks.[42] Extracts proved strongly molluscicidal.[217] Removal of leaf proteins could yield a nutritious food and a safer smoking product, with carotenoids as a by-product.

According to Hartwell, the plant, usually the leaves, is used in folk remedies for cancer, carcinomata, cirrhosis, indolent ulcers, scirrhi, and tumors.[4] Reported to be anodyne, anorexigenic, antidotal, CNS-stimulant, intoxicant, laxative, narcotic, parasiticidal, psychedelic, purgative, sedative, and vermifuge, tobacco is a folk remedy for asthma, backache, boils, catarrh, cholecystosis, colds, coughs, debility, dysentery, earache, epistaxis, flu, gastrosis, headache, head cold, lethargy, lumbago, malaria, menorrhagia, paralysis, pediculosis, puerperium, rheumatism, scabies, snakebite, sores, spasm, stomachache, tetanus, toothache, ulcers, and wounds.[32] Medicinally, leaves are sedative, narcotic, emetic, antispasmodic, and used for rheumatic swelling, skin diseases, for scorpion sting, and as a fish poison. As a medicinal agent, it produces great depression, emesis, and convulsions, sometimes in very moderate doses. Internally, for this reason, it is seldom used. As an ointment made by simmering the leaves in lard, it is employed in curing old ulcers and painful tumors.

To minimize tobacco pagination, I refer readers to References 1, 3, and 33, etc., for too many details on the chemistry of tobacco, the killer. Lung cancer is expected to kill 117,000 Americans in 1983, 7000 more than in 1982. On a more positive note, a recent OTA panel[218] evaluated the potential of tobacco as a source of leaf protein. Speaking for tobacco, S. G. Wildman's associates calculated that with 150 tons plant per hectare, they could get 13.2 MT white fibers for cigarette tobacco worth $8750, 4.5 MT decolorized insoluble protein, and starch equivalent to soybean meal worth $725, 600 kg Fraction 1 Protein equivalent to egg white worth $2400, 300 kg Fraction 2 Protein equivalent to soy protein worth $300, and 15 kg carotenoids for poultry worth $3750. Not necessarily speaking against tobacco, Telek told the same 1982 OTA[218] audience that, with suitable plant material, the yield/ha/ yr of leaf proteins can be at least four times higher than that of seed proteins. Kung and Tso[219] reported that Fraction 1 Protein contained in the amino acids *circa* 9% aspartic acid, 5.2% threonine, 3.1% serine, 11.5% glutamic acid, 5.1% proline, 10.3% glycine, 9.4% alanine, 8% valine, 1.2% methionine, 4.5% isoleucine, 8.9% leucine, 4.4% tyrosine, 4.1% phenylalanine, 6% lysine, 2.8% histidine, and 6.5% arginine. Leaves contain 0.6 to 0.9% alkaloids, including nicotine, nornicotine, anabasine, and anataline; roots also contain most of these alkaloids. Leaves also contain the aromatic nicotianin (tobacco camphor). Dry seeds contain 22.4% protein and 45.4% fat.[21]

Toxicity — Many interesting details are recounted by Mitchell and Rook.[6] Dermatitis has been attributed to dyes, fungicides, antibiotics, arsenic, coumarin derivatives, menthol, diethylene glycol, triacetin, and vanilla in cigarettes rather than the alkaloids themselves. Even cigarette paper made from flax can cause dermatitis. "Poisoning in human beings and livestock is not infrequent from intentional or accidental misuse of nicotine products containing it. The alkaloid is readily absorbed after either ingestion or inhalation or through scarified or intact skin and is rapidly fatal in small amounts."[14] Kingsbury quotes a story from the *Los Angeles Times* about a family that used wild tobacco as a potherb with one fatality. Horses left in tobacco barns overnight and hogs unleashed in tobacco fields have died from ingesting tobacco.[17]

242. *OCIMUM BASILICUM* L. (LAMIACEAE) — Sweet Basil, Garden Basil

Commonly cultivated as an aromatic herb, sweet basil produces an essential oil (*circa* 0.15%), used in cordials (e.g., Chartreuse), cosmetics, perfumes, soaps, and spices. The oil is sometimes used to adulterate patchouli. Placed in water to form a beverage, the seeds become coated with mucilage and can be used as a gruel. It is especially good in combination with parsley and savory. For a pungent liqueur, boil down sugared basil, parsley, and savory and add to gin; for a sweet liqueur, cut back on the parsley and add the basil and savory to vodka. Bloody Mary is better with basil and parsley. Leaves with a salty, clove-like taste used in cooking, especially in tomato recipes. Dried leaves used to flavor bean soup, dressing, eggplant, fish, minestrone, omelets, peas, pizza, potato, rice, sausage, shellfish, soups, spaghetti, and stews. Flowers said to be eaten in China. Basil gives a clove-like taste to omelets, salads, soups, and vegetables. It makes an interesting butter, particularly good with tomato sandwiches. Italian tomato paste cans often contain a leaf of basil. Good in onion and turtle soups. Basil rivals oregano for seasoning pizza. Interesting with baked apple and grilled cheese. Darrah[220] mentions other culinary and medicinal uses of basil. Basil oil possesses insecticidal and insect repellent properties, effective against house flies and mosquitoes.[221] In the Mediterranean region, basil was grown on unscreened windows to repel flies. Laid over tomatoes, basil is said to repel fruit flies. Basil is used in hair pomades by the Africans.[3] It is bactericidal against *Salmonella typhosa*. Chinese recommended basil plantings to keep down odors of manure; perhaps basil should be considered as an intercrop with tomatoes on farms fertilized with sewage sludge.[41]

Reported to be antispasmodic, alexiteric, anodyne, aphrodisiac, carminative, cyanogenetic, demulcent, diaphoretic, digestive, diuretic, expectorant, lactagogue, laxative, pectoral, refrigerant, sedative, stimulant, stomachic, and sudorific, basil is a folk remedy for alcoholism, anasarca, boredom, catarrh, cephalalgia, childbirth, cholera, colic, collapse, constipation, convulsion, cough, cramps, croup, deafness, delirium, depression, diarrhea, dropsy, dysentery, earache, enteritis, epilepsy, fever, flu, frigidity, gastroenteritis, gonorrhea, gout, gravel, halitosis, headache, hemiplegia, hiccup, hysteria, infection, inflammation, insanity, insect bites, labor, migraine, nausea, nephrosis, nerves, piles, paralysis, polyps, ringworm, snakebite, sinusitis, sores, sore throat, spasm, stings, stomach, throat, toothache, tumors, urinary ailments, wart, whooping cough, and worms.[32] An infusion of the plant is taken for halitosis, headache, and gout. Leaf juice allays irritation of the throat. It is applied to ringworm and snuffed for earache. In the Philippines, the leaves are poulticed onto fungal infections. Poultices of the seeds are applied to sores and sinuses.[127] Aqueous extracts of the seeds are active against Gram-positive bacteria and mycobacteria. In Salvador, placed in the ears to cure deafness.[9] Used in India as a nasal douche in myosis.[5] Leaves are used as a cataplasm for tumors. Hot baths with twigs are said to cure cancer of the stomach. The decoction from the whole plant is said to remedy nasal polyps.[4] A handful of leaves steeped in wine yields a digestive tonic. Seeds considered aphrodisiac, demulcent, diaphoretic, diuretic, refrigerant, and stimulant. Seeds given internally for constipation and piles, and used in poultices for sores and sinuses. In Malaya the leaf decoction may be administered after childbirth, and the juice may be taken if the menses are delayed. Expressed use of leaves said to expel worms and cure ringworm (a fungus). Seeds in water are administered to kidney ailments in India.

Contains eucalyptol, estragol, 1,8-cineol, eugenol, borneol, ocimene, geraniol, anethole, 10-cadinols, beta-carophyllene, alpha-terpinole, camphor, 3-octanone, methyl eugenol, safrol, sesquithujene, 1-epibicyclosequiphellandrine, linalool, and methyl chavicol.[29,33] Basil seeds yield a mucilage which on hydrolysis yields uronic acid, glucose, xylose, and rhamnose. Methylcinnamate may dominate the oil of some chemovars. The drying oil of the seed contains 7.0% palmitic, 0.2% stearic, 11.0% oleic, 60.0% linoleic, and 21% linolenic acids. The unsaponifiable fraction is said to contain beta-sitosterol, oleonolic, and ursolic acid. Per 100 g, the leaves are reported to contain 43 calories, 86.5 g H_2O, 3.3 g protein, 1.2 g fat, 7.0 g total carbohydrate, 2.0 g fiber, 2.0 g ash, 320 mg Ca, 38 mg P, 4.8 mg Fe, 12 mg Na, 429 mg K, 4500 µg β-carotene equivalent, 0.08 mg thiamine, 0.35 mg riboflavin, 0.80 mg niacin, and 27 mg ascorbic acid. Bowers and Nishida[222] isolated and identified two agents, juvocimene 1 and juvocimene 2, from the oil, with highly potent insect juvenile hormone activity. The compounds resulted in a dose-dependent inhibition of metamorphosis in *Oncopeltus fasciatus*. Per 100 g, ground basil is reported to contain 250 calories, 6.4 g H_2O, 14.4 g protein, 4.0 g fat, 61.0 g total carbohydrate, 17.8 g fiber, 14.3 g ash, 2113 mg Ca, 490 mg P, 42 mg Fe, 422 mg Mg, 34 mg Na, 3433 mg K, 5.8 mg Zn, 9375 IU vitamin A, 0.15 mg thiamine, 0.32 mg riboflavin, 6.9 mg niacin, and 61.2 mg ascorbic acid; of the amino acids, there are 211 mg tryptophan, 588 threonine, 588 isoleucine, 1078 leucine, 618 lysine, 202 methionine, 159 cystine, 733 phenylalanine, 432 tyrosine, 717 valine, 622 arginine, 287 histidine, 747 alanine, 1696 aspartic acid, 1565 glutamic acid, 690 glycine, 588 proline, and 561 mg serine.[39] There are 106 mg phytosterols.

Toxicity — Leaf juice said to be slightly narcotic.[1] Two carcinogens, safrole and estragole (methyl chavicol), are reported in some sweet basil oils.[29]

243. *OENANTHE PHELLANDRIUM* Lam. (APIACEAE) — Water Fennel, Water Dropwort

Poisonous plant, sometimes confused with edible relatives in the carrot family.

Reported to be alterative, carminative, diaphoretic, diuretic, emmenagogue, expectorant, poison, and sudorific, water fennel is a folk remedy for asthma, bronchitis, cancer, catarrh, consumption, dyspepsia, hemorrhoids, laryngitis, malaria, malignant ulcers, tuberculosis, tumors.[32,33] Roots are externally applied to hemorrhoids.[2] Used homeopathically as an expectorant, for digestive disorders and bladder ailments.[33]

Fruits contain 1 to 2.5% essential oil, about 80% of which is phellandrene,[2] *circa* 20% oil, 4% resin, 2 to 3% wax, galactan, mannan, and mucilaginous material.

Toxicity — In overdoses, the fruits produce vertigo, intoxication, and other narcotic effects. The root has caused fatalities when eaten by mistake. Early symptoms are stomach irritation, circulatory failure, cerebral disturbance, giddiness, convulsions, and coma. Fresh leaves may cause paralysis in cattle which eat them.[2]

244. *ORNITHOGALUM UMBELLATUM* L. (LILIACEAE) — Dove's Dung, Star of Bethlehem

Bulbs were used for food in Syria. In Dioscorides' day, the bulbs were commonly gathered, ground into meal, and mixed with flour to make bread. Modern Italians, in time of scarcity, eat the bulb.[38] Egyptians and Syrians store the bulbs for their pilgrimages to Mecca.[23] My favorite foraging book for use in the field, *Edible Wild Plants of Eastern North America*[13] treats the bulbs as edible, while my favorite poisonous plants text[14] reports that the bulbs have caused death in cattle in the U.S. Because the pretty stars open late in the day, they have been called "Sleepy Dick" or "Eleven O'clock Lady"[42] and they marked the morning 11 in Linnaeus' floral clock.[38]

Lebanese used the bulb for lymphatic ailments and recommended them in diets for debility.[23] They are considered diuretic[32] and emollient after roasting.[33] Bulbs of other Eastern species were used for cachexia, infections, parotitis, scabs, ulcers, and wasting disease. Used in Bach flower remedies "for all kinds of shock" and the "after effects of trauma."[71]

Although early reports of the gout medicine colchicine have been extricated from the credible literature, there are still reports of other toxins (e.g., convallotoxin, convalloside, and strophanthindin) in *Ornithogalum*.[38] According to *Hager's Handbook*, roots contain 0.04% convallotoxin, convalloside, rhodexin-A (sarmentogenin-alpha-L-rhamnoside), rodexoside (sarmentogenin-L-rhamnose-D-glucose), proteins, lipids, free fatty acids, unsaturated phospholipids, and mucilage.[33]

Toxicity — Reports of edibility are remarkable since chemical analysis shows that the entire plant is intensely poisonous. Grazing animals avoid it — or they may be poisoned.[38] Skeptical of the salubrity of dove's dung, I boiled a few bulbs from the dark green lawn weed, without salt, and then cautiously consumed one. It had a saponaceous quality, suggesting that I might be consuming a hemolytic saponin from a dangerous family. There was a bitter aftertaste. Then I salted a bulb which was a bit more palatable. I would need to be near starvation to consume more of these. I experienced a shortness of breath following the ingestion of only two bulbs. Symptoms of poisoning include salivation, flatulence, weakness, possibly leading to death.[33]

245. *PAEONIA OFFICINALIS* L. (RANUNCULACEAE) — Peony

Common ornamental. Roots were once celebrated for their medicinal value for cephalic and neuralgic disorders. Seeds and/or peony water was once used as a spice.[2] The seed oil has been extracted.

Reported to be alterative, antispasmodic, deobstruent, depurative, emetic, emmenagogue, poison, tonic, peony is a folk remedy for convulsions, diarrhea, dropsy, epilepsy, epistaxis, hemorrhage, intestines, liver ailments, lunacy, neuroses, pimples, pregnancy, puerperium, spasms, splenosis, stomach ailments, tumors, uterosis, varices, and wounds.[2] Used homeopathically for children's cramp, anal fissures, cystitis, and hemorrhoids.

Root contains 4 to 5% glucose, 8 to 14% saccharose, 14 to 20% starch, *circa* 2% metarabinic acid, 1% organic acids, tannic acid, calcium oxalate, 0.4% essential oil (with much peonol), benzoic acid and esters, glutamine, arginine, tannin, paeoniaflioureszine. Rootbark contains peoniide ($C_{15}H_{20}O_8$); the alkaloid peregrinine is uterotonic, hemostat, narrowing the renal capillaries. Flowers contain a red pigment, the anthocyanin, and tannin. Seeds also contain peregrinine, 23.6% fat, 1.13% resinic acid, 1.4% sugar, arabinic acid, 4% of the dye "peonia-brown", 11% protein, pectin; and, in the testa, peonifluoreszin.

Toxicity — Flowers and seed induce gastroenteritis with nausea, colic, and diarrhea.

246. *PANAX GINSENG* C. A. Mey. (ARALIACEAE) — Chinese or Asiatic Ginseng

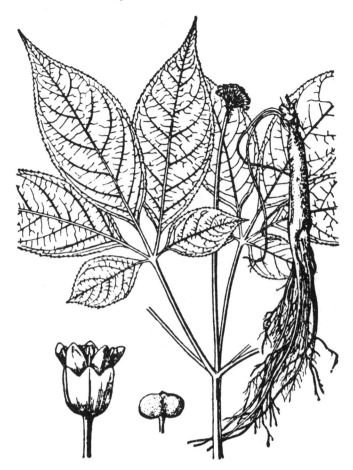

One of the most famous, and now rare, medicinal plants of China. Wild specimens may now command thousands of dollars. Leaves are considered emetic, expectorant. Roots are regarded as alterative, antiseptic, carminative, demulcent, stimulant, tonic; they are used for asthenic hemorrhages, cancer, cough, debility, dyspepsia, exhausting discharges, fever, malaria, nausea, polyuria, spermatorrhea, and are frequent in polyherbal prescriptions for diabetes mellitus. The plant is considered tonic. In spite of U.S. claims that American ginseng has no important medicinal virtues, Chinese pharmacologists insist that ginseng has positive action as a nerve and cardiac stimulant, increasing metabolism, retarding impotence, regulating blood pressure and blood sugar . . . "Indicated in asthma, dizziness, epistaxis, headache, nausea, parturition, rheumatism."[29] Of 42 ginseng recipes from the Pen Tsao Kang Mu, five contained ginseng along, two ginseng with mineral drugs, two with animal products, and 33 with plant drugs. Pure ginseng recipes were used for revitalization, thirst, swellings, and pains from dogbite, bites of poisonous centipedes, and stings of wasps and scorpions.[223] This would make one suspicious of the high prices Americans pay for Chinese ginseng, surely not for the bites of centipedes, dogs, and wasps. Some writers maintain that ginseng is mainly a tonic to increase the efficiency of mental and physical work.[224] In a rather extensive and expensive computer index,[32] ginseng is listed for folk use as alterative, amnesia, anemia, anodyne, anorexia, apertif, aphrodisiac, asthma, atherosclerosis, boils, bruises, cachexia, cancer, cardiotonic, carminative, convulsions, cough, debility, diabetes, diuretic, divination, dysentery, dysmenorrhea, dyspepsia, emetic, enterorrhagia, epilepsy,

epistaxis, estrogenic, expectorant, fatigue, fear, fever, forgetfulness, gastritis, gonado-
trophic, hangover, headache, heart, hematoptysis, hemorrhage, hyperglycemia, hypertension, hypotension, impotence, insomnia, intestines, longevity, malaria, menorrhagia, nausea,
nervine, neurasthenia, palpitations, polyuria, pregnancy, puerperium, rectocele, renitis,
rheumatism, sedative, shortbreath, sialogogue, sores, spermatorrhea, splenitis, stimulant,
stomachic, swelling, tonic, tranquilizer, and vertigo.[41]

Subcutaneous injections in young mice of ether extracts induce follicule arrector muscle
stimulation, the action like that of yohimbine. Ginseng saponin, at a dose of 10 μg/mℓ,
was radioprotective when present prior to gamma-irradiation. Constituents of Asiatic ginseng
increase adrenal capacity in stressed animals. Ginseng may improve radiation resistance as
part of a general improvement in stress response. Tong and Chao[225] even adduce evidence
in support of the ginseng longevity connection: "the cell density of the human amnion cells
grown in the medium containing crude aqueous extract of ginseng is greater than that of the
control due to the prolonged life span of the treated cells." If Rg_1 can promote mitosis in
some tissues in human body in vivo, and this seems likely according to the results reported
in rats and mice, perhaps it can improve the general resistance and regenerative condition
of a person, especially in the old one, by activating the metabolic processes and cell proliferation.[41] Ginseng extract is said to be mitogenic as well as antimitogenic, depending on
the dose used.[225] Ginseng has been shown to facilitate the mating behavior of male rats.[226]
Males under the influence of ginseng began ejaculation earlier and repeated the action more
often than controls.[226] A compound A ($C_6H_6O_3$ = 3-hydroxy-2-methyl-γ-pyrine = maltol)
was shown to be the antioxidant lying behind the "anti-aging activity of Korean ginseng."[227]
Panaxin stimulates the cerebellum system, heart, and blood vessels, panaxic acid invigorates
the heart and metabolism; panaquilon stimulates internal secretions; panacene benefits the
cerebrum and spinal cord, while ginsenin shows hypoglycemic activity. Both Chinese and
Russian investigators have demonstrated hypoglycemic and cardiotonic activity. Ginseng
has been given for arrhythmia with shock-like condition, hypotension, and shock. In patients
with hypotension and shock, oral administration of ginseng decoction or ginseng powder
seemed to strengthen myocardial contraction and elevate the blood pressure. Four-year-old
roots contained, per 100 g: 338 calories, 10.0% water, 12.2 g protein, 70.0 g carbohydrate,
4.2% fiber, 1.0 g fat, 2.6% ash. 100 IU vitamin A, 1.0 mg vitamin C, 0.10 mg vitamin
B-1, 0.108 mg vitamin B-2, 4.70 mg niacin, 234 mg calcium; 490 mg iron, 1.49 IU vitamin
E, 0.48 mg vitamin B-6, 0.0506 mg folic acid, 0.31 mcg vitamin B-12, 216 mg phosphorus,
5.0 mcg iodine, 98.0 mg magnesium, 1.04 mg zinc, 0.62 mg copper, 0.00772 mg biotin,
and 0.69 mg pantothenic acid.[228]

Toxicity — Emboden lists the plant with his narcotic tranquilizers.[54] JAMA recently
carried an article called "Ginseng Abuse Syndrome",[229] which failed to convince me that
moderate use of ginseng was harmful. The histamine-liberating property would indicate
caution by asthmatics or emphysema patients. "In 1958, the FDA gave panax ginseng the
'Generally Recognized as Safe (GRAS) status', but only as a tea infusion; the agency has
repeatedly confiscated other forms of the herb."[231] According to Baranov,[154] "Soviet physicians who have tested ginseng in clinics state that healthy people, under the age of 40,
should not use ginseng . . . patients have been found to be allergic to ginseng, showing
symptoms such as palpitation, insomnia, and pruritis. Ginseng overdose may cause similar
symptoms as well as heart pain, decrease in sexual potency, vomiting, hemorrhage, diathesis,
headache, and epistaxis; ingestion of a very large dose of ginseng preparations may even
cause death."

247. *PANAX NOTOGINSENG* (Burkill) Hoo & Tseng (ARALIACEAE) — Sanchi Ginseng, Tienchi Ginseng

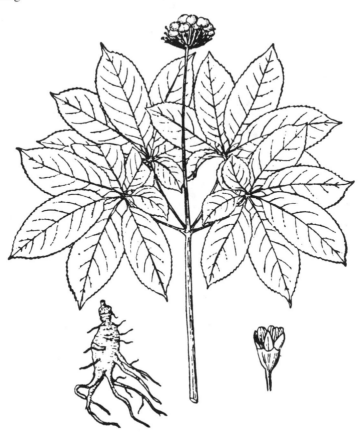

Being introduced into Chinese medicinal shops in the U.S. and Europe, sanchi ginseng, a rather expensive item, is suggested as an additive to chicken soup. I suspect it is too expensive for most Chinese to use in chicken soup. Perhaps the chicken soup is as valuable medicinally as the Sanchi but not nearly so expensive. One of its names translates ''mountain varnish'', because it is used to stick the edge of wounds together.[16]

The rhizome is used for contusions, epistaxis, hematochezia, hematuria, metrorrhagia; the tincture, rubbed on bruises, is astringent, discutient, styptic, tonic, vulnerary. The root is regarded as analgesic, antiphlogistic. Tierra classes it as the best treatment for hemorrhage, adding that it maintains normal body weight, prevents fatigue and alleviates stress,[4] normalizes blood pressure and heart rate, and improves circulation[29] and inflammation.

Liu and Staba,[232] comparing the ginsenosides with those of other ginseng plants and products, found the highest total ginsenoside concentration in sanchi ginseng (8.7 w/w%) with larger proportions of Rbg, Re, and Rgl than Rc, Rd, and F_{11}. Both (20S)-protopanaxadiol and (20S)-protopanaxatriol ginsensoides were detected. Sanchi saponins have sedative effects on mice. In vitro, sanchi saponins inhibit *Escherichia coli, Sporotrichum schenchii, Staphylococcus, Trichophyton gypseum*, and *T. tonsurans*. The anticancer compound, beta-sitosterol (perhaps ubiquitous), has been reported with daucosterol, a glycoside of quercitin and six saponins from root hairs.[41]

Toxicity — The LD_{50} for mice (subcutaneously) is 1667 mg/kg.

248. *PANAX QUINQUEFOLIUS* L. (ARA LIACEAE) — Ginseng, Sang

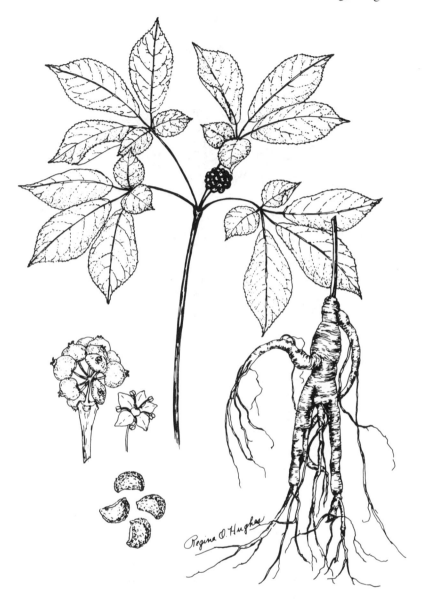

With few genuine uses in the U.S., ginseng is one of the biggest herbal exports, exceeding 35 million dollars worth per year recently. Ginseng is one of the most popular items, with aloe and jojoba, in natural cosmetics of all sorts. It is one of the main components in a tea I call root booster, with ginger, sarsaparilla, and sassafras. Recently, the FDA has suggested that it is legitimate to add water to ginseng (water then being an approved food additive), but not the ginseng to water (water is a food, but ginseng is not approved as a food additive or a drug).[233] This prompted a pretty witty ditty.[234]

"Sang's" Swan Song

According to their spokesman, water is a food,
Perpetrate a hoax man, water is real good,
The FDA, it say, hey, ginsang's also good,
Until it's put in water, and then it ain't no good.

So that's the way they play, hey, the good old FDA,
Ring along with dingdong, sing a song of "sang," hey,
Done sung the "sang's" swan song.

Add the water to the "sang" and then it's all OK,
Don't add the sang to water, warns the FDA,
To add the sang to water, you have to prove "sang" safe,
That costs a pretty dollar, 1 million and a half.

Let's not get real neuro, cats, nor friends, nor FDA,
Another group of bureaucrats, will sell the "sang" away,
List endangered species, put a price upon its head,
Then comes the bounty hunter, and bang! the "sang" is dead.

So that's the way they play, hey, the good, great, grand, ESSA,
Ring along with dingdong, sing a song of "sang," hey,
Done sung the "sang's" swan song.

A sad song for your soul son, endangered species we,
Water's foul and the "sang" is sung, can't drink no ginseng tea,
Water and "sang" are toxic, they've saved us from the worst,
FDA fed logic, we'll only die of thirst.*

Root used as a panacea (cure-all) by wealthy Chinese and other orientals, who consider it carminative, diuretic, stimulant, and tonic, Chinese sick chew the root to recover health. Healthy Chinese chew it to increase their vitality. It is said to remove both mental and bodily fatique, to cure lung disorders, dissolve tumors, and prolong life. Reportedly, it reduces blood sugar concentration and acts favorably on metabolism, the central nervous system, and on the endocrine secretions. Employed in the Orient in the treatment of anemia, diabetes, insomnia, neurasthenia, gastritis, and especially for sexual impotence, hence, used as an aphrodisiac. The American medical profession recognizes only its demulcent properties, although Appalachian folk uses include tonic and aphrodisiac. An infusion of the leaves is said to make a palatable tea. I can't tell where Parvati learned that Chicana midwives give it to prevent any infection nor that Indians used it as a female contraceptive under the name "tartar root".[48]

Nutritionally, American ginseng is probably similar to Oriental ginseng. According to *Hager's Handbook*, the herbage contains panasenoside, a kaempferol-3-glucogalactoside, kaempferol and trifolin, and something like panaxoside C. The roots contain a complex array of saponins, primary glycosides which are based on oleanolic acid or dammarol. Formulas of many of these are presented.[33] Additionally, ginseng contains the acetylene derivative panaxynol ($C_{17}H_{26}O$), a 1,9-*cis*-hepta-decadiene-4,6-diyn-3-ol, beta-elemene, beta-sitosterol and its glucoside, 0.05% essential oil, panacene ($C_{15}H_{24}$), a pyrrolidone, 5-peptides (D-glucose and D-fructose), 3.3% disaccharides (saccharose and maltose). Among vitamins 0.66 mm/100 g pantothenic acid, 0.92 μg/100 g biotin, vitamins B_1, B_2. B_{12}, nicotinic acid, oleic acid, 0.1 to 0.2% choline, citric, fumaric, malic, maleic, panaxic, and tartaric acids, traces of Mn, V, Cu, Co, and As, and 0.15% S.

Toxicity — The LD_{50} for mice (subcutaneously) is 1667 mg/kg.

* ©J. A. Duke, 1977. Following the 1st Herb Trade Association Meeting, Santa Cruz, California.

249. *PANCRATIUM TRIANTHUM* Herb. (AMARYLLIDACEAE) — Kwashi

Bulbous ornamental. Botswana bushmen prize it, not as an ornament, but to alter the mind, producing colorful and vivid hallucinations. Rather than eat the bulb, they cut it open and apply it to self-inflicted wounds on the forehead.

Other species, reported to be anodyne, emetic, pectoral, poison, purgative, are used in folk remedies for asthma, ear problems, fever, inflammation, orchitis, and splenosis.[32]

Toxicity — Toxic alkaloids of the tazettine type occur in the genus. Listed by Emboden as a narcotic hallucinogen.[54] Some species, perhaps even this one, contain one or more cardiac toxins.

250. *PAPAVER BRACTEATUM* Lindl. (PAPAVERACEAE) — Great Scarlet Poppy, Thebaine Poppy

A handsome ornamental perennial now of greater interest as a source of thebaine, a medicinally important alkaloid. Thebaine can be rather readily converted to codeine, a widely used analgesic and antitussive, but it is not so readily converted into the drug of abuse, heroin. Naloxone, an important narcotic antagonist derived from thebaine, is administered to infants borne of heroin addicts, lest the children experience withdrawal.[17] Watson et al.[236] showed that naloxone produced decreases in auditory hallucinations in some schizophrenic patients. Faden and Holaday report the use of naloxone to treat shock following acute blood loss in conscious rats.[237] Perhaps even more interesting is the report of the induction of copulatory behavior in sexually inactive rats by naloxone.[238] My Bihai driver in Iran told me that his grandfather, like many other shepherds, would smoke the "bracteum" (congealed latex) when he could not get opium (now a capital offense in Iran)! Whole seeds are used in the Kurdistan for baked goods and confectionary. Seeds can be used in lieu of opium poppyseed for food and as a source of oil. But in 1977, the *Chemical Marketing Reporter*[239] stated that the Drug Enforcement Agency Administrator announced that the agency would not authorize domestic commercial production.

Lack of citation in Hartwell[4] and Duke and Wain[32] might suggest that the great scarlet poppy is not used in folk remedies. However, Hartwell did report that the closely related. *P. orientale* was used for tumors of the heart.[4]

The seed contains 45 to 48% oil, rich in unsaturated fatty acids, 7.7 to 7.9% palmitic-, 2.2 to 2.5% stearic-, 8.6 to 9.2% oleic-, and 79.7 to 79.9% linoleic-acids.[240] Per 100 g, the seed meal is reported to contain 4.9 to 5.3 g H_2O, 32.7 to 42.0 g protein, 6.3 to 20.7 g fat, 14.3 to 14.7 g fiber, 7.5 to 8.2 g ash. Aspartic acid ran 9.2 to 9.5 g/16 g N; threonine, 3.4 to 3.6; serine, 4.0 to 4.1; glutamic acid, 20.1 to 20.4; proline, 3.7 to 3.9; glycine, 4.2 to 4.4; alanine, 3.8 to 4.0; valine, 4.6 to 5.0; isoleucine, 3.75 to 3.85; leucine, 6.2 to 6.3; tyrosine, 3.8 to 4.0; phenylalanine, 3.6 to 3.8; lysine, 3.9 to 4.0; histidine, 2.2 to 2.4; arginine, 9.9 to 10.1; methionine, 2.35 to 2.45; and cystine, 2.2 to 2.3 g/16 g N.[241] Duke[91] summarizes the reported alkaloids: acetylornithine, alpinigenine, bractamine, bractavine, bracteine, bracteoline, codeine, coptisine, isothebaine, morphine, nuciferine, oreophiline, orientalinone, oripavine, oxysanguinarine, papaverrubines B, D, and E, protopine, salutaradine, and thebaine, the most important alkaloid. Kettenes et al.[242] review the biological activities of thebaine, isothebaine, (−)-nuciferine, protopine, tetrahydropalmatine, coptisine, and alpinigenine. According to Morton,[17] the major alkaloids are thebaine ($C_{19}H_{21}NO_3$) and alpinigenine ($C_{22}H_{27}NO_6$). Fresh capsule latex may contain 0.125% thebaine; dried, 0.25 to 0.26%. Thebaine in dried roots ranges from 0.50 to 1.3%; in stems, 0.09 to 0.22%. None or only a little occurs in the leaves. In some strains, shoot concentration in young plants ranges from 233 to 808 ppm; in older plants, 437 to 1606 ppm. Alpinigenine occurs in all above-ground parts and may constitute 18% of the total alkaloids in the capsule latex.

Toxicity — Etorphine, or M99, one of the "Bentley Compounds" derived from thebaine, is much used to sedate large wild animals for scientific purposes.[17] Toxic activities of some of the alkaloids are discussed by Duke.[243] The dependence potential of thebaine is partially attributed to oripavine, one of the principal metabolites of thebaine. The analgesic potency of oripavine in mice is higher than that of thebaine and comparable to morphine.[244] At doses of 4 mg/kg rats show preconvulsive disorders, hyperirritability, tremor, muscle rigidity, motor impairment, and transient muscle contraction. At 5 mg/kg the rats died immediately in a persistent tonic seizure. Thebaine has an LD_{50} of 20 mg/kg in mice. In chicken embryos, thebaine is reported to induce pseudohyperfeminization.[245]

251. *PAPAVER SOMNIFERUM* L. (PAPAVERACEAE) — Opium Poppy, Poppyseed Poppy, Keshi

Opium is the air-dried milky exudation obtained from excised unripe fruits. It is extensively smoked as an intoxicant. Commercial products are called Turkey opium, Indian opium, Persian opium, Chinese opium, and Egyptian opium, and they differ in appearance and quality. Opium is largely used for manufacture of morphine, codeine, narcotine, laudanine, papaverine, and many other alkaloids. Codeine phosphate is the biggest selling opium derivative,[246] with industry sales averaging around 60,000 k/year, worth about $40 million at the going market price. One lab bought about half the codeine sold commercially in the U.S. Its "tylenol" with codeine was the biggest selling codeine analgesic in the U.S. in 1982, racking up sales of $72 million. In one survey, codeine showed up in 22.23% of the analgesic preparations in the U.S. Other opium derivatives being oxycodone (3.63%), dihydrocodeinone (0.54%), dihydromorphinone (0.45%), morphine (0.41%), oxymorphone (0.10%), and opium tinctures and extracts (0.16%).[68] Codeine appeared in more than one quarter of our antitussives and decongestants. Papaverine, ephedrine, and caffeine are three of one survey of 50 pure drugs (obtained from higher plants), those 3 being entirely prepared by synthesis. Papaverine is the fourth most common coronary vasodilator.[88] Opium is also the source of the toxic and extremely habit-forming narcotic heroin or dimorphine, prohibited in some countries. Seeds contain no opium, and are used extensively in baking and sprinkling on rolls and bread. Seeds are a good source of energy. They are also the source of a drying oil, used for manufacture of paints, varnishes, and soaps, and in foods and salad dressing. Oil cake is a good fodder for cattle. Seeds used for preparation of emulsions (white-seeded varieties preferred); the bluish-black varieties are generally used for baking. Stems used for straw. Lecithin has been extracted from poppy seed meal. Seedlings are eaten as a potherb in Iran. As the peony flowered poppy, the opium poppy is widely grown as an ornamental, even here in the U.S. where it is illegal to grow.

Regarded as analgesic, anodyne, antitussive, aphrodisiac, astringent, bactericidal, calmative, carminative, demulcent, emollient, expectorant, hemostat, hypotensive, hypnotic, narcotic, nervine, sedative, sudorific, tonic, poppy has been used in folk remedies for asthma, bladder, bruises, cancer, catarrh, cold, colic, conjunctivitis, cough, diarrhea, dysentery, dysmenorrhea, enteritis, enterorrhagia, fever, flux, headache, hemicrania, hypertension, hypochondria, hysteria, inflammation, insomnia, leucorrhea, malaria, mania, melancholy, nausea, neuralgia, otitis, pertussis, prolapse, rectitis, rheumatism, snakebite, spasm, spermatorrhea, sprain, stomachache, swelling, toothache, tumor, ulcers, and warts.[32,33,243] Hartwell[4] mentions opium as a remedy for such cancerous conditions as cancer of the skin, stomach, tongue, uterus, carcinoma of the breast, polyps of the ear, nose, and vagina; scleroses of the liver, spleen, and uterus; and tumors of the abdomen, bladder, eyes, fauces, liver, spleen, and uvula. The plant, boiled in oil, is said to aid indurations and tumors of the liver. The tincture of the plant is said to help cancerous ulcers. Smoking the plant is said to cure cancer of the tongue. The capsule decoction is said to cure uterine cancer. An injection of the seed decoction is also said to help uterine cancer. Egyptians claim to become more cheerful, talkative, and industrious following the eating of opium. When falling asleep, they have visions of "orchards and pleasure gardens embellished with many trees, herbs, and various flowers." Lebanese used their opium wisely; to quiet excitable people, to relieve toothache, headache, incurable pain, and for boils, coughs, dysentery, and itches. Algerians tamp opium into tooth cavities. Iranians use the seed for epistaxis; a paste made from *Linum*, *Malva*, and *Papaver* is applied to boils. Ayurvedics consider the seeds aphrodisiac, counstipating, and tonic, the fruit antitussive, binding, cooling, deliriant, excitant, and intoxicant, yet, aphrodisiac if freely indulged, the plant aphrodisiac, astringent, fattening, stimulant, tonic, and good for the complexion; in Unani medicine, the fruit is suggested as well for anemia, chest pains, dysentery, fever, but is correctly deemed hypnotic, narcotic, and, perhaps, harmful to the brain.[38] The plant provides a narcotic that induces sleep, a sleep so heavy that the person becomes insensible. When the Roman soldiers at Golgotha took pity on their prisoner on the cross, they added this poppy juice to the potion of sour wine. Its compounds are used in medicine as analgesic, anodyne, antispasmodic, hypnotic, narcotic, sedative, as respiratory depressants, and to relieve severe pain. Jewish authorities maintain that the plant and its stupefacience were well known among the Hebrews more than 2000 years ago. The Jerushalmi warns against opium eating. Although the seeds contain no narcotic alkaloids, urinalysis following their ingestion may suggest the morphine of a heroin addict's urinalysis.[38]

Seed is reported to contain 4.3 to 5.2% moisture, 22.3 to 24.4% protein, 46.5 to 49.1% ether extract, 11.7 to 14.3% N-free extract, 4.8 to 5.8% crude fiber, 5.6 to 6.0% ash, 1.03 to 1.45% calcium, 0.79 to 0.89% phosphorous, 8.5 to 11.1 mg/100 g iron, 740 to 1181 μg/100 g thiamine, 765 to 1203 μg/100 g riboflavin, 800 to 1280 μg/100 g nicotinic acid; carotene is absent. Minor minerals in the seeds include 6 μg/kg iodine, 29 mg/kg manganese, 22.9 mg/kg copper, 15.6 g/kg magnesium, 0.3 g/kg sodium, 5.25 g/kg potassium, and 130 mg/kg zinc; the seeds also contain 2.80% lecithin, 1.62% oxalic acid, 3.0 to 3.6% pentosans, traces of narcotine, and an amorphous alkaloid; and the enzymes diastase, emulsin, lipase, and nuclease. Poppyseed oil cakes were estimated to have 88 feed units per 100 kg, 27.5% digestible crude protein, and 25.6% digestible true protein. Another analysis showed 54% organic matter, 87% crude protein, 100% crude fat, 0% crude fiber, and 48% N-free extract. The opium alkaloids are rather numerous, Duke[97] having listed allocryptopine, apomorphione, berberine, codeine, codeinone, corytuberine, cryptopine, desmethylepiporphyroxine, glaudine, gnoscopine, hydroxycodeine, isoboldine, isocorypalmine, lanthopine, laudanidine, laudanine, laudanosine, 6-methylcodeine, morphine, narcotine, narcotoline, nornarceine, oxycryptopine, oxymorphine, oxynarcotine, palaudine, papaveraldine, papaveramone, papaverine, papaverubines B, C, D, and E, protopine, pseudomorphine, reticuline, roemerine, salutaridine, salutaridinol-1, sanguinarine, and thebaine.

Toxicity — Opium and its derivatives are responsible for many deaths each year in the U.S., especially when a high potency heroin gets through to a street user, accustomed to low potency adulterated dosages. Duke[243] discusses the toxicities of some of these alkaloids. Morton[17] adds that as little as 300 mg opium can be fatal to a human, though addicts can tolerate 2000 mg morphine over 4 hr. Death from circulatory and respiratory failure follows cold and clammy skin, weak and rapid pulse, pulmonary edema, and cyanosis. Pupil dilation may occur just before death.[17]

252. *PARIS QUADRIFOLIA* L. (TRILLIACEAE) — Herb Paris

Much used in traditional medicine back in the old days, more recently largely restricted to homeopathy. "It has been used as an aphrodisiac — the seeds and berries have something of the nature of opium."[2] Herb paris (tincture of fresh plant) is used as an antidote to arsenic and mercurial sublimate.[2]

Reported to be emetic, narcotic, poison, purgative, herb paris is a folk remedy for bronchitis, colic, cough, cramps, gout, insanity, neuralgia, palpitations, rabies, rheumatism, sores, spasms, tumors, and ulcers.[2,32,33] The seeds and plant juice are mixed to apply to inflammations and tumors. Russians prescribe the leaves for madness. The berry juice is used as a collyrium.[2] Used homeopathically for headache, neuralgia, and vertigo.[33]

Contains a glucoside called paradin which hydrolyzes to glucose and paridol (a sapogenin) and the saponin paristyphnine, L-asparagine, citric acid, and pectin.[33]

Toxicity — "Narcotic, in large doses producing colic, nausea, vomiting, tenesmus, headache, vertigo, delirium, convulsions, restlessness, profuse sweating, and dry throat. The drug should be used with great caution; overdoses have proven fatal to children and poultry."[2]

To the physician — *Hager's Handbook*[33] suggests mucilaginous antispasmodics for the gastroenteritis, sedatives or hypnotics for cramps and restlessness, maintenance of the circulatory and respiratory systems.[33]

253. *PASSIFLORA INCARNATA* L. (PASSIFLORACEAE) — Passionflower, Maypop

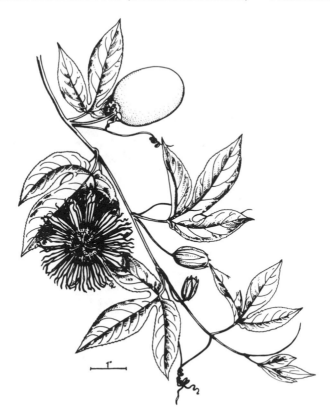

Fruit edible. One anonymous author suggests smoking passionflower leaves ''for mild but very relaxing high'' or as a tobacco substitute in trying to recover from nicotine addiction.

Reported to be cyanogenetic, narcotic, sedative, soporific, spasmolytic, passionflower is a folk remedy for colic, diarrhea, dysentery, dysmenorrhea, epilepsy, eruptions, insomnia, morphinism, neuralgia, neurosis, ophthalmia, piles, spasms.[32,33] Official in homeopathic medicine for insomnia and neurasthenia.[33]

Per 100 g, the fruit is reported to contain 111 calories, 72.5 g H_2O, 2.3 g protein, 3.3 g fat, 21.0 g total carbohydrate, 7.3 g fiber, 0.9 g ash, 14 mg Ca, 43 mg P, and 1.6 mg Fe.[21] Contains 0.025 to 0.032% total alkaloids, among them harman $C_{12}H_{10}N_2$, harmol $C_{12}H_{10}N_2O$, harmalol $C_{12}H_{12}N_2O$, harmine $C_{13}H_{12}N_2O$, harmaline $C_{13}H_{24}N_2O$. The steroid fraction contains 13% sitosterol, stigmasterol; *n*-nonacosane, Ca 0.25% gums. Flavone-C-glycosides include saponarin, vitexin, saponaretin; possibly, homo-orientine and orientine. The ''active principle'', has been called passiflorine, and is suggested to be somewhat similar to morphine. The alkaloid fraction has sedative activity on experimental mice.[33]

Toxicity — Narcotic, the ''drug is known to be a depressant to the motor side of the spinal cord, slightly reducing arterial pressure, though affecting circulation but little, while increasing the rate of respiration.''[2]

254 *PASSIFLORA QUADRANGULARIS* L. (PASSIFLORACEAE) — Granadilla

Fruits edible, and mixed with wine and sugar to make a sherbert. Green fruits are sometimes boiled as vegetable.

Reported to be anthelmintic, diuretic, and emetic. Small doses of the seed are used as a CNS sedative. Brazilians take the leaf decoction for worms; Venezuelans for skin disorders. Venezuelans use the rind as emollient and poultice. Trinidad natives take the leaf decoction for diabetes, dysuria, and hypertension. Guatemalans take the fruit as a stomachic vermifuge. Leaves steeped in almond oil and ethanol, and poulticed over hepatitis and applied to the temples for headache.[42]

Per 100 g, the fruit is reported to contain 20 to 77 calories, 80.0 to 94.4 g H_2O, 0.7 to 2.6 g protein, 0.2 to 1.9 g fat, 4.3 to 14.5 g total carbohydrate, 0.7 to 4.9 g fiber, 0.4 to 1.0 g ash, 9 to 14 mg Ca, 17 to 36 mg P, 0.6 to 0.8 mg Fe, 0.00 mg thiamine, 0.03 mg riboflavin, 3.8 mg niacin, and 15 to 20 mg ascorbic acid.[21] Seeds contain 81 calories, 78.4% H_2O, 1.9% protein, 1.3% fat, 17.6% total carbohydrate, 5.6% fiber, 0.8% ash, 9 mg Ca, 39 mg P, 2.9 mg Fe, 20 μg β-carotene, 0.00 mg thiamine, 0.12 mg riboflavin, 1.9 mg niacin, and 15 mg ascorbic acid per 100 g.

Toxicity — Root said to be very poisonous and a powerful narcotic.[2] Leaves, rind, and immature seeds may produce HCN.[42]

255. *PAULLINIA CUPANA* Kunth ex H.B.K. (SAPINDACEAE) — Guarana, Uabano, Brazilian Cocoa

Guarana is a dried paste, chiefly of crushed seeds, which may be swallowed, powdered, or made into a beverage. It is a popular stimulant in Brazil among natives who grate a quantity into the palm of hand, swallow it, and wash it down with water. Taste is astringent and bitterish, then sweetish. A refreshing guarana soft drink is made in Brazil similar to making the ordinary drink, but sweetened and carbonated. Odor is similar to chocolate. Cultivated by the Indians, the seed made into a paste, sold in two grades. Said to be used also in cordials and liqueurs. Tyler notes that Coca Cola® — Brazil uses guarana in a carbonated beverage it markets there.[37] Zoom®, a rather tasty beverage, has been promoted as a "cocaine" substitute.

A nervine tonic and stimulant, the drug owing its properties to caffeine. Used for cardiac derangements, headaches, especially those caused by menstrual or rheumatic derangements, intestinal disorders, migraine, and neuralgia.[27] Action is sometimes diuretic, and used for rheumatic complaints and lumbago. With words like aphrodisiac, diet, narcotic, and stimulant associated with guarana in the herbal literature, it is little wonder that the herb has excited curiosity among avante garde Americans. Promotional literature states that guarana outsells Coke® in Brazil, suggesting that Amazon natives sniff the powdered seeds, and stating, wrongly or rightly, that guarana decreases fatigue and curtails hunger. However, Latin Americans used the plant mainly as a stimulant and for treating chronic diarrhea.[29] Rose adds that guarana is also used for hangovers and menstrual headaches.[47]

Indians in South America also made an alcoholic beverage from the seeds along with cassava and water. Guarana contains guaranine, an alklaoid similar to theine of tea and caffeine of coffee; about 2.5 to 5% caffeine and 5 to 25% tannin, as catechutannic acid. An 800-mg tablet of Zoom® is said to contain *circa* 60 mg caffeine.[37] Adenine, catechin, choline, guanine, hypoxanthine, mucilage, resin, saponins, 8.5% tannin, theobromine, theophylline, timbonine, and xanthine are reported, in addition to the caffeine.[29]

Toxicity — Narcotic stimulant;[54] may be quite high in caffeine (Cheney[352] says the highest of any plant) and tannin. Dysuria often follows its administration. Has been approved for food use (§172.510).[29]

256. *PAULLINIA YOKO* Schultes & Killip (SAPINDACEAE) — Yoko

Employed as a stimulant by the Indians of southern Colombia and adjacent Ecuador and Peru.[247] Sap expressed from the stems makes a light chocolate-brown mixture when added to cold water. Yoko is taken at dawn by the Putumayo Indians who normally eat nothing til noon. One or two "cups" (*jicaras*), each dosage representing the expressed sap of *circa* 100 g rasped material, allay all sensations of hunger for at least 3 hr, yet supply muscular stimulation.

Reported to be anorexiac, febrifuge, intestional disinfectant, purgative, stimulant, tonic, yoko is a folk remedy for dysentery, fever, malaria, and stomachache.[32] Putumayo use the plant in large doses as an antimalarial febrifuge and in treating bilious disorders.[247]

Yoko bark contains 6.1% ash, 12.3% water, and 2.73% caffeine, also found in the inflorescence.[247]

257. *PAUSINYSTALIA JOHIMBE* (K. Schum.) (RUBIACEAE) — Yohimbe

Yohimbine (quebrachine, corynine, aphrodines), found in *P. johimbe* (also, in *Rauvolfia*), is an adrenergic blocking agent, which has been used in angina pectoris and arteriosclerosis, and was formerly used as a local anesthetic, mydriatic, and, especially in veterinary medicine, as an aphrodisiac.[20] Yohimbine, which causes hypotension, is reported to be a cardiovascular depressant, with hypnotic activity and a relatively high toxicity.[1] Tyler notes that some authors recommend snuffing yohimbine to obtain both stimulant and mild hallucinogenic effects.[37] Ajmaline has been reported to stimulate respiration and intestinal movements. Its action on systemic and pulmonary blood pressure is similar to serpentine's. "The drug dilates the blood vessels of the skin and mucous membranes and thereby lowers blood pressure. Its alleged aphrodisiac effects are attributed not only to this enlargement of blood vessels in the sexual organs but to increased reflex excitability in the sacral (lower) region of the spinal cord."[37] "At Queen's University in Kingston, Ont., urologist Alvaro Morales and a team of researchers have conducted a study of yohimbine's effect on 23 men with impotence related to physical problems, like diabetes. The men were given laboratory-synthesized yohimbine daily for eight to ten weeks. Ten improved, with six once again able to sustain erection and reach orgasm. The drug unexpectedly also relieves some of the numbness and prickling in the legs that frequently afflict diabetics. Unpleasant side effects were limited to temporary dizziness or gastrointestinal upset. The Canadian team is now beginning a two-year study of yohimbine vs. placebo in 120 men with impotence of organic or psychological origin. How yohimbine helps potency is a mystery. The chemical blocks or stimulates the release of adrenaline at nerve endings in different parts of the body. Researchers think that this action changes blood flow or the transmission of nerve impulses to genital tissue."[248]

Widely regarded in the counterculture as an aphrodisiac and not so widely accepted to lower blood pressure. The closely related *P. macroceras* has other reputed attributes. In Ghana, the Ivory Coast, and Upper Volta, a bark decoction is used for fevers and leprosy and the bark is chewed for coughs. Yohimbine is available commercially as an aphrodisiac, but claims for sexual stimulation have not been adequately supported.[11] Yohimbine is hallucinogenic in large doses,[11] but, I might add, a dangerous hallucinogen or aphrodisiac. Rose[47] states that yohimbe "causes a tingling sensation in the genitals."[47] According to *Hager's Handbook*, the herb is used as an aphrodisiac, for psychic impotence, for dysmenorrhea, for prostatitis with bladder complaints, and as a local anesthetic for eye, ear, and nose operations.[33]

The bark of both species contains yohimbine, and various configurations thereof, while that of yohimbe also contains ajmaline, alloyohimbine, corynanthine, corynantheine, dihydrocorynantheine, pseudoyohimbine, quebrachine, and tetrahydromethylcorynanthein. Yohimbine is available commercially as yohimbine hydrochloride combined with methyltestosterone and nux vomica; but the manufacturer's claim of relieving male impotence, as judged by the number of erections and orgasms per week achieved by patients, apparently deserves no great credence.[11] Tannin is also reported.

Toxicity — The FDA classifies this as an unsafe herb, containing "the toxic alkaloid, yohimbine (quebrachine), and other alkaloids." Yohimbine is a monoamine oxidase inhibitor, hence, users should avoid tyramine-containing foods (cheese, liver, red wines) and drugs containing phenylpropanolamine. Persons suffering diabetes, low blood pressure, and heart, kidney, and liver disorders should avoid the drug. According to Tyler, yohimbine may actually activate psychoses in schizophrenics.[37] Upon drinking a tea made by adding 6 to 10 teaspoons shaved yohimbe bark to 1 pt boiling water, the experimenter may experience lethargic debility of the limbs, restlessness, chills and shivers, with some nausea and vertigo.[51] The LD_{50} of corynanthine is 250 mg/kg subcutaneously in mice.

258. *PEGANUM HARMALA* L. (RUTACEAE) — Harmel, Syrian Rue, African Rue

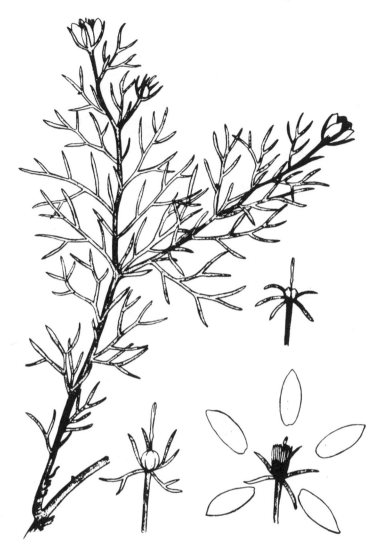

Seeds are source of the dye Turkish Red. According to Emboden, Nazi scientists used harmine as a truth serum during World War II.[54] Small doses of harmine stimulate the brain, but in excess it depresses the CNS. Hallucinations are produced by oral doses as small as 25 mg.[54] Egyptian studies[249] show that the extract is markedly fungicidal and bactericidal. The plant is also used as an insect repellant, pediculicide, and protisticide. Algerians extract the seed oil; other orientals use the seed as a spice. Unfortunately for the gullible, harmine, harmaline, and harmalol are powerful stimulants capable of inducing wild visions.[51]

Reported to be abortifacient, alterative, amebicidal, anodyne, aphrodisiac, diuretic, emetic, emmenagogue, intoxicant, lactagogue, narcotic, soporific, stimulant, sudorific, and vermifuge, harmal is a folk remedy for asthma, calculus, cancer, colic, dysmenorrhea, fever, gallstones, hiccup, hysteria, jaundice, laryngitis, malaria, neuralgia, parkinsonianism, prolapse of the womb, rheumatism, and urogenital ailments.[32,33] Seeds used by Greeks for eye ailments, diaphoretic, emmenagogue, vermifuge, sedative, and intoxicant. Used in India for syphilis, in N. Africa for fever.[53]

Per 100 g, the fresh forage is reported to contain 67.7 g H_2O, 5.7 g protein, 1.5 g fat, 16 g N-free extract, 5.7 g fiber, 3.4 g ash, 0.56% Ca, 0.06% P, 0.0009% Cu, and 250 mg thiamine. Egyptians sell the seed oil as the aphrodisiac Zit-el-Harmel. Dry seed contain 29.4 to 35.2% protein, 13.5 to 17.5% oil. Seeds contain harmine, harmaline, harmalol, harman, peganine, isopeganine, dipegene, vasicinone, and deoxyvasicinone.[80] According to Emboden, harmine, harmaline, and harmalol have produced sexual responses in rats under laboratory conditions, 5 mg harmine eliciting measurable sexual activity. Harmine is used in research on mental disease and encephalitis.[54] Harmine and harmaline elicit hallucinogenic effects in humans at doses of 4 mg/kg. Closely related to the harmala bases is 6-methoxytetrahydroharman, a natural hormone of the pineal body, hallucinogenic at doses of 1.5 mg/kg.

Toxicity — Narcotic hallucinogen.[54] Overdoses cause hallucinations, colored visions, tremors, salivation, nausea. Fatal doses in experimental animals are at *circa* 1500 mg/kg body weight. The principle narcotic effect is stimulation of the motor tracts of the cerebrum. In poisonous doses results in a depression of the central nervous system.[53] Seeds used as a vermifuge and amebicide in Asia.

259. *PERILLA FRUTESCENS* (L.) Britt. (LAMIACEAE) — Perilla, Wild Coleus, Beefsteak Plant

Leaves and seeds alike are eaten in the orient and are part of the Japanese national dish "shisho". The preserved seeds are eaten as the core of some Japanese confections. Leaves and seeds of var. *crispa* are also used as food coloring. After frost, the old foliage and flower heads made very potent additives to herbal teas. The mucilaginous seeds can be eaten as sprouts. The edible oil is used in making varnishes and lacquers, dyes and inks, and in the manufacture of synthetic leather, oil paper, linoleum, etc. It is also used for burning. The expressed cake makes a satisfactory fodder. A kilogram of dried herb yields 1 to 1.5 g volatile oil. Oil distilled from the herb, as opposed to the seed oil, is a flavoring agent used sparingly in oriental sauces and confections. It gives peculiar flavors to dentifrices, and is used to kill mildew and other germs. In Japan, its chief use was in preparing a sweetening agent. The seed oil (which contains antioxidant and antiseptic compounds) is a preservative utilized in making soy sauce.[16]

In folk medicine the antispasmodic, diaphoretic, sedative herb is prescribed for cephalic pulmonary and uterine trouble. Reported to be alexeteric, anodyne, antidotal, antiseptic, antitussive, carminative, diaphoretic, expectorant, febrifuge, pectoral, preventitive (cold), sedative, stomachic, and tonic, perilla is a folk remedy for asthma, bronchitis, cephalgia, chest, cholera, colds, cough, fish poison, flu, malaria, nausea, pregnancy, rheumatism, spasm, sunstroke, and uteritis.[32] In Korea the leaves are used for colds, cough, and dyspepsia.

Per 100 g, the seed is reported to contain 425 calories, 17.8 g H_2O, 15.7 g protein, 26.3 g fat, 37.0 g total carbohydrate, 28.0 g fiber, 3.2 g ash, 350 mg Ca, 33 mg P, 11.1 mg Fe, 10 μg β-carotene equivalent, 0.32 mg thamine, 0.11 mg riboflavin, 3.1 mg niacin, and 0 mg ascorbic acid. The amino acid composition of the seed protein is (in g/16 g N) arginine, 14.8; histidine, 2.5; leucine, 6.3; isoleucine, 4.3; lysine, 4.4; methionine, 1.4; phenylalanine, 5.1; theonine, 3.01; and valine, 6.0.[1] The seed cake, rich in proteins, contains 0.56% Ca, 0.47% P, and 6.14% N. The average feed value of the cake is 38.4% protein, 8.4% fat; 16.0% N-free extract, 20.9% N-free extract, 34.2% digestible protein, and 61.4% digestible nutrients. Apigenin and luteolin are the chief flavones in the seeds (1:1). They also occur in the leaves with nine flavone glycosides, the chief one being shishonin, and five anthocyanins including cyanadin 3.5 diglucoside and its esters with cinnamic acid derivatives.[251] The leaves contain an anthocyanin, perillanin chloride, which on hydrolysis yields probably delphinidin, protocatechuic acid, and glucose. The plant is reported to contain perillartine (perillaldehyde), about 2000 times as sweet as cane sugar (four to eight times as sweet as saccharin), hence, used as a sweeetening agent in Japan.[20] The volatile perilla oil also contains aldehyde antioxine (used in the tobacco industry), ten times sweeter than saccharin, but poisonous, not dissolving easily in water but disintegrating by heating. The volatile perilla oil also contains citral, 1-limonene, and alpha-pinene. The leaves, per 100 g, contain 42 calories, 86 g water, 3.4 g protein, 0.6 g fat, 8.0 g carbohydrate, 1.5 g fiber, 2.0 g ash, 197 mg Ca, 73 mg P, 6.7 mg Fe, 20 mg Na, 650 mg K, 4380 μg β-carotene equivalent, 0.07 mg thiamine, 0.32 mg riboflavin, 0.8 mg niacin, and 46 mg ascorbic acid. When dried, the edible seed contains 23.12% protein, 45.07% oil, 10.28% N-free extract, 10.28% fiber, and 4.64% ash.[1]

Toxicity — Perilla ketone is a potent pulmonary edemagenic agent for lab animals and livestock. Perilla ketone, egomaketone, and isoegonaketone are chemically closely related to the toxic ipomeanols of moldy sweet potatoes; intravenous doses of the compounds generally resulted in illness or death, while intraruminal injections had no effect. All resulted in pleural effusion and edema.[252,253] Okazaki et al.[254] report dermatitis on the hands of 20 to 50% of long-time workers with the plant. Patch tests were performed with results suggesting that this skin disease is associated with 1-perillaldehyde and perillalcohol contained in shiso oil. Patients responded negatively to patch tests, with perillic acid detected on cotton gloves once used by them. Guinea pig maximization tests were carried out with these shiso oil components. The experimental animals became sensitized with perillaldehyde and not with perillalcohol or perillic acid.

260. *PETROSELINUM CRISPUM* (Mill.) Nym. (APIACEAE) — Parsley

   Parsley is viewed as a healthful garnish, capable of masking foul odors. Leaves and roots, fresh or fried, serve as vegetable or condiment. In European cookery, it enters egg, fish, fowl, meat, shellfish, and soup dishes. It is useful in the preparation of aromatic vegetable tisanes. In France a mixture of parsley and shallot, finely chopped, is added as *persillade* toward the end of cooking a dish. It is important in bouquet garni, in butters and vinegars, and in *ravigote, sauces tartare, vinaigrette*, and *verte*. Mixed with bulgur wheat, it is an important middle eastern salad ingredient. Roots used as a soup ingredient. Fruits yield *circa* 20% fatty oil with up to 76% petroselenic acid. It has been suggested for lubricants, plastics, and synthetic rubbers. A newspaper account mentions parsley, licorice, hot pepper, and rose hips among a mixture of herbal species which lower the cholesterol content of eggs from chickens fed the herbal mixture.

   Parsley, pounded with snails, was applied to scrofulous swellings as an ointment. Bruised leaves are used like those of celandine, clover, comfrey, and violet, to dispel cancerous tumors.[2] It has also been used in uterine disorders. A strong decoction of the root is said to be good for congestion of the kidney, dropsy, gravel, jaundice, and stone. Parsley tea was once served the troops in the trenches suffering from dysentery. Bruised leaves are used to alleviate insect bites and to get rid of lice and skin parasites. Various parts of the plant have been used for tumors of the bladder, breast, eyes, liver, sinews, spleen, throat, and uvula; indurations of the bladder, kidney, liver, spleen, stomach, and uterus; and condylomata, warts, and whitlows.[4] Leung suggests that parsley does, in fact, have antimicrobial, hypotensive, laxative, and tonic properties.[29] Parsley herb tea is "used to treat gallstones."[29]

Leaves, stems, and fruits contain the glucoside apiin, which on hydrolysis yields apigenin, glucose, and apiose. Parsley seed oil contains 1-allyl-2,3,4,5-tetramethoxybenzene, apiole, myristicin, palmitic acid, petrosilane, α-pinene, and various aldehydes, ketones, and phenols. Parsley leaf oil contains apiole, ethanol, hex-3-en-l-yl acetate, and cis-3-hexen-l-ol, that of the "Curley Moss" cultivar up to 85% myristicin. A good source of calcium, iron, vitamin A, and C, the green leaves, per 100 g, contain 68.4% water, 5.9% protein; 1.0% fat; 19.7% carbohydrate, 1.8% fiber; 3.2% mineral matter, 390 mg Ca, 200 mg P, 17.9 mg Fe, 3200 IU vitamin A, 0.04 mg thiamine, 0.5 mg nicotinic acid, and 281 mg ascorbic acid. It contains the furocoumarin, bergapten, which may cause skin reactions in some people. Flavonoids include apiin, luteolin-7-apiolglucoside, apigenin 7-glucoside, and luteolin-7-diglucoside. Leung lists other ingredients as well.[29] Per 100 g, dried parsley is reported to contain 276 calories, 9.0 g H_2O, 22.4 g protein, 4.4 g fat, 51.7 g total carbohydrate, 10.3 g fiber, 12.5 g ash, 1.468 mg Ca, 351 mg P, 97.9 mg Fe, 249 mg Mg, 452 mg Na, 3805 mg K, 4.8 mg Zn, 23,340 IU vitamin A, 0.17 mg thiamine, 1.23 mg riboflavin, 7.9 mg niacin, 122 mg ascorbic acid, and 1.0 mg vitamin B_6.[89]

Toxicity — In one canning factory, the majority of female workers preparing parsley developed vesicular inflammation and purple discoloration of the skin and hands, followed by puberulent folliculitis and carbuncles. Oil of parsley in perfumery can also cause dermatitis.[6] The apiol in parsley, used for ague, nervous ailments, and formerly official in the U.S. as an antipyretic and emmenagogue, can be poisonous. In large doses the oleoresin of parsley (apiol, apiolin, and myristicin) produces giddiness and deafness, fall of blood pressure, and some slowing of the pulse and paralysis, followed by fatty degeneration of the liver and kidney, similar to that caused by myristicin.

261. *PEUMUS BOLDUS* Molina (MONIMIACEAE) — Boldo

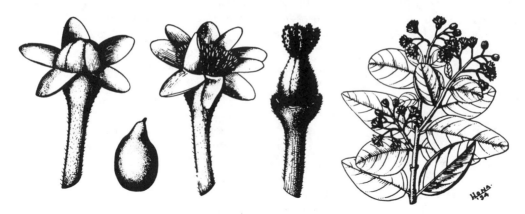

Leaves sometimes consumed as a medicinal tea in South America. The fruits are said to be eaten and leaves used as a spice in Chile. Bark used for tanning and dyeing fibers. Shredded young twigs and bark are boiled in 25 ℓ of water for 1 ℓ of wool or fiber (with *Persea*) to dye it yellow.[255] Wood used for charcoal. Though used as a unique fragrance compound, the oil and/or absolute of the leaves is not very popular.

The aromatic leaves are used as a mild diuretic, especially in liver ailments like jaundice. Said, also, to be anodyne, antiseptic, choleretic, hepatotonic, hypnotic, stimulant, tonic, and vermifuge. Used for urogenital inflammations, like gonorrhea, in Latin America. Elsewhere it is used for dyspepsia, gout, hepatosis, rheumatism, syphilis, and worms.[32,33] An infusion of fragrant leaves is taken for stomach and liver troubles.[9] Chileans "cure earaches" with the sap of the leaves. One or two small leaves in hot water, taken occasionally for liver ailment. Three shredded bark flakes with laurel twigs and burned surga (or toasted salt) used for stomach (south of Chile, Mapuche Indians). Bark mixed with picapica for cough.[255] Leaves, bruised and half roasted and sprayed with wine, are used to treat running sores and head colds. Warm baths with the leaf decoction are highly recommended for dropsy and rheumatism.[256]

The genus *Peumus* is reported to contain the toxins pachycarpine and terpineol. Leaves are said to contain the alkaloid boldine. Sparteine is also reported. The essential oil (2% of leaf) is stated to be chemically related to oil of *Chenopodium*. Grieve also lists the glucosides boldin or boldoglucin.[2] Ascaridol and flavonoids are also reported. Bruns and Kohler[257] found that the essential oil of boldo contained alpha-pinene (4.0%), camphene (0.6%), beta-pinene (0.8%), sabinene (0.8%), α-3-carene (0.5%), terpinolene (0.4%), limonene (1.6%), 1.8-cineole (16.0%), gamma-terpinene (1.0%), *p*-cymene (28.6%), 2-nonanone (0.4%), fenchone (0.8%), 1-methyl-4-isopropenyl-benzene (0.3%), camphor (0.6%), linalool (9.1%), bonyl acetate (0.2%), alpha-fenchol (0.09%), terpinen-4-ol (2.6%), alpha-terpineol (0.9%), cuminaldehyde (0.3%), farnesol — no isomer given (0.4%), ascaridol (16.1%), alpha-methyl ionone (0.4%), methyl eugenol (0.5%), alpha-hexyl cinnamaldehyde (0.4%), diethyl phthalate (0.3%), coumarin (0.5%), and benzyl benzoate (0.4%) (trace amounts of 2-tridecanone, beta-iso-methyl ionine, 2-heptanone, 2-octanone, 2-decanone, benzaldehyde, 2-undecanone, and myrtenal). According to Lawrence,[31] that "Burns and Kohler[257] found alpha-hexyl cinnamaldehyde and diethyl pthalate naturally occurring is very unlikely." He thought it more likely that the authors obtained adulterated oil from a commercial house. "The authors should know that it is not nice to fool mother nature."[31]

Toxicity — I would rank this, if forced to, as, perhaps, more poisonous than coffee or tea, but I would not be afraid to drink a tea made from the leaves. Many of the compounds listed above are toxic and/or carcinogenic. Mapuche Indians are a little fearful of the herb, believing it might curtail the vision.[255] Approval for food use only in alcoholic beverages (§172.510).[29] Boldine, when injected hyperdermically, paralyzes both motor and sensory nerves; also, the muscle fibers. Given internally in toxic doses, it causes excitement, exaggerates the reflexes and respiratory movement, causes cramps and convulsions, ending in death from centric respiratory paralysis, the heart beating after breathing stops.[2]

262. *PHASEOLUS LUNATUS* L. (FABACEAE) — Lima Bean, Sieva Bean, Butterbean

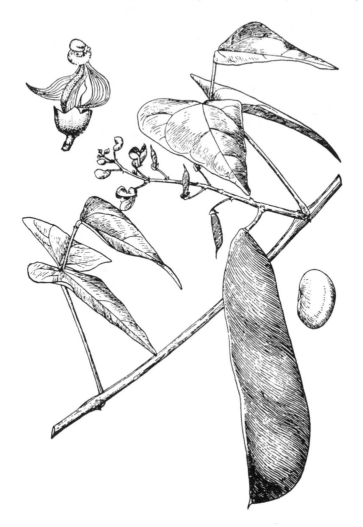

Lima bean is grown for green or dried shelled beans that are eaten cooked and seasoned or mixed with other vegetables or foods. Lima beans are marketed green or dry, canned or frozen. Green immature pods of some cultivars may be eaten cooked as a green vegetable. Sprouts are also eaten.

Yucatan Mayans apply the leaves to painful breasts presaging a chill and fever.[42] Seeds are astringent and are used as diet food in fevers. Mexicans crush the raw black seed to serve as an emetic.[42] A decoction of green pods, seeds, and stems reportedly has been used for Bright's disease, diabetes, dropsy, and eclampsia. In Java, seeds are poulticed onto the abdomen for stomachache.[40] Seeds also used for warts and tumors.[4]

Green lima beans contain moisture, 69.2%; N, 1.3%; ether extract, 0.3%; crude fiber, 0.5%; and ash, 1.5%; Ca, 9; P, 97; Fe, 1.3; carotene, 0.06; thiamine, 0.03; riboflavin, 0.09; nicotinic acid, 1.6; and ascorbic acid, 30.8 mg/100 g. Another report gives 123 calories, 67.5% moisture, 8.4 g protein, 0.5 g fat, 22.1 g total carbohydrate, 1.8 g fiber, 1.5 g ash. Raw dried mature seeds contain, per 100 g, 345 calories, 10.3% moisture, 20.4 g protein, 1.6 g fat, 64.0 g total carbohydrate, 4.3 g fiber, 3.7 g ash. Lima bean flour is reported to contain, per 190 g, 343 calories, 10.5% moisture, 21.5 g protein, 1.4 g fat, 63,0 g total carbohydrate, 2.0 g fiber, 3.6 g ash. The average amino acid composition is, per 16 g N,

11.9 g asparagine, 5.1 g threonine, 8.1 g serine, 14.9 g flutamic acid, 4.6 g proline, 4.4 g glycine, 4.7 g alanine, 5.8 g valine, 1.2 g methionine, 5.3 g isoleucine, 8.9 g leucine, 3.4 g tyrosine, 6.4 g phenylalanine, 3.2 g histidine, 7.5 g lysine, 6.3 g arginine, and 1.1 g cystine. Raw leaves contain (per 100 g edible portion) 8 calories, 97.2% moisture, 0.6 g protein, 1.7 g total carbohydrate, 0.5 g ash, 8 mg Ca, 36 mg P, 2.3 mg Fe. Cooked leaves contain, per 100 g, 112 calories, 70.5% moisture, 7.0 g protein, 0.2 g fat, 21.2 g total carbohydrate, 1.1 g fiber, 1.1 g ash, 36 mg Ca, 98 mg P, 0.7 mg Fe, 0.14 mg thiamine, 0.04 mg riboflavin, 0.9 mg niacin, and 18 mg ascorbic acid. Dried seeds contain 13.2% water, 20.0% protein, 1.5% fat, 58.0% carbohydrate, 3.7% crude fiber, and 3.4% ash. Seeds are reported to contain trypsin inhibitors and chymotrypsin inhibitors. The seeds may contain 0.62% lecithin, 0.09% cephalin, a papainlike protease, carotene oxidase, gum, and tannin.[40] Maturing seeds are reported to synthesize S-methylseleno cysteine. Silage made from the leaves and stems is reported to contain 27.3% DM, 3.3% protein, 2.1% digestible protein, 14.2% total digestible nutrients (nutritive ratio 5:8).

Toxicity — The root, considered poisonous by some, has been reported to cause colic, giddiness, nausea, prostration, purgation, and rapid pulse.[3] Addition of raw limas to the diets of rats is said to diminish the rate of growth and the apparent digestibility of protein and fat. Maya says that pods are poisonous to pigs.[42] Containing varying amounts of cyanogenetic glycosides, beans which have reverted to the wild or semicultivated forms should be avoided; ''the dark purple beans are the most toxic and have caused death.''[16] ''Cooking does not altogether destroy these compounds.''[11] Lima beans also contain a lectinic glycoprotein. Deeply colored red or black testas have, in the past, been associated with high levels of cyanogenic glucosides in the seed, but cyanogenic glucosides have been reduced to safe levels by selection in the U.S. and elsewhere. The supposed absence of HCN in white cultivars may represent wishful thinking. There seems to be no reliable correlation between seed color and cyanide content.[40] Some tropical limas may contain 0.3% HCN. The U.S. rejects beans containing more than 0.01%, while Canada stops imports with more than 0.02%. Plants containing 0.02% HCN or more are regarded as potentially dangerous to livestock. The small black lima of Puerto Rico contains as much as 0.3% and is said to have caused fatalities when ingested. Silage and wilted haulms may cause intoxication as well.[14]

263. *PHASEOLUS VULGARIS* L. (FABACEAE) — Bean, Kidney Bean, Haricot Bean

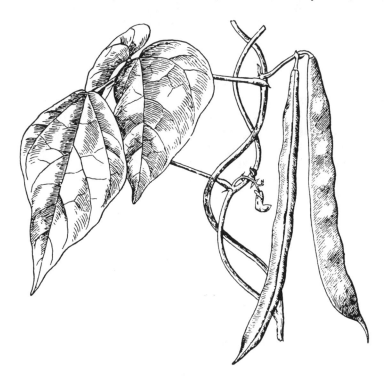

Widely cultivated in temperate and widely semitropical regions. In temperate regions green immature pods are cooked and eaten as a vegetable. Immature pods are marketed fresh, canned, frozen, whole, cut, or french-cut. Mature ripe beans, called navy beans, kidney beans, white beans, northern beans, or pea beans, are widely consumed. In lower latitudes, dry beans furnish a large portion of the protein needs of low- and middle-class families. In some parts of the tropics leaves are used as a pot-herb, and to a lesser extent the green-shelled beans are eaten. In Java, young leaves are eaten as a salad. After beans are harvested, straw is used for fodder.[40] Phaseollin from the seeds is fungicidal.[11] Leaves yield a yellow dye.[3]

Regarded as carminative, depurative, diaphoretic, digestive, diuretic, emmenagogue, febrifuge, and resolvent, beans are said to be used for acne, albuminuria, bladder, burns, cardiac, carminative, cold, depurative, diabetes, diarrhea, dropsy, dysentery, eczema, emollient, headache, hiccups, itch, kidney, nephritis, rheumatism, sciatica, scurvy, tenesmus, tumors, and warts.[32,33,40] *Hager's Handbook* mentions several other medicinal applications.[33]

Beans are highly nutritive, relatively low-cost protein food. Green snap beans contain 6.2% protein, 0.2% fat, and 63% carbohydrate. Analysis of a sample of dried beans marketed under the name "Rajmah" gave the following values per 100 g: moisture, 12.0; protein, 22.9; fat, 1.3; carbohydrates, 60.6; and minerals, 3.2%; Ca, 260, P, 410, and iron, 5.8 mg; 346 calories/100 g. The vitamin contents of the dried beans are thiamine, 0.6; riboflavin, 0.2; nicotinic acid, 2.5; and ascorbic acid, 2.0 mg/100. Analysis of dried beans from another source yielded (mg/100 g): Na, 43.2; K, 1160; Ca, 180; Mg, 183; Fe, 6.6; Cu, 0.61; P, 309; S, 166; and Cl, 1.8 mg/100 g. Beans also contain I(1.4 µg/100 g), Mn (1.8 mg/100 g), and arsenic (0.03 mg/100 g). Raw immature pods of green, and yellow or wax snap beans are reported to contain, per 100 g, 32 and 27 calories, 90.1 and 91.4 moisture, 1.9 and 1.7 g protein, 0.2 g fat, 7.1 and 6.0 g total carbohydrate, 1.0 g fiber, and 0.7 g ash, respectively. Raw pods of kidney beans contain (per 100 g edible portion) 150 calories,

60.4% moisture, 9.8 g protein, 0.3 g fat, 27.8 g total carbohydrate, 2.3 g fiber, 1.7 g ash, 59 mg Ca, 213 mg P, 3.6 mg Fe, 10 μg vitamin A, 0.38 mg thiamine, 0.12 mg riboflavin, 1.5 mg niacin, 7 mg ascorbic acid. Raw, dried mature seeds of white, red, and pinto beans are reported to contain, per 100 g, 340, 343, and 349 calories, 10.9, 10.4, and 8.3% moisture, 22.3, 22.5, and 22.9 g protein, 1.6, 1.5, and 1.2 g fat, 61.3, 61.9, and 63.7 g total carbohydrate, 4.3, 4.2, and 4.3 fiber, 3.9, 3.7, and 3.9 ash, respectively. Whole seeds of kidney beans contain (per 100 g) 86 mg Ca, 247 mg P, 716 mg Fe, 5 μg vitamin A, 0.54 mg thiamine, 0.19 mg riboflavin, 2.1 mg niacin, 3 mg ascorbic acid. Whole seeds cooked contain 141 calories, 68.0% moisture, 5.9 g protein, 5.7 g fat, 17.9 g total carbohydrate, 1.1 g fiber, 2.5 g ash, 46 mg Ca, 120 mg P, and 1.9 mg Fe. Raw leaves contain (per 100 g) 36 calories, 86.8% moisture, 3.6 g protein, 0.4 g fat, 6.6 g total carbohydrate, 2.8 g fiber, 2.6 g ash, 274 mg Ca, 75 mg P, 9.2 mg Fe, 3230 μg β-carotene equivalent, 0.18 mg thiamine, 0.06 riboflavin, 1.3 mg niacin, 110 mg ascorbic acid. After harvest, plants can be fed to cattle, sheep, and horses as a part of the roughage if fed with good hay. Comparable to corn or sorghum fodder in nutritive value, it contains moisture, 10.9; protein, 6.1; fat, 1.4; N-free extract, 34.1; fiber, 40.1; ash, 7.4; Ca, 1.7; P, 0.1; K, 1.0 digestible protein, 3.1; and total digestible nutrients, 45.2%. After pod removal, silage may be prepared from green vines. Dehydrated bean vine meal prepared from green plants after pod removal contains protein, 18.3; digestible protein, 12.3; and total digestible nutrients, 46.3%. Meal made from vines with mature leaves is inferior in quality. Leaves contain carotene (178.8 mg/100 g), thiamine, riboflavin, nicotinic acid, folic acid, and pantothenic acid. They also contain a quercetin glycoside. The hull is said to yield 0.13% rubber. The leaves are said to contain allantoin.[40]

Toxicity — The roots are reported to cause giddiness in human beings and animals. Seeds are reported to contain trypsin inhibitors and chymotrypsin inhibitors.[40] Negritos in the Philippines suspect the plant is poisonous.[16] Kidney beans contain a lectinic phytohemagglutinin (PHA). The PHA enhances the antitumor activity of cultured lymphocytes. Hungarian beans may contain 0.035% HCN while other estimates put it at 8 ppm.[3]

264. *PHORADENDRON SEROTINUM (RAF.)* M.C. Johnston (LORANTHACEAE) — Mistletoe

In eastern North America, this is the mistletoe so frequently used at Christmas. Leaves have considerable reputation as a home remedy.[37]

Oxytocic, the plant is said to be efficacious for arresting post-partum hemorrhage.[8] American mistletoe is believed to stimulate smooth muscles, causing a rise in blood pressure and increased uterine and intestinal contractions, as opposed to the European mistletoe which is condered antispasmodic, calmative, and hypotensive.[37]

Contains beta-phenylethylamine and tyramine.[14] Injected into test animals the toxic protein phoratoxin induced hypotension, slower, weaker heartbeats, and constriction of the blood vessels in the skin and skeletal muscles.[37] On a zero-moisture basis, the leaves (41.0% dry matter) contain 22.0% crude protein, 5.6% ether extract, 18.8% crude fiber, 5.9% ash, and 46.7% N-free extract.[21]

Toxicity — The berries contain toxic amines (tyramine) and proteins that may cause gastroenteritis "if eaten in large quantities.[11] Said to cause contact dermatitis as well.[11] Kingsbury[14] notes that the National Clearinghouse for Poison Control Centers has reported a fatality following ingestion of a tea brewed from the berries. Death occurred *circa* 10 hr after symptoms of acute gastroenteritis and cardiovascular collapse.

To the physician — Hardin and Arena[34] give the following suggestions in cases of poisoning: gastric lavage or emesis; supportive treatment; potassium, procainamide, quinidine sulfate or disodium salt of edetate.

265. *PHYSOSTIGMA VENENOSUM* Balf. (FABACEAE) — Ordeal Bean

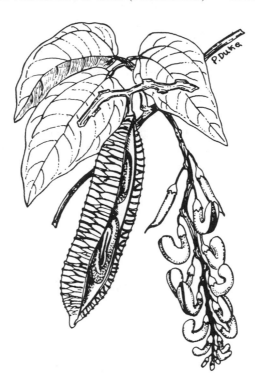

Lewis and Elvin-Lewis give an interesting account of Nigerian use of the plant to reveal and destroy witches and/or witchcraft. The suspect is given eight ground beans in water to drink. If guilty, his mouth shakes and mucus emanates from his nostrils, but if he is innocent, he lifts his hand and regurgitates. "If the poison continues to affect the suspect after he has established his innocence, he is given a concoction of excrement mixed in water that has been used to wash the external genitalia of a female. When a person dies from the ordeal the usual procedure is to remove the eyes and throw the body into the forest."[11] The leaves yield a dye used to stain gourds or wood black. Stems are split and made into mats. At one time physostigmine (eserine), which prolongs the effect of the transmission of nerve impulses to voluntary muscles by anticholinesterase action, was used to treat myasthenia gravis. Now it is more frequently used for the eye. More recently, Davis et al.,[258] noting that physostigmine improves long-term memory processes in humans, suggest further studies of the cholinomimetics in Alzheimer's disease. Bartus[259] has some comments on physostigmine that might fuel homeopathic flames. "The performance of the young monkeys treated with physostigmine was similar to that recently reported for young humans — no effects at low doses, some improvement at a restricted range of doses, and deficits at the highest dose." Getting variations with other age groups as well, Bartus concluded "physostigmine cannot easily or reliably be used as an agent for treating geriatric cognition."[259] Sometimes used as an antidote to atropine, curare, nicotine, and strychnine.[17]

Reported to be antidotal, myotic, parasiticidal, pediculicidal, raticidal, and sedative, the ordeal bean is a folk remedy for arthritis, bursitis, colic, diabetes, edema, fibrositis, glaucoma, myasthenia gravis, ophthalmia, psoriasis, rheumatism, spasms, tetanus.[32] Africans use the seed to kill mice and, with palm oil, to kill lice. Homeopaths use the drug for diabetes.[33]

Seeds contain 48% starch, 23% protein, and the alkaloids phystosterine (ereine), calabarine, eseranine, eseridine, physovenine, isophysostigmine, *N*-8-norphysostigmine.[17] Physostigmine salicylate has been administered intravenously for intoxication, due to antidepressants, antihistamines, and certain sedatives. It is also given as an antispasmodic in bursitis, fibrosis, and rheumatoid arthritis.[17] Eseridine is purgative, physostigmine a spinal sedative; calabarine a spinal stimulant (again showing my contention that if you study a plant hard enough you may find many antagonistic compounds. Cannot the homeostatic human body pull out the needed compound, rejecting the unneeded?); physovenine is myotic. The fatty oil contains behenic-, palmitic-, stearic-, oleic-, and linoleic acids; sitosterol, stigmasterol, stearines, and essential oil; trifolianol $C_{21}H_{34}O_2(OH)_2$ and calabarol $C_{23}H_{34}(OH)_3$.[33]

Toxicity — Extremely poisonous. The maximum number of beans eaten and survived is 35. Acts as a sedative on the spinal cord, paralyzing the legs and the heart (with large doses), and death by asphyxia.

266. *PHYTOLACCA AMERICANA L. (PHYTOLACCACEAE) — Poke, Pokeweed, Scoke*

Young shoots, 15 to 20 cm tall, used for greens or potherbs when thoroughly cooked and the water discarded. Plant becomes poisonous as it matures. Roots, leaves, and berries are poisonous. Poultry eat the berries; large quantities give the flesh an unpleasant flavor causing it to become purgative when eaten. Poke greens, called "poke salet", are commercially canned in northern Kentucky and southern Ohio. Pennsylvania Dutch use berries for ink and to color wines (the latter practice has been discontinued because of poisonous effects). Indians used the powdered root to treat cancer and early settlers applied the berry juice to cancerous skin ulcers. The juice, as an unguent, is said to alleviate cancer. The root, used in an ointment or decoction, is used to treat cancer and tumors. An ointment or a cataplasm derived from the leaf is said to aid cancer.[4] Leaf juice with gunpowder was one old cancer nostrum.[46] Pamunkey Indians drank tea of the berries for rheumatism.[11] Medicinally, poke root is sliced and dried, and considered by some a valuable remedy for catarrh, dysmenorrhea, dyspepsia, granular conjunctivitis, laryngitis, mumps, rheumatism, ringworm, scabies, syphilis, tonsillitis, and ulcers. Berries have a milder action and are relatively nonpoisonous except for children. In Appalachia dried fruits used as poultice on sores. Morton[46] reports dangerous treatments for hemorrhoids, topical application of macerated leaves, or repeated enemas with a strong leaf injection. Poke is alterative, cathartic, emetic, laxative, slightly narcotic, nervine, and stimulant. It is also a slow-acting emetic and purgative. In Spain a pomade made of roots is used to treat skin eruptions, skin rash, and ringworms. Decoction of roots used for drenching cattle. Often used to adulterate belladonna; said to help various types of headaches. I think Tyler's hyperbole might help spare another pokeweed incident. "Pokeweed is not therapeutically useful for anything."[37]

Hager's Handbook lists caryophyllene, phytolaccanin (betanine, which hydrolyzes to betanidine and isobetanidine), isobetanine, isoprebetanine, and prebetanine; salts of phytolaccic acid. Phytolaccin is the active principle in all parts of the plant. A sapogenin, phytolaccagenin acts as a powerful molluscicide and parasiticide. Per 100 g, the shoots are reported to contain 23 calories, 91.6 g H_2O, 2.6 g protein, 0.4 g fat, 3.7 g total carbohydrate, 1.7 g ash, 53 mg Ca, 44 mg P, 1.7 mg Fe, 5220 μg β-carotene equivalent, 0.08 mg thiamine, 0.33 mg riboflavin, 1.2 mg niacin, and 136 mg ascorbic acid. Gums, resins, tannins, and waxes are also reported.[27] The seed oils contain 12.5% saturated acids, 49% mono-acids, and 37.5% di-acids.

Toxicity — The FDA[62] classifies this as an herb of undefined safety, adding that it "contains an acidic steroid saponin. Emetic action is slow but of long duration. Narcotic effects have been observed." It has been employed internally in chronic rheumatism but is not therapeutically useful and is no longer prescribed. Overdoses have sometimes been fatal. Because poke is mitogenic, handlers should wear gloves.[11] The proteinaceous mitogen PWM may produce blood cell abnormalities when absorbed.[32] Dust of the dried root irritates the eye and induces sneezing.[6] The roots and seeds are especially toxic, due to triterpene saponins (especially phytolaccigenin).[11] Tierra says, "Poke root contains toxic mitogenic substances and therefore must be used in small quantities, not to exceed about one gram per day."[28] I think 1 g is too much! Hardin and Arena[34] recall attending to a 5-year-old girl who died from ingesting poke berries, crushed and added to water to simulate grape juice.

To the physician — For poisoning, they recommend gastric lavage or emesis and symptomatic and supportive treatment. "Children have died and adults have been hospitalized from the gastroenteritis, hypotension, and diminished respiration caused by eating pokeroot or the berries or leaves . . ."[37] In 1981[144] there was a mass poisoning of New Jersey campers from eating the young leaves, picked, boiled, drained, and reboiled, "a method that reputedly ensures the plant's edibility." Sixteen (31%) of the 51 interviewed met the case definition (vomiting accompanied by any three of the following: nausea, diarrhea, stomach cramps, dizziness, and headache). Of 21 ill persons (5 from a different camping group), 18 (86%) experienced nausea, 18 (86%) stomach cramps, 17 (81%) vomiting, 11 (52%) headache, 10 (48%) dizziness, 8 (38%) burning in the stomach or mouth, and 6 (29%) diarrhea. Persons became ill $\frac{1}{2}$ to 5 hr (mean 3) after eating the pokeweed. Symptoms lasted 1 to 48 hr, with a mean of 24 hr. Eighteen persons were seen in local emergency rooms or physicians' offices. Four of these were hospitalized for 24 to 48 hr for protracted vomiting and dehydration.[144]

267. *PICRASMA EXCELSA* (Sw.) Planch. (SIMAROUBACEAE) — Jamaican Quassia

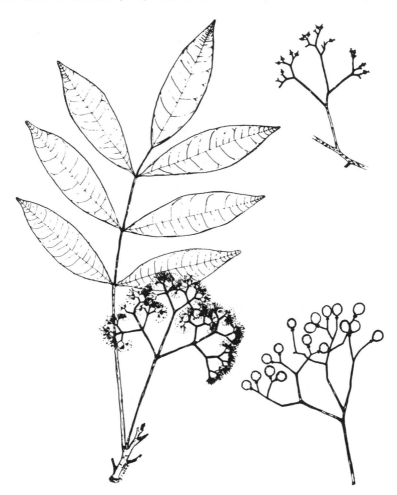

Extracts and quassin (purified mixture of bitter principles) used with alcoholic beverages, baked goods, bitters, candy, dairy products, desserts, gelatins, liqueurs, and puddings. Highest average maximum use level in nonalcoholic beverages is *circa* 75 ppm.

Reported to be digestive, insecticidal, narcotic, piscicidal, stomachic, tonic, and vermifuge, Jamaican quassia is a folk remedy for epithelioma and fever.[32]

Contains isoquassin, *circa* 50 times more bitter than quinine; also, 1.8% thiamine, beta-sitosterol, and beta-sitosterone. It is said, though, to contain no tannin.[29] Grieve mentions pectin.[2]

Toxicity — Quassia wood is slightly narcotic.[2]

268.   *PILOCARPUS* SPP. (RUTACEAE) — Jaborandi

The eighth edition of the *Merck Index* lists pilocarpine hydrochloride (trading at $13.00/ oz. in November 1979) as a parasympathomimetic and topical mitotic and as an antidote for atropine. Pilocarpine is a cholimimetic alkaloid, used widely in treating glaucoma. It acts directly on cholinergic receptor sites, thus, mimicking the action of acetylcholine. Intraocular pressure is thereby reduced, and despite its short-term action, it is the standard drug for initial and maintenance therapy in certain types of glaucoma. Pilocarpine stimulates secretions in the respiratory tract, as well as gastric, lacrimeal, salivary, and other glands, weakens heart action, accelerates pulse, increases peristalsis, and promotes uterine contraction. A single dose of pilocarpine may induce 250 to 450 g sweat. Small doses quench the thirst in fever or chronic renal ailments. Frequently used in hair preparations. In 1973 in the U.S., nearly four million (0.26% of total) prescriptions contained pilocarpine.[98]

Used in asthma, baldness, Bright's disease, catarrh, coronary, deafness, diabetes, dropsy, intestinal atony, jaundice, nausea, nephritis, pleurisy, prurigo, psoriasis, rheumatism, syphilis, and tonsillitis.[32] Induces sweat, salivation, perhaps even lactation. Leaf decoction applied externally to prevent baldness.

Tannic acid; volatile oils (including dipentene and other hydrocarbons). Three alkaloids prevail, pilocarpine $C_{11}H_{16}N_2O_2$, pilocarpidine, and isopilocarpine $C_{11}H_{16}N_2O_2$. Jaborine, in the leaves, may be antagonistic to pilocarpine. Pilosine, isopilosine, isopilocarpine, jaborandine, jaboric, and pilocarpic acid are also reported.

Toxicity — Leaves containing the teratogenic alkaloid pilocarpine are poisonous to cattle and donkeys. Jaborandi may irritate the stomach, causing nausea. An overdose may cause flushing, profuse sweating, salivation, nausea, rapid pulse, contracted pupils, diarrhea, perhaps even fatal pulmonary edema. The stomach should be evacuated and atropine given. The essential oil may be irritant to the skin. The oral LD_{50} of pilocarpine in rats is 911 mg/ kg body weight.

269. *PIMENTA DIOICA* (L.) Merr. (MYRTACEAE) — Allspice, Jamaica Pepper, Clove Pepper

Allspice of commerce is the dried unripe fruit, used as a condiment, in mixed spices as in pickles, ketchup, baked goods, and in flavoring sausages and curing meats. Allspice powder consists of whole ground dried fruit. In Jamaica, a local drink, "pimento dram", is made of ripe fruits and rum. A volatile oil, extracted from the spice and leaves, is used to flavor essences and perfumes, and as a source of eugenol and vanillin. Bahamians make a pleasing tea from the leaves. Saplings are used as walking sticks and umbrella handles.

Reported to be anodyne, antioxidant, bactericidal, carminative, fungicidal, stimulant, and stomachic, allspice is a folk remedy for corns, neuralgia, and rheumatism.[32] Jamaicans take the fruit decoction for colds, menorrhagia and stomachache; Costa Ricans take the leaf infusion as a carminative and stomachic, useful for diabetes; Guatemalans apply it externally for bruises and rheumatic pain; Cubans drink the refreshing tea as depurative, stimulant, and tonic.[42] Aromatic, carminative, and stimulant, the spice and oil are used medicinally for diarrhea, dyspepsia, and flatulence.

Dry, unripe fruits, the allspice of commerce, contains 2 to 5% essential oil with *circa* 35% eugenol, 40 to 45% eugenolmethylether, caryophyllene, (−)-ox-phellandrene, cineole, palmitic acid, fatty oils, resin, sugar, starch, malic acid, calcium oxalate, and tannin.[33] Purseglove et al.[64] compare the berry and leaf oils, mentioning, in addition to the above, cymene, carene, limonene, myrcene, ocimenes, pinene, sabinene, terpinenes, terpinolene, thujene, cadinenes, calamene, copaene, curcumene, elemene, gurgunene, humulenes, iso-caryophyllene, muurolenes, selinenes, etc. Per 100 g, ground allspice is reported to contain 263 calories (1099 kJ), 8.5 g H_2O, 6.1 g protein, 8.7 g fat, 72.1 g total carbohydrate, 21.6 g fiber, 4.6 g ash, 661 mg Ca, 113 mg P, 7 mg Fe, 135 mg Mg, 77 mg Na, 1044 K, 1 mg Zn, 540 IU vitamin A, 0.101 mg thiamine, 0.063 mg riboflavin, 2.86 mg niacin, and 39 mg ascorbic acid. There are 61 mg phytosterols.[89]

Toxicity — Eugenol, the principal constituent of leaves and fruits, is toxic in large quantities and causes contact dermatitis. Allspice itself is irritant to the skin.[6] Of 408 patients with hand eczema, 19 showed positive patch test reaction to allspice.

270. *PIMENTA RACEMOSA* (Mill.) J. W. Moore (MYRTACEAE) — Bayrum Tree, Bayberry, Wild Cinnamon

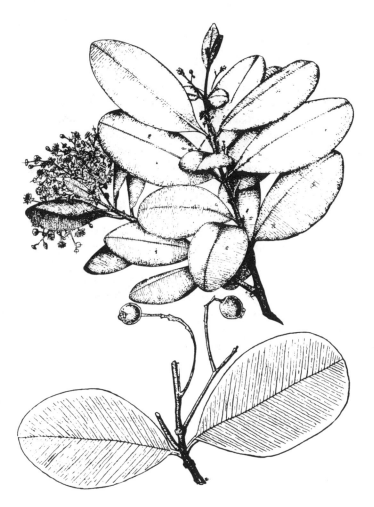

Leaves distilled for the spicy bay oil, used in perfumery and in the preparation of bay rum. Formerly leaves were distilled in rum and water; now oil is dissolved in alcohol or, in Dominica, only in water. Bay rum, with soothing and antiseptic qualities, also used in toilet preparations and as a hair tonic. Oil also used on a limited scale for flavoring foods, chiefly in table sauces. Bay rum is occasionally drunk. Wood is moderately hard and heavy, with a fine compact texture. A leaf held in the mouth is said to discourage smoking.[42,62] The dried green berries have a flavor with hints of cinnamon, clove, and nutmeg. They are used as a condiment. The hard heavy wood (specific gravity 0.9) is described as strong, tough, durable, and resistant to attack by dry wood termites. It is used for carpentry, firewood, and posts.

Reported to be anodyne, antiseptic, carminative, digestive, expectorant, and stomachic, the bay rum tree is a folk remedy for adenopathy, bruise, bug bites, cancer, chest colds, diarrhea, dyspepsia, dysuria, edema, elephantiasis, fever, flu, headache, incontinence, nausea, pleurisy, pneumonia, sore throat, spasms, stroke, varicosity, and vertigo.[32] Cubans decoct four seeds in a cup of water as a stimulant. Bahamians drink the hot leaf tea as a stimulant, using the cooled tea as a skin bracer; Jamaicans take the tea for cold and fever; Grenadans for diarrhea; Trinidad natives for chest colds, flu, pneumonia, and stroke. Leaves are pasted with macerated leaves of *Caesalpinia coriaria* to relieve toothache.[42]

The essential oil content of Puerto Rican bay leaves runs 1.0 to 3.4%, the highest occurring in regions of lower rainfall (*circa* 11 dm), the lowest in wetter areas (*circa* 22 dm annual precipitation). Main constiuents of the leaf oil are eugenol, alpha-pinene, myrcene, alpha-phellandrene, limonene, diterpene, cineol, citral, and chavicol.[42] McHale et al.[260] compared the typical bay oil with anise- and lemon-scented cultivars (formerly known as *P. acris* var. *citrifolia*). Eugenol constitutes 56.2% of the typical oil, chavicol 21.6%. These are present as methyl ethers in the anise-scented (43.1 and 31.6%). The lemon-scented oil contained 32.6% neral and 53.2% geranial and contained no phenols or phenol ethers. Lawrence[193] compared the bay rum oil with that of California bay and laurel bay.

Toxicity — Bayrum used in hair dressings has caused dermatitis of the face and scalp. Contact sensitivity has been reported for eugenol and phellandrene.[6]

271. *PIMPINELLA ANISUM* L. (APIACEAE) — Anise

Anise is grown primarily for the seeds, used for flavoring curries, sweets, confectionaries, and liqueurs, such as anisette. One good anise liqueur is made by stirring 6 spoons crushed aniseed (fennel seed may be substituted) in a quart of brandy. Anisette combines equal parts (2 spoons each) of rather equal-flavored anise, coriander, and fennel seed in a quart of sugared vodka. A pleasant way to take your vitamins would be with Farrel's Rose-Hip and Anise Liqueur, which can be simulated by boiling aniseed and rosehips down to a syrup in sugar water to add to vodka. For those not in a hurry, steeping in the vodka is favored over boiling, but weekend liqueur connoisseurs can achieve results quicker using the herbal syrup approach. Allasch is a Latvian Kummel with anise, almond, and caraway. Anesone is an anise-flavored cordial, sweeter and stronger than anisette. Ojen is a Spanish liqueur, high in alcohol and anise. Ouzo is a Greek anise liqueur. A French anise-based liqueur is called Pastis. Tres Castillos is a Puerto Rican anise-flavored liqueur.[261] Anise makes a nice ancillary ingredient to other more delicate liqueurs but should be used cautiously. The highest maximum levels for anise oil are *circa* 570 ppm in alcoholic beverages, *circa* 680 ppm in candy. Seeds yield an essential oil upon distillation, used in medicine, perfumery, soaps, and beverages. Anise oil, rather than licorice root, generally used to provide "licorice" flavoring used in baked goods, beverages, brandy, cordials, cough drops, in dentrifices as an antiseptic, gelatins, meat and meat products, puddings, etc., and as a photographic sensitizer. If you grow a good quantity of seed, they are useful for refreshing the breath. Teas and salads are embellished by the addition of small quantities of leaves. Anise odor sometimes used in England and U.S. as an artificial fragrance for ''drag hunting'' with fox hounds. In the old days, anise seed was valued against the evil eye and the bad breath. The condiment anise seed was taxed by Edward I to help repair London Bridge.[6] Powdered anise used to flavor horse and cattle feed. Oil of anise regarded as an excellent bait for mouse and rat traps, and as fish lure, said to be poisonous to pigeons. Ground seeds are used in sachets and have been smoked to promote expectoration. Anise oil, mixed with sassafras oil, is used against insects.[65] Anethole, anisaldehyde, d-carvone, and myristicin all have mild insecticidal properties.[65] The fungicidal oil is used in oil-painting china.[61]

Reported to be abortifacient, anodyne, antiseptic, antispasmodic, aperient, aphrodisiac, aromatic, balsamic, carminative, collyrium, diaphoretic, digestive, diuretic, expectorant, fungicidal, lactagogue, pectoral, stimulant, stomachic, sudorific, and tonic, anise is a folk remedy for asthma, bronchitis, cancer, cholera, colic, cough, dropsy, dysmenorrhea, epilepsy, gall bladder, halitosis, indigestion, insomnia, lice, migraine, nausea, nephrosis, neuralgia, purperium, scabies, stomach, and stones. Yucatan natives take a decoction of 3 to 4 g fruit per 160 cc water three to four times a day as a tonic and galactagogue.[42] Said to promote milk production as well as sleep in nursing mothers. Placed under the pillow, anise is supposed to ward off bad dreams.

Per 100 g, anise seed is reported to contain 337 calories (1412 kJ), 9.5 g H_2O, 17.6 g protein, 15.9 g fat, 50.0 g total carbohydrate, 14.6 g fiber, 7.0 g ash, 646 mg Ca, 440 mg P, 37 mg Fe, 170 mg Mg, 16 mg Na, 1441 mg K, and 5.3 mg Zn.[89] Fresh leaves, used in garnishes and salads, may contain 8 to 9 mg vitamin C per 100 g. Simple coumarins (6,7-furocoumarins) and acetylinic compounds have been reported.[262] Anethol-glycol, creosol, anethol, acetaldehyde, isoamylamine, umbelliferone, bergaptene, isopimpinellin, isobergaptene, and sphondin have been reported from *Pimpinella*. Kunzemann and Hermann[263] report the flavonoids: quercetin-3-glucoronide, rutin, luteolin 7-glucoside, isoorientin, isovitexin, and apigenin 7-glucoside. El-Moghazi et al.[264] add luteolin and luteolin 7-O-xyloside for the fruits.

Toxicity — Its major component, anethole, can cause dermatitis (erythema, scaling, and vesiculation). Anethole has two isomers, the cis isomer 15 to 38 times more toxic than the trans isomer.[29] Like fennel oil, anise oil contains compounds that can be aminated in vivo resulting in a series of three dangerous hallucinogenic amphetamines.[54] Anethole used to flavor toothpaste has produced contact sensitivity, according to reports.[6] Aniseed is reported to have caused cheilitis and stomatitis.[6] Anethole (1-methoxy-4-propenylbenzene) is a major component of the essential oils of bitter fennel, anise, star anise, et al.[15] Anethole is a moderate acute toxin with an oral LD_{50} of 2090 mg/kg in rats. Bitter fennel oil, with an oral LD_{50} of 4.52 mℓ/kg in rats, is irritant when applied to the skin. Rats fed a diet containing 0.25% anethole for 1 year showed no ill effects, while those receiving 1.0% anethole for 15 weeks had microscopic alterations of the hepatocytes. Anethole and the essential oils of anise and fennel significantly stimulated hepatic regeneration in rats.[117]

272. *PINUS ELLIOTTII* Engelm. (ABIETACEAE) — Slash Pine

A major source of pulp and tall oils in the deep south of the U.S. The wood is very hard, heavy, strong, coarse grained, and durable. It is used for construction and railroad ties.[8] According to Morton, terpin hydrate is the main synthetic product of turpentine used in pharmaceutical preparations, used as an expectorant in humans and for veterinary bronchitis. Other synthetics produced from pine are anethole, camphor, and DL-menthol. Tall oil rosin contains sterols, mainly sitosterol. Russians built a factory in 1968 to produce steroids from pine pulp extractives. They are also pioneering in the commercial production of vitamin A and E.[17] Synthetic materials derived from turpentine are used in perfumery and to impart flavors suggestive of cinnamon, citrus, lemongrass, licorice, nutmeg, peppermint, and spearmint. Menthol from turpentine is added to cigarettes and cosmetic and toilet products.

According to Morton,[17] turpentine has long been used internally for catarrh, chronic bowel inflammation, colds, gonorrhea, leucorrhea, rheumatism, and various urinary complaints, rheumatism, and ulcers. Pine tar has been used for many ailments in the past, but lately it is prescribed only for external use in chronic and parasitic skin diseases. It shows up in several of the drugs I have resorted to in futile efforts to cure psoriasis. (I have had better luck in substituting rice flour for wheat flour with its glutein.)

Leaves yield *circa* 0.3% of a balsam-scented oil compared to about 0.4% for longleaf pine. This leaf oil consists mostly of borneol, cadinene, camphene, and beta-pinene. The natural oleoresin exudate from the resin ducts contains *circa* 66% resin acids, 25% turpentine, 7% nonvolatiles, and 2% water. Turpentine from slash pine contains 1-*a*-pinene, while that from longleaf contains some *d*-pinene. Pinene is the main constituent of turpentine. Dipentene and other monocyclic terpenes constitute 5 to 8% of gum and refined sulfate turpentine, 15 to 20% of wood and crude sulfate turpentine. Camphene constitutes 4 to 8% of wood turpentine, and 0% of gum turpentine. Rosin consists mostly of diterpene resin acids of the abietic (abietic, neoabietic, palustric, and dehydroabietic) and pimaric types (pimaric, iso-pimaric, and sanaracopimaric). Pine tar contains turpentine, resin, guaiacol, creosol, meth-ylcreosol, phenol, phlorol, toluene, xylene, etc. Crude tall oil contains 40 to 60% resin acids, 40 to 55% fatty acids (mostly n-C_{18}, 75% monoenoic, and 25% dienoic, with traces of trienolic and saturates), and 5 to 10% neutral properties.[17]

Toxicity — Raised in the southern U.S. like me, Sam Page of the FDA tells me that as a child he was given oral doses (a couple of drops on the tongue) of turpentine as a mosquito repellant, an effective but dangerous application. My mom applied it to cuts and sores as a disinfectant, perhaps less dangerous.

273. *PIPER BETEL* L. (PIPERACEAE) — Betel Pepper, Betelvine, Pan Tambult

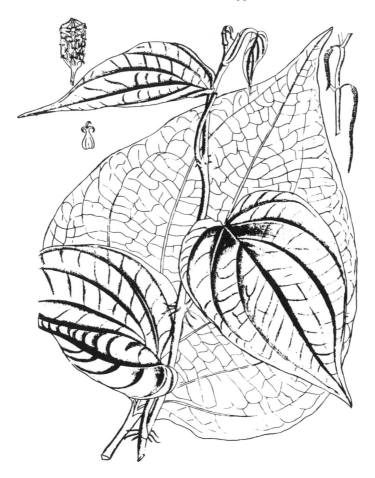

Betel leaf is an important article of daily consumption in Asia and Africa since ancient times, both for rich and poor. Leaves are used in wrapping pellets of betelnut and lime for use as a masticatory. Pellets are hot, acrid, aromatic, and astringent. They redden the saliva and blacken the teeth, and eventually corrode them. One astute observer[265] challenges this, sensing that Indonesians (including dentists) widely believe the converse to be true, i.e., that chewing the betel strengthens the teeth and prevents decay. Betel leaf chewed with the betelnut and lime acts as a gentle stimulant and exhilarant. Those accustomed to its use feel a sense of languor when deprived of it. Leaves and/or the essential oil therefrom are antiseptic and antioxidant. Heated with oils and fats, they check rancidity, effective, e.g., in coconut, groundnut, mustard, safflower, and sesame oils (due to hydroxy chavicol). The essential oil and leaf extracts possess activity against several Gram-positive and Gram-negative bacteria: *Bacillus subtilis* and *B. megaterium, Diplococcus pneumoniae, Escherichia coli, Erwinia carotovora, Micrococcus pyogenes, Proteus vulgaris, Pseudomonas solanaoearum, Salmonella typhosa, Sarcina lutea, Shigella dysenteriae, Streptococcus pyogens*, and *Vibrio comma*. Antiseptic activity is probably due to chavicol. Essential oil and leaf extracts also show antifungal activity against *Aspergillus niger, A. oryzae, Curvularia lunata*, and *Fusarium oxysporum*. Oil is lethal in about 5 min to *Paramoecium caudatum*, in dilutions of up to 1:10,000. It inhibits the growth of *Vibrio cholerae* in a dilution of 1:4000, *Salmonella typhosum* and *Shigella flexneri* in 1:3000, and *Escherichia coli para S.* and *Micrococcus pyogenes* var. *aureus* in 1:2000.[1]

The leaf and root, in oil, are used as a salve or ointment for hard tumors and scirrhi. A bolus, made of the leaf, is also used for cancer.[4] Leaves are poulticed onto boils, bruises, ulcers, and wounds. Asian Indians add the leaf juice to medications for maladies of the mucous lining of the mouth, nose, and stomach. They also use the leaves as a counterirritant to suppress milk secretion in mammary abscesses. Malayans poultice the leaf on ulcerated noses and apply them to the body after childbirth. The juice or infusion is dropped into the ear for wounds, the eyes as a collyrium. Chewing the leaves is believed to prevent dysentery, fever, and gastrosis.[16] Believing that the betel pepper is counterindicated in alcoholism, asthma, eye diseases, leprosy, poisoning, thirst, and unconsciousness, Ayurvedics prescribe it for bronchitis, elephantiasis, halitosis, inappetence, and ozena, believing the leaves to be anthelmintic, aphrodisiac, carminative, laxative, stomachic, and tonic. Yunani regard the leaf as styptic and vulnerary, prescribing it to improve the appetite and taste, to strengthen the teeth, as a tonic to the brain, heart, liver, and throat. Duke and Wain,[32] citing the betel leaf as a folk anodyne, antioxidant, antiseptic, aphrodisiac, astringent, bactericide, carminative, CNS-depressant, contraceptive, deobstruent, digestive, expectorant, fungicide, inhalant, malaria-preventative, masticatory, sialogogue, sterilant, stimulant, and stomachic, note that it is a folk remedy for adenopathy, asthma, bronchitis, bruises, cancer, catarrh, colic, congestion, consumption, cough, diptheria, edema, gastroenteritis, hepatosis, inflammation, intestines, laryngitis, madness, malaria, phthisis, rheumatism, satyriasis, snakebite, sores, sore throat, swellings, syphillis, thirst, tumors, venereal diseases, and wounds.

The active principle responsible for the stimulating effects upon the central nervous system is the alkaloid of the betel nut (see *Areca*); lime helps liberate the alkaloid(s). The essential oil in the leaf is reported to enhance the arecanut effects and to act synergistically on the CNS. Excessive indulgence in chewing for long periods is liable to produce dental caries, pyorrhea alveolaris, oral sepsis, dyspepsia, palpitation, neurosis, and slow cerebration. Chewing is reported to lead sometimes to carcinomatous growths in the mouth, but it is considered that the use of tobacco with pan may be responsible for it. The availability of calcium from the leaf may be poor, since nearly 94% is bound up with oxalic acid. There is considerable free oxalic acid which interferes with calcium utilization from other foods. Fresh leaves contain moisture, 85.4; protein, 3.1; fat, 0.8; carbohydrates, 6.1; fiber, 2.3; mineral matter, 2.3%; calcium, 230 mg; phosphorus, 40 mg; iron, 7 mg; ionizable iron, 3.5 mg; carotene (as vitamin A), 9600 I.U.; thiamine, 70 μg; riboflavin, 30 μg; nicotinic acid, 0.7 mg; and vitamin C, 5 mg/100 g. Betel leaves contain 3.4 μg/100 g of iodine. They have 0.26 to 0.42% potassium nitrate, the amount depending on the position of the leaf on the vine. The important constituents which determine the value of the leaf for chewing are the essential oil and the sugars. Betel leaves from Bombay contained reducing sugars (as glucose),1.4 to 3.2; nonreducing sugars (as sucrose), 0.6 to 2.5; total sugars, 2.4 to 5.6; starch, 1.0 to 1.2; essential oil, 0.8 to 1.8; and tannin, 1.0 to 1.3%. The leaf juice is acidic in nature; malic and oxalic acids have been reported. Diastase and catalase are among the enzymes. Alkaloids and glycosides are reportedly absent. Leaves contain significant amounts of all essential amino acids except lysine, histidine, and arginine which occur in traces. Large concentrations of asparagine are present, while glycine (in a combined state) and proline occur in good amounts; ornithine is present in traces. An aqueous diffusate of the bleached leaves contained: leucine, 18.3; phenylalanine, 14.2; alanine, 11.0; arginine, 2.4; threonine, 12.0; serine, 22.1; aspartic acid, 23.0; glutamic acid, 29.7; methionine, 13.5; valine, 3.8; tyrosine, 1.2; and y-aminobutyric acid, 20.2 mg/100 mℓ. The essential oil contains allyl catechol, 2.7 to 4.6; cadinene, 6.7 to 9.1; carvacrol, 2.2 to 4.8; caryophyllene, 6.2 to 11.9; chavibetol, nil to 1.2; chavicol, 5.1 to 8.2; cineole, 3.6 to 6.2; estragol, 7.5 to 14.6; eugenol, 26.8 to 42.5; and eugenol methyl ether, 8.2 to 15.8%.[1] Pyrocatechin is also reported.[6]

Toxicity — The oil shows a marked irritant action on the skin and mucous membrane.

It produces an inflammatory reaction when injected. In moderate doses, it appears to have antispasmodic action on involuntary muscle tissue, inhibiting excessive peristaltic movements of the intestines. It exhibits a depressant action on the central nervous system of mammals; lethal doses produce deep narcosis leading to death within a few hours.[1] Excessive use produces effects somewhat similar to those of alcoholic intoxication.

274. *PIPER METHYSTICUM* Forst. (PIPERACEAE) — Kava-Kava

Stimulant consumed with Polynesian religious rites, usually as a fermented liquor made from the upper portion of the rhizome. The natives believe it relaxes the mind and body, eases pain, and induces restful sleep. Kava prepared by chewing and fermenting is said to be narcotic. Unlike alcohol, the drug does not impair mental alertness.[11] The deep dreamless sleep following kava ingestion is not followed by hangover.[11] Tierra[28] speaks not of dreamless sleep, but of "a deep restful sleep with clear, epic-length dreams." According to Duve,[266] the chemical constituents possess anaesthetic, analgesic, anticonvulsive, antifungal, sleep-inducing, and spasmolytic properties. Dihydrokawain and dihydromethysticin are both said to have sedative effects.

Considered anesthetic, anodyne, antiseptic, aphrodiiac, aromatic, diuretic, expectorant, galactogogue, narcotic, sedative, sudorific, and tonic.[2,11] The cold water extract of the roots is considered stmulant and tonic.[11] Used in backache, bronchitis, chills, colds, coughs, debility, elephantiasis, gonorrhea, headache, leucorrhea, lungs, myalgia, nocturnal emissions, renitis, rheumatism, skin ailments, tuberculosis, urethritis, vaginitis.[32,33] The diuretic properties led to its use for bronchial and rheumatic ailments resulting from heart troubles, as well as gout. Homeopathically suggested for headache and neurasthenia.[33] Rose describes it as a lilac-scented douche for vaginitis which "gives a tingling type of sensation to the mucous membranes."[47]

Contains a greenish-yellow aromatic resin called kawine, and various alkaloids, with an abundance of starch.[2] The extract dihydromethysticin may cause exfoliative dermatitis. Methysticin, dihydrokawain, and yangonin are also reported.[11] Benzoic and cinnamic acids may have a local anesthetic action.[5] Also, contains yangonin, 11-methoxyangonin, demethoxyyangonin, methysticin, dihydromethisticin, kawain, dihydrokawain, 5-dihydroyangonin, and tetrahydroyangonin.[266]

Toxicity — Continued usage may cause inflammation of the body and eyes, resulting in ugly ulcers, parching, and peeling of the skin.[2] According to the FDA kava resin acts upon the spinal column.[62] Tierra suggests that "regular use of large doses will cause an accumulation of toxic substances in the liver."[28]

275. *PIPER NIGRUM* L. (PIPERACEAE) — Black Pepper

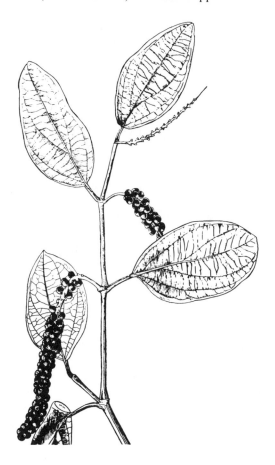

Black pepper is the unripe dried fruit; white pepper is obtained by removal of outer coating (pericarp). Both are available whole, cracked, coarsely ground, or finely ground. Both have numerous culinary uses, including seasoning and flavoring of soups, meats, poultry, fish, eggs, vegetables, salads, sauces, and gravies, and are employed commercially in preparation of processed meats of all kinds, soups, sauces, pickles, salad dressings, mayonnaise, and other foods. Pepper has been used as a condiment and medicine since times of Hippocrates. In the old days, peppercorns were used like currency to pay dowries, rents, taxes, and tolls.[64] Pepper oil, distilled from fruit, serves in perfumery. Piperine markedly increases the kill of houseflies as given by pyrethrin. It improves insecticidal activity like eucalyptus oil. Piperine is synergistically insecticidal to rice weevils and cowpea weevils.[41] Benzene extracts are markedly fungitoxic to mycelia and certain fungi.[267]

Medicinally, it is considered aromatic, carminative, febrifuge, rubefacient, and stimulant. Piperine is used in synthesizing heliotropine which is recommended as an antiseptic and antipyretic, and has been used for arthritic disorders, cholera, constipation, diarrhea, scarlatina, vertigo, paralytic disorders, and sometimes used as a gargle. Piperine should not be combined with astringents as it renders them inert. The root, in the form of ghees, powders, enemas, and balmes, is applied to abdominal tumors. The powdered fruit is said to remedy superfluous flesh. An electuary prepared from the seed is said to help hard tumors; a salve from the seed, eye indurations and internal tumors. The grain, with warm wine and egg, is used for indurations of the stomach. A poultice, made from the pepper plus salt and vinegar, is used for corns.[4] Suggested for urinary calculus. Poulticed for colic, headache, parturition, puerperium, and rheumatism. White pepper suggested for cholera, malaria, and stomachache; black for abdominal fullness, adenitis, cancer, cholera, cold, colic, diarrhea, dysentery, dysmenorrhea, dysuria, furuncles, headache, gravel, nausea, poisoning due to fish, mushrooms, or shellfish. Heavy dose of pepper with wild bamboo shoots said to produce abortion.[41]

Per 100 g, the fruit is reported to contain 255 calories, 10.5% H_2O, 11.0 g protein, 3.3 g fat, 64.8 g total carbohydrate, 13.1 g fiber, 4.3 g ash, 437 mg Ca, 173 mg P, 28.9 mg Fe, 44 mg Na, 1259 mg K, 114 μg β-carotene equivalent, 0.11 mg thiamine, 0.24 mg riboflavin, and 1.14 mg niacin. Pepper contains 2 to 4% volatile oil; and 5 to 9% piperine, piperidine, piperettine, and other minor alkaloids (piperyline, piperolein A, piperolein B, piperanine, etc.).[29] White pepper contains little volatile oil but has the same pungent principles and alkaloids as black pepper. Pepper oil contains a complex mixture of monoterpenes (70 to 80%), sesquiterpenes (20 to 30%). Major monoterpenes include alpha-thujene, alpha-pinene, camphene, sabinene, beta-pinene, myrcene, 3-carene, limonene, and beta-phellandrene. Sesquiterpenes include, mostly, beta-caryophyllene, some beta-bisabolene, beta-farnesene, ar-curcumene, humulene, beta-selinene, alpha-selinene, beta-elemene, alpha-cubebene, alpha-copaene, and sesquisabinene. Oxygenated components include linalool, 1-terpinen-4-ol, myristicin, nerolidol, safrol, beta-pinone, N-formalpiperidine, etc.[29] Purseglove et al. devote a $2^1/_2$-page table to the constituents of black pepper.[64] One pungent principle of pepper is piperine, present at levels of 2 to 6%. Piperine may be useful as an analeptic in barbiturate poisoning. It has a central stimulant action in frogs, mice, rats, and dogs. Piperine, at 1 mg/mℓ, decreased the contraction of isolated guinea pig ileum. When injected i.v. into dogs at 1 mg/kg, it decreased blood pressure and respiration rate. When given orally to rats at 100 mg/100 g, it showed slight febrifugal activity. Piperine interacts with nitrite in vitro under slightly acidic conditions at 37° to form carcinogenic nitrosamines. Piperine is a stimulant. It is mutagenic with Leptospira; with large doses a bactericidal effect is produced. Isolated piperine has an inhibiting effect on *Lactobacillus plantarum, Micrococcus specialis*, and two fecal microorganisms (*Escherichia coli* and *Streptococcus faecalis*.)[41]

Toxicity — Reviewing work on piperine, Buchanan[117] notes that black pepper is probably the most abundantly employed of the various methylenedioxybenzone-containing spices in the U.S. While containing some myristicin and safrole, pepper's main pungent flavoring compound is the piperine (trans, *trans*-5-[3,4-methylenedioxyphenyl-2,4-pentadienoic peridide]). Since piperine and other pepper alkaloids have chemical structures similar to that of the mutagenic urinary safrole metabolite, 3-piperidyl-1-(3′,4′-methylenedioxyphenyl)-1 propanone, pungent components of black pepper are sometimes suspected to be mutagenic and/or carcinogenic. Russians have suggested that the consumption of tea flavored with black pepper may have contributed to the unusually high incidence of esophagal cancer in the Aktibinsk region of the U.S.S.R. Like safrole, piperine stimulates hepatic regeneration in partially hepatocomized rats. Topical application of pepper extract to mice skin has increased the incidence of total malignant tumors. Reviewing the work on safrole, Buchanan concluded that it is the most thoroughly investigated methylenedioxybenzene derivative. Safrole also occurs in black pepper, basil,[10] as well as cinnamon leaf oil, cocoa, mace, nutmeg, parsley, and star anise oil. Safrole was banned in root beer. The oral LD_{50} for safrole in rats is 1950 mg/kg body weight, with major symptoms including ataxia, depression, and diarrhea, death occurring in 4 to 5 days. Ingestion of relatively large amounts of sassafras oil produces psychoactive and hallucinogenic effects persisting several days in humans. With rats, dietary safrole at levels of 0.25, 0.5, and 1% produced growth retardation, stomach and testicular atrophy, liver necrosis, and biliary proliferation and primary hepatomas. Also, reviewing research on myristicin, which occurs in nutmeg, mace, black pepper, carrot seed, celery seed, and parsley, Buchanan hypothesized that myristicin and elemicin can be readily modified in the body to amphetamines.

276. *PISCIDIA PISCIPULA* Sarg. (FABACEAE) — Jamaica Dogwood

Best known for its action as a fish poison, stupefying the fish in waters to which it has been added. Fruits from some species have been used as arrow poison. The plant was described as narcotic in 1844.[2] Has been used as a substitute for morphine and opium.[33] Extracts are insecticidal. The heavy, yellow-brown, close-grained wood is used for boat building, charcoal, and fuel.

Analgesic, diaphoretic, diuretic, emetic, hypnotic, mydriatic, narcotic (root said to smell like opium),[22] sedative, and sudorific, the plant is said to help asthma, dysmenorrhea, neuralgia, pertussis, spasm, and toothache. Root bark applied locally in toothache. Extracts have been used in alcoholism, asthma, bronchitis, delirium, dysmenorrhea, headache, hysteria, insomnia, neuralgia, pertussis and tuberculosis.[2,33,42] Bahamians bind crushed leaves to the head, the emanating fumes believed to alleviate headache. Jamaicans add the bark to their baths to alleviate backache and other pains. Yucatan natives use leaf preparations for cold, cough, skin diseases, and wounds, the bark extract as an antispasmodic, sedative, and sudorific. The bark extract is said to cure mange in dogs.[42]

Hager's Handbook lists jamaicin $C_{22}H_{18}O$, beta-sitosterol, cerotinic-, stearic-, malic-, succinic-, eretic-, tartaric-, and citronic-acid, tannin, resins, sumatrol, lisetine, piscerythrone, piscidone, millettone, isomellettone, and dehydromellettone. Rotenone, $C_{23}H_{22}O_6$, while piscicidal and insecticidal, has also shown anticancer activity against lymphocytic leukemia[10] and cell cultures of human epidermoid carcinoma of the nasopharynx.[4] Like so many other anticancer compounds, the rotenone is also said to be carcinogenic.[10] The aqueous extract of the bark contains piscidic acid and a bitter glucoside. It also is reported to contain an alkaloid. All parts, more especially the root bark, contain ichthyotoxic substances variously reported as *circa* 0.25% ichthynone ($C_{21}H_{14}O_5$), piscidin ($C_{29}H_{24}O_8$) or piscidic acid ($C_{11}H_{12}O_7$), and rotenone.[33,42]

Toxicity — It may cause gastric distress and nausea, overdoses "produce toxic effects",[2] numbness, tremors, salivation, sweating, etc.[33] The wood is said to be irritant and toxic to humans.[6]

277. *PISTACIA LENTISCUS* L. (ANACARDIACEAE) — Mastic

Cultivated primarily for the resin, used as a masticatory and a medicine. Also, used to harden gums and alleviate toothache. Chewed by oriental women as a breath freshener. Used for filling dental caries.[2] Gum used in surgical varnishes and plasters. Used in the manufacture of confectionery, liqueurs, and varnishes. The Greek liqueur mastiche is made from grape skins flavored with mastic.[38] The varnish is used for coating metals and paintings, for lithography, and for retouching negatives. Egyptians used mastic as an embalming agent. Sometimes used in incense; oil of mastic used in cosmetics.[6] Arabs use the seed oil for food and illumination. The wood and leaves burn green. Twigs are used in basketry.[38]

Frequently cited in the cancer folklore, the resin or juice from mastic is used for indurations or tumors of the anus, breast, liver, parotid, spleen, stomach, testicles, throat, and uterus. Regarded as analgesic, antitussive, aperitive, aphrodisiac, astringent, carminative, diuretic, expectorant, hemostatic, stimulant to the mucous membranes, stomachic, and sudorific. It is said to be used for apostemes, boils, cankers, carbuncles, cardiodynia, caries, catarrh, cholecystosis, cirrhosis, condylomata, debility, diarrhea, excrescences, gingivitis, gonorrhea, halitosis, hepatosis, leucorrhea, mastitis, phymata, sclerosis, stomach ailments, and tumors.[32,38] The leaves are used for blennorrhea and dysentery.

The plant contains a little volatile oil, 9% resin (soluble in alcohol and ether) and 10% resin insoluble in alcohol.[2] Mastic contains 90% resin, masticin, mastichic acid, and a bitter principle.[8] *Hager's Handbook* lists masticodienonic acid, isomasticodienoic acid, oleanolic acid, and tirucallol. Young leaves and/or twigs contain myricetin ($C_{15}H_{10}O_8$), quercetin, kaempferol, shikimic acid, lupeal, cycloartenol, beta-sitosterol, pinene, camphene, and terpene.[33] Seeds contain aucubin (rhinanthin), choline, and organic acids, and 0 to 0.022% plantease (a crystalline trisaccharide), much starch, and up to 22.08% of an edible oil.[42]

Toxicity — Classified by the FDA as an Herb of Undefined Safety: "Scarcely ever given internally except in certain cathartic pills."[62] Mastic can cause dermatitis.[6]

278. *PLANTAGO MAJOR* L. (PLANTAGINACEAE) — Plantain

Seeds once used for bird seed. Stripping the seeds (and husks) off the fruiting spikes, I have added them to milk and sugar as a poor man's branflakes (or Metamucil®)! Not only do the seeds exude a copious mucilage, they seem to have a rennet effect on the milk! Eaten by sheep, goats, and swine, and one of the favorite snacks of my rabbit, plantain leaves are said to be rejected by cattle and horses. I often add the leaves to my potherb potpourri. An old cosmetic embraced plantain, houseleeks, and lemon juice.

Widely touted, especially as Latin America's "llanten", plantain is prominently mentioned in folk cancer literature for cancer of the anus, breast, eyes, feet, gums, intestines, liver, mouth, parotids, sexorgans, throat, and uterus and for neuroblastoma, lacrymatory tumors, and nasal polyps.[4] Fresh leaves are poulticed onto herpes.[42] Decoctions of plantain entered nearly every old European nostrum.[2] Mentioned in "remedies" for asthma, bronchitis, bruises, cold, cough, convulsions, diarrhea, dropsy, dysentery, earache, enuresis, epilepsy, epistaxis, fever, gout, headache, hemoptysis, hemorrhoids, hepatitis, jaundice, kidneystone, lunacy, malaria, piles, otitis, puerperium, renitis, ringworm, shingles, sore throat, splenitis, strangury, syphilis, thrush, and toothache.[16,32,42] Perry recounts a Soviet report that refrigerated leaves produce a stimulant which, when extracted, may be injected subcutaneously for tubercular ulcers in the bronchial tubes, scrofula, and other tubercular and eye ailments. The extract stimulates the mucous membrane of the trachea and the secretory tissues of the digestive organs and slows and deepens respiration.[16] Chinese use the seeds as an antidiarrheic, antidysenteric, antipyretic, antirheumatic, demulcent, diaphoretic, and pectoral. They apply the leaves externally for sores, ulcers, and whitlow; bruised leaves to the cheek swollen from abscessed gums. Juice from heated leaves is used by Malayans to cleanse thrush in an infant's mouth.[16] Chinese also use the plant for asthma, bronchitis, cholecystosis, colic, cough, gonorrhea, hemorrhage, hemorrhoids, and malaria.[16] Seeds used for bowel ailment, dysentery and thrush. Poulticed onto boils, e.g., Barbadans poultice leaves onto boils or infections. Seeds used as a diuretic in Chinese medicine.[8] In both Europe and Russia, the root or strands from the leaf petiole inserted in the ear for a toothache.[11] Added to creams for back massages.[11] "The common plantain weed was formerly considered refrigerant, diuretic, deobstruent, and somewhat astringent. The ancients esteemed it highly but it is at present rarely used, except externally in domestic practice as a stimulant application to sores. The leaves are applied whole or bruised in the form of a poultice."[268] Rubbed onto insect and stinging nettle rashes. Liniment made from plantain and rose oil applied to gout of the extremities. Steinmetz[27] lists the herb as alterative, astringent, demulcent, deobstruent, diuretic, expectorant, and refrigerant.

Per 100 g, the leaves are reported to contain 61 calories, 81.4 g H_2O, 2.5 g protein, 0.3 g fat, 14.6 g total carbohydrate, 1.2 g ash, 184 mg Ca, 52 mg P, 1.2 mg Fe, 16 mg Na, 277 mg K, 2520 μg β-carotene equivalent, 0.28 mg riboflavin, 0.8 mg niacin, and 8 mg ascorbic acid.[21] Contains, also, aucubin, gum, mucilage, resin, and tannin.[27] *Hager's Handbook* adds quite a few compounds to the biochemical menu: allantoin, adenine, baicalein, baicalin, benzoic acid, chlorogenic acid, choline, cinnamic acid, ferulic acid, L-fructose, fumaric acid, gentisic acid, D-glucose, *p*-hydroxybenzoic acid, indicain, lignoceric acid, neochlorogenic acid, oleanolic acid, plantagonine, planteose, saccharose, salicylic acid, scutellarein, sitosterol, sorbitol, stachyose, syringic acid, tyrosol, ursolic acid, vanillic acid, and D-xylose. The seeds contain 18.8% protein, 19% fiber, 10 to 20% oil (with 37% oleic-, 25% linoleic-, 0.9% linolenic-acids), adenine, aucubin, choline, mucilage, plantenolic, and succinic acid.[16,33]

Toxicity — Classified by the FDA as an Herb of Undefined Safety.[62]

279. *PODOPHYLLUM PELTATUM* L. (PODOPHYLLACEAE) — Mayapple, American Mandrake

Though poisonous, like the leaves and roots, when green, the fully ripe fruit is edible and said to make a luscious marmelade and jelly. I have enjoyed the fruit pulp in July and August but never in May. By September aerial portions of the plant have usually disappeared (in Maryland). Fernald et al. mention a southern drink from mayapple juice, sugar, and Madeira.[13] Such a liqueur might be a useful addition to the folk medicine chest. Mayapple ade can be prepared in the same fashion as lemonade, from the ripe fruits. Back in 1925, mayapple root diggers earned less than 10¢ a pound. A Chinese correspondent, fearing cancer, offered me $50 for a pound, which my son easily could have supplied with an hours digging. However, that would have constituted a serious conflict of interest. And suppose she overdosed and killed herself! I see it advertised now for less than $10 a pound.[269]

Much interest in the *poisonous* mayapple stems from Hartwell's investigations, prompted by the fact that Maine's Penobscot Indians used mayapple in folk cancer treatments.[4,10] Mayapple root contains podophyllotoxin and two other lignans which were powerfully active against Sarcoma 37 in mice. Hence, the mayapple, recommended for cancerous tumors in folk medicine, contains several ingredients said to be useful in the treatment of cancers. Podophyllin, a resin from the plant, has been used to treat venereal warts.[210] This dangerous allergenic resin also exhibits antitumor activity. Mayapple is a dangerous plant that shows up in many folk remedies. North American Indians used the plant as an emetic and vermifuge. In small doses, it is used for skin disorders. Dissolved in alcohol it is used as a counterirritant. It is also recommended in amenorrhea, biliousness, dropsy, dysentery, dyspepsia, hepatitis, jaundice, liver, prostatitis, rheumatism, scrofula, and other disorders. Podophyllin is considered alterative, cholagogue, emetic, hydragogue, purgative, and bitter tonic. It has been tried in tinea capitis with equivocal results. It is described as an effective vermifuge, first stimulating, then killing Ascaris, It has given symptomatic relief in some skin allergies and inflammations. Papilloma and senile keratoses have responded satisfactorily. *Drugs of Choice*[270] states that condylomata acuminata (veneral warts), unless they have formed vegetating masses (giant condylomas), often yield to *Podophyllum*. Resistance or recurrences often mean urethral, vaginal, or anal involvement, which reservoirs must also be corrected. R_x126 in *Drugs of Choice* consists of podophyllum resin dissolved in a mixture of benzoin, aloe, storax, and tolu balsam, quite an herbal *potpourri*. Podophyllotoxin is one 1976 Drug of Choice for brain tumors and for Hodgkin's disease. The natural podophyllotoxins may produce myelosuppression, hair loss, and, rarely, vascular collapse. Persons handling the powdered rhizome may display conjunctivitis, keratitis, and ulcerative skin lesions.[6] Still I coated my lichen planus (identified by one doctor as psoriasis) with the green stain of podophyllum leaves by crushing them onto the lesions. The herbal treatment was no more, no less, effective than the prescribed treatment, which was applied to the lesions of the other leg.

The root contains 3.5 to 6% of podophyllum resin, which, in turn, contains 20% podophyllotoxin ($C_{22}H_{22}O_8$), and toxic principle yielding podophyllic acid ($C_{22}H_{24}O_9$) and picropodophyllin ($C_{22}H_{22}O_8$); 10% alpha-peltatin and 5% beta-peltatin. Several flavonoids (astragalin, isorhamnetin, kaempferol, kaempferol-3-glucoside, quercetin, and quercetin-3-galactoside) are also reported.[29] Albumen, calcium oxalate, gallic acid, gum, quercetin, starch, and volatile and fixed oils are also reported.

Toxicity — One herbal says its greatest action is on the liver and bowels, as a powerful hepatic and intestinal stimulant. In large doses it causes nausea and inflammation of the stomach and intestines which have proved *Fatal*.[2] Amerindians once used the young shoots for suicide. One woman is said to have died 31 hr after ingesting 300 mg podophyllum powder. Another died when podophyllum ointment was left too long ($>^1/_2$ hr) on a venereal wart of the vulva.[17]

To the physician — Hardin and Arena suggest gastric lavage or emesis, activated charcoal, and antidiarrheal agents to treat poisoning.

280. *POLYGONUM AVICULARE* L. (POLYGONACEAE) — Prostrate Knotweed

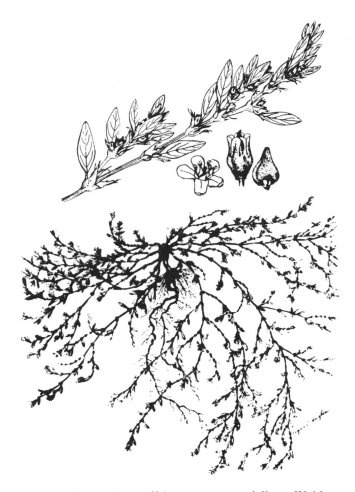

Weed, sometimes used to make tea,[16] known commercially as Weidemannscher tea and Homeriana tea.[3] Used as a substitute for ergot in Russia.[33] Bactericidal. Although regarded as a folk remedy for malaria, tests have proven negative.

Used in folk medicine for tumors and cancers, especially of the kidneys, mammaries, or stomach.[4] A mild astringent formerly employed as a vulnerary and styptic. Used in some countries of Europe as a home remedy for cancer, chancroids, diabetes, diarrhea, dysentery, dysuria, gonorrhea, gout, hemorrhoids, hypertension, malaria, nephrosis, night sweats, rheumatism, stones, tonsillitis, tuberculosis, and vaginitis.[32] Toward the end of the 19th century, Hemero tea derived from it was esteemed for asthma and bronchitis in Germany and Austria.[8] Said to be a good gargle and remedy for inflammatory diseases of the tissues.[2] Used for gonorrhea in Malaya.[16] In China, where considered anthelmintic, antidiarrheic, astringent, diuretic, and emollient the plant is employed for chancroid, cholera, eczema, fever, hemorrhoids, jaundice, menorrhagia, pruritis, stomachache, and sunstroke.[12,16] The herb has been classified anodyne, antiperiodic, antiseptic, astringent, cholagogue, demulcent, diuretic, emetic, emollient, expectorant, hemostat, laxative, tonic, vasoconstricting, vermifuge, and vulnerary.[27,32,33]

"Contains the glycosides, quercetin 3-arabinoside and avicularin" ($C_{20}H_{18}O_{11}$).[8] Also, contains calcium oxalate, polygonic acid, salicic acid, tannic acid, and a trace of volatile oil. Avicularin, catechin, delphinidin, hyperin, myricetin, querciten, quercitrin, and rutin are also reported.[12,33] Roots contain 0.35% oxymethylanthraquinone, twigs 0.2%, leaves 0.15%.[3]

Toxicity — Can cause contact dermatitis.[6] Classified by the FDA[62] as an Herb of Undefined Safety. The oral minimum lethal dose for the cat and rabbit is *circa* 20 mℓ/kg of herbal infusion, intravenous *circa* 2 mℓ.

281. *POLYGONUM MULTIFLORUM* Thunb. (POLYGONACEAE) — Fo-Ti, Climbing Knotweed

Fo-ti is a magical or medicinal or ornamental plant or weed, depending on your point of view. It is sold by many American herbalists. Tyler says, hopefully tongue-in-cheek: ''use of 50-year-old root preserves one's natural hair color; 100-year-old root helps one maintain a cheerful appearance; 150-year-old root causes new teeth to grow; 200-year-old root preserves one's youth and energy; and the 300-year-old product makes one immortal. Needless to say, very little (if any) of the truly ancient product is available.''[37]

Leaves and roots believed to tonify the liver and kidneys, fortify the blood, strengthen muscles and bones, and keep the hair black; used for anemia, backache, kneeache, lymphadenitis, neurasthenia, premature graying, and traumatic bruises. Stems considered deostruent, nerve calmant; for scabies and itching skin. Chinese decoct the root for cancer, constipation, insomnia, scrofula, and weakness of liver and spleen; they prescribe it for cold, piles, parturition, puerperium, and tumors. Considered one of China's great four herbal tonics (with Angelica, Lycium, and Panax), this is believed to strengthen the liver, kidney, bone marrow, tendons, and bones, maintaining youthfulness and general good health. Others rank fo-ti among the five major tonic herbs (with *Panax ginseng, Eleutherococcus senticosus, Angelica sinensis*, and *Glycyrrhiza galabra*). Tyler mentions that Europeans once used it for diabetes, a use not apparently in vogue in China.[37]

Extracts have shown antiprogestational, antipyretic, antitumor, and sedative effects. Two antitumor agents, emodin and rhein, occur, along with chrysophanic acid, cyrysophanic acid anthrone, and chrysophanol.

282. *PROSOPIS JULIFLORA* (Sw.) DC. (MIMOSACEAE) — Velvet Mesquite

Prosopis pods are among the earliest known foods of prehistoric man in the New World. Today flour products made from the pods are still popular, although only sporadically prepared, mostly by Amerindians. Pods are made into gruels, sometimes fermented to make a mesquite wine. The leaves can be used for forage. Providing good bee pasturage, also, nectar from mesquite yields a superior honey. The wood is used for parquet floors, furniture, and turnery items, fencepost, pilings, as a substrate for producing single-cell protein, but, most of all, for fuel (in World War II, Argentina fueled its industrial boilers and steam locomotives with *Prosopis* spp.). Toasted seeds are added to coffee. Bark, rich in tannin, is used for roofing in Columbia. The gum forms an adhesive mucilage, used as an emulsifying agent. Gum is used in confectionary and mending pottery. Roots contain 6 to 7% tannin, which might discourage Rhizobia.[63]

The juice is used in folk remedies for that cancerous condition Hartwell terms "superfluous flesh".[4] Reported to be cathartic, cyanogenetic, discutient, emetic, POISON, stomachic, and vulnerary, mesquite is a folk remedy for catarrh, colds, diarrhea, dysentery, excrescences, eyes, flu, headcold, hoarseness, inflammation, itch, measles, pinkeye, stomachache, sore throat, and wounds.[32] Pima Indians drank the hot tea for sore throat.[11] Aqueous and alcoholic extracts are markedly antibacterial. Yucatan Indians made a gargle from the bark and/or seeds for bronchitis, laryngitis, and pharyngitis and an enema for diarrhea and dysentery. Curacao natives used the leaves for eye inflammation and abscessed teeth, applying packs for breast pain. Mexicans use the gum for dysentery.

Per 100 g, the flower is reported to contain, on a dry weight basis, 21.0 g protein, 3.2 g fat, 65.8 g total carbohydrate, 15.5 g fiber, 10.0 g ash, 1,310 mg Ca, and 400 mg P. Leaves contain 19.0 g protein, 2.9 g fat, 69.6 g total carbohydrate, 21.6 g fiber, 8.5 g ash,

2080 mg Ca, and 220 g P. Fruits contain 13.9 g protein, 3.0 g fat, 78.3 g total carbohydrate, 27.7 g fiber, and 4.8 g ash. Seeds contain 65.2 g protein, 7.8 g fat, 21.8 g total carbohydrate, 2.8 g fiber, and 5.2 g ash, all on a zero-moisture basis.[273] However, Earle and Jones[271] report only 33.9% seed protein, 6.9% oil, and 30% ash, while Jones and Earle report 30.0% seed protein, 5.3% seed oil.[272] Another analysis of the fruit shows 14.35% water (hygroscopic), 1.64% oil, 16.36% starch, 30.25% glucose, 0.85% nitrogenous material, 5.81% tannin-like material, 3.5% mineral salts, and 27.24% cellulose. Mesquite gum readily hydrolyzes with dilute sulfuric acid to yield L-arabinose and D-galactose and 4-O-methyl-D-glucuronic acid at 4:2:1. Owing to the high content of arabinose, the gum is an excellent source of sugar. Roots contain 6.7% tannin, bark 3 to 8.4%, and dry wood 0.9%. The alkaloids 5-hydroxytryptamine and tryptamine are reported from this species.[274]

Toxicity — Thorns from mesquite, on penetrating the eye, cause more inflammation than expected from the physical injury. The irritation may be due to waxes. Injection of cerotic acid is destructive to the eye.[6] (Still Amerindians applied the leaves for conjunctivitis.) Using the wood in a fireplace has caused dermatitis, as has working with seasoned wood. The gum has irritant properties. Reports on cattle toxicity vary. Ingestion over long periods of time will result in death in cattle.[11] The pollen may cause allergic rhinitis, bronchial asthma, and/or hypersensitivity pneumonitis.[11] Kingsbury[14] goes into some detail on mesquite poisoning in cattle, including cases where autopsies showed pods and seeds in the rumen 9 months after the cattle could have ingested them. Mesquite poisoning may induce a permanent impairment of the ability to digest cellulose.[14] Felker and Bandurski[275] also provide interesting detail. If *Prosopis* pods are the sole food source for cattle, *circa* 1% becomes sick, and some die with a compacted pod ball in the rumen. Death is attributed to high sugar content repressing the rumen-bacterial cellulose activity. Mesquite feeding to pigs was promising during the first 4 weeks, deteriorating thereafter, perhaps due to phytohemagglutinins and trypsin inhibition. Feeding trials with sheep show a 15% higher protein disgestibility coefficient for mesquite pods than for alfalfa hay. Trypsin inhibition has been demonstrated. Contains isorhamnetin 3-glucoside, apigenin, 6,8-diglycoside, and traces of quercitin 3′,3diOMe, leutolin 3′-OMe, and apigenin diglycoside.[274] According to Morton, the gum is irritant and causes dermatitis in susceptible people. Flowers may cause respiratory irritation.[42]

283. *PRUNUS ARMENIACA* L. (ROSACEAE) — Apricot, Chinese Almond, Apple

Apricots are cultivated for the fruit, eaten fresh out of hand, or dried, made into conserves or alcoholic beverages. Kernels produce a sweet edible oil sometimes used as substitute for almond oil. Chinese almonds are the seed kernels of several sweet varieties of apricot, used for almond cookies, eaten salted and blanched, or made into gruel or flour. Afghans also use the seeds as almonds. Bitter apricot kernel is highly toxic because of prussic acid present. Expressed oil, known as persic oil or apricot oil, is used as a pharmaceutical vehicle; it is obtained by the same process as bitter almond oil. Pit shells have been used to prepare activated charcoal, via destructive distillation.[38]

An apple a day keeps the doctor away. In Biblical days, Solomon said "comfort me with apples for I am sick."[38] Surely they didn't mean apricot pits in the Garden of Eden. As Milton[276] says, "The fruit of that forbidden tree whose mortal taste brought death into the world, and all our woe." The pits do contain laetrile-like compounds which can cure or kill, depending on dosage. Considered antidotal, antitussive, aphrodisiac, cyanogenetic, demulcent, emollient, expectorant, preventitive, sedative, tonic, vermifuge, vulnerary, the apricot is an element in folk medicines for anemia, asthma, bronchitis, catarrh, cold, constipation, cough, eyes, fertility (female), fever, heart, hemorrhage, inflammation, laryngitis, puerperium, rheumatism, spasm, swellings, thirst, and tumors. The fruit is said to be a folk remedy for cancer. Medicinally, a paste, obtained by crushing the kernel, is used for inflammation of the eyes and is considered antispasmodic, demulcent, pectoral, sedative, vulnerary, and anthelmintic. Ginger and licorice combined with kernels make a confection used as a tussic and expectorant remedy. Another concoction made by fermentation is used as a prophylactic and tonic. Decoction of kernels made into a beverage is used for cough, asthma, and catarrhal ailments. Kernel juice is used for hemorrhages. In Chinese medicine, fruit of bitter almond is useful in heart disease.[38] In Korea, the expectorant kernel is used to treat dry throat.

Laetrile, according to some laetrile advocates, is broken down by the enzyme, beta-glucosidase, to release toxic cyanide. The enzyme is believed to be prevalent in tumor tissue but scarce elsewhere. Triggered by laetrile advocates, the FDA banned laetrile from interstate commerce in 1971. In a 1980 clinical study by the NCI, it was concluded that laetrile and natural products containing it, such as apricot pits, were "ineffective as a treatment for cancer."[37] Per 100 g, the fruit is reported to contain 38 to 58 calories, 83.4 to 85.6 g H_2O, 0.5 to 1.2 g protein, 0.4 to 0.6 g fat, 9.0 to 14.1 g total carbohydrate, 1.1 to 1.5 g fiber, 0.6 to 1.2 g ash, 13 to 30 mg Ca, 24 to 34 mg P, 0.7 to 1.1 mg Fe, 1 mg Na, 218 mg K, 1340 to 3145 μg β-carotene equivalent, 0.02 to 0.04 mg thiamine, 0.05 to 0.06 mg riboflavin, 0.4 to 0.7 mg niacin, and 6 to 10 mg ascorbic acid. Fruits also contain lycopine, quercetin, rutin, isoquercitrin, benzyl alcohol, epoxydihydrolinalool, linalool, isobutyric acid, capronic acid, and lactone.[33] WOI reports xylose, glucose, fructose, sucrose, sorbitol, meso-inositol, etc. among the sugars and polyols. The fruits also contain pectic substances (1% as calcium pectate). Malic and citric are said to be the principal acids, with tartaric, quinic, and succinic also reported. The volatile essence contains myrcene, limonene, *p*-cymene, terpinoleine, *trans*-2-hexenol, α-terpeneol, geranial, geraniol, 2-methylbutyric acid, acetic acid, linalool, etc.[1] The edible portion (93% of fruit) of dried Indian apricots contains, per 100 g: 19.4% H_2O, 1.6 g protein, 0.7 g ether extract, 2.1 g fiber, 73.4 g other carbohydrates, 2.89 ash, 110 mg Ca, 70 mg P, 4.6 mg Fe, 98 IU vitamin A, 0.22 mg thiamine, 2.3 mg nicotinic acid, 2 mg ascorbic acid, and 306 calories.[1] The cake left after oil extraction from the seed contains too much amygdalin, yielding 0.06% HCN. Used as a fertilizer, it contains 6.64% N, 2.2% P_2O_5 and 1.14% K_2O. The seed contains, per 100 g, 549 calories, 8% water, 29% protein, 47.3% fat, 12.9% total carbohydrate, 3.0 g fiber, 2.8 g ash, 140 mg Ca, 276 mg P, 4.4 mg Fe, 716 mg K, 0 mg β-carotene equivalent, 0.14 mg thiamine, 0.49 mg riboflavin, and 1.6 mg niacin. The seed oil (*circa* 40%) contains 62 to 80% oleic and 14 to 80% linoleic acid, 0.02% squalene, beta-sitosterol, campesterol, and the pangamic acid vitamin, useful in the treatment of pachymeningitis.[33] An Indian sample showed 1.1% myristic, 3.5% palmitic, 2.0% stearic, 73.4% oleic, and 20% linoleic acid.[1]

Toxicity — *Poison:* a double kernel is said to be enough to kill a man (deaths are reported).[3] If eaten in excess, fruit is believed to harm the bones and muscles, to promote blindness and falling hair, to numb mental facilities, and to injure parturient women. Apricot pits vary in their amygdalin or laetrile "content which may reach 8%, but the kernels of some wild varieties contain 20 times as much as those of cultivated varieties of apricot. Serious cases of poisoning, especially among children, have been reported as a result of eating quantities of these seeds."[37]

284. *PRUNUS LAUROCERASUS* L. (ROSACEAE) — Cherry-Laurel

Evergreen tree or shrub. Leaves are macerated in water and distilled to yield cherry laurel water and standardized to contain 0.1% HCN, which is used as a flavoring and ophthalmic ingredient. This can be redistilled to yield oil of cherry laurel. The most active essence is reserved for perfumery.[27] Crushed leaves have been used to prepare the entomologist's killing bottle. Bruised leaves, like those of almond and peach, will deodorize pots and pans of balsam of copaiba or clove oil odors, if grease has first been cut with alcohol.[2] Fruits are said to be edible but the seeds should be expectorated. The wood is used in carpentry and lathe work.

The leaves and roots are used in folk remedies for scirrhous tumors and ulcerated cancers.[4] Reported to be anodyne, antispasmodic, antitussive, aromatic, cyanogenetic, narcotic, poison, sedative, and tonic, cherry laurel is a folk remedy for asthma, cancer, cough, dyspepsia, nausea, pertussis, and tumors.[4,32,33] Homeopathically used for cyanosis, dry throat, pertussis, and spasms.[33]

Seeds contain 0.034% HCN and 25 to 30% fatty oil, the oil containing 9.9% palmitic; 1.7% stearic-; 1.8% myristic-, 73.4% oleic-, and 13.2% linoleic-acid. Protein content runs *circa* 31%, fat content to nearly 45%.[21] Leaves contain 0.12 HCN, younger leaves containing more than older leaves. Hence, fresh leaves give aroma of bitter almond when mashed.

Leaves yield 0.05% essential oil with 82.7 to 91.2% benzaldehyde, 1.4 to 2.8% HCN; also, probably some *d*-mandelonitrile and benzyl alcohol. 0.1% HCN translated to 100 mg HCN per 100 g fresh leaves. Here are levels reported[1,14] for various *Prunus* species: *P. armeniaca* pits, 275 mg/100 g; *P. laurocerasus* leaves, 100 mg/100g; *P. melanocarpa* fresh leaves, 368; *P. pensylvanica* fresh leaves, 91; wilted leaves, 143; *P. persica* peach pits, 164; fresh leaves, 66; *P. serotina* fresh leaves, 212; *P. virginiana* fresh leaves, 143; wilted leaves, 243. Plants containing 0.02% HCN or more are generally considered potentially dangerous to livestock. The U.S. rejects for humans lima beans containing more than 10 mg/100 g, Canada more than 20 mg/100 g (some tropical forms of lima beans contain 300 mg HCN/ 100 g). Leaves contain the glucoside prulaurasine and prunasine ($C_{14}H_{17}NO_6$). According to Grieve's *Herbal*, a drop of sulfuric acid will keep cherry laurel water unchanged for 2 years.[2]

Toxicity — Narcotic.[2] For cyanide symptoms and treatments, see *Malus*. Leaves can kill cattle that might ingest them.

285. *PTYCHOPETALUM OLACOIDES* Benth. (OLACACEAE) — Potency Wood, Muira Puama

The "aphrodisiac" drug, muira puama, comes from the stems of *P. olacoides* and *P. uncinatum*, while the roots are considered a panacea.[24] The drug is administered by mouth, either as a powder, an alcoholic extract, or a decoction (extract formed by boiling in water). An alternative method of obtaining the aphrodisiac effect is to bathe the genitals with a concentrated decoction. It is applied locally for rheumatism and muscle paralysis.[37]

Reported to be apertif, aphrodisiac, CNS-stimulant, nervine, and tonic, potency wood is a folk remedy for dyspepsia, neuralgia, paralysis, and rheumatism.[32,33] The drug has a long history of use in Brazil as a powerful aphrodisiac and nerve stimulant. It is an ingredient in a number of proprietary remedies and folk medicines for sexual impotence. Muira puama is also touted for dyspepsia, menstrual irregularities, rheumatism, and paralysis caused by poliomyelitis, and as a general tonic and aperitif.

Hager's Handbook reports 0.05% "muirapuamin", 0.5% alkaloid, 0.4% fat, 0.6% phlobaphene, 0.6% alpha-resinic acid, 0.7% beta resinic acid.[33] Chemically, muira puama contains 0.4 to 0.5% of a mixture of esters, two thirds of which is behenic acid, lupeol, and beta-sitosterol; in the remaining portion, campesterol and other fatty acids (arachinic-, lignoceric-, uncosanic-, tricosanic-, and pentacosanic-acids) replace the behenic acid. Other more-or-less routine plant constituents are volatile oil, resin, fat, tannin, and various fatty acids.

Toxicity — "None of the constituents in this drug is known to exhibit any pronounced physiological activity. This, plus the lack of any reported clinical testing of muira puama, causes us to review its reported effects with considerable skepticism. Until such tests have been carried out, no claims of efficacy or safety can be substantiated, and we must conclude at this point that potency wood is instead impotent."[37]

286. *QUASSIA AMARA* L. (SIMAROUBACEAE) — Surinam Quassia, Bitterwood

Bark has been used as a febrifuge and insecticide. Recently, the extracts and quassin, a mixture of bitter principles, have been used in bitters, liqueurs, and nonalcoholic beverages, candy, baked goods, desserts, gelatins, and pudding. The highest average maximum use level is reported at below 75 ppm in nonalcoholic beverages.[29] Central Americans store clothing in boxes made from the wood, said to repel moths.[22] Grieve mentions a rather drastic cure for alcoholism: ''Quassia with sulphuric acid acts as a cure for drunkeness, by destroying the appetite for alcoholics.''[2]

Reported to be apertif, depurative, insecticidal, laxative, pediculicide, stomachic, tonic, and vermifuge, quassia is a folk remedy for carcinoma, debility, dyspepsia, fever, hepatosis, hyperglycemia, malaria, snakebite, and spasms.[32] Mexicans take the bark infusion rectally for intestinal parasites. Costa Ricans use the root decoction for diabetes, diarrhea, and fever; Brazilians for diarrhea, dysentery, dyspepsia, flatulence, and gonorrhea. Surinamese steep the bark or twigs in liqueurs and drink the bitters to prevent fever; Guyanese steep it in alcohol for a tonic. Panamanians take it for fever, hepatosis, and snakebite.[42] Homeopathically used for debility, diarrhea, dyspepsia, fever, hepatosis, jaundice. Quassimarin is said to have antileukemic properties.[29]

Contains quassin, $C_{22}H_{28}O_6$, quassinol, 18-hydroxyquassin, and neoquassin, bitter principles said to be 50 times more bitter than quinine, and two more quassinoids, quassinacin, and simalikalactone D.[29] *Hager's Handbook* adds beta-sitosterol, isoquassin ($C_{22}H_{28}O_6$), *circa* 0.25% alkaloids. The root bark contains quassin, a volatile oil, malic acid, gallic acid, calcium tartrate, and potassium acetate.[1]

Toxicity — Narcotic.[2] GRAS § 172.510.[29]

287. *QUILLAJA SAPONARIA* Molina (ROSACEAE) — Soaptree, Soapbark

Bark used for washing clothes in South America, contains *circa* 10% saponin; used in manufacture of commercial saponins; used in mineral water industry, shampoo liquids, etc. as a foaming agent.[20] Has been suggested as a substitute for senega and sarsaparilla. Used in mouthwash, shampoos, and toothpowders.[33] As a shampoo — 1 part bergamot, 5 parts powdered quillaja, and 20 parts alcohol, has been recommended to promote the growth of hair. Also, used extensively as a foaming agent in cocktail mixes and root beers. Highest maximum use level averages around 0.01% for such beverages. Also, reported in baked goods, candy, frozen dairy desserts, gelatins, and puddings.[29]

Formerly used internally for bronchitis. Used externally as a detergent and local irritant. Has been suggested for aortic disease with hypertrophy, its efficacy depending on the diminished action of the cardiac ganglia and muscle.[2] According to Leung,[29] the herb is used for vaginal douches, itchy scalp, dandruff, skin sores, athlete's foot, and cough.

Contains quillaic acid, quillaja saponin (relatively stable sternuatatory foaming agent), galactose, sucrose, tannin.[20,33] Calcium oxalate (11%) present in bark.[2]

Toxicity — Quillaja saponin (sapotoxin) is very poisonous,[2] with digitalis recommended as an antidote. Quillaja has been approved (§ 172.510) for food use; powdered quillaja bark has highly local irritant and sternutatory properties.[29] It is expectorant and depresses the heart and respiration. The saponin is reported to be ''too strongly hemolytic and irritating in the gastrointestinal tract to be used internally.'' Severe toxic effects due to large doses include liver damage, gastric pain, diarrhea, hemolysis of red blood corpuscles, respiratory failure, convulsions, and coma.

288. *RANUNCULUS BULBOSUS* L. (RANUNCULACEAE) — Bulbous Buttercup

Roots, poisonous raw, are said to be edible when boiled. Pigs eat them raw. Yet it is said that people who consumed bulbs boiled in chicken broth suffered severe gastritis.[19] "The round bulbs which have wintered over are in the spring surprisingly mild and sweet, with no more pungency than a somewhat strong radish, and when thoroughly dried they lose essentially all their acridity and become very sweet."[13] Beggars use the vesicant latex to make their sores more pitiable. Protoanemonin is active against both Gram-positive and Gram-negative bacteria, and inhibits *Candida, Escherichia,* and *Staphylococcus.*[1]

The plant, especially the roots, has been used for cancers, especially of the breast, cervix, etc., corns, warts, and wens, from Chile to California.[4] Roots raise blisters with more safety and less pain than Spanish Fly, and are applied like it to the joints in gout. The sternutatory juice has been used to clear up headache. Leaves applied as a vesicant to rheumatism of the wrist and rheumatic neuralgia. The tincture in wine is used for shingles and sciatica. Infusion of the root is said to have cured nursing sore mouth. Homeopathic tinctures are used for such diverse ailments as alcoholism, breast pain, chilblains, corns, delirium tremens, diarrhea, dyspnea, eczema, epilepsy, flu, foot pain, gastralgia, gout, hayfever, herpes, hiccup, hydrocele, jaundice, liver pain, meningitis, neuralgia, nyctalopia, ovaries, pemphigus, pleuritis, pleurodynia, rheumatism, spinal irritation, and, fortunately for me, writer's cramps.[30]

Reported to contain ranunculin, protoanemonin, anemonin, and labenzyme. Leaves said to contain 1-caffeoylglucose.[33]

Toxicity — Millspaugh recalls the weird and, perhaps, dubious tale of a sailor who inhaled the fumes of the burning plant and then had his first attack of epilepsy which returned in 2 weeks, accompanied by cachexia, headache, gout, and finally death. Contains *circa* 2% (dry weight) protoanemonin. This species is blamed, rightly or wrongly, with poisoning hogs, resulting in blindness. Children who ate several of the bulbs are said to have been poisoned.[14] A 5-year old who took the buttercups to bed with her suffered vesicles and bullae of the lower limbs.[16]

289. *RAUVOLFIA SERPENTINA* (L.) Benth. (APOCYNACEAE) — Rauvolfia, Chandra, Sarpaganda

Rauvolfia is cultivated for medicinal uses of the 30 alkaloids, especially reserpine, found in the root. Drug used mostly in treating high blood pressure, also for insomnia, hyperglycemia, hypochondria, mental disorders, and certain forms of insanity. Its great value lies in its not requiring to be administered in critical dosages, rare side effects (recently stated to be carcinogenic and teratogenic), nonhabit-forming, without withdrawal symptoms or contraindication. Also used for hypertension, neuropsychiatry, gynecology and geriatrics. Emboden notes that under the name "sarpaganda" it was used by holy men such as Mahatma Gandhi to promote states of meditation and introspection. In experimental animals reserpine is said to interrupt the vaginal cycle, inhibit ovulation, induce false pregnancy, inhibit androgenic secretions of the male gonads, and decrease hypertrophy of the testes. In laboratory rats, the bacteriostatic alkaloid lessens skin graft rejections.[1] Rescinnamine is second in importance as a hypotensive drug and deserpidine nearly as important as a hypotensive and sedative. Serpentinine is a weak hypotensive, sarpagine has only fleeting effects on blood pressure and pressure amines, ajmaline stimulates respiration and intestinal movement, rauvolfinine is hypertensive, while yohimbine is hypotensive, a cardiovascular depressant, and hypnotic.[1] According to Morton, ajmalinine is hypotensive, ajmalicine is a CNS-depressant, and chandrine is antiarrhythmic.[17] There is such a great demand for the alkaloids and the raw drug that India is taking steps to increase production to 100 to 150 MT root per year. In 1973, *circa* 25 million prescriptions containing reserpine were sold in the U.S. (1.45% of all prescriptions and nearly six million (0.38% of total) containing *Rauvolfia* extracts).

Hindus uses the root for fever, dysentery, and other painful intestinal ailments, e.g., cholera, gastritis, etc., and for snakebite. In Ayurvedic medicine it was used for worms, ulcers, scorpion sting, and snakebite. Leaves have been used to remove opacities of the cornea. In Java it is regarded as vermifuge. Holy men used it for meditation, and witch doctors used it to calm the mentally disturbed.

The total alkaloid content ranges from about 1.5 to 2.4% and is concentrated in the root bark, latex vessels, and secretory cells. Some of the more often reported alkaloids are ajmalicine, ajmaline, ajmalinine, alloyohimbine, chandrine, corynanthine, deserpidine, iscajmaline, neoajmaline, papaverine, raunatine, raunolinine, rauwolscine, rescinnamine, reserpiline, reserpine, reserpinine, reserpoxidine, sarpagine, serpertine, serpentinine, serpine, serpinine, thebaine (probably erroneous?), and yohimbine. Reserpine not only tranquilizes, but lowers blood pressure, taking several days for early effects to be felt, several weeks before total effects are realized. Emboden speculates that the reserpine in the body is transformed into secondary active substances. "Remarkable cures have been achieved in cases of acute schizophrenia, migraine headaches, high blood pressure, and withdrawal from opiates."[54]

Toxicity — Listed by Emboden as a narcotic tranquilizer,[54] often leading to bizarre dreams. Reserpine, the active tranquilizing ingredient, recently got bad press as a carcinogen, far exceeding in its publicity what the public received by the press retraction. Side effects of *Rauvolfia* preparations include drowsiness, bradycardia, salivation, nausea, hypergastric secretions, nasal congestion, and diarrhea, and certain endocrine disorders (fertility depression, inhibition of menstruation, feminization, and impairment of sexual function in males). Mental depression has often been severe enough to lead to suicide.

290. *RAUVOLFIA TETRAPHYLLA* L. (APOCYNACEAE) — Pinque-Pinque

Roots are pulverized and made into tranquilizing infusions.[54] Deserpidine (canascine, recanescine) is extracted commercially from the roots for use as hypotensive and tranquilizer.[42] Bark is inserted in teeth to cause them to disintegrate. Purple juice used for dye and ink in Central America.[17]

Reported to be parasiticidal, avicidal, and poisonous, pinque-pinque is a folk remedy for skin ailments, syphilis, and wounds.[32] Mexicans use the roots for erysipelas, gingivitis, sore throat, stomatitis, ulcers, and wounds. West Indians use the bark for skin ailments and syphilis. Yucatanese use the latex as an antiedemic, cathartic, collyrium (or for granulated eyelids), diuretic, emetic, and expectorant. Central Americans use the plant for fever, malaria, skin ailments, and snakebite.[22,32,42] Crushed roots used for mange in dogs.[17]

The alkaloid extract, especially yohimbine, is hypotensive, but only in cases of high blood pressure.[33] Rauvolscine, the alkaloid that is the major element in the entire plant, is hypotensive but not sedative.[42] Guatemalan investigators have isolated two alkaloids, chalchupine A and B, from the plant.[22] *Hager's Handbook* lists alstonine, ajmalicine, ajmaline, aricine, corynanthine, deserpidine, isoreserpiline, renoxydine, reserpiline, reserpine, reserpinine, sarpagine, serpentine, yohimbine, alpha- and beta-yohimbine, pseudoyohimbine, canembrine (raunescine), 19-methyl- (or allo-) yohimbine, raujemidine, isoreserpinine ($C_{22}H_{26}N_2O_4$), isoraunescine ($C_{31}H_{36}N_2O_8$), pseudoreserpine ($C_{32}H_{38}N_2O_9$). Morton gives formulas for many of these compounds.[17] The Conservation Foundation Letter[277] says "only 10% of the pharmaceutical compounds obtained from plants can be commercially synthesized at competitive costs." Reserpine, for example, can be commercially prepared from natural sources for as little as \$1/g but synthesizing it costs twice as much.

Toxicity — Listed as a narcotic tranquilizer,[54] fruits toxic, sap may cause dermatitis. Said to have caused fatalities;[11] causes stomatalgia, pharyngeal constrictions, intense thirst, enteritis, often violent, acute burning sensation, bloody diarrhea, nausea, vomiting, convulsions, cold hands and feet, death in extreme cases.[17]

291. *RHAMNUS PURSHIANUS* DC (RHAMNACEAE) — Cascara Sagrada, Cascara Buckthorn

Bitter cascara extract has been used in liqueurs, the debittered extract in baked goods, ice creams, and soft drink. Drug of the bark commonly employed as a purgative, strong when fresh, milder with age (e.g., 3 years). Usually sold in quills. Used veterinarily, especially for constipated dogs. Flowers an important bee forage in California. Fruits eaten raw or cooked but can cause a transient reddening of the skin.[17] Useful in sunscreens.[211] Nearly 2.5 million prescriptions (0.16% of all prescriptions sold in U.S. in 1973) sold in 1973 contain cascara extracts.[98]

Juice of the plant has been used for cancer and tumors, even in Kentucky.[4] It contains the quinone emodin,[2] which has shown activity in the lymphocytic leukemia and Walker carcinosarcoma tumor systems.[10] Used as a laxative and tonic, in gallstones and liver ailments. Tierra classifies cascara as valuable "whenever there are hemorrhoids associated with poor bowel function."[28] Duke and Wain classify it as aperient, diuretic, emetic, laxative, poison, and purgative.[32]

Primary active ingredients are anthracenes, 10 to 20% O-glycosides, based on emodins, 80 to 90% C-glycosides (like aloin) including barbaloin and deoxybarbaloin (yields cascarosides), emodin, oxanthrone, aloe-emodin, and chrysophanol. It also contains rhamnol (cinchol, cupreol, quebrachol), linoleic-, myristic-, and syringic-acids.[33] Contains, also, resins, fat, starch, glucose, malic, and tannic acid. Bark said to contain a glucoside which yields crysophanic acid on hydrolysis.[2] Also, contains cascarin and purshianin.[8] Dry seeds yield 6.7 to 25.4% protein, 13.4 to 56.9% oil, and 1.3 to 2.3% ash.[21]

Toxicity — Emodin can cause dermatitis.[6] Habitual use may induce chronic diarrhea with weakness from excessive loss of potassium.[17] Over a long period, melanin will pigmentize the mucous membranes of the colon. Approved by the FDA for food use (§ 172.510).[29]

292. *RHEUM OFFICINALE* Baill. (POLYGONACEAE) — Chinese Rhubarb, Canton Rhubarb, Shensi Rhubarb

Chinese rhubarb is used as a flavor component in alcoholic bitters and nonalcoholic beverages, baked goods, candy, frozen dairy desserts, gelatins, and puddings. The highest average maximum use level is 0.05% in alcoholic beverages and baked goods.[29] Grown as a medicinal plant in the Orient, particularly China. Aloe-emodin has shown anticancer activity in the lymphocytic leukemia and Walker carcinosarcoma tumor systems.[10] According to *Hager's Handbook,* rhein and emodin inhibit the further growth of mammary carcinomata and Ehrlich-ascites-carcinomata in mice.[33]

Dried rhizome and roots with the periderm removed are used medicinally as antiseptic, aperient, astringent, cathartic, hypotensive, laxative, stomachic, and colic. Said to be antidiarrheal in small doses, purgative in large. Compound powder is powdered rhubarb with magnesium oxide and ginger. The drug has been known to the Chinese for centuries, and may have reached Europe as early as the first century A.D. Russia maintained monopoly of the drug from 1650 to nearly 1800 by bringing high quality drug overland from China. Chinese use the herb for cancer, constipation, diarrhea, dropsy, fever, headache, jaundice, malaria, sores, thermal burns, and toothache.[16,29]

Medicinal rhubarb contains, per 100 g, 16 calories, 94.9% water, 0.5% protein, 0.1% fat, 3.8% total carbohydrates, 0.7% fiber, 0.7% ash, 51 mg Ca, 25 mg P, 0.5 mg Fe, 10 μg vitamin A activity, 0.1 mg thiamine, 0.02 mg riboflavin, 0.1 mg niacin, and 10 mg ascorbic acid.[21] Drug contains anthraquinone compounds, as rhein, chrysophanol, erythro-eretin, emodin, aloe-emodin, and emodin monomethyl ether, glucogallin, rheotannic acid, rhabarberon, tetrasin, and methylchrysophanic acids, as well as gallic acid, catechin, and 7% calcium oxalate. The volatile oil contains chrysophanic acid, other anthraquinones, diisobutyl phthalate, cinnamic acid, phenylpropionic acid, and ferulic acid. Chrysophanol is passed on in the milk by nursing mothers causing a laxative effect.[33] Plant contains 1.3% rutin.[59]

Toxicity — Hemorrhoid patients and expectorant or puerperal mothers are advised to avoid it.[16] Leaves of most rhubarb species are poisonous if ingested. Stomach pains, nausea, vomiting, weakness, dyspnea, burning of mouth and throat, internal bleeding, coma, even death can occur from eating rhubarb leaves.

To the physician — Hardin and Arena suggest gastric lavage or emesis with lime water, chalk, or calcium salts, calcium gluconate, parenteral fluids, and supportive treatments.[34] Cleared by the FDA for food use (§ 172.510).[29]

293. *RHUS TOXICODENDRON* L. (ANACARDIACEAE) — Poison Ivy

Bees can make nontoxic honey from this evil plant, sometimes cultivated foolishly for fall foliage. Dutch have planted poisonous *Rhus* to stablize their dikes. Birds eat the fruits, and herbivores, except man, eat the leaves, apparently with impunity. One of the reasons I went into botany was an early childhood ethnobotanical application of poison oak (or was it ivy) as a substitute for toilet paper. Still, 40 years later I was not sensitized at all, while my father-in-law was hospitalized from smoke of poison ivy. Poisonous as it is, I have harvested and sent some to homeopathic colleagues in India. As an intense dermatogen, it is not often used in allopathic medicine. I have been asked in the past how long a dead poison ivy retained its poison. According to Gillis,[278] herbarium specimens several centuries old have induced dermatitis in sensitive people who handled them. Hence, I suspect that long dead logs of *Rhus* can cause dermatitis if burned.

Poison ivy is a folk remedy for cancer, carcinoma, coronary insufficiency, eczema, epistaxis, headache, infections, malaria, pain, rheumatism, scrofula, typhus, vaginitis, and warts.[4,32] Cherokee used the decoction as an emetic. Chippewa used the plant to prevent infection and heal wounds. Delaware compounded the roots in salves for sore or swollen glands. Fox poulticed the pounded roots onto swellings. Houma took the leaf decoction as a rejuvenation or tonic. Following up on this lead perhaps, the USP listed the leaves of *R. toxicodendron* as a stimulant and narcotic. Potawatomi poulticed pounded roots onto swellings. Some say that drinking impatiens tea serves as a preventitive, while others keep frozen tea on hand to rub the dermatitis if the preventitive did not work. "The idea that American Indians chewed a leaf of poison-ivy to confer immunity is a myth that has never been documented."[278] I would prefer drinking the milk of a goat (but not milking her, as the poison is transmitted on animal hairs) that has eaten poison oak as the folkloric preventative.

Contains urushiol and ureshenol (3-[8-pentadecenyl]-brenzcatechin), the latter probably the worst dermatogen (leaves contain 3.3%, the twigs, 1.6%, the unripe fruits, 3.6%). Also present: the flavone derivatives myricetin, quercetin, kaempferol, fisetin ($C_{15}H_{10}O_6$), up to 25% tannins, gallotannic acid, gallotannic methylester, and rhamnose. Fruits contain heneicosandicarbonic acid ($C_{23}H_{44}O_4$), the fruit walls with myristic- and palmitic-acids as the chief fatty acids. The seed oil contains oleic and lineolic acids.[33] The poisons are not volatile and cannot be contracted "out of the air", but tiny droplets (nonvolatilized) may be transmitted in the smoke of burning poison ivy. Dry seeds of *R. radicans* contain 10.6% protein and 22.4% fat.[21]

Toxicity — The syndrome has been known since the days of primitive Indian cultures. Natural immunity is originally present in all persons, but is reduced to one or another threshold of sensitivity by contact with the poison. A dermatitis is manifested by reddened and itchy skin in mild cases to blisters which exude serum in severe cases. Mucous and alimentary canal membranes may be affected; serious gastric upset, even death, may result from ingestion of leaves or fruits. Eating a leaf, contrary to many old wives' tales, does *not* confer immunity. Infection or other complications may bring about death on occasion even when only the skin is affected.[14]

To the physician — Kingsbury suggests topical treatment with patent lotions and creams. This treatment is symptomatic only; there is no known cure. In severe cases, physicians administer ACTH or cortisone derivative under carefully controlled conditions. Injections should generally be avoided since they produce sporadic results, often disastrous — especially when administered during an attack.[14] Indian herbal "remedies" for poison oak and ivy included *Astragalus nitidus* (Cheyenne), *Fagus grandifolia* (Rappahannock), *Grindelia robusta* (Pacific Coast Indians), *Hedeoma pulegioides, Impatiens balsamifera* (Meskwaki, Potawatomi), and *Lepidium viriginicum* (Menominee), as well as *Atropa, Bryonia, Cornus, Grindelia, Pycnanthemum,* or *Verbena.*[43]

294. *RHYNCHOSIA PYRAMIDALIS* (Lam.) Urban (FABACEAE) — Piule, Virility Vine

In Salvador, the stems and leaves are used for washing clothes, the red and black seeds for bracelets and necklaces. For a brief period in the 1950s the West Indies were pushing bits of the stem as an aphrodisiac called "pega palo". Small pieces were advertised for sale from the Dominican Republic at fantastic prices, to be soaked in rum, to be drunk as an aphrodisiac.[22] Farnsworth pointed this out and added, "In 1961, a well-known medical journal published a scientific report pointing out that in a controlled clinical study of fifty impotent male patients, ranging in age from twenty-five to ninety, 82% of the group treated with an extract of this plant experienced increased libido during and after treatment."[279] In two cases the wife became pregnant while her husband, formerly incapacitated, was sexually active under the pega palo treatment. Farnsworth published[147] that "a group of Texas realtors supposedly paid the sum of $5,000,000 to a Mexican physician for rights to market this plant here in the U.S. as an aid to restoring sexual vigor and vitality to males . . . Interested parties are said to have purchased pieces of pega palo vine for as much as $15.00 . . . Such a large amount of the material was being sold that the U.S. Government, in order to protect the health of its citizens, placed a restriction on the importation of pega palo to the U.S."

Reported to be aphrodisiac, narcotic, and poison, piule is a folk remedy for coughs.[32] Dry seeds contain *circa* 19% protein and 1% fat.

Toxicity — Classed as a narcotic hallucinogen.[54]

295. *RICINUS COMMUNIS* L. (EUPHORBIACEAE) — Castorbean

Castorbean seeds yield a fast-drying, nonyellowing oil, used mainly in industry and medicines. The oil is used to coat fabrics and other protective coverings, to manufacture high-grade lubricants, transparent typewriter and printing inks, to dye textiles (when converted into sulfonated castor oil or Turkey-red oil; for cotton fabrics with alizarine), to preserve leather, and produce "Rilson", a polyamide nylon-type fiber. Dehydrated oil, an excellent drying agent comparing favorably with tung oil, is used in paints and varnishes. Hydrogenated oil is utilized in the manufacture of waxes, polishes, carbon paper, candles, and crayons. "Blown oil" is used for grinding lacquer paste colors, and when hydrogenated and sulfonated used for preparation of ointments. Castor oil pomace, the residue after crushing, is used as a high-nitrogen fertilizer. Although it is highly toxic due to the ricin, a method of detoxicating the meal has now been found, so that it can safely be fed to livestock. Stems are made into paper and wallboard.[61]

Considered anodyne, antidote, aperient, bactericide, cathartic, cyanogenetic, discutient, emetic, emollient, expectorant, insecticide, lactagogue, larvicidal, laxative, POISON, purgative, tonic, and vermifuge, castor or castor oil is a dangerous ingredient in folk remedies for abscess, anasarca, arthritis, asthma, boils, burns, cancer, carbuncles, catarrh, chancre, cholera, cold, colic, convulsions, corns, craw-craw, deafness, delirium, dermatitis, dogbite, dropsy, epilepsy, erysipelas, fever, flu, gout, guineaworm, headache, inflammation, moles, myalgia, nerves, osteomyelitis, palsy, parturition, prolapse, puerperium, rash, rheumatism, scald, scrofula, seborrhea, skin, sores, stomachache, strabismus, swellings, toothaches, tuberculosis, tumors, urethritis, uteritis, venereal disease, warts, whitlows, and wounds. The oil and seed have been used as folk remedies for: warts, cold tumors, indurations of the abdominal organs, whitlows, lacteal tumors, indurations of the mammary gland, corns, moles, etc. Castor oil is a cathartic and has labor-inducing properties. Ricinoleic acid has served in contraceptive jellies. Ricin, a toxic protein in the seeds, acts as a blood coagulant. Oil used externally for dermatitis and eye ailments. Seeds, which yield 45 to 50% of a fixed oil, also contain the alkaloids ricinine and toxalbumin ricin, and considered purgative, counterirritant in scorpion sting and fish poison. Leaves applied to the head to relieve headache and as a poultice for boils.[32,33,63]

Per 100 g, the leaves are reported to contain, on a zero-moisture basis, 24.8 g protein, 5.4 g fat, 57.4 g total carbohydrate, 10.3 g fiber, 12.4 g ash, 2670 mg Ca, and 460 mg P. The seed contains 5.1 to 5.6% moisture, 12.0 to 16.0% protein, 45.0 to 50.6% oil, 3.1 to 7.0 N-free extract, 23.1 to 27.2% crude fiber, and 2.0 to 2.2% ash. Seeds are high in phosphorus, 90% in the phytic form. The castor oil consists principally of ricinoleic acid with only small amounts of dihydroxystearic, linoleic, oleic, and stearic acids. The unsaponifiable matter contains beta-sitosterol. The oilcake from crushing whole seeds contains 9.0% moisture, 6.5% oil, 20.5% protein, 49.0% crude fiber and carbohydrates, and 15.0% ash. The manural value is 6.6% N, 2.6% P_2O_5, and 1.2% K_2O.[1] There are 60 mg/kg uric acid and 7 ppm HCN in the seed. The seeds contain a power lipase, employed for commercial hydrolysis of fats; also, amylase, invertase, maltase, endotrypsin, glycolic acid, oxidase, ribonuclease, and a fat-soluble zymogen. Sprouting seeds contain catalase, peroxidase, and reductase.

Toxicity — The seeds contain 2.8 to 3% toxic substances, 2.5 to 20 seeds killing a man, 4 a rabbit, 5 a sheep, 6 an ox, 6 a horse, 7 a pig, 11 a dog, but 80 for cocks and ducks. The principal toxin is the albumin, ricin. However, it produces antigenic or immunizing activity, producing in small doses an antitoxin analogous to that produced against bacteria.[63]

296. *RIVEA CORYMBOSA* Hall. F. (CONVOLVULACEAE) — Snakeplant

Seeds used by Amerindians as a hallucinogen, used in divination and witchcraft. Aztec priests mixed the seeds with ashes of poisonous insects, tobacco, and some live insects as a body rub before sacrifices "to make them fearless of every danger . . . Those who used it were said to have communication with the devil, believe in the owl and suck blood, among other things."[54] Flowers, rich in nectar, may provide $1/3$ of the total Cuban honey crop.[42] Honey used to ferment an intoxicating beverage. Latex used to coagulate rubber.[42] Seeds have shown antibiotic activity.

Reported to be anodyne, aphrodisiac, carminative, contraceptive, diuretic, hallucinogenic, narcotic, psychomimetic, snakeplant is a folk remedy for chills, dislocations, flatulence, fracture, gout, ophthalmia, pain, paralysis, parturition, rheumatism, sores, swellings, syphilis, and tumors.[42,52] Seeds are ground up and applied to gout.[52] In San Juan del Rio, the seed infusion with lemon juice is taken on the last day of menstruation as a contraceptive. Yucatanese boil the flowers and leaves for urinary calculus or obstruction. Cubans infuse the leaves, roots, and stems to stimulate labor in parturition. Mexicans bathe swollen limbs in the leaf decoction, and rub paralyzed limbs with a hot mescal mixture of powdered seeds and leaves. They rub a seed ointment onto rheumatism to relieve the pain.[42]

Dry seeds contain *circa* 18% protein, 9% fat. Seeds contain 0.056% alkaloids, with 18.3% (+)-lysergic acid amine (ergine), 33.7% isoergine, penniclavine, 6.9% chanoclavine, 2.7% elymoclavine, 3.7% ergometrine, 3.3% ergometrinine, lysergol, tryptophane, tubicoryne ($C_{27}H_{48}O_{11}$), which on dehydrolysis gives phenanthrene and 1,7-dimethylphenanthrene. Active principles include ergine, isoergine, and minor alkaloids.[54] Hofmann concludes that the glucoside turbicoryn has nothing to do with the psychotropic activity.[182] However, another glucoside $C_{28}H_{46}O_{12}$ is a CNS-stimulant.[33]

Toxicity — Excessive use said to lead to madness.[52] Classified as a narcotic hallucinogen.[54] Honey may cause headache and vertigo.[42]

297.   *ROBINIA PSEUDOACACIA* L. (FABACEAE) — Black Locust, False Acacia

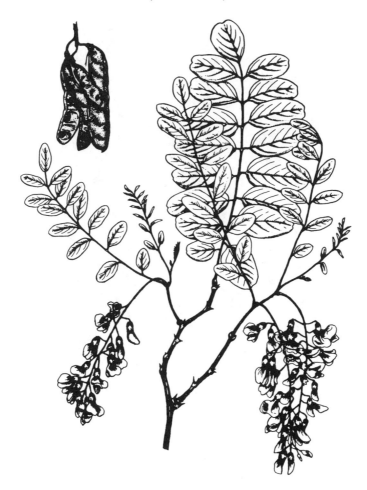

According to Grieve, this is "one of the most valuable timber trees of the American forest, where it grows to a very large size."[2] It was one of the first trees introduced into England from America, and is cultivated as an ornamental tree in the milder parts of Britain. It is great for posts, but one of the hardest of American woods and very difficult to work (640 to 800 kg/m³, 15% moisture). Amerindians and Asian Indians report that the seeds are edible. Watt and Breyer-Brandwijk suggest that the seeds left in fruits hanging on the trees may still be edible, after processing.[3] Frankly, I am reluctant to experiment with this as a food source. Seeds are also suggested as a coffee substitute.[3] *Robinia*, dangerous as it is, can serve as a vegetable rennet. The essential oil from the flowers has been used as a spice, in sherberts and toilet waters. Wood is suitable for agricultural implements, tool handles, shoe lasts, sports goods, dowels and pins for insulators on telephone and telegraph wires, tree nails, boat ribs, brackets, sleepers, and sills. It is used also for light construction, gates, wagon hubs, cart wheels, shipbuilding, furniture, and turnery work. Burrs from the trees provide attractive wood for tabletops and music cabinets. Robinetin is a strong dyestuff yielding with different mordants different shades similar to those obtained with fisetin, quercetin, and myricetin; with aluminum mordant, it dyes cotton to a brown-orange shade.[1] Country folk in the U.S. once macerated the leaves in water as a fly killer.[63]

Reported to be astringent, cholagogue, diuretic, emetic, emollient, laxative, *poison,* protisticidal, purgative, sedative, tonic, and viricidal, black locust is a folk remedy for dyspepsia and spasms.[32] Cherokee used the plant as an emetic and for toothache. Chinese cook the flowers with meat to make a broth used to treat eye diseases. Homeopaths use it for constipation, duodenal and ventricular ulcers, and hyperacidity.[33]

Per 100 g, the seed is reported to contain 17.0 to 25.5 g protein, 3.0 to 3.3 g fat, 35.0 to 46.5 g N-free extract, 17.2 to 39.0 g fiber, 6.1 to 7.5 g ash, 1290 to 1500 mg Ca, 0.26 to 0.32 mg P. Higher values for protein (43.2%) and fat (10.8%) are reported.[21] The predominant flavonoids in the heartwood are dihydrorobinetin (17.6%), robinetin (3,3',4',5',7-pentahydroxyflavone, 8%), leucorobinitinidine 7.3',4',5'-tetrahydroxyflavan-3,4-diol (6.2%), and robtin (1.5%). *Hager's Handbook* gives other percentages.[33] Other flavonoids present in the heartwood are liquiritigenin, robtein, fustin, butin, butein, fisetin, 7.3',4'-trihydroxy-flavan-3,4-diol, and 2',4',4-trihydroxy chalkone. Bark also contains a glucoside robinitin (3%), syringin, tannin (up to *circa* 7.0%), a toxalbumin (phasin), resins, phytosterols, some coloring matter, and an unidentified, unstable alkaloid. Inner bark is reported to contain amygdalin and urease. Leaves, considered antispasmodic and laxative, prescribed in digestive disorders, are poisonous to chicken. Leaves contain a coloring matter acacetin (apigenin-4'-methyl ether). Apigenin-7-bioside, apigenin-7-trioside, and indican have also been reported. Leaves contain a volatile oil (0.01%) and carotene (209 mg/100 g). Hexene-3-ol[1] and *trans*-2-hexenal have been identified in the oil, the latter toxic to ciliates such as *Paramoecium*. Flowers are powerfully diuretic due to the glycoside, robinin (kaempferol-7-1-rhamnosido-3'-robinobioside, 4.4%). Flowers also contain 1-asparagine, a volatile oil, and wax. The oil contains methyl anthranilate, linalool, alpha-terpineol, benzaldehyde, benzylalcohol, alpha-terpineol, farnesol, heliotropin (nerol, piperonal), indole, an aldehyde or ketone having a peach-like odor, and traces of pyridine-like bases. Flowers contain cardiac glycosides.[33] Seeds contain: moisture, 10.3 to 11.5; crude protein, 38.8 to 39.5; fat, 10.2 to 11.0; N-free extract, 20.4 to 23.0; crude fiber, 12.9 to 13.6; ash, 4.0 to 4.7; calcium (CaO), 0.19; and phosphorus (P_2O_5), 1.65%.[63] Seeds contain the sugars sucrose, raffinose (traces), and stachyose, the amino acids arginine and glutamic acid, and canavanine. Roots are rich in asparagine and are also reported to contain robin.[1]

Toxicity — Bark, leaves, and roots are reported to be toxic due to the presence of a toxalbumin, robin (1.6% in the bark). Toxic symptoms are suggestive of those associated with belladonna poisoning. Phasin is also very poisonous.[33] Occasional cases of poisoning are on record in which boys have chewed the bark and swallowed the juice, the principal symptoms being dryness of the mouth and throat, burning pain in the abdomen, hypersomnia, dilation of the pupils, nausea, diarrhea, sometimes bloody, vertigo, muscular twitches, paleness, stupor, and shock; excessive quantities causing, also, weak and irregular heart action.[2,33,36,56] Resistance to rot may be due to the wood containing 4% taxifolin, an isomer of dihydroquercetin, or dihydrorobinetin, a growth-inhibitor of wood-destroying fungi. The flower is said to contain the antitumor compound benzaldehyde. Some have classified the honey as toxic, others as the best of honeys.

To the physician — Hardin and Arena suggest that fatalities are due to black locust, recommending gastric lavage or emesis, with symptomatic and supportive treatment, keeping the urine alkaline.[34]

298. *ROSMARINUS OFFICINALIS* L. (LAMIACEAE) — Rosemary

Rosemary is grown primarily for the leaves, used ground or whole as a condiment in lamb dishes, stews, or soups. Oil of rosemary, because of its persistent, woody odor, is used in wine, cosmetics, soaps, perfumes, deodorants, and hair tonics. A perennial to 2 m tall, this evergreen blue-flowered shrub is native to the Mediterranean. As a camphoraceous spice, it is used with chicken, duck, fish, lamb, pork, rabbit, shellfish, soups, stews, and veal. Among vegetables, it is used with eggplant or potatoes, or zucchini, but sparingly. Some cultivars are adaptable to bonsai and hanging gardens. Rosemary is found in herbal baths, herb biscuits, bouquets, potpourris, weddings, and even funerals as a symbol of fidelity. It is a cordial, reputed to stimulate the blood, eyes, hair, and memory. Placing a pan of water with crushed rosemary and juniper on the radiator can improve the aroma of the place. The French are said to burn the mixture of rosemary and juniper berries. The oil is used in cosmetics, deodorants, liniments, lotions, soaps, and tonics, as well as liqueurs and medicinals.[29] Rosemary lotion is said to stimulate hair growth and prevent baldness. Used as an insecticide in Latin America. It is a fragrant moth repellent. The oil is bactericidal, fungicidal, and protisticidal. Rosemary makes good bee forage. Rosemary extracts are antioxidant, perhaps due to carnosic- and labiatic-acid.[29] Inatani et al.[280] indicate a strong antioxidant activity for an odorless and tasteless lactone known as rosmanol.

The leaf, in distilled water, is said to alleviate cancer, and the seed, powdered for an ointment, to alleviate indurations. The essential oil in wine is said to help cancers.[4] In Central America, rosemary figures in a concoction (with "Artemisia maritima"?) for fertility control, which may produce temporary sterility or possibly interfere with the implantation of the ovum or affect menstruation (*Artemisia* contains thujone).[3] Rosemary oil is used, perhaps dangerously, for headache and tardy menstruation.[1] Rosemary tea is said to cause temporary sterility in females, but I would not count on it. Rosemary, steeped in vodka, makes a satisfactory liqueur, called Hungary Water. I would count on it, having made it. It was once massaged onto gout of the hands and legs. Fresh leaves are cooked in wine to poultice on hemorrhoids.[42] As a massage ointment, rosemary spirits are said to prevent premature baldness. A cold mixture of dried rosemary mixed with borax is said to be a good dandruff shampoo. Mixed with lavender (10 g) in alcohol, 20 g rosemary is used for alopecia, bruises, and sprains.[42] "A tisane of Rosemary will cure a nervous headache and has a beneficial effect on the brain. Its constant use will greatly improve a bad memory."[2] The tea is also good for colds, colic, dyspepsia, headcolds, and nervous tension.[29] Yucatan natives use rosemary for dysmenorrhea, dyspepsia, neuralgia, and rheumatism.[42] Cubans use it for fever, hoarseness, and insomnia. Rubbed together with coltsfoot leaves, rosemary is smoked to treat asthma, lungs, and throat conditions. Brazilians douche with the decoction for leucorrhea, and take the decoction for asthma, bronchitis, cough, and hysteria. Medicinally, rosemary is variously considered aperient, astringent, carminative, diaphoretic, dissolvent, diuretic, ecbolic, emmenagogue, febrifuge, nervine, stimulant, stomachic, and tonic. In Latin America it is used also for cardosis, diarrhea, edema, nausea, nephrosis, neuralgia, polyuria, rheumatism, stomatosis, toothache, and vertigo.[42]

Per 100 g, the dried herb is reported to contain 331 calories, 9.3 g H_2O, 4.9 g protein, 15.2 g fat, 64.1 g total carbohydrate, 17.6 g fiber, 6.5 g ash, 1280 mg Ca, 70 mg P, 29.2 mg Fe, 220 mg Mg, 50 mg Na, 955 mg K, 3.2 mg Zn, 3128 IU vitamin A, 0.5 mg thiamine, 1.0 mg niacin, and 61 mg ascorbic acid.[21] Plant tops contain up to 3.9% ursolic acid. The diuretic, glycollic acid, has also been isolated from the tops.[3] Tannins are also reported. Oil contains alpha – thujene and alpha-pinene (12.5%), camphene (4.0%), beta-pinene (1.3%), delta-3-carene and myrcene (1.3%), alpha-terpinene (0.4%), limonene (3.0%), 1,8-cineole (47.0%), gamma-terpinene (0.4%), *p*-cymene (1.8%), copanene (0.6%), camphor and linalool (10.7%), bornyl acetate (0.9%), terpinen-4-ol (1.3%), caryophyllene (4.9%), delta-terpineol (0.9%), borneol, alpha-humulene and alpha-terpineol (4.0%), gamma-muurolene (3.1%), beta-bisabolene and carvone (4.0%), and ar. curcumene (0.2%). Trace amounts of sabinene, terpinolene, alpha-p. dimethyl styrene, alpha-fenchyl alcohol, verbenone, ledene, delta-cadinene, gamma-cadinene, alpha-selinene, cubenene, p. cymen-8-ol, *trans*-anethol, calamenene, calocorene, caryophyllene oxide, methyl eugenol, humulene epoxide I, humulene epoxide II, -corocalene, carvacrol, thymol, and cadalene also occur.[27,31,33,65] Flavonoids include apigenin, diosetin, diosmin, genkwanin, genkwanin-4'-methyl ether, hispidulin, luteolin, 6-methoxygenkwanin, 6-methoxyluteolin, 6-methoxyluteolin-7-glucoside, 6-methoxyluteolin-7-methyl ether, and methyl ether. Diosmin is said to be stronger than rutin in decreasing capillary permeability and fragility. O–O–N trimethylrosmaricine stimulates smooth muscles in vitro and has moderate analgesic activity.

Toxicity — Bath preparations containing the oil can cause erythema. Toiletries containing the oil can cause dermatitis in hypersensitive individuals.[6]

299. *RUMEX CRISPUS* L. (POLYGONACEAE) — Yellow Dock

Amerindians cooked the seeds in a gruel. Young leaves are cooked and eaten by various ethnic groups. The plant is sold for "herb tea" in the U.S. Herbal extracts may inhibit *Escherichia, Salmonella,* and *Staphylococcus.*[3] Rumicin ($C_{14}H_{10}O_4$) is discutient, ectoparasiticidal, and rubefacient.

Used by Paiutes and Shoshones for bruises, burns, swelling, and venereal disease; by Navajo for cutaneous and oral sores; by Tarahumara for skin lesions, especially sores on legs and feet.[281] Africans treat anthrax with a decoction of *Rumex* and *Teucrium.* Used as an internal remedy for scrofula, in herpes, the herb may cause nausea. Asian Indians use root juice for toothache, powdered roots for gingivosis and as a dentifrice. The rhizome is used for laryngitis.[3] Brazilians apply root decoction externally in adenopathy, internally as an antiscorbutic, depurative, febrifuge, and tonic.[42] The herb is also used for anemia, anthrax, cancer, diarrhea, eczema, fever, itch, leprosy, malaria, rheumatism, ringworm, syphilis, tuberculosis, urticaria.[3,32,33] Taiwanese use *Rumex* in vapor baths to relieve the eyes and ulcers.[16] Contradictorily, Chinese employ the rhizome as a laxative, while Australians, Americans, and Europeans use it for diarrhea and as a laxative. The tannins work as astringents; the anthraquinones as laxatives. Could the human body select from the herbal concoction the medicinal element it needed, avoiding the element it did not need? Perhaps the answer lies in homeostasis.

Contains anthraquinones, brassidinic acid, calcium oxalate, catechol tannins (5%), chrysophanic acid, emodin, oxalic acid, potassium oxalate, resin, rumicin, tannin. Leaves contain 92.6% water, 21 calories, 1.5% protein, 0.3% fat, 4.1% total carbohydrate, 0.9% fiber, 1.5% ash, 74 mg Ca, 56 mg P, 5.6 mg Fe, 1.38 mg vitamin A, 0.06 mg thiamine, 0.08 mg riboflavin, 0.4 mg nicotinic acid, and 30 mg ascorbic acid per 100 g.[21] As late as 1977 chrysarobin was used topically for fungal infections, psoriasis, ringworm, and other skin disorders, being one of the most effective agents for psoriasis. Rumex also contains tannins, which are effective (but hepatotoxic) when used topically for burns and other exudative conditions, like inflammatory diarrhea. Tannic acid has long been an astringent for diarrhea. The secretory activity and the transudation of fluids in the gut is hindered, and the underlying mucose may be protected from the effects of irritants in the bowel. When used topically for burns, the same mechanism prevails. In 1925, tannic acid treatment was advocated for burns, and for almost 20 years it was a favored local application for injured tissues. Unfortunately, tannic acid is hepatotoxic when absorbed in large quantities, causing liver damage in some tannic-acid treatments.[281] If watered with an iron carbonate solution, the rhizome may segregate 1.5% iron in organic combination. Such rhizome in 1- to 3-g doses has been given in anemia.[3] Containing more (2.17%) anthraquinones than rhubarb (1.42%), *Rumex* is pharmacologically more active than rhubarb.[1]

Toxicity — Overdoses of the root extract may cause diarrhea, nausea, and polyuria. Plants have caused fatalities in animals, perhaps due to nitrate and/or oxalate poisoning. Rumicin is said to lie behind dermatitis and gastric disturbances caused by the plant.[1,3]

300. *RUMEX HYMENOSEPALUS* Torr. (POLYGONACEAE) — Canaigre

Useful as a tannin source for dyeing wool and tanning leather. For a while, canaigre was promoted as a cheap substitute for ginseng. Cheap it is, like most *Rumex*. A ginseng substitute is another matter. According to Tyler, "to promote canaigre as a kind of American ginseng is a recent deceptive practice, probably due to the high prices commanded by the ginseng of today."[37]

Although the promotional literature makes another panacea out of canaigre, most American herbals report no medicinal uses whatsoever.[37] Granted, there are many medicinal uses for other species of *Rumex* high in tannin and anthraquinone, and many of these attributes might accrue to canaigre. *Hager's Handbook* adds that leucoanthocyanin fractions have antitumor activity.[33]

Contains 18 to 35% tannin and smaller amounts of anthraquinone (1.2% chrysophanic acid, 0.3% physcion emodin), 8 to 12% sugar, 25 to 40% starch and resin.[37] Canaigre does not contain any of the active panaxoside-like saponin glycosides responsible for ginsengs physiological activities.[37] Roots contain leucocyanidine, leucodelphinidine, leucopelargon-idine, and polymers of flavan-3,4-diole.[33] No ginsenosides were found,[232] although they have been reported.

Toxicity — According to Tyler,[37] "because of its high tannin content, the root may have considerable carcinogenic potential."

301. *RUTA GRAVEOLENS* L. (RUTACEAE) — Rue, Garden Rue, German Rue

Fresh or dried leaves are used sparingly to season such beverages and foods as cheese, meat, vegetable juice, salads, stews, vegetables, and wine. Eaten in Italian salads, where said to preserve the eyesight.[2] Ethiopians grind the fruits with capsicum to make a hot sauce.[38] They also use the leaves as a rennet[38] and with coffee leaves to make a coffee-like beverage. Fresh herbs contain about 0.06% of a volatile oil, for flavoring, as an aromatic in perfumes, in soaps, and toilet preparations. Rue water, sprinkled about the house, was supposed to kill fleas, and still believed to repel insects. Arabs chew the plant and add it to suspect drinking water. Affirmed by the FDA as GRAS (rue § 184.1698; oil or rue § 184.1699), oil of rue is said to be used as a flavor component (e.g., coconut type) in "most" major food products, including alcoholic (bitters, vermouth) and nonalcoholic beverages, baked goods, candy, frozen dairy desserts, gelatin, and pudding. Average maximum use levels are reported below 0.01%. The herb itself shows up in baked goods, candy, frozen dairy desserts, and nonalcoholic beverages. Average maximum use levels are below 0.0002% (2 ppm).[29] The rue alkaloid, arborinine, has abortive, antiinflammatory, antihistaminic, and spasmolytic properties. The furocoumarins, bergapten and xanthotoxin, have spasmolytic effects on smooth muscles and have phototoxic properties useful in treating psoriasis. Hirudicidal, nematicidal, and vermicidal activity is attributed to 2-undecanone, the major component of rue oil. Fagarine, graveolinine, and skimmianine are weakly abortive and spasmolytic.[33] Rutin, first isolated from rue, is best known for its ability to decrease capillary permeability and fragility. It is also said to be a cancer preventitive, i.e., inhibiting tumor formation on mouse skin by the carcinogen benzopyrene, and protective against irradiation damage. Rutin is also useful to counteract edema, atherogenesis, thrombogenesis, inflammation, spasms, hypertension. It was once official in the U.S. for arteriosclerosis, hypertension, diabetes, and allergic manifestations. It is suggested that it may be useful for stroke prevention.

Pliny is said to have mentioned 84 remedies containing rue. Warm fomentations of the oil are folk remedies for liver indurations. The leaf is said to alleviate scirrhus and scleroma of the uterus, cancer of the mouth, tumors, and warts. Expressed juice of the leaves is dropped, lukewarm, into the ears for earache.[42] The seed is said to alleviate warts and tumors.[4] In old herbals, recommended for ague, cough, croup, earache, epilepsy, gout, headache, hysteria, nightmare, sciatica, and vertigo. In Poland, used as an aphrodisiac and choleretic. Herb used medicinally as a bitters, an aromatic stimulant, antispasmodic, ecbolic, and emmenagogue, in suppression of the menses, for gas pains and colic, hysteria, and amenorrhea. In Chinese medicine rue, regarded as an emmenagogue, hemostat, intestinal antispasmodic, sedative, uterine stimulant, and vermifuge, is used for bug bite, cold, fever, rheumatism, snakebite, and toothache.[28,29] In Latin America it is used as an abortifacient and emmenagogue, and to treat epilepsy, fever, headache, heart conditions, measles, parturition, scarlet fever, and worms.[42]

Rue oil contains up to 90% or more of two ketones, methyl nonyl ketone and methyl heptyl ketone. Several alkaloids (arborinine, gamma-fagarine, graveoline, graveolinine, kokusaginine, 6-methoxydictamine, rutacridone, skimmianine, pinene, limonene, cineole, methyl–n–heptyl carbinol, methyl-n-nonyl carbinol, salicylic acid,[7] 2-heptanone, 2-octanone, 2-nonanol, 2-undecanol, undecyl-2-acetate, anisic acid, phenol, guaiacol, bergapten, xanthotoxin, rutamarin, psoralen, isoimperatorin, pangelin, and rutarin. Caprinic, caprylic, oenanthylic, and plagonic acids and rutin are also reported.[2,7,29] Plant yields 5-methoxypsoralen and 8-methoxypsoralen. Dry seeds of *R. chalepensis* contain 26.4% protein and 33.2% fat.[21]

Toxicity — An acro-narcotic poison in excessive doses internally; externally irritant and rubefacient. Oil, very dangerous, has been said to cause abortion. Handling of the foliage, flowers, and/or fruit can produce burning, erythema, itching, and vesication. Can cause dermatitis and photodermatitis. Taken internally, rue oil can cause severe stomach pain, vomiting, exhaustion, confusion, and convulsion, even fatality. According to Tierra,[28] "if there are any adverse symptoms of overdose a small amount of goldenseal root will act as an antidote . . . because rue has emmenagogue properties, it should be avoided by all pregnant women."[28]

302. *SALVIA DIVINORUM* Epl. and Jativa (LAMIACEAE) — Pipiltzintzintli ''Salvia of the Seers'', Hojas de la Pastora

A decoction made by steeping about 50 leaves in water overnight is drunk (and the leaves, chewed) by Amerindians and their California cultural counterparts as an hallucinogen (said to induce vivid dreams, dancing colors, clairvoyance, and telepathic insight).[61] As many as 100 leaves given a patient may induce a trance-like state in 15 min.[54] Fifty leaves in alcohol were prescribed for nonalcoholics, 100 for those habituated to alcohol. Wassen[252] describes his personal experiences with the herb. Macate Indians regard several labiates as hallucinogenic: *S. divinorum* known as ''la hembra''; *Coleus pumile*, of European origin, as ''el mache'', and ''el rene'' and ''el ahijado'', both forms of *C. blumei*. Indians are said to insist that the others are psychotropic, too.

Reported to be emetic, hallucinogenic, psychotropic.[32,33] Perhaps, because of its instability, the active ingredient has not been identified. The effect is weak and of short duration.[33]

Toxicity — Classed as a narcotic hallucinogen.[54]

303. *SALVIA OFFICINALIS* L. (LAMIACEAE) — Sage

Dried leaves of this evergreen perennial woody subshrub are best known as a spice for flavoring sausages, stuffings, soups, and some canned vegetables. Fresh leaves are also used in herb butters, cheeses, liqueurs, pickles, salads, and vinegars. Fresh leaves make a good dentifrice. The infusion applied to the scalp is said to darken the hair as well as stimulate hair growth and strengthen the hair. Chinese were said to prefer sage tea to their own true tea. An ale, made of sage, betony, scabious, spikenard, squinnette, and fennel seed, was said to be quite wholesome. Sage, speedwell, and wood betony is said to make an excellent breakfast tea. Italians nibble on sage to preserve the health. Sage butter and bread is said to be a satisfying meal in itself. Various combinations of sage, onion, butter, and lemon make excellent relishes for meat dishes. Sage oil is used in perfumery, for deodorants, and in insecticidal preparations.[1] Tumors (''apples'') on sage, caused by an insect puncture, are preserved in honey and eaten. Leaves are pickled in vinegar and eaten.[61]

The powdered leaf is a folk remedy for gum cancer. The leaf, pounded with salt and vinegar to form a plaster, is said to help cancer of the mouth, brow, and other parts. Decoctions and fomentations of the leaf and flower are said to remedy tumors.[4] Decoction of sage, in wine or hot water, used for toothache. Plant used to treat hyperhidrosis. Sage, containing estrogenic substances, has been prescribed since antiquity for female disorders.

Strong sage tea is said to dry up the mammaries when one wishes to wean.[1] A sage gargle is recommended for bleeding gums, sore throat, and tonsillitis. The tea is a good tonic and recommended for kidney, lethargy, liver, lungs, measles, nerves, and phthisis. The oil is applied in cases of rheumatic pain. An oil distilled from the plant is said to be a violent epileptiform convulsant, resembling the essential oils of absinth, nutmeg, and wormwood. The dried leaves have been smoked to treat asthma. Smelled for some time, it is said to cause intoxication and giddiness.

The oil contains cis-2-methyl-3-methylenehept-5-ene (0.1%), alpha-thujene and methyl isovalerate (0.3%), alpha-pinene (1.5%), camphene (3.3%), beta-pinene (1.4%), myrcene (1.2%), alpha-terpinene (0.2%), limonene (2.2%), 1,8-cineole and cis-ocimene (9.3%), gamma-terpinene (0.5%), p. cymene (0.8%), terpinolene (0.4%), alpha thujone (29.1%), beta-thujone (5.5%), camphor (26.3%), linalool (0.5%), bornyl acetate (1.9%), terpinen-4-ol (4.0%), caryophyllene (0.4%), alloaromadendrene (0.3%), sabinyl acetate and alpha-humulene (4.4%), alpha-terpineol and sabinol (3.4%), gamma- and delta-cadinene (0.1%), calamenene (0.1%), caryophyllene oxide (1.1%), and trace amounts of trans-2-methyl-3-methylene-hept-5-ene, trans-ocimene, trans-3-hexenal, trans-allo-ocimene, alpha-p.dimethylstyrene, beta-bourbonene, alpha-gurjunene, isocaryophyllene, alpha-maaliene, beta-copaene, aromadendrene, delta-terpineol, borneol, and p.cymen-8-ol in S. officinalis oil. Trace amounts of 1-octen-3-ol, alpha-corocalene, selina-5, 11-diene, and ledene have also been identified.[193] Per 100 g, the dried herb is reported to contain 315 calories, 8.0% H_2O, 10.6 g protein, 12.7 g fat, 60.7 g total carbohydrates, 18.0 g fiber, 8.0 g ash, 1652 mg Ca, 91 mg P, 28.1 mg Fe, 428 mg Mg, 11 mg Na, 1070 mg K, 4.7 mg Zn, 5900 IU vitamin A, 0.75 mg thiamine, 0.34 mg riboflavin, 5.7 mg niacin, and 32 mg ascorbic acid. Saturated fatty acids total 7.03 g, monounsaturated 1.87 g, and polyunsaturated 1.76 g.[89] Leaves contain 3% tannin, fumaric, malic, and oxalic acids, picrosalvin, saponins, pentoses, a wax, and KNO_3.[1] Seeds contain 18% protein and an oil (used as a bonding agent in oil paints) which contains 14.2% oleic acid, 29.2% linoleic, 34.7% linolenic, and 12.0% saturated acids.[1]

Toxicity — Cheilitis and stomatitis follow some cases of sage tea ingestion.[6]

304. *SALVIA SCLAREA* L. (LAMIACEAE) — Clary, Cleareye, Muscatel Sage

Dried leaves are used occasionally for flavoring in cookery. Leaves and flowers yield an essential oil, oil of clary sage, used in perfumery and in flavorings, combining a delicate fragrance with a powerful odor. Oil and absolute are used in colognes, creams, detergents, lotion, perfume, soap, and tobacco, with leaves up to 0.8% in perfumes. In Germany it is added to wines to give them a musky bouquet (e.g., muscatel). Either flowers alone, or the whole plant with the woody stalks removed, are used to extract the oil. High grade oil has a typical odor, reminiscent of ambergris, which blends well with lavender, bergamot, or jasmine. The oil is used in beverages, baked goods, candy, gelatins, liqueurs, puddings, relishes, and spices, with highest use of around 155 ppm for the oil in alcoholic beverages.[29] After the essential oil has been distilled, the residue is used as a source of sclareol, used like its derivative sclareolide for flavoring tobacco. Sclareolide is used in producing an ambergris substitute.[29]

The plant is used as antiseptic, antispasmodic, astringent, balsamic, emmenagogue, stimulant, stomachic, and for the treatment of cancer, catarrh, debility, dyspepsia, headache, kidney diseases and digestive, ophthalmic, and uteral disorders. Flowers and seeds have also been used medicinally in a limited way for treatment of inflammation of the eyes. The mucilage of the seed is said to be a folk remedy for tumors or swellings.[4] It is also used for removing dust particles from the eye.[28] The leaf, with vinegar or honey, is said to alleviate felons.[4] Fresh leaves used for kidney stones.

Dry seeds, with 20.7% protein and 31.8% fat, contain beta-thujone, linalyl-acetate, and beta-pinene; 5 days after germination, beta-thujone, alpha-pinene, cineol, borneol, beta-pinene, and terpene acid are present. Volatile oil (0.2 to 0.5%) of fully developed plants contains 30% linalool, 2% aldehyde, and 3% pinene and secondary alcohol. The dominant element of the essential oil is linalool, with alpha- and beta-pinene, beta-phellandrene, citranellol, limonene, *n*-pentanol, 3-octanol, furfurol, terpinolene benzaldehyde, delta 3-carene, *n*-nonanal, cineole, *p*-cymene, cadinene, caryophyllene, beta-ocimene, *cis*- and *trans*-alloocimene, gamma-terpinene, acetic-, propionic-, butyric-, and valerianic-acid esters, sclareol, beta-sitosterol, ursolic- and oleanolic-acid, and tannin. Seeds, fatty oil with glycerides of caprylic- (0.92%), palmitic- (7.05%), stearic- (2.82%), oleic- (21.71%), linoleic- (16.82%), linolenic- (50.66%), arachinic-, behenic-, lignoceric- and cerotinic acid.[33]

Toxicity — Surely the essential oil is toxic in quantity (GRAS § 182.10 and 182.20).[29]

305. *SAMBUCUS CANADENSIS* L. (CAPRIFOLIACEAE) — Elderberry

Fruit not very tasty when fresh, but flavor improves with drying. Dried berries simmered in water with sugar and lemon make excellent summer drink. Also, used in pies, cobblers, and jellies. Elderberry wines and cordials have been made for centuries. They may be substituted in any recipe calling for blackberries. Fruit is excellent source of vitamin C. Elderberry flowers (Elder Blow) used in cooked products by adding to batter for pancakes, waffles, and muffins, or the entire flower cluster, excluding the tough stem, may be dipped into fritter batter and fried in deep hot fat, or used plain with meal or with whipped cream or granulated sugar as a dessert. Elder Blow wine is something special, delicious, with a beautiful pale yellow color; flowers also used for tea. Flower extracts are used in perfumery. Unopened flower buds are pickled and used as substitute for capers. Young shoots from which pith has been extracted used for whistles and an Italian musical instrument called the Zampogna. Pith used for pith-balls for electrical experiments, and in soups to thicken them. New shoots may be cooked and served like asparagus; older green parts are reported to be poisonous. Leaves yield a green dye, and bruised leaves have been used in summer for keeping flies away. Bark yields a black dye. Pith is removed from stems to make nesting sites for solitary bees, especially near alfalfa fields where their presence aids in pollination. Many game birds eat the berries.[62]

Leaves are poulticed to relieve pain and promote healing of bruises and sprains. Dried flowers with mint leaves used in dyspepsia. The flower influsion is alterative and laxative. Juice of fruit with honey makes an effective cough syrup, this mixture along with sumac extract (fruit of *Rhus glabra*) makes a healing gargle for treating sore throat. Inner bark used in preparing ointments. Meskwaki women used elderberries to assist in parturition.[11] Considered alterative, aperient, carminative, cathartic, cyanogenetic, depurative, diuretic, emetic, excitant, hydragogue, intoxicant, laxative, poison, refrigerant, stimulant, sudorific, elderberry is used in folk remedies for abrasions, asthma, bronchitis, bruises, burns, cancer, chafing, cold, dropsy, epilepsy, fever, gout, headache, neuralgia, psoriasis, rheumatism, skin ailments, sores, sore throat, swelling, syphilis, toothache.[32]

Per 100 g, the fruit is reported to contain 72 calories, 79.8 g H_2O, 2.6 g protein, 0.5 g fat, 16.4 g total carbohydrate, 7.0 g fiber, 0.7 g ash, 38 mg Ca, 28 mg P, 1.6 mg Fe, 300 mg K, 360 μg β-carotene equivalent, 0.07 mg thiamine, 0.06 mg riboflavin, 0.5 mg niacin, and 36 mg ascorbic acid. The leaf is reported to contain 47.8 g H_2O, 10.2 g protein, 2.1 g fat, 37.9 g total carbohydrate, 15.7 g fiber, and 2.0 g ash. Most of the chemical work has been done in the European elder, *S. nigra* L., according to Leung who gives a long list of compounds.[29] Seeds contain 10.6% protein, 20.1% oil (with 5.8% palmitic-, 2.8% stearic-, 0.6% arachidic-, 4.0% oleic-, 53.0% linoleic-, and 34% linolenic-acid).[1,33] Leaves contain 3.5% rutin, immature flowers 5.2%, and mature flowers 3.0%.[59] Bark contains baldrianic acid. Young shoots contain the iridoid morroniside (0.1% in April, 0.05% in May).[33]

Toxicity — Because partly of HCN content, leaves have been approved for use in alcoholic beverages only provided HCN does not exceed 25 ppm (0.0025%) in the flavor (§ 182. and 182.20).[29] According to Lewis and Elvin-Lewis, children using pea shooters made of elderberry stems may be poisoned by alkaloids and cyanide. As little as 60 mg cyanide has caused death in man. The largest dose from which man has recovered is 150 mg.[11]

To the physician — Hardin and Arena suggest emesis or gastric lavage and treatment for cyanide poisoning (see Malug).[34]

306. *SANGUINARIA CANADENSIS* L. (PAPAVERACEAE) — Bloodroot

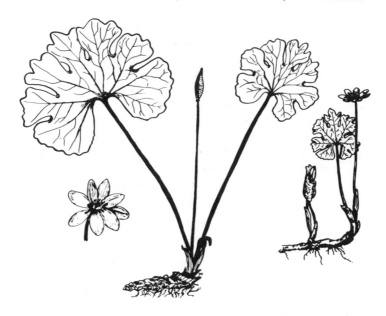

As alternative source to some opium alkaloids, bloodroot has a very limited market in some American herbal shops. Bloodroot is extensively used in cough remedies, almost always with other herbs like spikenard root, balm of gilead bud, white pine, and wild cherry bark, as in Compound White Pine Syrup and other compound concoctions.[29] Indians once stained their bodies with the bloody preparation, with the intent of frightening their enemies. Used also for dyeing cloth. Captain John Smith reported that the Indian maidens selected to cohabit with him painted their bodies with bloodroot. Protopine can induce brachycardia experimentally. Sanguinarine, which can induce glaucoma in experimental animals, is anticancer, antiseptic, and locally anesthetic. Once used by the Amerindians to aid in divination.[18] New England Indians squeezed juice from the root into maple sugar lump as a lozenge for sore throats. Flowers have been used in herbal tea.[27] Some Indians chewed the root and spat the juice onto skin burns. The extract is active against tuberculosis.

Root or juice used for cancer and tumors, especially of the breast, ear, nose, skin, and uterus, also, the "ethmoid sinuses",[4] and for fungoid growths and fleshy excrescences.[2] Highly regarded by some old-fashioned Indian and colonial "medics" for cancer. Taken with bayberry as a snuff for nasal polyps. The sulphate of the alkaloid berberine is active in B1, KB, and PS-131 tumor systems.[10] Chelerythrine and sanguinarine exerted a therapeutic action on Ehrlich carcinoma in mice and necrotize sarcoma 37.[11] Bloodroot with zinc chloride paste is said to help external tumors. Lewis and Lewis[11] summarize several studies of the relationships between bloodroot and cancer. Kloss recommends the herb for adenoids, nasal polyps, sore throat, and syphilis, suggesting a strong injection (anal?) of bloodroot tea for piles.[44] Rappahannocks used the root infusion for rheumatism.[11] In small doses, bloodroot is said to stimulate the digestive organs, acting as a stimulant and tonic; in large doses, an arterial sedative. It was used by the Amerindians in asthma, bronchitis, laryngitis, pertussis, and other respiratory problems. Also, used for gastrointestinal catarrh and hepatomegaly with jaundice. Homeopaths and others recommend bloodroot in alcoholism, anemia, aphonia, arthritis, asthma, breast tumor, bronchitis, burns, cancer, catarrh, chest pain, cold, cough, croup, deafness, diptheria, dysmenorrhea, dyspepsia, ear polyps, eczema, edema, fever, flushes, frigidity, climacteric gleet, glossitis, gout, granular lids, headache, hemoptysis, influenza, keratitis, laryngitis, liver ailments, migraine, neuralgia, ophthalmia, pertussis, pharyngitis, phthisis, physometra, pneumonia, poisoning, polyps, pregnancy, pyrosis, quinsy, rheumatism, rhinosis, scarlatina, shoulder, smell, stomach, syphilis, tinnitus, tumor, typhoid, nausea, whitlow, and whooping cough.[4,30,32,33] Roots applied to dental caries to relieve toothache.[11] Regarded as alterative, aphrodisiac, diaphoretic, diuretic, emetic, emmenagogue, expectorant, febrifuge, laxative, narcotic, prophylactic, rubefacient, sedative, sternutatory, stimulant, and vermifuge.[4,27,32,33]

Contains the alkaloids alpha-allocryptopine, beta-allocryptopine, berberine, chelerythrine, chelilutine, chelirubine, coptisine, homochelidonine, oxysanguinarine, protopine, pseudo-chelerythrine, sanguidimerine, sanguilutine, sanguinarine (*circa* 1%), and sanguirubine. Rhizome also contains a reddish resin and much starch. The seed contains 0.1 and 28% fatty oil. Leaves contain 0.08% alkaloid,[33] the rhizome 4.7%, the roots 1.8%. Citric and malic acids, gum, and porphyroxin are reported.[27]

Toxicity — Classified by the FDA[62] as an unsafe herb, containing the "Poisonous alkaloid sanguinarine, and other alkaloids." Chelerythrine and saguinarine disturb mitosis. Dust of the root irritant to the mucous membranes, "The root is violently toxic."[18] Jim Duke experienced tunnel vison when he took a small bite of rhizome. Urine analyses subsequently suggested the presence of opium-like alkaloids! In toxic doses bloodroot causes burning in the stomach, intense thirst, paralysis, vomiting, faintness, vertigo, collapse, and intense prostration with dimness of eyesight.[3,33]

To the physician — For poisonings, Hardin and Arena[34] suggest gastric lavage or emesis with symptomatic treatments. *Hager's Handbook* suggests warm milk for the nausea,[33] cardiac stimulants for impending collapse.[33]

307. *SANTALUM ALBUM* L. (SANTALACEAE) — Sandalwood

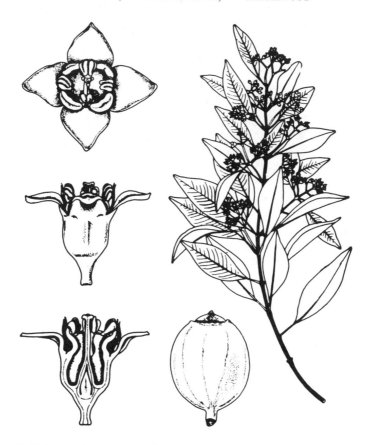

Sandalwood oil is widely used as a fragrance ingredient in creams, detergents, incenses, lotions, perfumes, and soaps. Also, used as a flavor component in beverages, baked goods, candies, liqueurs, gelatins, and puddings, with reported average maximum use levels generally below 0.001%. The oil has a wide array of folk medicinal applications. Chinese import the wood for furniture, saving the scraps for medicine.[16] In Malaya the leaves are smoked with those of *Mimusops* (perhaps, strictly medicinal, for asthma). The oil has diuretic and urinary antiseptic properties.[29] Sandalwood is one of the finest woods for carving, next only to ivory for intricate workmanship. The endosperm of the seed is edible.[1]

Reported to be anodyne, antiseptic, astringent, carminative, diaphoretic, diuretic, expectorant, febrifuge, stimulant, stomachic, sandalwood is a folk remedy for acne, boneache, bronchitis, cholera, cystitis, diarrhea, dysentery, dyspepsia, elephantiasis, enteralgia, gastralgia, gleet, gonorrhea, headache, hiccups, nausea, skin ailments, splenosis, stomachache, tumors, urethritis, and urogenital ailments.[16,32,33] The seed is used for skin diseases.

Sandalwood oil contains 90% alpha- and beta-santalols. Minor constituents include *circa* 6% sequiterpene hydrocarbons (mostly alpha- and beta-santalenes and epi-beta-santalene; small amounts of alpha- and beta-curcumenes, possibly beta-farnesene, and dendrolasin), dihydro-beta-agarofuran, santene, teresantol, borneol, teresantalic acid, tricycloekasantalal, santalone, and santanol, among others. Alpha-santalol and beta-santalol account for most of the odor of sandalwood oil.[29] Demole et al. blame 2-furfuryl pyrrole.[284] Seeds yield 50 to 55% of a dark red viscid fixed oil containing stearolic acid (9-octadecynoic acid) and santalbic acid (11-octadecen-9-ynoic acid). Oil of the fruit contains santalbic-, palmitic-, oleic-, and linoleic-acid; the fruit contains betulic acid, beta-sitosterol, glucose, fructose, and sucrose. Fresh leaves yield a wax, the unsaponifiable fraction containing *circa* 75% wax, made up largely of *n*-triacontanol, *n*-octacosanol, palmitone, and *d*-10-hydroxy palmitone.[1]

Toxicity — The oil is irritating to mouse and rabbit skin. Still, it has been approved for food use (§ 172.510). Santalol can cause dermatitis in sensitive individuals.[29]

308. *SANTOLINA CHAMAECYPARISSUS* L. (ASTERACEAE) — Lavender-Cotton

Ornamental shrub, foliage gray, flowers yellow, sometimes planted as an insect repellent. Foliage brought into the house as a moth repellant.[8] Essential oil occasionally used in perfumery.[2] Extracts of flower, leaf, and root active against Gram-positive bacteria.[1]

Considered analgesic, antispasmodic, bactericide, digestive, fungicidal, emmenagogue, stimulant, stomachic, vermifuge, and vulnerary. Once considered a snakebite remedy. Arabs are said to use the juice of the plant as a collyrium. Used also as a children's vermifuge.[2]

The flowers contain *circa* 0.86% essential oil, which, in turn, contains thujene, pinene, phellandrene, myrcene, limonene, p-cymol, acetone, cryptone, phellandral, menth-1-en-4-ol, alpha-terpineol, patuletin (6-methoxyquercetin), luteolin, caffeic acid, vanillic acid, *o*- and *p*-cumaric acids, and catechin;[33] these data may, however, apply to an *Artemisia*. The root contains the *cis*- and *trans*-isomers of 2-[4-(5-formyl-2-thienyl)but-3-yn-1-enyl]-furan.[33] Resin and tannin are also reported

Toxicity — Classified by the FDA[62] as an Herb of Undefined Safety: "Reputed anthelmintic. Stated to have long been used popularly against the round worm in Scotland."

309. *SARCOSTEMMA ACIDUM* (Roxb.) Voigt. (ASCLEPIADACEAE) — Soma

The sap is said to alleviate thirst, but is also used to produce an insecticide and intoxicant.[16] Emboden evokes evidence suggesting that this might be the Ayurvedic soma.[54] Indians use the plant to keep termites out of cane fields.[1]

Reported to be antidotal, depurative, emetic, insecticidal, intoxicant, and poison.[32] The dried stem is emetic,[54] the root infusion used as an antidote for rabid dogbites.[1]

Contains malic acid, succinic acid, reducing sugar, sucrose, traces of tannin, an alkaloid, a phytosterol, alpha- and beta-amyrins, lupeol and lupeol acetate, and beta-sitosterol. The milky latex contains 4.1% caoutchouc; the coagulum contains 16% caoutchouc, 68.1% resins, and 15.9% insolubles. Treatment of the waxy material from the exudate with sulfur, formaldehyde, and phthalic anhydride yields a composite resin. The waxy product can be used as a plasticizer for reclaimed rubber (from waste rubber). Condensed with glycol, the waxy product yields glyptal type of resins useful for the lacquer and varnish industries.

Toxicity — Listed as a narcotic inducing hallucinations and giddiness.[54] "Active constituents of the plant are said to be toxic to man."[1]

310. *SASSAFRAS ALBIDUM* (Nutt.) Nees (LAURACEAE) — Sassafras

Dried leaves constitute the file of file gumbo in Creole cooking. Two or three leaves in a glass of water yield a mucilaginous beverage. Oil of sassafras applied externally as a pediculicide, and to relieve stings and bites. Once used in dentistry to disinfect root canals. Once used to make "Godfrey's Cordial", a mixture of opium and sassafras. Used as a source of artificial heliotrope. Oil also used to flavors dentifrices, masticatories, mouthwashes, soaps, candies, root beers, and "sassaparillas", as well as tobaccos.[8] Twigs used for cleaning the teeth. Wood and bark furnish a yellow dye.[2] Pith of sassafras was once official in the U.S. as a mucilaginous demulcent, used for eye inflammation. I enjoy very much a tea made from sassafras roots and sumach berries. My 100-year old grandmother was not apparently frightened by the sassafras-cancer scare. Pioneers boiled sassafras in maple sap. Sassafras is said to be antagonistic to the narcotic effects of alcohol. South Carolina blacks make a soup from young sassafras leaves with *Viola palmata* and *V. septemloba*.[46] Sassafras extracts show very good activity against Ancylostoma and Strongyloides.[285]

Teas, made from the oil or root bark, have been applied to cancer, corns, osteosarcomas, tumors, and wens.[4] Sassafras tea used in Appalachia as a diaphoretic and diuretic for bronchitis, gastritis, and indigestion, and to slow down the milk of nursing mothers. South Carolina blacks give it to children to "bring out the measles."[46] Tea also used for arthritis, acne, bronchitis, catarrh, dropsy, dysmenorrhea, dysentery, fever, gleet, gonorrhea, gout, hypertension, kidney trouble, mumps, nephrosis, ophthalmia, pneumonia, respiratory ailments, rheumatism, skin trouble, syphilis, and typhus.[2,32,46] The herb is alterative, anodyne, antiseptic, aromatic, carminative, depurative, diaphoretic, demulcent, diuretic, emmenagogue, stimulant, and sudorific.[28,32] Externally, sassafras has been used as a rubefacient on bruises, rheumatism, sprains, and swellings. Lewis and Elvin-Lewis[11] cite an interesting cancer "cure", based on a carcinogen, that echos some of the cancer-preventitive suggestions recently emanating from congress and the NIH: "Let him drink Sassafras Tea every Morning, live temperately, upon light and innocent Food; and abstain entirely from strong liquor. The Way to prevent this Calamity, is, to be very sparing in eating Pork, to forbear all salt, and high season'd Meats, and life chiefly upon the Garden, The Orchard, and the Hen-House (cancer cure in Virginia — 1734).''

The volatile oil contains *circa* 80% safrole, some anethole, apiole, asarone, camphor, caryophyllene, coniferaldehyde, copaene, elemicin, eugenol, 1-menthone, 5-methoxyeugenol, myristicin, phellandrene, pinene, piperonylacrolein, sesquiterpenes, thujone. The alkaloids (*circa* 0.02%) include boldine, cinnamolaurine, isoboldine, norboldine, norcinnamolaurine, and reticuline. Two lignins, sesamin and desmethoxyaschantin, are reported along with sitosterol. Gum, mucilage, resin, tannin, and wax also occur.[27]

Toxicity — Safrole, having produced liver tumors in rats, has, like oil of sassafras, been banned because of its purported hepatotoxicity. Oil of sassafras is said to produce dermatitis in sensitive individuals. Morton[46] adds that the decoction makes pimples come out on arms and body if the root is used before it is aromatic. Heliotropin, a safrole derivative, used in cosmetics, has been reported to cause dermatitis.[6] Safrole (chemically like myristicin and asarone) is suspected of being hallucinogenic in large doses (carcinogenic and hepatotoxic).[11] A teaspoonful of oil produced vomiting, dilated pupils, stupor, and collapse in a young man. Its use has caused abortion.[2] Safrole-free sassafras extract (§ 172.580) and safrole-free sassafras leaves and extracts (§ 172.510) are approved for food use. Safrole, sassafras, and sassafras oil are prohibited from use in foods.[29] Safrole is also reported from basil, black pepper, mace, and nutmeg. After quoting a nice verse from the ''Spring Ode'' by Donald Robert Perry Marquis:

Fill me with sassafras, nurse,
And juniper juice!
Let me see if I'm still any use!

Tyler[37] gives some pessimistic data on sassafras. ''As a result of research conducted in the early 1960s, safrole was recognized as a carcinogenic agent in rats and mice . . . No one really knows just how harmful it is to human beings, but it has been estimated that one cup of strong sassafras tea could contain as much as 200 mg of safrole, more than four times the minimal amount believed hazardous to man if consumed on a regular basis . . . Recent studies have shown that even safrole-free sassafras produced tumors in two-thirds of the animals treated with it. Apparently other constituents in addition to safrole are responsible for part of the root bark's carcinogenic acitivity.''[39] Benedetti et al.,[286] however, detected no 1'-hydroxysafrole, the metabolite considered responsible for safrole's carcinogenicity, when small amounts of safrole were given by mouth to human volunteers, as opposed to rats. Tyler notes that the doses of safrole were very small (maximum 1.655 mg), perhaps resulting in the nonappearance of 1-hydroxy-safrole in the human's urine. Perhaps the high tannin content may be synergistic. Reviewing work on safrole, Buchanan[117] concluded that it is the most thoroughly investigated methylenedioxybenzene derivative. The major flavoring constituent in sassafras root bark, safrole, was identified as a ''low grade hepatocarcinogen.'' It was banned in root beer, and the FDA in 1976 banned interstate marketing of sassafras for sassafras tea. The oral LD_{50} for safrole in rats is 1950 mg/kg body weight, with major symptoms including ataxis, depression, and diarrhea, death occurring in 4 to 5 days. Ingestion of relatively large amounts of sassafras oil produced psychoactive and hallucinogenic effects persisting several days in humans. With rats, dietary safrole at levels of 0.25, 0.5, and 1% produced growth retardation, stomach and testicular atrophy, liver necrosis, biliary proliferation, and primary hepatomas.[117]

311. *SATUREJA MONTANA* L. (LAMIACEAE) — Winter Savory, White Thyme, Spanish Savory

Winter savory is cultivated as a spicy herb, used fresh or dried, as seasoning in poultry or with green beans. Fresh flowering plants yield oil, used in flavoring condiments, relishes, soups, sausages, canned meats, and in spicy table sauces. According to Lawrence, the value of savory oil is in its high carvacrol content and its fresh, spicy phenolic notes, reminiscent of oregano and thyme.[193] Dried herb mixed with crumbs to bread meat and fish.

Savory is sometimes added to massage creams. Spinal massage with herbal creams (savory, plantain, peppermint, cow parsnip, celandine), along with foot baths and vaginal douches, are believed by some European natural healers to work miracles with the frigid female.[11] Oil is also an efficient antidiuretic because of carvacrol present, 35 to 40% in wild plants, up to 65% in cultivated plants. Considered antioxidant, aperitive, astringent, bactericidal, carminative, digestive, expectorant, fungicidal, laxative, sedative, stimulant, stomachic, sudorific, tonic, and vermifuge, winter savory is used for catarrh, colic, otitis, sclerosis, and spasms.[32]

Leaves contain ursolic acid. Dried savory (*S. hortensis*) herbage contains, per 100 g, 272 calories, 9.0 g water, 6.7 g protein, 5.9 g fat, 68.7 g total carbohydrate, 15.3 g fiber, 9.6 g ash, 2132 mg Ca, 140 mg P, 37.9 mg Fe, 24 mg Na, 1051 mg K, 3078 μg β-carotene, 0.37 mg thiamine, and 4.08 mg niacin.[21] Fresh flowering herb yields 0.15 to 0.23% of an orange yellow essential oil containing 28 to 65% carvacrol, 27% *p*-cymene, 14% dipentene, and 10% of an unidentified alcohol.[1] Leung, probably reporting on dried savory, reports 1.6% volatile oil, containing, in addition to the above, thymol, alpha- and beta-pinene, limonene, cineole, borneol, alpha-terpineol, oleanolic-, and ursolic-acid.[29] *Hager's Handbook* adds phenol, *p*-cymene, terpene, carvone, perilla alcohol, and methadiene-1,8(9)-ol-(7) ($C_{10}H_{16}O$).[33] San Martin et al.[287] examined the chemical composition from 12 different areas in Spain. From four areas, the lab-distilled oils had a normal or classical composition: carvacrol (60 to 75%), thymol (1 to 5%), *p*-cymene (10 to 20%), gamma-terpinene (2 to

10%), and small quantities of linalool, terpinen-4-ol, alpha-terpineol, and their acetates. One oil contained carvacrol (41.5%), thymol (0.5%), *p*-cymene (15%), linalool (24%), and alpha-terpineol (7.5%). Two other oils were found to contain *p*-cymene as the major constituent (32.5 and 39%), along with varying amounts of carvacrol (21.5 and 1.5%), thymol (0 and 3.5%), gamma-terpinene (13.5 and 1%), linalool (9.0 and 7.0%), alpha-terpineol (9.5 and 6%), and terpinen-4-ol (trace and 7%). The other five oils contained linalool as the major compound (24.3 to 54.1%), with varying amounts of alpha-pinene, camphene, beta-pinene, myrcene, limonene, gamma-terpinene, *p*-cymene, thujanol-4 (sabinene hydrate), linalyl acetate, terpinen-4-ol, terpinen-4-yl acetate, alpha-terpineol, alpha-terpinyl acetate, geranyl acetate, geraniol, thymol, and carvacrol. It can be seen from these results that Spanish savory oil, which is obtained from *S. montana*, can vary considerably depending upon which chemical race of the plant is used for essential oil production.[193,287]

Toxicity — § GRAS 182.10 and 182.20.[29]

312. *SCHINUS MOLLE* L. (ANACARDIACEAE) — Peppertree, Peruvian Mastic Tree

Berries used in syrups, vinegar, beverage (Peru), wine (Chile), and as a pepper substitute, when berries are dried and ground up (Africa). Leaves used as a dye, have a peppery odor when crushed. Leaves so rich in oil that if fragments are placed in water, the oil is expelled with such force that pieces twist and jerk as if by spontaneous motion. Resin exudes from trees, known as American mastic. Plants make a good hedge, and are often planted for ornamental purposes along streets, for which purpose they are not well suited as they spread too much and branch too close to ground. Has been used as an adulterant of black pepper.[3] Chileans make a vinegar-like brew by steeping mashed berries in water. Spaniards macerate husked fruits in water, steeping for 3 days to yield 35% alcohol. The plant shows up in religious artifacts or idols among some Chilean Amerindians. Wood, used as a fuel,[63] seems to be immune to termites.[3]

The balsam, resin, or bark decoction are used in folk remedies for tumors and warts[4] from Mexico to Peru and Brazil. Reported to be astringent, balsamic, collyrium, diuretic, emmenagogue, masticatory, piscicide, purgative, stomachic, tonic, viricidal, and vulnerary, peppertree is a folk remedy for amenorrhea, apostemes, blenorrhagia, bronchitis, cataracts, dysmenorrhea, gingivitis, gonorrhea, gout, ophthalmia, rheumatism, sores, swellings, tuberculosis, ulcers, urethritis, urogenital and venereal disorders, warts, and wounds.[32,33] The juice of the leaves was employed in ophthalmia and rheumatism; the decoction used in washing prolapsed uterus. The bark is employed in Colombia for diarrhea, hemoptysis, and rheumatism.[57] Mexicans steep the berries in pulque for 3 days to make the beverage called capalote.

Fruits contain 3.4 to 5.2% essential oil with alpha- and beta-phellandrene, beta spathulene, D-limonene, silvestrene, alpha- and beta-pinene, perillaldehyde, carvacrol, myrcene, camphene, *o*-ethyl phenol, *p*-cymene, *p*-cymol, etc.[33,288] The bark contains *circa* 23% tannin, and, like *S. terebinthifolius,* is said to be a source of Aroeira resin, with *circa* 55% resin, 40% gums, etc. Leaves contain 0.2 to 1.% essential oil with phellandrene and carvacrol, myricetin, quercetin, kaempferol, leucodelphinidin, lignoceric acid, and beta-sitosterol.[33] Dry seeds contain 8.1 to 8.5% protein and 7.7 to 10.6% oil.[21] Analyses in Egypt, where the fruits are used as substitutes for black pepper, showed 6% moisture, 5.47% ash, 3.31% acid insoluble ash, and 15.5% crude fiber.[289]

Toxicity — Children are frequently poisoned from eating the fruits, leading to nausea, diarrhea, gastroenteritis, headache, and lassitude.[3]

313. *SCHINUS TEREBINTHIFOLIUS* Raddi (ANACARDIACEAE) — Brazilian Pepper-tree, Florida Holly

According to the *Wall Street Journal*[290] much recent material sold here at more than $125/kilo as "pink peppercorns" or "red peppercorns" is, in fact, nothing more than fruits of the Brazilian pepper plant. This tree was introduced to Florida as an ornamental, but now many Floridians regret the introduction. Said to be a good honey plant, making the plant more appealing to beekeepers than to those who are allergic to the plant. Reports on the marginal pulping potential are included in an interesting account by Morton.[291] Goats graze the foliage with impunity but cattle may develop enteritis. Stem is source of a resin called Balsamo de Missiones.[8]

The balsam and bark, fruits, and leaves are used in folk remedies for tumors, especially of the foot, in Brazil.[4] Reported to be antiseptic, aphrodisiac, astringent, bactericidal, stimulant, tonic, and viricidal, Brazilian peppertree is a folk remedy for atony, bronchitis, bruises, chills, diarrhea, gout, hemoptysis, rheumatism, sciatica, sores, swellings, syphilis, tendonitis, tumors, ulcers, and wounds.[3,32] The leaf and fruit, both said to possess antibiotic activity, have been used to bathe sores and wounds. Brazilians poultice the dried leaves on ulcers. The bark decoction is added to the bath water of rheumatic and sciatic patients. Macerated root juice is applied to contusions and/or ganglionic tumors. Leaves are used to bathe sores and wounds.[3] Homeopathically, the plant is used for arthritic pain, chills, diarrhea, gout, hemoptysis, intestinal weakness, lymphatic swellings, sexual inertia, skin complaints, and tumors.

Fruits contain the triterpenes terebinthone, $C_{30}H_{46}O_3$, and schinol, $C_{30}H_{50}O_3$. Aroeira resin from the bark contains *circa* 55% resin, 40% gum, and some essential oils, tannins, saponins, and fatty acids. The resin also contains cardol ($C_{21}H_{39}O_2$). Leaves contain myricetin, quercetin, kaempferol, leucocyanindine, triacontane, beta-sitosterol, masticodienonic acid, and 3 alpha-hydroxymasticodienoic acid.[33] Seeds contain 25 to 45% essential oil, composed mostly of phellandrene, and 8 to 11% dark green fatty oils. USDA reports 10.8% protein and 32.2% fatty oil in the dry seeds.[21]

Toxicity — Can wreak havoc on the human digestive system (with such aftereffects as vomiting, diarrhea, and hemorrhoids) . . . extremely upset stomach . . . violent headaches, swollen eyelids, and shortness of breath . . ."Birds are said to become intoxicated when they eat the 'peppers' and fish die in ponds bordered by the Brazilian pepper plant."[290] Horses may develop fatal colic upon ingesting the berries.[291]

314. *SCHOENOCAULON OFFICINALE* A. Gray (LILIACEAE) — Sabadilla (Anglicized),
Cebadilla (Spanish)

According to Roark, the seed has been used as an insecticide since the 16th century,
mostly for the destruction of lice on man and domestic animals. Following a war-induced
shortage of rotenone in 1939, sabadilla was tested (and sold) for control of hemipterous
insects, e.g., squash bugs, chinch bugs, harlequin bugs, and Lygus bugs. (Around 1945,
annual imports exceeded 50 MT, mostly from Venezuela.) To increase the toxicity, powdered
seeds are heated in kerosene or some other solvent to about 150°C. If the seed were destined
for dusting powders, it was heated without the solvents. According to the *Wealth of India,*
the alkaloidal mixture, especially cevadine, is a powerful insecticide against hair lice, thrips,
and a variety of pests of horticultural crops and vegetables. Toxicity of the seed powder to
houseflies is increased by incorporating pyrethrum as a synergist.[1] "Veratrine can be safely
used on food crops even shortly before harvest, since it decomposes rapidly on exposure to
sunlight. It has no residual effect."[1]

Reported to be cathartic, emetic, *Poison,* sternutatory, and vermifuge, cebadilla is a folk
remedy for arthritis, cancer, gout, hypertension, inflammation, neuralgia, pediculosis, and
rheumatism.[2,32] Homeopathically used as a mucal and neural stimulant, for angina, grippe,
headache, hysteria, migraine, neurasthenia, etc.[33]

Seeds contain 4(1 to 6)% steroid alkaloids which work rather like veratrine, composed
mostly (*circa* 75%) of cebadine (an angelic acid ester of veracevine) ($C_{32}H_{49}NO_9$) and 25%
veratridine (a veratric acid ester of veracevine) ($C_{36}H_{51}NO_{11}$). Vanilloylcevine is hypotensive.[1] Also, cevacine ($C_{29}H_{45}NO_9$), vanillylveracevine ($C_{29}H_{45}NO_{11}$), veracevine ($C_{27}H_{43}NO_8$),
neosabadine ($C_{27}H_{43}NO_8$), sabadine ($C_{29}H_{47-49}NO_8$), hydroalkamine S ($C_{27}H_{45}NO_8$), vera-
grinine ($C_{31}H_{53-57}NO_{13}$), sabatine ($C_{29}H_{47-49}NO_8$). The seed contains 9 to 20% oil composed
mostly of palmitic and oleic acids, phytosterols, wax, resin, and angelic-, acetic-, ceva-
dic-, chelidonic-, tiglic-, vanillic-, and veratric-acids.[2,33]

Toxicity — Cebadilla is an acrid, drastic emeto-cathartic; in overdoses preparations produce numbness and tingling, followed by anesthesia and coldness. Large doses paralyze
heart action and respiration. Internal use is extremely dangerous.[2] The oral LD_{50} in rats is
5000 mg/kg.

315. *SCLEROCARYA CAFFRA* Sond. (ANACARDIACEAE) — Marula

Fruits (or kernels, or both) edible, yet said to serve as an insecticide.[33] Zulu use the fruits to destroy ticks.[3] Bark contains 10.7 to 20.5% tannin, hence, may be used as tanning agent. The tannin-rich gum is the source of a fixative for dyes. The seed oil is used for food and industrial purposes. Zulu women crush the kernels, boil them down to an oily residue, which is applied to skin shirts. Pedi use the ground-up kernels for making a porridge, the embryo as a condiment, and the leaf as a relish. The edible kernel yields an edible oil.[3] Fruits used to brew a highly intoxicating beverage, so potent "that a man is not allowed to bear arms having drunk it" lest he harm his neighbor.[54] The timber, fairly durable, is used for canoes, furniture, and structures.

Reported to be astringent, marula is a folk remedy for diarrhea, dysentery, malaria, proctitis.[3,32] The bark decoction is used for diarrhea, dysentery, and malaria, and to clean out wounds. The leaf juice is applied to gonorrhea.[33] Europeans in South Africa take the bark decoction both for the cure and prevention of malaria (but experiments have not confirmed antimalarial activity.)[3] Zulu use the bark decoction to prevent gangrenous rectitis. Fruits are believed to serve both as an aphrodisiac and contraceptive for females. (African cattle, having partaken of too much fruit, have been observed to become both agressive and infertile.)[3,32,33]

Per 100 g, the fruit is reported to contain 30 calories, 91.7 g H_2O, 0.5 g protein, 0.1 g fat, 7.5 g total carbohydrate, 0.5 g fiber, 0.2 g ash, 6 mg Ca, 19 mg P, 0.1 mg Fe, 0.03 mg thiamine, 0.05 mg riboflavin, 0.2 mg niacin, and 68 mg ascorbic acid. The seed is reported to contain 604 calories, 3.9 g H_2O, 24.6 g protein, 57.5 g fat, 9.2 g total carbo-hydrate, 2.7 g fiber, 4.8 g ash, 143 mg Ca, 1248 mg P, 0.4 mg Fe, 0.04 mg thiamine, 0.12 mg riboflavin, 0.7 mg niacin. Bark contains 3.5 to 10% tannin, leaves 20% tannin, a trace of alkaloids, and 10% gum. Fruits contain citric and malic acid, sugar, and 54 mg vitamin C per 100 g. Seed oil (53 to 60%) contains circa 55 to 70% oleic acid.[33] The pattern of the amino acids (particularly rich in arginine, aspartic acid, and glutamic acid) in the mean differ only slightly from that in human milk and eggs.

Toxicity — Listed as a narcotic hallucinogen(?) by Emboden.[54] In 1972, a flurry of newspaper articles heralded the propensity of pachyderms to get pickled on the fruit. Ele-phants, baboons, monkeys, warthogs, and humans may overindulge[292] in Kruger Park (South Africa).

316. *SCOPOLIA CARNIOLICA* Jacq. (SOLANACEAE) — Scopolia

Used in America for years in manufacturing "belladonna" plasters. In 1900, scopolamine and morphine were mixed to produce an anesthesia, the "Twilight Sleep", used alone or before chloroform or ether. This caused loss of memory, including that of pain (but mortality was high).[21] Has been used as a sleep inducer for children.[33]

Reported to be a cerebral sedative useful as a folk remedy for colic, cramps, drug addiction, epilepsy, gout, hysteria, insomnia, nymphomania, paralysis, rheumatism.[32,33] Hyoscine predominates along with scopolamine, often yielding more total alkaloid than belladonna. The total alkaloid content of the rhizome is 0.3 to 0.8% (highest during flowering and fruiting), up to 0.4% L-hyoscyamine, to 0.03% atropine, and traces of scopolamine, scopine, tropine, cuscohygrine, pseudotropine, 3-α-tigloyloxytropane, scopoline, scopoletine, chlorogenic acids. Leaves contain chlorogenic- and caffeic-acid, rutin, and aesculetin.[33]

Toxicity — Narcotic;[2] overdoses of scopolamine have proved fatal. Intoxication may cause disorientation and delirium similar to atropine poisoning, with or without somnolence, mydriasis, accelerated pulse rate, dry mouth, laryngeal paralysis. Where there is difficult breathing strychnine may be used. It is better not to give drugs for the relief of the delirium, but, if very active, small doses of paraldehyde and bromides may be employed.

317. *SCUTELLARIA LATERIFLORA* L. (LAMIACEAE) — Virginian Skullcap

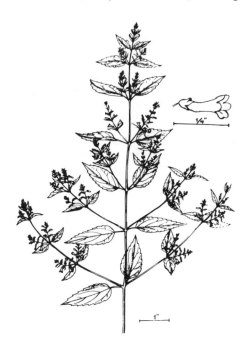

On sale as a medicinal herb in the American herb industry and elsewhere as a bitter tonic.[33]

Anaphrodisiac, antispasmodic, astringent, diaphoretic, febrifuge, and tonic, skullcap was once said to be the finest of nervines;[2] also, with a great reputation for hydrophobia. In commenting on its toxicity, the FDA said:[62] "As destitute of medicinal properties as a plant may be. Probably of no remedial value. At one time used as a remedy in hydrophobia. Formerly used in nervous diseases." Herbals, however, commend it as useful in chorea, convulsions, epilepsy, headache, hiccups, hypertension, hysteria, insomnia, nervous headache, nervousness, neuralgia, palsy, rickets, snakebite, and wakefullness. Tierra,[28] calling skullcap a safe and reliable nerve sedative, suggests that it is useful for drug and alcohol withdrawal symptoms. Said to increase activity of the kidneys. Indians are said to have used it to eliminate a mother's afterbirth, to promote menstruation, and to treat diarrhea and heart disease. With pennyroyal in tea, skullcap is said to serve successfully in female cramps, especially severe cramps due to menstruation suppressed by colds.[30] Homeopaths prescribe the tincture for ardor urinea, brain irritation, chorea, delirium tremens, dentition, flatulence, headache, hiccups, hydrophobia, hysteria, night terrors, sleeplessness, and tobacco heart. Chinese use skullcap (mostly S. *baikalensis*) as a nervine, sedative, stimulant, and tonic to treat convulsions, cramps, epilepsy, heart conditions, and rheumatism, and to reduce fever, pain, and to expel tapeworm.[30] Perhaps you can detect a contradiction in the following two quotes from Kloss:[44] "Skullcap as a substitute for quinine is more effective and is not harmful as quinine is" (page 313), and "Nonpoisonous herbs will do everything for which the allopath gives cadmium, mercury, antitoxin, serums, vaccines, insulin, strychnine, digitalis, and all the poisonous drug preparations, and nonpoisonous herbs do not leave any bad after effects." (p. 281)

Contains a volatile oil, scutellarin, and a bitter glucoside, yielding scutellarein on hydrolysis. Also contains tannin, lignin, resin, fat, sugar, cellulose, and a bitter principle.[2,33]

Toxicity — Classified by the FDA[62] as an Herb of Undefined Safety. Overdoses of the tincture cause giddiness, stupor, confusion of mind, twitchings of the limbs, intermission of the pulse, and other symptoms indicative of epilepsy.[2]

318. *SELENICEREUS GRANDIFLORUS* Britt. and Rose (CACTACEAE) — Night-Blooming Cereus

Ornamental cactus. Elsewhere it has quite a reputation for coronary problems.[33] Europeans use the extract or tincture for angina pectoris, bladder irritability, cardiac neuralgia, cardiac palpitation, kidney congestion, nervous headache, and prostatic diseases. Cubans use the stem juice as a vermifuge and vesicant, the flower-stem infusion as antirheumatic and cardiotonic.[42] The sap is used in middle America for cystitis, dropsy, dyspnea, and rheumatism.[33] Homeopaths use it for angina pectoris, endocarditis, myocarditis, and stenocardy.[33]

According to Morton,[42] the stems and flowers contain the alkaloid cactine which has digitalis-like activity. Working with fresh material, Petershofer-Halbmayer et al.[293] detected only one alkaloid hordenine (*N,N*-dimethyl-4-hydroxy-beta-phenethylamine) and concluded that so-called "cactine" must be identical with hordenine. Contains betacyanin, isorhamnetin-3-glucoside. Flowers contain narcissin ($C_{28}H_{32}O_{16}$), rutin, cacticine ($C_{22}H_{22}O_{12}$), kaempferitrin, grandiflorine ($C_{20}H_{18}O_{10}$), hyperoside, isorhamnetin-3-beta-(galactosyl)-rutinoside, and isorhamnetin-3-beta-(xylosyl)-rutinoside.

Toxicity — Said to contain digitalis-like compounds. Fresh juice burns the mouth, causing nausea, vomiting, and diarrhea.

319. *SENECIO AUREUS* L. (ASTERACEAE) — Liferoot, Squaw Weed, Golden Groundsel

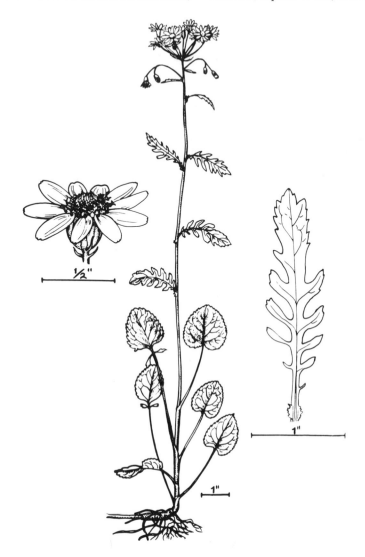

Parvati describes this as one of the most certain and safe "emmenagogics".[48] It is used by European homeopaths as well. Where I came from, it was nothing more than an attractive weed.

Reported to be anodyne, astringent, diaphoretic, diuretic, emmenagogue, expectorant, febrifuge, pectoral, stimulant, tonic, uterotonic, and vulnerary, squaw weed is a folk remedy for circulation, consumption, diarrhea, dysmenorrhea, dysuria, gravel, hemorrhage, leucorrhea, nephrosis, neurosis, parturition, and stone.[2,32,33,44] Amerindians used it an an antihemorrhagic, abortivant, and vulnerary.[19] Later it was recommended as an ergot-substitute, "as an excellent drug to control pulmonary hemorrhage, generally as a diuretic, pectoral, diaphoretic, tonic, and a substance to be thought of in various forms of uterine trouble." Catawba Indians used the tea to hasten labor and check the pains of parturition.[11]

Hager's Handbook[33] lists senecionine and other alkaloids (senecifoline, senecine), as well as resin. Tannin is also reported.[19]

320. *SERENOA REPENS* (Bartel.) Small (ARECACEAE) — Saw Palmetto

Indians used the seed for food. Flowers a good source of honey. Fruit used as an aromatic in cognac.[33]

Fruits have been used as a folk remedy for tumors.[4] Morton[50] quotes some of the herbalist's claims: "Palmetto Berries are of great service in cold in the head, irritated mucous membrane of the throat, nose, and air passages, and in numerous other conditions . . . The reputed effect is primarily rejuvenation, . . . The fruit is a nutritive tonic, diuretic, sedative. It is highly recommended in all wasting diseases, as it has a marked effect upon all the glandular tissues, increasing flesh rapidly and building up to strength. Should be used in atrophy of testes, mammae, etc." Unfortunately or not, *Hager's Handbook*[33] is almost as laudatory, stating that the herb is aphrodisiac, stimulating bladder, prostate, testicles, ovaries, and uterus. Wagner and Flachsbarth report an antiphlogistic principle.[294] Indians used an infusion for stomachache and dysentery. Inner bark of the trunk poulticed onto bugbite, snakebite, and ulcers. Dried fruits believed useful for indigestion, respiratory infections, and catarrhal irritation. Used for underdeveloped breasts.[33] Steinmetz[27] lists the herb as aphrodisiac, diuretic, sedative, stimulant, tonic, and "one of the most nourishing medicines we have." According to Hutchens,[30] "Serenoa is of great service for colds in the head, irrated mucous membrane of the throat, nose, and air passages, and chronic bronchitis of lung, asthma. Of use in renal conditions and diabetes." Also, suggested for epididymitis and cystitis.

Contains volatile oil, fixed oil, glucose, *circa* 63% of free acids, 37% ethyl esters.[2] Dry fruits contain beta-sitosterol (said to have both aphrodisiac and anticancer activity).[4,10,15] On alkaline hydrolysis, the shrubs yield 61.8% *p*-oxybenzoic acid, 0.6% p-oxybenzaldehyde, 1.5% vanillic acid, 0.3% vanillin, 0.6% acetovanillone, 1.0% syringic acid, 0.8% syringaldehyde, 0.9% acetosyringone, and 1.9% ferulic acid. In the seed oil there is stearic acid and the glycerides of capric-, lauric-, myristic-, palmitic-, and oleic-acids. The fruit contains carotene, lipase, tannin, resin, *circa* 28.2% invert sugar, mannitol; the dried fruit contains 0.0189% beta-sitosterol and 0.022% beta-sitosterol-D-glucoside; anthranilic acid and three flavonoids. Beta-sitosterol probably lies behind the counterculture claims that saw palmetto berries will enlarge the breasts. There are relatively high concentrations of free and bound sitosterols in dry berries. Injected into immature female mice, beta-sitosterol is estrogenic. Still, the saw palmetto extract is only 1/10,000 as potent as estradiol. Pure beta-sitosterol is less than 1/10 as potent. Further, the studies injected the sitosterols, which are poorly absorbed in the gastrointestinal tract. Since beta-sitosterol is not very soluble in water, herb teas would not contain much in solution, so Tyler concludes, perhaps correctly (but his conclusion could readily be tested analytically), "a cup of saw palmetto tea contains about as much real estrogenic activity as a cup of hot water."[37] The fruit flesh contains about 1.5% palmetto oil, up to 63% free fatty acids and caproic-, caprylic-, capric-, lauric-, palmitic-, and oleic-acids, and up to 38% of their ethyl esters.[33]

Toxicity — Classified by the FDA[62] as an Herb of Undefined Safety: "Saw Palmetto berries yield about 1.5% of oil composed of free fatty acids and ethyl esters of these acids. Claimed to exert a stimulant action upon mucous membrane of genitourinary tract. Has been used in chronic and subacute cystitis and in cases of enlarged prostate in old men. Reputed aphrodisiac."

321. *SIMMONDSIA CHINENSIS* (Link) C. Schneid. (BUXACEAE) — Jojoba

Simmondsia is unique among plants in that its seeds contain an oil which is a liquid wax. Oil of *Simmondsia* is obtained by expression or solvent extraction. It is light yellow, unsaturated, of unusual stability, remarkably pure, and need not be refined for use as a transformer oil or as a lubricant for high-speed machinery or machines operating at high temperatures. The oil does not become rancid, is not damaged by repeated heating to temperatures over 295°C or by heating to 370°C for 4 days; the color is dispelled by heating for a short time at 285°C, does not change in viscosity appreciably at high temperatures, and requires little refining to obtain maximum purity. Since *Simmondsia* oil resembles sperm whale oil both in composition and properties, it should serve as a replacement for the applications of that oil. With the American ban on spermwhale oil, a market of nearly 60 tons oil per year must be filled by substitutes.[11] Jojoba oil could be one of the most effective.[296] Further, the oil can be easily hydrogenated into a hard white wax, with a melting point of about 73 to 74°C, and is second in hardness only to carnauba wax. The oil is a potential source of both saturated and unsaturated long-chain fatty acids and alcohols. It is also suitable for sulfurization to produce lubricating oil and a rubber-like material suitable for use in printing ink and linoleum. The residual meal from expression or extraction contains 30 to 35% protein and is acceptable as a livestock food. Seeds were said to be palatable and were eaten raw or parched by Indians. They may also be boiled to make a well-flavored drink similar to coffee, hence the name coffeeberry. It is an important browse plant in California and Arizona, the foliage and young twigs being relished by cattle, goats, and deer, hence the name goatnut.

Indians of Baja California highly prized the fruit for food and the oil as a medicine for cancer and kidney disorders. Indians in Mexico use the oil as a hair restorer. Reported to be emetic, jojoba is a folk remedy for cancer, colds, dysuria, eyes, head, obesity, parturition, poison ivy, sores, sore throat, warts, and wounds. Seri Indians applied jojoba to head sores and aching eyes. They drank jojoba-ade for colds and to facilitate parturition.[32,63]

I was amazed, in searching through files on jojoba, to find no conventional proximate analysis. It was not even included in two of my most treasured resources, *Hager's Handbook*[33] and the *Wealth of India*.[1] Perhaps this is due to the relative novelty of interest and the unique situation that the seed contains liquid wax rather than oil, sort of unusual for the conventional analyses. Verbiscar and Banigan[295] approximated a proximate analysis, some of which follows: per 100 g, the seed is reported to contain 4.3 to 4.6 g H_2O, 14.9 to 15.1 g protein, 50.2 to 53.8 g fat, 24.6 to 29.1 g total carbohydrate, 3.5 to 4.2 g fiber, and 1.4 to 1.6 g ash. USDA analyses show 15.6% protein and 45.6% fat.[21] The amino acid composition of deoiled jojoba seed meal is 1.05 to 1.11% lysine, 0.49% histidine, 1.6 to 1.8% arginine, 2.2 to 3.1% aspartic acid, 1.1 to 1.2% threonine, 1.0 to 1.1% serine, 2.4 to 2.8% glutamic acid, 1.0 to 1.1% proline, 1.4 to 1.5% glycine, 0.8 to 1.0% alanine, 1.1 to 1.2% valine, 0.2% methionine, 0.8 to 0.9% isoleucine, 1.5 to 1.6% leucine, 1.0% tyrosine, 0.9 to 1.1% phenylalanine, 0.5 to 0.8% cystine and cysteine, and 0.5 to 0.6% tryptophane. Detailed analyses of the wax esters, free alcohols, and free acids are reported in NAS.[297] Per 100 g jojoba meal, there is 1.4 g lysine, 0.6 g histidine, 1.9 g arginine, 2.6 g aspartic acid, 1.3 g threonine, 1.3 g serine, 3.2 g glutamic acid, 1.5 g proline, 2.4 g glycine, 1.1 g alanine, 0.6 g cystine, 1.5 g valine, 0.1 g methionine, 0.9 g isoleucine, 1.8 g leucine, 1.1 g tyrosine, and 1.2 g phenylalanine. The two major flavonoid constituents of the leaves are isorhamnetin 3-rutinoside (narcissin) and isorhamnetin 3,7-dirhamnoside. Seed meals (27% protein, 12% fiber, 3.9% ash, and 3.3 mcal/kg) from seven locations averaged 6610 μg/g K, 136 Na, 1410 Mg, 372 Ca, 46 Fe, 21 Mn, 16 Zn, 10 Cu, 3.8 Ni, and 0.7 μg/g Co.[299]

Toxicity — The acute oral LD_{50} for crude johoba oil to male albino rats is higher than 21.5 mℓ/kg body weight. Strains of *Lactobacillus acidophilus* can ameliorate this toxicity. Rats injected with jojoba oil were similar to controls injected with olive oil. Guinea pigs treated topically with refined jojoba wax for prolonged periods show normal growth and no histopathological changes in the internal organs. Seeds contain 2.25 to 2.34%; seed hulls, 0.19%; core wood, 0.45%; leaves, 0.19 to 0.23%; twigs, 0.63 to 0.75%, and inflorescence, 0.22%; simmondsin is a demonstrated, appetite-depressant toxicant. Three related cyano-methylenecyclohexyl glucosides have also been isolated from the seed meal. Swingle[299] found problems using jojoba meal as a protein source in livestock feed. Feeder steers ate significantly less feed when it was supplemented with 10% untreated jojoba meal as compared to cottonseed meal. They were not so quick to digest feed and gain weight.

322. *SMILAX ARISTOLOCHIIFOLIA* Mill. (LILIACEAE) — Mexican Sarsaparilla

Once widely used in root beer, sarsaparilla extracts are used now (up to 13 ppm) in baked goods (up to 0.2%), beverages, candies, and frozen dairy desserts. Sarsaparilla is advertised as an aphrodisiac in the U.S. Such ads have resulted in sarsaparilla's inclusion in the tea I called the "Root Booster" composed of roots of ginger, ginseng, sarsaparilla, and sassafras, all reputed aphrodisiacs. Sarsaparilla saponins are reported to facilitate body absorption of other drugs.[42]

Reported to be depurative, diaphoretic, sudorific, tonic, Mexican sarsaparilla is a folk remedy for arthritis, cancer, dyspepsia, eczema, fever, gonorrhea, leprosy, nephrosis, rash, rheumatism, scrofula, skin ailments, syphilis, and wounds.[29,32,33] New Guinea natives apply macerated bark from the base of the stem of a *Smilax* species to toothache.

Speaking of several sarsaparilla species,[29] Leung notes that the plants contain steroids (sarsasopogenin, smilagenin, sitosterol, stigmasterol, and pollinastanol) and their glycosides (saponins) including sarsasaponin (parillin), smilasaponin (smilacin), sarsaparilloside, sitosterol glucoside, etc. Other constituents present include starch, resin, cetyl alcohol, and a trace of a volatile oil.[20,29]

Toxicity — In unusually large doses, the saponins could possibly be harmful. (Approved for food use; § 172.510.[29])

323. *SOLANUM DULCAMARA* L. (SOLANACEAE) — Bittersweet, Bitter Nightshade, Felonwood

Shepherds once hung it around the necks of sheep suspected of being under the influence of the "evil eye". Sold as a dangerous drug. Shoots said to be eaten by muskrats, the berries in winter by grouse and pheasant. In Russia, the plant is used for the synthesis of progesterone, cortisone, and other hormones.

Widely used folklorically for felons, tumors, and warts, bittersweet or alcoholic extracts of rhizomes and roots, etc. possess significant antitumor activity. Recommended by Galen (AD150) for tumors, cancer, and warts, woody nightshade has proven out. Beta-solamarine has proven active against Sarcoma 180. Extracts also inhibit Walker carcinosarcoma-256 in rats.[1,4,10] Recommended for asthma, arthritis, bladder ailments, bronchitis, cancer, catarrh, dysmenorrhea, eczema, excrescences, fever, hepatitis, jaundice, leprosy, malaria, pertussis, rheumatism, scrofula, skin ailments, splenitis, swellings, tumors, ulcers, warts, whooping cough.[16,33] Considered alterative, analgesic, anaphrodisiac, anodyne, antidotal, cardiotonic, depurative, diaphoretic, diuretic, emetic, expectorant, hypnotic, laxative, narcotic, refrigerant, resolvent, sedative, stimulant, sudorific, tonic, and toxic. The infusion of dried branches is said to be analgesic and sedative. Homeopaths recommend a tincture of fresh green stems and leaves, gathered just before flowering, for adenitis, angina, aphonia, asthma, bladder, blepharophthalmia, catarrh, cholera, colic, crusta lactea, cystitis, diarrhea, dropsy, dysentery, dysuria, emaciation, enteritis, exostosis, gastritis, glossitis, hemorrhage, hemorrhoids, hay fever, incontinence, meningitis, myalgia, myelitis, neuralgia, ophthalmia, paralysis, pemphigus, psoriasis, rheumatism, scarlatina, scrofula, stammering, stiff neck, tenesmus, tibial pain, tongue, tonsillitis, tumor, typhoid, urticaria, warts, and whooping cough.[30,33]

Bittersweet contains the alkaloid solanine, which acts narcotically, and the glucoside dulcamarine. Glycoalkaloids present in the plant include: alpha-, beta-, and gamma-soladulcine (aglycone, soladulcidine [solasodan-3-ol], and alpha-, beta-, and gamma-solamarine (aglycone, delta-5-tomatiden-3 beta-ol). An isomer of gamma-solamarine, viz. gamma-solamarine, and a derhamnosyl derivative of alpha-solamarine, named delta-solamarine, have been identified. Soladulcine is tetraoside of soladulcidine, the sugars identified in the hydrolysate of the glycoalkaloid being D-xylose, L-rhamnose, D-galactose, and D-glucose. Alpha-solamarine is a trisaccharide of tomatiden-3 beta-ol, the sugar components being D-glucose, D-galactose, and L-rhamnose; the sugar components of beta- and gamma-solamarines are, respectively, D-glucose and L-rhamnose (2 mol); and D-glucose and L-rhamnose. The presence of solanine, solasonine, and solamargine in the plant has also been reported. Yamogenin, tigogenin, and diosgenin are present; the first two compounds occur in high concentrations in the inflorescence of the plant. From the roots, 15 alpha-hydroxy-soladulcidine, 15 alpha-hydroxysolasodine, 15 alpha-hydroxytomatidine, and 15 alpha-hydroxytomatidenol have been isolated. Green and yellowing fruits contain a higher percentage of glycoalkaloids than the ripe (red) fruits. As the fruit ripens, the glycoalkaloids and their aglycones tend to disappear, while the nitrogen-free sapogenins remain. Wild plants contain soladulcidine, while cultivated forms contain either soladulcidine or 5-tomatiden-3-beta-ol, or both.[1] Fruits contain lycopene. A monohydroxy lycopene, lycoxanthin ($C_{40}H_{56}O$), and a dihydroxy lycopene, lycophyll ($C_{40}H_{56}O_2$), have been identified. Resins, saponins, and tannins are also reported.[27] Leaves contain the galactosides from cholesterol, brassicasterol, campesterol, stigmasterol, and beta-sitosterol and their palmitic-acid esters.[33] Willuhn and May[300] identified from tissue cultures the 4,4-dimethylsterols cycloartenol, cycloartanol 24-dihydrolanosterol, and 24-methylenecycloartanol, and the sterols cholesterol, 24-methylenecholesterol, campesterol, stigmasterol, isofucosterol, and sitosterol. The main fatty acids of the petrolether soluble lipids of the callus are palmitic-, linoleic-, and linolenic-acid. On

p. 99 of the same journal, Willuhn et al. report still other compounds from the seeds, e.g., tigogenin, diosgenin, soladulcidine, solasodine, 31-norlanosterol, lophenol, obtusifoliol, gramisterol, and citrostadienol.

Toxicity — The FDA classifies this as an unsafe poisonous herb, containing the toxic glycoalkaloid solanine; also, solanidine and dulcamarin.[62] According to Hardin and Arena, the symptoms of solanine poisoning are headache, convulsions, cyanosis, stomachache, subnormal temperature, paralysis, dilated pupils, vertigo, vomiting, diarrhea, speech difficulties, shock, circulatory and respiratory depression, loss of sensation, and death.

To the physician — Hardin and Arena recommend "Gastric lavage or emesis; symptomatic, support respiration, paraldehyde (2 to 10 mℓ. I.M.)."[34] Treat as you would atropine poisoning.[33]

324. *SOLANUM NIGRUM* L. (SOLANACEAE) — Wonderberry, Black Nightshade, Prairie Huckleberry

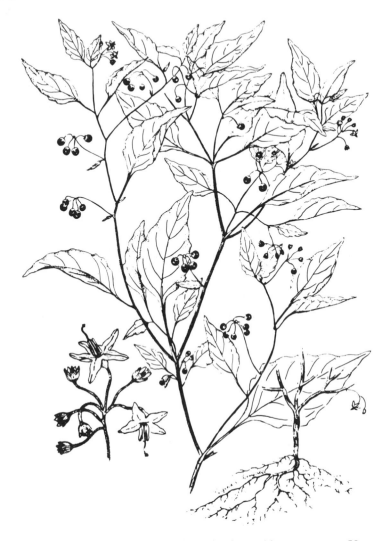

Fruits of some strains are eaten in pies and used for making preserves. Young shoots and leaves are cooked as a potherb. The black berries and whole plant are used for food in Java; however, in Holland, plants are regarded as extremely poisonous. Like many other species, this appears to have poisonous and innocuous strains, perhaps indistinguishable to the taxonomist and almost certainly liable to confuse the lay person. Like many other *Solanums*, this could be a source of the steroidal starter, diosgenin.

The juice, prepared in various manners, is said to be a folk remedy for tumors and cancer, as is the case with the leaf. A poultice of the root is said to remedy scirrhous tumors of the rectum. Medicinally, many parts of the plant are used; berries are considered diuretic and are used for eye diseases, fevers, and hydrophobia; juice of the plant is considered cathartic, diuretic, hydragogue, laxative, alterative, emollient, and is used for chronic enlargement of the liver, in blood-spitting, piles, and dysentery. Ayurvedics, considering the berries alterative, aperitif, aphrodisiac, diuretic, laxative, and tonic, use them for asthma, bronchitis, dysentery, erysipelas, eye ailments, fever, heart ailments, hiccups, inflammation, itch, leucoderma, ratbite, urinary discharges, and vomiting, believing that the berries facilitate delivery and favor conception.[26] Yunani only seem to use the fruits for fever, inflammation, and thirst, the laxative seed for giddiness, gonorrhea, inflammation, and thirst, the leaves for headache and rhinosis, and the root bark, not to be given to pregnant women, for eye, ear, nose, and throat ailments, fever, griping, hepatitis, and neck ulcers.[26] Bengalis use the berry for diarrhea, eye disease, fever, and hydrophobia. Regarded as cathartic, diuretic, and hydragogue in Bombay, the juice is prescribed in rather large, perhaps, dangerous doses (280 to 240 g) for chronic liver enlargement. In India's northwestern provinces, juice is used for dysentery, hemoptysis, and piles. Konkanese use the shoots for chronic skin diseases, like psoriasis, apparently with some success.[26] In Japan, a decoction of the plant is regarded as narcotic and antispasmodic. Young shoots are used for skin diseases and psoriasis. Leaves are an official drug in India, Africa, China, Japan, and the Philippines, and in Java are used for sore eyes in chickens. Homeopathically used for asthma, cramps, and rheumatism. On account of the solanine, used for convulsive coughs, neuralgia, and rheumatism.[33]

Per 100 g, the leaves are reported to contain 39 to 45 calories, 85.0 to 87.8 g H_2O, 3.2 to 5.0 g protein, 0.4 to 1.0 g fat, 6.4 to 8.9 g total carbohydrate, 1.1 to 2.2 g fiber, 1.6 to 1.8 g ash, 199 to 216 mg Ca, 54 to 88 mg P, 0.3 to 9.9 mg Fe, 460 to 3660 μg β-carotene equivalent, 0.12 to 0.18 mg thiamine, 0.05 to 0.24 mg riboflavin, 1.0 to 1.3 mg niacin, and 24 to 61 mg ascorbic acid.[21] Vitamin C values in India range from 11 to 40 mg for the leaves, in Pakistan 158 to 186 mg; stems 24 to 27 mg, fruits 47 to 59 mg/100 g.[1] Fruits also contain glucose and fructose (15 to 20%), beta-carotene. Seeds, comprising nearly 10% of the weight of the fresh fruits, contain 17.5% protein on a dry weight basis and 21.5% oil. The component fatty acids are 46.6% linoleic, 49.7% oleic, 1.8% palmitic, and 1.9% stearic.[1] Sitosterol and cholesterol are also reported. Fruits also contain diosgenin and ti-gonenin. Roots, shoots, and mature fruits are low in alkaloids, but the green fruits contain solanine $C_{45}H_{73}NO_{15}$ which can be separated into alpha-solanine ($C_{45}H_{73}NO_{15}$), beta-solanine ($C_{36}H_{36}NO_{11}$), gamma-solanine ($C_{33}H_{53}NO_6$), alpha-chaconine ($C_{45}H_{73}NO_{14}$), beta-chaconine ($C_{39}H_{63}NO_{10}$), and gamma-chaconine ($C_{33}H_{53}NO_6$). Solasodine, solasonine, solamargine, beta-solamargine and alpha-beta-solansodamine, -(L-rhamnosyl-D-glucosyl)-solasoidine, -solanigrine gitogenin, traces of saponins, 7 to 10% tannin.[1,33] Khanna and Rathore[301] examined diploid, tetraploid, and hexaploid populations and found significant amounts of diosgenin (0.4 to 1.2%) and solasodine (0.09 to 0.65%), the maximum occurring in the diploid.

Toxicity — Mortality or severe poisoning has been described for cattle, chickens, ducks, horses, sheep, and swine.[14] Solanine in doses of 200 to 400 mg will induce, in humans, gastroenterosis, tachycardia, dyspnea, vertigo, sleepiness, lethargy, twitching of the extremities, cramps.[33] Solanine is also said to exhibit teratogenic properties.[302] Other symptoms for poisoning by the plant include diarrhea, mydriasis, panic, excitation, coma, hyperthermia, later dazed state, paralysis, rarely fatality due to respiratory difficulty and hypothermia (more normally due to respiratory paralysis). Black nightshade can contain up to 2.5% N as NO_3 and can cause nitrate toxicity. In cattle acute nitrate toxicity leads to death, but chronic toxicity may lead to decreased milk yield, abortion, and impaired vitamin A and iodine nutrition. The proposed LD_{50} is 160 to 224 mg/NO_3/kg.[302]

325. *SOLANUM TUBEROSUM* L. (SOLANACEAE) — Potato

Roots are one of the temperate staples, eaten boiled, baked, fried, stewed, etc. Surplus potatoes are used for fodder and alcohol, and chemurgic applications. The flour can be used for baking. Potato starch is used to determine the diastatic value of starch. Boiled with weak sulphuric acid, potato starch is changed into glucose, fermented into alcohol, to yield "British Brandy". Ripe potato juice is excellent for cleaning cottons, silks, and woolens.[3] Potato wastes are taken to be 3/17ths of tuber weight, while the moisture content of the haulms (not normally gathered) are figured at 77.5%. The residue potential is calculated by multiplying tuber production by 0.2. Processing wastes at factories or at home, where peels are removed, are calculated by multiplying production by 0.1 to 0.2. Rotten potatoes and the $33^{1}/_{3}\%$ waste in the potato chip industry can be converted to butanol with a greater energy content than ethanol. Twenty percent butanol can be added to regular gas and up to 40% to diesel as extenders without engine modification. Germans developed the energetic "Fuselol" from potato alcohol.[3] Root and leaf diffusates of growing potato plants possess cardiotonic activity. Dried ethanol extracts of above-ground parts show marked hypotensive and myotropic action and a spasmolytic and soothing effect on intestinal musculature. Ethanol extracts of leaves have antifungal properties, active against *Phytophthora infestans*. Leaves, seeds, and tuber extracts show antimicrobial activity against Gram-positive and Gram-negative bacteria.

Reported to be alterative, aperient, bactericide, calmative, diuretic, emetic, lactagogue, potato is a folk remedy for burns, corns, cough, cystitis, fistula, prostatitis, scurvy, spasms, tumors, and warts. The mealy flour of baked potato is oiled and applied to frostbite.[2] The tea, made from the peels of the tuber, is said to be a folk remedy for tumors. The boiled tuber is said to remedy corns. The powdered tuber, with copper sulfate, is said to remedy callused fistulas. Europeans tie raw potatoes behind the ears for delirium.

Per 100 g, the tubers contain 76 to 82 calories, 77.7 to 79.8 g H_2O, 1.7 to 2.8 g protein, 0.1 to 0.2 g fat, 17.1 to 18.9 g total carbohydrate, 0.4 to 0.6 g fiber, 0.9 to 1.6 g ash, 7 to 13 mg Ca, 50 to 53 mg P, 0.6 to 1.1 mg Fe, 3 to 7 mg Na, 396 to 407 mg K, traces to 25 μg β-carotene equivalent, 0.07 to 0.11 mg thiamine, 0.03 to 0.04 mg riboflavin, 1.3 to 1.6 mg niacin, and 18 to 21 mg ascorbic acid. Mineral elements present are (mg/100 g): Mg, 20; Na, 11.0; K, 247; Cu, 0.21; S, 37.0; and Cl, 16.0; small quantities of iodine (11 μg/kg), Mn, and Zn are also present. Potato is among the richest foods in potassium, poorest in sodium. The more important sugars are sucrose, glucose, and fructose; some galactose, melibiose, raffinose, stachyose, planteose, myoinositol, maltotriose, manninotriose, galactinol, trigalactosyl glycerol, digalactosyl glycerol, glucosyl myoinositol, ribosylglucose, xylosyl-glucose, arabinosylglucose. Nonstarch polysaccharides include hemicellulose, cellulose, and pectic substances. The pectin content varies from 1.8 to 3.3%. The pectic substance consists of anhydrogalacturonic acid (51%) and polysaccharides (49%), composed of rhamnose (6%), fucose (0.6%), arabinose (5.6%), xylose (1.8%), and galactose (86%). The amino acid composition of potato globulin is as follows: arginine, 6.0; histidine 2.2; lysine, 7.7; tryptophan, 1.6; phenylalanine, 6.6; cystine, 2.1; methionine, 2.3; threonine, 5.9; and valine, 6.1%. The protein is somewhat deficient in sulfur amino acids and probably, also, histidine. It is rich in lysine. Potato contains, also, gamma-aminobutyric acid, alpha-aminobutyric acid, beta-alanine, and methionine sulphoxide. Other nitrogen compounds include: glutathione, choline, acetyl choline, trigonelline, cadaverine, adenine, hypoxanthine, and allantoin. Potato contains a phenolase, also called phenol oxidase, polyphenol oxidase, catecholase, and tyrosinase, which oxidizes phenols. The vitamins present in potato are, per 100 g edible material: vitamin A, 40 IU; thiamine, 0.1 mg; riboflavin, 0.01 mg; nicotinic acid, 1.2 mg; vitamin C, 17 mg; and choline, 100 mg. A small quantity of folic acid (total, 7.4 μg/100 g; free acid, 3.0 μg/100 g) is also present.[1] Linoleic acid is the predominant (41.3% of the total) acid in potato fat; other acids present are palmitic, 24.9; linolenic, 19.4; oleic, 6.4; stearic, 5.4; and myristic, 0.6%; two unidentified acids, and a few hydroxylated fatty acids are also reported to be present. Cholesterol, sigmasterol, and beta-sitosterol are present in the unsaponifiable fraction. Organic acids present in the tuber (excluding ascorbic acid, amino acids, and fatty acids) are lactic, succinic, oxalic, malic, tartaric, hydroxymalonic, citric, isocitric, aconitic, alpha-ketoglutaric, phytic, caffeic, quinic, and chlorogenic acids. Citric acid is present also in the stems and leaves. Tannins are localized in the suberized tissue of potato and are also present in potato leaves (*circa* 3.2%). Potato seeds contain 2-flavonol glycosides, kaempferol-3-diglucoside-7-rhamnoside, and kaempferol-3-triglucoside-7-rhamnoside. The flavonols, myricetin and quercitin, are also present. Volatiles from cooked potatoes include: hydrogen sulphide, acetaldehyde, methanethiol, acrolein, acetone, ethanethiol, dimethylsulphide, iso- and n-butyraldehyde, isovaleraldehyde, methylisopropylketone, etc.[1] Fresh potato tops may be used as feed for cattle and sheep. Analysis gave (dry-matter basis): total N, 1.82 to 2.30; crude fat, 3.06 to 4.63; crude fiber, 15.36 to 23.67; N-free extract, 40.46 to 50.51; and ash, 15.97 to 22.28; digestibility of protein, 60%.

Toxicity — Although the foliage is considered poisonous, some African tribes used the tip as a potherb, while others, like Mauritians, extract the green parts as a narcotic. Solanine is one toxic ingredient in the green tuber and sprouts. The "green fruit has caused fatalities"[3] . . . "potato with a green discoloration as a result of exposure to the sun, contains solanine and has been known to cause fatal poisonings."[3] There are records of severe solanine poisoning in 60 persons in Cyprus, with one death, from eating green potato shoots collected about the time of flowering and boiled *circa* $\frac{1}{2}$ hr before eating. The shoots contained *circa* 27 to 49 mg solanine per 100 g. Animals fed large residues of raw or cooked potato or "distiller's-slop" develop a disorder known as potato eruption. Symptoms in mild intoxications include a slight rise in temperature, anorexia, constipation, stiff gait, salivation, lacrimation, all preceding a vesicular inflammation on the lower part of the limbs.[3] Persons harvesting, handling, or peeling potatoes may develop allergy or urticaria.[6] Sir Walter Raleigh is said to be the first to plant the potato, near Cork, U.K., but knowing little about it, he ate the berries. Discovering their narcotic nature, he ordered his plants uprooted.[2]

326. *SOLIDAGO VIRGAUREA* L. (ASTERACEAE) — European Goldenrod, Woundwort

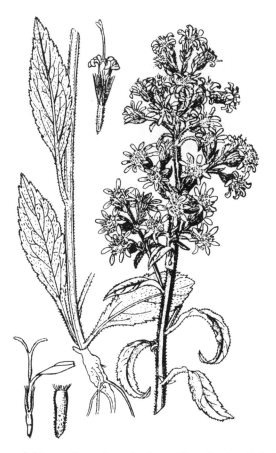

Leaves and flowers yield a yellow dye. An ingredient in the Swiss Vulnerary *faltrank*. Leaves sometimes used as a tea substitute.[1]

Used by Syrians and others for tumors, especially mammary tumors.[4] In powdered form the herb is used to cicatrize old ulcers. Herb decoction recommended as a gargle, while the powdered root is used for diarrhea and dysentery.[1] Chinese used the seed as a carminative and decoagulant and for treating cholera, diarrhea, dysmenorrhea, hemorrhages, and wounds.[16] Taiwanese use the plant for cholecystosis and nephrosis.[16] Said to be good for amenorrhea, arthritis, asthma, bladderstones, cholera, diarrhea, diptheria, dropsy, dysmenorrhea, eczema, edema, fever, flu, gastritis, gravel, headache, indigestion, kidneystones, malaria, measles, menorrhagia, nephritis, rheumatism, sore throat, spasms, uremia, whooping cough.[1,2,32,33] Antispasmodic, aromatic, astringent, carminative, diaphoretic, digestive, diuretic, expectorant, febrifugal, hemostat, lithontriptic, nervine, stimulant, sudorific, tonic, and vulnerary.[1,2,33] Homeopathically suggested for nephrolithiasis, albuminuria, nephritis, gout, hypertrophy of the prostate, and cystitis.[33]

Essential oil, mucilage, resin, saponin, and tannin (catechins) are reported. The plant produced maculopapular dermatitis with itching and burning for workers making hay for 3 to 12 hr. The irritant is said to be carried by the pollen.[6] Fresh root contains 0.8% matricariaester and inulin; kaempferol rhamnoside, quercitin, quercitrin, rutin, and astragalin occur in the leaves; flowers contain cyanidin-3-glucosylglucoside, isoquercitrin, kaempferolrhamnoglucoside, caffeic-, chlorogenic-, hydroxycinnamic-, and quinic-acids.[1]

Toxicity — Case of poisonings are reported for grazing livestock in North America.[33]

327. *SOPHORA SECUNDIFLORA* (Ortega) Lag (FABACEAE) — Mescal Bean

The dark red beans are said to be hallucinogenic,[51] used in ritual by Amerindians. The "red bean dance" of Mexico and Texas was based on this hallucinogenic bean. Necklaces made from the beans were believed to ward off evil.

Reported to be emetic, hallucinogenic, insecticidal, intoxicant, narcotic, poison, stimulant, mescal bean is a folk remedy for earache.[32] Comanche and Kickapoo used the seed infusion for ear and eye problems, perhaps reflecting the doctrine of signatures.

Per 100 g, the dry seed is reported to contain 15.6 to 22.4 g protein, 13.4 to 25.2 g fat, 73.1 to 78.0 g total carbohydrate, 15.1 to 31.1 g fiber, 3.2 g ash, 500 to 1990 mg Ca, 140 mg P, 1150 mg K.[21] The psychoactive nature is attributed to a "toxic pyridine called cytisine."[51] Schultes and Farnsworth[80] note that the beans, containing cytisine, are capable of causing death by asphyxia.

Toxicity — Classed as a narcotic hallucinogen.[54] Seeds and flowers are very poisonous.[1] Can cause nausea and convulsions, and has caused death due to respiratory failure.[1] One seed, thoroughly chewed, is sufficient to kill a child.

To the physician — Hardin and Avena suggest emesis or gastric lavage and symptomatic treatment.[34]

328. *SPIGELIA MARILANDICA* L. (LOGANIACEAE) — Pinkroot

Evergreen herb with an attractive wild flower; source of a dangerous drug. Used by Amerindians as a dangerous anthelminthic. Listed in *Merck Index*[20] as a vermifuge. In Appalachia, a tea made from the leaves is used to aid digestion. Root was used in old-fashioned malaria remedies.[18] Once used, also, for children's fever, not necessarily associated with worms. If used as a children's vermifuge, it should be followed by a saline aperient. Once recommended for convulsions. Steinmetz[27] lists the herb as cardiac, narcotic, sedative, soporific, tonic, toxic, and vermifuge. Used homeopathically for mania and strabismus.

Spigeline, tannin, resin, bitter principle, volatile oil, wax, fat, mucilage, albumen, myricin, and lignin are recorded.[2]

Toxicity — Has been used for poisoning humans, the toxic effects similar to those of strychnine.[11] Effects of spigeline are rather like those of coniine, lobeline, and nicotine. If not followed by a saline aperient, even proper doses may lead to disturbed vision (dimness), dizziness, muscular spasms, twitching eyelids, dilated pupils, facial spasms, and increased heart activity. In larger doses, circulation and respiration are depressed and muscular power diminished; there have been fatalities in children.[2,56]

329. *STACHYS OFFICINALIS* (L.) Trevisan (LAMIACEAE) — Betony

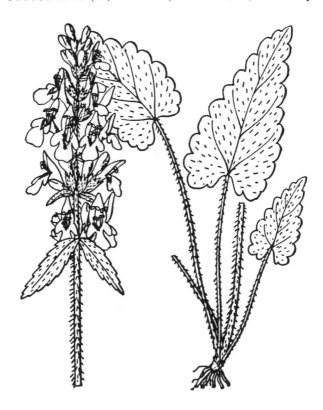

Ornamental perennial herb, once carefully planted in churchyards and hung around the neck to ward off evil spirits. A weak leaf infusion is said to be a satisfactory substitute for tea (1 pt of boiling water over an ounce of dried herbs).[2] The dried herb has been smoked like tobacco (more medicine than pleasure, with coltsfoot and eyebright for headache). Once an ingredient of Rowley's British Herb Snuff (also for headache). Fresh leaves used to dye wool a fine yellow. Leaves sometime used as a spice with meat. The tea is used as well for a gargle or mouthwash, good for the gums, mouth, or throat.[37] "The only known utility of the drug is its astringent action, due to the tannins, which makes it effective in treating diarrhea and various irritations of the mucous membranes. In normal usage, betony should not cause any notable side effects."[37]

The plants (or leaves) are used in folk remedies for genital tumors, sclerosis of the spleen and liver, morbid granulations, and wens.[4] Though claimed to cure 47 ailments in medieval times, headache was one of its prime targets. Folk remedy for ague, bladderstone, catarrh, cancer, colds, convulsions, cough, dropsy, dyspepsia, epidemic epilepsy, epistaxis, gout, headache, heartburn, hematoptysis, hysteria, kidney stones, nephrosis, neuralgia, neurosis, palpitation, palsy, sores, splenitis, stomachache, and toothache.[32,33] Used homeopathically for asthma and debility. In Denmark employed as a nervine tonic. Steinmetz calls it a brain tonic.[27]

The herbage contains *circa* 0.5% betaine with much betonicine (achillein, [−] oxystachydrine) ($C_7H_{13}NO_3$) and a little (±)-stachydrine ($C_7H_{13}NO_2$), and turicine (+)-oxystachydrine ($C_7H_{13}NO_3$). Also, 0.5% 4,1-caffeic acid, about 15% tannins, choline, and other alkaloids with traces of essential oil. May also include chlorogenic-, neochlorogenic-, 4-caffeic-, and rosmarinic-acids; harpagide. A hypotensive glycoside is also reported.

Toxicity — Overdosing may result in excessive irritation of the stomach.

330. *STELLARIA MEDIA* (L.) Vill. (CARYOPHYLLACEAE) — Chickweed

More regarded as a weed than a useful plant, chickweed does find its way into a few pots for food or folk medicine. Obese Americans like myself might profit by substituting a gram of chickweed for a gram of chicken fat. Seed has been used as birdfood (said to produce gastrointestinal problems if lambs eat too much). Cosmetologists might view (with wholesome scepticism) the suggestions in Culpepper's herbal that the juice corrects various facial imperfections.

INTERIOR'S SUPERIOR CRITERIA

In weeding my garden
My arteries hardened
When spoke to me the Park Ranger
"Poison Ivy and Dodder,
Though good for goat fodder,
Are close kin to species in danger

"You may kill any weed
Just as long as you heed
And don't hurt an endangered species
But if you're in doubt
You can't rip it out"
Until your Park Ranger, he sees

It's not only illegal
To shoot at the eagle
Who's eating the chickens your family needs.
It's also immoral
To engage in the quarrel
What's better, the chickens or chickweeds

"Chickweed's in danger
Says Interior's Ranger
"And you can't hurt an endangered species
Without identification"
There's procrastination
Until your Ranger, he sees, or she sees. . .or fe-fi-fo-fum — fee sees!*

*   J. A. Duke, 1975.

Reported to be demulcent, depurative, diuretic, emmenagogue, expectorant, galactagague, poison, refrigerant, chickweed is a folk remedy for asthma, blood disorders, cancer, conjunctivitis, constipation, cough, dyspepsia, eczema, elephantiasis, erysipelas, fever, fractures, hemorrhoids, hoarseness, hydrophobia, infection, inflammation, obesity, ophthalmia, skin ailments, sores, spasms, swellings, tuberculosis, and urogenital ailments.[16,32,37] Ainu use the plant to treat bruises. Chinese use a sugar infusion for epistaxis.[16] The plant is poulticed onto abscesses, boils, ulcers, and other ailments. Homeopaths use it for arthritis, gout, and rheumatism.

Per 100 g, the leaves are reported to contain 91.7 g H_2O, 1.2 g protein, 0.2 g fat, 5.3 g total carbohydrate, 1.7 g fiber, 1.6 g ash, 0.02 mg thiamine, 0.14 mg riboflavin, 0.51 mg niacin, and 375 mg ascorbic acid.[21] Seeds contain 4.8 to 5.9% oil, *circa* 18% protein,[21] 0.15 to 0.55% ascorbic acid. Leaves contain 0.1 to 0.15% vitamin C, rutin, and various acids (octadecatetraenic and linolenic), alcohol and esters (hentriacontanol, cerylcerotate).

Toxicity — Grazing animals have experienced nitrate poisoning.

331. *STILLINGIA SYLVATICA* (L.) Mull. (EUPHORBIACEAE) — Queen's Root

Medicinal plant of southeastern U.S.[2] Reported to be alterative, cathartic, depurative, diuretic, emetic, expectorant, laxative, sialogogue, and tonic, queen's root is a folk remedy for bronchitis, cancer, croup, dysmenorrhea, hematosis, laryngitis, leucorrhea, puerperium, scrofula, skin ailments, syphilis.[32,33] Homeopaths use in secondary and tertiary syphilis.

Per 100 g, one species is reported to contain, on a zero-moisture basis, 21.7 g protein, 3.2 g fat, 63.4 g total carbohydrate, 13.9 g fiber, 11.7 g ash, 310 mg Ca, 240 mg P, 2590 mg K.[21] Fresh root contains stillingine and resin,[56] 3 to 4% essential oil, a glycoside, fatty oils, resinic acid, silvacrol, tannin, starch, calcium oxalate. Seeds contain *circa* 30% fatty oil.[33] The poisonous principle in queen's delight, *S. treculeana* (Muell. Arg.) Johnst., is hydrocyanic acid.[14]

Toxicity — Overdoses may cause vertigo, burning mouth, throat, and gastrointestinal tract, diarrhea, nausea, vomiting, cough, depression, dysuria, aches and pains, itching and eruptions, perspiration, and fatigue.[56] Sap from the root inflames the skin and produces swellings.[33]

332. *STRYCHNOS NUX-VOMICA* L. (LOGANIACEAE) — Nux-Vomica, Strychnine

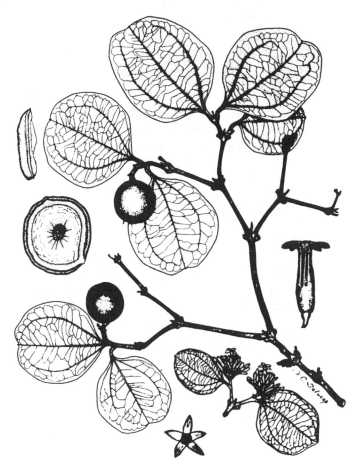

Brucine has been used to denature alcohol for cosmetics and perfumery. Nux-vomica is important as a source of the medicinal alkaloids brucine and strychnine. "Strychnine causes the sense organs to become more sensitive, it stimulates the visceral organs and the urogenital tract, it is ingested to treat neurasthenia and as an aphrodisiac."[16] Strychnine increases the reflex excitability of the spinal cord and the medullary centers; therapeutic doses produce a tonic effect on the alimentary canal, and a limited amount of respiratory and vasomotor stimulation. In spite of its toxicity, strychnine or nux-vomica finds its way into at least two "aphrodisiacs" on the market, one with yohimbine hydrochloride and methyl-testosterone, the other with caffeine and theophylline.[11] Back in 1927, after a long battle with predators like badgers, coyotes, skunks, snakes, and weasles, Californians had really cut down on them. Coupled with bumper crops, the lack of predators led to a mouse explosion, seemingly unprecedented, slickenening the highways, infesting bed covers. The U.S. Biological Survey's Pied Piper, S. E. Piper, set to work with 40 tons of strychnine-laced alfalfa. "Within a month, the mouse population had plummeted, and human beings had once more subdued the animal kingdom."[304] Had the mice known about the 5.4% methyl eugenol in California bay oil, they could have saved themselves; it prevents the death of mice treated with lethal convulsant doses of strychnine.[29] One avant garde publication states that the seeds "have occasionally been ingested in minute quantities for their stimulating effect upon the spinal cord and cerebrum" . . ."It has been found to enhance the learning process."[51] Wood is used for agricultural tools.

Shoots are used in folk remedies for tumors in India, the seed for cancerous sores in China.[4] Reported to be abortifacient, aphrodisiac, CNS-stimulant, laxative, nervine, POISON, stimulant, stomachic, tonic, and vasoconstrictor, nux-vomica is a folk remedy for ague, alcoholism, amblyopia, anemia, arthritis, asthma, cancer, cholera, chorea, colic, cynanche, diptheria, dyspepsia, dysentery, epilepsy, fever, itch, jaundice, laryngitis, larygeal paralysis, lumbago, malaria, myocarditis, neuralgia, neurasthenia, ophthalmia, paralysis, polio, rheumatism, snakebite, sores, sore throat, swellings, tumors, typhoid, and ulcers.[16,32,33] Also, used in homeopathy. The seed is prescribed as bitter stomachic, nerve tonic, and spinal stimulant for abdominal enlargements, ague, fever, laryngitis, laryngoparalysis, and other throat ailments. Seed powder blown into the throat for cynache. Introduced to the vagina for abortion in China. Kills parasites, heals feverish stomach, draws small wounds together.[41]

Per 100 g, the fruits of related "edible" species are reported to contain *circa* 72 calories, 70 to 80 g H_2O, 1.6 g protein, 0.6 g fat, 17.1 g total carbohydrate, 0.6 g fiber, 1.0 g ash, 28 mg Ca, 42 mg P, 0.7 mg Fe, 0.11 mg thiamine, 0.17 mg riboflavin, 1.9 mg niacin, and 18 mg ascorbic acid (data mostly from *Strychnos spinosa*).[21] Nux-vomica is said to contain *circa* 1.2% strychnine, 1.6% brucine (the sulphate of which was selling for about \$170/kg, December 31, 1982), struxine, vomicine, alpha- and beta-colubrine, loganin, and chlorogenic acid.[17] *Hager's Handbook* adds stigmasterol, cylcoartenol, beta-amyrin, choline, a mucilage which hydrolizes to 5:2:1:1 galactose, mannose, xylose, arabinose. Seed contains 4 to 5% oil with 0.9% $C_8C_{10}C_{14}$, 12.6% C_6, 6.6% C_{18}, 7% C_{20}, 1.7% C_{22}, 62% oleic-, and 9% linoleic-acids.[33] Leaves contain 1.6% brucine + strychnine, 0.025% strychnicine.[1]

Toxicity — Even *Herbal Highs* handles this one carefully, "the margin between useful and dangerous (usually lethal) doses is very narrow."[51] Fatal to man at doses of 30 to 90 mg.[1,11] Toxic doses cause characteristic tetanus, spasmodic respiration, violent changes in blood pressure. Death occurs from asphyxia and the paralysis which follows the stimulations.[41]

To the physician — Treatment should be aimed to prevent convulsions (rectal sodium bromide, oral chloroform or barbituates, chloroform inhalation) and to support respiration (endotracheal intubation followed by curariiform drugs in small doses). Gastric lavage to remove any residual poison. A 1:1000 potassium permanganate solution is an effective antidote.[1]

333. *STYRAX BENZOIN* Dryander (STYRACACEAE) — Benzoin, Sumatra Benzoin, Styrax

Benzoin is a gum exuded when the trunk is wounded. The gum is used in the pharmacopeias. In the U.S. Sumatra benzoin (*S. benzoin* and *S. paralleoneurus*) is more customarily used in pharmaceutical preparations, Siam benzoin (*S. tonkinensis* et al.) in the flavor and fragrance industries, e.g., in alcoholic beverages, baked goods, beverages, candy, desserts, gelatins, and puddings, the highest average maximum level *circa* 0.014% in baked goods and candies. It is used also as an incense, and preservative for fats.[29] Perhaps its most important pharmaceutic role is in the "Compound Benzoin Tincture", which also contains aloe, balsam tolu, storax, etc.[29] Benzoin is sometimes added to cigarettes as a flavoring.

Reported to be antiseptic, carminative, deodorant, disinfectant, diuretic, expectorant, insecticidal, sedative, stimulant, and vulnerary, benzoin is a folk remedy for arthritis, bronchitis, cancer, catarrh, cold sores, colic, cough, herpes, laryngitis, ringworm, shingle, spermatorrhea, and wounds.[4,17,32,38] Benzoin vapors are expectorant, the tincture antiseptic.[29] Malayans use it for cracked feet and circumcisions. Under the name "virgins milk" it has been used to heal cracked nipples; also, in feminine hygiene.

Contains 70 to 80% coniferyl cinnamate, cinnamyl cinnamate (styracin), and coniferyl benzoate; *circa* 10% free cinnamic acid; some benzoic acid, traces of benzaldehyde, vanillin and styrene. Another analysis suggests that the benzoin contains *circa* 90% resinous matter composed mainly of sumaresinoic acid and coniferyl cinnamate, with 10 to 20% benzoic acid and 10 to 30% cinnamic acid; other constituents being 2 to 3% phenylpropyl cinnamate, 1% vanillin, traces of cinnamyl cinnamate, styrene, and benzaldehyde.[29]

Toxicity — Benzoin is regarded as moderately toxic.[29] Heated benzoin gives off a white vapor which may induce coughing; powdered resin may induce sneezing. Approved for food use (§ 172.510).[29]

334. *SYMPHYTUM PEREGRINUM* Ledeb. (BORAGINACEAE) — Comfrey, Russian Comfrey, Quaker Comfrey

Comfrey is mainly cultivated as a green forage, and can be used with maize to make ensilage. Comfrey is difficult to conserve because of the low amounts of carbohydrates. Here in Maryland it is one of the last green snacks in fall and the first in spring for my vegetarian goats! It is also used as a green vegetable, some cooking it in two waters and serving it with chives. Proofreading this on the first day of November, I decided to try it again, and prefer this potherb mixture to pure spinach. Only small and medium leaves are used. They may be boiled or steamed for 2 min, and then eaten in salads or with a sauce. Without saying so herself Rose says, "Comfrey can be said to be the world's fastest protein builder and an absolute must in any vegan's diet as it is one of the few plants that can produce vitamin B12 from the cobalt in the soil."[49] Comfrey is an ingredient in various herbal, pharmaceutical, and cosmetic preparations such as creams, eyedrops, hair products, lotions, and ointments.[29] It is said to be antiinflammatory. The mucilage is thought to soften the skin when used in baths. "Whatever healing properties comfrey may have are probably caused by its content of allantoin, an agent which promotes cell proliferation."[37] "As a cosmetic and bath herb, with continuous use, it regenerates aging skin."[47]

Plant contains allantoin, used in some face creams, or to cure scours in pigs and calves or give a bloom to horses. For humans, it is reported to be good for asthma, whooping cough, stomach and duodenal ulcers, and lung ailments. Gerard's herbal is quoted in the *New Scientist*[305] on multicolored comfrey, variously known as Abraham, Isaac and Jacob, Pigweed, Suckers, or Church Bells. "The slimie substances of the roots made in a posset of ale and given to drink against the pain in the back gotten by any violent motion such as wrastling or overmuch use of women, doth in foure or five days perfectly cure the same, although involuntarie flowing of seed in men be gotten thereby." Comfrey is said to be alterative, astringent, demulcent, emollient, expectorant, hemostat, nutritive, and vulnerary. The root decoction is used as a mouthwash or gargle for asthma, bleeding gums, hoarseness, sore throat, and stomatitis. It is also used for arthritis, bronchitis, bloody urine, cholecystitis, cough, dysentery, dysmenorrhea, enterorrhagia, gallstones, gastritis, gout, hematochezia, hematuria, hemoptysis, hepatosis, internal ulcers, leucorrhea, metrorrhagia, rheumatism, scrofula, tonsilitis, and ulcers of the kidney.[32,47] In a vague reference Hutchens[30] quotes, "Numerous uncontradicted reports of lung cancer cured where all other means have failed and in which the sole treatment consisted of infusion made from the whole green plant and, even in some instances, of infusion made from the powder of the entire plant." Interestingly, comfrey does contain beta-sitosterol which shows anticancer activity against the Lewis Lung Carcinoma Adenocarcinoma 755 and Walker Carcinosarcoma 256 tumor systems. Homeopathically, comfrey is prescribed for abscess, bone cancer, breast, enlarged glands, eye pain, fracture, gunshot wounds, hernia, menstrual arrest, sexual abscess, sprains, and wounds.[30] Comfrey and alfalfa increase the activity of aminopyrine N-demethylase. Allantoin, present in comfrey leaves and roots, is said to be a cell proliferant, making the edges of wounds grow together and healing sores. Rose has some rather positive comments: "Eat the raw leaves daily for an ulcer or use a powder of the leaves in soup. As a douche it will cleanse the vagina and cure the whites. A leaf compress can be used for sore, swollen breasts."[47] Parvati adds that "Chicana midwives apply this mending herb as a poultice for vaginal tears."[48] For yeast infection (vaginitis) Parvati recommends a douche with *circa* 4 tablespoons of comfrey, goldenseal, and sage, 2 tablespoons of acidophilus, and 2 tablespoons cidar vinegar in 1 qt water. Steep the herbs for 15 to 20 min, strain, and add the acidophilus and vinegar. "Douche remembering to use light pressure so as not to push any organisms up into your urethral canal and bladder."[48] I would not give any of these herbs a completely clean bill of health.

Per 100 g, comfrey is said to contain 0.5 mg thiamine, 1.0 mg riboflavin, 5.0 nicotinic acid, 4.2 mg pantothenic acid, 0.07 mg vitamin B12 (rare in vegetarian diet), 28,000 IU vitamin A, 100 mg vitamin C, 30 mg vitamin E, and 0.18 mg allantoin. So-called Russian comfrey (84.93% water) analyzed on a zero-moisture basis 22.73% crude protein, 5.39% ether extract, 4.22% N-free extract, 21.25% ash, 5.35% silica, 2.02% Ca, and 0.57% P. Based on 16 analyses, prickly comfrey ranged from 83.1 to 89.8% H_2O, and on a zero-moisture basis 16 to 23% protein, 1.0 to 2.8% ether extract, 11.9 to 18.2% crude fiber, 12.4 to 22.1% ash, and 43.2 to 49.1% N-free extract. The root contains 0.60 to 2.55% allantoin and about 0.3% alkaloids (symphytine, echimidine), lithospermic acid (said to be antigonadotrophic), 29% mucopolysaccharide (of glucose and fructose). There is a gum consisting of L($-$)-xylose, L-rhamnose, L-arabinose, D-mannose, and D-glucuronic acid. Also, reported are 2.4% pyrocatechol tannins, 0.65% carotene, glycosides, isobauerenol, beta sitosterol, stigmasterol, steroidal saponins, triterpenoids (e.g., isobanerenol), choline, consolidine, silicic acid, lasiocarpine, viridiflorine, echinatine, and heliosupine-N-oxide.[33] "Comfrey (*Symphytum officinale* L.) has considerable tannin in the leaves and the infusion of the dried leaves is very astringent. Comfrey root possesses 2.4% tannin . . .used in Germany for tanning leather. The above-ground plant contains the alkaloid *lasiocarpine* . . . consolidine and the N-oxide of heliosupine. Additional alkaloids have been found in the root . . . The acid fraction of an aqueous extract of the plant has shown antigonadotrophic activity in mice."[50] Tannin-containing extracts from several plants produced either sarcomas or liver tumors.[10] Tannic acid is hepatotoxic and several workers report that tannic acid or tannin extracts from plants were oncogenic in animals. *Lasiocarpine* is reported as an oncogenic compound. Based on these cancer treatment reports,[10] one might conclude that comfrey contains two oncogenic or tumor-inducing substances.

Toxicity — ''Both root and leaf of this plant have been shown to be carcinogenic in rats when fed in concentrations of as little as 0.5% and 8% of their diet, respectively. Eight pyrrolizidine alkaloids have been identified in the closely related Russian comfrey, *Symphytum* × *uplandicum* Nym. Tests have shown that these combined alkaloids cause chronic hepatotoxicity in rats and raise concern regarding human consumption of the plant.''[37] I really prefer my chives without the comfrey, even if the comfrey had an absolutely clean bill of health. I fear I am not competent to weigh the risks of tannin and lasiocarpine against the benefits of allantoin.

335. *SYMPLOCARPUS FOETIDUS* (L.) Nutt. (ARACEAE) — Skunk Cabbage

Used for food, after elaborate preparation, by the Amerindians. Seeds said to be narcotic.[33] Reported to be anodyne, antispasmodic, diaphoretic, emetic, emmenagogue, expectorant, narcotic, poison, sedative, sialogogue, stimulant, styptic, sudorific, skunk cabbage is a folk remedy for asthma, catarrh, cancer, chorea, convulsions, cough, dropsy, epilepsy, headache, hemorrhage, hysteria, itch, parturition, pregnancy, rheumatism, ringworm, snakebite, skin, sores, spasms, splinters, swellings, toothache, wounds, and worms.[32,33]

Contains starch, gum-sugar, fatty oil, essential oil, tannin, iron. Leaves contain *n*-hydroxytryptamine (the leaf extract is hemolytic). The root contains crystals of calcium oxalate.

Toxicity — Narcotic[2] — large doses cause nausea, vomiting, headache, vertigo, and dimness of vision. Contact with the plant has been reported to cause intolerable itching and inflammation of the skin.[6]

336. *SYZYGIUM AROMATICUM* (L.) Merr. & Perry (MYRTACEAE) — Cloves, Clavos

Cloves of commerce are the dried, unexpanded flower-buds with a lower nail-shaped portion consisting of the calyx-tube enclosing the upper half of the ovary, the four calyx teeth surrounding the unopened globular petals and stamens. Cloves are used as a condiment or spice in cordials, soups, sauces, tobaccos, masticatories, curries, pickles, preserves, desserts, cakes, and puddings. In Indonesia, cloves are used to make special cigarettes which crackle when burning. Cloves have been used in both alcoholic and nonalcoholic beverages, e.g., Benedictine and cola. Ground cloves enter many spice mixtures, curry powders, pumpkin-pie spice, and sausage seasonings. Whole or ground cloves are used in sachets, pomanders, and potpourris. Clove oil, clove-stem oil, and clove-bud oil, obtained by steam distillation, are used in soaps, insect repellents, perfumes, mouthwashes, medicines, and as an antiseptic. They contain eugenol, which is important in the manufacture of synthetic vanilla. Eugenol is mixed with zinc oxide and used as the temporary filling to disinfect root canals from which the pulp has been removed prior to permanent restoration, mixed with catnip, ground cloves with sassafras are applied as a poultice to aching teeth.[11] Oil also used as a clearing agent in biomicroscopy. It is bactericidal. Takechi and Tanaka report the antiviral substance, eugeniin, from the buds.[306]

Reported to be analgesic, anesthetic, anodyne, antidotal, antioxidant, antiperspirant, antiseptic, bactericidal, carminative, deodorant, digestive, disinfectant, rubefacient, stimulant, stomachic, tonic, and vermifuge, cloves is a folk remedy for abdominal problems, callus, cancer, caries, cholera, cough, diarrhea, dyspepsia, enterosis, gastritis, heart, hernia, hiccups, nausea, parturition, polyps, sores, spasm, sterility, toothache, uteropathy, warts, worms, and wounds.[4,32,33] In China, crushed flower buds have been used for nasal polyps, in Malaya for callous ulcers, in California for warts.[4] Sold in oriental bazaars as a carminative and stimulant, and to relieve the irritation of sore throat. Clove oil is widely used for toothache. Clove oil is locally irritant and stimulates peristalsis. A powerful antiseptic, perhaps dangerously so, it has been applied as a local anesthetic for toothache. Has been used as an expectorant in bronchitis and phthisis. As an aromatic, powdered cloves or an infusion thereof has been given for emesis, flatulence, languid indigestion, and dyspepsia.

Per 100 g, the ground cloves are reported to contain 323 calories, 6.9 g H_2O, 6.0 g protein, 20.1 g fat, 61.2 g total carbohydrate, 9.6 g fiber, 5.9 g ash, 646 mg Ca, 105 mg P, 8.7 mg Fe, 264 mg Mg, 243 mg Na, 1102 mg K, 1.0 mg Zn, 530 IU vitamin A, 0.12 mg thiamine, 0.27 mg riboflavin, 1.46 mg niacin, 80.1 mg ascorbic acid, 0 mcg vitamin B_{12}, and 2.56 mg phytosterols. Campesterol, crataegol acid, sitosterols, and stigmasterol are reported from clove bud extracts. From an essential-oil point of view, dried cloves contain 4.0 to 10.0% nonvolatile ether extract, 5.0 to 8.0% total ash, 0.1 to 0.6% acid-insoluble ash, not less than 15% volatile oil, less than 8% moisture, and 10 to 13% crude fiber.[193] Trace element composition is 0.7 to 0.8% Ca, 0.1% P, 0.02 to 0.3% Na, 1.0 to 1.3% K, 90 to 150 ppm Fe, <0.05 ppm Co, 0.25 to 0.3 Mg, <0.2 ppm Pb, <0.5 ppm F, 0.2 ppm V, <0.04 ppm Se, <0.05 ppm Cd, <0.05 ppm Hg, 80 to 160 ppm Al, 20 to 30 ppm Ba, 30 to 60 ppm Sr, 10 to 40 ppm B, 2 to 5 ppm Cu, 4 to 25 ppm Zn, 170 to 410 ppm Mn, and 3 to 5 ppm.[193] Many compounds have been identified in the essential oil, among them 2-heptanone, ethyl hexanoate, 2-heptanyl acetate, 2-heptanol, 2-nonanone, methyl octanoate, ethylooctanoate, 2-nonanyl acetate, alpha-cubenene, copaene, 2-nonanol, linalool, 2-undecanone, caryophyllene, methyl benzoate, ethyl benzoate, alpha-humulene, methyl chavicol, styralyl acetate, alpha-terpinyl acetate, alpha-amorphene, benzyl acetate, alpha-muurolene, carvone, zonarene, gamma-carinene, phenyl ethyl acetate, calamenene, *trans*-anethole, benzyl alcohol, calacorene, caryophyllene oxide, methyl eugenol, 2-heptanyl benzoate, humulene epoxide, cinnamaldehyde, benzyl tiglate, ethyl cinnamate, eugenol, 10-alpha-cadinol, eugenyl acetate, chavicol, and humulen-7-ol.

Toxicity — Clove bud oil is reported to have an oral LD_{50} of 2650 mg/kg body weight in rats (equaling that of the major ingredient, eugenol, which sensitizes some people, causing contact dermatitis).

337. *TABEBUIA* SPP. (BIGNONIACEAE) — Pao D'Arco

Timbers of many species are used, though "axe-breakers". Sawdust of *T. avellanedae* is resistant to termites, perhaps due to deoxylapachol. Recently, there has been a flurry of interest in "herbal tea" made from ipe roxo, lapacho, and/or pao d'arco.

According to Hartwell,[4] the tea of the bark is used in folk remedies for adenocarcinoma (pancreas), cancer of the esophagus, head, intestines, lung, prostate, and tongue, Hodgkin's disease, leukemia, and lupus.[4] Reported to be alexiteric, analgesic, anodyne, antidotal, diuretic, and fungicidal, *Tabebuia* species show up in folk remedies for boils, chlorosis, diarrhea, dysentery, enuresis, fever, pharyngitis, snakebite, syphilis, and wounds.[32] The pao d'arco and ipe roxo popularity stems to such species as *T. avellanadae, T. heptophylla, T. impetiginosa,* and *T. ipe.* Many Latin American species are used in folk medicines. According to Morton,[42] the leaf decoction of *T. heterophylla* is used in the Bahamas for backache, dysuria, gonorrhea, and toothache; the bark decoction for enuresis and incontinence; leaves are boiled with other species, *Bourreria, Bursera, Cassytha, Guettarda,* and/or *Smilax,* as an aphrodisiac. Costa Ricans take the bark decoction of *T. rosea* for colds, fever, and headache; the flowers, leaves, and shoots for snakebite. Guatemalans give the bark decoction to dogs to protect them from rabies; Mexicans take the root decoction for anemia, the leaf-and-bark decoction for fever.[42] *T. serratofolia* has been used traditionally for cancer in Colombia, the activity probably due to quinone. The following "quote" got me in trouble: "American herbal medicine experts Dr. James Duke, of the National Institute of Health, and Dr. Norman Farnsworth, of the University of Illinois, confirm Martin's claims 'Taheebo undoubtedly contains a substance found to be highly effective against cancers.' Farnsworth told GLOBE." I quote from the *Globe* newspaper as it was printed (September 15, 1981, *Globe*).[307] Shortly thereafter, *Tabebuia* was being promoted under the bold headlines, "1,000-year-old Inca cancer cure works". In the PR material the three words "Farnsworth told GLOBE" has been deleted. I imagine that Dr. Farnsworth was as badly represented by the quote as I was, but note how the omission of three little words made it look as though Duke and Farnsworth were being quoted, not Martin. Perhaps similar quotes may be taken from this book, but neither I, the editors, CRC Press, NCI, or the USDA are endorsing any herbal remedy for cancer. I am merely compiling information on biological activity of some plants and plant-derived compounds. Even though some folk cancer plants do contain biologically active compounds with antitumor compounds, we do not endorse herbal medication. Lapachol was listed recently by Perdue[308] among the most important antitumor agents from plants. There are many folk uses reported in the popular press. Californians tell me in 1983 that they are drinking pao d'arco for fungal infections and applying the decoction (one spoon soaked in a quart of water overnight) locally to *Candida*. They report favorable results. Personally, I would not hesitate to apply such a decoction or drink so dilute a tea. But, if there are data to show long-term effects, positive or negative, I do not know them. Xyloidone, as noted, is active against *Candida*.

According to *Hager's Handbook*,[33] the wood of *T. ipe* contains *circa* 3.7% lapachol ($C_{15}H_{14}O_3$); the essential oil ranges from 0.55 to 1.49% (predominantly sesquiterpenes), resin 3.3 to 4.5%, waxy matter with ceryl alcohol and lignoceric acid 0.95 to 1.18%, lactone bitter substances 0.85 to 1.4%, glycosidol bitter substances 0.025 to 0.042%, 12.2 to 17.8% tannin, yielding protocatechuic acid, and 3 to 4% acids and neutral saponins; wood also contains napthaquinones and anthraquinones. The latter might explain its folk usages for psoriasis. According to Prakash and Singh,[309] the stembark of *T. pentaphylla* contains lapachol, nonacosane, dehydrotectol, and beta-sitosterol, the roots hexacosane, dehydrotectol, beta-sitosterol, and oleanolic acid. Some species contain xyloidone, which is active against *Brucella* and *Candida*. Much of the activity may be traced to the lapachol. Lapachol has shown antimalarial activity in animals.[309] It is known to uncouple oxidative phosphorylation. Lapachol, active against Gram-positive and acid-fast bacteria, as well as fungi, was once of great interest to the NCI. Lapachol was found to have anticancer activity. Filed in 1967, lapachol was dropped from NCI investigations because of therapeutic inactivity. Reported by Hartwell[4] from *Stereospermum suaveolens*, lapachol occurs in several species of *Tabebuia*, e.g., *T. rufescens* and *T. serratifolia*. Old correspondence from Hartwell suggests that lapachol has also been isolated from woods of taigu, greenheart, lapacho, mao, ipedo campo, ipeamarillo, ipe tabaco, mostly species of *Bignonia*, *Tabebuia*, and *Tecoma*. In my "Phytotoxin Tables",[15] I cited *Avicennia*, *Bassia*, *Bignonia*, *Paratecoma*, *Tabebuia*, *Tecoma*, and *Tectona* with questionable citations for *Adenanthera*, *Andira*, and *Intsia*. Brazilian species of *Tabebuia*, known as pao d'arco and lapacho, have gotten big press sporadically over the last 50 years as cancer remedies. Thus, *Stereospermum* and *Tabebuia*, classical sources of lapachol, both have folk histories as "cancer remedies". I have received unsolicited testimonials from "recovered patients" supposedly cured by pao d'arco. Recently, I submitted an unvouchered sample of ipe roxo ("*Tecoma curialis*") to the NCI for screening. This is also labeled pao d'arco herbal tea. There are no cancer claims on the label, but the pao d'arco has gotten enough press to generate a North American interest in lapachol, dropped by NCI because of "no therapeutic effect." Hartwell[4] notes that lapachol was carried into clinical trial because "of its high Walker 256 activity even when given orally." Lack of toxicity permitted large oral doses but sufficiently high blood levels could not be obtained to show a therapeutic effect. While lapachol was inactive against L1210 leukemia, the sodium salt of lapachol was active.

Toxicity — At high oral doses, lapachol causes nausea, vomiting, and a reversible prolongation of the prothrombin time, the latter effect possibly due to lapachol's structural similarity to vitamin K. Clinical studies were discontinued not only because of anticoagulant effects, but also because therapeutic plasma levels of >30 µg/mℓ of lapachol could not be achieved without encountering toxicity.[310]

338. *TABERNANTHE IBOGA* Baill. (APOCYNACEAE) — Iboga

The yellowish root is the hallucinogen, iboga, an African narcotic of growing social importance, especially in Gabon and vicinity.[58] In their very interesting "Ethnomedical, Botanical and Phytochemical Aspects of Natural Hallucinogens",[80] Shultes and Farnsworth "note that of more than 200 species of hallucinogenic plants only two are legally prohibited from use in the United States by Federal Law — Cannabis sativa and Tabernanthe Iboga."

Reported to be apertif, aphrodisiac, CNS-stimulant, hallucinogenic, stimulant, and tonic, iboga is a folk remedy for convalescence, debility, fever, hypertension, neurasthenia.[32,33,58]

Indole alkaloids comprise up to 6% of the dried root. Ibogaine ($C_{20}H_{26}N_2O$), the main one, is a cholinesterase inhibitor, stimulating appetite and digestion, as well as hypotension,[58] a strong CNS-stimulant.[80] Other alkaloids include ibogamine ($C_{19}H_{24}N_2$), tabernanthine ($C_{20}H_{26}N_2O$), ibagamine ($C_{21}H_{28}N_2O_2$), iboluteine ($C_{20}H_{26}N_2O_2$), desmethoxyiboluteine ($C_{19}H_{24}N_2O$), hydroxyindoleninibogamine ($C_{19}H_{24}N_2O$), hydroxyindoleninibogaine ($C_{20}H_{26}N_2O_2$), ibochine ($C_{20}H_{24}N_2O_2$), iboxygaine ($C_{20}H_{26}N_2O_2$), kimvuline ($C_{20}H_{26}N_2O_2$), kisantine ($C_{21}H_{28}N_2O_3$), gabonine ($C_{21}H_{28}N_2O_4$), and voacangine ($C_{22}H_{28}N_2O_3$). Tannin is also reported.[33]

Toxicity — Crediting the plant with ibogaine and ibogaine-like alkaloids, Emboden lists it as a narcotic hallucinogen.[54] The unworldly hallucination often results only with doses large enough to cause death. Ibogaine causes visual and other hallucination, often associated with severe anxiety and apprehension, leading in toxic doses to convulsions, paralysis, and, finally, arrest of respiration.[58] Brachycardia and hypotension may result as well.

To the physician — The alkaloids are said to be antagonistic to reserpine.[33]

339. *TANACETUM VULGARE* L. (ASTERACEAE) — Tansy

Supposed by the ancients to impart immortality, the herb was used for embalming. As a strewing herb, tansy was dedicated to the Virgin Mary. Mixed with elderberry leaves, tansy is supposed to repel flies. Russians use powdered tansy as an insecticide. Tansy extracted in distilled water deters feeding of some lepidopterous larvae.[311] If meat is rubbed with tansy, flies won't bother it. Oil of tansy rubbed on the skin is supposed to repel insects. Tansy is one old cancer folk remedy that showed some activity in the NIH cancer program. Although poisonous, tansy finds its way into omelets, puddings, and herbal teas. Fresh tender young leaves are used sparingly as a spice to flavor an omelet, a baked fish, or in meat pie. Tansy "tea" is a bitter beverage brewed from either fresh or dried leaves and tops, and is said to have a calming effect on the nerves. Leafy tips are used in the preparation of cosmetics and ointments. The essential oil has been used in perfumery. "A tea made from the leaves and flowering tops is probably sufficiently dilute to permit ingestion."[11] Cows and sheep eat tansy, but horses, goats, and hogs are said to refuse it. My rabbit and goats certainly turned it down. Said to repel deer, e.g., if planted around a poisonous yew tree, tansy can be planted to repel ants as well as flies. Tyler's stance is not quite so positive. "Since far more effective insect repellants are readily available, there is no real reason to use tansy for anything; well perhaps there is just one. Tansy is used as a flavoring agent in certain alcoholic beverages, including Chartreuse, but the resulting product must be thujone-free."[37] It is the source of green dye. In Maine, tansy water is poured into milk before making the curds of tansy cheese. Sometimes dangerously mixed in with green salads and potherbs, cheeses, salad dressings, omelets. Once used as a pepper substitute. The flowers and foliage do well in dry bouquets. Oil is of some use in the perfume industry.

Medicinally, tansy leaves and flowers have stimulating and tonic properties. An oil obtained from the flowers is used primarily as an anthelmintic. The unguent made from the leaves is said to be a folk remedy for tumors in the tendons. The root is said to be good for gout, yet a tea of dried flowers and seeds is used in Scotland. Russians steep the flowers in vodka for stomach and duodenal ulcers. Leaves, placed in the shoes, are said to cure ague. It is reportedly used for amenorrhea, bruises, burns, cholecystosis, cold, dropsy, epilepsy, fever, freckles, gout, hepatosis, hysteria, kidney problems, nerves, rheumatism, sores, spasms, sprains, swellings, tuberculosis, and is considered abortifacient, antibiotic, anthelmintic, antioxidant, antiseptic, ascaricide, bactericide, cordial, diaphoretic, emmenagogue, narcotic, nervine, pediculicide, poison, pulicide, sedative, stimulant, stomachic, sudorific, tonic, vermifuge, and vulnerary. Hot fomentations are recommended for bruises, freckles, inflammation, leucorrhea, palpitation, sciatica, stomachache, sunburn, swellings, toothache, and tumors. Homeopaths prescribe tinctures for abortion, amenorrhea, chorea, dysmenorrhea, epilepsy, eyes, hydrophobia, labial abscess, paralysis, strabismus, and worms.

Fresh tansy contains 0.12 to 0.18% volatile oil of highly variable chemical composition. Some races contain the ketone beta-thujone as the major constituent (to 95%) of the poisonous essential oil; others are devoid of it. Artemisia ketone, borneol, camphor, 18-cineole, *cis*-chrysanthenyl acetate, isopinocamphone, isothujone, piperitone, gamma-terpinene, umbellulone, and unidentified terpenes have also been reported as major ingredients of the essential oil.[31,32] Citric acid, probably of universal distribution, and tartaric acid occur in the leaves. Gallic acid, gum, mucilage, resin, and tannins are also reported. The acetylenic compound, pontica epoxide, is reported, and 1-viburnitol (wrongly called 1-quercitol). Appendino et al.[313] report the new hydroperoxysesquiterpene lactone, crispolide. Amines were sought but undetected in the flowers. Dry seed contain 28.9% protein and 26.5% fat.[21]

Toxicity — The oil is quite toxic and should be used only with extreme caution. Ten drops of the oil could be lethal. Keindorf and Keindorf[314] report hyperestrogenism in cattle. Tansy is said to be prohibited for botanical dealers, and cannot be sold as a dried herb by mail. Causing some cases of contact dermatitis, tansy yields the potentially allergenic sesquiterpene lactones, arbusculin-A and tanacetin.[6] Symptoms of internal tansy poisoning include rapid and feeble pulse, severe gastritis, violent spasms, and convulsions.

To the physician — Hardin and Arena[34] suggest gastric lavage or emesis and symptomatic treatment.

340. *TARAXACUM OFFICINALE* Weber ex Wigg. (ASTERACEAE) — Dandelion

Potherb sometimes eaten raw in salads, but often blanched like endive and used as a green; frequently cooked with salt pork or bacon to enhance the flavor. One recipe calls for a hot cream sauce of bacon, sugar, salt, cornstarch, egg, cream, and vinegar poured over coarsely chopped dandelion leaves.[315] Roots are sometimes pickled. Flowers used to make a wine. Ground, roasted roots used for dandelion coffee, and sometimes mixed with real coffee, like chicory. As late as 1957 more than 50 tons dandelion roots were imported to the U.S. for medicinal purposes.[315] Dried leaves are an ingredient in many digestive or diet drinks and herb beers. Birds like the seeds and pigs devour the whole plant. Goats eat the leaves, but sheep, cattle, and horses do not care for it. Dandelion has also been used as a source of latex.[62] Leaf extracts are phototoxic and weakly antibiotic against *Candida albicans* and *Saccharomyces cerevisiae*.[316]

Citing dandelion as alterative, aperient, apertif, bactericidal, cholagogue, depurative, diuretic, intoxicant, lactagogue, laxative, stimulant, stomachic, and tonic, Duke and Wain[32] and *Hager's Handbook*[33] list dandelion in folk remedies for abscess, bruises, cancer, caries, catarrh, cholecystosis, dyspepsia, eczema, gastritis, gout, heart ailments, heartburn, hemorrhoids, hepatosis, inappetence, jaundice, nephrosis, rheumatism, sclerosis, skin ailments, snakebite, splenosis, swellings, tumors, and warts.[32,33] Roots have been used medicinally as a simple bitter (as in the Lydia M. Pinkham tonic) and milk laxative. Roots, considered collyrium, depurative, diuretic, aperient, and tonic, are used as a remedy for chronic disorders of kidney and liver, gallstones, piles, warts,[317] and are supposed to be useful for anemia, diabetes, dyspepsia, gout, hepatosis, jaundice, and rheumatism. Experimentally hypoglycemic.[11] Chinese regard the whole plant as alterative, stomachic, sudorific, and tonic, believing it useful for abscesses, boils, snakebites, sores, ulcers, even dental caries, internal injuries, itch, and scrofula. They poulticed macerated plants and distillers grain on abscesses or cancers of the breast.[16]

Nutritional analyses of the leaves reveal, per 100 g, 45 calories, 85.6% water, 2.7 g protein, 0.7 g fat, 9.2 g carbohydrate, 1.6 g fiber, 1.8 g ash, 187 mg Ca, 66 mg P, 3.1 mg Fe, 76 mg Na, 297 mg K, 8400 μg β-carotene equivalent, 0.19 mg thiamine, 0.26 mg riboflavin, and 35 mg ascorbic acid. (Fresh leaves may contain 73 mg; fresh root contains

7 mg, dried root 0.3 mg.) Root contains taraxacin, inulin, glutin, gum, and potash. The latex contains ceryl alcohol, glycerin, tartaric acid, 0.1% caoutchouc, taraxasterol, and esters of acetic and higher fatty acids.[33] Plants also contain the phytosterols taraxasterol, homo-taraxasterol, and saponin, androsterol, homoandrosterol, cluytianol, cerylic alcohol, arabinose, and beta-hydroxyphenyl-acetic acid,[2,16] cerotic-, linoleic-, linolenic-, melissic-, oleic-, and palmitic-acids.[33] Spanish roots contain 0.13 per caoutchouc. Flowers contain beta-amyrin, beta-sitosterol, lutein, taraxanthine, taraxiene, flavoxanthin, vitamin B_2, arnidiol ($C_{30}H_{50}O_2$), and faradiol ($C_{30}H_{50}O_2$). Pollen contains cycloartenol, cycloartanol, 31-norcycloartanol, and pollinastanol.[33] "In spring, it contains mannite, or mannitol, a substance used as a base for pills, as a treatment for hypertension and coronary insufficiency, and in the manufacture of radio condensers and percussion caps."[315]

Toxicity — According to "Jake's Page",[315] "there is, experts aver, nothing toxic about a dandelion."[315] Contact sensitivity to this species has been reported. Mitchell and Rook[6] dig up one old case from *The Lancet* 1881 (not seen): "Two patients who had 'weed dermatitis' showed positive patch test reactions to the plant; the patients were also contact sensitive to Chrysanthemum and Olearia."[6] I feel that it is a relatively safe herb, about to take over the world.

> Back in the old days, the beasts and the birds
> Bugs and the bass talked, even herbs had their words
> They lived with the redman, they got along well
> But overpopulation turned their heaven to hell
> Too many redmen killed the big game for meat
> While smaller game perished 'neath too many feet
> So the animals met where their many paths crossed
> Making plans to regain, their paradise lost.
>
> Bears took the matters in their own paws
> But they couldn't shoot bows without clipping their claws
> They took to the mountains leaving humans behind
> They abandoned their problem and they found peace of mind
> The deer too convened and they talked of revenge
> If the redman kept on with his flesheating binge
> Rheumatism they invented, to bring down despair
> On the hunter who slew them, without first a prayer.
>
> Like the deer other species, contrived a disease
> To plague their tormentor 'til he came to his knees
> Diseases and ailments, a thousand and one
> 'Till it looked like the human race had been run
> The animals surely knew what they were about
> Many meat eaters came down with crippling gout
> The list of diseases, the redman's defeat
> From cancer to the colon of those who eat meat.
>
> Green plants respected red man's adventures
> And they offered a thousand and twenty one cures
> Man trod more lightly, and ate much less meat
> And medicine made from the plants at his feet.
>
> So daddy won't you show me to the Cherokee teepee
> Down by the green valley where paradise lies
> I'm sorry my son but you're asking too lately
> Mr. Watt's second cousin's done built a high rise.
> Where once the green trees were kissed by the sunrise
> There's a highrise 'tween the sunrise and the smog in your eyes
> All the other flow'rs got twisted by the herbicide squirt
> The last dandelions laughing, man's bitter dessert.*

* ©J. A. Duke, 1982.

341. *TEPHROSIA VIRGINIANA* (L.) Pers. (FABACEAE) — Devil's Shoe String

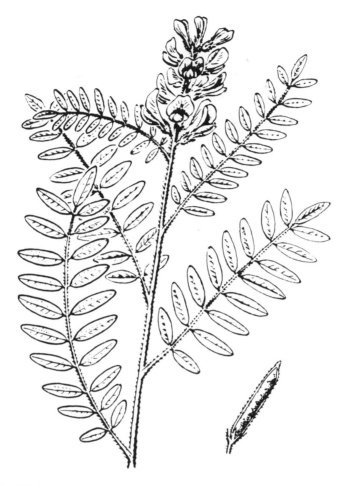

Ornamental wild flowers, sometimes used to poison fish. Rotenone, the insecticidal and piscicidal compound in *Tephrosia,* has shown anticancer activity in the lymphocytic leukemia and nasopharyngeal tumor systems. Like so many other antitumor agents, rotenone is also described as carcinogenic.[10]

Duke and Wain mention folk uses for alopecia, cholecystosis, cough, syphilis, and worms.[32] Roots considered anthelmintic, cathartic, diaphoretic, fortificant, laxative, piscicidal, stimulant, and tonic.[8,32] Creek Indians used the roots for consumption, bladder trouble, and loss of manhood. Root decoction used by Cherokees as children's tonic. The root contains deguelin, dehydrorotenone ($C_{23}H_{20}O_6$), rotenone, and tephrosin.[33]

Toxicity — The crude plant preparation or rotenone derived from it may cause dermatitis, conjuntivitis, and rhinitis.[6] Wood and seed reported toxic.[56] Deguelin, rotenone, and tephrosin can cause paralysis and death,[33] but they are not so toxic to man as they are to fish and insects.

342. *THEOBROMA CACAO* L. (STERCULIACEAE) — Chocolate, Cacao, Cocoa

Cacao seeds are the source of commercial cocoa, chocolate, and cocoa butter. Fermented seeds are roasted, cracked, and ground to give a powdery mass from which fat is expressed. This is the cocoa from which a popular beverage is prepared. In the preparation of chocolate, this mass is mixed with sugar, flavoring, and extra cocoa fat. Milk chocolate incorporates milk, in addition. Cocoa butter is used in confections and in manufacture of tobacco, soap, and cosmetics. Cocoa butter has been described as the world's most expensive fat, used rather extensively in the emollient "bullets" used for hemorrhoids.

Reported to be antiseptic, diuretic, ecbolic, emmenagogue, parasiticide, and *Poison,* cacao is a folk remedy for alopecia, burns, cough, dry lips, eyes, fever, listlessness, malaria, nephrosis, parturition, pregnancy, rheumatism, snakebite, and wounds.[32] Cocoa butter is applied to wrinkles in the hope of correcting them.[29] Cuna Indians apply the flowers in a remedy for worms of the eye.[60] The antitumor activity of the rootbark is attributed to its tannin content. Venezuelans apply cocoa butter to burns, cracked lips, genitals, and sore breasts to relieve irritation.[42]

Per 100 g, the seed is reported to contain 456 calories, 3.6 g H_2O, 12.0 g protein, 46.3 g fat, 34.7 g total carbohydrate, 8.6 g fiber, 3.4 g ash, 106 mg Ca, 537 mg P, 3.6 mg Fe, 30 μg β-carotene equivalent, 0.17 mg thiamine, 0.14 mg riboflavin, 1.7 mg niacin, and 3 mg ascorbic acid. Cocoa contains over 300 volatile compounds, including esters, hydrocarbons, lactones, monocarbonyls, pyrazines, pyrroles, and others. The important flavor components are said to be aliphatic esters, polyphenols, unsaturated aromatic carbonyls, pyrazines, diketopiperazines, and theobromine. Cocoa also contains about 18% proteins (*circa* 8% digestible); fats (cocoa butter); amines and alkaloids, including theobromine (0.5 to 2.7%), caffeine (*circa* 0.25% in cocoa; 0.7 to 1.70 in fat-free beans, with forasteros containing less than 0.1% and criollos containing 1.43 to 1.70%), tyramine, dopamine, salsolinol, trigonelline, nicotinic acid, and free amino acids; tannins; phospholipids; etc. Cocoa butter contains mainly triglycerides of fatty acids that consist primarily of oleic, stearic, and palmitic acids. Over 73% of the glycerides are present as monounsaturated forms (oleopalmitostearin and oleodistearin), the remaining being mostly diunsaturated glycerides (palmitodiolein and stearodiolein), with lesser amounts of fully saturated and triunsaturated (triolein glycerides). Linoleic acid levels have been reported to be up to 4.1%. Also present in cocoa butter are small amounts of sterols and methylsterols; sterols consist mainly of beta-sitosterol, sigmasterol, and campesterol, with a small quantity of cholesterol. In addition to alkaloids (mainly theobromine), tannins, and other constituents, cocoa husk contains a pigment that is a polyflavone glucoside with a molecular weight of over 1500; this pigment is claimed to be heat and light resistant, highly stable at pH 3 to 11, and useful as a food colorant; it was isolated at a 7.9% yield.[29]

Toxicity — Sutton[318] reports the collapse and death of a 3-year-old bitch that had eaten a 250-g package of cocoa. Post-mortem examination revealed congestion of lungs, liver, kidney, and pancreas, and petechial and ecchymotic hemorrhage of the thymus, all compatible with acute circulatory failure. The stomach contained high concentrations of theobromine and/or caffeine. Though used cosmetically, cocoa butter has been reported to have allergenic and comedogenic properties in animals. Plant reported to contain small quantities of safrole, the compound which led the FDA to ban sassafras. Cocoa extracts are GRAS (§182.20).[29]

343. *THEVETIA PERUVIANA* (Pers.) K. Schum (APOCYNACEAE) — Lucky Nut, Yellow Oleander

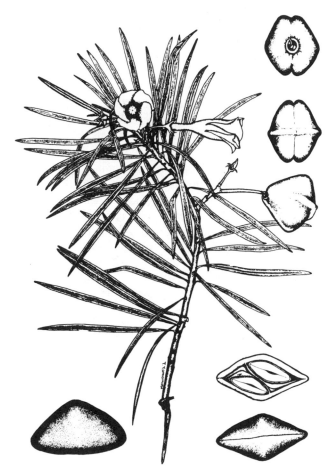

Ornamental but very dangerous. Seeds used as a fish poison, insecticide, and bactericide. They have also been used for homicide and suicide.[56] Latex has been used in arrow poisons.[10] Black pulp of the fruits is said to be eaten in Ghana. Burkill hints that dried leaves, poisonous though they are, may sometimes be smoked in Java.[5] Oblong seeds strung into necklaces. The solvent-extracted seed oil is said to be free of cardiac glycosides and appears suitable for soap-making and edible purposes. Perry hopes that the cardiac glycosides could serve as substitutes for digitalis.[16]

Poisonous latex applied to toothache in Yucatan, the leaf decoction for fever and malaria.[42] In the Philippines the bark and in Puerto Rico the seed are used as an emetic. Cubans treat insomnia with a floral infusion.[42] In Bengal the seed is used as an abortifacient. Bark used as febrifuge in the Philippines for malaria. Kernel oil has been applied externally to skin ailments. Seeds employed as a purgative in dropsy and rheumatism. Seed decoction used for hemorrhoids.[3] Regarded as abortifacient, alexiteric, cardiotonic, emetic, febrifuge, fumitory, insecticide, piscicide, poison, purgative, the plant is used in folk remedies for coronary insufficiency, dropsy, fever, heart conditions, malaria, piles, rheumatism, skin ailments, tachycardia, toothache, and tumors.[32,33]

Leaves poulticed onto tumors in Latin America, may contain the anticancer cardenolide glycoside cerberin (active in KB, LL, and SA systems), and the triterpene ursolic acid (active in PS 125 system). Active principles of the kernel are thevetin, thevetoxin, and neriifolin, all toxins with digitalis-like effects. One might conceivably obtain rubber as a by-product of glycoside extraction (leaf contains 0.44% rubber, stem 0.19%, whole plant 0.23%). Latex contains 9.7 to 13.3% caoutchouc, 67.4 to 69.7% resin, and 17.0 to 22.9% insolubles. Seeds yield 57 to 67% of a nondrying oil, used in India for skin ailments. The fatty acid composition is 17.1% palmitic, 11.8% stearic, 0.4% arachidic, 64.3% oleic.

Toxicity — Extremely poisonous plant. "Many deaths result from the use of the plant in folk medicine."[56] On Oahu, this is considered the most frequent plant cause of fatal or dangerous poisoning in man. Most human fatality has resulted from misuse of the plant, or substances derived from it as medicine by natives.[14] Symptoms include numbness and burning in the mouth, dry throat, diarrhea, vomiting, dilated pupils, slow, irregular heartbeat, high blood pressure, occasional convulsions, heart failure, and death. One to two seeds have killed children.

To the physician — Gastric lavage or emesis, atropine, symptomatic, and supportive treatment.[56] The thevetin is the superstitious ingredient separating the gringo doctors' hemorrhoid treatment from the Venezuelan "witch doctor" who hangs thevetia fruits or seeds from his belt while inserting cocoa butter suppositories rectally three times a day. Pulverized seed with bits of radish and lemon juice is said to be emetic and litholytic.[42]

344. *THYMUS VULGARIS* L. (LAMIACEAE) — Common Thyme

Used to flavor or scent butters, cheeses, chowders, fish, liqueurs, meats, olives, onions, perfumes, pickles, poultry, sauces, soaps, soups, stews, stuffings, tomatoes, and vinegars. Some cooks (definitely not French) say "use thyme almost as freely as salt." It is an ingredient of the liqueur Benedictine. Thyme is one of the better honey plants. If the bees won't make your thyme honey add hot thyme tea to other honey types. Sheep who feed on thyme are said to develop an especially delicate flavor. Currently it is popular as a tisane, mixed with such herbs as mint and rosemary. Thyme butter spread over steak before cooking is good. Dried flowers, like lavender, are said to preserve linen from insects, scenting them at the same time. In sufficient dose, carvacrol and thymol have tracheal relaxant properties. Thymol and thyme oil act as secretomotoric and disinfectant agent. Application of thyme extracts causes an important increase of mucus secretion of the bronchii. The antispasmodic action of thyme oil has been demonstrated, with the phenols generally regarded as the active agents. This may be true for the secretomotoric, secretolytic, and antiseptic effects. One author has suggested, though, that thyme oil increases the antitussive action of unknown, water-soluble components. Others report spasmolytic activity of fresh–plant extracts compared to inactivity in dried plants.

An ointment derived from the plant is a folk remedy for indurations and warts. Thyme tea (1 teaspoon fresh leaves steeped in 1 cup boiling water) said to be good for headache. The hot aqueous extract is said to cure tumors of the digestive tract. The juice, with vinegar, is said to cure tumors and cancers. Thyme has reportedly been used in bronchitis, catarrh, colic, diabetes, fever, gout, laryngitis, leprosy, rheumatism, scarlet fever, sciatica, sore throat, spleen disorders, uterine disorders, warts, and whooping cough.[32] It is considered antiseptic, antispasmodic, carminative, and tonic. Smoked like tobacco, it is used for digestion, drowsiness, and headache. Burning thyme is supposed to repel insects. Thymol, the oil of thyme, is an antiseptic and deodorant, used externally and internally. It has been used as a meat preservative, and is used to treat burns, eczema, hookworm, psoriasis, ringworm, and worms. It is used as a fungicide to prevent mildew, and as an ingredient in toothpastes. It is used as a diffusible stimulant in case of collapse. It is said to cause mental excitement. It has been used externally as a rubefacient counterirritant. Thymol may cause abortion, coloring the urine green. Said to be anodyne, antiseptic, antitussive, apertif, carminative, demulcent, depurative, diaphoretic, digestive, diuretic, expectorant, fungicide, nervine, pectoral, rubefacient, sedative, stimulant, and vermifuge. Thyme is a reputed folk remedy for anemia, asthma, bad breath, bronchitis, bruises, callosities, cancer, catarrh, colds, colic, cough, cramps, debility, diabetes, diarrhea, dysmenorrhea, fever, flatulence, gastritis, gastroenteritis, gingivitis, gout, headache, indurations of heart, indigestion, leucorrhea, lung, melancholy, nerves, neuralgia, sciatica, sclerosis, skin, snakebite, sore, sore throat, spasm, sprain, stomach, stomatitis, tumors (of the liver, spleen, and digestive tract), warts, whooping cough, and worms.[32,33]

The thyme of commerce originates from either *T. vulgaris* L. or *T. zygis* L., wild thyme originates from *T. serphyllum* L., and Moroccan thyme originates from *T. satureioides* Cass. Essential oils from the four species contain alpha-pinene, camphene, beta-pinene, myrcene, alpha-phellandrene, limonene, 1,8-cineole, p. cymene, linalool, linalyl acetate, bornyl acetate terpinen-4-ol, alpha-terpinyl acetate, alpha-terpineol, borneol, citral (neral, geranial, or both?), geraniol, thymol, and carvacrol. In *T. vulgaris,* thymol and terpinen-4-ol were major compounds; in *T. zygis* and *T. serphyllum,* thymol, linalool, and linalyl acetate were major compounds, while in *T. satureioides* thymol, borneol, and alpha-terpineol were major compounds.[193] Per 100 g, ground thyme is reported to contain 276 calories, 7.8 g H_2O, 9.1 g protein, 7.4 g fat, 63.9 g total carbohydrate, 18.6 g fiber, 11.7 g ash, 1890 mg Ca, 201 mg P, 123.6 mg Fe, 220 mg Mg, 55 mg Na, 814 mg K, 6.2 mg Zn, 3800 IU vitamin A,

0.51 mg thiamine, 0.4 mg riboflavin, 4.94 mg niacin, 0 cholesterol, and 163 mg phytosterols. Saturated fatty acids totaled 2.73 g, monounsaturated 0.47 g, and polyunsaturated 1.19 g. Per 100 g, there are 186 mg tryptophan, 252 threonine, 468 isoleucine, 430 leucine, 207 lysine, 274 methionine and cystine, 482 phenylalanine and tyrosine, and 502 mg valine.[89] Seeds yield 37% of a drying oil consisting mainly of linolenic, linoleic, and oleic acids. Triterpenoid saponins, flavones, ursolic acid (1.5%), caffeic acid, tannins, and resins also occur in the herb.[33]

Toxicity — Oil of thyme is itself quite poisonous. Thymol has caused dermatitis in dentists, and, when used in toothpaste, chelitis and glossitis. Oil of thyme, in bath preparations, has been reported to cause hyperemia and severe inflammation, even in the director himself of the Platonic Academy of Herbalism.

345. *TILIA EUROPAEA* L. (TILIACEAE) — Linden Tree (America), Lime Tree (Europe)

French households keep stocks of the dried flowers for making "Tilleul". Honey from the flowers is one of the world's best flavored, used almost exclusively in liqueurs and medicine. According to Tyler, "Authorities generally agree that linden flower tea is both a pleasant-tasting beverage and a useful diaphoretic." Most of the wooden floral carvings and figurines by Grinley Gibbons for St. Paul's Cathedral and Windsor Castle were made of the wood, one of the lightest of the broadleaf trees of Europe, very resistant to worm damage. Bark or bast widely used for basketry and cordage. Sugar can be obtained from the sap, drawn off in spring. Foliage is eaten by cattle, fresh or dry. They contain "a saccharine matter having the same composition as the manna of Mount Sinai."[2] The seed oil resembles olive oil.[33]

Reported to be diaphoretic, diuretic, hemostat, nervine, sedative, spasmolytic, stomachic, tranquilizer, linden is a folk remedy for burns, cardosis, catarrh, cholecystosis, dyskinesia, dyspepsia, epilepsy, gallstones, gastritis, headache, hepatosis, hysteria, nausea, nephrosis, neuralgia, palpitations, rheumatism, sciatica, sores, spasms.[32,33] Homeopathically used for enuresis, hemorrhage, incontinence, leucorrhea, metritis, ophthalmia, prolapsed uterus, rheumatism, urticaria, and uterosis.[33] The charcoal is recommended for dyspepsia; it was once regarded so effective that an epileptic need only sit under a linden tree to effect a cure.[37] Russians apply the bark to wounds.[33]

Per 100 g, the leaf is reported to contain 12.0 g H_2O, 16.2 g protein, 2.9 g fat, 57.5 g total carbohydrate, 13.2 g fiber, and 11.4 g ash.[21] The fruit is reported to contain 68.1 g H_2O, 3.6 g protein, 3.0 g fat, 23.2 g total carbohydrate, 13.5 g fiber, and 2.1 g ash. Leaves contain linarin and the glycoside tiliacine, sugar, starch, lipids, phytosterols, resinic acids, tannins, phlobaphene, a yellow xanthophyll pigment, vitamin C, and beta-amyrin. The bark contains taraxerol ($C_{30}H_{50}O$), tiliadine ($C_{21}H_{32}O_2$), vanillin, and mucilage. The wood contains beta-sitosterol, stigmasterol, stigmastanol, saccharose, and fatty acid esters of lineolic-, palmitic-, oleic-, and linolenic-acids. The fruits contain 58% fatty oil, phytosterols, tannin, sugar, asparagine, glutamic-acid, serine, glycine, alanine, tyrosine, valine, and leucine. Flowers contain mucilage, tannin, quercetin-3-gluco-7-rhamnoside, kaempferol-3-gluco-7-rhamnoside, quercetin-rhamnoxyloside, kaempferol-3,7-dirhamnoside, isoquercitrin, astragalin, quercitrin, afzelin, tiliroside ($C_{30}H_{26}O_{13}$), *p*-cumaric-, chlorogenic-, and caffeic-acids, and essential oil containing docosane, eucosane, eugenol, farnesol, farnesylacetate, geraniol, geranylacetate, heneicosane, hentriacontane, heptacosane, hexacosane, linalool, nonadecane, octacosane, octadecane, pentacosane, 2-phenylethanol, 2-phenylethylbenzoate, 2-phenylethylphenylacetate, tetracosane, triacontane, and tricosane. Also, listed are citral, citronellal, citronellol, limonene, linalylacetate, nerol, nerolidol, alpha-pinene, terpineol, and vanillin. Flowers also contain the free amino acids cysteine, cystine, and phenylalanine, and all tissues alanine, isoleucine, leucine, and serine. Gums, hesperidin, saponin, sugar, and tocopherol are also reported.[33]

Toxicity — Using old flowers may induce narcotic intoxication.[2] Too frequent use of linden flower tea may result in heart damage.[37] Those with cardiac problems would do well to avoid using the drug.[37]

346. *TRICHOCEREUS PACHANOI* Britton and Rose (CACTACEAE) — Achuma, San Pedro Aguacolli

Used by Andean Indians as an intoxicant, the cactus is cut up into small pieces and boiled down. Healers believe that the hallucinations induced by the cactus facilitate their divination of the cause and cure of an ailment.[52]

Said to be emetic, hallucinogenic, and psychedelic, achuma is a folk remedy for enteritis, gastritis, pneumonia, and sterility.[32] Used by Andean curanderos in the diagnosis and cure of illness, employed in magic and folk medicine. Mixed with another cactus, *Neoraimondia macrostibas*, *Brugmansia*, *Isotoma*, and *Pedilanthus* for the hallucinogenic drink called *cimora*.[80] Mescaline is apparently the active principle.[54]

Toxicity — San Pedro is listed as a narcotic hallucinogen.[54]

347. *TRIFOLIUM PRATENSE* L. (FABACEAE) — Red Clover, Pavine Clover, Cowgrass

4 cm

Extensively grown for pasturage, hay, and green manure. As a forage crop, it is excellent for livestock and poultry. In comparison with alfalfa, red clover has about two thirds as much digestible protein, slightly more total digestible nutrients, and slightly higher net energy value. The best approximation to vegetable bouillion I have made consisted of red clover and chicory flowers, boiled vigorously with wild onion or chives. Red-clover flowers are reported to possess antispasmodic, estrogenic, and expectorant properties. The solid extract is used in many food products usually at less than 20 ppm, but in jams and jellies it may be 525 ppm.

Said to be used for: alterative, antiscrofulous, antispasmodic, aperient, athlete's foot, bronchitis, burns, cancer, constipation, diuretic, expectorant, gall bladder, gout, liver, pertussis, rheumatism, sedative, skin, sores, tonic, ulcers. Flowers have been used as a sedative.[3] Russians recommend the herb for bronchial asthma. Chinese take the floral tea as an expectorant.[16] Kloss recommends that every family "stash" red clover blossoms, gathered in summer, and dried on paper in shade. "Use this tea in place of tea and coffee and you will have splendid results." This is one of Kloss' diets that doesn't offend me. I have enjoyed pink clover tea with about half melilot. Kloss devotes pages to the anticancer activity of the floral tea, one remedy not yet tested by the National Cancer Institute.[320] Kloss also recommends the clover for bronchitis, leprosy, pertussis, spasms, and syphilis. Recently, there has been a flap over Jason Winters' tea containing red clover and chaparral and some unidentified secret spice, sold at rather high prices as a "cancer cure".[320]

Green forage of red clover is reported to contain 81.0% moisture, 4.0% protein, 0.7% fat, 2.6% fiber, 2.0% ash. Hay of red clover contains 12.0% moisture, 11.8% protein, 2.6% fat, 27.2% fiber, and 6.4% ash. Seeds are reported to contain trypsin inhibitors and chymotrypsin. Estrogenic disorders have been reported in cattle grazing largely on red clover, apparently due to activity of the isoflavones-formononetin, biochanin A, and to some small extent daidzein and genistein. Protein content varies from 14 to 28% and total digestible nutrients from 65 to 88%. More nutritional data are located in the *Handbook of Legumes of World Economic Importance*.[40] Flowers contain salicylic acid, *p*-coumaric acid, isorhamnetin ($C_{16}H_{12}O_7$), glucosides including a phytosterol glucoside trifolianol, trifolin ($C_{22}H_{22}O_{11}$), trifolitin ($C_{16}H_{10}O_6$), rhamnose ($C_{16}H_{12}O_5$), isotrifolin ($C_{22}H_{22}O_{11}$), pratol ($C_{15}H_8O_2[OH]OCH_3$), pratensol ($C_{17}H_9O_2[OH]_3$), genestein, coumestrol, etc.,[3] *trans*- and *cis*-cloramide (L-dopa conjugated with *trans*- and *cis*-caffeic acids), phaseolic acid, and pterocarpan phytoalexins (in response to fungal or viral infection).[29]

348. *TRIGONELLA FOENUM-GRAECUM* L. (FABACEAE) — Fenugreek

Widely cultivated as a condiment crop, fenugreek seeds, containing coumarin, are used in curries and soups. Plant is used to make horse hair shiny. Powdered seeds are used locally for a yellow dye. Harem women are said to eat roasted fenugreek seed to attain buxomness. Mixed with cottonseed, the seed increases the flow of milk in cows, but imparts the fenugreek aroma to milk. Plant serves as a potherb; used to flavor imitation maple syrup. In North Africa it is mixed with breadstuffs. In Greece, raw or boiled seeds are eaten with honey. The celery-scented oil is used in butterscotch, cheese, licorice, pickle, rum, syrup, and vanilla flavors. Used also in cosmetics, hair preparations, and perfumery. Indians grow the plant as a forage, considered a good soil renovator. Fenugreek is a source of diosgenin, used in the synthesis of hormones. As with other chemurgic crops, a large percentage of the crop must be thrown into the pot to extract a small percentage (1 to 2% on a dry weight basis) of pharmaceutical (diosgenin). While the ''pot is boiling'', proteins, fixed oils, oleoresins, e.g., coumarin, mucilages, and/or gums, might be extracted. Organic residues might be used for biomass fuels or manures, inorganic residues for ''inorganic'' chemical fertilizers. The husk of the seed might be removed for its mucilage, with the remainder partitioned into oil, sapogenin, and protein-rich fractions.[46] Seed mucilage (*circa* 45%) could be prepared from the marc left after extraction of the fixed oil (used as a lactagogue). Its relatively high viscosity makes it a good emulsifying agent to be used in pharmaceutical and food industries. Due to its neutral ionic properties, it is compatable with other drugs or compounds sensitive to acids.[321]

Mucilaginous seeds regarded as having tonic, emollient, and vermifugal properties, and used for oral ulcers, chapped lips, and stomach irritation. Indian women believe the seeds promote lactation. Seeds carminative, tonic, used for diarrhea, dyspepsia, rheumatic conditions. Crushed leaves taken internally for dyspepsia. Said to be used for aphrodisiac, astringent, bronchitis, demulcent, diuretic, emmenagogue, emollient, expectorant, fever, fistula, glands, gout, lactogogue, neuralgia, restorative, sciatica, skin, sore, sore throat, tonic, tuberculosis, tumor, wounds. Chinese use the seed for abdominal pain, chilblains, cholecystosis, fever, hernia, impotence, hypogastrosis, nephrosis, and rheumatism.[16] Malayans poultice the seeds onto burns and use them for chronic coughs, dropsy, hepatomegaly, and splenomegaly.[16] Ecbolic and spermicidal activity has been reported.[322]

Seeds contain 6.2% moisture, 23.2% protein, 8.0% fat, 9.8% fiber, 26.3% mucilaginous material, and 4.3% ash. Whole grain is reported to contain (per 100 g edible portion): 369 calories, 7.8% moisture, 28.2 g protein, 5.9 g fat, 54.5 g total carbohydrate, 8.0 g fiber, 3.6 g ash, 220 mg Ca, 358 mg P, 24.2 mg Fe, 55 µg β-carotene equivalent, 0.32 mg thiamine, 0.30 mg riboflavin, 1.5 mg niacin, and 274 mg tryptophane. The protein is characterized by low levels of S-amino acids and high levels of lysine and tryptophane. Flour contains 375 calories, 9.9% moisture, 25.5 g protein, 8.4 g fat, 53.1 g total carbohydrate, 7.1 g fiber, 3.1 g ash, 213 mg Ca, 270 mg P, 32.4 mg Fe, 0.06? mg thiamine, 0.05? mg riboflavin, and 1.5 mg niacin. Raw leaves contain 35 calories, 87.6% moisture, 4.6 g protein, 0.2 g fat, 6.2 total carbohydrate, 1.4 g fiber, 1.4 g ash, 150 mg Ca, 48 mg P/100 g.[40] A hypoglycemic betain, trigonelline, a methylbetaine of nicotinic acid, was considered responsible for the hypoglycemic activity of fenugreek. However, Israeli researchers found that coumarin and nicotinic acid were the main hypoglycemic constituents of all compounds isolated from the active fraction of the seed. Trigonelline exerted a less pronounced though more persistent activity. Singhal et al.,[323] feeding diets containing 50% fenugreek seed for 2 weeks to normal and hypercholesterolemic rats, significantly lowered serum cholesterol levels by 42 and 58%, respectively. Khurana et al.[324] describe 3,4,7,trimethyl coumarin from the stems.

Toxicity — Coumarins and estrogen can be toxic in overdoses. Seeds are also reported to contain trypsin inhibitors and chymotrypsin inhibitors.

349. *TRILISA ODORATISSIMA* (J.F. Gmel.) Cass (ASTERACEAE) — Deertongue, Deer's Tongue

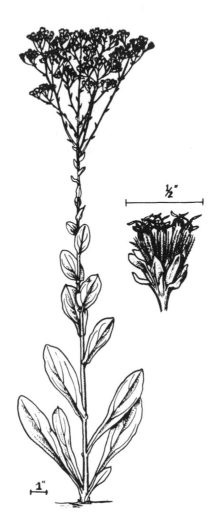

Leaves used to flavor pipe and cigar tobacco and cigarettes. (American market alone estimated at one million pounds in 1974.) Deertongue extracts serve as fixatives or fragrance components in perfumes and cosmetics (creams, detergents, lotions, soaps).[29] Also, used as a substitute for tonka beans and as a moth repellant.[33]

Said to be demulcent, diaphoretic, diuretic, febrifuge, stimulant, sudorific, and tonic, deertongue is used in malaria, neuroses, and pertussis.[32]

Heavy with coumarin crystals, especially on the upper leaf surface, the plant contains about 1.6% coumarin. Other volatiles include dihydrocoumarin, 2,3-benzofuran, terpenes, straight-chain aldehydes, and ketones, plus nearly 100 other identified items. Nonvolatiles include triterpenes (e.g., beta-amyrin, lupeol, lupenone, 11-oxo-beta-amyrin, 11-oxo-alpha-amyrin) and sesquiterpenes (eudesmin and epieudesmin). Fresh leaves are high in *cis-* and *trans-o*-hydroxycinnamic acids, the former, perhaps, converted to coumarin in curing.[29]

Toxicity — Coumarins may cause hemorrhage and liver damage. Still, coumarins are reportedly effective in reducing high-protein edema, especially lymphedema.[29]

350. *TURNERA DIFFUSA* Willd. (TURNERACEAE) — Damiana

Mexicans drink the leaf teas and use the leaves to flavor liqueurs, especially the Guadalajara liqueur called ''Damiana''.[42] Leung mentions its use in alcoholic and nonalcoholic beverages, baked goods, candies, frozen dairy desserts, gelatins and pudding, with use levels up to 0.125% crude damiana in baked goods. Siegel notes[55] that damiana is advertised as producing a marihuana-like euphoria lasting 60 to 90 min, adding ''such effects are little more than anecdotal at this time.'' One source suggests smoking the leaves and drinking the tea simultaneously.[51]

Reported to be aphrodisiac, CNS-depressant, CNS-stimulant, diuretic, expectorant, laxative, purgative, stimulant, tonic, damiana is a folk remedy for amaurosis, catarrh, cholecystosis, cold, cough, diabetes, dysentery, dysmenorrhea, dyspepsia, enuresis, headache, impotence, infections, inflammations, nephrosis, neurosis, orchitis, paralysis, renosis, spermatorrhea, stomachache, syphilis.[32,33] The tea made from the leaves has been considered a mild aphrodisiac and tonic.[51,57] Yucatanese use the leaf and flower decoction for asthma and bronchitis. Bahamians give the leaf decoction for enuresis. Cubans take the decoction as aphrodisiac, diuretic, and stimulant. Homeopaths use for impotence, migraine, neurasthenia, and sterility.[33]

Contains 0.5 to 1% essential oil, with a low boiling fraction containing mainly 1.8-cineole, alpha- and beta-pinenes, *p*-cymene, and a higher boiling fraction consisting primarily of thymol and sesquiterpenes (α-copaene, δ-cadinene, and calamenene. Also, contains 8% chlorophyll, *circa* 3.5% tannin, about 7% alkaloid, 13.5% gum, 15% protein, 6.5% resin, 6% starch, sugar, fatty oils, and traces of acids. Leaves are said, also, to contain arbutin. The antitumor compounds beta-sitosterol, gonzalitosin I (5-hydroxy-7,3',4'– trimethoxyflavone), resin, and a bitter substance termed damianin.[29]

Toxicity — Reported to contain cyanogenic glycosides, tannins, and the medicinal compound arbutin. Approved for food use (§ 172.510).[29]

351. *TUSSILAGO FARFARA* L. (ASTERACEAE) — Coltsfoot, Coughwort, Horse-Hoof

Silky seed once used for stuffing pillows and mattresses. Young leaves occasionally used in soups, and older leaves consumed as a vegetable or as a "tea". In England it is used with buckbean, eyebright, wood betony, rosemary, thyme, lavender, and rose petals. Leaves and flowers used in preparing coltsfoot candy, and as a tobacco substitute. Gibbons[39] gives recipes for coltsfoot cough drops, cough syrup, smoking mixture, and tea.

The root, with wine, is a folk remedy for liver indurations. Bruised leaves are applied as a poultice to scrofulous tumors.[4,44] The flower, in distilled water, is said to be a cure for cancer. Chinese dig next year's flower bud in winter, let it airdry, unwashed, then shake the half-dry herb to remove the dirt, allowing it then to dry further. Such buds, demulcent and lung-tonic according to the Chinese, are prescribed for asthma, hemoptysis, and lung cancer. Coltsfoot is also used in the Orient for apoplexy, cough, influenza, lung congestion, and phthisis.[16] Suggested as antihistamine, antitussive, bactericide, collyrium, demulcent, diuretic, emollient, expectorant, pectoral, spasmolytic, styptic, sudorific, coltsfoot occurs in folk "remedies" for apoplexy, asthma, bronchitis, catarrh, cold, cough, diarrhea, dyspepsia, fever, hoarseness, mucous congestion, neuroses, ophthalmia, phthisis, rheumatism, scrofula, and tumor. According to the FDA, "The only therapeutic value the leaves possess is a demulcent effect due to their mucilage. They were used in a pectoral tea called Species Pectorales, which also contains five other herbs." According to Grieve's *Herbal*,[2] it was one of the most popular of cough remedies. Dried leaves used medicinally in cough medicines for colds and bronchial catarrh, or smoked with other herbs for asthma and coughs. Smoke from the herb is said to be anticholinergic and antihistaminic. Powdered leaves snuffed up the nostrils are said to be "excellent for nasal obstruction and headache."[44] For lung congestion or consumption, coltsfoot is combined with elderflowers, groundivy, horehound, and marshmallow. Drug is demulcent, diuretic, emollient, expectorant, pectoral, spasmolytic, sudorific, and tonic. When dried, all parts have a bitter mucilaginous taste. The root was used for scrofula. Homeopaths prescribe the tincture of the whole plant for corpulence and plethora. Kloss recommends the roots and leaves for ague, asthma, bronchitis, catarrh, consumption, cough, fever, inflammation, piles, stomach troubles, and swellings.[44]

Coltsfoot, collected or cultivated as source of drug, contains an acrid essential oil, a bitter glucoside, a resin, and gallic-, malic-, and tartaric-acids. The flowers contain phytosterols (stigmasterol, taraxasterol, sitosterol), a dihydride alcohol, faradial,[2] choline, inulin, tannic acid, tannin, tussilagin, essential oil, KNO_3, and 8.2% (dry basis) mucilage containing 14.7% glucose, 23.6% galactose, 30.1% fructose, 21.1% arabinose, 10.5% xylose, and 5.8% uronic acid. Leaves contain essential oil, up to 17% tannin, 8.2% mucilage (rhamnose, L-fucose, 2-O-methyl-D-xylose), inulin, sugar, glycosides, choline, paraffin, stigmasterol, taraxasterol, beta-amyrin, sitosterol, calendol, kaempferol, and quercetin.[33] Roder et al. describe the new pyrrolizidine alkaloid tussilagine.[325]

Toxicity — Classified by the FDA[62] as an Herb of Undefined Safety. Tyler recounts recent studies of the dried young flowers showing small traces (0.015%) of hepatotoxic pyrrolizidine alkaloid senkirkine. Rats fed diets with >4% coltsfoot developed cancerous tumors of the liver. Senkirkine has also been found in the leaves.[37] Rosberger et al.[326] report, also, the hepatotoxic pyrrolizidine alkaloid senecionine in preblooming material.

352. *ULMUS RUBRA* Muhl. (ULMACEAE) — Slippery Elm

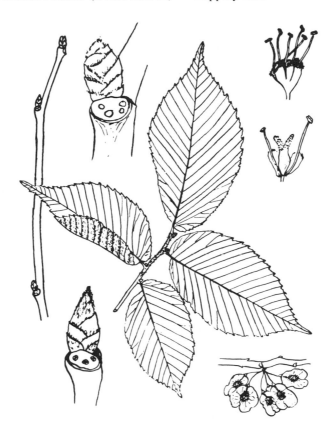

Inner bark used medicinally. *U. rubra* was an official drug of the U.S. pharmacopeia. It was used by the American Indians for making canoes, sides of wigwams, baskets, kettles, and related items, and as fiber and caulking.[45] Indians of the Missouri River Valley cooked the bark with buffalo fat to prevent rancidity. Hence, they were using it as an antioxidant. Ground bark has been mixed with milk as a nutrient for infants and invalids.[2,8] Slippery elm bark is very nutritional and has formed the basis of many patent foods.[2] Wood used for pulpwood, charcoal, and lumber.[63]

The bark (or root) is used in folk remedies for cancer, carcinomas, felons, tumors, and whitlows.[4] Reported to be antiseptic, demulcent, diuretic, emollient, expectorant, vulnerary, preventative, laxative, the bark is a folk remedy for abscesses, diarrhea, dysentery, urinary inflammations, inflammation, burns, fever, parturition, tumor, wounds.[32] Dried inner bark is considered demulcent and emollient. Powdered bark was once used as a soothing poultice. The infusion was given to relieve gastric and intestinal inflammation. It was powdered and made into lozenges for throat irritation.[17] Powdered bark is used in the form of a poultice to remove discoloration from blackened eyes and to treat cold sores; it was commonly used by North American Indians in the treatment of skin diseases and wounds.[11] Antiherpetic and antisyphilitic uses are noted by Millspaugh.[19] The bark is cleansing, healing, and strengthening.[2] Taken as an infusion, it is said to be effective against gastritis, gastric catarrh, mucous colitis, and lung disturbances (bronchitis, bleeding, consumption, pleurisy, and coughs). Because of its mucilaginous character, it is very helpful in removing excess mucous from the body. It has also been found helpful in typhoid fever and diseases of the organs. As an injection, it has been used for diarrhea, constipation, dysentery, and urinary tract diseases; and for expulsion of *Ascarides* worms and tapeworms. It is considered one of the best poultices and the uses attributed to it are many, including the reduction of swollen glands (mixed with milk and brewer's yeast), for severe rheumatism and gout (mixed with vinegar and bran), and to arrest the spread of gangrene. A pinch of the powder is said to stop tooth decay (if used at onset) and allay toothache.[2] Indians used mashed bark to ease removal of lead from gunshot wounds. To assist in childbirth, a tea from the roots, bark, or sap was often used. Bricklin[327] mentions the use of slippery elm bark with jimsonweed for treating eczema, and gives an old miner's remedy of sucking stocks of the bark with a little kerosene added to prevent coal dust from sticking to the throat. According to Dioscorides, used in a bath, slippery elm heals broken bones. Potawatomi Indians used the bark for cramps and eye inflammations, and a splinter of the bark was used to pierce boils.[45]

The wood contains compounds like cholesterol, campesterol, beta-sitosterol, citrostadienol, dolichol, and the sesquiterpene 2-hydroxy-5-isopropyl-6-methoxy-8-methyl-3-naphthaldehyde, 5-isopropyl-3,8-dimethyl-2-napthol, and 2-hydroxy-5-isopropyl-8-methyl-5,6,7,8-tetrahydro-3-napthaldehyde.[33] The bark contains pentoses, methylpentoses, and hexoses, which, after hydrolysis, give galactose and traces of glucose and fructose, two polyuronides, galacturonic acid, L-rhamnose, and D-galactose. In addition to mucilage, the bark contains (twin crystals of) calcium oxalate.[2]

Toxicity — The pollen is allergenic.[210]

353. *UMBELLULARIA CALIFORNICA* (Hook. and Arn.) Nutt. — California Bay, California Laurel, California Sassafras

Aromatic evergreen tree, sometimes grown as ornamental.[36] Indians and early Spanish settlers employed the leaves as a condiment, and as ingredients in counterirritants and insecticides. Crushed leaves were packed around game immediately after the kill, to (1) keep away flies and (2) improve the flavor.[7] Wood valued for fine carpentry. According to Hardin and Arena, the berries are eaten raw or roasted.

California Bay is reported to be a folk remedy for colic, diarrhea, headache, rheumatism, and stomachache.[3,32]

Bark and leaves contain a pungent volatile oil, the stem yielding 0.56 to 1.13%, leaves up to 4.69%, the chief constituent being umbellulone (40 to 60%), a ketone $C_{10}H_{14}O$; eugenol, 1-alpha-pinene, cineole, safrole, methyl eugenol, formic acid. Buttery et al.[328] found the volatile oil of the leaves to contain 39% umbellulone, 19% 1-8-cineole, 7.6% alpha-terpineol, 6.2% terpinen-4-01, 6% sabinene, 4.7% alpha-pinene, and 5.4% 3,4-dimethoxyallylbenzene. Twenty-six other compounds were characterized. Lawrence[193] compares the chemical menus from California Bay, Grecian Bay, and Jamaica Bay, devoting two pages of tables to *Umbellularia*.

Toxicity — Produces irritation of the mucous membrane of some persons who handle it.[6] Inhalation of the oil causes severe headache[7] and raises the blood pressure. According to MacGregor et al.,[329] the acute oral toxicity to mice of the oil was greater than that of *Laurus* and *Pimenta*, due primarily but not entirely to umbellulone. DMAB (3,4-dimethoxyallylbenzene) produced sedation at low doses and a reversible narcosis at higher doses. DMAB prevented the death of mice treated with lethal convulsant doses of stychnine. Once I read a report of a fatality attributed to the California Bay, but I can no longer find the reference.

354. *UNCARIA GAMBIR* (Hunter) Roxb. (RUBIACEAE) — Gambir, Gambier, Pale Catechu

Gambir is widely used as a tanning material and is employed medicinally as an astringent. Gambir is the dried, aqueous extract prepared from the leaves and twigs. The drug contains catechutannic acid (22 to 50%) which resembles the tannin in kino (*Pterocarpus marsupium* Roxb.) and *Krameria*. Gambir is also used as a mordant in dyeing. In Malaya, it is used for chewing with the betel leaf; some believe it causes the reddening of the mouths of betel chewers. It is often planted in Malaya as well for the aromatic flowers.

Gambir is a powerful astringent, used in diarrhea. Spasmolytic activity has been reported for an unnamed alkaloid or base occurring in gambir. In Malaya, gambir is applied to burns and scurf. In Borneo, it is applied externally for lumbago and sciatica. In Johore, an infusion of fresh leaves and young shoot is taken for diarrhea and dysentery and used as a gargle for sore throat.[16]

According to the *Wealth of India,* gambir contains 22.50% catechutannic acid, 7 to 33% catechin, 3 to 5% mineral matter, small amounts of (+)-epicatechin, catechu-red, quercetin, fixed oil, wax, sugar, starch, cellulose, weak acids (probably 2,4 dihydrobenzoic and adipic) and their salts, and a mixture of fluorescent alkaloids (at least in commercial Italian samples). The alkaloid mixture (*circa* 0.05%) consists of gambirtannine ($C_2H_{18}N_2O_2$), oxogambirtannine ($C_{21}H_{16}N_2O_3$), and 3-14-dihydrogambirtannine. Gambirtannine has the skeleton of yohimbine, a reputedly aphrodisiac alkaloid that has received a lot of press lately. The *Dictionary of Chinese Traditional Medicine* also indexes roxburghines A, B, C, D, and E, rhynchophylline, isorhynchophylline, mitraphylline, rotundifoline, protocatechu tannins, and pyrogallic tannins.[41] *Hager's Handbook*[33] adds gambirine, ourouparin, dihydrocorynanthein, and 11-methoxyyohimbine, giving structure for these and other compounds, also reporting the active compound rutin. Pyrocatechol (up to 30%) has also been reported.

355. *URGINEA MARITIMA* (L.) Baker (LILIACEAE) — Squill, Sea Onion

Main function is to poison rats, mice, and other rodents.[20] The drug scilla represents the inner bulb scales of the white strain. While the rat poison is drawn from the red strain,[8] an article in *Organic Gardening* suggests it might be edible, but I would prefer to think of it as toxic. Balbaa et al.[33] note that the decreased hazard of red squill to other organisms than rats is because red squill induces emesis in other organisms while rats cannot vomit. Scilliroside is active, making methanol extracts of red squill effective as hair tonics in treating chronic seborrhea and dandruff. White squill is used as an expectorant in some cough preparations.[29]

The bulb is a popular old ''remedy'' for cancer, tumors, and indurations, e.g., of the eyes, liver, parotid, spleen, and viscera as well as calluses, corns, fungous flesh, felons, and warts.[4] Scilliglaucosidin has shown activity in KB cancer systems.[10] Formerly used as a cardiotonic, diuretic, emetic, expectorant.[20] As a diuretic, it is used for dropsy due to renal or cardiac ailments. In the 6th century B.C., Pythagoras invented Oxymel of Squill, used for coughs. Considered useful in asthma, bronchitis, catarrh, croup, and whooping cough. Leaves are crushed and applied to bruises, burns, cuts, and ulcerating sores. Homeopathically suggested for cardiac insufficiency with edema and arrhyrhmia, congestive, bronchitis, meteorism, and stomachache.[33]

Contains glucoscillaren A (scillarenin + rhamnose + 2 glucose), scillaren A (scillarenin + rhamnose + glucose), proscillaridin A (scillarenin + rhamnose), scillaridin A, scilliglaucoside, scilliphaeoside, glucoscilliphaeoside, scillicyanoside, scillicoeloside, scillazuroside, scillicryptoside.[20] Flavonoids (dihydroquercitin, dihydroquercitin-4′-monoglucoside, isoorientin, isovitexin, orientin, quercitin, scoparin, taxifolin, vitexin) and stigmasterol are also reported. Phytosterol and calcium oxalate also present.

Toxicity — Bulbs contain cardiotonic glucosides, but human poisonings are rare.[11] Fresh bulbs contain a vesicant liquid. Scillitin, the active diuretic and expectorant principle, is the main toxic element. In poisonous doses, squill produces violent gastrointestinal and genitourinal inflammation, nausea, purging, pain, dullness, convulsions, and stupor, a marked temperature drop, slowed circulation, and sometimes death.[2]

356. *URTICA DIOICA* L. (URTICACEAE) — Stinging Nettle, Common Nettle, Greater Nettle

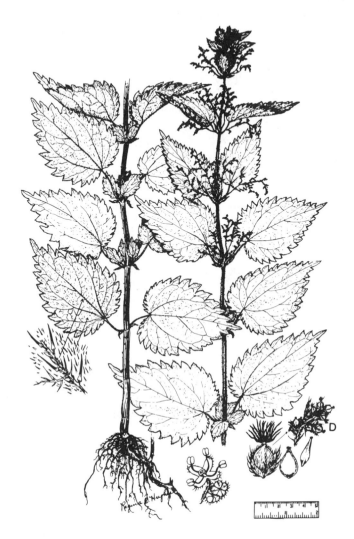

Young tops, gathered when about 15 cm high, can be used as a spring green vegetable, usually in the form of a puree, but their rather earthy flavor is not liked by some. In Scotland, nettles are combined with leeks or onions, broccoli or cabbage, and rice, boiled in a muslin bag, and served with butter or gravy. Nettle beer and nettle tea are made by some people. Dried nettles can be fed to livestock and poultry, but few animals will eat the living plants. I have eaten our species, raw and cooked. Swedes and Russians cultivate nettles as fodder crops. Alcoholic extracts of nettle, camomile, thyme, and burdock have been used in hair and scalp preparations. Russians use the nettle as a green pigment in confectionary. Nettles have served as a commercial source of chlorophyll.[37] Nettle fiber can be used for making textiles and paper, as has been done during wartime or when other fibers are not available. Roman soldiers, when they got too far north, flegellated themselves with nettle to keep warm. The fiber is similar to hemp or flax and can be used for fine or coarse materials.

Nettle is cited as a folk remedy for cancerous ulcers, inflamed and/or edematous tumors, and, specifically, cancer of the breast, ear, face, feet, lungs, joints, mouth, nostrils, spleen, stomach, womb, etc.[4] In Russia, leaves enter the preparation "Alochol", used for chronic hepatitis, cholangitis, cholecystitis, and habitual constipation.[30] Russians also recommend the herb for bronchial asthma.[332] Roots and seed are prescribed as a vermifuge. Clinical experiments are said to have confirmed the utility of the herb as a hemostatic. It is also used in anemia, asthma, bronchitis, constipation, diabetes, diarrhea, dropsy, dysentery, dysmenorrhea, dyspnea, epistaxis, gastritis, gravel, hematoptysis, headache, jaundice, malaria, menorrhea, nephritis, neuralgia, palsy, paralysis, pertussis, piles, rheumatism, sciatica, and tuberculosis, and as a hair tonic. Algerians powder nettle and jasmine for gonorrhea.[38] Roots are diuretic. Juice of the plants is used as an external irritant. Decoction of plant is anthelmentic, antiseptic, astringent, depurative, diuretic, emmenagogue, rubefacient, and vasoconstrictor. Homeopaths prescribe a tincture of the flowering plant for agalactia, beestings, burns, colic, dysentery, erysipelas, erythema, gout, gravel, hemorrhage, lactation, leucorrhea, malaria, menorrhagia, phlegmasia, preventing calculus, renitis, rheumatism, sore throat, splenitis, uremia, urticaria, vertigo, whooping cough, and worms.

Dried, young preflowering plants contain, per 100 g, 30.4 g crude protein, 3.4 g fat, 10.3 g cellulose, 39.6 g N-free extract, 16.3 g ash, 2970 mg Ca, 680 mg P, 32.2 mg Fe, 650 mg Mg, 3450 mg K, 140 mg Na, 4.3 mg Mn, 540 mg S, 680 mg Si, 270 mg Cl, and 20.2 mg beta-carotene. Oil from the seeds contained 11.5% oleic-, 73.7% linoleic-, 1.7% linolenic-acid, and *circa* 7.0% saturated acid (mainly palmitic), 4.5% glycerol, and 1.6% unsaponifiable material. Fresh plant material contained 80 μg vitamin B_1 per 100 g and 15.7 mg chlorophyll. Betaine, choline, and lecithin occur in the leaves. Carbonic, formic, and silicic acids are also reported, with phytosterins and tannin.

Toxicity — The sting is due to acetylcholine, histamine, and 5-hydroxytryptamine. Toxic effects from drinking nettle tea have been recorded: gastric irritation, burning sensation of the skin, edema, and urine suppression. Juice contains lecithin and is an antidote for the nettles own sting, and will quickly relieve the stinging sensation when rubbed on the affected spot. Juice of dock (*Rumex*) or touch-me-not (*Impatiens*) which usually grows nearby has the same beneficial action. Rosemary, sage, or mint leaves rubbed on nettle stings also relieve the itching. Classified by the FDA[62] as an Herb of Undefined Safety.

357. *VALERIANA OFFICINALIS* L. (VALERIANACEAE) — Valerian

Largely used as a drug, especially a sedative (formerly termed narcotic). Valerian vale-potriates have CNS-depressant and anticonvulsant activities in lab animals. Once a spice and perfume, still occasionally used in perfuming soaps. Anglo-Saxons used valerian as a salad, which sometimes took over its corner of the garden, and Scots added it to broths, meats, and pottage. According to Leung, the extracts and essential oil are components in beers, liqueurs, root beers, candy, frozen dairy desserts, baked goods, gelatins and puddings, meat, and meat products. Valerian is said to attract rats (the Pied Piper is rumored to have secreted valerian on his person). Valerian with CNS-depressant activities is reported to have antispasmodic and equalizing (sedative in states of agitation and stimulant in fatigue) activ-ities. The valepotriates are mainly responsible for the CNS-depressant and antispasmodic effects. Other activities include hypotensive in experimental animals; antibacterial, antidi-uretic, protective against experimental liver necrosis. An ethanol extract of valerian is said to have antidandruff properties.[29]

Valerian is regarded as a powerful nervine antispasmodic, as anodyne, bactericide, car-minative, CNS-depressant, hypnotic, nervine, sedative, stimulant, stomachic, sudorific, tonic, tranquilizer. Its sedative virtues are used in convulsions, fever, hypochondriasis, hysteria, insomnia, neuralgia, and St. Vitus' dance; said to strengthen the eyesight. Once used as an anticonvulsant in epilepsy. Also, used for cardiac palpitation, catarrh, cholera, cold, flu, menoxenia, neurasthenia, numbness, polyps, rheumatism, sores, spasms, trauma, worms, and wounds.[32,33] Culpepper prescribed it with aniseed, licorice, and raisins for coughs. Bruised valerian, applied to the forehead, was supposed to alleviate headache. Indochinese use the roots for dyspepsia, inflammation, and toothache.[16]

An oil in the subepidermal layers of the root contains acetic, formic and valeric acids, borneol, bornyl formate, bornyl acetate, bornyl butyrate, bornyl isovalerianate, camphene, pinene, and two alkaloids called chatinine and valerianine.[16] The important active compounds are iridoid compounds called valepotriates (valtrate, valtrate isovaleroxyhydrin, acevaltrate, valechlorine, etc.), didrovaltrates (didrovaltrate, homodidrovaltrate, deoxydodidrovaltrate, homodeoxydodidrovaltrate, isovaleroxyhydroxydidrovaltrate, etc.), and isovaltrates (isovaltrate, 7-epideacetylisovaltrate, etc.); valtrate and didrovaltrate are the major valepotriates. Also, contains valerosidatum (an iridoid ester glycoside) and a volatile oil (0.5 to 2%) containing, e.g., bornyl acetate and isovalerate (major compounds), caryophyllene, alpha- and beta-pinenes, valeranone, beta-ionone, eugenyl isovalerate, isoeugenyl isovalerate, patchouli alcohol, valerianol, borneol, camphene, beta-bisabolene, ledol, isovaleric acid, terpinolene, etc. Valerian also contains the alkaloids actinidine, valerianine, valerine, chatinine, etc. Other constituents include choline (*circa* 3%), methyl 2-pyrrolyl ketone, chlorogenic acid, and caffeic acid; beta-sitosterol; tannins; gums; etc.[29] One anonymous writer describes valeric acid as a very potent tranquilizer.[51] Dry seeds contain 19.4 to 19.9% protein and 30.0 to 34.4% fat.[21]

Toxicity — The compound α-methylpyrryl ketone is believed to be narcotic.[16] Approved for food use by the FDA § 172.510.[29]

358. *VANILLA PLANIFOLIA* Andr. (ORCHIDACEAE) — Vanilla

Pods the source of commercial vanilla. Vanilla is said to alleviate the sweet tooth of obese people fearing dental caries, the vanilla extract cutting back on the sugar requirement for fresh fruit salads in reducing diets, the catechin said to curb caries. Vanilla extract is used to flavor beverages, cakes, chocolates, confections, custards, ice creams, and puddings, and in perfumes, sachet powders, and soap. Highest maximum use level is nearly 10,000 ppm for vanilla in baked goods.[29] Vanilla has served in poison baits for fruitflies, grasshoppers, and melon beetles.[1]

Reported to be aphrodisiac, carminative, stimulant, and vulnerary. Venezuelans use the pods against fevers and spasms. Yucatanese use vanilla extract (pod steeped in alcohol) as aphrodisiac and stimulant; Argentinans as an antispasmodic, aphrodisiac, or emmenagogue.[42] In Palau, vanilla is used for dysmenorrhea, fever, and hysteria.[16] Vanillin is choleretic.[33] Vanilla is said to inhibit caries, probably because of the catechin.

Per 100 g, the beans are reported to contain 25.9 to 30.9 g H_2O, 2.6 to 4.9 g protein, 4.7 to 6.7 g fat, 30.5 to 32.9 g N-free extract, 7.1 to 9.1 g sugar, 15.3 to 19.6 g fiber, 4.5 to 9.7 g ash. Purseglove et al. dedicate more than four pages of tabulations to the chemistry of vanilla.[64] Cured pods contain anisic acid, anisaldehyde, glucovanillin, vanillic acid, and vanillin.[42] *Hager's Handbook* adds vanillyl alcohol, protocatechualdehyde, protocatechuic acid, *p*-hydroxybenzaldehyde, piperonal, anisalcohol, balsam, sugar (15% glucose and fructose, 35% saccharose) enzyme, fatty oil (glycerides of oleic-, palmitic-, and stearic-acid), tannin, resin, mucilage, essential oil, citric-, malic-, oxalic-, and tartaric-acids.[33] The aroma compounds include *p*-hydroxybenzylalcohol, acetaldehyde, diacetyl, furfurol, 2.5-methyl-furfurol, benzaldehyde, acetophenone, acetic acid, isobutyric-, caproic-, isovalerianic-, benzoic-, and anisic-acid, guaiacol, *p*-cresol, *n*-capric acid, *n*-caprylic acid, benzyl benzoate, etc.[33]

Toxicity — The calcium oxalate crystals in the plant may cause dermatitis.[1] Workers with vanilla may exhibit dermatitis, headache, and insomnia, all symptoms of vanillism.[64] Several toxic compounds are present in minor quantity. GRAS § 182.10, 182.20, and 169.3.[29]

359. *VERATRUM VIRIDE* Ait. (LILIACEAE) — American Hellebore

A "Major Medicinal Plant",[17] valued throughout the 18th and 19th centuries as an analgesic in treating painful diseases, epilepsy, convulsions, pneumonia, peritonitis, and as a cardiac sedative.[17] Starting in the 1950s there was an upsurging interest in the plant, and in 1979 rumor has it that one company was willing to pay $1.50/lb for 100,000 lb, possibly because of its hypotensive properties. The ester alkaloids reduce both systolic and diastolic pressure, slow the heart rate, and stimulate peripheral blood flow in the kidneys, liver, and extremities. Currently, largely reserved for hypertensive toxemia during pregnancy and the pulmonary edema which arises in acute hypertensive crises. Powdered rhizomes or some of its alkaloids are used in some insecticidal preparations.[17] Over a million prescriptions containing veratrum extracts were sold in the U.S. in 1973.[98]

Tinctures have been suggested for cancers, e.g., breast tumors.[4] Powdered rhizome and tincture thereof employed for respiratory afflictions, convulsions, mania, neuralgia, and headache.[1] Infusions gargled for sore throat and tonsillitis. Veratrine, a pale grey amorphous powder, is used externally as an analgesic and parasiticide. It can be used as a counterirritant in neuralgia. Also, said to be useful in arteriosclerosis and interstitial nephritis. Russians steep the herb in vodka for painful rheumatism and sciatica. Described as abortifacient, analgesic, anodyne, arteriosedative, cardiosedative, CNS-sedative, decongestant, deobstruent, diaphoretic, diuretic, emetic, febrifuge, nervine, sedative, spasmolytic, sternutatory, sudorific, tranquilizer, American hellebore has been prescribed for asthma, backache, cancer, cholera, cold, consumption, convulsion, croup, dandruff, dyspepsia, epilepsy, fever, gout, headache, herpes, hypertension, inflammation, mania, miscarriage, neuralgia, pertussis, pneumonia, puerperal fever, rheumatism, scarlet fever, sciatica, scrofula, shingles, sore throat, swellings, tonsillitis, toothache, tumors, typhoid, and wounds.[32,33] Homeopaths prescribe the tincture of the fresh root, gathered in autumn, for amaurosis, amenorrhea, apoplexy, asthma, bunions, cecal inflammation, chilblains, chorea, congestion, convulsions, diplopia, diaphragmitis, dysmenorrhea, erysipelas, esophageal spasm, fever, headache, heart, hiccups, hyperpyrexia, influenza, malaria, measles, meningitis, myalgia, orchitis, pneumonia, proctalgia, puerperal convulsion, puerperal mania, sleep, spinal congestion, splenic congestion, sunstroke, typhoid, and uteral congestion.

Alkaloids in the rhizome are placed in three groups: group A — alkamines (esters of the steroidal bases) with organic acids, including germidine and germitrine, most valued therapeutically, also, cevadine, neogermitrine, neoprotoveratrine, protoveratrine, and veratridine; group B — (glucosides of the alkamines), mainly pseudojervine and veratrosine; group C — (alkamines), germine, jervine, rubijervine, and veratramine.[17] Muscle relaxation has been induced in cats by administration of protoveratine with phenyldiguanide and 5-hydroxytryptamine.

Toxicity — All parts, especially the rhizome, are highly toxic. Young animals, especially lambs, have died from leaf ingestion, and poultry have succumbed to eating leaves or seeds. All veratrum species are said to be irritant. Several of the alkaloids cause watering of the mouth, nausea, diarrhea, stomachache, general paralysis, and spasms, with severe poisoning leading to shallow breathing, slower pulse, lower temperature, convulsions, and possibly death.

To the physician — For poisoning, gastric lavage or emesis is recommended followed by activated charcoal, atropine, and/or hypotensive drugs.[34]

360. *VERBENA OFFICINALIS* L. (VERBENACEAE) — Vervain, Verbena

Used in some liqueurs, and rather popular of old, because of a reputed aphrodisiac quality. The essential oil may find some use in perfumery. Verbenalin hastens blood coagulation. Perry mentions its use as an insecticide.[16] The extracts are analgesic and antiphlogistic.[33]

Widely noted (Mexico, Brazil, China) as a folk remedy for cancer and tumors, especially of the neck, parotids, scrotum, spleen, and viscera, for indurations of the veins and for polypus.[4] Fresh leaves have been used as a rubefacient in rheumatism and to promote wound healing. Once used for eye disease. Tierra describes it as excellent for fever, especially combined with boneset.[28] Has been used at one time or another for anemia, bronchitis, calculus, colds, cramps, dropsy, dysuria, eczema, edema, fever, headache, hemorrhoids, inflammation, insomnia, malaria, metrorrhagia, neuralgia, pertussis, pleurisy, tympanites, ulcers, and uterosis. The herb is said to be anthelmintic, antiscorbutic, antispasmodic, aphrodisiac, astringent, detersive, diaphoretic, diuretic, emetic, emmenagogue, expectorant, febrifugal, lactagogue, nervine, purgative, sudorific, and tonic.[16,32] Used homeopathically for stones, as diuretic, and emmenagogue.[33]

Plant contains the glucoside, verbenalin ($C_{17}N_{24}O_{10}$), which on hydrolysis with emulsin, also present, and yields glucose and verbenalol ($C_{11}H_{14}O_5$). The essential oil contains citral, geraniol, limonene, terpenes, terpene alcohols, and verbenone.[12] Leaves contain adenosine and beta-carotene. Stachyose has been found in the roots (2%) and stems (1.3%).[1] There is the iridoid glycoside, hastatoside ($C_{17}H_{24}O_{11}$).[33] Another glucoside verbenin stimulates the flow of milk; in small doses it stimulates, in large doses inhibits the sympathetic nerve endings of the epidermal mucous glands of the heart and blood vessels and or the intestine and salivary glands. A peculiar tannin and tannic acid are also present with invertin and saponin.

Toxicity — Classified by the FDA[62] as an Herb of Undefined Safety: ''Herb is an astringent vulnerary. Leaves a substitute for Chinese tea. Listed in 21 CFR 121.1163 under Vervain. European for use in alcoholic beverages only.'' Suspected of poisoning cattle in Australia. With frogs, small doses of verbenalin stimulate the uterus, but high doses paralyze the CNS, following stupor, clonic, and tetanic convulsions.

361. *VINCA MINOR* L. (APOCYNACEAE) — Periwinkle

Often planted in the temperate zone as an evergreen ornamental ground cover. Sometimes used in garlands and Christmas wreaths. Of interest as a temperate source of drugs. The alkaloid vincamine is used clinically in eastern Europe for treating arterial hypertension, and as a sedative (perhaps due to the reserpine content). It is said to be effective against headache, migraine, and neurogenic tachycardia. Once used in magic, a favorite flower for making charms and love philters. It was believed to exorcize evil spirits.

In old herbals, periwinkle was used for tumors of the uvula. Reserpine shows anticancer activity in the CA, SA, and WA tumor systems; beta-sitosterol in the CA, LL, and WA systems; ursolic acid in the PS system.[10] Acid, astringent, bactericide, carminative, collyrium, depurative, diuretic, emetic, hemostat, lactagogue, sedative, spasmolytic, and tonic, the periwinkle is used for bleeding, catarrh, diarrhea, dysentery, eczema, fits, hypertension, hysteria, menorrhagia, nervous disorders, phthisis, piles, tumors, and nightmares.[32,33] Periwinkle used to be tied around cramped limbs as a remedy. Bruised leaves with lard used for piles. Used homeopathically for hemorrhagia. Flowers said to be purgative when fresh. Bruised leaves are stuffed into the nostrils to stop nose bleed.[2] Leaves are a bitter astringent used against dysentery and hemorrhoids.[11] Aqueous extracts injected in cats are hypotensive. Given orally, the plant lowers arterial hypertension. Vincamine has a weak antagonistic action on the pressor activity of adrenaline. Infusion of the herb used as a gargle and mouthwash and the extract as a vermifuge.[1] The plant is applied to scalp ailments. While the literature available to me suggests that this plant is lactagogue, Parvati suggests its usage to control overabundant milk and to dry up milk when weaning. She recommends it for vaginal hemorrhage, with motherwort for bloated congestion with a late period, and as a douche with chaparral, oatstraw, and slippery elm for trichomoniasis.[48]

Many alkaloids are reported from this species: 1,2-dehydroaspidospermidine, eburnamenine, eburnamine, eburnamonene, epivincamine, 14-epivincamine, isoburnamine, isovincarnine, methoxyminovincine, 16-methoxy-20-oxo-1-vin-cadifformine, methoxyvincamine, *N*-methylaspidospermidine, minorine, minovincine, minovincinine, minovine, perivincine, pubescine, quebrachamine, reserpine, strictamine, vincadifformine, vincadine, vincamidine, vincamine (0.062 to 0.168% in cultivars, 0.033 to 0.073 in wild), vincaminine, vincaminoridine, vincaminorine, vincarorine, vincine, vincinine, vincoridine, vincorine, vinine, vinoxine, and vintsine. The alkaloidal content of the leaves, according to Hungarian analyses, varied from 0.11 to 7.06% (dry matter), and vincamine content from 0.02 to 1.75% (also dry matter). The total alkaloidal content of the roots from different countries varied from 1.24 to 1.98%. Nonalkaloidal constituents include a number of acids, 1-glutamate carboxylase (leaves), 1-bornesitol and the enzyme forming the same, dambonitol, ornol, phenols, rubber (2.0 to 2.2% in the plant; 1.1% in the leaves), ursolic acid (0.24 to 1.34% in the plant; 0.14 to 3.7% in the leaves), beta-sitosterol, saponin, 3-beta-D-glucosyloxy-2-hydroxybenzoic acid (leaves, an amorphorus and bitter but odorless glucoside called vincoside, 1%), flavonoid glycosides including robinin (present in the flowers, 0.4%) and delphinidin-3,5-diglycoside.[1] The decoction contains tannin. The leaf ash yields the glucoside C-beta-*d*-glucosyloxy-2-hydroxybenzoic acid as well as ursolic acid. Flowers yield 0.4% robinoside. Leaves contain the anticancer agent beta-sitosterol.

Toxicity — Although clearly a biologically active species belonging to a dangerous family, this periwinkle has been little cited as an allergic or toxic plant!

362. *VIROLA CALOPHYLLA* Warb. (MYRISTICACEAE) — Epena, Parica, Yakee

Bark made into an hallucinogenic snuff.[52] Highly narcotic, it was taken only by witch doctors to diagnose and treat disease, and for divination, magic, and prophecy.[58]

Reported to be narcotic.[32] The active hallucinogenic principles are *N,N*-dimethyltryptamine, *N*-monomethyltryptamine, and 5-methoxy-*N,N*-dimethyltryptamine.[54]

Toxicity — Classed as a narcotic hallucinogen.[54] Mixed with Theobroma, *Virola* snuff is powerful, causing intoxication which, it is reported, has occasionally led to death among witch doctors.[58]

363. *VISCUM ALBUM* L. (LORANTHACEAE) — European Mistletoe

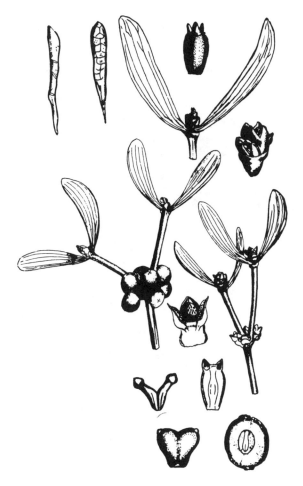

According to Keough,[333] recent research shows that extract of mistletoe, long a folk remedy for cancer, is effective against the growth of cancer cells. A tiny amount prolongs the lives of leukemic mice. Cancer cells treated with mistletoe extract were unable to form new colonies in laboratory cultures. In Germany and Austria, a derivative is used in post-operative treatment following many types of cancer.[333] According to Morton,[50] "It is known to possess narcotic, hypotensive, antispasmodic, tumor-inhibiting and thymus-stimulating activity." The tumor-inhibiting activity of the drug is due to the presence of a basic protein complex. Several protein fractions showing marked cancerostatic effects in extremely low doses (0.006 to 0.05 mg/kg) have been isolated from the press juice of the plant; some of the fractions were specific for malignant HeLa cells. Tumor-inhibiting activity of the fractions was independent of the toxicity.[1] Iscador is a natural fermentation product of plant juice of viscum, and is used "in human cancer therapy." It appears to be a good adjuvant for the induction of antibody formation and delayed hypersensitivity. Protein fractions induce splenomegaly and thymus hyperplasia in tumor-bearing mice all the while inhibiting tumor growth.[334] In spite of reputed toxicity, mistletoe is cut for winter feed for cattle; the fruits are used in decorated confectionary; the starch-containing stems are dried, ground, and mixed with rye to make bread in times of scarcity.[1] Berries once a source of bird lime. The Druids believed that mistletoe protected its possessors from all evil. Tea from this plant was sold in "health food stores" at about 15¢ an ounce in 1977.[335] Perhaps it is popular because it was believed to be useful in cases of sterility in the older herbals.

Dried leaves steeped in wine are given for nervous complaints. Regarded as antiseptic, antispasmodic, astringent, cardiac, cardiotonic, digestive, diuretic, emetic, lactagogue, narcotic, nervine, purgative, stimulant, and vasodilator, mistletoe has been recommended for such things as amenorrhea, apoplexy, arthrosis, asthma, arteriosclerosis, cancer, chilblains, chorea, convulsions, delirium, epilepsy, heart disease, hemorrhage, hepatosis, hypertension, hypertomy, hysteria, lumbago, malaria, metrorrhagia, nervous debility, neuralgia, neuritis, otitis, piles, St. Vitus disease, spasms, splenomegaly, spondylosis, tumors, typhoid, ulcers, urinary disorders, uterosis, varicose veins, and vertigo. The juice of the berries is applied externally to sores and ulcers. The tea is compressed onto varicose veins. In Europe the drug (derived from leaves, stems, and berries) enters several proprietary hypotensive drugs. It is also used for hepatic and splenic enlargements, and in hemorrhoids and lumbago.[1] Koreans make a tonic tea for colds, lumbago, rheumatism, and weak muscles. Chinese use the dried inner portions of the stem as antispasmodic, carminative, hypotensive, laxative, lactogogue, and sedative, believing it promotes hair growth, alleviates liver congestion, stimulates the kidneys, strengthens bones, quiets the pregnant womb.[16]

Said to contain aromatic vasopressor amines (sympathomimetic amines).[11] The active part, the resin viscin, is accompanied by mucilage, tannin, a fixed oil, inositol, xanthophyll, sugar, and starch. Ripe berries contain 750 mg vitamin C per 100 g; leaves, 75 mg. Carotenoids in the leaves include alpha- and beta-carotenes and lutein. The hypotensive action of the drug is attributed to acetylcholine, histamine, gamma-aminobutyric acid, and flavones. Oral efficacy of these items is doubtful.[1] Freshly prepared juice contains arginine, asparagine, cysteic acid, hydroxylysine, and 1-kynurenine.[16] Leaves and twigs contain beta-amyrin, lupeol, oleanolic acid, tyramine, beta-phenylalanine, acetylcholine, ceryl alcohol, mannitol, quercitin, inositol, glucose, arabinose, rhamnose, caffeic-acid, sinapic-acid, quercetin-3-rhamnoside, quercetin-3-arabinoside, syringin, myristic acid, flavoyedorinin A and B.

Toxicity — Fatalities have been reported following ingestion of the berries. Children frequently suffer epileptiform convulsions following ingestion of the berries. "Contains viscotoxin, a mixture of toxic proteins which apparently affect RNA and DNA synthesis. Lethal doses of Viscum proteins produce anemia and hemorrhage in the liver, intestinal hemorrhage and fatty degeneration of the thymus in experimental animals."[50] The LD_{50} i.p. in mice of the plant juice was 32 mg (dry weight) of juice per kilogram body weight. Fermentation renders the plant juice less toxic.

364. *WITHANIA SOMNIFERUM* (L.) Dunal (SOLANACEAE) — Ashwagandha

Seeds are used as a vegetable rennet to coagulate milk.[2] Tender shoots are said to serve as a vegetable in India. Seeds are used as a masticatory. Withaferin A shows antiarthritic, antibiotic, antimitotic, antitumor, and fungicidal activity. Withaferinile is antimitoic. The drug is sedative, hypnotic, laxative, and diuretic, fruits analgesic, diuretic, proteolytic, and sedative.

Reported to be abortifacient, amebicide, anodyne, bactericide, contraceptive, diuretic, emmenagogue, fungicide, narcotic, pediculicide, poison, sedative, spasmolytic, tonic, ashwagandha is a folk remedy for adenopathy, anthrax, arthritis, asthma, bronchitis, cancer, candida, cold, cough, cystitis, debility, diarrhea, dropsy, dyspepsia, erysipelas, fever, furuncle, gynecopathy, hiccups, hypertension, inflammation, lumbago, marasmus, nausea, piles, proctitis, psoriasis, rheumatism, ringworm, scabies, senility, smallpox, sores, syphilis, tuberculosis, tumors, typhoid, uterosis, and wounds.[3,32,33] Steeped in warm castor oil, the leaves are applied to carbuncles, inflammations, and swellings.[2] Bruised berries are rubbed onto ringworm.[1] Lesotho natives take the root decoction for colds and chills. The bark infusion is taken for asthma and applied topically to bedsores. Zulu give an enema of decorticated root to feverish infants. Tanganyikans use the root as a sexual stimulant and ecbolic. Masai use the leaf juice for conjunctivitis. Alcoholic preparations of the plant have been used in alcoholism, emphysema, and pulmonary tuberculosis.

The roots contain anahygrine, meso-anaferine, cuscohygrine, isopelletierine, hygrine, tropine, pseudotropine, 3 alpha-tigloyloxytropane, choline, withasomnine, 0.1% saccharose, 0.02% beta-sitosterol, somniferine ($C_{12}H_{16}N_2$), withanine ($C_{44}H_{80}N_2O_{12}$), withaninine, nicotine, ipuranol, hentriacontane, fatty oils, essential oils, withanolide, scopoletin.[3,33] The leaves contain withaferin A ($C_{28}H_{38}O_6$) and withanolide ($C_{28}H_{38}O_6$) glycosides, reducing sugars, somnitol, withanone ($C_{24}H_{32}O_5$), glycine, cystine, glutamic acid, alpha-alanine, proline, and tryptophane.[33] Chlorogenic acid, condensed tannins, and flavonoids are also reported, along with somnisol ($C_{32}H_{43}O_6OH$), somnitol ($C_{33}H_{44}O_5[OH]_2$) and withanic acid ($C_{29}H_{45}O_6COOH$).

Toxicity — Narcotic, sedative, and tranquilizer;[54] suspected of poisoning stock.[1] The berry, sometimes eaten by children, causes severe gastrointestinal upset.[3]

365. *ZIZIPHUS SPINA-CHRISTI* (L.) Willd. (RHAMNACEAE) — Syrian Christthorn, Thorns

Planted as a hedge, this thorn is highly successful, for it forms an almost impenetrable fence. Arabs use it to keep goats and cattle off the fields. The Arabs revere the tree as sacred, planting it for shade.[37] If, however, it is grown with crops, cattle will eat it as herbage. A good liquor is also obtained from the fruit. Fruits, eaten by Arabs, have the tase of dried apples; if not grapes, the Arabs gathered apples of "thorns". Some think this is the thorn used to make Christ's cruel corona. The bark can be used as a source for tannin. The hard, heavy, termite-proof wood is used in African carpentry.

Said to be anodyne, astringent, demulcent, depurative, emollient, laxative, pectoral, refrigerant, stomachic, and tonic, this thorn is said to be used for toothaches and tumors, even as a mouthwash. Lebanese believe that all parts of the plant are medicinal, the powdered seeds with lemon juice for liver complaints, the flower infusion as an eyewash and febrifuge, the boiled bark for venereal disease, the cathartic raw root juice for arthritis and rheumatism, the fruits for bronchitis, coughs, and tuberculosis.[30] In Curacao, the leaf decoction is used for colds and hypertension.[42] Perhaps the antitumor activity may be attributed to the presence of beta-sitosterol, a proven antitumor agent.[10]

The antitumor estrogen beta-sitosterol has been reported. Morton lists the peptide alkaloids amphibine A, E, and F, and mauritine A and C.[42] Relatively dry fruits, per 100 g, contain 314 calories, 9.3% H_2O, 4.8% protein, 0.9% fat, 80.6% total carbohydrate, 4.4% ash, 140 mg Ca, 3.0 g Fe, 0 μg vitamin A, 0.04 mg thiamine, 0.13 mg riboflavin, 3.7 mg niacin, and 30 mg ascorbic acid.[21]

Table 1
365 MEDICINAL HERBS: TOXICITY RANKING AND PRICELIST[37,47]

Score[a]	Scientific name	Price (source)[b]	Common name
3-2-X	Abelmoschus moschatus	22(A)	Musk okra
0-X-X	Abrus precatorius		Jequerity
3-1-X	Acacia farnesiana	11(A)	Cassie
3-1-X	Acacia senegal	14(A), 17(T)	Gum arabic
2-X-X	Acalypha indica		Indian acalypha
3-1-3	Achillea millefolium	12(A), 6(B), 11(T)	Yarrow
0-X-X	Acokanthera schimperi		Arrowpoison tree
0-0-X	Aconitum napellus		Monkshood
1-0-1	Acorus calamus	14(A), 39(T)	Sweet flag
1-X-X	Actaea pachypoda		Baneberry
1-X-X	Actinidia polygama		Cat powder
1-X-X	Adhatoda vasica		Malabar nut
1-X-X	Aesculus hippocastanum	14(A)	Horse chestnut
0-X-X	Aesthusa cynapium		Fools parsley
3-1-3	Agave sisalana		Mescal
2-1-X	Agrimonia eupatoria	12(A), 12(T)	Agrimony
1-X-X	Agrostemma githago		Corncockle
2-X-X	Ajuga reptans		Bugleweed
0-X-X	Alchornea floribunda		Niando
1-X-X	Aletris farinosa	28(T)	Unicorn root
1-X-X	Aleurites moluccana		Candlenut
2-1-X	Alnus glutinosa	13(A)	Alder
3-2-3	Aloe barbadensis	20(A), 22(T)	Aloe
2-1-X	Aloysia triphylla	22(A), 11(B), 23(T)	Lemon verbena
1-X-X	Anadenathera colubrina		Vilca
1-X-X	Anadenathera peregrina		Niopo
2-1-X	Ananas comosus		Pineapple
3-X-X	Anaphalis margaritacea	35(A), 21(T)	Pearly everlasting
1-X-X	Andira araroba		Goa
1-X-X	Andira inermis		Cabbagebark
1-0-X	Anemone pulsatilla		Pasque flower
3-3-X	Anethum graveolens	30(A), 18(B), 19(T)	Dill
2-1-1	Angelica archangelica	26(A), 18(B), 36(T)	Angelica
1-X-1	Angelica polymorpha	118(B), 275(T)	Dong quai
2-3-X	Apium graveolens	8(A), 6(T)	Celery
1-X-X	Apocynum androsaemifolium		Bitter root
1-X-X	Apocynum cannabinum		Indian hemp
1-3-X	Aquilegia vulgaris		Columbine
1-X-X	Arbutus unedo		Strawberry tree
2-3-2	Arctium lappa	20(A), 14(B), 23(T)	Burdock
2-1-3	Arctostaphylos uva-ursi	13(A), 14(T)	Bearberry
2-X-X	Areca catechu	12(T)	Betel-nut
1-X-X	Argemone mexicana		Prickly poppy
1-X-X	Argyreia nervosa		Woodrose
1-X-X	Arisaema triphyllum		Jack-in-the-pulpit
2-X-X	Aristolochia serpentaria	146(A)	Snakeroot
1-0-3	Arnica montana	47(A), 60(T)	Mountain tobacco
1-0-X	Artemisia abrotanum	22(A), 22(T)	Southernwood
1-0-1	Artemisia absinthium	10(A), 13(T)	Wormwood
2-3-X	Artemisia dracunculus	97(A), 29(B), 61(T)	Tarragon
1-0-X	Artemisia vulgaris	11(A), 6(B), 12(T)	Mugwort
1-X-X	Asclepias syriaca		Common milkweek
0-0-X	Atropa bella-donna		Belladonna
1-0-X	Banisteriopsis caapi		Caapi
2-X-2	Barosma betulina	17(A), 10(B), 18(T)	Buchu
1-0-3	Berberis vulgaris	14(A)	Barberry

Table 1 (continued)
365 MEDICINAL HERBS: TOXICITY RANKING AND PRICELIST[37,47]

Score[a]	Scientific name	Price (source)[b]	Common name
0-X-X	Blighia sapida		Akee
3-1-2	Borago officinalis	15(A), 9(B)	Borage
0-X-X	Bowiea volubilis		Climbing onion
0-X-X	Brunfelsia uniflorus		Manaca
1-X-X	Buxus sempervirens		Boxwood
1-X-X	Caladium bicolor		Heart-of-Jesus
1-X-X	Calea zacathechichi		Mexican calea
3-3-2	Calendula officinalis	16(T)	Marigold
1-X-X	Calliandra anomala		Cabeza de angel
1-X-X	Calotropis procera		Giant milkweek
2-X-2	Camellia sinensis		Tea
1-X-X	Cananga odorata		Cananga
2-0-X	Cannabis sativa		Marijuana
2-X-X	Capsicum annuum	14(A)	Chili
3-3-3	Carica papaya	12(A), 7(B), 13(T)	Papaya
1-X-3	Cassia angustifolia	10(A), 8(T)	Indian senna
1-X-3	Cassia senna	10(A)	Alexandrian senna
1-X-X	Catha edulis		Khat
1-X-X	Catharanthus lanceus		Lance-leaf periwinkle
1-X-X	Catharanthus roseus		Madagascar periwinkle
1-X-1	Caulophyllum thalictroides	8(B), 16(T)	Blue cohosh
1-3-1	Centella asiatica	12(A), 7(B), 11(T)	Gotu kola
3-1-3	Chamaemelum nobile	59(A), 29(B)	Roman chamomile
0-X-X	Chamaesyce hypericifolia		Spurge
1-1-X	Chelidonium majus	16(A), 17(T)	Celandine
1-1-X	Chenopodium ambrosioides	31(A), 22(T)	Wormseed
1-X-X	Chionanthus virginica		Fringe tree
2-1-X	Chrysanthemum cinerariifolium		Pyrethrum
3-1-X	Chrysanthemum parthenium		Feverfew
0-X-X	Cicuta maculata		Water hemlock
1-0-2	Cimicifuga racemosa	20(A), 7(B), 14(T)	Black cohosh
2-2-X	Cinchona sp.	23(A)	Quinine
1-X-X	Cineraria aspera		Mohodu
2-2-X	Cinnamomum camphora	88(A)	Camphor
2-3-X	Cinnamomum verum	14(A), 41(T)	Cinnamon
0-X-X	Citrullus colocynthis		Colocynth
1-0-X	Clematis vitalba		Traveler's joy
2-X-2	Coffea arabica		Coffee
2-3-2	Cola acuminata	12(A), 7(B), 22(T)	Kola nuts
2-3-2	Cola nitida	12(A), 22(T)	Kola
0-X-X	Colchicum autumnale		Autumn crocus
1-X-X	Coleus blumei		Nene
0-0-X	Conium maculatum		Poison hemlock
0-0-X	Consolida ambigua		Larkspur
1-0-X	Convallaria majalis	17(A), 25(T)	Lily-of-the-valley
0-X-X	Coriaria thymifolia		Shanshi
2-X-X	Cornus florida		Dogwood
1-2-3	Crataegus oxyacantha	12(A), 7(B), 12(T)	Hawthorn
2-3-X	Crocus sativus	2280(B), 1925(T)	Saffron
1-3-X	Croton eleuteria	28(T)	Cascarilla
0-X-X	Cryptostegia grandifolia		Rubber vine
0-X-X	Cycas revoluta		Sago cycas
2-0-X	Cypripedium calceolus	80(A), 83(T)	Yellow ladyslipper
1-0-1	Cytisus scoparius	11(A), 12(T)	Scotch broom
1-3-X	Daemonorops draco	40(A), 50(T)	Dragon's blood

Table 1 (continued)
365 MEDICINAL HERBS: TOXICITY RANKING AND PRICELIST[37,47]

Score[a]	Scientific name	Price (source)[b]	Common name
0-X-X	Daphne mezereum	14(A)	Mezereon
0-0-X	Datura candida		Borrachero
0-0-X	Datura innoxia		Thornapple
0-0-X	Datura metel		Unmatal
0-0-X	Datura stramonium		Jimsonweed
2-X-X	Daucus carota		Carrot
1-X-X	Dieffenbachia seguine		Dumbcane
0-0-X	Digitalis purpurea		Digitalis
1-X-X	Dioscorea composita	14(A)	Barbasco
2-1-X	Dipteryx odorata	62(A), 72(T)	Tonka bean
2-0-X	Dryopteris filix-mas	21(T)	Male fern
0-X-X	Duboisia myoporoides		Corkwood
1-X-X	Duranta repens		Golden dewdrop
1-X-X	Elaeophorbia drupifera		Dodo
3-3-2	Eleutherococcus senticosus	44(A)	Spiny ginseng
1-1-1	Ephedra gerardiana	11(A)	Pakistani ephedra
1-1-1	Ephedra nevadensis	8(B), 10(T)	Mormon tea
1-2-2	Equisetum arvense	13(A), 13(T)	Field horsetail
1-2-2	Equisetum hyemale	13(A), 7(B), 13(T)	Shavegrass
1-X-X	Erythrina fusca		Coral bean
0-X-X	Erythrophleum suaveolens		Sassybark
1-0-X	Erythroxylum coca		Coca
1-0-X	Eschscholzia californica		California poppy
2-1-X	Eucalyptus spp.	12(A), 8(B), 13(T)	Eucalyptus
1-X-X	Euonymus atropurpureas	56(A), 69(T)	Wahoo
2-X-2	Eupatorium perfoliatum	11(A), 10(T)	Boneset
0-X-X	Euphorbia lathyris		Mole plant
0-X-X	Euphorbia pulcherrima		Poinsettia
0-X-X	Euphorbia tirucalli		Aveloz
2-1-0	Euphrasia officinalis	37(A), 22(B), 34(T)	Eyebright
2-3-X	Ferula assa-foetida	60(A)	Asafetida
2-3-X	Ferula sumbul		Sumbul
3-3-X	Filipendula ulmaria	15(A), 25(T)	Meadowsweet
3-3-3	Foeniculum vulgare	10(A), 7(B), 7(T)	Fennel
1-1-X	Frangula alnus	17(A), 9(B), 20(T)	Buckthorn
3-3-X	Galium odoratum	16(A), 14(T)	Woodruff
1-0-X	Garcinia hanburyi		Mangosteen
1-1-X	Gaultheria procumbens	26(A), 23(T)	Wintergreen
0-X-X	Gelsemium sempervirens		Yellow jessamine
0-X-X	Genista tinctoria		Dyer's broom
2-2-3	Gentiana lutea	27(A), 24(T)	Gentian
2-1-X	Geranium maculatum	21(A), 28(T)	Cranesbill
2-X-X	Glechoma hederacea	25(T)	Ground ivy
2-X-X	Gleditsia triacanthos		Honey locust
0-X-X	Gloriosa superba		Glory lily
1-3-2	Glycyrrhiza glabra	22(A), 8(B), 30(T)	Licorice
1-X-X	Gossypium barbadense	28(T)	Sea island cotton
1-X-X	Gossypium hirsutum	28(T)	Upland cotton
1-1-X	Grindelia spp.	25(T)	Rosinweed
2-3-3	Hamamelis virginiana	17(A), 19(T)	Witch-hazel
2-X-2	Harpagophytum procumbens	34(A), 22(B)	Devil's claw
2-X-2	Hedeoma pulegioides	8(B)	Pennyroyal
1-X-X	Hedera helix		Ivy
1-X-X	Heimia salicifolia		Sinicuichi
1-1-X	Heliotropium europaeum		Heliotrope

Table 1 (continued)
365 MEDICINAL HERBS: TOXICITY RANKING AND PRICELIST[37,47]

Score[a]	Scientific name	Price (source)[b]	Common name
0-0-X	Helleborus niger	8(A)	Christmas rose
3-2-2	Hibiscus sabdarriffa	11(T)	Roselle
0-X-X	Hippomane mancinella		Manchineel
1-X-X	Homalomena sp.		Homalomena
0-X-X	Hoslundia opposita		Kamyuye
2-3-2	Humulus lupulus	20(A), 7(B), 22(T)	Hops
0-X-X	Hura crepitans		Sandbox tree
2-3-2	Hydrangea arborescens	20(A), 20(T)	Seven barks
2-1-1	Hydrangea paniculata		Peegee
1-3-2	Hydrastis canadensis	92(A), 29(B), 79(T)	Goldenseal
0-0-X	Hyoscyamus niger		Henbane
1-1-3	Hypericum perforatum	12(A), 15(T)	St. John's wort
2-X-X	Ilex opaca		Holly
2-3-2	Ilex paraguariensis	8(B), 19(T)	Mate
1-3-X	Illicium verum	18(A), 6(B)	Star anise
2-X-X	Indigofera tinctoria	28(A), 25(T)	Indigo
1-X-X	Ipomoea purga	31(A), 37(T)	Jalap root
1-X-X	Ipomoea violacea		Tlitliltzen
2-X-X	Iris versicolor		Blue flag
0-X-X	Jatropha curcas		Physic nut
0-X-X	Jatropha gossypiifolia		Tua-tua
1-3-2	Juniperus communis	20(A), 8(B), 16(T)	Juniper
1-0-X	Juniperus sabina		Sabine
1-X-X	Justicia pectoralis		Bolek hena
2-3-X	Kaempferia galanga		Maraba
0-X-X	Kalmia latifolia		Mountain laurel
0-X-X	Laburnum anagyroides		Golden chain
1-X-X	Lachnanthes tinctoria		Redroot
2-0-2	Lactuca virosa		Lettuce
1-X-X	Lagochilus inebrians		Intoxicating mint
1-X-X	Lantana camara		Lantana
2-2-0	Larrea tridentata	10(A), 12(T)	Creosotebush
0-X-X	Latua pubiflora		Latua
2-1-X	Laurus nobilis	20(A), 5(B), 8(T)	Bay
2-3-X	Lavandula angustifolia	30(A), 12(B)	Lavender
2-1-X	Lawsonia inermis	58(T)	Henna
1-0-X	Ledum palustre	25(T)	Marsh tea
1-X-X	Leonotis leonurus		Dagga
2-X-X	Leonurus cardiaca	12(T)	Motherwort
1-X-X	Ligustrum vulgare		Privet
1-0-1	Lobelia inflata	25(A), 28(T)	Indian tobacco
1-X-X	Lobelia tupa		Tupa
0-X-X	Lolium temulentum		Darnel
2-1-X	Lophophora williamsii		Peyote
2-X-X	Lycopersicon esculentum		Tomato
1-X-3	Mahonia aquifolia	19(A), 20(T)	Oregon grape
3-X-X	Malus sylvestris		Apple
3-2-X	Malva rotundifolia		Dwarf mallow
0-X-X	Mandragora officinarum	18(A)	Mandrake
1-X-X	Manihot esculenta		Cassava
3-3-X	Maranta arundinacea	13(A), 4(B), 7(T)	Arrowroot
1-X-X	Marsdenia reichenbachii	14(A)	Condurango
3-1-3	Matricaria chamomilla	22(A), 18(T)	German chamomile
3-2-2	Medicago sativa	10(A), 5(B)	Alfalfa
2-2-X	Melaleuca leucadendron		Cajeput
0-X-X	Melia azedarach		Chinaberry

Table 1 (continued)
365 MEDICINAL HERBS: TOXICITY RANKING AND PRICELIST[37,47]

Score[a]	Scientific name	Price (source)[b]	Common name
2-2-X	Melilotus officinalis	25(T)	Sweetclover
1-X-X	Menispermum canadense		Moonseed
2-3-1	Mentha pulegium	13(A), 8(B), 13(T)	Pennyroyal
0-X-X	Mercurialis annua		Annual mercury
0-X-X	Methystichodendron amesianum		Culebra
1-X-X	Mimosa hostilis		Jurema
1-X-X	Mirabilis multiflora		Soksi
2-X-X	Mitchella repens	29(A), 26(T)	Partridgeberry
1-X-X	Mitragyna speciosa		Katum
1-X-X	Momordica charantia		Balsam pear
2-3-2	Myrica cerifera	31(A), 14(B), 26(T)	Bayberry
1-2-X	Myristica fragrans	16(A), 6(B), 17(T)	Nutmeg
1-1-X	Myroxylon balsamum var. pereirae		Balsam of Peru
0-0-X	Narcissus tazetta		Daffodil
3-X-2	Nepeta cataria	16(A), 8(B), 15(T)	Catnip
0-0-X	Nerium oleander		Oleander
0-X-X	Nicotiana glauca		Tree tobacco
0-X-X	Nicotiana rustica		Aztec tobacco
0-X-X	Nicotiana tabacum		Tobacco
2-3-X	Ocimum basilicum	28(A), 12(B), 12(T)	Basil
0-X-X	Onenanthe phellandrium		Water fennel
1-X-X	Ornithogalum umbellatum		Dove's dung
0-0-X	Paeonia officinalis	57(A), 14(T)	Peony
3-3-2	Panax ginseng	528(A), 528(T)	Oriental ginseng
3-3-2	Panax notoginseng		Sanchi ginseng
3-3-2	Panax quinquefolius	598(T)	American ginseng
0-X-X	Pancratium trianthum		Kwashi
1-0-X	Papaver bracteatum		Scarlet poppy
1-0-X	Papaver somniferum	14(A), 9(T)	Opium poppy
1-X-X	Paris quadrifolia		Herb paris
2-3-3	Passiflora incarnata	10(A), 10(T)	Passionflower
2-X-X	Passiflora quadrangularis		Granadilla
2-3-2	Paullinia cupana	70(A), 53(B), 62(T)	Guarana
2-X-X	Paullinia yoko		Yoko
1-0-1	Pausinystalia johimbe	28(T)	Yohimbe
1-X-X	Peganum harmala		Harmel
2-X-X	Perilla frutescens		Perilla
2-3-2	Petroselinum crispum	28(A), 12(B), 21(T)	Parsley
0-X-X	Peumus boldus	11(A), 11(T)	Boldo
2-X-X	Phaseolus lunatus		Lima bean
3-X-X	Phaseolus vulgaris		Bean
1-0-1	Phoradendron serotinum		American mistletoe
0-X-X	Physostigma venenosum		Ordeal bean
0-X-0	Phytolacca americana	19(A)	Pokeweed
0-3-X	Picrasma excelsa		Jamaican quassia
1-0-X	Pilocarpus spp.	15(B), 33(T)	Jaborandi
2-3-X	Pimenta dioica	13(A), 7(B), 9(T)	Allspice
2-X-X	Pimenta racemosa		Bayrum tree
2-3-X	Pimpinella anisum	15(A), 10(B)	Anise
1-X-X	Pinus elliottii		Slash pine
1-X-X	Piper betel		Betel pepper
1-0-X	Piper methysticum	30(A), 33(B), 37(T)	Kava-kava
1-X-X	Piper nigrum	12(A), 5(B), 7(T)	Black pepper
1-X-X	Piscidia piscipula		Jamaica dogwood
2-3-X	Pistacia lentiscus		Mastic

Table 1 (continued)
365 MEDICINAL HERBS: TOXICITY RANKING AND PRICELIST[37,47]

Score[a]	Scientific name	Price (source)[b]	Common name
3-1-X	Plantago major	15(A), 8(B), 16(T)	Plantain
1-0-X	Podophyllum peltatum	17(T)	Mayapple
2-X-X	Polygonum aviculare	13(T)	Knotweed
2-X-2	Polygonum multiflorum	97(A), 11(B), 121(T)	Fo-ti
2-1-X	Prosopis juliflora		Mesquite
1-3-1	Prunus armeniaca		Apricot
1-X-X	Prunus laurocerasus		Cherry-laurel
1-3-1	Ptychopetalum olacoides	18(A), 28(T)	Muira puama
0-3-X	Quassia amara	12(A), 13(T)	Surinam quassia
2-3-X	Quillaja saponaria		Soaptree
2-X-X	Ranunculus bulbosus		Bulbous buttercup
1-X-X	Rauvolfia serpentina		Sarpaganda
1-X-X	Rauvolfia tetraphylla		Dinque pinque
1-1-X	Rhamnus purshianus	13(A), 7(B), 14(T)	Cascara sagrada
2-X-X	Rheum officinale	13(A), 15(T)	Chinese rhubarb
1-X-X	Rhus toxicodendron		Poison Ivy
1-X-X	Rhynchosia pyramidalis		Piule
0-X-X	Ricinus communis		Castor
1-X-X	Rivea corymbosa		Snakeplant
1-X-X	Robinia pseudoacacia		Black locust
2-2-3	Rosmarinus officinalis	10(A), 4(B), 8(T)	Rosemary
1-X-X	Rumex crispus	23(A), 6(B), 21(T)	Yellow dock
1-X-0	Rumex hymenosepalus		Canaigre
2-0-2	Ruta graveolens	14(B), 14(T)	Rue
2-0-X	Salvia divinorum		Holy sage
3-3-2	Salvia officinalis	24(A), 8(B), 12(T)	Sage
2-1-X	Salvia sclarea	20(A)	Clary sage
1-1-X	Sambucus canadensis	17(A)	Elderberry
1-X-X	Sanguinaria canadensis	24(A), 27(T)	Bloodroot
2-2-X	Santalum album	21(A), 22(T)	Sandalwood
1-X-X	Santolina chamaecyparissus		Lavender-cotton
1-X-X	Sarcostemma acidum		Soma
2-3-1	Sassafras albidum	36(A), 11(B), 28(T)	Sassafras
2-X-X	Satureja montana	11(T)	Winter savory
1-X-X	Schinus molle		Peruvian peppertree
1-X-X	Schinus terebinthifolius		Brazilian peppertree
0-X-X	Schoenocaulon officinale		Sabadilla
1-X-X	Sclerocarya caffra		Marula
0-X-X	Scopolia carniolica		Scopolia
2-X-2	Scutellaria lateriflora	28(A), 21(T)	Virginian scullcap
2-1-X	Selenicereus grandiflorus		Night-blooming cereus
1-X-1	Senecio aureus	11(A)	Life root
2-3-2	Serenoa repens	18(A), 14(B), 18(T)	Saw palmetto
2-X-2	Simmondsia chinensis		Jojoba
2-3-2	Smilax aristolochiifolia	31(A), 14(B), 28(T)	Sarsaparilla
1-X-X	Solanum dulcamara		Bittersweet
1-X-X	Solanum nigrum		Nightshade
2-X-X	Solanum tuberosum		Potato
3-X-X	Solidago virgaurea	10(A), 15(T)	European goldenrod
1-X-X	Sophora secundiflora		Mescal bean
0-X-X	Spigelia marilandica	50(T)	Pinkroot
2-3-2	Stachys officinalis		Betony
2-X-2	Stellaria media	19(A), 9(B), 20(T)	Chickweed
0-0-X	Stillingia sylvatica	37(A), 39(T)	Queen's delight

Table 1 (continued)
365 MEDICINAL HERBS: TOXICITY RANKING AND PRICELIST[37,47]

Score[a]	Scientific name	Price (source)[b]	Common name
0-X-X	Strychnos nux-vomica		Nux-vomica
1-3-X	Styrax benzoin	14(A), 9(B)	Benzoin
2-3-1	Symphytum peregrinum	20(A), 14(T)	Comfrey
1-X-X	Symplocarpus foetidus		Skunk cabbage
2-3-X	Syzygium aromaticum	57(A), 22(B)	Clove
1-X-X	Tabebuia sp.	15(A)	Pao d'arco
0-0-X	Tabernanthe iboga		Iboga
2-2-2	Tanacetum vulgare	11(A), 8(B), 11(T)	Tansy
3-3-2	Taraxacum officinale	25(A), 9(B), 26(T)	Dandelion
1-X-X	Tephrosia virginiana	14(A)	Devil's shoe string
2-3-2	Theobroma cacao		Cocoa
0-X-X	Thevetia peruviana		Luckynut
2-3-X	Thymus vulgaris	14(A), 6(B)	Thyme
2-3-3	Tilia europaea	30(A), 14(B), 9(T)	Linden
1-X-X	Trichocereus pachanoi		Achuma
3-2-2	Trifolium pratense	18(A)	Pink clover
2-X-2	Trigonella foenum-graecum	6(A), 4(B), 8(T)	Fenugreek
2-3-X	Trilisa odoratissima	20(A), 21(T)	Deer's tongue
2-3-2	Turnera diffusa	16(A), 8(B), 13(T)	Damiana
2-3-1	Tussilago farfara	9(B), 12(T)	Coltsfoot
2-3-X	Ulmus rubra	19(A), 15(B), 22(T)	Slippery elm
1-X-X	Umbellularia californica		California bay
2-X-X	Uncaria gambir		Gambir
0-0-X	Urginea maritima	27(T)	Sea onion
3-1-2	Urtica dioica	18(A), 10(T)	Stinging nettle
2-3-3	Valeriana officinalis	9(B), 16(T)	Valerian
2-2-X	Vanilla planifolia	6(B), 218(T)	Vanilla
0-0-X	Veratrum viride		American hellebore
3-3-X	Verbena officinalis	18(A), 14(T)	Verbena
0-3-X	Vinca minor	24(A), 18(T)	Periwinkle
1-X-X	Virola calophylla		Epena
1-0-1	Viscum album	15(A), 16(T)	European mistletoe
0-X-X	Withania somniferum		Ashwagandha
2-X-X	Ziziphus spina-christi		Christthorn

[a] My rating in first column, Rose's in second, Tyler's in third after my interpretation of their books.

   0 = very dangerous; Duke wouldn't drink any cups of it; Rose gave three asterisks as hazardous; Tyler gave it minus for safety and efficacy.

   1 = more dangerous than coffee; Duke wouldn't be afraid to drink one cup containing 10 g herb steeped; Rose gave two stars; Tyler gave negative in the safety column.

   2 = as dangerous as coffee; Duke wouldn't be afraid to drink two cups a day; Rose gave one star; Tyler gave it a plus in the safety column, tempered with a minus somewhere.

   3 = Duke considers it safer than coffee and wouldn't hesitate, for health reasons, to drink three cups a day; Rose gave it no negative asterisks; Tyler gave it only positive ratings.

   X = Rating not available to me.

[b] Rounded to nearest dollar per kilogram (divide by 2 to get rough estimate of price per pound). Prices quoted are 1983 prices.

   A = Aphrodisia, 282 Bleeker Street, New York, New York, 10014.

   B = Mark Blumenthal, Wholesale Suggested Price, 1983. Sweethardt Herbs, Austin, Texas, 78725.

   T = Tommy Wolfe's "Tea Here Now" Catalog No. 3, Smile Herb Shop, 4908 Berwyn Road, College Park, Maryland, 20740.

Table 2
TOXINS: THEIR TOXICITY AND DISTRIBUTION IN PLANT GENERA

Chemical	Toxicity[a,b]	Plant genera
Abobioside	ivn cat LD_{50}, 0.699 mg	*Adenium*
Abomonoside	ivn cat LD_{50}, 679 µg	*Adenium*
Acetaldehyde	orl rat LD_{50}, 1,930 mg	*Brassica, Carpinus. Carum, Cinnamomum, Citrus, Foeniculum, Lycopersicon, Mentha, Nicotiana, Pimpinella, Pyrus, Quercus, Rosa, Rosmarinus, Sorbus*
Acetic acid	orl hmn TDLo, 1,470 µg GIT	Widespread, both free and combined
	ihl hmn TDLo, 816 ppm/3 min IRR	
	orl rat LD_{50}, 3,310 mg	
	orl mus LD_{50}, 4,960 mg	
Acetone	orl rbt LD_{50}, 5,300 mg	*Coriandrum, Erythroxylum, Hevea, Lycopersicon, Manihot, Phaseolus,·Pogostemon*
	orl dog LDLo, 8,000 mg	
	ihl man TCLo, 12,000 ppm/4 hr CNS	
	ihl hmn TCLo, 500 ppm EYE	
Acetophenone	orl rat LD_{50}, 900 mg	*Camellia, Cistus, Corchorus, Iris, Populus, Stillingia, Urtica*
(hypnone)	ipr mus LDLo, 200 mg	
Acetylcholine	orl mus LD_{50}, 3,000 mg	*Artocarpus, Capsella, Crataegus, Digitalis, Laportea, Solanum, Urtica, Viscum*
(choline acetate ester)		
α-Acetyldigitoxin	orl cat LD_{50}, 250 µg	*Digitalis*
	ivn cat LD_{50}, 514 µg	
β-Acetyldigitoxin	orl gpg LD_{50}, 50 mg	*Digitalis*
	ivn cat LD_{50}, 476 µg	
α-Acetyldigoxin	orl cat LD_{50}, 200 µg	*Digitalis*
	orl gpg LD_{50}, 3,300 µg	
	ivn cat LD_{50}, 466 µg	
c-Acetyldigoxin	ivn cat LD_{50}, 430 µg	*Digitalis*
α-Acetylgitoxin	orl gpg LD_{50}, 40 mg	*Digitalis*
Acolongifloroside E	ivn cat LD_{50}, 259 µg	*Acokanthera*
Aconitine	ipr rat LDLo, 250 µg	*Aconitum*
	ivn dog LDLo, 350 µg	
Adenine	orl rat LD_{50}, 755 mg	*Beta, Camellia, Coffea, Pachyrrhizus. Saccharum*
	ipr rat LD_{50}, 198 mg	
	ipr mus LD_{50}, 335 mg	
Adipic acid	orl rat LDLo, 3,600 mg	*Beta*
	orl mus LD_{50}, 1,900 mg	
Adonitoxin	ivn cat LDLo, 191 µg	*Adonis*
Ajmalicine	orl hmn TDLo, 43 µg PSY	*Catharanthus, Corynanthe, Mitragyna, Rauwolfia, Stemmadenia, Tonduzia*
	orl mus LD_{50}, 400 mg	
	orl rbt LD_{50}, 500 mg	
Ajmaline	ipr mus LDLo, 75 mg	*Aspidosperma, Rauwolfia, Tonduzia*
Albitocin	orl gpg LD_{50}, 19 mg	*Albizzia*
	ipr rat LD_{50}, 0.8 mg	
Allyl alcohol	orl mus LD_{50}, 96 mg	*Martynia*
	orl dog LDLo, 43 mg	
	orl rbt LDLo, 53 mg	
	ihl hmn TCLo, 25 ppm TWA 2 ppm	
Allyl benzyl piperidine (HCl)	ipr mus LD_{50}, 68 mg	*Voacanga*
Allyl isothiocyanate	orl rat LD_{50}, 148 mg	*Armoracia, Brassica, Thlaspi*
(mustard oil)	ipr mus LDLo, 4 mg	
Amboside	ivn cat LD_{50}, 827 µg	*Strophanthus*
o-Aminobenzoic acid (anthranilic acid)	orl rat LD_{50}, 4,620 mg	In bacteria but probably not in higher plants
	orl rat LDLo, 16 Dg/24 WI NEO	higher plants
o-Aminobenzoil acid, methyl ester (methyl anthranilate)	orl rat LD_{50}, 2,910 mg	
	orl mus LD_{50}, 3,900 mg	
	orl gpg LD_{50}, 2,780 mg	
γ-Aminobutyric acid (butyric acid, 4-amino)	ivn mus LD_{50}, 2,748 mg	May be universal in plants
Ammonia	orl rat LD_{50}, 350 mg	*Amorphophallus, Arum, Dracunculus, Nicotiana, Sauromatum*
	ihl hmn LCLo, 10,000 ppm/3 hr	
	ihl hmn TCLo, 20 ppm	
Amyl alcohol	orl rat LD_{50}, 3,030 mg	*Mentha, Origanum*
(pentyl alcohol)	orl mus LD_{50}, 200 mg	
	ipr rat LDLo, 490 mg	
Amyris oil	orl rat LDLo, 4,540 mg	*Amyris*

Table 2 (continued)
TOXINS: THEIR TOXICITY AND DISTRIBUTION IN PLANT GENERA

Chemical	Toxicity[a,b]	Plant genera
Anabasine	orl rat LDLo, 10 mg	*Anabasis, Aniba, Duboisia, Nicotiana, Sophora,*
	orl dog LDLo, 50 mg	*Zinnia*
Andromedol	ipr mus LD_{50}, 908 μg	Ericaceae
Andromedotoxin	ipr mus LD_{50}, 1.28 mg	Ericaceae
Anemonin	ipr mus LD_{50}, 150 mg	*Aconitum, Anemone, Caltha, Clematis, Ranunculus*
Anethole	orl rat LD_{50}, 2,090 mg	*Backhousia, Canarium, Clausena, Foneniculum,*
		Illicium, Magnolia, Ocimum, Piper
	orl mus LD_{50}, 3,050 mg	
	orl gpg LD_{50}, 2,160 mg	
Anhaline	orl dog LDLo, 2,000 mg	*Acacia, Avena, Hordeum, Lophophora, Mammillaria,*
(hordenine)	scu rat LDLo, 1,000 mg	*Oryza, Panicum, Phalaris, Sorghum, Trichocereus,*
	ivn dog LDLo, 275 mg	*Zea*
16-Anhydrodigitalinum		*Digitalis*
verum		
Anhydrogitalin	orl cat LDLo, 880 μg	*Digitalis*
Anhydroperiplogenone	ivn cat LD_{50}, 1 mg	*Pachycarpus, Xysmalobium*
p-Anisaldehyde	orl rat LD_{50}, 1,510 mg	*Acacia, Agastache, Erica, Illicium, Magnolia,*
	orl gpg LD_{50}, 1,260 mg	*Mimosa, Pelea, Protium, Vanilla*
Anthraquinone	orl rat TDLo, 90 g/90 DC NEO	*Acacia, Quebrachia*
	orl rat TDLo, 2,740 mg/61 WI NEO	
Antiarin	ivn rbt LDLo, 1 mg	*Antiaris, Antiaropsis, Ogcodeia*
Apiole	scu mus LDLo, 1,000 mg	*Apium, Crithmum, Licaria, Ocotea, Petroselinum,*
	scu dog LDLo, 500 mg	*Piper*
Apoatropine	orl mus LD_{50}, 160 mg	*Atropa*
	ipr mus LD_{50}, 14 mg	
L-Arabinose	ivn dog LD_{50}, 5,000 mg	*Aloe, Prosopis*
(pectinose)		
Arecoline	scu mus LDLo, 65 mg	*Areca*
Aribine	ipr mus LD_{50}, 50 mg	*Sickingia*
(harman)	scu rbt LDLo, 200 mg	
Aristolochine	ivn man LDLo, 3 mg/2 DI	*Aristolochia*
(aristolochic	ipr cat LDLo, 40 mg	
acid)	ipr rbt LDLo, 1,500 μg	
Asarone	ipr gpg LD_{50}, 275 mg	*Acorus, Asarum, Daucus, Orthodon, Piper*
Ascaridole	orl rat LDLo, 250 mg	*Chenopodium*
	skn mus TDLo, 25 g/42 WI NEO	
Ascorbic acid	ivn mus LD_{50}, 518 mg	Widespread
(vitamin C)		
Asiaticoside	skn mus TDLo, 400 mg/52 WI NEO	*Centella, Hydrocotyle*
Aspidospermine	ipr mus LDLo, 40 mg	*Aspidosperma, Vallesia*
Astrophyllin	scu mus LDLo, 824 mg	*Astrocasia*
	ivn mus LD_{50}, 1,080 mg	
Atractyloside	ims rat LD_{50}, 431 mg	*Atractylis*
Atropine	orl hmn TDLo, 100 μg PSY	*Atropa, Datura, Duboisia, Hyoscyanus,*
	orl rat LD_{50}, 622 mg	*Mandragora, Scopolia, Solandra, Solanum*
	orl mus LD_{50}, 400 mg	
	orl rbt LDLo, 1,450 mg	
	orl gpg LD_{50}, 1,100 mg	
Benzaldehyde	orl rat LD_{50}, 1,300 mg	*Acacia, Cananga, Dianthus, Eugenia, Hyacinthus,*
	orl gpg LD_{50}, 1,000 mg	*Melaleuca, Rosa, Rubus, Ruta*
	scu rat LDLo, 5,000 mg	
Benzoic acid	orl rat LD_{50}, 3,040 mg	*Aniba, Cananga, Cinnamomum, Eugenia, Gaultheria,*
	orl mus LD_{50}, 2,370 mg	*Globularia, Myroxylon, Vaccinium, Xanthorrhea*
Benzyl alcohol	orl rat LD_{50}, 1,230 mg	*Acacia, Cananga, Dianthus, Eugenia, Hyacinthus*
	orl mus LD_{50}, 1,580 mg	*Jasminum, Narcissus, Polianthes, Viola*
	orl rbt LD_{50}, 1,940 mg	
Benzyl cyanide	orl rat LD_{50}, 270 mg	*Codonocarpus, Lepidium, Leptactina, Tropaeolum*
(phenyl,	orl mus LD_{50}, 78 mg	
acetonitrile)		
Berbamine	ipr rat LDLo, 500 mg	*Atherosperma, Berberis, Mahonia, Pycnarrhena,*
		Stephania
Betaine	ivn mus LD_{50}, 830 mg	*Atriplex, Beta, Cichorium, Glycine, Gossypium,*
		Helianthus, Lycium, Orthosiphon, Parthenium,
		Scopolia
Bicuculline	ipr mus LDLo, 25 mg	*Adlumia, Corydalis, Dicentra, Fumaria*
Biflorine	ipr rat LDLo, 100 mg	*Adlumia, Arctomecon, Argemone, Bocconia,*
(fumarine,		*Chelidonium, Corydalis, Cysticapnos,*
protopine)		*Dactylicarpus, Dendromecon, Dicentra,*
		Dicranostigma, Eschscholtzia, Fumaria, Glaucium,

Table 2 (continued)
TOXINS: THEIR TOXICITY AND DISTRIBUTION IN PLANT GENERA

Chemical	Toxicity[a,b]	Plant genera
		Hunnemannia, Hypecoum, Maccleaya, Meconella, Meconopsis, Nandina, Papaver, Platycapnos, Platystemon, Pteridophyllum, Roemeria, Romneya, Sanguinaria, Sarcocapnos, Stylophorum
Borneol	orl rbt LDLo, 2,000 mg	Angelica, Asarum, Blumea, Boswellia, Cinnamomum, Citrus, Coriandrum, Cymbopogon, Dryobalanops, Elettaria, Eucalyptus, Kaempferia, Lavandula, Lindera, Matricaria, Myristica, Rosmarinus, Salvia, Thymus, Valeriana, Zingiber
Boroside D	ivn cat LD_{50}, 0.11 mg	Bowiea
Bracken	orl rat TDLo, 3,050 g/17 WC CAR	Pteris
	orl ctl TDLo, 495 g/68 WC NEO	
Brucine	orl hmn LDLo, 30 mg	Strychnos
	orl rat LD_{50}, 1 mg	
	orl rbt LD_{50}, 4 mg	
Bufotenine	ivn hmn TDLo, 2 mg	Desmodium, Lespedeza, Phalaris, Piptadenia, Virola
	ipr mus LD_{50}, 290 mg	
Bulbocapnine	scu mus LD_{50}, 195 mg	Corydalis, Dicentra, Fumaria
Butyl alcohol	ihl hmn TCLo, 25 ppm	Mentha
	orl rat LD_{50}, 790 mg	
	orl rbt LDLo, 4,250 mg	
	TWA 100 ppm	
Butyraldehyde	ihl hmn TCLo, 580 mg/m³	Acacia, Artemisia, Camellia, Carpinus, Eucalyptus,
	orl rat LD_{50}, 2,490 mg	Lavandula, Melaleuca, Monarda, Morus, Quercus, Raphanus
Butyric acid	orl rat LD_{50}, 2,940 mg	Eucalyptus?, Sapindus?
	orl mus LDLo, 500 mg	
	orl rbt LDLo, 3,600 mg	
Caffeine	orl hmn LDLo, 192 mg	Annona, Camellia, Cereus, Coffea, Cola, Combretum,
(guaranine)	orl rat LD_{50}, 192 mg	Davilla, Erodium, Gallesia, Genipa, Guazuma,
	orl mus LD_{50}, 620 mg	Harrisia, Helicteres, Herrania, Ilex, Leocereus,
	orl cat LDLo, 125 mg	Maytenus, Neea, Oldenlandia, Paullinia,
	orl rbt TDLo, 1,500 mg	Pilocereus, Piriqueta, Pleiocarpa, Sterculia,
	(1–15 D preg TER)	Theobroma, Trichocereus, Turnera, Urginea,
	orl mus TDLo, 650 mg	Villaresia,
	(6–18 D preg TER)	
Cajeput oil	orl rat LD_{50}, 387 mg	Melaleuca
Cajeputol	orl rat LD_{50}, 2,480 mg	Melaleuca
(epoxymenthane)	scu mus LDLo, 50 mg	
Calamus oil	orl rat LD_{50}, 777 mg	Acorus
Calotropin	orl cat LDLo, 120 mg	Asclepias, Calotropis, Pergularia
Camphor	ipr rat LDLo, 900 mg	Achillea, Acorus, Alpinia, Aristolochia,
	scu mus LDLo, 2,200 mg	Artemisia, Chenopodium, Chrysanthemum,
	TWA 2 ppm	Cinnamomum, Curcuma, Lavandula, Lippia,
		Matricaria, Meriandra, Ocimum, Prunella,
		Rosmarinus, Salvia, Sassafras, Tanacetum
Camptothecin	ivn hmn TDLo, 2,500 µg/7 DI BLD	Camptotheca
(Na-salt)		
Cannabinol, tetrahydro	ipr mus LD_{50}, 125 mg	Cannabis
Cannabinol, (delta 8, tetrahydro)	orl rat LD_{50}, 860 mg	Cannabis
Cannabinol, (delta 9, tetrahydro)	orl rat LD_{50}, 666 mg	Cannabis
	orl mus LD_{50}, 482 mg	
	orl dog LDLo, 525 mg	
Capuarine (capauridine)	ipr rat LDLo, 500 mg	Corydalis
Capric acid (decanoic acid)	ivn mus LD_{50}, 129 mg	Citrus, Cuphea, Cymbopogon, Lavandula, Litsea, Neolitsea, Ruta, Sassafras, Ulmus, Zelkova
Caproic acid (hexanoic acid)	orl rat LDLo, 3,000 mg	Camellia, Chrysanthemum, Cinnamomum, Cocos, Laurus, Lavandula, Mentha, Prunus, Ruta
Caprylic acid (octanoic acid)	ivn mus LD_{50}, 600 mg	Astrocaryum, Cinnamomum, Citrus, Cocos, Cuphea, Cymbopogon, Mentha, Myristica, Reseda, Roystonea, Spartium, Ulmus, Zelkova
Capsaicin	ivn cat LDLo, 1,600 µg	Capsicum
Caraway oil	orl rat LD_{50}, 3,500 mg	Carum
	skn rbt LD_{50}, 1,780 mg	
Carbazole (class of alkaloid)	ipr mus LD_{50}, 200 mg	Clausenia, Glycosmis, Murraya
Carene	orl rat LD_{50}, 4,800 mg	Chenopodium, Citrus, Dacrydium, Eucalyptus, Illicium, Kaempferia, Pinus, Piper

Table 2 (continued)
TOXINS: THEIR TOXICITY AND DISTRIBUTION IN PLANT GENERA

Chemical	Toxicity[a,b]	Plant genera
Carthamoidine	ivn mus LD_{50}, 68 mg	*Senecio*
Carvacrol	orl rat LD_{50}, 810 mg	*Carum, Cinnamomum, Mentha, Nepeta, Ocimum,*
	orl rbt LDLo, 100 mg	*Origanum, Orthodon, Ruta, Satureja, Schinus,*
	orl rbt LDLo, 2,700 mg	*Thymus, Zea*
1-Carveol	orl rat LD_{50}, 3,000 mg	*Carum*
Carvone	orl rat LD_{50}, 1,640 mg	*Anethum, Carum, Chrysanthemum, Citrus, Cympobogon,*
	orl gpg LD_{50}, 766 mg	*Eucalyptus, Lavandula, Lindera, Lippia, Litsea,*
		Mentha, Nigella, Orthodon, Satureja
Cassia oil	orl rat LD_{50}, 2,800 mg	*Cinnamomum*
	skn rbt LD_{50}, 320 mg	
Catechol (see Pyrocatechol)		
Cedarleaf oil	orl rat LD_{50}, 830 mg	*Thuja*
Cephaeline (HCl)	ipr rat LD_{50}, 10 mg	*Alangium, Bothriospora, Capirona, Cephaelis,*
		Ferdinandusa, Hillia, Psychotria,
		Remijia, Tocoyena
Cerberin	ivn cat LD_{50}, 370 µg	*Cerbera, Tanghinia, Thevetia*
Cevadine	ivn mus LD_{50}, 1 mg	*Schoenocaulon, Veratrum*
	scu rbt LDLo, 500 µg	
	scu gpg LDLo, 1 mg	
Cevine	ipr rat LD_{50}, 67 mg	*Schoenocaulon, Veratrum*
	ivn mus LD_{50}, 87 mg	
Ceylon cinnamon oil	orl rat LD_{50}, 4,160 mg	*Cinnamomum*
Cheiranthum	ivn cat LDLo, 44 µg	*Cheiranthus*
Cheirotoxin	ivn cat LDLo, 119 µg	*Cheiranthus*
Chelidonin	scu rat LDLo, 300 mg	*Chelidonium, Dicranostigma, Glaucium, Stylophorum*
	scu mus LDLo, 300 mg	
	scu rbt LDLo, 300 mg	
	scu gpg LDLo, 300 mg	
	scu frg LDLo, 300 mg	
Chlorophyll	ipr mus LD_{50}, 400 mg	All green plants
	ivn mus LD_{50}, 285 mg	
	ivn gpg LDLo, 80 mg	
Cholesterol	scu mus TDLo, 600 mg/9 WI CAR	*Calendula, Dioscorea, Haplopappus, Helianthus,*
	orl rbt TDLo, 84 g/42 DI NEO	*Hypochoeris, Ilex, Phoenix, Populus,*
		Solanum, Tamnus
Choline	ipr rat LD_{50}, 400 mg	*Acorus, Adonis, Artemisia, Atropa, Caltha,*
		Canavalia, Cannabis, Capsella, Cephaelis, Cicer,
		Cichorium, Coffea, Combretum, Convallaria, Daucus,
		Dictamnus, Digitalis, Doryphora, Erodium,
		Eupatorium, Fagus, Genista, Glycine, Gossypium,
		Helianthus, Humulus, Hyoscyamus, Hyssopus,
		Lactuca, Lagenaria, Lens, Leonurus, Linum, Luffa,
		Magnolia, Marrubium, Medicago, Morus, Olea,
		Orthosiphon, Oryza, Pachyrrhizus, Panicum,
		Pennisetum, Phaseolus, Pimpinella, Pluchea,
		Scopolia, Strophanthus, Taraxacum, Trigonella,
		Valeriana, Vicia, Withania
Chrysarobin	ipr mus LDLo, 4 mg	*Andira, Cassia, Rumex*
Cicutoxin	orl cat LDLo, 7 mg	*Cicuta*
Cinchonidine	ipr rat LD_{50}, 206 mg	*Cinchona, Remijia, Strychnos*
Cinchonine	ipr rat LD_{50}, 152 mg	*Cinchona, Remijia*
Cineole (1-8)	orl rat LD_{50}, 2,480 mg	*Achillea, Aglaia, Alpinia, Artemisia, Blumea,*
(eucalyot-1)	scu mus LDLo, 50 mg	*Cinnamomum, Comptonia, Curcuma, Cyperus,*
		Elletaria, Eucalyptus, Eugenia, Illicium, Laurus,
		Lavandula, Lippia, Litsea, Luvunga, Melaleuca,
		Mentha, Michelia, Myrothamnus, Mytrus, Ocimum,
		Pimenta, Piper, Rosmarinus, Ruta, Salvia,
		Zingiber
Cinnamaldehyde	orl gpg LD_{50}, 1,160 mg	*Cassia, Cinnamomum, Hyacinthus, Lavandula,*
	orl rat LD_{50}, 2,220 mg	*Narcissus, Pogostemon*
	ipr mus LD_{50}, 200 mg	
	par mus LDLo, 200 mg	
Cinnamyl alcohol	orl rat LD_{50}, 2,000 mg	*Cinnamomum, Hyacinthus, Narcissus, Populus,*
		Xanthorrhoea
Citral	orl rat LD_{50}, 4,960 mg	*Aglaia, Backhousia, Camellia, Citrus, Cymbopogon,*
		Eucalyptus, Lavandula, Leptospermum, Lippia,
		Melissa, Ocimum, Pelargonium, Phebalium, Pimenta,
		Piper, Prunus, Rosa, Sassafras, Thymus, Zingiber

Table 2 (continued)
TOXINS: THEIR TOXICITY AND DISTRIBUTION IN PLANT GENERA

Chemical	Toxicity[a,b]	Plant genera
Citric acid	ipr rat LD_{50}, 884 mg ipr mus LD_{50}, 961 mg ivn mus LD_{50}, 42 mg ivn rbt LD_{50}, 330 mg	*Ananas, Angelica, Asperula, Berberis, Beta, Bryophyllum, Chelidonium, Citrus, Digitalis, Drosera, Fagus, Fragaria, Galium, Garcinia, Helianthus, Hibiscus, Ilex, Kleinia, Lactuca, Lolium, Malus, Mammea, Medicago, Morus, Musa, Myrtus, Nicotiana, Oryza, Papaver, Passiflora, Plantago, Pyrus, Ribes, Sclerocarya, Sedum, Spinacia, Tamarindus, Tanacetum, Vitis*
Citronella oil	orl mam LDLo, 1,000 mg	*Cymbopogon*
Citronellol	ims mus LD_{50}, 4,000 mg	*Backhousia, Calythrix, Camellia, Cinnamomum, Citrus, Cymbopogon, Lippia, Melissa, Pelargonium, Rosa, Xanthorrhoea*
Clove oil	orl rat LD_{50}, 3,720 mg	*Syzygium*
Cocaine	orl rbt LDLo, 126 mg ivn mus LDLo, 30 mg ivn rat LD_{50}, 18 mg	*Erythroxylum, Gleditsia?[c]*
Codeine	orl mus LD_{50}, 250 mg ivn rbt LD_{50}, 34 mg scu rat LD_{50}, 500 mg ipr mus LD_{50}, 200 mg scu mus LDLo, 241 mg	*Argemone?, Eschscholzia?, Humulus?, Papaver*
Colchamine	ivn rat TDLo, 5,700 µg NEO ipr mus LDLo, 75 mg unk rbt TDLo, 500 µg (preg) TER	*Colchicum, Gloriosa, Merendera*
Colchiceine	orl hmn LDLo, 43 µg ipr mus LD_{50}, 84 mg	*Colchicum, Merendera*
Colchicine	orl hmn LDLo, 186 µg 4 D skn mus TDLo, 144 mg/14 W NEO ipr mus TDLo, 2,500 mg MUT ivn ham TDLo, 10 mg (14 D preg) TER	*Androcymbium, Asphodelus, Bulbine, Bulbocodium, Camptorrhiza, Colchicum, Crocus, Dipidax, Gloriosa, Hemerocallis, Iphigenia, Littonia, Lloydia, Merendera, Muscari, Narcissus, Ornithoglossum, Sandersonia, Tofieldia, Tulipa, Veratrum*
Coniine (cicutine)	scu mus LDLo, 75 mg	*Aethusa, Amorphophallus, Arisarum, Arum, Caladium, Conium, Humulus, Parietaria, Punica, Sarcolobus*
Convallamarin	scu mus LDLo, 600 mg ipr rat LD_{50}, 3,400 µg	*Convallaria*
Convallatoxin	ivn rat LDLo. 38 mg ivn frg LDLo, 300 µg	*Antiaris, Convallaria, Ornithogalum*
Convallatoxol	ivn cat LD_{50}, 87 µg	*Antiaris, Convallaria*
Convalloside	ivn cat LDLo, 215 µg	*Convallaria, Ornithogalum*
Coramine	orl hmn TDLo, 18 mg PSY orl rbt LD_{50}, 650 mg scu rat LD_{50}, 240 mg ivn dog LDLo, 175 mg	*Corydalis*
Corchoroside A	ivn cat LD_{50}, 77 µg	*Castilla, Corchorus, Erysimum*
Coriamyrtin	orl mus LD_{50}, 1 mg scu rat LDLo, 1 mg ivn rat LDLo, 700 µg	*Coriaria*
Coriander oil	orl rat LD_{50}, 4,130 mg	*Coriandrum*
Corlumine	ipr rat LDLo, 25 mg	*Corydalis, Dicentra*
Corynantheine	ivn mus LDLo, 1,400 µg	*Corynanthe, Pausinystalia, Pseudocinchona*
Costus oil	orl rat LD_{50}, 3,400 mg	*Costus*
Cotoin	scu frg LDLo, 224 mg	*Aniba, Nectandra, Rudgea*
Coumarin	skn mus TDLo, 1,800 mg NEO orl mus LD_{50}, 196 mg orl rat LD_{50}, 293 mg orl gpg LD_{50}, 202 mg	Wide distribution, *Achylis, Alyxia, Asperula, Ceratophyllum, Chrysophyllum, Cinnamomum, Dipteryx, Ficus, Hemidesmus, Herniaria, Laserpitium, Lavandula, Levisticum, Liatris, Macrosiphonia, Melilotus, Pastinaca, Peristrophe, Petroselinum, Phoenix, Prunus, Rhinacanthus, Ruta, Stenolobium, Tabebuia, Talauma, Verbascum, Vitis*
p-Cresol	orl rat LD_{50}, 1,454 mg orl mus LD_{50}, 861 mg	*Acacia, Camellia, Citrus, Gnaphalium, Jasminum, Ledum, Lilium, Pimpinella*
p-Cresol methyl ether	orl rat LD_{50}, 1,920 mg	*Cananga*
Croton oil	ipr mus LDLo, 1 mg	*Croton*
Cryogenine	ipr rat LD_{50}, 55 mg	*Decodon, Heimia*
Cuachichicine (HCl)	ivn mus LDLo, 10 mg ivn cat LDLo, 10 mg	*Garrya*

Table 2 (continued)
TOXINS: THEIR TOXICITY AND DISTRIBUTION IN PLANT GENERA

Chemical	Toxicity[a,b]	Plant genera
Cucurbitacin E (elaterin)	ipr mus LD_{50}, 2 μg	*Bryonia, Citrullus, Cucurbita, Ecballium, Echinocystis, Kedrostis, Lagenaria, Luffa, Peponium, Telfairia*
Cumene	orl rat LD_{50}, 1,400 mg	*Cuminum*
Cumin oil	orl rat LD_{50}, 2,500 mg	*Cuminum*
Cuminic alcohol	orl rat LD_{50}, 1,020 mg	*Cinnamomum, Cuminum, Eucalyptus, Lavandula, Ledum*
Cuminic aldehyde (cuminaldehyde)	orl rat LD_{50}, 1,390 mg skn rbt LD_{50}, 2,800 mg	*Acacia, Aegle, Artemisia, Cicuta, Cinnamomum, Cuminum, Lavandula, Pectis, Peumus, Ruta, Zanthoxylum*
Curare	ivn dog LD_{50}, 1,200 μg ivn rbt LD_{50}, 1,300 μg	*Strychnos*
Curarine	scu dog LD_{50}, 500 μg	*Strychnos*
Cyanocobalamin (vitamin B_{12})	ipr mus LDLo, 3 mg	*Medicago, Symphytum*
Cycasin	orl rat LDLo, 400 mg/2 DC CAR orl mus TDLo, 300 mg CAR orl ham TDLo, 100 mg CAR	*Cycas*
Cyclopamine	orl dom TDLo, 800 mg (8 D preg) TER	*Veratrum*
Cymarin	ivn cat LDLo, 130 μg	*Adonis, Antiaris, Apocynum, Castilla, Pachycarpus, Strophanthus*
Cymarol	ivn cat LDLo, 99 μg	*Antiaris, Strophanthus*
p-Cymene	orl hmn TDLo, 86 mg CNS orl rat LD_{50}, 4,750 mg ipr gpg LDLo, 2,162 mg	*Aegle, Boswellia, Chenopodium, Cinnamomum, Citrus, Coriandrum, Croton, Cuminum, Eschscholtzia, Eucalyptus, Eupatorium, Illicium, Ledum, Litsea, Melaleuca, Myristica, Nigella, Origanum, Satureja, Thymus*
Cytisine	orl mus LD_{50}, 101 mg	*Anagyris, Argyrolobium, Baptisia, Cladrastis, Coronilla, Cytisus, Euchresta, Genista, Gleditsia, Laburnum, Lotus, Piptanthus, Senecio, Sophora, Spartium, Templetonia, Thermopsis, Ulex*
Daphe	ipr mus LDLo, 500 mg	*Daphne*
Damascenini (HCl)	orl mus LD_{50}, 1,800 mg	*Nigella*
Decamine	orl hmn LD_{50}, 80 mg orl rat TDLo, 25 mg (6–15 D preg) TER orl mus LD_{50}, 368 mg	
Decanal	orl rat LD_{50}, 3,730 mg	*Acacia, Cassia, Cinnamomum, Citrus, Coriandrum, Iris, Lavandula*
Decanoic acid (capric acid)	ivn mus LD_{50}, 129 mg	*Citrus, Cuphea, Cymbopogon, Lavandula, Litsea, Neolitsea, Ruta, Sassafras, Ulmus, Zelkova*
Decyl alcohol	orl rat LD_{50}, 4,720 mg ihl mus LC_{50}, 4,000 mg/m³	*Citrus, Prunus, Simmondsia*
Demecolcine	ivn rat TDLo, 6 mg	*Colchicum, Merendera*
Derris	orl rat LDLo, 100 mg orl dog LDLo, 100 mg orl rbt LDLo, 200 mg orl gpg LDLo, 75 mg	*Derris*
Deserpidine	ipr mus LD_{50}, 60 mg orl mus LD_{50}, 500 mg	*Rauwolfia, Tonduzia*
Diacetyl (butanedione)	orl rat LD_{50}, 1,580 mg orl gpg LD_{50}, 990 mg	*Carum, Fagraea, Polyalthia*
Dichroine	orl mus LD_{50}, 2,740 μg ivn mus LD_{50}, 10 mg	*Dichroa*
Diethylamine	ivn rat TDLo, 5,720 μg 1 YI CAR scu rat TDLo, 0.5 mg (12–15 D preg) TER	*Arum, Sauromatum*
Digitalis glucoside (uzara)	orl dog LDLo, 180 mg ivn dog LDLo, 4 mg	*Digitalis*
Digitoxin	orl rat LD_{50}, 100 μg orl cat LD_{50}, 0.18 mg	*Digitalis*
Digitoxoside	ivn cat LDLo, 400 μg	*Digitalis*
Digoxin	orl gpg LD_{50}, 3,500 μg orl cat LD_{50}, 200 μg	*Digitalis*
Dihydrokawain	orl mus LD_{50}, 920 mg ipr mus LD_{50}, 325 mg ipr rbt LD_{50}, 350 mg	*Piper*
Dimethylamine	orl rat LD_{50}, 698 mg orl mus LD_{50}, 316 mg orl rbt LD_{50}, 240 mg orl gpg LD_{50}, 240 mg TWA 10 ppm	*Arum, Courbonia, Crataegus, Dracunculus, Heracleum, Sauromatum*

Table 2 (continued)
TOXINS: THEIR TOXICITY AND DISTRIBUTION IN PLANT GENERA

Chemical	Toxicity[a,b]	Plant genera
Dimethyl	ims man TDLo, 1 mg CNS	*Acacia, Banisteriopsis, Desmodium, Lespedeza,*
tryptamine	ipr mus LD_{50}, 110 mg	*Mimosa, Petalostylis, Phalaris, Piptadenia,*
(indole)	ivn mus LD_{50}, 43 mg	*Prestonia*
Dipentene (see Limonene)		
Divaricoside	ivn cat LD_{50}, 165 µg	*Strophanthus*
Dodecyl alcohol	ipr rat LD_{50}, 800 mg	*Citrus, Furcraea, Ligusticum, Rhamnus*
Donaxine (gramine)	ipr rat LDLo, 250 mg	*Acer, Arundo, Desmodium, Hordeum, Lupinus,*
	ipr mus LD_{50}, 122 mg	*Phalaris*
Dopa	orl rat LD_{50}, 582 mg	*Astragalus, Baptisia, Cytisus, Euphorbia, Lupinus,*
	orl rbt LDLo, 950 mg	*Mucuna, Musa, Portulaca, Vicia*
	orl man TDLo, 5,733 mg	
Dopamine	orl wmn TDLo, 300 µg/D	*Hermidium, Mucuna, Musa*
	(1st 7 W preg) TER	
	orl rat LD_{50}, 165 mg	
	orl mus LD_{50}, 145 mg	
Echubioside	ivn cat LD_{50}, 290 µg	*Adenium*
Echujin	ivn cat LD_{50}, 300 µg	*Adenium*
Elaidic acid	ipr mus LDLo, 512 mg	*Delphinium, Physalis, Tribulus*
Emetine	ipr mus LD_{50}, 12 mg	*Alangium, Borreria, Bothriospora, Capirona,*
	orl gpg LDLo, 20 mg	*Cephaelis, Ferdinandusa, Hillia, Hybanthus,*
		Manettia, Psychotria, Remijia,
		Richardia, Tocoyena
Enteramine (see Serotonin)		
Ephedrine	orl mus LD_{50}, 400 mg PSY	*Aconitum, Catha, Ephedra, Moringa, Roemeria,*
	orl rbt LDLo, 590 mg	*Sida, Taxus*
	scu hmn TDLo, 357 µg	
	scu rat LD_{50}, 300 mg	
Ergine (see Lysergamide)		*Argyreia, Ipomoea, Rivea*
Ergometrine	scu rat LDLo, 500 µg	*Argyreia, Ipomoea*
	ivn rat LDLo, 7,500 µg	
	ivn gpg LDLo, 80 mg	
Ergosterol	orl dog LDLo, 4 mg	*Citrus, Hevea, Ilex, Lactuca, Nicotiana, Triticum*
(ergocalciferol)	ipr dog LDLo, 10 mg	
(vitamin D_2)	ivn dog LDLo, 5 mg	
	ims dog LDLo, 5 mg	
Erysodine (HCl)	orl mus LD_{50}, 155 mg	*Erythrina*
	scu mus LD_{50}, 100 mg	
Erysopine (HCl)	orl mus LD_{50}, 18 mg	*Erythrina*
	scu mus LD_{50}, 15 mg	
Erysothiopine (Na)	scu mus LD_{50}, 76 mg	*Erythrina*
Erythraline (HBr)	orl mus LD_{50}, 80 mg	*Erythrina*
	scu mus LD_{50}, 92 mg	
Erythramine (HBr)	scu mus LD_{50}, 104 mg	*Erythrina*
β-Erythroidine	ipr mus LD_{50}, 24 mg	*Erythrina*
	ivn rbt LD_{50}, 8,600 µg	
Esdragole	orl rat LD_{50}, 1,820 mg	*Artemisia, Feronia, Foeniculum, Illicium, Ocimum,*
(estragole)	orl mus LD_{50}, 1,250 mg	*Persea, Pimenta, Pimpinella*
Eserine (see Physostigmine)		
Estrone	imp rat TDLo, 16 mg NEO	*Elaeis, Malus, Phoenix, Punica*
	imp gpg TDLo, 2 mg NEO	
	imp ham TDLo, 100 mg CAR	
	par mus TDLo, 30 mg (11—16 D preg) TER	
Ethyl alcohol	orl chd LDLo, 2,000 mg	*Angelica, Anthriscus, Castanea, Citrus,*
	orl man TDLo, 50 mg	*Eucalyptus, Fragaria, Heracleum, Lycopersicon,*
	orl mus LDLo, 220 mg	*Mentha, Nicotiana, Pastinaca, Rubus*
	orl dog LDLo, 5,500 mg	
	orl cat LDLo, 6,000 mg	
	orl rbt LD_{50}, 6,300 mg	
	orl gpg LD_{50}, 5,560 mg	
Ethylamine	orl rat LDLo, 400 mg	*Amorphophallus, Arum, Bryonia, Crataegus,*
(ethanamine)	ihl rat LCLo, 3,000 ppm/4 hr	*Dracunculus, Erodium, Sambucus, Sauromatium*
(ethyl mercaptan)	skn rbt LD_{50}, 390 mg	
Eucalyptol (see Cineole)		
Eucalyptus oil	orl rat LD_{50}, 4,440 mg	*Eucalyptus*
Eugenol	orl rat LDLo, 500 mg	*Acorus, Alpinia, Cananga, Cannabis, Cinnamomum,*
	orl mus LD_{50}, 3,000 mg	*Citrus, Cryptocarya, Cymbopogon. Dicypellium*
	orl gpg LD_{50}, 2,130 mg	*Eugenia, Geum, Levisticum, Laurus, Lavandula,*
		Majoranum, Michelia, Myristica, Nicotiana,
		Ocimum, Pimenta, Pimpinella, Piper, Pogostemon,
		Polianthes, Prunus, Reseda, Syzygium

Table 2 (continued)
TOXINS: THEIR TOXICITY AND DISTRIBUTION IN PLANT GENERA

Chemical	Toxicity[a,b]	Plant genera
Eugenol methyl ether	orl rat LD_{50}, 1,179 mg ipr mus LD_{50}, 540 mg ivn mus LD_{50}, 112 mg	*Acacia, Acorus, Asarum, Atherosperma, Cananga, Cinnamomum, Hyacinthus, Melaleuca, Pimenta, Piper, Rosa*
Evonoside	ivn cat LDLo, 839 µg	*Euonymus*
Fenchone (norbornanone)	orl rat LD_{50}, 4,400 mg	*Blumea, Foeniculum, Lavandula, Prunella, Thuja*
Fennel oil	orl rat LD_{50}, 3,120 mg	*Foeniculum*
Ferulic acid	par mus LDLo, 1,200 mg	*Elodea, Ferula, Lycopersicon*
Ficin	orl rat LD_{50}, 5,000 mg orl rbt LD_{50}, 5,000 mg orl rbt LD_{50}, 5,000 mg orl gpg LD_{50}, 5,000 mg ivn mam LDLo, 50 mg	*Ficus*
Folic acid (vitamin M)	ipr mus LD_{50}, 100 mg ivn mus LD_{50}, 239 mg	*Gossypium, Hordeum, Juglans, Lactuca, Lathyrus, Lens, Lycopersicon, Madhuca, Malus, Morus, Musa*
Formaldehyde	orl wmn LDLo, 36 mg orl rat LD_{50}, 800 mg orl gpg LD_{50}, 260 mg scu rat LDLo, 1,300 mg/65 WI NEO TWA 3 ppm	*Achillea, Beta, Cymbopogon, Humulus, Monarda*
Formic acid	orl rat LD_{50}, 1,210 mg orl mus LD_{50}, 1,100 mg orl dog LD_{50}, 4,000 mg ivn mus LD_{50}, 145 mg	*Arnica, Artemisia, Cinnamomum, Cistus, Citrus, Daucus, Digitalis, Eriodictyon, Humulus, Malus, Mentha, Myristica, Nicotiana, Ricinus, Triticum, Urtica, Valeriana*
Fructose [fructopynanose (β-D)]	scu mus TDLo, 5,000 mg NEO	Widespread
Fumaric acid	ipr mus LD_{50}, 200 mg	General? *Acer, Angelica, Arnica, Capsella, Carduus, Corydalis, Dicentra, Fumaria, Glaucium, Helianthus, Myrrhis, Oldenlandia, Phaseolus, Ricinus, Senecio, Silybum*
Funtumidene	ivn mus LD_{50}, 21 mg	*Funtumia*
Funtumine	ivn mus LD_{50}, 30 mg	*Funtumia*
Furfural	orl rat LD_{50}, 127 mg orl mus LD_{50}, 425 mg orl dog LD_{50}, 2,300 mg orl rbt LD_{50}, 928 mg orl gpg LD_{50}, 541 mg TWA 5 ppm	*Angelica, Atractylodes, Cananga, Carum, Cinnamomum, Citrus, Cymbopogon, Gossypium, Hordeum, Ilex, Juglans, Lavandula, Mentha, Myristica, Pimenta, Syzygium, Zea*
Furfuryl alcohol	orl rat LD_{50}, 275 mg ihl rat LD_{50}, 233 ppm 4 hr orl mus LD_{50}, 40 mg ivn rbt LD_{50}, 650 mg TWA 50 ppm	*Coffea, Syzygium*
Gallic acid	orl rat LD_{50}, 5,000 mg	*Acacia, Arctostaphylos, Areca, Astilbe, Bergenia, Camellia, Cornus, Digitalis, Juniperus, Lawsonia, Malus, Mangifera, Myrica, Phyllanthus, Polygonum, Prunus, Psidium, Punica, Pyrus*
Garryfoline (acetate)	ivn mus LD_{50}, 16 mg	*Garrya*
Garryine (HCl)	ivn mus LD_{50}, 13 mg	*Garrya*
Gaultheria oil	par rat TDLo, 1,500 mg (9—11 D preg) TER orl hmn TDLo, 170 mg orl rat LD_{50}, 887 mg orl mus LD_{50}, 1,110 mg orl dog LD_{50}, 2,100 mg orl rbt LD_{50}, 1,300 mg orl gpg LD_{50}, 1,060 mg	*Gaultheria*
Gelsemine	scu rbt LDLo, 500 µg ivn rbt LDLo, 800 µg	*Gelsemium, Mostuea*
Gentisic acid	orl rat LD_{50}, 800 mg	75 Dicot, 7 monocot families
Geraniol	orl rat LD_{50}, 3,600 mg ims mus LD_{50}, 4,000 mg ivn rbt LD_{50}, 50 mg	*Bursera, Camellia, Cinnamomum, Citrus, Darwinia, Elsholtzia, Eucalyptus, Eugenia, Humulus, Jasminum, Lavandula, Lippia, Mangifera, Melissa, Michelia, Myristica, Pelargonium, Polianthes, Prunus, Thymus, Zingiber*
Geranium oil	skn rbt LD_{50}, 2,500 mg	*Pelargonium*
Germerine	orl rat LD_{50}, 30 mg scu rat LD_{50}, 3,700 µg scu cat LDLo, 500 µg scu rbt LDLo, 2 mg	*Veratrum*

Table 2 (continued)
TOXINS: THEIR TOXICITY AND DISTRIBUTION IN PLANT GENERA

Chemical	Toxicity[a,b]	Plant genera
Gitalin	scu mus LDLo, 29 mg	*Digitalis*
Gitorin	ivn cat LD_{50}, 0.43 mg	*Digitalis*
Gitoxin	orl cat LDLo, 880 μg	*Digitalis*
Glaucarubin	orl hmn TDLo, 3 mg GIT	*Perriera, Simaruba*
Glyceraldehyde	ipr rat LD_{50}, 2,000 mg	Universal?
Glycerol	orl gpg LD_{50}, 7,750 mg	Universal?, *Olea, Phoenix, Theobroma*
Gnoscopine (see Narcotine)		
Gossypol	orl rat LD_{50}, 2,315 mg	*Gossypium*
	orl pig LD_{50}, 550 mg	
Gossyverdurin	orl rat LD_{50}, 660 mg	*Gossypium*
Gramine	ipr rat LDLo, 250 mg	*Acer, Arundo, Desmodium, Hordeum, Lupinus,*
(donaxine)	ipr mus LD_{50}, 122 mg	*Phalaris*
Guaiacol	orl rat LD_{50}, 725 mg	*Acer, Apium, Betula, Cannabis, Coffea, Citrus,*
		Guaiacum, Nicotiana, Pandanus, Ruta
Guanidine	orl rbt LDLo, 500 mg	*Beta, Brassica, Galega, Glycine, Vicia, Zea*
	scu rbt LDLo, 500 mg	
	scu mus LDLo, 500 mg	
Guanine (HCl)	ipr mus LDLo, 200 mg	*Aralia*?
Guaranine (see Caffeine)		
Halostachyine (HCl)	ivn rbt LDLo, 100 mg	*Halostachys*
Harmaline	scu rat LDLo, 120 mg	*Banisteriopsis, Peganum*
	scu rbt LDLo, 100 mg	
	ivn rbt LDLo, 20 mg	
Harman	ipr mus LD_{50}, 50 mg	*Calligonum, Carex, Neuboldia, Passiflora,*
	scu rbt LDLo, 200 mg	*Sickingia, Symplocos, Tribulus*
Harmine	ivn hmn TDLo, 2 mg PSY	*Banisteriopsis, Passiflora, Peganum, Sickingia,*
		Tribulus, Zygophyllum
Heliotrine	ivn rat LD_{50}, 274 mg	*Heliotropium*
	par rat LDLo, 150 mg (preg) TER	
Helleborein	ivn cat LDLo, 1,900 μg	*Helleborus*
Heptanal	orl mus LD_{50}, 500 mg	*Cananga, Gaultheria, Hyacinthus, Ricinus*
Heptane	ihl hmn TCLo, 1,000 ppm/6 min CNS	*Pinus, Pittosporum*
Heptanoic acid	orl mus LD_{50}, 160 mg	*Acacia, Acorus, Arnica, Viola*
	ivn mus LD_{50}, 1,200 mg	
Heptyl alcohol	orl rat LD_{50}, 3.25 mg	*Hyacinthus, Litsea*
	orl mus LD_{50}, 1,500 mg	
	orl rbt LD_{50}, 750 mg	
Hesperidin	ipr mus LD_{50}, 1,000 mg	*Citrus*
Heteroxanthine	ukn frg LDLo, 10 mg	*Beta*
Hexadecanol	skn rbt LD_{50}, 2,600 mg	*Ambrosia, Dorema, Ipomoea, Loranthus, Smilax,*
		Sorghum
Hexanal	orl rat LD_{50}, 4,890 mg	*Quercus*
(acpronaldehyde)	ihl rat LDLo, 2,000 ppm/4 hr	
Hexanol	orl rat LD_{50}, 720 mg	*Camellia, Fragaria, Lavandula, Litsea,*
(hexyl alcohol)	orl mus LD_{50}, 1,950 mg	*Pelargonium, Salvia*
	skn rbt LD_{50}, 3,100 mg	
Himandrine	ivn rat LD_{50}, 34 mg	*Galbulimima*
Hinokitiol	scu gpg LDLo, 500 mg	*Thuja*
Histamine	scu gpg LD_{50}, 0.5 mg	*Amorphophallus, Beta, Chelidonium, Chenopodium,*
	ivn gpg LD_{50}, 180 μg	*Citrus, Cucumis, Cyclamen, Delphinium, Drosera,*
	ivn dog LDLo, 50 mg	*Erodium, Lamium, Laportea, Lycopersicon, Mimosa,*
	scu rat LDLo, 250 mg	*Nepenthes, Pinguicula, Plantago, Salsola, Salvia,*
	scu mus LDLo, 600 mg	*Sarracenia, Silybum, Spinacia, Trifolium, Urera,*
		Urtica, Viscum
Ho leaf oil	orl rat LD_{50}, 3,270 mg	
Hordenine (see Anhaline)		
Hydantoin	orl man TDLo, 15 g	*Beta*
	orl hmn TDLo, 1 mg PSY	
	orl rat TDLo, 1,500 mg CAR	
	ipr rat TDLo, 100 mg	
	(7—10 D preg) TER	
Hydrastine (HCl)	orl rat LD_{50}, 1,000 mg	*Corydalis, Hydrastis, Meconopsis*
	scu rat LD_{50}, 1,270 mg	
Hydrocyanic acid	orl hmn LDLo, 570 μg	*Acacia, Acalypha, Acanthospermum, Adenia, Adenium,*
	orl mus LD_{50}, 3,700 μg	*Ageratum, Agrimonia, Agropyron, Ailanthus,*
	orl dog LDLo, 4 mg	*Albizia, Allium, Alocasia, Alternanthera,*
		Amaranthus, Annona, Anthephora, Aristida,
		Artocarpus, Asplenium, Atriplex, Avena, Bauhinia,
		Berkheyopsis, Beta, Boehmeria, Boscia, Brabejum,
		Brassica, Cadaba, Caesalpinia, Calodendrum,

Table 2 (continued)
TOXINS: THEIR TOXICITY AND DISTRIBUTION IN PLANT GENERA

Chemical	Toxicity[a,b]	Plant genera
Hydrocyanic acid (continued)		*Canavalia, Cardiospermum, Carica, Cassia, Castalis, Castanospermum, Ceiba, Centella, Centaurea, Cercocarpus, Chaenomeles, Chenopodium, Chironia, Chloris, Chrysanthemum, Chrysopogon, Cinnamomum, Cirsium, Cissus, Citrus, Citrullus, Coix, Colocasia, Combretum, Commelina, Corchorus, Cotoneaster, Cotyledon, Courbonia, Crotalaria, Cryptolepis, Cucurbita, Cydonia, Cymbopogon, Cynanchum, Cynodon, Cyperus, Dactyloctenium, Datura, Daucus, Dichapetalum, Dichrostachys, Digitaria, Dimorphotheca, Dipcadi, Diplolophium, Dodonaea, Dolichos, Duranta, Eleusine, Entada, Epaltes, Eragrostis, Eriobotrya, Erythrophleum, Eschscholtzia, Eucalyptus, Euphorbia, Eustachys, Fagopyrum, Florestina, Galenia, Gibberella, Gliricidia, Glyceria, Glycine, Grevillea, Hertia, Hevea, Hibiscus, Holcus, Hydrangea, Hydnocarpus, Indigofera, Ipomoea, Jatropha, Kalanchoe, Kedrostis, Kiggelaria, Lagenaria, Lagurus, Lantana, Lasiosiphon, Lathyrus, Lepidium, Leucaena, Linum, Lolium, Lotononis, Lotus, Luffa, Lupinus, Macadamia, Malus, Mangifera, Manihot, Medicago, Melica, Melilotus, Melolobium, Mercurialis, Mollugo, Moringa, Mundulea, Myrsine, Nasturtium, Nerium, Nicotiana, Ochna, Ocimum, Olinia, Oryza, Osteospermum, Pangium, Panicum, Pappea, Paspalum, Passiflora, Pennisetum, Persea, Phaseolus, Physalis, Piper, Poa, Pogonarthria, Polygonum, Portulaca, Prunus, Psidium, Pteridium, Pygeum, Pyracantha, Pyrus, Rhynchelytrum, Ricinus, Saccharum, Salsola, Sambucus, Sarcostemma, Setaria, Silybum, Solanum, Sorbaria, Sorghum, Spiraea, Sporobolus, Stephanorossia, Sterculia, Suckleya, Synadenium, Tagetes, Tamarindus, Tephrosia, Terminalia, Thalictrum, Themeda, Thesium, Tithonia, Tribulus, Trifolium, Triglochin, Urginea, Vernonia, Vicia, Viola, Vitex, Xanthium, Ximenia, Zea, Zygophyllum*
Hydrohydratinine (HCl)	par mus LDLo, 150 mg	*Corydalis*
Hydroquinone	orl rat LD_{50}, 370 mg	*Arbutus, Bergenia, Hydrangea, Pimpinella, Protea, Pyrus, Rhododendron, Rubus, Vaccinium,*
	ipr rat LD_{50}, 170 mg	
	skn mus TDLo, 800 mg NEO	
	orl mus LD_{50}, 400 mg	*Xanthium*
	TWA 2 mg/m³	
Hydroxysenkirkine	ipr rat TDLo, 300 mg CAR	*Crotalaria*
Hyoscyamine	ivn mus LD_{50}, 95 mg	*Anthocercis, Atropa, Datura, Duboisia, Hyoscyamus, Lactuca, Mandragora, Physochlaina, Scopolia, Solandra, Solanum*
Hypoglycine (cyclopropanealanine)	ipr rat TDLo, 30 mg (1–6 D preg) TER	*Blighia*
Hypoxanthine	ipr mus LDLo, 750 mg	*Lupinus, Solanum*
Imperatorin	par mus LDLo, 600 mg	*Aegle, Ammi, Angelica, Pastinaca, Peucedanum, Prangos, Ruta*
Indican	scu mus TDLo, 250 mg	*Asclepias, Bletia, Calanthe, Crotalaria, Echites, Epidendrum, Indigofera, Isatis, Phaius, Polygala, Polygonum, Robinia, Wrightia*
	scu mus TDLo, 500 mg 26 WI CAR	
Indole	orl rat LD_{50}, 1,000 mg	*Cheiranthus, Chimonanthus, Citrus, Jasminum, Narcissus, Robinia*
	skn mus TDLo, 480 mg	
	scu mus LD_{50}, 225 mg	
	scu mus TDLo, 1,000 mg (20 WI NEO)	
Inositol-(meso) (myo-inositol)	ipr mus LDLo, 500 mg	*Alnus, Cotinus, Juglans, Phaseolus, Quercus, Rhus, Viburnum, Viscum, Zea*
Ionone (irisone)	orl rat LD_{50}, 4,590 mg	*Boronia, Lawsonia, Saussurea*
Isatidene (retrorsine N oxide)	orl rat TDLo, 1,044 mg/87 W CAR	*Erechtites, Senecio*
	ipr rat TDLo, 225 mg/9 W CAR	
	ivn mus LD_{50}, 835 mg	
Isoamyl alcohol	orl rat LDLo, 1,300 mg	*Eucalyptus, Fragaria, Lavandula, Mentha, Pelargonium, Rubus, Solanum*
	ipr rat LDLo, 813 mg	
	orl rbt LDLo, 4,250 mg	

Table 2 (continued)
TOXINS: THEIR TOXICITY AND DISTRIBUTION IN PLANT GENERA

Chemical	Toxicity[a,b]	Plant genera
Isobutyl alcohol	skn rbt LD_{50}, 3,970 mg TWA 100 ppm orl rat LD_{50}, 2,460 mg ihl rbt LDLo, 8,000 ppm 4 hr orl rbt LDLo, 3,750 mg	Anthemis
Isobutyraldehyde	skn rbt LD_{50}, 4,240 mg TWA 100 ppm orl rat LD_{50}, 2,810 mg ihl rat LCLo, 8,000 ppm/4 hr	Acacia, Camellia, Datura, Morus, Nicotiana, Raphanus
Isobutyric acid	orl rat LD_{50}, 280 mg skn rbt LD_{50}, 500 mg	Anthemis, Arnica, Artemisia, Ceratonia, Cinnamomum, Croton, Daucus, Eucalyptus, Euryops, Laurus, Lavandula, Peucedanum, Prunus, Seseli
Isochaksine	ipr rat LD_{50}, 56 mg	Cassia
Isoeugenol	orl rat LD_{50}, 1,560 mg orl gpg LD_{50}, 1,410 mg	Cananga, Cinnamomum, Leptactinia, Michelia, Myristica, Nectandra, Prunus
Isopentyl alcohol (see Isoamyl alcohol)		
Isosafrole	orl rat LD_{50}, 1,340 mg orl mus LD_{50}, 2,470 mg orl mus TDLo, 61 g/81 WIC CAR ipr mus LDLo, 256 mg	Cananga, Illicium, Ligusticum, Murraya
Isovaleraldehyde		Camellia, Cinnamomum, Citrus, Cymbopogon, Eucalyptus, Glycine, Helichrysum, Lavandula, Mentha, Monarda, Quercus, Raphanus
Isovaleric acid	orl rat LDLo, 3,200 mg ivn mus LD_{50}, 1,120 mg	Achillea, Angelica, Artemisia, Calotropis, Cinnamomum, Citrus, Croton, Eucalyptus, Humulus, Lavandula, Levisticum, Lippia, Malus, Melaleuca, Mentha, Musa, Nicotiana, Pelargonium, Prunus, Rosmarinus, Ruta, Theobroma, Valeriana, Viburnum
Jacobine	ivn mus LD_{50}, 77 mg NEO	Crotalaria, Senecio
Jervine	ivn mus LD_{50}, 9,300 µg	Amianthium, Veratrum, Zigadenus
Juglone	orl mus LD_{50}, 2,500 µg ipr mus LD_{50}, 3 mg	Carya, Juglans, Platycarya, Pterocarya
Khellin	orl rat LD_{50}, 80 mg orl mus LD_{50}, 400 mg	Ammi
Lactose	scu mus TDLo, 50 mg/DC NEO ivn rbt LDLo, 1,500 mg	Achras, Ceratonia, Forsythia, Lippia, Mimusops, Oldenlandia, Pouteria, Sapindus
Lapachol	orl rat LDLo, 1,200 mg orl mus LD_{50}, 487 mg	Adenanthera?, Andira?, Avicennia, Bassia, Bignonia, Intsia?, Paratecoma, Tabebuia, Tecoma, Tectona
Larasha	ims rat TDLo, 2,250 mg 61 WI NEO	Citrus
Lasiocarpine	orl rat LD_{50}, 150 mg ipr rat LD_{50}, 78 mg ipr rat TDLo, 470 mg 56 WI CAR	Cynoglossum, Heliotropium, Lappula, Symphytum
Lasiocarpine N-oxide	ipr rat LD_{50}, 600 mg	Heliotropium
Lauric acid (n-dodecanoic acid)	ivn mus LD_{50}, 131 mg	Acrocomia, Actinodaphne, Areca, Astrocaryum, Attalea, Cinnamomum, Cocos, Elaeis, Erisma, Hyphaene, Irvingia, Laurus, Litsea, Neolitsea, Ruta, Salvadora, Umbellularia
Lemon oil	orl rat LD_{50}, 2,840 mg	Citrus
Leptoside	ivn cat LD_{50}, 1.9 mg	Strophanthus
Leucaenine	orl rat TDLo, 350 mg (preg) TER	Leucaena, Mimosa
Leucomycin	ivn mus LD_{50}, 650 mg	Artemisia
Leurocristine (vincristine)	ivn rat LD_{50}, 1.3 mg ipr rat LD_{50}, 1,300 µg unk rat TDLo, 250 µg (9 D preg TER)	Catharanthus
Limonene (dipentene)	orl rat LDLo, 4,600 mg	Anethum, Apium, Baeckea, Barosma, Boswellia, Canarium, Carum, Chenopodium, Cinnamomum, Citrus, Coriandrum, Croton, Cuminum, Cymbopogon, Dacrydium, Daucus, Elettaria, Eucalyptus, Foeniculum, Humulus, Hyptis, Illicium, Lavandula, Lippia, Litsea, Mangifera, Medicago, Melaleuca, Mentha, Murraya, Myristica, Nicotiana, Ocimum, Origanum, Perilla, Pilocarpus, Pimenta, Piper, Pittosporum, Premna, Prunus, Psidium, Salvia, Seseli, Siler, Solidago, Valeriana

Table 2 (continued)
TOXINS: THEIR TOXICITY AND DISTRIBUTION IN PLANT GENERA

Chemical	Toxicity[a,b]	Plant genera
Linalool	orl rat LD_{50}, 2,790 mg	*Acacia, Artemisia, Atalantia, Bursera, Camellia, Cananga, Cinnamomum, Citrus, Coriandrum, Cymbopogon, Eschscholtzia, Eucalyptus, Humulus, Jasminum, Lavandula, Majorana, Melissa, Mentha, Morus, Myristica, Ocimum, Origanum, Pandanus, Prunus, Salvia, Sassafras, Satureja, Thymus, Zingiber*
Lobeline	ipr mus LD_{50}, 39.9 mg scu mus LD_{50}, 37 mg	*Lobelia*
Longilobine (retrorsine)	ivn mus LD_{50}, 78 mg	*Crotalaria, Erechtites, Senecio*
Loturine (see Harman)		
Lunarine	ivn mus LDLo, 62 mg	*Lunaria*
Lupinidene (see Sparteine)		
Lupulone (humulone)	orl rat LD_{50}, 1,800 mg ims rat LD_{50}, 330 mg orl mus LD_{50}, 1,500 mg orl rbt LDLo, 1,000 mg orl gpg LD_{50}, 130 mg	*Humulus, Lupulus*
Lysine	ivn mus LD_{50}, 181 mg	Universal? *Lupinus, Pisum, Robinia, Vicia*
Maleic acid	orl rat LD_{50}, 708 mg skn rbt LD_{50}, 1,560 mg	
Malic acid	orl rat LDLo, 1,600 mg	*Acer, Ananas, Artemisia, Asperula, Berberis, Beta, Bryophyllum, Capsicum, Coriandrum, Cydonia, Datura, Daucus, Drosera, Ficus, Fragaria, Fraxinus, Helianthus, Hibiscus, Hippophae, Ilex, Juniperus, Lactuca, Lolium, Lycopersicon, Madhuca, Malus, Medicago, Morus, Musa, Myrtus, Nicotiana, Oryza, Papaver, Passiflora, Prunus, Pyrus, Rhus, Ribes, Ricinus, Rubus, Rumex, Sambucus, Sclerocarya, Sedum, Sempervivum, Solanum, Sorbus, Tamarindus, Vaccinium, Vitis*
Malonic acid	orl rat LD_{50}, 1,310 mg ipr mus LD_{50}, 300 mg	*Anthriscus, Anthyllis, Apium, Astragalus, Avena, Bunias, Colutea, Helianthus, Hordeum, Lotus, Lupinus, Medicago, Melilotus, Ononis, Phaseolus, Sophora, Stemona, Thermopsis, Trifolium, Trigonella, Triticum*
Maltol (larixic acid)	orl rat LD_{50}, 1,440 mg orl mus LD_{50}, 550 mg orl rbt LD_{50}, 1,620 mg orl gpg LD_{50}, 1,410 mg orl ckn LD_{50}, 3,720 mg	*Abies, Cichorium, Corydalis, Larix*
Maltose	ipr mus LDLo, 200 mg scu mus TDLo, 1,750 g/50 WC NEO	*Aconitum, Bassia, Bryophyllum, Ceratonia, Chrysactinia, Daucus, Glycine, Hygrophila, Lathyrus, Lippia, Malus, Mercurialis, Musa, Oryza, Panax, Phaseolus, Schizopepon, Trifolium, Tropaeolum, Typha, Umbilicus*
Mandelonitrile	scu mus LDLo, 23 mg orl rat LD_{50}, 116 mg	*Cotoneaster, Eucalyptus, Prunus, Pteris, Sambucus*
Mansonin	ivn cat LD_{50}, 0.15 mg	*Mansonia*
Melilotin (hydrocoumarin)	orl rat LD_{50}, 1,460 mg orl gpg LD_{50}, 1,760 mg	*Melilotus*
Menadione (vitamin K_3)	orl mus LDLo, 400 mg ipr mus LD_{50}, 50 mg orl rbt LDLo, 230 mg	
Menthol	orl rat LD_{50}, 3,180 mg scu rat LDLo, 2,000 mg	*Mentha, Thymus*
dl-Menthol	orl rat LD_{50}, 3,300 mg orl cat LDLo, 800 mg ipr rat LD_{50}, 710 mg	*Calamintha, Hedeoma, Hyptis, Mentha, Pycnanthemum*
Mercurialin (see Menthylamine)		
Mesaconic acid	ipr mus LDLo, 500 mg	*Brassica*
Mescaline (phenethylamine, trimethoxy)	orl hmn TDLo, 5 mg ivn hmn TDLo, 7 mg orl mus LD_{50}, 880 mg ivn gpg TDLo, 450 μg 8 D preg TER	*Gymnocalycium, Lophophora, Opuntia, Trichocereus*
Methanethiol (methyl mercaptan)	ihl rat LCLo, 10,000 ppm scu mus LD_{50}, 2.4 mg	*Brassica, Camellia, Coprosma, Lasianthus, Paederia, Raphanus*

Table 2 (continued)
TOXINS: THEIR TOXICITY AND DISTRIBUTION IN PLANT GENERA

Chemical	Toxicity[a,b]	Plant genera
Methanol	orl hmn LDLo, 340 mg orl hmn TDLo, 100 mg EYE ihl hmn TCLo, 300 ppm CNS orl mus LDLo, 420 mg orl dog LDLo, 6,300 mg orl mky LDLo, 7,000 mg orl rbt LDLo, 4,750 mg TWA 200 ppm	*Angelica, Anthriscus, Boehmeria, Camellia, Carum, Gossypium, Heracleum, Lycopersicon, Mentha, Pastinaca, Pimenta, Syzygium* *Vitex*
Methoxalen (see Xanthotoxin)		
Methylamine (mercurialin)	scu mus LDLo, 2,500 mg scu frg LDLo, 2,000 mg TWA 10 ppm	*Acorus?, Amorphophallus, Arum, Atropa, Beta, Chaerophyllum, Conium, Delphinium, Dracunculus, Heracleum, Iris, Leptotaenia, Lilium, Mentha, Mercurialis, Nicotiana, Nuphar, Philadelphus, Sambucus, Sauromatium, Stapelea, Staphylea, Thalictrum, Veratrum, Viburnum*
Methyl anthranilate	orl rat LD_{50}, 2,910 mg orl mus LD_{50}, 3,900 mg orl gpg LD_{50}, 2,780 mg	*Cananga, Citrus, Jasminum, Kaempferia, Michelia, Polianthes*
S-Methyl-1-cysteine	ipr mus LDLo, 500 mg	*Astragalus, Phaseolus*
Methyl salicylate	orl hmn LDLo, 170 mg orl rat LD_{50}, 887 mg orl mus LD_{50}, 1,110 mg orl dog LD_{50}, 2,100 mg orl rbt LD_{50}, 1,300 mg orl gpg LD_{50}, 1,060 mg scu gpg LDLo, 1,500 mg par rat TDLo, 1,500 mg (9–11 D preg TER)	*Acacia, Alphitonia, Alsodeia, Alstonia, Ardisia, Atalantia, Baccaurea, Barringtonia, Betula, Bignonia, Bridelia, Calycanthus, Camellia, Cananga, Carallia, Ceanothus, Chenopodium, Chilocarpus, Clematis, Cocculus, Comesperma, Conocephalus, Cryptolepis, Cyclostemon, Dendrocalamus, Diospyros, Elaeocarpus, Erythroxylum, Eugenia, Ficus, Fragaria, Garuga, Gaultheria, Glycosmis, Homalium, Hunteria, Hydnocarpus, Lindera, Linociera, Maba, Macaranga, Marsdenia, Meliosma, Metrosideros, Monotropa, Murraya, Myristica, Nyctanthes, Nyctocalos, Paliuris, Parinari, Photinia, Platea, Polianthes, Polygala, Prunus, Ribes, Rubus, Ryparosa, Scolopia, Sideroxylon, Stiftia, Styphelea, Symplocos, Syzygium, Tabebuia, Taraktogenos, Thunbergia, Turpinia, Vernonia, Viburnum, Vitis, Xanthophyllum*
Milloside	ivn cat LD_{50}, 1.3 mg	*Roupellina*
Monocrotalic acid	ivn rat LD_{50}, 581 mg ivn mus LD_{50}, 606 mg	*Crotalaria*
Monocrotaline	orl rat LD_{50}, 66 mg scu rat LDLo, 60 mg ipr rat TDLo, 5 mg CAR ivn rat LD_{50}, 92 mg ivn mus LD_{50}, 261 mg	*Crotalaria*
Morphine	orl mus LD_{50}, 745 mg ipr mus LD_{50}, 480 mg scu mus LD_{50}, 360 mg ivn mus LD_{50}, 199 mg scu rat LDLo, 420 mg orl rat LD_{50}, 600 mg	*Argemone?, Eschscholtzia?, Humulus?, Papaver*
Myristic acid (tetradecanoic acid)	ivn mus LD_{50}, 43 mg	*Calendula, Cinnamomum, Citrus, Cocos, Corylus, Croton, Cyperus, Elaeis, Erisma, Guizotia, Helianthus, Irvingia, Levisticum, Macadamia, Moringa, Myrica, Myristica, Myrtus, Phyllanthus, Pimpinella, Pycnanthus, Ruta, Salvadora, Virola*
Myristicin	orl cat LDLo, 570 mg	*Anethum, Carum, Cinnamomum, Levisticum, Myristica, Oenanthe, Orthodon, Pastinaca, Petroselinum, Peucedanum*
Napthalene	orl chd LDLo, 100 mg orl rat LD_{50}, 1,780 mg ipr mus LDLo, 150 mg scu rat TDLo, 3,500 mg TWA 150 mg/98 DI NEO	*Iris, Oryza, Saussurea, Syzygium*
Narcotine	ipr rat LD_{50}, 750 mg	*Brassica, Citrus, Lycopersicon, Papaver, Rauwolfia, Solanum, Strychnos*
Neogermitrine	ipr mus LD_{50}, 510 µg	*Veratrum, Zigadenus*
Neriin	scu mus LDLo, 95 mg	*Nerium*
Niacin (nicotinic acid)	scu rat LDLo, 4,000 mg orl mus LD_{50}, 5,000 mg scu mus LDLo, 4,000 mg	Ubiquitous, *Zinnia*

Table 2 (continued)
TOXINS: THEIR TOXICITY AND DISTRIBUTION IN PLANT GENERA

Chemical	Toxicity[a,b]	Plant genera
Nicotinamide (vitamin B_3) (vitamin PP)	orl rat LD_{50}, 3,500 mg scu rat LD_{50}, 1,680 mg scu mus LDLo, 1,800 mg	
Nicotine	orl hmn LDLo, 1 mg orl rat LD_{50}, 53 mg orl mus LD_{50}, 3 mg orl dog LD_{50}, 9,200 μg orl pgn LD_{50}, 75 mg orl dck LD_{50}, 75 mg	*Acacia, Aesculus, Asclepias, Cannabis, Cyphomandra, Duboisia, Echeveria, Eclipta, Equisetum, Erythroxylum, Herpestis, Juglans, Mucuna, Prunus, Sempervivum, Solanum, Urtica, Withania*
Nitrosodimethyl amine	ihl rat TCLo, 37 mg CAR ipr rat TDLo, 30 mg NEO unk mus TDLo, 13 mg (preg TER) orl rat LD_{50}, 26 mg orl dog LDLo, 20 mg	*Solanum*
Nonanoic acid (pelargonic acid)	orl rat LD_{50}, 3,200 mg ivn mus LD_{50}, 224 mg	*Artemisia, Citrus, Cuphea, Eremocitrus, Iris, Lavandula, Litsea, Mentha, Pelargonium, Rhus*
Nonlyl alcohol (1-nonanol)	orl rat LDLo, 1,400 mg	*Cinnamomum, Citrus, Eremocitrus*
Noradrenaline (norepinephrine)	ivn rat LD_{50}, 0.10 mg orl mus LD_{50}, 20 mg scu mus LD_{50}, 7,600 μg ivn mus LD_{50}, 3,700 mg ivn rbt LDLo, 750 μg	*Citrus, Musa, Portulaca, Prunus, Solanum*
D-Nornicotine	ipr rat LDLo, 6 mg	*Duboisia, Nicotiana, Salpiglossis, Zinnia*
L-Nornicotine	ipr rat LDLo, 23.5 mg	
DL-Nornicotine	ipr rat LDLo, 10.5 mg	
Norpseudoephedrine (cathine)	scu mus LD_{50}, 275 mg	*Catha, Ephedra, Maytenus*
Noscapine	ivn mus LD_{50}, 83 mg	*Lycopersicon*
Nutmeg oil (expressed)	orl rat LD_{50}, 3,640 mg	*Myristica*
Nutmeg oil (volatile)	orl rat LD_{50}, 2,620 mg	*Myristica*
Odorobioside K	ivn cat LD_{50}, 2.3 mg	*Nerium*
Odoroside K	ivn cat LD_{50}, 4.74 mg	*Nerium*
Oleic acid	ivn mus LD_{50}, 230 mg ivn cat LDLo, 50 mg scu rbt TDLo, 3,120 mg 52 WI NEO	Major component of oilseeds, *Myrica*
Origanum oil	orl rat LD_{50}, 1,850 mg skn rbt LD_{50}, 320 mg	*Origanum*
Orcinol (methyl resorcinol)	orl rat LD_{50}, 844 mg orl mus LD_{50}, 772 mg orl rbt LD_{50}, 2,400 mg orl gpg LD_{50}, 1,687 mg	Leaf extracts of Ericaceae
Ouabain	ivn rat LD_{50}, 14 mg ivn dog LDLo, 54 μg ivn cat LDLo, 120 μg ivn rbt LDLo, 100 μg ims gpg LDLo, 220 μg	*Acokanthera, Strophanthus*
Oxalic acid	orl hmn LDLo, 700 mg orl dog LDLo, 1,000 mg TWA 1 mg/m³	*Abelmoschus, Allium, Anacardium, Ananas, Benincasa, Beta, Brassica, Camellia, Capsicum, Chenopodium, Cicer, Citrus, Colocasia, Coriandrum, Daucus, Eleusine, Ficus, Glycine, Halogeton, Ipomoea, Juniperus, Lactuca, Lens, Lolium, Lycopersicon, Mangifera, Manihot, Medicago, Mentha, Momordica, Musa, Nicotiana, Oryza, Oxalis, Pastinaca, Pennisetum, Phaseolus, Pisum, Prunus, Raphanus, Rheum, Rumex, Solanum, Sorghum, Spinacia, Trichosanthes, Triticum, Vigna, Vitis, Zea*
Oxyacanthine	ipr rat LDLo, 250 mg	*Berberis, Mahonia, Michelia*
Pachycarpine (D-sparteine)	ivn mus LD_{50}, 26 mg	*Aconitum, Ammodendron, Ammothamnus, Anagyris, Baptisia, Chelidonium, Cytisus, Genista, Hevea, Leontice, Leptorhabdos, Lupinus, Peumus, Sophora, Thermopsis*
Palmitic acid	ivn mus LD_{50}, 57 mg	Common in oilseeds, *Acacia, Aconitum, Apium, Daucus, Equisetum, Hydnocarpus, Levisticum, Myrica, Pimenta, Ricinus, Ruta, Sterculia, Vetiveria*

Table 2 (continued)
TOXINS: THEIR TOXICITY AND DISTRIBUTION IN PLANT GENERA

Chemical	Toxicity[a,b]	Plant genera
Pantothenic acid (Ca-salt) (vitamin B$_5$)	ipr rat LD$_{50}$, 820 mg scu rat LD$_{50}$, 3,400 mg ipr mus LD$_{50}$, 920 mg	*Gossypium, Hordeum, Juglans, Lathyrus, Lens, Linum, Lolium, Lycopersicon, Madhuca, Malus, Medicago, Oryza, Passiflora, Persea, Saccharum*
Papaverine	ipr rat LD$_{50}$, 64 mg orl mus LD$_{50}$, 350 mg scu mus LDLo, 160 mg ivn mus LD$_{50}$, 32 mg	*Papaver, Rauwolfia*
Parasorbic acid	scu rat TDLo, 64 mg/32 W CAR orl rat TDLo, 90 mg/64 WC NEO ipr mus LD$_{50}$, 750 mg	*Sorbus*
Parsley oil	orl rat LD$_{50}$, 3,960 mg	*Petroselinum*
Paulioside	ivn cat LD$_{50}$, 710 μg	*Roupellina*
Pectin	unk mus LDLo, 1,800 mg	Universal in higher plants?, *Cephaelis, Citrullus, Croton, Daucus, Diospyros, Feronia, Flacourtia, Fortunella, Fragaria, Hibiscus, Hordeum, Ipomoea, Juniperus, Lactuca, Lagenaria, Linum, Lupinus, Lycopersicon, Malus, Medicago, Moringa, Morus, Musa, Myristica, Opuntia, Oryza, Papaver, Passiflora, Phaseolus, Physalis, Phyllanthus, Ribes, Tamarindus, Taraxacum, Taxus*
Pelargonic acid (see Nonanoic acid)		
Pelletierine (punicine)	ivn rbt LDLo, 40 mg	*Punica, Sedum, Withania*
Pellotine	scu rat LDLo, 90 mg	*Lophophora*
Pennyroyal oil	orl rat LD$_{50}$, 400 mg	*Mentha*
Pentadecanoic acid	ivn mus LD$_{50}$, 54 mg	*Calendula, Zanthoxylum*
Peppermint oil	orl rat LD$_{50}$, 4,441 mg ipr rat LD$_{50}$, 819 mg	*Mentha*
Peru balsam oil	orl rat LD$_{50}$, 2,360 mg	*Myroxylon*
Petaline (chloride)	ivn mus LDLo, 3 mg ivn rbt LDLo, 16 mg	*Leontice*
Peucedanin	orl mus LD$_{50}$, 315 mg	*Peucedanum*
Phellandrene	No toxicity data	*Aegle, Anethum, Angelica, Artemisia, Boswellia, Bupleurum, Bursera, Caesalpinia, Canarium, Cinnamomum, Citrus, Coriandrum, Cuminum, Curcuma, Cymbopogon, Eucalyptus, Foeniculum, Haplopappus, Illicium, Laurus, Lavandula, Ledum, Magnolia, Mangifera, Melaleuca, Mentha, Monodora, Myroxylum, Oenanthe, Pelargonium, Pimenta, Piper, Sassafras, Schinus, Skimmia, Zingiber*
Phenethyl alcohol	orl rat LD$_{50}$, 1,790 mg orl mus LD$_{50}$, 800 mg skn rbt LD$_{50}$, 790 mg	*Citrus*
Phenethylamine	orl rat LDLo, 800 mg ipr rat LDLo, 100 mg ipr mus LD$_{50}$, 366 mg	*Acacia, Phoradendron*
Phenol	orl hmn TDLo, 14 mg GIT skn mus TDLo, 4,000 mg/20 WI CAR orl rbt LDLo, 420 mg orl hmn LDLo, 140 mg TWA (skn) 5 ppm	*Artemisia, Camellia, Nicotiana, Reseda, Ribes, Ruta, Salix*
Phorbol	ipr mus TDLo, 400 mg 25 W CAR	*Croton*
Phthalic acid	orl rat LDLo, 4,600 mg ipr mus LD$_{50}$, 1,670 mg	*Papaver*
Physostigmine	orl mus LD$_{50}$, 4,500 μg scu mus LDLo, 750 μg scu rat LDLo, 2,192 μg	*Dioclea, Hippomane, Mucuna, Physostigma, Vicia*
Picropodophyllin	ipr mus LD$_{50}$, 280 mg	Artefact?
Picrotoxin	orl mus LD$_{50}$, 15 mg ipr mus LD$_{50}$, 7 mg ivn mus LD$_{50}$, 2,440 μg ipr rat LD$_{50}$, 5,600 μg ivn rat LD$_{50}$, 1,960 μg	*Anamirta, Cocculus, Tinomiscium*
Pilocarpine	ims hmn TDLo, 140 mg CNS orl rat LD$_{50}$, 911 mg ipr mus LD$_{50}$, 500 mg ivn rbt LDLo, 175 mg	*Pilocarpus*

Table 2 (continued)
TOXINS: THEIR TOXICITY AND DISTRIBUTION IN PLANT GENERA

Chemical	Toxicity[a,b]	Plant genera
Pimenta leaf oil	orl rat LD_{50}, 3,600 mg	*Pimenta*
Piperidine	orl rat LD_{50}, 400 mg	*Cannabis, Conium, Lupinus, Nicotiana,*
	ihl rat LCLo, 4,000 ppm/4 hr	*Petrosimonia, Piper, Psilocaulon*
	skn rbt LD_{50}, 320 mg	
Piperonal	orl rat LD_{50}, 2,700 mg	*Cinnamomum, Doryphora, Eryngium, Piper,*
(heliotropin)	ipr rat LDLo, 1,500 mg	*Polianthes, Robinia, Spiraea, Vanilla, Viola*
Plicatic acid	ihl hmn TDLo, 2,500 g/m³	
Podophyllic acid	orl mus LD_{50}, 899 mg	*Podophyllum*
	scu mus LD_{50}, 700 mg	
Podophyllum	scu rat LDLo, 18 mg	*Podophyllum*
	orl mus LD_{50}, 68 mg	
	scu mus LD_{50}, 58 mg	
Podophyllotoxin	scu rat LD_{50}, 8 mg	*Callitris, Diphylleia, Juniperus, Linum,*
	orl mus LD_{50}, 90 mg	*Podophyllum*
1,2-Propanediamine	orl rat LD_{50}, 2,230 mg	*Arum, Sauromatum*
	skn rbt LD_{50}, 500 mg	
1-Propanethiol	orl rat LD_{50}, 1,790 mg	
(propyl	ihl rat LC_{50}, 7,300 ppm 4 hr	*Allium*
mercaptan)	ipr rat LD_{50}, 515 mg	
	ihl mus LC_{50}, 4,010 ppm 4 hr	
Propionaldehyde	orl rat LDLo, 800 mg	*Lycopersicon*
	orl mus LDLo, 800 mg	
Propionic acid	orl rat LD_{50}, 1,510 mg	Widespread, *Cinnamomum, Lavandula, Pimpinella*
	orl mus LD_{50}, 1,370 mg	
	orl rbt LD_{50}, 1,900 mg	
Propylamine	orl rat LDLo, 570 mg	*Camphorosma*
	ihl rat LC_{50}, 2,310 ppm 4 hr	
	skn rbt LD_{50}, 560 mg	
Protopine	ipr rat LDLo, 100 mg	*Adlumia, Arctomecon, Argemone, Bocconia,*
		Chelidonium, Corydalis, Cysticapnos, Dactylicapnos,
		Dendromecon, Dicentra, Dicranostigma,
		Eschscholtzia, Fumaria, Glaucium, Hunnemannia,
		Hypecoum, Maclaeya, Meconella, Meconopsis,
		Nandina, Papaver, Platycapnos, Platystemon,
		Pteridophyllum, Roemeria, Romneya, Sanguinaria,
		Sarcocapnos, Stylophorum
Protoveratrine	orl rat LD_{50}, 5 mg	*Veratrum*
	scu rat LD_{50}, 600 μg	
	ipr mus LD_{50}, 370 μg	
	ivn mus LD_{50}, 48 μg	
Pseudoephedrine	orl mus LDLo, 116 mg	*Ephedra, Roemeria, Taxus*
	scu rat LDLo, 650 mg	
	scu rbt LDLo, 400 mg	
Pterophine	ivn mus LD_{50}, 58 mg	*Senecio*
Pulegone	ipr mus LD_{50}, 150 mg	*Agastache, Bystropogon, Calamintha, Hedeoma,*
		Mentha, Micromeria, Origanum, Poliomintha,
		Pycnanthemum, Satureja
Pyrethrin I	orl rat LD_{50}, 1,200 mg	*Chrysanthemum*
	unk mam LD_{50}, 960 mg	
Pyrethrin II	orl rat LD_{50}, 1,200 mg	*Chrysanthemum*
Pyrethrum	orl rat LD_{50}, 200 mg	*Chrysanthemum*
	orl mam LD_{50}, 250 mg	
Pyridine	orl rat LD_{50}, 891 mg	*Atropa?, Haplopappus*
	ihl rat LC_{50}, 4,000 ppm 4 hr	
	orl gpg LDLo, 4,000 mg	
	ipr gpg LDLo, 870 mg	
Pyrocatechol	orl rat LD_{50}, 3,890 mg	*Allium, Ampelopsis, Beta, Betula, Citrus,*
	scu rat LDLo, 300 mg	*Eucalyptus, Fragaria, Garcinia, Paullinia,*
	orl rbt LDLo, 1,000 mg	*Platanus, Populus, Psidium, Psorospermum,*
	orl gpg LDLo, 550 mg	*Raphanus*
Pyrogallol	unk man LDLo, 120 mg	
	orl rat LD_{50}, 787 mg	
	orl dog LDLo, 25 mg	
	orl rbt LD_{50}, 1,600 mg	
Pyrrolidine	orl rat LD_{50}, 300 mg	*Daucus, Nicotiana*
	ihl mus LC_{50}, 1,300 mg/m³/2 hr	
Quercitin	orl rat LD_{50}, 161 mg	*Acacia, Aesculus, Allium, Arctostaphylos, Aster,*
	orl mus LD_{50}, 159 mg	*Bauhinia, Calluna, Capparis, Cornus, Cosmos,*
	scu mus LD_{50}, 97 mg	*Crataegus, Crocus, Dryobalanops, Dipterocarpus,*
	ivn rbt LD_{50}, 100 mg	*Erigeron, Euphoria, Euphorbia, Fatsia, Fragaria,*

Table 2 (continued)
TOXINS: THEIR TOXICITY AND DISTRIBUTION IN PLANT GENERA

Chemical	Toxicity[a,b]	Plant genera
		Fuchsia, Gossypium, Helinis?, Hypericum, Illicium, Iris, Malus, Marsdenia, Moringa, Nolana, Podophyllum, Polygonum, Prunus, Psidium, Quercus, Rhamnus, Ribes, Rosa, Shorea, Solidago, Thespesia, Trifolium, Vaccinium, Viola, Vitis, Zea
Quercitrin (flavone)	ipr mus LDLo 200 mg	*Aesculus, Agrimonia, Aleurites, Bauhinia, Camellia, Ceratostigma, Cercidiphyllum, Citrus, Crataegus, Crusea, Engelhardtia, Erigeron, Eucalyptus, Eucryphia, Forsythia, Fraxinus, Houttuynia, Humulus, Illicium, Lathyrus, Leucaena, Lycopersicon, Malus, Neolitsea, Nicotiana, Plumbago, Prunus, Psittacanthus, Quercus, Rhododendron, Ribes, Rosa, Solidago, Stuartia, Thuja, Tilia, Vaccinium, Vicia, Vitis*
Quinaldine	orl rat LD_{50}, 1,230 mg skn rbt LD_{50}, 1,870 mg	*Galipea*
Quinazoline	skn mus TDLo, 4,000 mg/YI NEO	*Dichroa?*
Quinic acid	scu mus LD_{50}, 10,000 mg	*Aconitum, Angelica, Arctostaphylos, Cinchona, Daucus, Eucalyptus, Illicium, Linum, Malus, Medicago, Nicotiana, Pistacia, Prunus, Rosa, Terminalia, Vaccinium*
Quinidine	orl rat LD_{50}, 1,000 mg orl mus LD_{50}, 594 mg ipr rat LDLo, 174 mg ipr mus LD_{50}, 190 mg	*Cinchona, Coutarea, Enantia, Remijia, Strychnos*
Quinine	orl wmn TDLo, 20 mg (4–5 W preg) TER orl rbt LDLo, 800 mg orl gpg LDLo, 300 mg unk gpg TDLo, 200 mg preg TER	*Cinchona, Cornus, Coutarea, Enantia, Ladenbergia, Picrolemma, Remijia, Strychnos*
Quinoline	orl rat LD_{50}, 460 mg ipr mus LDLo, 64 mg skn rbt LD_{50}, 540 mg	*Citrus*
Raton	scu rat TDLo, 31 g/61 WI NEO	*Gliricidia*
Red squill	orl rat LD_{50}, 200 mg	*Scilla*
Red thyme oil	orl rat LD_{50}, 4,700 mg	*Thymus*
Rescinnamine	orl mus LD_{50}, 1,420 mg	*Rauwolfia, Tonduzia*
Reserpine	orl hmn TDLo, 14 μg PSY ims hmn TDLo, 357 μg PSY unk rat TDLo, 1,500 μg (9–10 D preg) TER	*Alstonia, Aspidosperma, Bleekeria, Excavatia, Ochrosia, Rauwolfia, Tonduzia, Vallesia, Vinca, Voacanga*
Retronecine	ivn rat LD_{50}, 1,311 mg ivn mus LD_{50}, 634 mg	Usually combined. *Amsinckia, Brachyglottis, Crotalaria, Echium, Emilia, Erechtites, Heliotropium, Petasites, Senecio, Trichodesma, Tussilago*
Retinol (vitamin A)	orl rat TDLo, 55 mg (4–6 D preg) TER ipr mus TDLo, 180 mg (preg) TER	Widespread?
Retrorsine	orl rat TDLo, 30 mg CAR ipr rat TDLo, 150 mg CAR ivn rat LD_{50}, 38 mg ivn mus LD_{50}, 59 mg	*Crotalaria, Erechtites, Senecio*
Riboflavine (vitamin B_2, vitamin G)	ipr rat LD_{50}, 560 mg scu LD_{50}, 5,000 mg	Universal?
Ricin	orl hmn TDLo, 2 mg orl rat LDLo, 100 mg	*Ricinus?*
Ricinoleic acid	scu rbt TDLo, 3,120 mg /52 WI NEO	*Agonandra, Argemone, Cassia, Cephalocroton, Ricinus, Solanum, Vitis, Wrightia*
Riddelline	unk rat TDLo, 209 mg/52 W NEO ivn mus LD_{50}, 105 mg	*Crotalaria, Senecio*
Rotenone	orl rat LD_{50}, 132 mg ipr rat LD_{50}, 2,800 μg ipr rat TDLo, 91 mg/42 DI NEO ipr mus LD_{50}, 2,800 μg TWA 5 mg/m^3	*Cracca, Derris, Lonchocarpus, Milletia, Mundulea, Pachyrhizus, Piscidia, Spatholobus, Tephrosia*
Rubijervine	ivn mus LD_{50}, 70 mg	*Veratrum*

Table 2 (continued)
TOXINS: THEIR TOXICITY AND DISTRIBUTION IN PLANT GENERA

Chemical	Toxicity[a,b]	Plant genera
Rutin (sophorin) (vitamin P)	ivn mus LD_{50}, 950 mg	Widespread. *Abutilon, Acacia, Acanthopanax, Aesculus, Aleurites, Ammi, Artemisia, Asparagus, Atropa, Baptisia, Bauhinia, Begonia, Betula, Boehmeria, Boenninghausenia, Brassica, Bryophyllum, Buniza, Bupleurum, Caccinia, Calotropis, Camellia, Capparis, Capsella, Crataegus, Cyclamen, Daviesia, Eschscholtzia, Eucalyptus, Eupatorium, Fagopyrum, Festuca, Ficus, Firmiana, Forsythia, Fothergilla, Galium, Globularia, Gossypium, Grevillea, Hedera, Hemidesmus, Heracleum, Herniaria, Heterostemma, Hibiscus, Hippophae, Hydrangea, Hyoscyamus, Leonurus, Leptadenia, Leucadendron, Liatris, Limonium, Lithospermum, Lycopersicon, Magnolia, Marsdenia, Menyanthes, Nardosmia, Nerium, Nicotiana, Oenanthe, Osyris, Paliurus, Pastinaca, Phaseolus, Phoenix, Photinia, Phrygilanthus, Phytolacca, Platanus, Polygonum, Protea, Prunus, Pyrus, Rhamnus, Rheum, Rhododendron, Rivea, Ruta, Salix, Sambucus, Senecio, Solanum, Solidago, Sophora, Tephrosia, Tulipa, Turbina, Tussilago, Viola, Vinca*
Ryania	orl rat LD_{50}, 750 mg orl mus LD_{50}, 550 mg orl dog LD_{50}, 150 mg orl rbt LD_{50}, 650 mg orl gpg LD_{50}, 2,500 mg	*Ryania*
Safrole (shikimol)	orl rat LD_{50}, 1,950 mg orl rat TDLo, 180 g/2 YC CAR orl mus LD_{50}, 2,350 mg orl mus TDLo, 132 g/81 WIC CAR orl rbt LDLo, 1,000 mg	*Asarum, Beilsmiedia, Cananga, Chenopodium, Cinnamomum, Doryphora, Eremophila, Heterotropa, Illicium, Licaria, Magnolia, Myristica, Nectandra, Nemuaron, Ocimum, Ocotea, Piper, Protium, Sassafras, Schinus, Zieria*
Sage oil	orl rat LD_{50}, 2,600 mg	*Salvia*
Sali	scu rat TDLo, 31 g/61 WI NEO	*Heliotropium*
Salicin (salicylamide)	orl rat LD_{50}, 1,890 mg orl mus LD_{50}, 1,400 mg orl rbt LDLo, 3,000 mg unk mky TDLo, 1,500 mg/18—26 preg TER	*Populus, Salix, Spiraea, Viburnum*
Salicylaldehyde	scu rat LDLo, 1,000 mg	*Ceanothus, Cinnamomum, Cordia, Filipendula, Homalium, Nicotiana, Prunus, Rauwolfia*
Salicylic acid	orl rat LD_{50}, 891 mg orl dog LDLo, 450 mg orl rbt LDLo, 1,300 mg	*Arctostaphylos, Betula, Calendula, Cananga, Cinnamomum, Cucurbita, Gaultheria, Jasminum, Matricaria, Mentha, Ophthalmoblapton, Polygala, Populus, Primula, Quercus, Ruta, Salix, Spiraea, Viola*
Sanguinarine	ipr rat LD_{50}, 18 mg	*Ammodendron, Argemone, Bocconia, Chelidonium, Corydalis, Dicentra, Dicranostigma, Eschscholtzia, Glaucium, Hunnemannia, Hylomecon, Hypecoum, Macleaya, Meconopsis, Papaver, Platystemon, Robinia, Romneya, Sanguinaria, Sapindus, Scabiosa, Stylomecon, Stylophorum*
L-Santonin	scu mus LDLo, 250 mg	*Artemisia*
Saponin	orl mus LDLo, 3,000 mg scu mus LDLo, 800 mg ivn mus LDLo, 1,000 mg ivn cat LDLo, 46 mg ivn rbt LDLo, 40 mg	Very widespread. *Abrus, Acacia, Achras, Achyranthes, Acorus, Actaea, Aegiceras, Aegle, Aesculus, Agave, Agrostemma, Akebia, Albizia, Aleurites, Allium, Alnus, Alstonia, Amaranthus, Anabasis, Anagallis, Anemone, Aporusa, Aralia, Artocarpus, Asparagus, Aster, Astragalus, Atriplex, Avena, Bacopa, Balanites, Barringtonia, Bassia, Beta, Bixa, Blighia, Butyrospermum, Calendula, Camellia, Canarium, Caryocar, Cassia, Castanospermum, Calophyllum, Celastrus, Centella, Cephaelis, Cestrum, Chenopodium, Chrysophyllum, Cicer, Cimicifuga, Cinchona, Citrullus, Citrus, Clematis, Corchorus, Cornus, Coronilla, Cucumis, Cucurbita, Cydonia, Dictamnus, Digitalis, Dioscorea,*

Table 2 (continued)
TOXINS: THEIR TOXICITY AND DISTRIBUTION IN PLANT GENERA

Chemical	Toxicity[a,b]	Plant genera
Saponin (continued)		*Diospyros, Ecballium, Echinocystis, Entada, Eriobotrya, Eryngium, Eucalyptus, Eugenia, Euphorbia, Euphoria, Fagus, Ficus, Furcraea, Glinus, Glycine, Glycyrrhiza, Gratiola, Guaiacum, Hedera, Helianthus, Hepatica, Hibiscus, Hypericum, Ilex, Impatiens, Jacquinia, Kochia, Lagenaria, Lathyrus, Lens, Leontice, Lippia, Liquidambar, Lotus, Luffa, Lupinus, Maclura, Madhuca, Manihot, Manilkara, Matricaria, Medicago, Melaleuca, Menyanthes, Micromeria, Mimusops, Mollugo, Momordica, Mora, Morus, Nephelium, Nicotiana, Nigella, Ocimum, Olea, Ononis, Pachyrhizus, Panax, Panicum, Phaseolus, Phyllanthus, Phytolacca, Pisum, Pithecellobium, Plantago, Platanus, Plumeria, Polygala, Polyscias, Primula, Prunella, Randia, Raphanus, Ricinus, Rosa, Salvia, Sanicula, Sansevieria, Sapindus, Saponaria, Schinus, Sesbania, Solidago, Sophora, Spartium, Spergularia, Strophanthus, Stryphnodendron, Swartzia, Telfairia, Terminalia, Tetragonia, Thymus, Trifolium, Trillium, Vaccinium, Vangueria, Viburnum, Viscum, Vitis, Zizyphus*
Sarmentoside A	ivn cat LD_{50}, 100 μg	*Strophanthus*
Sarmutoside	ivn cat LD_{50}, 0.48 mg	*Strophanthus*
Sceleratine	ivn mus LD_{50}, 139 mg	*Senecio*
Scillaren	orl rbt LDLo, 900 μg	*Urginea*
	scu rbt LDLo, 700 μg	
	scu rat LDLo, 10 mg	
Scillaren A	ivn cat LD_{50}, 170 μg	*Urginea*
Scillaren B	ivn cat LDLo, 140 μg	*Urginea*
Scillarenin	ivn cat LDLo, 0.16 mg	*Urginea*
Scilliroside	orl rat LD_{50}, 430 μg	*Urginea*
Scopolamine	orl hmn TDLo, 14 μg CNS	*Atropa, Datura, Duboisia, Hyoscyamus, Loranthus, Mandragora, Physochlaina, Scopolia, Utricularia*
	scu hmn TDLo, 2 μg CNS	
	scu mus LD_{50}, 1,700 mg	
	ivn mus LD_{50}, 163 mg	
Sebacic acid	ipr mus LD_{50}, 500 mg	*Ipomoea*
Senecionine	ivn mus LD_{50}, 64 mg	*Brachyglottis, Crotalaria, Emilia, Erechtites, Petasites, Senecio, Tussilago*
	ivn ham LD_{50}, 61 mg	
Seneciphylline	ivn rat LDLo, 60 mg	*Crotalaria, Erectites, Senecio*
	ivn mus LDLo, 50 mg	
	ivn gpg LDLo, 50 mg	
Serotonine	ipr rat LD_{50}, 117 mg	*Ananas, Carica, Lycopersicon, Musa, Persea, Prunus, Solanum*
	scu rat LD_{50}, 117 mg	
	ipr mus LD_{50}, 868 mg	
Serpentine	ipl mus TDLo, 400 mg NEO	*Catharanthus, Rauwolfia, Vinca*
Shikimic acid	orl mus TDLo, 4,000 mg NEO	Ubiquitous? *Acer, Aesculus, Akebia, Anogeissus, Aruncus, Aucuba, Caesalpinia, Carpinus, Cercidiphyllum, Cinnamomum, Cleyera, Corylus, Cyclobalanopsis, Cydonia, Dahlia, Distylium, Eleamus, Eriobotrya, Eucalyptus, Euonymus, Euptelea, Gardenia, Geranium, Hedera, Helianthus, Houttinia, Hypericum, Illicium, Kadsura, Lagenaria, Lespedeza, Ligustrum, Liquidambar, Liriodendron, Lithocarpus, Lithraea, Lolium, Lycopersicon, Macleaya, Magnolia, Mahonia, Malus, Mammea, Medicago, Michelia, Myrica, Patrinia, Phlox, Pieris, Pistacia, Platanus, Platycarya, Pteridium, Pyrus, Quercus, Ribes, Rubia, Rubus, Salix, Sapindus, Sarcandra, Saxifraga, Schima, Schinus, Solanum, Syringa, Terminalia, Ternstroemia, Thalictrum, Viburnum, Vicia, Vitex, Vitis, Weigela*
	ipr mus TDLo, 40 mg NEO	
	ipr mus LD_{50}, 1,000 mg	
Solanine	ipr mus LD_{50}, 32 mg	*Capsicum, Lycopersicon, Physochlaina, Solanum*
	ipr rbt LDLo, 20 mg	
	ivn rbt LDLo, 20 mg	
Sorbic acid	unk hmn LDLo, 5,000 mg	*Pyrus, Sorbus*
	scu rat TDLo, 2,600 mg/65 W CAR	
Sorsaka	scu rat TDLo, 31 g/61 WI NEO	*Annona*

Table 2 (continued)
TOXINS: THEIR TOXICITY AND DISTRIBUTION IN PLANT GENERA

Chemical	Toxicity[a,b]	Plant genera
Sparteine (lupinidine)	scu mus LDLo, 120 mg scu rbt LDLo, 100 mg ivn rbt LDLo, 30 mg	*Aconitum, Adenocarpus, Ammodendron, Ammothamnus, Anagyris, Baptisia, Chelidonium, Cytisus, Genista, Hovea, Leptorhabdos, Lupinus, Peumus, Piptanthus, Sarothamnus, Solanum, Sophora, Spartium, Templetonia, Thermopsis*
Sprintillamine	scu mus LDLo, 1 mg ivn rbt LDLo, 5 mg par frg LDLo, 6 mg	*Helleborus*
Sprintillin	scu mus LDLo, 1 mg	*Helleborus*
Squill	orl rat LD_{50}, 125 mg	*Urginea*
Stearic acid (octadecanoic acid)	ivn rat LD_{50}, 22 mg ivn mus LD_{50}, 23 mg ivn cat LDLo, 5 mg	*Citrus, Croton, Myrica*
Stearyl alcohol		*Ambrosia, Piper*
Stroboside	ivn cat LDLo, 0.25 mg	*Roupellina*
Strophanthidin	ivn cat LDLo, 280 μg ivn rbt LDLo, 110 μg	*Adonis, Antiaris, Apocynum, Castilla, Cheiranthus, Convallaria, Corchorus, Erysimum, Mansonia, Ornithogalum, Pachycarpus, Periploca, Strophanthus, Syrenia*
Strophanthin	ivn cat LDLo, 160 μg ivn rbt LDLo, 200 μg	*Strophanthus*
Strychnine	orl hmn LDLo, 30 mg orl rat LD_{50}, 16 mg orl dog LDLo, 1,100 μg orl cat LDLo, 750 μg	
Styrene	ihl hmn LCLo, 10,000 ppm 30 min ihl hmn TCLo, 376 ppm CNS orl rat LD_{50}, 5,000 mg orl mus LD_{50}, 316 mg TWA 100 ppm	*Liquidambar, Xanthorrhea*
Succinic acid	scu frg LDLo, 2,000 mg	General? *Artemisia, Atropa, Erodium, Gossypium, Helianthus, Phaseolus, Pyrus, Saccharum, Sedum, Vitis*
Tanghiniferin	ivn cat LD_{50}, 0.9 mg	*Tanghinia*
Tanghinin	ivn cat LD_{50}, 352 μg	*Tanghinia*
Tannic acid	orl mus LDLo, 2,000 mg scu mus LDLo, 75 mg scu rat TDLo, 4,450 mg/17 WI CAR	Widespread. *Acacia, Acorus, Agrimonia, Alnus, Aralia, Arctium, Arctostaphylos, Aralia, Areca, Arnica, Artemisia, Asperula, Berberis, Betula, Camellia, Castanea, Catharanthus, Cephaelis, Cinnamomum, Coffea, Corchorus, Coriandrum, Cornus, Cuminum, Cydonia, Diospyros, Drimys, Eucalyptus, Eugenia, Eupatorium, Euphoria, Ficus, Fragaria, Galium, Garcinia, Geum, Glycyrrhiza, Helianthus, Hibiscus, Humulus, Hydnocarpus, Hyssopus, Ilex, Ipomoea, Lavandula, Lawsonia, Lespedeza, Leucaena, Litchi, Lycopus, Madhuca, Majorana, Malus, Mangifera, Marrubium, Melia, Morus, Musa, Myrica, Myristica, Nephelium, Papaver, Passiflora, Peltophorum, Pimpinella, Phyllanthus, Phytolacca, Polygonum, Prunus, Punica, Rheum, Rosmarinus, Rubus, Rumex, Ruta, Salvia, Sassafras, Schinopsis, Sclerocarya, Symphytum, Syzygium, Tanacetum, Taraktogenos, Terminalia, Tussilago, Veronicastrum, Vitis*
Tartaric acid	orl dog LDLo, 5,000 mg	*Cichorium, Helianthus, Hibiscus, Musa, Persea, Rumex, Tamarindus, Vitis*
Tecomine	orl mus LD_{50}, 300 mg	*Tecoma*
Terpineol	orl rat LD_{50}, 4,300 mg	*Acacia, Artemisia, Asarum, Bursera, Cinnamomum, Citrus, Cuminum, Cymbopogon, Dryobalanops, Elettaria, Hypericum, Illicium, Jasminum, Laurus, Levisticum, Lippia, Melaleuca, Origanum, Pelargonium, Peumus*
Thebaine	scu mus LD_{50}, 117 mg	*Papaver, Rauwolfia?, Strychnos*
Theobromine	orl hmn TDLo, 26 mg CNS orl cat LD_{50}, 200 mg unk frg LDLo, 5 mg	*Camellia, Coffea, Cola, Guazuma, Herrania, Ilex, Paullinia, Sterculia, Theobroma*
Theophylline	orl hmn TDLo, 9 mg CNS scu rat TDLo, 100 mg/17 D preg TER	*Camellia, Ilex, Paullinia*

Table 2 (continued)
TOXINS: THEIR TOXICITY AND DISTRIBUTION IN PLANT GENERA

Chemical	Toxicity[a,b]	Plant genera
	orl mus LD_{50}, 600 mg	
	orl dog LDLo, 290 mg	
	orl cat LDLo, 100 mg	
	orl rbt LDLo, 350 mg	
Thevetin	ivn cat LDLo, 1,240 µg	*Thevetia*
Thiamine	scu mus LD_{50}, 301 mg	Widespread
Thiouracil	orl mus TDLo, 310 g/74 WC NEO	*Brassica*
	orl rbt LDLo, 3,700 mg	
Thujone	ipr rat LDLo, 120 mg	*Artemisia, Lavandula, Lippia, Salvia, Tanacetum, Thuja*
Thyme oil	orl rat LD_{50}, 2,840 mg	*Thymus*
Thymol	orl rat LD_{50}, 980 mg	*Carum, Lavandula, Monarda, Nepeta, Ocimum,*
	orl mus LD_{50}, 1,800 mg	*Origanum, Satureja, Thymus*
	orl gpg LD_{50}, 880 mg	
Thymoquinone	skn mus TDLo, 50 mg/DI NEO	*Callitris, Monarda*
Tomatine	orl rat LDLo, 800 mg	*Lycopersicon*
Tonka absolute	orl rat LD_{50}, 1,380	*Dipteryx*
Tridecanoic acid	ivn mus LD_{50}, 130 mg	*Cocos, Iris*
Tridecanol	orl rat LDLo, 4,750 mg	*Musa*
Trigonelline	scu rat LDLo, 5,000 mg	*Acacia, Astragalus, Avena, Canavalia, Cannabis, Coffea, Cucurbita, Dahlia, Dichapetalum, Dictamnus, Glycine, Ilex, Lupinus, Medicago, Mirabilis, Morus, Oryza, Pisum, Sambucus, Scorzonera, Solanum, Stachys, Strophanthus, Trigonella*
Trimethylamine	ipr mus LDLo, 75 mg	*Acorus, Amorphophallus, Aristolochia, Arnica, Beta,*
	ivn mus LD_{50}, 90 mg	*Chenopodium, Chaerophyllum, Chrysosplenium,*
	scu mus LDLo, 1,000 mg	*Clematis, Cornus, Cotyledon, Crataegus,*
	scu rbt LDLo, 800 mg	*Dracunculus, Fagus, Gossypium, Heracleum, Maerua,*
	ivn rbt LDLo, 400 mg	*Menyanthes, Mercurialis, Nicotiana, Prunus,*
	scu frg LDLo, 2,000 mg	*Pyrus, Rhagodia, Sauromatium, Sorbaria, Sorbus, Spathiphyllum, Spiraea, Taraxacum, Viburnum*
Tropacocaine	ivn rat LDLo, 20 mg	*Erythroxylum*
	ivn cat LDLo, 18 mg	
	scu rbt LDLo, 400 mg	
	ipr gpg LDLo, 170 mg	
Tryptophane	orl rat TDLo, 1,100 mg/DC NEO	*Eleusine, Fagopyrum, Glycine, Helianthus, Ipomoea, Juglans, Linum, Luffa, Lupinus, Lycopersicon, Manihot, Medicago, Moringa, Murraya, Musa, Nicotiana, Nigella, Panicum, Papaver, Paspalum, Peltophorum, Pennisetum*
Tubocurarine	scu mus LD_{50}, 560 µg	*Chondrodendron*
	ipr mus LD_{50}, 500 µg	
	ivn mus LD_{50}, 180 µg	
	ivn rbt LD_{50}, 200 µg	
Tylocrebrine	orl rat LD_{50}, 65 mg	*Ficus, Tylophora*
	ivn rat LD_{50}, 32 mg	
	orl mus LD_{50}, 50 mg	
	ivn mus LD_{50}, 42 mg	
Tyramine	ipr mus LDLo, 800 mg	*Acacia, Capsella, Carduus, Citrus, Crinum, Cytisus,*
	scu mus LDLo, 225 mg	*Erodium, Lophophora, Phoradendron, Psittacanthus,*
	ivn mus LD_{50}, 260 mg	*Silybum, Viscum*
	unk mus LDLo, 2,200 mg	
	scu cat LDLo, 30 mg	
	ivn rbt LD_{50}, 300 mg	
Undecanoic acid	ivn mus LD_{50}, 140 mg	*Artemisia, Cocos, Iris, Thymus*
Undecylenic acid	orl rat LD_{50}, 2,500 mg	
	orl mus LD_{50}, 2,300 mg	
	ipr mus LD_{50}, 960 mg	
Valeraldehyde	orl rat LD_{50}, 3,200 mg	*Carpinus, Eucalyptus, Lavandula, Melaleuca, Ocotea, Quercus, Syzygium*
Valeric acid	ivn mus LD_{50}, 1,290 mg	*Ananas, Angelica, Andropogon, Atractylis, Boronia,*
	orl mus LD_{50}, 500 mg	*Cananga, Carthamus, Croton, Erythraea, Ferula,*
	scu mus LD_{50}, 3,590 mg	*Humulus, Laurus, Lavandula, Mentha, Sambucus, Valeriana, Viburnum*
Vanillin	orl rat LD_{50}, 1,580 mg	*Cymbopogon, Gossypium, Ilex, Liquidambar,*
	orl rbt LDLo, 3,000 mg	*Myroxylon, Nigritella, Ruta, Styrax, Syzygium*
	orl gpg LD_{50}, 1,400 mg	

Table 2 (continued)
TOXINS: THEIR TOXICITY AND DISTRIBUTION IN PLANT GENERA

Chemical	Toxicity[a,b]	Plant genera
Veatchin (HCl)	ivn mus LD_{50}, 13 mg	*Garrya*
	ivn cat LDLo, 2 mg	
Veratridine	ipr rat LD_{50}, 3,500 μg	*Veratrum*
	ipr mus LD_{50}, 1,350 μg	
	ivn mus LD_{50}, 420 μg	
Veratrine	orl rat LDLo, 25 mg	*Sarracenia, Schoenocaulon*
	orl rat LD_{50}, 4,000 mg	
	ipr mus LD_{50}, 7.5 mg	
	ivn mus LD_{50}, 420 μg	
Verbenol	ipr mus LDLo, 125 mg	*Boswellia*
Verbenone	ipr mus LDLo, 250 mg	*Boswellia, Lippia*
Veriloid	orl rat LD_{50}, 12 mg	*Veratrum*
	ipr rat LD_{50}, 1,690 μg	
	orl mus LD_{50}, 4,500 μg	
	orl rbt LD_{50}, 18 mg	
Vinblastine	unk rat TDLo, 250 μg/8	*Catharanthus*
	D preg TER	
	ivn mus LD_{50}, 17 mg	
	ivn ham TDLo, 250 μg	
	/8 D preg TER	
Vincamine	scu rat LD_{50}, 1,000 mg	*Catharanthus*
Vincristine (see Leurocristine)		
Visnadin	orl mus LD_{50}, 2,240 mg	*Ammi*
	ivn rbt LDLo, 50 mg	
	ivn dog TDLo, 20 mg	
Xanthine	ipr mus LD_{50}, 500 mg	*Beta, Camellia, Lupinus, Vicia*
	scu rat TDLo, 3,600 mg	
	(18 WI) NEO	
	imp mus TDLo, 80 mg NEO	
	unk frg LDLo, 20 mg	
Xylenol	orl mus LD_{50}, 1,070 mg	*Betula*
Yohimbine	orl mus LDLo, 25 mg	*Alstonia, Aspidosperma, Catharanthus, Corynanthe,*
(Aphrodine)	ivn mus LDLo, 16 mg	*Ladenbergia, Pouteria, Pseudocinchona,*
	scu dog LDLo, 20 mg	*Rauwolfia, Vinca*
	scu rbt LDLo, 50 mg	
	scu frg LD_{50}, 34 mg	

[a] All toxicity data derived from NIOSH, 1975.[353] Unless otherwise indicated, all gram, milligram, and microgram entries are expressed as per kilogram body weight.

[b] The following list of abbreviations is taken from NIOSH, 1975.[353] BLD, blood effects; C, continuous; Ca, calcium; CAR, carcinogenic effects; chd, child; ckn, chicken; CNS, central nervous system effects; ctl, cattle; D, day; dck, duck; d, dog; dom, domestic animal; EYE, eye effects; frg, frog; GIT, gastrointestinal tract effects; gpg, guinea pig; ham, hamster; hmn, human; I, intermittent; ihl, inhalation; ims, intramuscular; ipr, intraperitoneal; IRR, irritant effects; ivn, intravenous; LC_{50}, lethal concentration 50% kill; LCLo, lowest published lethal concentration; LD_{50}, lethal dose 50% kill; LDLo, lowest published lethal dose; mam, mammal (species unspecified); mky, monkey; mus, mouse; MUT, mutagenic effects; NEO, neoplastic effects; orl, oral; par, parental; PNS, peripheral nervous system effects; pgn, pigeon; preg, pregnant; PSY, psychotropic; rbt, rabbit; scu, subcutaneous; skn, skin; TCLo, lowest published toxic concentration; TDLo, lowest published toxic dose; TER, teratogenic effects; TWA, time-weighted average (usually applies to air standards); unk, unreported; W, week; wmn, woman; Y, year.

[c] ? = Reports are very dubious.

Table 3
HIGHER PLANT GENERA AND THEIR TOXINS

Genus[a]	Family	Toxin
Abelmoschus	Malvaceae	Oxalic acid
Abies	Abietaceae	Maltol
Abrus[b]	Fabaceae	Saponin
Abutilon	Malvaceae	Rutin
Acacia[b]	Mimosaceae	Anhaline, anisaldehyde, anthraquinone, benzaldehyde, benzyl alcohol, butyraldehyde, cresol, cuminic aldehyde, decanal, dimethyl tryptamine, eugenol methyl ether, gallic acid, heptanoic acid, hydrocyanic acid, indole, isobutyraldehyde, linalool, methyl salicylate, nicotine, palmitic acid, phenethylamine, quercitin, rutin, saponin, tannic acid, terpineol, trigonelline, tyramine
Acalypha	Euphorbiaceae	Hydrocyanic acid
Acanthopanax[b]	Araliaceae	Rutin
Acanthospermum	Asteraceae	Hydrocyanic acid
Acer[b]	Aceraceae	Donaxine, fumaric acid, guaiacol, hypoglycine, inositol, malic acid, shikimic acid
Achillea[b]	Asteraceae	Camphor, cineole, formaldehyde, isovaleric acid
Achras	Sapotaceae	Lactose, saponin
Achyranthes	Amaranthaceae	Saponin
Acokanthera	Apocynaceae	Acolongifloriside, ouabain
Aconitum[b]	Ranunculaceae	Aconitine, anemonin, ephedrine, maltose, pachycarpine, palmitic acid, quinic acid, sparteine
Acorus[b]	Araceae	Asarone, calamus oil, camphor, choline, eugenol, eugenol methyl ether, heptanoic acid, methylamine, saponin, tannic acid, trimethylamine
Acrocomia	Arecaceae	Lauric acid
Actaea[b]	Ranunculaceae	Saponin
Actinea[b]	Asteraceae	
Actinodaphne	Lauraceae	Lauric acid
Adenanthera	Mimosaceae	Lapachol
Adenia	Passifloraceae	Hydrocyanic acid
Adenium[b]	Apocynaceae	Abobioside, abomonoside, echubioside, echujin, hydrocyanic acid
Adenocarpus	Fabaceae	Sparteine
Adlumia	Fumariaceae	Bicuculline, biflorine, protopine
Adonis[b]	Ranunculaceae	Adonitoxin, choline, cymarin, cymarol, strophanthidin
Aegiceras	Myrsinaceae	Saponin
Aegle	Rutaceae	Cuminic aldehyde, cymene, imperatorin, phellandrine, saponin
Aesculus[b]	Hippocastanaceae	Nicotine, quercitin, quercitrin, rutin, saponin, shikimic acid
Aethusa[b]	Apiaceae	Coniine
Agastache	Lamiaceae	Anisaldehyde, pulegone
Agave[b]	Agavaceae	Saponin
Ageratum	Asteraceae	Hydrocyanic acid
Aglaia	Meliaceae	Cineole, citral
Agonandra	Opiliaceae	Ricinoleic acid
Agrimonia	Rosaceae	Hydrocyanic acid, quercitrin, tannic acid
Agropyron[b]	Poaceae	Hydrocyanic acid
Agrostemma[b]	Caryophyllaceae	Saponin
Agrostis[b]	Poaceae	
Ailanthus	Simaroubaceae	Hydrocyanic acid
Akebia	Lardizabilaceae	Saponin, shikimic acid
Alangium	Nyssaceae	Cephaelin, emetine
Albizia	Mimosaceae	Albitocin, hydrocyanic acid, saponin
Aleurites[b]	Euphorbiaceae	Quercitrin, rutin, saponin
Allemanda	Apocynaceae	
Allium[b]	Alliaceae	Hydrocyanic acid, oxalic acid, propanethiol, pyrocatechol, quercitin, saponin
Alnus	Betulaceae	Inositol, saponin, tannic acid
Alocasia	Araceae	Hydrocyanic acid
Aloe	Liliaceae	Arabinose
Aloysia[b]	Verbenaceae	
Alphitonia	Rhamnaceae	Methyl salicylate
Alpinia	Zingiberaceae	Camphor, cineole, eugenol
Alsodeia	Violaceae	Methyl salicylate

Table 3 (continued)
HIGHER PLANT GENERA AND THEIR TOXINS

Genus[a]	Family	Toxin
Alstonia	Apocynaceae	Methyl salicylate, reserpine, saponin, yohimbine
Alternanthera	Amaranthaceae	Hydrocyanic acid
Alyxia	Apocynaceae	Coumarin
Amaranthus[b]	Amaranthaceae	Hydrocyanic acid, saponin
Amaryllis[b]	Amaryllidaceae	
Ambrosia[b]	Asteraceae	Hexadecanol, stearyl alcohol
Amianthium[b]	Liliaceae	Jervine
Ammi[b]	Apiaceae	Imperatorin, khellin, rutin, visnadin
Ammodendron	Fabaceae	Pachycarpine, sanguinarine
Ammothamnus	Fabaceae	Pachycarpine
Amorphophallus	Araceae	Ammonia, coniine, ethylamine, histamine, methylamine, trimethylamine
Ampelopsis	Vitaceae	Pyrocatechol
Amsinckia[b]	Boraginaceae	Retronecine
Amyris	Rutaceae	Amyris oil
Anabasis	Chenopodiaceae	Anabasine, saponin
Anacardium[b]	Anacardiaceae	Oxalic acid
Anagallis	Primulaceae	Saponin
Ananas	Bromeliaceae	Citric acid, malic acid, oxalic acid, serotonine, valeric acid
Anagyris	Fabaceae	Cytisine, pachycarpine, sparteine
Anamirta	Menispermaceae	Picrotoxin
Andira	Fabaceae	Chrysarobin, lapachol
Androcymbium	Liliaceae	Colchicine
Andromeda[b]	Ericaceae	
Andropogon	Poaceae	Valeric acid
Anemone	Ranunculaceae	Anemonin, saponin
Anethum	Apiaceae	Carvone, limonene, myristicin, phellandrene
Angelica	Apiaceae	Borneol, citric acid, ethyl alcohol, fumaric acid, furfural, imperatorin, isovaleric acid, methanol, phellandrene, quinic acid, valeric acid
Aniba	Lauraceae	Anabasine, benzoic acid, cotoin
Anogeissus	Combretaceae	Shikimic acid
Annona	Annonaceae	Caffeine, hydrocyanic acid, sorsaka
Anthemis[b]	Asteraceae	Isobutyl alcohol, isobutyric acid
Anthephora	Poaceae	Hydrocyanic acid
Anthocercis	Solanaceae	Hyoscyamine
Anthriscus	Apiaceae	Ethyl alcohol, malonic acid, methanol
Anthyllis	Fabaceae	Malonic acid
Antiaris	Moraceae	Antiarin, convallatoxin, convallatoxol, cymarin, strophanthidin
Apium[b]	Apiaceae	Apiole, guaiacol, limonene, malonic acid, palmitic acid
Apocynum	Apocynaceae	Cymarin, strophanthidin
Aporusa	Euphorbiaceae	Saponin
Arachis[b]	Fabaceae	
Aralia[b]	Araliaceae	Guanine, saponin, tannic acid
Arbutus	Ericaceae	Hydroquinone
Arctium	Asteraceae	Tannic acid
Arctomecon	Papaveraceae	Biflorine, protopine
Arctostaphylos	Ericaceae	Gallic acid, quercitin, quinic acid, salicylic acid, tannic acid
Ardisia	Myrsinaceae	Methyl salicylate
Areca	Arecaceae	Arecoline, gallic acid, lauric acid, tannic acid
Argemone[b]	Papaveraceae	Biflorine, codeine (?), morphine (?), protopine, ricinoleic acid, sanguinarine
Argyreia	Convolvulaceae	Ergometrine
Argyrolobium	Fabaceae	Cytisine
Arisaema	Araceae	
Arisarum	Araceae	Coniine
Aristida	Poaceae	Hydrocyanic acid
Aristolochia	Aristolochiaceae	Aristolochine, camphor, trimethylamine
Armoracia[b]	Brassicaceae	Allyl isothiocyanate
Arnica[b]	Asteraceae	Formic acid, fumaric acid, heptanoic acid, isobutyric acid, tannic acid, trimethylamine

Table 3 (continued)
HIGHER PLANT GENERA AND THEIR TOXINS

Genus[a]	Family	Toxin
Artemisia	Asteraceae	Butyraldehyde, camphor, choline, cineole, cuminic aldehyde, esdragole, formic acid, inositol, isobutyric acid, isovaleric acid, leucomycin, linalool, malic acid, nonanoic acid, phellandrene, phenol, rutin, santonin, succinic acid, tannic acid, terpineol, thujone, undecanoic acid
Artocarpus[b]	Moraceae	Acetylcholine, hydrocyanic acid, saponin
Arum[b]	Araceae	Ammonia, coniine, diethylamine, dimethylamine, ethylamine, methylamine, propanediamine
Aruncus	Rosaceae	Shikimic acid
Arundo	Poaceae	Donaxine
Asarum[b]	Aristolochiaceae	Asarone, borneol, eugenol methyl ether, safrole, terpineol
Asclepias[b]	Asclepiadaceae	Calotropin, indican, nicotine
Asimina	Annonaceae	
Asparagus[b]	Liliaceae	Rutin, saponin
Asperula	Rubiaceae	Citric acid, coumarin, malic acid, tannic acid
Asphodelus	Liliaceae	Colchicine
Aspidium[b]	Aspidiaceae	
Aspidosperma	Apocynaceae	Ajmaline, aspidospermine, reserpine, yohimbine
Asplenium	Aspleniaceae	Hydrocyanic acid
Aster[b]	Asteraceae	Quercitin, saponin, selenium
Astilbe	Saxifragaceae	Gallic acid
Astragalus[b]	Fabaceae	Dopa, malonic acid, methyl cysteine, saponin, selenium, trigonelline
Astrocaryum	Arecaceae	Caprylic acid, lauric acid
Astrocasia	Euphorbiaceae	Astrophyllin
Atalantia	Rutaceae	Linalool, methyl salicylate
Atherosperma	Monimiaceae	Berbamine, eugenol methyl ether
Atractylis	Asteraceae	Atractyloside, valeric acid
Atractylodes	Asteraceae	Furfural
Atriplex[b]	Chenopodiaceae	Betaine, hydrocyanic acid, saponin, selenium
Atropa[b]	Solanaceae	Apoatropine, atropine, choline, hyoscyamine, methylamine, pyridine, rutin, scopolamine, succinic acid
Attalea	Arecaceae	Lauric acid
Aucuba	Cornaceae	Shikimic acid
Avena[b]	Poaceae	Anhaline, ferulic acid, hydrocyanic acid, malonic acid, saponin, trigonelline
Avicennia	Avicenniaceae	Lapachol
Baccaurea	Euphorbiaceae	Methyl salicylate
Baccharis[b]	Asteraceae	
Backhousia[b]	Myrtaceae	Anethole, citral, citronellol
Bacopa	Scrophulariaceae	Nicotine, saponin
Baeckea[b]	Myrtaceae	Limonene
Bahia[b]	Asteraceae	Hydrocyanic acid
Baileya[b]	Asteraceae	
Balanites	Balanitaceae	Saponin
Bambusa	Poaceae	Hydrocyanic acid
Banisteriopsis	Malpighiaceae	Dimethyl tryptamine, harmaline, harmine
Baptisia[b]	Fabaceae	Cytisine, dopa, pachycarpine, rutin, sparteine
Barbarea[b]	Brassicaceae	
Barosma	Rutaceae	Limonene
Barringtonia	Barringtoniaceae	Methyl salicylate, saponin
Bauhinia	Caesalpiniaceae	Hydrocyanic acid, quercitin, quercitrin, rutin
Begonia	Begoniaceae	Rutin
Beilschmiedia	Lauraceae	Safrole
Benincasa	Cucurbitaceae	Oxalic acid
Berberis	Berberidaceae	Berbamine, citric acid, malic acid, oxyacanthine, tannic acid
Bergenia	Saxifragaceae	Gallic acid, hydroquinone
Berkheyopsis	Asteraceae	Hydrocyanic acid
Berula[b]	Apiaceae	
Beta[b]	Chenopodiaceae	Acetaldehyde, acetone, adenine, adipic acid, betaine, citric acid, ethyl alcohol, ferulic acid, formaldehyde, guanidine, heteroxanthine, histamine, hydantoin, hydrocyanic acid, malic acid, methylamine, oxalic acid, pyrocatechol, saponin, trimethylamine, xanthine
Betonica[b]	Lamiaceae	

Table 3 (continued)
HIGHER PLANT GENERA AND THEIR TOXINS

Genus[a]	Family	Toxin
Betula	Betulaceae	Guaiacol, methyl salicylate, pyrocatechol, rutin, salicylic acid, tannic acid, xylenol
Bidens[b]	Asteraceae	
Billia[b]	Hippocastanaceae	Hypoglycine A
Bignonia	Bignoniaceae	Lapachol, methyl salicylate
Bixa	Bixaceae	Saponin
Bleekeria	Apocynaceae	Reserpine
Bletia	Orchidaceae	Indican
Blighia[b]	Sapindaceae	Hypoglycine, saponin
Blumea[b]	Asteraceae	Borneol, cineole, fenchone
Bocconia[b]	Papaveraceae	Protopine, sanguinarine
Boehmeria	Urticaceae	Hydrocyanic acid, methanol, rutin
Boenninghausenia	Rutaceae	Rutin
Boronia[b]	Rutaceae	Ionone, valeric acid
Boscia	Capparaceae	Hydrocyanic acid
Boswellia	Burseraceae	Borneol, cymene, limonene, phellandrene, verbenol, verbenone
Bothriospora	Rubiaceae	Emetine
Bowiea	Liliaceae	Boroside D
Brabejum	Proteaceae	Hydrocyanic acid
Brachyglottis[b]	Asteraceae	Retronecine, senecionine
Brassica[b]	Brassicaceae	Acetaldehyde, allyl isothiocyanate, guanidine, hydrocyanic acid, mesaconic acid, methanethiol, narcotine, oxalic acid, rutin
Bredemeyera	Polygalaceae	Methyl salicylate
Bridelia	Euphorbiaceae	Methyl salicylate
Bromus[b]	Poaceae	
Bryonia[b]	Cucurbitaceae	Cucurbitacin E, ethylamine
Bryophyllum	Crassulaceae	Citric acid, malic acid, maltose, rutin
Bulbine	Liliaceae	Colchicine
Bulbocodium	Liliaceae	Colchicine
Bunias	Brassicaceae	Malonic acid, rutin
Bupleurum	Apiaceae	Phellandrene, rutin
Bursera	Burseraceae	Linalool, phellandrene, terpineol
Butyrospermum	Sapotaceae	Saponin
Buxus[b]	Buxaceae	
Bystropogon	Lamiaceae	Pulegone
Caccinia	Boraginaceae	Rutin
Cadaba	Capparaceae	Hydrocyanic acid
Caesalpinia[b]	Caesalpiniaceae	Hydrocyanic acid, phellandrene, shikimic acid
Caladium[h]	Araceae	Coniine
Calamintha	Lamiaceae	Menthol, pulegone
Calamagrostis[b]	Poaceae	
Calanthe	Orchidaceae	Indican
Calendula	Asteraceae	Cholesterol, myristic acid, pentadecanoic acid, salicylic acid, saponin
Calla[b]	Araceae	
Calligonum	Polygonaceae	Harman
Callitris	Cupressaceae	Podophyllotoxin, thymoquinone
Calluna	Ericaceae	Quercitin
Calodendrum	Rutaceae	Hydrocyanic acid
Calophyllum	Clusiaceae	Saponin
Calotropis	Asclepiadaceae	Calotropin, isovaleric acid, rutin
Caltha[b]	Ranunculaceae	Anemonin, choline
Calycanthus[b]	Calycanthaceae	Methyl salicylate
Calycophyllum	Rubiaceae	Saponin
Calytrix	Myrtaceae	Citronellol
Camelina	Brassicaceae	Ferulic acid
Camellia	Theaceae	Acetophenone, adenine, butyraldehyde, caffeine, caproic acid, citral, citronellol, cresol, gallic acid, geraniol, hexanol, isobutyraldehyde, isovaleraldehyde, linalool, methanethiol, methanol, methyl salicylate, oxalic acid, phenol, quercitrin, rutin, saponin, tannic acid, theobromine, theophylline, xanthine
Camphorosma	Chenopodiaceae	Propylamine
Camptorrhiza	Liliaceae	Colchicine
Camptotheca	Nyssaceae	Camptothecin

Table 3 (continued)
HIGHER PLANT GENERA AND THEIR TOXINS

Genus[a]	Family	Toxin
Cananga	Annonaceae	Benzaldehyde, benzoic acid, benzyl alcohol, cresol, methyl ether, eugenol, eugenol methyl ether, furfural, heptanal, isoeugenol, iso-safrole, linalool, methyl anthralinate, methyl salicylate, safrole, salicylic acid, valeric acid
Canarium	Burseraceae	Anethole, limonene, phellandrene
Canavalia[b]	Fabaceae	Choline, hydrocyanic acid, trigonelline
Cannabis[b]	Cannabinaceae	Cannabinols, choline, eugenol, guaiacol, nicotine, piperidine, trigonelline
Capirona	Rubiaceae	Cephaeline, emetine
Capparis	Capparidaceae	Quercitin, rutin
Capsella	Brassicaceae	Acetylcholine, choline, fumaric acid, rutin, tyramine
Capsicum	Solanaceae	Capsaicin, malic acid, oxalic acid, solanine
Carallia	Rhizophoraceae	Methyl salicylate
Cardaria[b]	Brassicaceae	
Cardiospermum	Sapindaceae	Hydrocyanic acid
Carduus[b]	Asteraceae	Fumaric acid, tyramine
Carex[b]	Cyperaceae	Harman
Carica	Caricaceae	Hydrocyanic acid, serotonine
Carpinus	Betulaceae	Acetaldehyde, butyraldehyde, shikimic acid, valeraldehyde
Carthamus	Asteraceae	Valeric acid
Carum	Apiaceae	Acetaldehyde, caraway oil, carvacrol, carveol, carbone, diacetyl, furfural, limonene, methanol, myristicin, thymol
Carya	Juglandaceae	Juglone
Caryocar	Caryocaraceae	Saponin
Cassia[b]	Caesalpiniaceae	Chrysarobin, cinnamaldehyde, decanal, hydrocyanic acid, isochaksine, ricinoleic acid, saponin
Castalis	Asteraceae	Hydrocyanic acid
Castanea	Fagaceae	Ethyl alcohol, tannic acid
Castanospermum	Fabaceae	Hydrocyanic acid, saponin
Castilla	Moraceae	Corchoroside, cymarin, strophanthidin
Castilleja[b]	Scrophulariaceae	Selenium
Catha	Celastraceae	Ephedrine, norpseudoephedrine
Catharanthus	Apocynaceae	Ajmalicine, leurocristine, serpentine, tannic acid, vinblastine, vincamine, yohimbine
Ceanothus	Rhamnaceae	Methyl salicylate, salicylaldehyde
Ceiba	Bombacaceae	Hydrocyanic acid
Celastrus	Celastraceae	Saponin
Centaurea[b]	Asteraceae	Hydrocyanic acid
Centaurium[b]	Gentianaceae	Hydrocyanic acid
Centella	Apiaceae	Asiaticoside, hydrocyanic acid, saponin
Cephaelis	Rubiaceae	Cephaeline, choline, emetine, pectin, saponin, tannic acid
Cephalanthus[b]	Rubiaceae	
Cephalocereus	Cactaceae	Caffeine
Cephalocroton	Euphorbiaceae	Ricinoleic acid
Ceratonia	Caesalpiniaceae	Isobutyric acid, lactose, maltose
Ceratophyllum	Ceratophyllaceae	Coumarin
Ceratostigma	Plumbaginaceae	Quercitrin
Cerbera	Apocynaceae	Cerberin
Cercidiphyllum	Cercidiphyllaceae	Quercitrin, shikimic acid
Cercocarpus	Rosaceae	Hydrocyanic acid
Cereus	Cactaceae	Caffeine
Cestrum[b]	Solanaceae	Saponin
Chaenomeles	Rosaceae	Hydrocyanic acid
Chaerophyllum	Apiaceae	Methylamine, trimethylamine
Cheilanthes[b]	Sinopteridaceae	
Cheiranthus	Brassicaceae	Cheiranthum, cheirotoxin, indole, strophanthidin
Chelidonium[b]	Papaveraceae	Biflorine, chelidonin, citric acid, histamine, pachycarpine, protopine, sanguinarine, sparteine
Chenopodium[b]	Chenopodiaceae	Ascaridole, carene, cymene, histamine, hydrocyanic acid, limonene, methyl salicylate, oxalic acid, safrole, saponin, trimethylamine
Chilocarpus	Apocynaceae	Methyl salicylate
Chloris	Poaceae	Hydrocyanic acid
Chondrodendron	Menispermaceae	Tubocurarine

Table 3 (continued)
HIGHER PLANT GENERA AND THEIR TOXINS

Genus[a]	Family	Toxin
Chrysopogon	Poaceae	Hydrocyanic acid
Chrysanthemum	Asteraceae	Caproic acid, carbone, hydrocyanic acid, pyrethrin, pyrethrum
Crysophyllum	Sapotaceae	Coumarin, saponin
Chrysosplenium	Saxifragraceae	Trimethylamine
Chrysothamnus[b]	Asteraceae	
Cicer	Fabaceae	Choline, oxalic acid, saponin
Cichorium	Asteraceae	Betaine, choline, maltol, tartaric acid
Cicuta[b]	Apiaceae	Cicutoxin, cuminic aldehyde
Cimicifuga	Ranunculaceae	Saponin
Cinchona	Rubiaceae	Cinchonidine, cinchonine, quinic acid, quinidine, quinine, saponin
Cinnamomum	Lauraceae	Acetaldehyde, benzoic acid, borneol, caproic acid, caprylic acid, carvacrol, cassia oil, cineole, cinnamaldehyde, cinnamyl alcohol, citronellol, coumarin, cuminic alcohol, cuminic aldehyde, cymene, decanal, eugenol, eugenol methyl ether, formic acid, furfural, geraniol, hydrocyanic acid, isobutyric acid, isoeugenol, isovaleraldehyde, isovaleric acid, lauric acid, limonene, linalool, myristic acid, myristicin, nonlyl alcohol, phellandrene, piperonal, propionic acid, safrole, salicylaldehyde, salicylic acid, shikimic acid, tannic acid, terpineol
Cirsium[b]	Asteraceae	Hydrocyanic acid
Cissus[b]	Vitaceae	Hydrocyanic acid
Cistus	Cistaceae	Acetophenone, formic acid
Citrullus	Cucurbitaceae	Cucurbitacin E, hydrocyanic acid, pectin, saponin
Citrus[b]	Rutaceae	Acetaldehyde, benzaldehyde, borneol, capric acid, caprylic acid, carene, carvone, citral, citric acid, citronellol, cresol, cymene, decanal, decanoic acid, decyl alcohol, dodecyl alcohol, ergosterol, ethyl alcohol, formic acid, furfural, geraniol, guaiacol, hesperidin, histamine, hydrocyanic acid, indole, isovaleraldehyde, isovaleric acid, lauric acid, lemon oil, limonene, linalool, methyl anthralinate, myristic acid, narcotine, nonanoic acid, nonlyl alcohol, noradrenaline, oxalic acid, phellandrene, phenethyl alcohol, pyrocatechol, quercitrin, quinoline, saponin, stearic acid, terpineol, tyramine
Cladrastis	Fabaceae	Cytisine
Clausena	Rutaceae	Anethole, carbazole
Clematis[b]	Ranunculaceae	Anemonin, methyl salicylate, saponin, trimethylamine
Cleome[b]	Capparaceae	
Cleyera	Theaceae	Shikimic acid
Cocculus	Menispermaceae	Methyl salicylate, picrotoxin
Cochlearia[b]	Brassicaceae	
Cocos	Arecaceae	Caproic acid, caprylic acid, lauric acid, myristic acid, tridecanoic acid, undecanoic acid
Codonocarpus	Gyrostemonaceae	Benzyl cyanide
Coffea	Rubiaceae	Adenine, caffeine, choline, furfuryl alcohol, guaiacol, tannic acid, theobromine, trigonelline
Coix	Poaceae	Hydrocyanic acid
Cola	Sterculiaceae	Caffeine, theobromine
Colchicum[b]	Liliaceae	Colchamine, colchiceine, colchicine, demecolcine
Colocasia[b]	Araceae	Hydrocyanic acid, oxalic acid
Colutea	Fabaceae	Malonic acid
Comandra	Santalaceae	Selenium
Combretum	Combretaceae	Caffeine, choline, hydrocyanic acid
Commelina	Commelinaceae	Hydrocyanic acid
Comptonia	Myricaceae	Cineole, quercitin
Conium[b]	Apiaceae	Coniine, methylamine, piperidine
Convallaria[b]	Liliaceae	Choline, convallamarin, convallatoxin, convallatoxol, convalloside, strophanthidin
Convolvulus[b]	Convolvulaceae	
Conyza[b]	Asteraceae	
Coprosma	Rubiaceae	Methanethiol
Corchorus	Tiliaceae	Acetophenone, corchoroside, hydrocyanic acid, saponin, strophanthidin, tannic acid
Cordia	Boraginaceae	Salicylaldehyde

Table 3 (continued)
HIGHER PLANT GENERA AND THEIR TOXINS

Genus[a]	Family	Toxin
Coriandrum	Apiaceae	Acetone, borneol, coriander oil, cymene, decanal, limonene, linalool, malic acid, oxalic acid, phellandrene, tannic acid
Coriaria	Coriariaceae	Coriamyrtin
Cornus	Cornaceae	Gallic acid, quercitin, quinine, saponin, tannic acid, trimethylamine
Coronilla[b]	Fabaceae	Cytisine, saponin
Corydalis[b]	Fumariaceae	Bicuculline, biflorine, bulbocapnine, caoaurine, coramine, corlumine, fumaric acid, hydrohydratinine, maltol, protopine, sanguinarine
Corylus	Corylaceae	Myristic acid, shikimic acid
Corynanthe	Rubiaceae	Ajmalicine, corynantheine, yohimbine
Cosmos	Asteraceae	Quercitin
Costus	Costaceae	Costus oil
Cotinus	Anacardiaceae	Inositol
Cotoneaster	Rosaceae	Hydrocyanic acid, mandelonitrile
Cotyledon	Crassulaceae	Hydrocyanic acid, trimethylamine
Coutarea	Rubiaceae	Quinidine, quinine
Cracca	Fabaceae	Rotenone
Crataegus	Rosaceae	Acetylcholine, dimethylamine, ethylamine, quercitin, quercitrin, rutin, trimethylamine
Crinum[b]	Amaryllidaceae	Tyramine
Crithmum	Apiaceae	Apiole
Crocus	Iridaceae	Colchicine, quercitin
Crotalaria[b]	Fabaceae	Hydrocyanic acid, hydroxysenkirkine, indican, jacobine, longilobine, monocrotaline, retronecine, retrorsine, riddelline, senecionine, seneciphylline
Croton[b]	Euphorbiaceae	Croton oil, cymene, isobutyric acid, isovaleric acid, limonene, myristic acid, pectin, phorbol, stearic acid, valeric acid
Crusea	Rubiaceae	Quercitrin
Cryptocarya	Lauraceae	Eugenol
Cryptolepis	Periplocaceae	Hydrocyanic acid, methyl salicylate
Cryptostegia[b]	Periplocaceae	
Cucumis[b]	Cucurbitaceae	Histamine, saponin
Cucurbita	Cucurbitaceae	Cucurbitacin E, hydrocyanic acid, salicylic acid, saponin, trigonelline
Cuminum	Apiaceae	Cumene, cumin oil, cuminic alcohol, cuminaldehyde, cymene, limonene, phellandrene, tannic acid, terpineol
Cuphea	Lythraceae	Capric acid, caprylic acid, decanoic acid, nonanoic acid
Cupressus[b]	Cupressaceae	
Curcuma	Zingiberaceae	Cineole, phellandrene
Cuscuta[b]	Cuscutaceae	
Cycas[b]	Cycadaceae	Cycasin
Cyclamen[b]	Primulaceae	Histamine, rutin
Cyclobalanopsis	Fagaceae	Shikimic acid
Cydonia	Rosaceae	Hydrocyanic acid, malic acid, saponin, shikimic acid, tannic acid
Cymbopogon	Poaceae	Borneol, capric acid, caprylic acid, carbone, citral, citronella oil, citronellol, decanoic acid, eugenol, formaldehyde, furfural, hydrocyanic acid, isovaleraldehyde, limonene, linalool, phellandrene, terpineol, vanillin
Cynanchum[b]	Asclepiadaceae	Hydrocyanic acid
Cynodon[b]	Poaceae	Hydrocyanic acid
Cynoglossum[b]	Boraginaceae	Lasiocarpine
Cyperus	Cyperaceae	Cineole, hydrocyanic acid, myristic acid
Cyphomandra	Solanaceae	Nicotine
Cysticapnos	Fumariaceae	Biflorine, protopine
Cytisus[b]	Fabaceae	Cytisine, dopa, pachycarpine, sparteine, tyramine
Dacrydium	Podocarpaceae	Carene, limonene
Dactylicapnos	Fumariaceae	Biflorine, protopine
Dactyloctenium	Poaceae	Hydrocyanic acid
Dahlia	Asteraceae	Shikimic acid, trigonelline
Daphne[b]	Thymelaeaceae	Daphne
Darwinia	Myrtaceae	Geraniol
Datisca[b]	Datiscaceae	
Datura[b]	Solanaceae	Atropine, hydrocyanic acid, hyoscyamine, isobutyraldehyde, malic acid, scopolamine
Daubentonia[b]	Fabaceae	

Table 3 (continued)
HIGHER PLANT GENERA AND THEIR TOXINS

Genus[a]	Family	Toxin
Daucus[b]	Apiaceae	Acetone, asarone, choline, ethyl alcohol, formic acid, hydrocyanic acid, isobutyric acid, limonene, malic acid, maltose, oxalic acid, palmitic acid, pectin, pyrrolidine, quinic acid
Daviesia	Fabaceae	Rutin
Davilla	Combretaceae	Caffeine
Decodon	Lythraceae	Cryogenine, decamine
Delonix	Caesalpiniaceae	Hydrocyanic acid
Delphinium[b]	Ranunculaceae	Elaidic acid, histamine, methylamine
Dendrocalamus	Poaceae	Methyl salicylate
Dendromecon	Papaveraceae	Biflorine, protopine
Derris	Fabaceae	Derris, rotenone
Descurainia[b]	Brassicaceae	
Desmodium	Fabaceae	Bufotenine, dimethyl tryptamine, donaxine
Dicentra[b]	Fumariaceae	Bicuculline, biflorine, bulbocapnine, corlumine, fumaric acid, protopine, sanguinarine
Dichapetalum[b]	Dichapetalaceae	Hydrocyanic acid, trigonelline
Dichroa	Hydrangeaceae	Dichroine, quinazoline?
Dicranostigma	Papaveraceae	Biflorine, chelidonin, protopine, sanguinarine
Dictamnus	Rutaceae	Anethole, choline, saponin, trigonelline
Dicypellium	Lauraceae	Eugenol
Dieffenbachia[b]	Araceae	Oxalic acid
Digitalis	Scrophulariaceae	Anhydrogitalin, choline, citric acid, digitalis, digitoxin, digitoxiside. digoxin, formic acid, gallic acid, gitalin, gitorin, gitoxin, saponim
Digitaria	Poaceae	Hydrocyanic acid
Dimocarpus	Sapindaceae	Quercitin, saponin, tannic acid
Dimorphotheca	Asteraceae	Hydrocyanic acid
Dioclea	Fabaceae	Physostigmine
Dioscorea	Dioscoreaceae	Cholesterol, saponin
Diospyros	Ebenaceae	Methyl salicylate, pectin, saponin, tannic acid
Dipcadi	Liliaceae	Hydrocyanic acid
Diphylleia	Podophyllaceae	Podophyllotoxin
Dipidax	Liliaceae	Colchicine
Diplolophium	Apiaceae	Hydrocyanic acid
Dipterocarpus	Dipterocarpaceae	Quercitin
Dipteryx	Fabaceae	Coumarin, tonka
Distylium	Hammamelidaceae	Shikimic acid
Dodonaea	Sapindaceae	Hydrocyanic acid
Dolichos	Fabaceae	Hydrocyanic acid
Dorema	Apiaceae	Hexadecanol
Doryphora	Atherospermataceae	Choline, piperonal, safrole
Dracunculus	Araceae	Ammonia, dimethylamine, ethyl amine, methyl amine, trimethylamine
Drimys	Winteraceae	Tannic acid
Drosera	Droseraceae	Citric acid, histamine, malic acid
Drymaria[b]	Caryophyllaceae	
Dryobalanops	Dipterocarpaceae	Borneol, quercitin, terpineol
Dryopteris[b]	Aspidiaceae	
Drypetes	Euphorbiaceae	Methyl salicylate
Duboisia	Solanaceae	Anabasine, atropine, hyoscyamine, nicotine, nornicotine, scopolamine
Duranta[b]	Verbenaceae	Hydrocyanic acid
Durio	Bombacaceae	Propanethiol
Ecballium	Cucurbitaceae	Cucurbitacin E, saponin
Echeveria	Crassulaceae	Nicotine
Echinochloa[b]	Poaceae	
Echinocystis	Cucurbitaceae	Cucurbitacin, saponin
Echites	Apocynaceae	Indican
Echium[b]	Boraginaceae	Retronecine
Eclipta	Asteraceae	Nicotine
Eleagnus	Elaeagnaceae	Shikimic acid
Elaeis	Arecaceae	Estrone, lauric acid, myristic acid
Elaeocarpus	Elaeocarpaceae	Methyl salicylate
Elettaria	Zingiberaceae	Borneol, cineole, limonene, terpineol
Eleusine[b]	Poaceae	Hydrocyanic acid, oxalic acid, tryptophane
Elodea	Hydrocharitaceae	Ferulic acid

Table 3 (continued)
HIGHER PLANT GENERA AND THEIR TOXINS

Genus[a]	Family	Toxin
Elsholtzia	Lamiaceae	Cymene, geraniol, linalool
Elymus[b]	Poaceae	
Emilia	Asteraceae	Retronecine, senecionine
Enantia	Annonaceae	Quinidine, quinine
Engelhardtia	Juglandaceae	Quercitrin
Entada	Mimosaceae	Hydrocyanic acid, saponin
Epaltes	Asteraceae	Hydrocyanic acid
Ephedra	Ephedraceae	Ephedrine, norpseudoephedrine, pseudoephedrine
Epidendrum	Orchidaceae	Indican
Equisetum[b]	Equisetaceae	Nicotine, palmitic acid
Eragrostis[b]	Poaceae	Hydrocyanic acid
Erechtites	Asteraceae	Isatidene, longilobine, retronecine
Eremocitrus	Rutaceae	Nonanoic acid, nonlyl alcohol
Eremophila	Myoporaceae	Safrole
Eremosemium[b]	Chenopodiaceae	Selenium
Erica	Ericaceae	Anisaldehyde
Erigeron	Asteraceae	Quercitin, quercitrin
Eriobotrya	Rosaceae	Hydrocyanic acid, saponin, shikimic acid
Eriodictyon	Hydrophyllaceae	Formic acid
Erisma	Vochysiaceae	Lauric acid, myristic acid
Erodium	Geraniaceae	Caffeine, choline, ethylamine, histamine, succinic acid, tyramine
Ervatamia	Apocynaceae	
Eryngium	Apiaceae	Piperonal, saponin
Erysimum	Brassicaceae	Strophanthidin
Erythraea	Gentianaceae	Valeric acid
Erythrina	Fabaceae	Erysodine, erysopine, erysothiopine, erythraline, erythramine, erythroidine
Erythronium[b]	Liliaceae	
Erythrophleum	Mimosaceae	Hydrocyanic acid
Erythroxylum	Erythroxylaceae	Cocaine, methyl salicylate, nicotine, tropacocaine
Eschscholzia	Papaveraceae	Biflorine, codeine, hydrocyanic acid, morphine, protopine, rutin, sanguinarine
Eucalyptus[b]	Myrtaceae	Benzaldehyde, borneol, butyraldehyde, butyric acid, carene, carvone, cineole, citral, cuminic alcohol, cymene, ethyl alcohol, eucalyptus oil, gallic acid, geraniol, hydrocyanic acid, isoamyl alcohol, isobutyric acid, isoprene, isovaleraldehyde, isovaleric acid, limonene, linalool, mandelonitrile, phellandrene, pyrocatechol, quercitrin, quinic acid, rutin, saponin, shikimic acid, tannic acid, valeraldehyde
Euchresta	Fabaceae	Cytisine
Eucryphia	Eucryphiaceae	Quercitrin
Eugenia	Myrtaceae	Benzoic acid, cineole, eugenol, geraniol, methyl salicylate, saponin, tannic acid
Euonymus[b]	Celastraceae	Evonoside, shikimic acid
Eupatorium[b]	Asteraceae	Cymene, rutin, tannic acid
Euphorbia[b]	Euphorbiaceae	Dopa, gallic acid, hydrocyanic acid, quercitin, saponin
Euptelea	Eupteleaceae	Shikimic acid
Euryops	Asteraceae	Isobutyric acid
Eustachys	Poaceae	Hydrocyanic acid
Excavatia	Apocynaceae	Reserpine
Fagopyrum[b]	Polygonaceae	Hydrocyanic acid, rutin, tryptophane
Fagraea	Potaliaceae	Diacetyl
Fagus[b]	Fagaceae	Choline, citric acid, saponin, trimethylamine
Fatsia	Araliaceae	Quercitin
Ferdinandusa	Rubiaceae	Cephaeline, emetine
Feronia	Rutaceae	Esdragole, pectin
Ferula	Apiaceae	Ferulic acid, valeric acid
Festuca[b]	Poaceae	Rutin
Ficus	Moraceae	Coumarin, ficin, malic acid, methyl salicylate, oxalic acid, rutin, saponin, tannic acid, tylocrebrine
Filipendula	Rosaceae	Salicylaldehyde
Firmiana	Sterculiaceae	Rutin
Flacourtia	Flacourtiaceae	Pectin
Florestina[b]	Asteraceae	Hydrocyanic acid

Table 3 (continued)
HIGHER PLANT GENERA AND THEIR TOXINS

Genus[a]	Family	Toxin
Flourensia[b]	Asteraceae	
Foeniculum	Apiaceae	Acetaldehyde, anethole, esdragole, fenchone, fennel oil, limonene, phellandrene
Forsythia	Oleaceae	Lactose, quercitrin, rutin
Fortunella	Rutaceae	Pectin
Fothergilla	Hamamelidaceae	Rutin
Fragaria	Rosaceae	Citric acid, ethyl alcohol, hexanol, isoamyl alcohol, malic acid, methyl salicylate, pectin, pyrocatechol, quercitin, tannic acid
Fraxinus	Oleaceae	Malic acid, quercitrin
Fritillaria[b]	Liliaceae	
Fuchsia	Onagraceae	Quercitin
Fumaria	Fumariaceae	Biflorine, fumaric acid, protopine
Funtumia	Apocynaceae	Funtumidene, funtumine
Furcraea	Agavaceae	Dodecyl alcohol, saponin
Galanthus[b]	Amaryllidaceae	
Galbulimima	Himantandraceae	Himandrine
Galega	Fabaceae	Guanidine
Galenia	Aizoaceae	Hydrocyanic acid
Galipea	Rutaceae	Quinaldine
Galium	Rubiaceae	Citric acid, rutin, tannic acid
Gallesia	Phytolaccaceae	Caffeine
Garcinia	Clusiaceae	Citric acid, pyrocatechol, tannic acid
Gardenia	Rubiaceae	Shikimic acid
Garrya	Garryaceae	Cuachichinene, garryfoline, garryine, veatchin
Garuga	Burseraceae	Methyl salicylate
Gaultheria	Ericaceae	Benzoic acid, gaultheria oil, heptanal, methyl salicylate, salicylic acid
Gelsemium[b]	Loganiaceae	Gelsemine
Genipa	Rubiaceae	Caffeine
Genista[b]	Fabaceae	Cytisine, pachycarpine, sparteine
Geranium	Geraniaceae	Shikimic acid
Geum	Rosaceae	Eugenol, tannic acid
Gladiolus[b]	Iridaceae	
Glaucium	Papaveraceae	Biflorine, chelidonin, fumaric acid, sanguinarine
Glechoma[b]	Lamiaceae	
Gleditsia[b]	Caesalpiniaceae	Cocaine, cytisine
Glinus	Aizoaceae	Saponin
Gliricidia	Fabaceae	Hydrocyanic acid, raton
Globularia	Globulariaceae	Benzoic acid, rutin
Gloriosa[b]	Liliaceae	Colchamine, colchicine
Glyceria[b]	Poaceae	Hydrocyanic acid
Glycine[b]	Fabaceae	Betaine, choline, guanidine, hydrocyanic acid, isovaleraldehyde, maltose, oxalic acid, saponin, trigonelline, tryptophane
Glycosmis	Rutaceae	Carbazole, methyl salicylate
Glycyrrhiza	Fabaceae	Saponin, tannic acid
Gnaphalium[b]	Asteraceae	Cresol
Gossypium[b]	Malvaceae	Betaine, choline, folic acid, furfural, gossypol, gossyverdurin, methanol, pantothenic acid, quercitin, rutin, succinic acid, trimethylamine, vanillin
Grabowskia	Solanaceae	Myristic acid
Gratiola[b]	Scrophulariaceae	Saponin
Grevillea	Proteaceae	Hydrocyanic acid, rutin
Grindelia[b]	Asteraceae	Selenium
Guaiacum	Zygophyllaceae	Guaiacol, saponin
Guazuma	Sterculiaceae	Caffeine
Guizotia	Asteraceae	Myristic acid
Gutierrezia[b]	Asteraceae	Selenium
Gymnocalycium	Cactaceae	Mescaline
Gymnocladus[b]	Caesalpiniaceae	
Haemanthus[b]	Amaryllidaceae	
Halogeton[b]	Chenopodium	Oxalic acid
Halostachys	Chenopodiaceae	Halostachyine
Hamamelis	Hamamelidaceae	Isoprene

Table 3 (continued)
HIGHER PLANT GENERA AND THEIR TOXINS

Genus[a]	Family	Toxin
Haplopappus[b]	Asteraceae	Cholesterol, phellandrene, pyridine, selenium
Harrisia	Cactaceae	Caffeine
Hedeoma	Lamiaceae	Menthol, pulegone
Hedera[b]	Araliaceae	Rutin, saponin, shikimic acid
Heimia	Lythraceae	Cryogenine
Helenium[b]	Asteraceae	
Helianthus[b]	Asteraceae	Betaine, cholesterol, choline, citric acid, fumaric acid, malic acid, malonic acid, myristic acid, saponin, shikimic acid, succinic acid, tannic acid, tartaric acid, tryptophane
Helichrysum	Asteraceae	Isovaleraldehyde
Helicteres	Sterculiaceae	Caffeine
Heliotropium[b]	Boraginaceae	Heliotrine, lasiocarpine, lasiocarpine *N*-oxide, retronecine, sali
Helleborus[b]	Ranunculaceae	Helleborein, sprintillamine, sprintillin
Hemerocallis	Liliaceae	Colchicine
Hemidesmus	Periplocaceae	Coumarin, rutin
Hepatica	Ranunculaceae	Saponin
Heracleum	Apiaceae	Dimethylamine, ethyl alcohol, methylamine, rutin, trimethylamine
Hermidium	Nyctaginaceae	Dopamine
Herniaria	Caryophyllaceae	Coumarin, rutin
Herrania	Sterculiaceae	Caffeine, theobromine
Hertia	Asteraceae	Hydrocyanic acid
Heterostemma	Asclepiadaceae	Rutin
Hevea	Euphorbiaceae	Acetone, ergosterol, hydrocyanic acid, pachycarpine
Hibiscus	Malvaceae	Citric acid, hydrocyanic acid, malic acid, pectin, rutin, saponin, tannic acid, tartaric acid
Hilaria[b]	Poaceae	
Hillia	Rubiaceae	Cephaeline, emetine
Hippomane[b]	Euphorbiaceae	Physostigmine
Hippophae	Elaeagnaceae	Malic acid, rutin
Holcus[b]	Poaceae	Hydrocyanic acid
Hordeum[b]	Poaceae	Anhaline, donaxine, ferulic acid, folic acid, furfural, malonic acid, pantothenic acid, pectin
Houttuynia	Saururaceae	Quercitrin
Hovea	Fabaceae	Sparteine
Humulus	Cannabinaceae	Choline, codeine(?), coniine, formaldehyde, formic acid, geraniol, isovaleric acid, limonene, linalool, lupulone, morphine (?), quercitrin, tannic acid, valeric acid
Hunnemannia	Papaveraceae	Biflorine, sanguinarine
Hunteria	Apocynaceae	Methyl salicylate
Hura[b]	Euphorbiaceae	
Hyacinthus[b]	Liliaceae	Benzaldehyde, benzyl alcohol, cinnamaldehyde, cinnamyl alcohol, eugenol methyl ether, heptanal, heptyl alcohol
Hybanthus	Violaceae	Emetine
Hydnocarpus	Flacourtiaceae	Hydrocyanic acid, methyl salicylate, palmitic acid, tannic acid
Hydrangea[b]	Hydrangeaceae	Hydrocyanic acid, hydroquinone, rutin
Hydrastis	Hydrastidaceae	Hydrastine
Hylomecon	Papaveraceae	Sanguinarine
Hydrocotyle	Apiaceae	Asiaticoside
Hygrophila	Acanthaceae	Maltose
Hymenoxys[b]	Asteraceae	
Hyoscyamus[b]	Solanaceae	Atropine, choline, hyoscyamine, rutin, scopolamine
Hypecoum	Papaveraceae	Biflorine, sanguinarine
Hypericum[b]	Hypericaceae	Quercitin, saponin, shikimic acid, terpineol
Hyphaene	Arecaceae	Lauric acid
Hyptis	Lamiaceae	Limonene, menthol
Hypochoeris	Asteraceae	Cholesterol
Hyssopus	Lamiaceae	Choline, tannic acid
Ilex	Aquifoliaceae	Caffeine, cholesterol, citric acid, ergosterol, furfural, malic acid, saponin, theobromine, theophylline
Illicium	Illiciaceae	Anethole, anisaldehyde, carene, cineole, cymene, esdragole, isosafrole, limonene, phellandrene, quercitin, quercitrin, quinic acid, safrole, shikimic acid, terpineol, trigonelline, vanillin

Table 3 (continued)
HIGHER PLANT GENERA AND THEIR TOXINS

Genus[a]	Family	Toxin
Impatiens	Balsaminaceae	Saponin
Indigofera[b]	Fabaceae	Hydrocyanic acid, indican
Intsia	Mimosaceae	Lapachol
Iphigenia	Liliaceae	Colchicine
Ipomoea[b]	Convolvulaceae	Acetaldehyde, acetone, ergometrine, ethyl alcohol, hexadecanal, hydrocyanic acid, methanol, oxalic acid, pectin, propionaldehyde, sebacic acid, tannic acid, tryptophane
Iris[b]	Iridaceae	Acetophenone, decanal, methylamine, nonanoic acid, quercitin, tridecanoic acid, undecanoic acid
Irvingia	Irvingiaceae	Lauric acid, myristic acid
Isatis	Brassicaceae	Indican
Jacquinia	Theophrastaceae	Saponin
Jasminum	Oleaceae	Benzyl alcohol, cresol, geraniol, indole, linalool, methyl anthranilate, salicylic acid, terpineol
Jatropha[b]	Euphorbiaceae	Hydrocyanic acid
Juglans	Juglandaceae	Folic acid, furfural, inositol, juglone, nicotine, tryptophane
Juncus[b]	Juncaceae	
Juniperus[b]	Cupressaceae	Gallic acid, malic acid, oxalic acid, pectin, podophyllotoxin
Kadsura	Schisandraceae	Shikimic acid
Kaempferia	Zingiberaceae	Borneol, methyl anthranilate
Kalachoe	Crassulaceae	Hydrocyanic acid
Kallstroemia[b]	Zygophyllaceae	
Kalmia[b]	Ericaceae	
Karwinskia	Rhamnaceae	
Kedrostis	Cucurbitaceae	Cucurbitacin E, hydrocyanic acid
Kiggelaria	Flacourtiaceae	Hydrocyanic acid
Kleinia	Asteraceae	Citric acid
Kochia[b]	Chenopodiaceae	Saponin
Laburnum[b]	Fabaceae	Cytisine
Lachnanthes[b]	Haemodoraceae	
Lactuca[b]	Asteraceae	Choline, citric acid, ergosterol, folic acid, hyoscyamine, malic acid, oxalic acid, pectin
Ladenbergia	Rubiaceae	Quinine, yohimbine
Lagenaria	Cucurbitaceae	Choline, curcurbitacin E, hydrocyanic acid, pectin, saponin, shikimic acid
Lagurus	Poaceae	Hydrocyanic acid
Lamium[b]	Lamiaceae	Histamine
Lantana[b]	Verbenaceae	Hydrocyanic acid
Laportea	Urticaceae	Acetylcholine, histamine
Lappula	Boraginaceae	Lasiocarpine
Larix	Abietaceae	Maltol
Laserpitium	Apiaceae	Coumarin
Lasianthus	Rubiaceae	Methanethiol
Lasiosiphon	Thymelaeaceae	Hydrocyanic acid
Lathyrus[b]	Fabaceae	Folic acid, hydrocyanic acid, maltose, pantothenic acid, quercitrin, saponin
Laurus	Lauraceae	Capric acid, caproic acid, cineole, eugenol, isobutyric acid, lauric acid, phellandrene, terpineol, valeric acid
Lavandula	Lamiaceae	Borneol, butyraldehyde, camphor, capric acid, caproic acid, carvone, cineole, cinnamaldehyde, citral, coumarin, cuminic alcohol, cuminic aldehyde, decanal, decanoic acid, eugenol, fenchone, geraniol, hexanol, isoamyl alcohol, isobutyric acid, isovaleraldehyde, isovaleric acid, limonene, linalool, nonanoic acid, phellandrene, propionic acid, tannic acid, thujone, thymol, valeraldehyde, valeric acid
Lawsonia	Lythraceae	Gallic acid, ionone, tannic acid
Ledum[b]	Ericaceae	Cresol, cuminic alcohol, cymene, phellandrene
Lens	Fabaceae	Choline, folic acid, oxalic acid, pantothenic acid, saponin
Leocereus	Cactaceae	Caffeine
Leontice	Leonticaceae	Pachycarpine, petaline, saponin
Leonurus	Lamiaceae	Rutin
Leptactina	Rubiaceae	Benzyl cyanide, isoeugenol
Leptadenia	Asclepiadaceae	Rutin
Leptorhabdos	Scrophulariaceae	Pachycarpine, sparteine

Table 3 (continued)
HIGHER PLANT GENERA AND THEIR TOXINS

Genus[a]	Family	Toxin
Leptospermum	Myrtaceae	Citral
Leptotaenia	Apiaceae	Methylamine
Lespedeza[b]	Fabaceae	Bufotenine, dimethyl tryptamine, shikimic acid, tannic acid
Leucadendron	Proteaceae	Rutin
Leucaena[b]	Mimosaceae	Hydrocyanic acid, leucaenine, quercitrin, tannic acid
Leucothoe[b]	Ericaceae	
Levisticum	Apiaceae	Coumarin, eugenol, isovaleric acid, myristic acid, myristicin, palmitic acid, terpineol
Liatris	Asteraceae	Coumarin, rutin
Licaria	Lauraceae	Apiole, safrole
Ligusticum	Apiaceae	Dodecyl alcohol, isosafrole
Ligustrum[b]	Oleaceae	Shikimic acid
Lilium	Liliaceae	Cresol, methylamine
Limonium	Plumbaginaceae	Rutin
Lindera	Lauraceae	Borneol, carvone, methyl salicylate
Linociera	Oleaceae	Methyl salicylate
Linum[b]	Linaceae	Choline, hydrocyanic acid, pantothenic acid, pectin, podophyllotoxin, quinic acid, tryptophane
Lippia[b]	Verbenaceae	Camphor, carbone, cineole, citral, citronellol, geraniol, isovaleric acid, lactose, limonene, maltose, saponin, terpineol, thujone, verbenone
Liquidambar	Altingiaceae	Saponin, shikimic acid, styrene, vanillin
Liriodendron	Magnoliaceae	Shikimic acid
Lithocarpus	Fagaceae	Shikimic acid
Lithospermum	Boraginaceae	Rutin
Lithraea	Anacardiaceae	Shikimic acid
Litsea	Lauraceae	Capric acid, carvone, cineole, cymene, decanoic acid, hexanal, lauric acid, limonene, nonanoic acid
Littonia	Liliaceae	Colchicine
Lloydia	Liliaceae	Colchicine
Lobelia[b]	Lobeliaceae	Lobeline
Lolium[b]	Poaceae	Citric acid, hydrocyanic acid, malic acid, oxalic acid, pantothenic acid, shikimic acid
Lomatia	Proteaceae	Juglone
Lophophora	Cactaceae	Anhaline, mescaline, pellotine, tyramine
Loranthus	Loranthaceae	Hexadecanol, scopolamine
Lotonotis	Fabaceae	Hydrocyanic acid
Lotus[b]	Fabaceae	Cytisine, hydrocyanic acid, malonic acid, saponin
Luffa	Cucurbitaceae	Choline, cucurbitacin E, hydrocyanic acid, saponin, tryptophane
Lunaria	Brassicaceae	Lunarine
Lupinus[b]	Fabaceae	Dopa, hydrocyanic acid, hypoxanthine, lycine, malonic acid, pachycarpine, pectin, piperidine, saponin, sparteine, trigonelline, tryptophane, xanthine
Luvunga	Rutaceae	Cineole
Lycium[b]	Solanaceae	Betaine
Lycopersicon[b]	Solanaceae	Acetaldehyde, acetone, ethyl alcohol, ferulic acid, folic acid, histamine, malic acid, methanol, narcotine, noscopine, oxalic acid, pantothenic acid, pectin, propionaldehyde, quercitrin, rutin, serotonine, shikimic acid, solanine, tomatine, tryptophane
Lycopodium[b]	Lycopodiaceae	
Lycopus	Lamiaceae	Tannic acid
Lygodesmia[b]	Asteraceae	
Lyonia[b]	Ericaceae	
Macadamia	Proteaceae	Hydrocyanic acid, myristic acid
Macaranga	Euphorbiaceae	Methyl salicylate
Machaeranthera[b]	Asteraceae	Selenium
Macleaya	Papaveraceae	Biflorine, sanguinarine, shikimic acid
Maclura[b]	Moraceae	Saponin
Macrosiphonia	Apocynaceae	Coumarin
Macrozamia[b]	Cycadaceae	
Madhuca	Sapotaceae	Folic acid, malic acid, panthothenic acid, saponin, tannic acid
Maerua	Capparidaceae	Dimethylamine, hydrocyanic acid, trimethylanine
Magnolia	Magnoliaceae	Anethole, anisaldehyde, choline, phellandrene, rutin, safrole, shikimic acid

Table 3 (continued)
HIGHER PLANT GENERA AND THEIR TOXINS

Genus[a]	Family	Toxin
Mahonia	Berberidaceae	Berbamine, shikimic acid
Malus[b]	Rosaceae	Citric acid, estrone, folic acid, formic acid, gallic acid, hydrocyanic acid, isovaleric acid, malic acid, maltose, pantothenic acid, pectin, phloroglucinol, quercitrin, quinic acid, shikimic acid, tannic acid
Malva[b]	Malvaceae	
Mammea	Clusiaceae	Citric acid, shikimic acid
Mammillaria	Cactaceae	Anhaline
Mandragora[b]	Solanaceae	Atropine, hyoscyamine, scopolamine
Manettia	Rubiaceae	Emetine
Mangifera[b]	Anacardiaceae	Gallic acid, geraniol, hydrocyanic acid, limonene, oxalic acid, phellandrene, tannic acid
Manihot	Euphorbiaceae	Acetone, hydrocyanic acid, oxalic acid, saponin, tryptophane
Manilkara	Sapotaceae	Saponin
Mansonia	Sterculiaceae	Mansonin, strophanthidin
Marrubium	Lamiaceae	Choline, tannic acid
Marsdenia	Asclepiadaceae	Methyl salicylate, quercitin, rutin
Martynia	Martyniaceae	Allyl alcohol
Matricaria	Asteraceae	Borneol, camphor, salicylic acid, saponin
Maytenus	Celastraceae	Caffeine, norpseudoephedrine
Meconella	Papaveraceae	Biflorine
Meconopsis	Papaveraceae	Biflorine, sanguinarine
Medicago[b]	Fabaceae	Choline, citric acid, hydrocyanic acid, limonene, malic acid, malonic acid, oxalic acid, pantothenic acid, pectin, quinic acid, saponin, shikimic acid, trigonelline, tryptophane
Melaleuca	Myrtaceae	Benzaldehyde, butyraldehyde, cajeputol, cineole, cymene, eugenol methyl ester, isovaleric acid, limonene, phellandrene, saponin, terpineol, valeraldehyde
Melia[b]	Meliaceae	Tannic acid
Melica	Poaceae	Hydrocyanic acid
Melilotus[b]	Fabaceae	Coumarin, hydrocyanic acid, malonic acid, melilotin
Meliosma	Sabiaceae	Methyl salicylate
Melissa	Lamiaceae	Citral, citronellol, geraniol, linalool
Melolobium	Fabaceae	Hydrocyanic acid
Menispermum[b]	Menispermaceae	
Mentha	Lamiaceae	Acetaldehyde, amyl alcohol, caproic acid, caprylic acid, carvacrol, carvone, cineole, ethyl alcohol, formic acid, furfural, isoamyl alcohol, isovaleraldehyde, isovaleric acid, limonene, linalool, menthol, methanol, methylamine, nonanoic acid, oxalic acid, pennyroyal oil, peppermint oil, phellandrene, pulegone, salicylic acid, valeric acid
Menyanthes[b]	Menyanthaceae	Rutin, saponin, trimethylamine
Menziesia[b]	Ericaceae	
Mercurialis[b]	Euphorbiaceae	Hydrocyanic acid, maltose, methylamine, trimethylamine
Merendera	Liliaceae	Colchamine, colchiceine, colchicine, demecolcine
Meriandra	Lamiaceae	Camphor
Metopium[b]	Anacardiaceae	
Metrosideros	Myrtaceae	Methyl salicylate
Michelia	Magnoliaceae	Cineole, eugenol, geraniol, isoeugenol, methyl anthranilate, shikimic acid
Micromeria	Lamiaceae	Pulegone, saponin
Milletia	Fabaceae	Rotenone
Mimosa	Mimosaceae	Anisaldehyde, dimethyl tryptamine, histamine, leucaenine
Mimusops	Sapotaceae	Lactose, saponin
Mirabilis	Nyctaginaceae	Trigonelline
Mitragyna	Rubiaceae	Ajmalicine
Modiola[b]	Malvaceae	
Mollugo	Aizoaceae	Hydrocyanic acid, saponin
Momordica[b]	Cucurbitaceae	Oxalic acid, saponin
Monarda	Lamiaceae	Butyraldehyde, formaldehyde, isovaleraldehyde, thymol
Monodora	Annonaceae	Phellandrene
Monotropa	Monotropaceae	Methyl salicylate
Montia[b]	Portulacaceae	
Mora	Caesalpiniaceae	Saponin

Table 3 (continued)
HIGHER PLANT GENERA AND THEIR TOXINS

Genus[a]	Family	Toxin
Moringa	Moringaceae	Ephedrine, hydrocyanic acid, myristic acid, pectin, quercitin, tryptophane
Morus	Moraceae	Butyraldehyde, choline, citric acid, isobutyraldehyde, linalool, malic acid, pectin, saponin, tannic acid, trigonelline
Mucuna[b]	Fabaceae	Dopa, dopamine, nicotine, physostigmine, serotonine
Mundulea	Fabaceae	Hydrocyanic acid, rotenone
Murraya	Rutaceae	Carbazole, isosafrole, limonene, methyl salicylate, tryptophane
Musa	Musaceae	Citric acid, dopa, dopamine, folic acid, isovaleric acid, malic acid, maltose, noradrenaline, oxalic acid, pectin, serotonine, tannic acid, tartaric acid
Muscari	Liliaceae	Colchicine
Myrica	Myricaceae	Gallic acid, myristic acid, oleic acid, palmitic acid, shikimic acid, tannic acid
Myristica	Myristicaceae	Borneol, caprylic acid, cymene, eugenol, formic acid, furfural, geraniol, isoeugenol, limonene, linalool, methyl salicylate, myristic acid, myristicin, pectin, safrole, tannic acid
Myrothamnus	Myrothamnaceae	Cineole
Myroxylon	Fabaceae	Benzoic acid, peru balsam oil, phellandrene, vanillin
Myrrhis	Apiaceae	Fumaric acid
Myrsine	Myrsinaceae	Hydrocyanic acid
Myrtus	Myrtaceae	Cineole, citric acid, malic acid, myristic acid
Nandina	Nandinaceae	Biflorine
Narcissus[b]	Amaryllidaceae	Benzyl alcohol, cinnamaldehyde, cinnamyl alcohol, colchicine, indole
Nardosma	Asteraceae	Rutin
Nasturtium	Brassicaceae	Hydrocyanic acid
Neea	Nyctaginaceae	Caffeine
Nectandra	Lauraceae	Cotoin, isoeugenol, safrole
Nemuaron	Atherospermaceae	Safrole
Neolitsea	Lauraceae	Capric acid, decanoic acid, lauric acid, quercitrin
Nepenthes	Nepenthaceae	Histamine
Nepeta[b]	Lamiaceae	Carvacrol, thymol
Nephelium	Sapindaceae	Saponin, tannic acid
Nerine[b]	Amaryllidaceae	
Nerium[b]	Apocynaceae	Hydrocyanic acid, neriin, odorobioside, odoroside, rutin
Nicotiana[b]	Solanaceae	Acetaldehyde, ammonia, anabasine, citric acid, ergosterol, ethyl alcohol, eugenol, formic acid, guaiacol, hydrocyanic acid, isobutyraldehyde, isovaleric acid, limonene, malic acid, methylamine, nicotine, nornicotine, oxalic acid, phenol, piperidine, pyrrolidine, quercitrin, quinic acid, rutin, salicylaldehyde, saponin, trimethylamine, tryptophane
Nigella[b]	Ranunculaceae	Carbone, cymene, damascenini, saponin, tryptophane
Nigritella	Orchidaceae	Vanillin
Nolana[b]	Nolanaceae	Quercitin
Nuphar	Nymphaeaceae	Methylamine
Nyctanthes	Verbenaceae	Methyl salicylate
Nyctocalos	Bignoniaceae	Methyl salicylate
Nymphaea[b]	Nymphaeaceae	
Ochna	Ochnaceae	Hydrocyanic acid
Ochrosia	Apocynaceae	Reserpine
Ocimum	Lamiaceae	Anethole, camphor, carvacrol, cineole, citral, esdragole, eugenol, hydrocyanic acid, limonene, linalool, safrole, saponin, thymol
Ocotea	Lauraceae	Apiole, safrole, valeraldehyde
Oenanthe[b]	Apiaceae	Myristicin, phellandrene, rutin
Oldenlandia	Rubiaceae	Caffeine, fumaric acid, lactose
Olea	Oleaceae	Choline, glycerol, saponin
Olinia	Oliniaceae	Hydrocyanic acid
Onoclea[b]	Aspidiaceae	
Ononis[b]	Fabaceae	Malonic acid, saponin
Opuntia	Cactaceae	Mescaline, pectin
Origanum	Lamiaceae	Carvacrol, cymene, eugenol, limonene, linalool, origanum oil, pulegone, tannic acid, terpineol, thymol
Ornithogalum[b]	Liliaceae	Convallatoxin, convalloside, strophanthidin
Ornithoglossum	Liliaceae	Colchicine

Table 3 (continued)
HIGHER PLANT GENERA AND THEIR TOXINS

Genus[a]	Family	Toxin
Orthodon	Lamiaceae	Asarone, carvacrol, carvone, myristicin
Orthosiphon	Lamiaceae	Betaine, choline
Oryza	Poaceae	Anhaline, choline, citric acid, hydrocyanic acid, malic acid, maltose, naphthaline, oxalic acid, pantothenic acid, pectin, trigonelline
Osteospermum	Asteraceae	Hydrocyanic acid
Osyris	Santalaceae	Rutin
Oxalis[b]	Oxalidaceae	Oxalic acid
Oxytenia[b]	Asteraceae	
Oxytropis[b]	Fabaceae	Selenium
Pachycarpus	Asclepiadaceae	Anhydroperiplogenone, cymarin, strophanthidin
Pachyrhizus	Fabaceae	Adenine, choline, rotenone, saponin
Paederia	Rubiaceae	Methanethiol
Paeonia[b]	Paeoniaceae	
Paliuris	Rhamnaceae	Methyl salicylate, rutin
Panax	Araliaceae	Maltose, saponin
Pandanus	Pandanaceae	Guaiacol, linalool
Pangium	Flacourtiaceae	Hydrocyanic acid
Panicum[b]	Poaceae	Anhaline, choline, saponin, tryptophane
Papaver[b]	Papaveraceae	Biflorine, citric acid, codeine, malic acid, morphine, narcotine, papaverine, pectin, sanguinarine, tannic acid, thebaine, tryptophane
Pappea	Sapindaceae	Hydrocyanic acid
Paratecoma	Bignoniaceae	Lapachol
Parietaria	Urticaceae	Coniine
Parinari	Chrysobalanaceae	Methyl salicylate
Paris[b]	Trilliaceae	
Parkinsonia[b]	Caesalpiniaceae	
Parthenium	Asteraceae	Betaine
Parthenocissus[b]	Vitaceae	
Paspalum[b]	Poaceae	Hydrocyanic acid, tryptophane
Passiflora	Passifloraceae	Citric acid, harman, harmine, hydrocyanic acid, malic acid, pantothenic acid, pectin, tannic acid
Pastinaca[b]	Apiaceae	Coumarin, ethyl alcohol, imperatorin, methanol, myristicin, oxalic acid, rutin
Patrinia	Valerianaceae	Shikimic acid
Paullinia	Sapindaceae	Caffeine, pyrocatechol, theobromine, theophylline
Pausinystalia	Rubiaceae	Corynantheine
Pectis	Asteraceae	Cuminic aldehyde
Pedicularis[b]	Scrophulariaceae	
Peganum[b]	Zygophyllaceae	Harmaline, harmine
Pelargonium	Geraniaceae	Citral, citronellol, geraniol, hexanol, isoamyl alcohol, isovaleric acid, nonanoic acid, phellandrene, terpineol
Pelea	Rutaceae	Anisaldehyde
Peltophorum	Caesalpiniaceae	Tannic acid, tryptophane
Pennisetum	Poaceae	Choline, hydrocyanic acid, oxalic acid, tryptophane
Penstemon[b]	Scrophulariaceae	
Peponium	Cucurbitaceae	Cucurbitacin
Perilla	Lamiaceae	Limonene
Periploca	Periplocaceae	Strophanthidin
Peristrophe	Acanthaceae	Coumarin
Perriera	Simaroubaceae	Glaucarubin
Persea[b]	Lauraceae	Anethole, esdragole, hydrocyanic acid, serotonine, tartaric acid
Petalostylis	Caesalpiniaceae	Dimethyl tryptamine
Petasites	Asteraceae	Retronecine, senecionine
Petroselinum	Apiaceae	Apiole, coumarin, myristicin
Petrosimonia	Chenopodiaceae	Piperidine
Peucedanum	Apiaceae	Imperatorin, isobutyric acid, myristicin
Peumus	Monimiaceae	Pachycarpine, terpineol
Phaius	Orchidaceae	Indican
Phalaris[b]	Poaceae	Anhaline, bufotenine, dimethyl tryptamine, donaxine
Phaseolus[b]	Fabaceae	Acetone, choline, fumaric acid, hydrocyanic acid, inositol, malonic acid, maltose, methyl-cysteine, oxalic acid, pectin, rutin, saponin, succinic acid
Phebalium	Rutaceae	Citral

Table 3 (continued)
HIGHER PLANT GENERA AND THEIR TOXINS

Genus[a]	Family	Toxin
Philadelphus	Philadelphaceae	Methylamine
Philodendron[b]	Araceae	
Phlox	Polemoniaceae	Shikimic acid
Phoenix	Arecaceae	Cholesterol, coumarin, estrone, glycerol, rutin
Phoradendron[b]	Loranthaceae	Phenethylamine, tyramine
Photinia	Rosaceae	Methyl salicylate, rutin
Phyrygilanthus	Loranthaceae	Rutin
Phyllanthus[b]	Euphorbiaceae	Gallic acid, myristic acid, pectin, saponin, tannic acid
Physalis[b]	Solanaceae	Elaidic acid, hydrocyanic acid, pectin
Physochlaina	Solanaceae	Hyoscyamine, solanine
Physostigma	Fabaceae	Physostigmine
Phytolacca[b]	Phytolaccaeae	Rutin, saponin, tannic acid
Picrolemma	Simaroubaceae	Quinine
Pieris[b]	Ericaceae	Shikimic acid
Pilocarpus	Rutaceae	Limonene
Pimenta	Myrtaceae	Cineole, citral, esdragole, eugenol, eugenol methyl ether, limonene, methanol, palmitic acid, phellandrene
Pimpinella	Apiaceae	Acetaldehyde, anethole, choline, cresol, esdragole, eugenol, furfural, hydroquinone, myristic acid, propionic acid, tannic acid
Pinguicula	Lentibulariaceae	Histamine
Pinus[b]	Abietaceae	Carene, citric acid, fumaric acid, heptane, malonic acid
Piper	Piperaceae	Anethole, apiole, asarone, carene, cineole, citral, dihydrokawain, eugenol, eugenol methyl ether, hydrocyanic acid, limonene, phellandrene, piperidine, piperonal, safrole, stearyl alcohol
Piptadenia	Mimosaceae	Bufotenine, dimethyl tryptamine
Piptanthus	Fabaceae	Cytisine, sparteine
Piriqueta	Turneraceae	Caffeine
Piscidia	Fabaceae	Rotenone
Pistacia	Anacardiaceae	Quinic acid, shikimic acid
Pisum[b]	Fabaceae	Oxalic acid, saponin
Pithecellobium	Mimosaceae	Saponin
Pittosporum	Pittosporaceae	Heptane, limonene
Plagiobothrys[b]	Boraginaceae	
Plantago	Plantaginaceae	Citric acid, histamine, saponin
Platanus	Platanaceae	Pyrocatechol, rutin, saponin, shikimic acid
Platea	Icacinaceae	Methyl salicylate
Platycapnos	Fumariaceae	Biflorine
Platycarya	Juglandaceae	Juglone, shikimic acid
Platystemon	Papaveraceae	Biflorine, sanguinarine
Pleiocarpa	Apocynaceae	Caffeine
Plumbago	Plumbaginaceae	Quercitrin
Plumeria	Apocynaceae	Saponin
Poa[b]	Poaceae	Hydrocyanic acid
Podophyllum[b]	Podophyllaceae	Podophyllic acid, podophyllotoxin, quercitin
Pogonarthria	Poaceae	Hydrocyanic acid
Pogostemon	Lamiaceae	Acetone, cinnamaldehyde, eugenol
Poikilospermum	Urticaceae	Methyl salicylate
Polianthes	Agavaceae	Benzyl alcohol, eugenol, geraniol, methyl anthranilate, methyl salicylate, piperonal
Poliomintha	Lamiaceae	Pulegone
Polyalthia	Annonaceae	Diacetyl
Polygala[b]	Polygalaceae	Indican, methyl salicylate, salicylic acid, saponin
Polygonatum	Liliaceae	Chrysarobin
Polygonum[b]	Polygonaceae	Gallic acid, hydrocyanic acid, indican, quercitin, rutin, tannic acid
Polyscias	Araliaceae	Saponin
Populus	Salicaceae	Acetophenone, cholesterol, cinnamyl alcohol, isoprene, pyrocatechol, salicin, salicylic acid
Portulaca[b]	Portulacaceae	Dopa, hydrocyanic acid, noradrenalin
Pouteria	Sapotaceae	Lactose, yohimbine
Prangos	Apiaceae	Imperatorin
Premna	Verbenaceae	Limonene
Prestonia	Apocynaceae	Dimethyl tryptamine

Table 3 (continued)
HIGHER PLANT GENERA AND THEIR TOXINS

Genus[a]	Family	Toxin
Primula	Primulaceae	Salicylic acid, saponin
Prosopis[b]	Mimosaceae	Arabinose
Protea	Proteaceae	Hydroquinone, rutin
Protium	Burseraceae	Anisaldehyde, safrole
Prunella	Lamiaceae	Camphor, fenchone, saponin
Prunus[b]	Rosaceae	Caproic acid, citral, coumarin, decyl alcohol, eugenol, gallic acid, geraniol, hydrocyanic acid, isobutyric acid, isoeugenol, isovaleric acid, limonene, linalool, malic acid, mandelonitrile, methyl salicylate, nicotine, noradrenaline, oxalic acid, quercitin, quercitrin, quinic acid, rutin, serotonine, tannic acid, trimethylamine
Psathyrotes[b]	Asteraceae	
Pseudocinchona	Rubiaceae	Corynantheine, yohimbine
Psidium	Myrtaceae	Gallic acid, hydrocyanic acid, limonene, pyrocatechol, quercitin
Psilocaulon	Aizoaceae	Piperidine
Psittacanthus	Loranthaceae	Quercitrin, tyramine
Psoralea[b]	Fabaceae	
Psorospermum	Clusiaceae	Pyrocatechol
Psychotria	Rubiaceae	Cephaeline, emetine
Pteridium[b]	Dennstaedtiaceae	Hydrocyanic acid, shikimic acid
Pteridophyllum	Pteridophyllaceae	Biflorine
Pteris[b]	Pteridaceae	Mandelonitrile
Pterocarya	Juglandaceae	Juglone
Pulsatilla[b]	Ranunculaceae	
Punica	Punicaceae	Coniine, estrone, gallic acid, pelletierine, tannic acid
Pycnanthemum	Lamiaceae	Menthol, pulegone
Pycnarrhena	Menispermaceae	Berbamine
Pygeum	Rosaceae	Hydrocyanic acid
Pyracantha	Rosaceae	Hydrocyanic acid
Pyrus[b]	Rosaceae	Acetaldehyde, citric acid, gallic acid, hydroquinone, malic acid, rutin, shikimic acid, sorbic acid, succinic acid, trimethylamine
Quassia	Simaroubaceae	Glaucarubin
Quercus[b]	Fagaceae	Acetaldehyde, butyraldehyde, isoprene, isovaleraldehyde, quercitin, quercitrin, valeraldehyde
Rafinesquia[b]	Asteraceae	
Randia	Rubiaceae	Saponin
Ranunculus[b]	Ranunculaceae	Anemonin
Raphanus[b]	Brassicaceae	Acetaldehyde, acetone, butyraldehyde, ethyl alcohol, isobutyraldehyde, isovaleraldehyde, methanethiol, methanol, oxalic acid, propionaldehyde, pyrocatechol, saponin
Rauvolfia	Apocynaceae	Ajmalicine, ajmaline, deserpidine, narcotine, rescinnamine, salicaldehyde, serpentine, yohimbine
Remijia	Rubiaceae	Cephaeline, cinchonidine, cinchonine, emetine, quinidine, quinine
Reseda	Resedaceae	Caprylic acid, eugenol, phenol
Reverchonia[b]	Euphorbiaceae	
Rhagodia	Chenopodiaceae	Trimethylamine
Rhamnus[b]	Rhamnaceae	Crysarobin, dodecyl alcohol, quercitin, rutin
Rheum[b]	Polygonaceae	Oxalic acid, rutin, tannic acid
Rhinacanthus	Acanthaceae	Coumarin
Rhododendron[b]	Ericaceae	Hydroquinone, quercitrin, rutin
Rhus[b]	Anacardiaceae	Inositol, malic acid, nonanoic acid
Rhynchelytrum	Poaceae	Hydrocyanic acid
Ribes	Grossulariaceae	Citric acid, malic acid, methyl salicylate, pectin, phenol, quercitin, quercitrin, shikimic acid
Richardia	Rubiaceae	Emetine
Ricinus[b]	Euphorbiaceae	Formic acid, fumaric acid, heptanal, hydrocyanic acid, malic acid, palmitic acid, ricin, ricinoleic acid, saponin,
Rivea	Convolvulaceae	Rutin
Robinia[b]	Fabaceae	Chrysarobin, indican, indole, lysine, piperonal, sanguinarine
Roemeria	Papaveraceae	Biflorine, ephedrine, pseudoephedrine
Romneya	Papaveraceae	Biflorine, sanguinarine
Rosa	Rosaceae	Acetaldehyde, benzaldehyde, citral, citronellol, eugenol methyl ether, quercitin, quercitrin, quinic acid, saponin
Rosmarinus	Lamiaceae	Acetaldehyde, borneol, camphor, cineole, isovaleric acid, tannic acid

Table 3 (continued)
HIGHER PLANT GENERA AND THEIR TOXINS

Genus[a]	Family	Toxin
Roupellina	Apocynaceae	Milloside, paulioside
Roystonea	Arecaceae	Caprylic acid
Rubia	Rubiaceae	Shikimic acid
Rubus[b]	Rosaceae	Benzaldehyde, ethyl alcohol, hydroquinone, isoamyl alcohol, malic acid, methyl salicylate, shikimic acid, tannic acid
Rudbeckia[b]	Asteraceae	
Rudgea	Rubiaceae	Cotoin
Rumex[b]	Polygonaceae	Chrysarobin, malic acid, oxalic acid, tannic acid, tartaric acid
Ruta[b]	Rutaceae	Benzaldehyde, capric acid, caproic acid, caprylic acid, cineole, coumarin, cuminic aldehyde, decanoic acid, guaiacol, imperatorin, isovaleric acid, lauric acid, myristic acid, palmitic acid, phenol, rutin, salicylic acid, tannic acid, vanillin
Ryparosa	Flacourtiaceae	Methyl salicylate
Saccharum	Poaceae	Adenine, hydrocyanic acid, pantothenic acid, succinic acid
Salix	Salicaceae	Estriol, phenol, rutin, salicin, salicylic acid, shikimic acid
Salpiglossis	Solanaceae	Nornicotine
Salsola[b]	Chenopodiaceae	Histamine, hydrocyanic acid
Salvadora	Salvadoraceae	Lauric acid, myristic acid
Salvia[b]	Lamiaceae	Borneol, camphor, cineole, hexanol, histamine, limonene, linalool, saponin, tannic acid, thujone
Sambucus[b]	Caprifoliaceae	Ethylamine, hydrocyanic acid, malic acid, mandelonitrile, rutin, trigonelline, valeric acid
Sandersonia	Liliaceae	Colchicine
Sanguinaria[b]	Papaveraceae	Biflorine, sanguinarine
Sanicula	Apiaceae	Saponin
Sansevieria	Agavaceae	Saponin
Sapindus	Sapindaceae	Lactose, sanguinarine, saponin, shikimic acid
Saponaria[b]	Caryophyllaceae	Saponin
Sarcandra	Chloranthaceae	Shikimic acid
Sarcobatus[b]	Chenopodiaceae	
Sarcocapnos	Fumariaceae	Biflorine
Sarcolobus	Asclepiadaceae	Coniine
Sarcostemma	Asclepiadaceae	Hydrocyanic acid
Sarothamnus	Fabaceae	Sparteine
Sarracenia	Sarraceniaceae	Histamine, veratrine
Sartwellia[b]	Asteraceae	
Sassafras	Lauraceae	Camphor, capric acid, citral, decanoic acid, linalool, phellandrene, safrole, tannic acid
Satureja	Lamiaceae	Carvone, cymene, linalool, pulegone, thymol
Sauromatum	Araceae	Ammonia, dimethylamine, ethylamine
Saussurea	Asteraceae	Ionone, napthalene
Saxifraga	Saxifragaceae	Shikimic acid
Scabiosa	Dipsacaceae	Sanguinarine
Schima	Theaceae	Shikimic acid
Schinopsis	Anacardiaceae	Tannic acid
Schinus	Anacardiaceae	Phellandrene, safrole, saponin
Schisandra	Schisandraceae	Borneol, camphor, cineole, cymene, limonene
Schizopepon	Cucurbitaceae	Maltose
Schoenocaulon	Liliaceae	Cevadine, cevine, veratrine
Scilla[b]	Liliaceae	
Scirpus[b]	Cyperaceae	
Sclerocarya	Anacardiaceae	Citric acid, malic acid, tannic acid
Scolopia	Flacourtiaceae	Methyl salicylate
Scopolia[b]	Solanaceae	Atropine, betaine, choline, hyoscyamine, scopolamine
Scorzonera	Asteraceae	Trigonelline
Secale[b]	Poaceae	Ferulic acid
Sedum[b]	Crassulaceae	Citric acid, malic acid, pelletierine, succinic acid
Sempervivum	Crassulaceae	Malic acid, nicotine
Senecio[b]	Asteraceae	Carthamoidine, cytisine, fumaric acid, isatidene, jacobine, longilobine, retronecine, retrorsine, riddlelline, rutin, sceleratine, senecionine, seneciphylline
Sesbania[b]	Fabaceae	Saponin

Table 3 (continued)
HIGHER PLANT GENERA AND THEIR TOXINS

Genus[a]	Family	Toxin
Seseli	Apiaceae	Isobutyric acid, limonene
Setaria[b]	Poaceae	Hydrocyanic acid
Shorea	Dipterocarpaceae	Quercitin
Sida	Malvaceae	Ephedrine
Sideroxylon	Sapotaceae	Methyl salicylate
Siler	Apiaceae	Limonene
Silybum[b]	Asteraceae	Fumaric acid, histamine, hydrocyanic acid, tyramine
Simira	Rubiaceae	Harman, harmine
Simmondsia	Simmondsiaceae	Decyl alcohol
Sinapis[b]	Brassicaceae	
Sium[b]	Apiaceae	
Skimmia	Rutaceae	Phellandrene
Smilax	Smilacaceae	Hexadecanol
Solandra[b]	Solanaceae	Atropine, hyoscyamine
Solanum[b]	Solanaceae	Acetylcholine, atropine, cholesterol, hydrocyanic acid, hyoscyamine, hypoxanthine, isoamyl alcohol, malic acid, narcotine, nicotine, nitrosodimethylamine, noradrenaline, oxalic acid, ricinoleic acid, rutin, serotonine, solanine, sparteine, trigonelline
Solidago[b]	Asteraceae	Limonene, quercitin, quercitrin, rutin, saponin
Sonchus[b]	Asteraceae	
Sophora[b]	Fabaceae	Anabasine, cytisine, malonic acid, pachycarpine, rutin, saponin
Sorbaria	Rosaceae	Hydrocyanic acid, trimethylamine
Sorbus	Rosaceae	Acetaldehyde, malic acid, parasorbic acid, sorbic acid, trimethylamine
Sorghum[b]	Poaceae	Anhaline, hexadecanol, hydrocyanic acid, oxalic acid
Spartium	Fabaceae	Caprylic acid, cytisine, saponin, sparteine
Spathiphyllum	Araceae	Trimethylamine
Spatholobus	Fabaceae	Rotenone
Spergularia	Caryophyllaceae	Saponin
Spinacia[b]	Chenopodiaceae	Citric acid, histamine, oxalic acid
Spiraea	Rosaceae	Hydrocyanic acid, piperonal, salicin, salicylic acid, trimethylamine
Sporobolus	Poaceae	Ferulic acid, hydrocyanic acid
Stachys[b]	Lamiaceae	Trigonelline
Stanleya	Brassicaceae	Selenium
Stapelia	Asclepiadaceae	Methylamine
Staphylea	Staphyleaceae	Methylamine
Stellaria[b]	Caryophyllaceae	
Stemmadenia	Apocynaceae	Ajmalicine
Stemona	Stemonaceae	Malonic acid
Stephania	Menispermaceae	Berbamine
Stephanorossia	Apiaceae	Hydrocyanic acid
Sterculia	Sterculiaceae	Caffeine, hydrocyanic acid, palmitic acid, theobromine
Stewartia	Theaceae	Quercitrin
Stifftia	Asteraceae	Methyl salicylate
Stillingia[b]	Euphorbiaceae	Acetophenone
Stipa[b]	Poaceae	
Strelitzia[b]	Strelitziaceae	
Strophanthus	Apocynaceae	Amboside, choline, cymarin, cymarol, divaricoside, leptoside, ouabain, strophanthidin, strophanthin, trigonelline, saponin, sarmentoside, sarmutoside
Strychnos	Loganiaceae	Brucine, cinchonidine, curare, curarine, narcotine, quinidine, thebaine
Stryphnodendron	Mimosaceae	Saponin
Stylophorum	Papaveraceae	Biflorine, chelidonin, sanguinarine
Styphelia	Epacridaceae	Methyl salicylate
Styrax	Styracaceae	Vanillin
Suckleya[b]	Chenopodiaceae	Hydrocyanic acid
Suriana	Surianaceae	Rutin
Swartzia	Fabaceae	Saponin
Symphytum	Boraginaceae	Cyanocobalamin, lasiocarpine, tannic acid
Symplocarpus[b]	Araceae	
Symplocos	Symplocaceae	Harman, methyl salicylate
Synadenium	Euphorbiaceae	Hydrocyanic acid
Syrenia	Brassicaceae	Strophanthidin

Table 3 (continued)
HIGHER PLANT GENERA AND THEIR TOXINS

Genus[a]	Family	Toxin
Syzygium	Myrtaceae	Eugenol, furfural, furfuryl alcohol, methanol, methyl salicylate, napthalene, tannic acid, valeraldehyde, vanillin
Tagetes	Asteraceae	Hydrocyanic acid
Tamarindus	Caesalpiniaceae	Citric acid, hydrocyanic acid, malic acid, pectin, tartaric acid
Tamus	Dioscoreaceae	Cholesterol
Tanacetum[b]	Asteraceae	Camphor, citric acid, tannic acid, thujone
Tanghinia	Apocynaceae	Cerberin, tanghiniferin, tanghinin
Taraxacum	Asteraceae	Choline, pectin, trimethylamine
Taxus[b]	Taxaceae	Ephedrine, pectin, pseudoephedrine
Tecoma	Bignoniaceae	Coumarin, lapachol
Tectona	Verbenaceae	Lapachol
Telfairia	Cucurbitaceae	Cucurbitacin E, saponin
Templetonia	Fabaceae	Cytisine
Tephrosia	Fabaceae	Hydrocyanic acid, rotenone, rutin
Terminalia	Combretaceae	Hydrocyanic acid, saponin, shikimic acid, tannic acid
Ternstroemia	Theaceae	Shikimic acid
Tetradymia[b]	Asteraceae	
Tetragonia[b]	Tetragoniaceae	Saponin
Thalictrum	Ranunculaceae	Hydrocyanic acid, methylamine, shikimic acid
Thelypodium[b]	Brassicaceae	
Themeda	Poaceae	Hydrocyanic acid
Theobroma	Sterculiaceae	Caffeine, glycerol, isovaleric acid, theobromine
Thermopsis[b]	Fabaceae	Cytisine, malonic acid, pachycarpine
Thesium	Santalaceae	Hydrocyanic acid
Thespesia	Malvaceae	Quercitin
Thevetia[b]	Apocynaceae	Cerberin, thevetin
Thlaspi[b]	Brassicaceae	Allyl isothiocyanate
Thuja	Cupressaceae	Fenchone, quercitrin, thujone
Thunbergia	Acanthaceae	Methyl salicylate
Thymus	Lamiaceae	Borneol, carvacrol, citral, geraniol, linalool, menthol, saponin, thymol, undecanoic acid
Tilia	Tiliaceae	Quercitrin
Tinomiscium	Menispermaceae	Picrotoxin
Tithonia	Asteraceae	Hydrocyanic acid
Tocoyena	Rubiaceae	Cephaeline, emetine
Tofieldia	Liliaceae	Colchicine
Tonduzia	Apocynaceae	Ajmalicine, deserpidine, rescinnamine, reserpine
Toxicodendron[b]	Anacardiaceae	
Tribulus[b]	Zygophyllaceae	Harman, harmine, hydrocyanic acid
Trichocereus	Cactaceae	Anhaline, caffeine, mescaline
Trichodesma	Boraginaceae	Retronecine
Trichosanthes[b]	Cucurbitaceae	Oxalic acid
Trifolium[b]	Fabaceae	Histamine, hydrocyanic acid, malonic acid, maltose, quercitin, saponin
Triglochin[b]	Juncaginaceae	Hydrocyanic acid
Trigonella	Fabaceae	Choline, malonic acid, trigonelline
Trillium	Trilliaceae	Saponin
Triticum[b]	Poaceae	Ergosterol, ferulic acid, formic acid, malonic acid, oxalic acid
Tropaeolum	Tropaeolaceae	Benzyl cyanide, maltose
Tulipa	Liliaceae	Colchicine, rutin
Turbina	Convolvulaceae	Rutin
Turnera	Turneraceae	Caffeine
Turpinia	Staphyleaceae	Methyl salicylate
Tussilago	Asteraceae	Retronecine, rutin, senecionine, tannic acid
Tylophora	Asclepiadaceae	Tylocrebrine
Typha[b]	Typhaceae	Maltose
Ulex	Fabaceae	
Ulmus	Ulmaceae	Capric acid, caprylic acid, decanoic acid
Umbellularia	Lauraceae	Lauric acid
Umbilicus	Crassulaceae	Maltose
Urera	Urticaceae	Histamine
Urginea[b]	Liliaceae	Caffeine, hydrocyanic acid, scillaren, scillarenin, scilliroside

Table 3 (continued)
HIGHER PLANT GENERA AND THEIR TOXINS

Genus[a]	Family	Toxin
Urtica	Urticaceae	Acetophenone, acetylcholine, formic acid, histamine, nicotine, serotonine
Utricularia	Lentibulariaceae	Scopolamine
Vaccinium	Ericaceae	Benzoic acid, hydroquinone, malic acid, quercitin, quercitrin, quinic acid, saponin
Valeriana[b]	Valerianaceae	Borneol, choline, formic acid, isovaleric acid, limonene, valeric acid
Vallesia	Apocynaceae	Reserpine
Vangueria	Rubiaceae	Saponin
Vanilla	Orchidaceae	Anisaldehyde, piperonal
Veratrum[b]	Liliaceae	Cevadin, cevine, colchicine, cyclopamine, germerine, jervine, methylamine, neogermitrine, rubijervine, veratridine, veriloid
Verbascum	Scrophulariaceae	Coumarin
Verbesina[b]	Asteraceae	
Veronia[b]	Asteraceae	Hydrocyanic acid, methyl salicylate
Veronicastrum	Scrophulariaceae	Tannic acid
Vetiveria	Poaceae	Palmitic acid
Viburnum	Caprifoliaceae	Inositol, isovaleric acid, methylamine, methyl salicylate, salicin, saponin, shikimic acid, trimethylamine, valeric acid
Vicia[b]	Fabaceae	Choline, guanidine, hydrocyanic acid, lysine, physostigmine, quercitrin, shikimic acid, xanthine
Vigna	Fabaceae	Oxalic acid
Viguiera[b]	Asteraceae	
Villaresia	Celastraceae	Caffeine
Vinca	Apocynaceae	Reserpine, rutin, serpentine
Viola	Violaceae	Benzyl alcohol, heptanoic acid, hydrocyanic acid, yohimbine
Virola	Myristicaceae	Bufotenine, myristic acid
Viscum[b]	Loranthaceae	Histamine, inositol, saponin, tyramine
Vitex	Verbenaceae	Hydrocyanic acid, methanol, shikimic acid
Vitis	Vitaceae	Citric acid, coumarin, malic acid, methyl salicylate, oxalic acid, quercitin, quercitrin, ricinoleic acid, saponin, shikimic acid, succinic acid, tannic acid, tartaric acid
Voacanga	Apocynaceae	Reserpine
Weigela	Caprifoliaceae	Shikimic acid
Wislizenia[b]	Capparaceae	
Wisteria[b]	Fabaceae	
Withania	Solanaceae	Choline, nicotine, pelletierine
Wrightia	Apocynaceae	Indican, ricinoleic acid
Xanthium[b]	Asteraceae	Hydrocyanic acid, hydroquinone
Xanthocephalum[b]	Asteraceae	
Xanthophyllum	Xanthophyllaceae	Methyl salicylate
Xanthorrhoea	Xanthorrhoeaceae	Benzoic acid, cinnamyl alcohol, citronellol, styrene
Xanthosoma[b]	Araceae	
Ximenia	Olacaceae	Hydrocyanic acid
Xysmalobium	Asclepiadaceae	Anhydroperiplogenone
Zamia[b]	Zamiaceae	
Zanthoxylum	Rutaceae	Cuminic aldehyde, pentadecanoic acid
Zea[b]	Poaceae	Anhaline, carvacrol, furfural, guanidine, hydrocyanic acid, inositol, oxalic acid, quercitin
Zelkova	Ulmaceae	Capric acid, caprylic acid, decanoic acid
Zephyranthes[b]	Amaryllidaceae	
Zieria	Rutaceae	Safrole
Zigadenus[b]	Liliaceae	Jervine, neogermitrine
Zingiber	Zingiberaceae	Borneol, cineole, citral, geraniol, linalool, phellandrene
Zinnia	Asteraceae	Anabasine, niacin, nornicotine
Ziziphus	Rhamnaceae	Saponin
Zygophyllum	Zygophyllaceae	Harmine, hydrocyanic acid

[a] A generic entry not followed by the name of a toxin indicates that the genus contains toxic species but that the toxin is not yet identified.

[b] Indexed by Kingsbury, J. M., Poisonous Plants of the United States and Canada, Prentice-Hall, Englewood Cliffs, N.J., 1974. With permission.

Table 4
PHARMACOLOGICALLY ACTIVE PHYTOCHEMICALS

Abrine: antiinflammatory, antiophthalmic
Acacetin: antiinflammatory
Acerin: phagicide, viricide
Acer saponin P: antitumor
Acer saponin Q: antitumor
3β-Acetoxynorerythrosuamine: antitumor
2'-Acetylglaucarubinone: antitumor
Achyranthine: lower blood pressure, heart rate, dilate blood vessels in animal, diuretic, purgative
Acnistin: antitumor
Acobioside A: antitumor
Acofrioside: antitumor
Acolongifloroside G: antitumor
Acolongifloroside H: antitumor
Acolongifloroside K: antitumor
Aconine: febrifuge, gastric anesthetic
Aconitine: cardiotoxic, febrifuge, gastric anesthetic
Acoric acid: anticonvulsant
Acoschimperoside P: antitumor
Acoschimperoside Q: antitumor
Acospectoside A: antitumor
Acovenoside A: antitumor
Acovenoside B: antitumor
Acridone: antitumor
Acronycine: antitumor
Actein: hypotensive
Adenine: antiviral, granulocytopenia
Adynerin: antitumor
Aescin: antiinflammatory, antitumor
Affinin: insecticide
Agrimol: antimalarial
Agrimorphol: taenicide
Agropyrene: germicide
Ailanthinone: antitumor
Ailanthone: antitumor
Ajmalicine: antidiuretic, CNS-depressant
Ajmalinine: hypotensive
Alantolactone: bactericide, hypotensive, hyperglycemic, vermifuge, antitumor
Albizziagenin: anthelmintic, expectorant
Alginine: like cocaine
Alkannin: antitumor
Alkannin β,β-dimethylacrylate: antitumor
Alkannin monacetate: antitumor
Allamandicin: antitumor
Allamandin: antitumor, protisticidal
Allantoin: antiinflammatory, suppurative
Allicin: antitumor, bactericide, fungicide, germicide, hypoglycemic, hypocholesterolemic, insecticide, larvicide
Allisatin: antiinflammatory, bactericide, fungicide
Allocryptopine: antiarrhythymic, analgesic, soporific
Allyl propyl disulphide (APDS): hypoglycemic
Allyl isothiocyanate: counterirritant
Aloe-emodin: anticancer, antiseptic
Aloin: laxative
Alpha-angelicalactone: antitumor

Alphadichroine: antimalarial
Alvaxanthone: larvicide, piscicide
Amabain: cardiotonic, diuretic
Ambrosin: antitumor
Amentoflavon: antibradykinin
Americanin: hepatotropic
Aminophylline: antianginal
Amygdalin: antiinflammatory
α-Amyrin: antitumor
Anacardic acid: antitumor, bactericide, molluscicide, nematicide
Anacardol: antitumor
Anagyrine: antiedemic, antiarrhythmic, cathartic, diuretic, oxytocic
Andrographolide: bactericide
Anemonin: antitumor, bactericide
Anethole: carminative, expectorant, gastric stimulant, insecticide, lactagogue
Angelic acid: sedative
16-Anhydrogitoxigenin: carcinostatic
Anisaldehyde: insecticide
Anisic acid: antiseptic, antirheumatic
Anisodamine: antispasmodic, anticholinergic
Anisodine: analgesic, antimigraine, antispasmodic
Anopterine: antitumor
Anthorhododendrin: antibronchitic
Apigenin: antispasmodic
Apiole: antipyretic, diuretic, emmenagogue, insecticide
Apoatropine: antispasmodic
Apocannoside: antitumor, cardiac
Apocyanin: cholagogue
Apocynin: cardiac
Arborinine: antihistamic, antiinflammatory, ecbolic, spasmolytic
Arbutin: antibacterial, diuretic
Arctiopicrine: antitumor
Ardisinol: antitubercular
Arecoline: anthelmintic, cholinergic, mitotic, parasympathomimetic
Arginine: diuretic
Aristolochic acid: antiseptic, antitumor, bactericide
Armillarisan A: anticholecystitic
Aromaticin: antitumor
Artarine: antimalarial, bactericide, trypanocide
Arteannuin: antimalarial
Artemisiifolin: antitumor
Artemisinine: antimalarial
Arternol: hypertensive, vasoconstrictor
Asarinin: antitubercular
Asarone: anticonvulsant, CNS-depressant, hypothermic, psychoactive
Ascaricidal: anthelmintic
Ascaridiole: anthelmintic: but neoplastic
Asclepin: cardiotonic, digitalic
Asiaticoside: antifertility, antitubercular, antilepric, carcinogenic

Table 4 (continued)
PHARMACOLOGICALLY ACTIVE PHYTOCHEMICALS

Asparagusic acid: nematicide

Asperillin: antitumor

Asperuloside: antiinflammatory

Astragalan: immunostimulant

Athonin: bactericide

Atractylodin: bactericide, fungicide

Atropine: analgesic, anthydrotic, anticholinergic, antispasmodic, antisialogogue, bronchodilator, mydriatic, psychoactive

Aucubin: antibacterial

Autumnolide: antitumor

Azadirachtin: antifeedant

Azulene: antiinflammatory, antipyretic

Baccharin: antitumor

Baicalin: diuretic

Baileyin: antitumor

Barbaloin: antitubercular

Bebeerine: antimalarial

Benzaldehyde: anesthetic, antipeptic, antispasmodic, antitumor

Benzoic acid: antifungal, antipyretic, antiseptic, expectorant, vulnerary

Benzoin: anesthetic, antiseptic

Benzoylcatalpol: antihepatotoxic

Benzyl benzoate: scabicide, pediculicide

Benzyl isothiocyanate: antitumor, bactericide, fungicide

Berbamine: hypotensive, leukocytotic

Berberine: amebicide, antibacterial, anticonvulsant, antidiarrheal, antiinflammatory, astringent, candidicide, carminative, collyrium, febrifuge, fungicide, hemostatic, herbicide, immunostimulant, sedative, trypanocidal, trypanosomicide, uterotonic

Berberine sulfate: antitumor

Bergapten: antipsoriac, antiinflammatory, antihistamine, spasmolytic

Bergenin: antitussive

Bersaldegenin 3-acetate: antitumor

Bersaldegenin 1,3,5-orthoacetate: antitumor

Bersamagenin 1,3,5-orthoacetate: antitumor

Berscillogenin: antitumor

Bersenogenin: antitumor

Betaine: antimyoatrophic (in combination with glycocyamine)

Betulin: antitumor

Betulinic acid: antitumor

Bhilawanol: antitumor

Bicuculline: convulsant

Biflorine: antibiotic, fungicide

Bilobalide: antibacterial

Biochanin A: estrogenic

α-Bisabolol: antiinflammatory, antipeptic, antiphlogistic, antiseptic, antitubercular, spasmolytic

Bisabolol: antiinflammatory, antiulcer, bactericide, fungicide

Bisaboloxide: spasmolytic

Biscatechin: antitumor

4,8″-Biscatechin: antitumor

Bishydroxycoumarin: anticoagulant

Boldine: diuretic

Bouvardin: antineoplastic

Bromelain: antiedemic, antiinflammatory, nematicide, proteolytic

Bruceantarin: antitumor

Bruceantin: antitumor

Bruceantinol: antitumor

Bruceine: antitumor

Bruceine B: antitumor

Brucine: circulatory stimulant

Bryophylline: antibacterial, antiseptic

Bufotalidin acetate: antitumor

Bulbocapnine: antidotal, cardiotonic, cataleptic, hypotensive

Burseran: antitumor

Cactine: digitalic

Caffeic acid: antitumor, choleretic, hepatotropic

Caffeine: cardiac stimulant, CNS-stimulant, respiratory stimulant, viricide

Cajanone: antiseptic

Calabarine: spinal-stimulant

Calophyllolide: antiinflammatory

Calotropin: antitumor, digitalic

Camphor: anesthetic, antipruritic, antiseptic, carminative, stimulant

Camptothecin: antileukemic

Canavanine: bactericide, fungicide

Canescine: hypotensive, tranquilizer

Cannabidiol: antibiotic

Cannabidiolic acid: antibiotic

Cannabigerol: antibiotic

Cannabigerolic acid: antibiotic

Capillin: antifungal

Capsaicin: diaphoretic, sialogogue

Carnosine: antiinflammatory

Carpaine: cardiodepressant, CNS depressant

Carvacrol: anthelmintic, antiseptic, fungicide, tracheal relaxant

Carveol: CNS stimulant

Carvone: CNS stimulant, carminative, insecticide

Casimiroedine: antitumor, carcinostatic

Cassaidine: antiseptic, cardiotonic

Cassaine: antiseptic, cardiotonic, convulsant

Cassamine: antiseptic, cardiotonic

Catechin: anticariogenic, antihepatotoxic, antioxidant, antiulcer

Catharanthine: diuretic, hypoglycemic

Cathine: anorexic

Cathinone: analgesic, anorexic, antinoriceptive

Caulosaponin: oxytocic

Cecropin: cardiotonic, diuretic

Celsioside C: antitumor

Centaureidin: antitumor

Cephaline: antiprotozoic

Cepharanthine: antihepatitic, carcinostatic, radioprotective

Cerberin: antitumor

Table 4 (continued)
PHARMACOLOGICALLY ACTIVE PHYTOCHEMICALS

Cesalin: antitumor

Chaksin: antiinflammatory

Chalepensin: molluscicide

Chamazulene: anodyne, antiinflammatory, antiphlogistic, antispasmodic, germicide, vulnerary

Chamic acid: antibiotic, fungicide

Chamillin: spasmolytic

Chamissonin: antitumor

Chandrine: antiarrhythmic

Chaparrinone: antitumor

Chapuzina: CNS-sedative

Charantin: hypoglycemic

Chaulmoogric acid: antileprous

Cheirolin: bactericide, fungicide

Chelerythrine: antitumor, fungistat

Chelidimerine: antitumor

Chelidonine: smooth muscle depressant

Chimaphilin: urinary antiseptic, bacteriostatic

Chlorogenic acid: choleretic

Chlorogenin: antiinflammatory

Chlorophorin: fungicide

Choline: lipotropic

Chrysarobin: antifungal, keratolytic, psoriasis

Chrysophanic acid: fungicide

Chrysophanol: antimicrobial, cathartic

Chymopapain: proteolytic

Cineole: bronchitis, expectorant, insectifuge, laryngitis, pharyngitis, rhinitis

Cincassiol D4: antiallergenic

Cinnamaldehyde: antipyretic, insecticide, sedative

Cinnamic acid: anesthetic, anthelmintic

Cinnzelanin: insecticide

Cinnzelanol: insecticide

Cissampareine: antitumor

Citrantin: antifertility

Citriodorol: antibiotic

Citrulline: diuretic

Cocaine: anesthetic, bactericide

Cocsulinine: antitumor

Codeine: anesthetic, antitussive

Colchicine: antiarthritic, anticancer, antiinflammatory, gout

Colchicoside: antiinflammatory

Colocynthin: purgative

Colubrinol: antitumor

Colubrinol acetate: antitumor

Columbamine: bactericide, CNS-depressant, hypotensive

Conessine hydrochloride: antitumor

Conoduramine: antineoplastic

Conodurine: antineoplastic

Convallotoxin: digitalic

Coptisine: antiinflammatory

Coptisine chloride: antitumor

Coriatin: antischizophrenic

Cornerine: cardiac

Coroglaucigenin: antitumor

Coronopilin: antitumor

Corydine: anticancer, CNS depressant

Corynantheidine: hypotensive, sympatholytic

Corynantheine: sympatholytic

Corynanthine: aphrodisiac, sympatholytic

Corytuberine: analeptic, anticancer

Costunolide: antitumor

Coumarin: antitumor, antiinflammatory, hypoglycemic

Coumestrol: estrogenic

Creosol: expectorant

Creosote: expectorant

Crepin: bactericide

Cresol: antiseptic

Crinamine: antitumor

Crocin: choleretic

Crotepoxide: antitumor

Cryogenine: anticholinergic, antiinflammatory, antispasmodic, tranquilizer, vasodilator

Cryptoaescin: antiinflammatory

Cryptograndoside: cardiac

Cryptoleptine: antimicrobial, antipyretic, hypotensive

Cryptomeridiol: antispasmodic

Cryptopine: uterotonic

Cryptopleurine: antitumor

Cryptowolline iodide: antitumor

Cubebin: urinary stimulant

Cucurbitacins: antihepatotoxic, antitumor, cathartic

Cucurbitacin glycoside: antitumor

Cucurbitin: antischistosomiac

Cucurbitocitrin: hypotensive

Cumene: narcotic

Cupressuflavone: antibradykinin

Curcumin: antiedemic, antiinflammatory, bactericide, cholagogue, fungicide

Curcumol: antitumor

Curdione: antitumor

Curine: myorelaxant

L-Curine: antitumor

Cusparine: antispasmodic

Cyclamin: antitumor

Cycleadrine: antitumor

Cycleanine: antiinflammatory

Cycleanine dimethobromide: hypotensive

Cycleanorine: antitumor

Cycleapeltine: antitumor

Cycloartanol: antiinflammatory

Cycloprotobuxine: antitumor

Cyclosadol: antiinflammatory

Cymarin: antitumor, myocardial stimulant

Cynarin: choleretic, antilipidermic

Cytisine: antiinflammatory, psychoactive

Daidzein: antispasmodic, estrogenic

Damasceine: antiinflammatory

Damsin: antitumor

Daphnetin: analgesic

Datiscacin (Cucurbitacin R): antitumor

Datiscoside (Cucurbitacin D dehydroepirhamnoside): antitumor

Daucosterol (β-sitosterol glucoside): antitumor

Table 4 (continued)
PHARMACOLOGICALLY ACTIVE PHYTOCHEMICALS

Dauricine: antiinflammatory, myorelaxant

Dauricinoline: myorelaxant

Dauricoline: myorelaxant

Daurinoline: myorelaxant

Daurisoline methyl bromide: myorelaxant

Deacetylconfertiflorin: antitumor

Deacetyleupaserrin: antitumor

Decarbomethoxytetrahydrosecamine: antiseptic

Dehydroailanthinone: antitumor

Dehydroailanthion: antitumor

Dehydroanhydropicropodophyllin: antitumor

Dehydrobruceantarin: antitumor

Dehydrobruceantin: antitumor

Dehydrobruceantol: antitumor

Dehydrocoryantheol: antiseptic

3-Dehydronobilin: antitumor

Delavaconitine: anesthetic

Demecolcine: antineoplastic

3'-Demethylpodophyllotoxin: antitumor

Deoxyharringtonine: antitumor

Deoxypodophyllotoxin: antitumor

Deserpidene: hypotensive, tranquilizer

Desglucomusennin: antitumor

Desglucouzarin: antitumor

Deslanoside: digitalic

Desmethoxyangonin: anesthetic, antipyretic, antiinflammatory, myorelaxant

5'-Desmethoxy-β-peltatin A methyl ether: antitumor

3-Desmethylcolchicine: antineoplastic

N-Desmethylthalidasine: antitumor

N-Desmethylthalistyline: antiseptic, hypotensive

Desmethyltylophorinine: antitumor

Desoxypodophyllotoxin: antitumor

Dhelwangine: bactericide, fungicide

Diallyl disulfide: germicidal, hypochloesterolemic,
hypoglycemic, insecticide, larvicide

Diallyl trisulfide: germicide, hypocholesterolemic, hypoglycemic, insecticide, larvicide

Dianethole: estrogenic

(−)-Dicentrine: antitumor

ℓ-Dicentrine: analgesic, sedative

Dichroine: schizonticide

Dicumarol: bactericide

Diginatin: digitalic

Digitoxin: antitumor, cardiac tonic, dropsy, stimulant

Diglucoacoshimperoside N: antitumor

Diglucoacoshimperoside P: antitumor

Digoxin: digitalic

Dihydrocucurbitacin B: antitumor

Dihydrohelenalin: analgesic, antibiotic,
antiinflammatory

Dihydrokawain: anesthetic, antiinflammatory, antipyretic, myorelaxant

Dihydromethysticin: anesthetic, antiinflammatory, antipyretic, myorelaxant

25,26-Dihydrophysalin C: antitumor

25,26-Dihydrophysalin D: antitumor

Dihydrovaltrate: anticancer, sedative

5,7-Dihydroxy-8-methoxy-2-methylcromone:
antitumor

3,3'-Diindolymethant: antitumor

Dillapiol: insecticide

Dimethoxyallylbenzene: sedative

(+)-Dimethylisolariciresinol-2α-xyloside: antitumor

Dimethyltryptamine: hallucinogenic

Dimethyltubocurarine: anticholinergic

Dioscorine: analeptic, CNS-stimulant

Diosgenin: antiinflammatory, estrogenic

Dipterin: vasopressor

Dircin: antileukemic

Ditetrahydropalmatine: analgesic, sedative

Dopa: antiparkinsonian

Dopamine: adrenergic, antihypotensive, anti-
lactagogue, antiparkinsonian

Dracorhodin: antiseptic

Dracorubin: antiseptic

Dregamine: antiinflammatory

Dulcitol: antitumor

Dumbcain: proteolytic

Ecdysterone: analgesic

Ecgonine: anesthetic

Edpetiline: antiinflammatory

Elephantin: antitumor

Elephantopin: antitumor

Ellagic acid: antitumor

Ellipticine: antineoplastic

Embelin: teniacide

Emetine: amebicide, anticancer, antiinflammatory, antiprotozoic, emetic, expectorant

Emodin: antimicrobial, antitumor, cathartic, purgative

Emodin (aloe-emodin): antitumor

En-yn-dicycloether: antispasmodic, antiinflammatory,
antianaphylactic, germicide

Ephedrine: adrenergic, antiepileptic, antitussive, bronchodilator, cardiostimulant, hypertensive, mydriatic,
vasopressor

3-Epiberscillogenin: antitumor

Epicatechin: antiinflammatory, hepatotropic

2-Epicucurbitacin B: antitumor

10-Epieupatoroxin: antitumor

Epigallocatechin: antioxidant

Epinephrine (adrenalin): adrenergic, mydriatic

Epipodophyllotoxin: carcinostatic

Epitulipinolide: antitumor

Eremantholide A: antitumor

Ergometrine: hemostat, uterotonic

Eriodictyol: expectorant

Eriofertopin: antitumor

Erioflorin: antitumor

Erioflorin acetate: antitumor

Erioflorin methacrylate: antitumor

Eriolangin: antitumor

Eriolanin: antitumor

Erucic acid: antitumor

Table 4 (continued)
PHARMACOLOGICALLY ACTIVE PHYTOCHEMICALS

Erythrophlamine: antiseptic, cardiotonic
Erythrophleine: antiseptic, cardiotonic, hemolytic, uterotonic, vasodilator
Esculin: antiinflammatory
Eseridine: purgative
Ethyl gallate: bactericide
Eucalyptol: bactericide, expectorant
Eugarzasadone: amebicide
Eugeniin: viricide
Eugenol: analgesic, antiseptic, anesthetic, fungicide, larvicide
Eugenol acetate: antispasmodic
Eupachlorin: antitumor
Eupachlorin acetate: antitumor
Eupachloroxin: antitumor
Eupacunin: antitumor
Eupacunolin: antitumor
Eupacunoxin: antitumor
Eupaformonin: antitumor
Euparotin: antitumor
Euparotin acetate: antitumor
Eupaserrin: antitumor
Eupatin: anticancer
Eupatocunin: antitumor
Eupatocunoxin: antitumor
Eupatofolin: antitumor
Eupatorietin: anticancer
Eupatoroxin: antitumor
Eupatundin: antitumor
Fabacein: antitumor
Fagarine: antiarrhythymic, ecbolic, spasmolytic
Fagaronine: antitumor
Falcarindiol: antibiotic
Farrerol: antitussive
Fastigilin B: antitumor
Fastigilin C: antitumor
Febrifugine: antimalarial
Ferulic acid: antitumor, choleretic, hepatotropic
Fetidine: antiinflammatory
Ficin: proteolytic, vermicide
Filicin: anthelmintic
Filmarone: taenifuge
Florilenalin: antitumor
Foetidin: hypoglycemic
Formononetin: estrogenic
N-Formyldesacetylcolchicine: antitumor
Fraxin: antiinflammatory
Fritilline: relax smooth muscles
Fritillarine: local anesthetic, hypertension
Fritimine: relax smooth muscle, lower blood pressure
Fulvine: antitumor
Fulvoplumierin: bactericide
Furfural: fungicide, germicide, insecticide
Furocoumarin: phototoxic
Gaillardilin: antitumor
Gaillardin: antitumor
Galanthamine: for myasthenia gravis

Galegine: hypoglycemic
Galipine: antispasmodic
Galipinine: antispasmodic
Gallic acid: antitumor, astringent, bacteristatic, styptic
Gambirinine: hypotensive
Gamma aminobutyric acid: hypotensive
Gammadichroine: antimalarial
Geiparvarin: antitumor
Gelsemine: CNS stimulant, analgesic
Geneserine: parasympathomimetic
Genistein: estrogenic
Genkwadaphnin: antileukemic
Gentianidine: antiinflammatory
Gentianine: antiinflammatory, antipsychotic
Gentiopicrin: antimalarial, larvicidal
Ginkgolide A: antibacterial
Ginkgolic acid: antituberculic
Ginsenin: antidiabetic
Gitogenin: antiinflammatory
Gitogenin galactoside: antitumor
Gitoxigenin: carcinostatic
Glabrol: bactericide
Glaucine: antitussive
Glaucarubinone: antitumor
Glaucarubolone: antitumor
Glaziovine: carcinostatic
Gleditschine: stupefacient
Glucotropaeoline: antibiotic
Glycollic acid: diuretic
Glyestrone: estrogenic
Glycyrrhetic acid: antibacterial, antiinflammatory, antitussive
Glycyrrhetinic acid: antiaddisonian, antiarthritic, antiinflammatory, antirheumatic
Glycyrrhizin: antiinflammatory
Gnidicin: antitumor
Gossypol: antitumor
Gramine: vasopressor
Graveolinine: ecbolic, spasmolytic
Guaiacol: antituberculic, expectorant
d-Guatambuine: antitumor
Guggulsterone: anticholesterolemic
Guttiferin: bactericide
Gymnemic acid: hypoglycemic
Harmine: antiparkinsonian, aphrodisiac, CNS-stimulant
Harmaline: antiparkinsonian, CNS-stimulant
Harmalol: CNS-stimulant
Harpagide: antiarthritic, antiinflammatory
Harpagoside: analgesic, antiinflammatory
Harringtonine: antitumor
Hecogenin glycoside: antitumor
Hederasaponin C: antitumor
Helenalin: antitumor
Helanalin acetate: antitumor
Heliamine: antitumor
Heliettin: anticonvulsant

Table 4 (continued)
PHARMACOLOGICALLY ACTIVE PHYTOCHEMICALS

Heliotrine: antitumor
Hellebrigenin 3-acetate (Bufotalidin acetate):
 antitumor
Hellebrigenin 3,5 diacetate: antitumor
Hellebrin: antitumor
Hemerocallin: antischistosomiac
Heraclenin: antiinflammatory
Hernandezine: antiseptic
Hesperidin: vasopressor
Himbacine: antispasmodic
Hirucin: fungicide
Hispaglabridin: bactericide
Hispidulin: antitumor
Holacanthone: antitumor
Holaphylline: antiinflammatory
Homoatropine: mydriatic
Homocarnosine: antiinflammatory
Homoharringtonine: antineoplastic
Homophleine: analgesic, antiseptic, cardiotonic,
 hemolytic
Honokiol: anticariogenic, bactericide
Humulone: germicide, spasmodic
Hydnocarpic acid: antileprous
Hydrastine: bactericide, hypotensive, sedative
Hydroquinone: antiseptic, astringent
Hydroxyacanthine: hypotensive
Hydroxyanisole: antitumor
16 β-Hydroxybersaldegenin 1-acetate: antitumor
16 β-Hydroxybersaldegenin 3-acetate: antitumor
16 β-Hydroxybersaldegenin 1,3,5-orthoacetate:
 antitumor
16 β-Hydroxybersamagenin 1,3,5-orthoacetate:
 antitumor
10-Hydroxycamptothecin: antitumor
3-Hydroxydamsin: antitumor
12-Hydroxydaphnetoxin: antitumor
9-Hydroxyparthenolide: antitumor
5-Hydroxy-7,8,3',4'-penta-methoxyflavone: antifungal
Hydroxyperezone monoangelate: laxative
Hydroxytoluene: antitumor
4-8-Hydroxywithanoline E: antineoplastic
Hymenoflorin: antitumor
Hyoscyamine: analgesic, anticholinergic, antispasmodic, antivinous, bronchodilator, mydriatic, psychoactive, sedative
Hypericin: antidepressant, tonic, tranquilizer
Hyperin: hypotensive
Hyperoside: antiinflammatory, diuretic, reduce capil-
 lary fragility, viricide
Hypoxysin: oxytocic
Hyrcanoside: antitumor
Ibogaine: antidepressant, aperitif, CNS-stimulant,
 digestive, hypotensive
Ibogaline: CNS-stimulant
Iboxygaine: CNS-stimulant
Imperatorin: anticonvulsant, antiinflammatory, for
 vitiligo

Imperialine: relax smooth muscle
Indicine-N-oxide: antineoplastic
Indole-3-acetonitrile: antitumor
Indole-3-carbinol: antitumor
Ingenol dibenzoate: antitumor
Inokosterone: analgesic
Inulin: antidiabetic
Ipolearoside: antitumor
Iridin: aperient, diuretic
Iridoid valepotriates: sedative
Irigenin: cholinergic
Irisquinone: antitumor
Isobaccharin: antitumor
Isobruceine B: antitumor
Isochondodendrine: antitumor
Isochondodedrine hydrochloride: analgesic
Isocucurbitacin B: antitumor
Isocucurbitacin B (2-Epicucurbitacin B): antitumor
Isodonal: antitumor, bactericide
Isogaillardin: antitumor
Isoharringtonine: antitumor
Isoliquiritin: antiinflammatory
Isomaniferin: antitussive, expectorant
Isopatrinene: sedative
Isoplumericine: antitumor
Isoquercitin: diuretic
Isotetrandrine: antiinflammatory, hypotensive
Isotomine: cardiac
Ivalin: antitumor
Ivasperin: antitumor
Jacaranone: antitumor
Jaligonic acid: antiinflammatory, diuretic
Jateorrhizine: antiinflammatory, CNS-stimulant,
 hypotensive
Jatropham: antitumor
Jatrophone: antitumor
Juglone: antitumor, fungicide, herbicide
Justicidin: antitumor
Kaempferitrin: antiinflammatory
Kaempferol: antiinflammatory, diuretic, natriuretic
Kaempferol 4'-0-glucoside: hypotensive
Kaempferol-7-glucoside: antiinflammatory
Kawain: anesthetic, antiinflammatory, antipyretic,
 myorelaxant
Karanjin: antifeedant
Khellin: anthelmintic, antianaphylactic, antiarterios-
 clerotic, antiasthmatic, antidiabetic, antispasmodic,
 antitussive, antiulcerogenic, bronchodilator,
 vasodilator
Kolanone: antimicrobial
Kuwanon G: hypotensive
Kuwanon H: hypotensive
Lagochin: sedative
Lanatoside: antiarrhythmic
Lanatoside A: antitumor
Lanatoside B: antitumor
Lanatoside C: antitumor

Table 4 (continued)
PHARMACOLOGICALLY ACTIVE PHYTOCHEMICALS

Lantanine: hypothermic
Lapachol: antitumor, bactericide, fungicide
Lasiocarpine: antitumor
Lasiodiplodin: antileukemic
Laudanine: convulsant
Laudanosine: convulsant
Lawsone: antitumor, antispasmodic, bactericide, fungicide
Leucoanthocyanadins: anticoagulant, antiinflammatory, cardiotonic, hypotensive, vasodilating
Leucodelphinidin: antiinflammatory
Leurosine: antineoplastic, hypoglycemic
Leurosine base: antitumor
Leurosine sulfate: antitumor
Liatrin: antitumor
Licurizid: antiinflammatory
Lignin: antitumor
Linalool: anticonvulsant, antimicrobial, spasmolytic
Lipiferolide: antitumor
Liquiritic acid: antiinflammatory
Liquiritin: antiinflammatory
Liquiritigenin: antiinflammatory
Liquiritone: antiinflammatory
Liriodenine: antitumor
Lithospermic acid: antigonadotrophic, antithyroid, cardiotonic
Lobelanine: emetic
Lobelanidine: emitic
Lobeline: antibronchitic, antiasthmatic, diuretic, expectorant, respiratory stimulant
Lupeol: antitumor
Lupinine: antiinflammatory, antiarrhythymic, cathartic, diuretic, oxytocic
Lupulin: germicide, spasmolytic
Lupulone: germicide, spasmolytic
Luteolin: antiinflammatory, antispasmodic, antitussive
Luteolin-7-glucoside:
Lycorine: anticancer, antiviral
Maakinine: pediculicide
Macin: proteolytic
Macluroxanthone: larvicide, piscicide
Madecassoid: antiinflammatory
Magnoflorine: hypotensive
Magnolol: bactericide, anticarigenic
Maltol: antioxidant
Mammein: pediculicide
Marrubiin: expectorant
Margaspidin: antiinflammatory
Matrine: antitumor
Matteucinol: antitussive
Maysenine: antitumor
Maysine: antitumor
Maytanacine: antitumor
Maytanbutine: antitumor
Maytanprine: antileukemic, antitumor
Maytansine: antineoplastic
Maytansinol: antitumor

Maytanvaline: antitumor
Medicagol: fungicide
Melampodinin A: antineoplastic
Menadione: coagulant
Menthol: anesthetic, antipruritic, counterirritant
Mescaline: psychoactive
Mesembrine: sedative
Methothalistyline: hypotensive
9-Methoxycamptothecin: antitumor
10-Methoxycamptothecin: antitumor
5-Methoxycanthin-6-one: antitumor
Methoxydihydronitidine: antitumor
9-Methoxyellipticine: antitumor
Methoxyharringtonine: antitumor
Methoxynapthoquinone: fungicide
Methuenine: anticholinergic
0-Methyl-atheroline: carcinostatic
4″-0-Methylcurine dimethochloride: myorelaxant
4÷-0-Methylcurine dimethochloride: myorelaxant
N-Methyldemecolcine: antitumor
Methyl eugenol: sedative
0-Methylfagaronine: antitumor
Methylhydroquinone: antibronchitic, antienteritic
Methyl salicylate: analgesic, antiinflammatory, antipyretic, antirheumatic
0-Methylthalibrine: antiseptic
0-Methylthalicberine: hypotensive
0-Methylthalmethine: antiseptic
4-Methylthiocanthin-6-one: antitumor
Methysticin: anesthetic, antiinflammatory, antipyretic, myorelaxant
Mexicanin I: antitumor
Mezerein: antitumor
Micromelin: antitumor
Mikanolide: antitumor
Mimosine: depilatory
Mirificin: estrogenic
Mitragynine: anesthetic
Molephantin: carcinostatic
Molephantinin: carcinostatic
Monocrotaline: antitumor
Montanic acid-monoglyceride: antitumor
Moracenin D: hypotensive
Morellin: bactericide
Moringinine: cardiotonic, hypertensive
Morphine: analgesic, euphoric, stimulant
Mucuanain: proteolytic
Mulberrofuran C: hypotensive
Multiradiatin: antitumor
Mycorinematairesinol monoglucoside: cathartic
Myricadiol: mineral corticoid
Myricitrin: bactericide, paramebicide, spermicide
Myricoside: antifeedant
Myristicin: diuretic, hallucinogen, insecticide
Myrsine saponin: antitumor
Myrtilin: hypoglycemic
Mucilage: demulcent

Table 4 (continued)
PHARMACOLOGICALLY ACTIVE PHYTOCHEMICALS

Narciclasin: viristatic
Narcotine: antitussive
Nardol diastereomer: antitumor
Naringin: antiinflammatory
Neoliquiritin: antiinflammatory
Neoisoliquiritin: antiinflammatory
Neopseudoephedrine: CNS-stimulant
Nepetalactone: insect repellant
Neriifolin: antitumor, digitalic
Nerin: cardiac
Nevadensin: antitubercular, antitussive, expectorant
Nicotine: insecticide
Nicotinic acid: hypoglycemic
Nigellone: antiasthmatic
Nimbin: spermicide
Nitidine chloride: antitumor
Nobiletin: antifungal, cholinergic
Nootkatin: fungicide
Norcassaidine: antiseptic, cardiotonic
Norcassamidine: antitumor
Nordihydroguaiaretic acid: antioxidant, bactericide
Norepinephrine: adrenergic
Norerythrostachamine: antitumor
Normaysine: antitumor
Normaytancyprine: antileukemic
Nornuciferine: antispasmodic
Norwedelolactone: estrogenic
Noscapine: antitussive, myorelaxant
Nuciferine: antispasmodic
Obamegin: antitumor
Oleandomycin: antibiotic
Oleandrigenin: carcinostatic
Oleandrigenin 3-rhamnoside: antitumor
Oleandrin: antitumor, cardiac
Oleanolic acid: antihepatitic
Oleuropein: hypotensive
Olivacine: antitumor
Opposide: antitumor
Oridonin: antitumor, bactericide
Oripavine: analgesic
Ortho-hydroxycinnamic acid: antitumor
Osajin: antioxidant
Ouabagenin: antitumor
Ouabain: antiarrhythymic, cardiotonic
Ovatifolin acetate: antitumor
Oxopurpureine: carcinostatic
Oxyacanthine: antitumor, bactericide
Oxymatrine: antiasthmatic, antitumor
Oxynitidine: antitumor
Oxysparteine: antiarrhythmic, cathartic, diuretic, oxytocic
Oxytylocrebrine: antitumor
Paeoniflorin: antiinflammatory, antispasmodic, antidiuretic, hypotensive
Paeonol: antiinflammatory
Palmatine: analgesic, antiinflammatory, CNS-depressant, hypotensive

Panacen: analgetic, tranquilizer
Panax saponin A
Panaxin: cardiotonic, cerebrotonic
Papain: antiedemic, antiinflammatory, proteolytic
Papaverine: antianginal, antiasthmatic, anticholinergic, antispasmodic, myorelaxant
Para-hydroxycinnamic acid: antitumor
Parasorbic acid: bactericide, protozoacide
Parillin: antitumor
Parthenin: adrenergic-blocker, antitumor, carcinostatic, CNS-depressant
Parthenolide: antitumor
Passiflorine: sedative
Patrine: sedative
Patuletin: antispasmodic
Paucin: antitumor
Pectins: antitussive, antihypercholesterolemic
Peimine: hypotensive
Peimisidine: hypotensive
Peimunine: like atropine, relaxes smooth muscles
Pelletierine: anthelmintic, antioxyuriasic
Peltatin: antitumor, purgative
α-Peltatin: antitumor
β-Peltatin: antitumor
β-Peltatin A methyl ether: antitumor
cis-1,8-Pentadecadiene: antitumor
1-Pentadecene: antitumor
1,2,3,4,6-Pentagalloylglucose: viricide
Pentamethyl quercetin: cholinergic
Peregrinine: hemostat, uterotonic
Perezone: laxative
Periplocin: cardiac
Periplocymarin
Perivine: antineoplastic
Peruvoside: digitalic
Petiline: antiinflammatory
Phantomolin: carcinostatic
Phaseolin: fungicide
Phenethyl isothiocyanate: antitumor
Phenylheptatriyne: bactericide, fungicide
Phloretin: bactericide
Phoratoxin: hypotensive
Phorbol 12-tiglate 13-decanoate: antitumor
Phyllanthoside: antineoplastic
Physalin B: antitumor
Physalin D: antitumor
Physcion: antimicrobial, cathartic
Physostigmine: antiglaucomic, myotic, spinal sedative
Physovenine: myotic
Phytonadione: prothrombogenic
Picrotoxin: anticonvulsant, antidote, antiepileptic, stimulant
Pilocarpine: antiglaucoma, cholinergic, diaphoretic, myotic, sialogogue
Pilocereine: antitumor
Pinene: expectorant
Pinguinain: parasiticide, proteolytic

Table 4 (continued)
PHARMACOLOGICALLY ACTIVE PHYTOCHEMICALS

Pinnatifidin: antitumor
Pinosylvine: bactericide, fungicide
Piperine: analeptic, bactericide, insecticide
Piperitone: antiasthmatic
Platycodin: analgesic, antitussive, expectorant
Plenolin: antitumor
Plumbagin: bactericide, fungicide, uterotonic, vesicant
Plumericine: antitumor, fungicide
Podolide: antitumor
Podophyllic acid: antimitotic
Podophyllin: cholagogue, emetic, purgative
Podophyllotoxin: anticancer, antimitotic
Podophyllotoxin glucoside: antitumor
Polygodiol: antibiotic, candicide
Pomiferin: antioxidant
Ponicidine: antitumor
Primulic acid: mollusicide
Pristimerin: bactericide
16-Propionylgitoxigenin: carcinostatic
Proscillaridin A: antitumor
Protoanemonine: antimicrobial, bactericide
Protopine: analgesic, convulsant, hypotensive, sedative
Protoveratrine: hypotensive
Provincialin: antitumor
Proximol: antilithic, antispasmodic, diuretic
Pseudoephedrine: antiasthmatic, decongestant
Pseudoivalin: antitumor
Pseudolycorine: antileukemic
Psilostachyin A: carcinostatic
Psorospermin: antileukemic
Pterigospermine: bactericide
Pterygospermin: bactericide, fungicide
Pulchellin: antitumor
Pulchellin E: antitumor
Pyrocatechol: antiseptic
Quassimarin: antileukemic
Quassin: antiamebic
Quercitin: antiinflammatory, antispasmodic
Quercitrin: antispasmodic, diuretic, vasopressor, viricide
Quercitroside: choleretic
Quinghaosu: antimalarial
Quinine: antimalarial, antipyretic, cardiodepressant, oxytocic
Quinidine: antiarrhythymic, cardiodepressant, oxytocic
Quinovic acid: carcinostatic
Radiatin: antitumor
Raphanin: antibacterial, antifungal
Rapine: fungicide
Rauwolscine: hypotensive
Rescinnamine: antihypertensive, sedative, tranquilizer
Reserpine: antiadrinergic, anticonvulsant, antitumor, hypotensive, sedative, tranquilizer
3-Rhamnoside: carcinostatic
Rhein: antimicrobial, antitumor, cathartic, proteinase-inhibitor

Rhodexin B: antitumor
Rhomotoxin: antihypertensive, antitachycardia
Robinin: diuretic
Robustanol: antimalarial
Rorifon: antitussive, expectorant
Rosmanol: antioxidant
Rotenone: antitumor, insecticide, piscicide
Rumicin: discutient, parasiticidal, rubefacient
Ruscogenin: antiinflammatory
Rutin: antiatherogenic, antiedemic, antiinflammatory, antithrombogenic, hypotensive, spasmolytic, vasopressor
Rutoside: choleretic
Safrole: anesthetic, antiseptic
Saikogenin: antiedematic
Saikosaponin: antihepatotoxic
Saikosides: analgesic, antiinflammatory, antihistaminic, antipyretic, antitussive, antiulcer, cardiotonic, hypotensive
Sakuranetin: antiseptic
Salicin: antipyretic, analgesic, antirheumatic
Salicyclic acid: antipyretic, analgesic, antirheumatic
Saligenin: analgesic
Sanguidimerine: antitumor
Sanguinarine: anesthetic, antiseptic, antitumor, fungicide, sialogogue
Santonin: anthelminic
Saponaria saponin: antitumor
Sarmentosine: antihepatotoxic
Schizandrin: antihepatotoxic
Schizandrol: antihepatotoxic
Scillaren A: antitumor
Scillarenin: antitumor
Scilliglaucosidin: antitumor
Scilliroside: antitumor, antiseborrheic
Scillitin: diuretic, expectorant
Scolymoside: choleretic
Scoparoside: diuretic
Scopolamine: analgesic, anticholinergic, antiinflammatory, antispasmodic, bronchodilator, psychoactive, sedative
Scopoletin: antispasmodic
Selovicin: bactericide
6 α-Senecioloxychaparrinone: antitumor
Senecionine: antitumor
Senecionine-N-oxide: antitumor
Sennoside: laxative
Serotonin: oxytocic
Sesbanine: antileukemic
Shikonin: antibacterial
Silymarin: antihepatotoxic
Simmondsin: appetite depressant
Sinicuichine: tranquilizer
Sinomenine: analgesic, antiinflammatory, sedative
β-Sitosterol: antihypercholesterolemic, antiprostatic, antiprostatadenomic, estrogenic

Table 4 (continued)
PHARMACOLOGICALLY ACTIVE PHYTOCHEMICALS

β-Sitosterol-D-glycoside: hypoglycemic
β-Sitoterol glucoside: antitumor
Skimmianine: ecbolic, spasmolytic
Solamargine: antifungal, molluscicide
β-Solamarine: antitumor
Solanine: antiasthmatic, antibronchitic, antiepileptic
Solanocapsine: antibacterial
Solapalmatenine: antitumor
Solapalmatine: antitumor
Solaplumbin: antitumor
Solasodine: antiandrogenic, antiinflammatory,
  antispermatogenic
Solasodine base: antitumor
Solasodine hydrochloride: antitumor
Solasodine rhamnoside: antitumor
Solasonine: molluscicide
Somalin: antitumor
Somniferine: hypnotic
Sophocarpine: antitumor
Sophorajaponicin: antitumor
Sophoradochromene: antiulcers
Sparteine: antiinflammatory, cathartic, diuretic,
  oxytocic
Speciociliatine: analgesic, antitussive
Spectabiline: antitumor
Spirochin: analgesic, antipyretic, antiseptic
Stachydrine: cardiotonic
Steganacin: antitumor
Steganangin: antitumor
Steganol: antitumor
Steganone: antitumor
Stenocarpine: analgesic
epi-Stephanine: antitumor
Stephavanine: antitumor
Strophanthidin: antitumor
Strophanthin: antiarrhythmic
κ-Strophanthoside: antitumor
Strychnine: stimulant
Supinine: antitumor
Synephrine: decongestant
Tageretin: antifungal
Tamaulipin A: antitumor
Tamaulipin B: antitumor
Tannins: antidiarrhetic, bactericide, viricide
Tanshinone: antiseptic
Taspine: antiinflammatory
Taxifolin: antiinflammatory, antitumor
Taxol: antitumor
Tectoridin: antiinflammatory
Tectorigenin: antiinflammatory
Ternocide: antiinflammatory
α-Terthienyl: nematocide
Terpineol: antiallergenic, antiasthmatic, antitussive,
  bacteriostatic, expectorant
Tetrahydropalmatine: analgesic, sedative, tranquilizer
Tetrahydroxystilbene: fungicide
Tetramethylpyrazine: bactericide

Tetrandrine: analgesic, antiarrhythmic, antiinflamma-
  tory, antimicrobial, antimitotic, antitumor,
  hypotensive
D-Tetradrine: antitumor
DL-Tetradrine: antitumor
L-Tetradrine: antitumor
α-Tetraol: antitumor
Thalcimene: antiinflammatory
Thalfine: antiseptic
Thalfinine: antiseptic
Thalfoetidine: antiinflammatory
Thalibrine: antiseptic
Thalicarpine: antitumor
Thalidasine: antitumor, hypotensive
Thalidezine: antiseptic
Thaligosidine: antiseptic
Thaligosine: antiseptic
Thaligosinine: antiseptic
Thalirabine: antiseptic
Thaliracebine: antiseptic, hypotensive
Thalirugidine: antiseptic
Thalirugine: antiseptic
Thalisopine: antiarrhythmic
Thalistine: antiseptic
Thalistyline: antiseptic, hypotensive
Thalistyline methiodide: antiseptic
Thalmine: antiinflammatory
Thalmirabine: antiseptic
Thalrugosaminine: antiseptic, hypotensive
Thalsimine: antiinflammatory, antitussive
Thebaine: analgesic, CNS-stimulant
Theobromine: diuretic, stimulant, vasodilator
Theophylline: antiasthmatic, bronchodilator, stimulant
Thevetin: digitalic
Thevetoxin: digitalic
Thiocyanic acid: bactericide
Thujaplicin: fungicide
α-Thujaplicin: antitumor
Thymol: bactericide, fungicide, larvicide, tracheal re-
  laxant, vermicide
Tigloidine: CNS-sedative, antiparkinsonian
Tigogenin glycoside: antitumor
Tocopherol: antioxidant, antisterility
Tomatine: antiinflammatory, bactericide, fungicide,
  molluscicide
Trans-4-propenylveratrole: CNS-depressant
Trewiasine: antitumor
Tricanthine: cardiodepressant
Trichocarpin: fungicide
Tricin: antioxidant, estrogenic
Triethylrutoside: choleretic
Trigonelline: hypoglycemic
3',5,7-Trihydroxy-3,4'-dimethoxyflavone: antitumor
Trimethylrosmaricine: analgesic, myostimulant
Tripdiolide: antineoplastic
Triptolide: antitumor
Triptonide: antitumor

Table 4 (continued)
PHARMACOLOGICALLY ACTIVE PHYTOCHEMICALS

Tryptamine: vasopressor

Tuberosin: antistaphylococal, antitubercular, fungicide

Tuberosine: antiinflammatory, protisticide

Tubocurarine: anticonvulsant, histaminic, vagolytic

Tubulosine: antitumor

Tulipinolide: antitumor

Tuliposide A: antiallergic

Tyramine: vasopressor

Tylocrebine: antitumor

Tylophoridine: antitumor

Tylophorine: antiallergenic, antiasthmatic, antirhinitic, antitumor

Tylophorinine: antitumor

Uliginosin: antibacterial

Umbelliferone: antifungal

Uresin: bactericide

Ursolic acid: antitumor, diuretic

1-Usnic acid: antitumor

Uvaol: antitumor

Uvaretin: antitumor

Uzarigenin: antitumor

Valeric acid: hypotensive, tranquilizer

Valerinic acid: spasmolytic

Valeranone: sedative

Vanilloylcatalpol: antihepatotoxic

Vanilloylcevine: hypotensive

Vasicine: expectorant, oxytocic, abortifacient

Verbenalin: coagulant, parasympathicomimetic, uterotonic

Vernodalin: antitumor

Vernolepin: antitumor

Vernolide: antitumor

Vernomenin: antitumor

Vernomygdin: antitumor

Verticilline: hypertensive, relax smooth muscles

Verticine: hypertensive, paralyze sensory nerves and respiration

Vicemine: hypotensive

Vinaline: bactericide, fungicide

Vinblastine: antileukemic, oncolytic

Vincristine: antileukemic, oncolytic

Vindoline: hypoglycemic

Vindolinine: hypoglycemic

Vinleurosine: antitumor, oncolytic

Vinrosidine: oncolytic

Voacamine: antineoplastic

Voacorine: antitumor

Withacnistin: antitumor

Withaferin: antitumor, bactericide, viricide

Withaferin A: antiarthritic, antibiotic, antimitotic, antitumor, bacteriostatic

Withaferinile: antimitotic

Withanolide E: antifeedant, antineoplastic

Wogonin: antibacterial, diuretic

Xanthotoxin: antiinflammatory, antihistaminic, antipsoriac, spasmolytic, for vitiligo

Xanthoxyletin: anticonvulsant

Xyloidone: candidicide

Yangonin: anesthetic, antiinflammatory, antipyretic, myorelaxant

Yiamoloside B: fungistat

Yohimbine: aphrodisiac

Yuanhuacine: abortifacient, antileukemic

Zaluzanin C: antitumor

Zygophyllin: antiinflammatory

Table 5
PROXIMATE ANALYSES OF CONVENTIONAL PLANT FOODS
(per 100g)

Food item	H₂O (g)	Cal	Prot (g)	Fat (g)	Total Carb (g)	Fiber (g)	Ash (g)	Ca (mg)	P (mg)	Fe (mg)	Na (mg)	K (mg)	β-Car (μg)	Thia (mg)	Rib (mg)	Nia (mg)	Vit C (mg)
Vegetable (aerial)																	
Eggplant	92.4	25	1.2	.2	5.6	.9	.6	12	26	.7	2	214	6	.05	.05	.6	5
Okra	88.9	36	2.4	.3	7.6	1.0	.8	92	51	.6	3	249	312	.17	.21	1.0	31
Pepper	93.4	22	1.2	.2	4.8	1.4	.4	9	22	.7	13	213	252	.08	.08	.5	128
Squash	94.0	19	1.1	.1	4.2	.6	.6	28	29	.4	1	202	246	.05	.09	1.0	22
Tomato	93.5	22	1.1	.2	4.7	.5	.5	13	27	.5	3	244	540	.06	.04	.7	23
Av. (APB)[a]	92.4	25	1.4	.2	5.4	.9	.6	31	31	.6	4	224	271	.08	.09	.8	42
Av. (ZMB)[b]	0	331	18.5	2.6	71.4	11.9	7.9	410	410	7.9	53	2,963	3,585	1.06	1.19	10.6	556
Leafy vegetable (cultivated)																	
Beet greens	90.9	24	2.2	.3	4.6	1.3	2.0	119	40	3.3	130	570	3,660	.10	.22	.4	30
Cabbage	92.4	24	1.3	.2	5.4	.8	.7	49	29	.4	20	233	78	.05	.05	.3	47
Chard	91.1	25	2.4	.3	4.6	.8	1.6	88	39	3.2	147	550	3,900	.06	.17	.5	32
Kale	87.5	38	4.2	.8	6.0	1.3	1.5	179	73	2.2	75	378	5,340	—	—	—	125
Spinach	90.7	26	3.2	.3	4.3	.6	1.5	93	51	3.1	71	470	4,860	.10	.20	.6	51
Av. (APB)	90.5	27	2.7	.38	5.0	.96	1.46	106	46	2.4	89	440	3,568	.08	.16	.45	57
Av. (ZMB)	0	289	28.4	4.0	54.7	10.5	15.8	1,116	484	25.3	937	4,633	37,571	.84	1.68	4.74	600
Shoots																	
Asparagus	91.7	26	2.5	.2	5.0	.7	.6	22	62	1.0	2	278	540	.18	.20	1.5	33
Bamboo	91.0	27	2.6	.3	5.2	.7	.9	13	59	.5	—	533	12	.15	.07	.6	4
Mungbean	88.8	35	3.8	.2	6.6	.7	.6	19	64	1.3	5	223	12	.13	.13	.8	19
Pokeweed	91.6	23	2.6	.4	3.7	—	1.7	53	44	1.7	—	—	5,220	.08	.33	1.2	136

Soybean	86.3	46	6.2	1.4	5.3	.8	.8	48	1.0	67	—	—	48	.23	.20	.8	13
Av. (APB)	89.9	31	3.5	.5	5.2	.6	.9	31	1.1	59	4	345	1,166	.15	.19	1.0	41
Av. (ZMB)	0	307	34.6	5.0	51.5	5.9	8.9	307	10.9	584	40	3,416	11,543	1.48	1.88	9.9	405

Greens (wild)

Amaranth	86.9	36	3.5	.5	6.5	1.3	2.6	267	3.9	67	—	411	3,660	.08	.16	1.4	80
Dandelion	85.6	45	2.7	.7	9.2	1.6	1.8	187	3.1	66	76	397	8,400	.19	.26	—	35
Dock	90.9	28	2.1	.3	5.6	.8	1.1	66	1.6	41	5	338	7,740	.09	.22	.5	119
Lambsquarters	84.3	43	4.2	.8	7.3	2.1	3.4	309	1.2	72	—	—	6,960	.16	.44	1.2	80
Purslane	92.5	21	1.7	.4	3.8	.9	1.6	103	3.5	39	—	—	1,500	.03	.10	.5	25
Av. (APB)	88	35	2.84	.54	6.48	1.34	2.1	186.4	2.66	57	40.5	382	5,652	.11	.24	.9	68
Av. (ZMB)		292	23.7	4.5	54.0	11.2	17.5	1,553	22.2	475	337	3,183	47,433	.9	2.0	7.5	567

Fruit

Apples	84.4	58	.2	.6	14.5	1.0	.3	7	.3	10	1	110	54	.03	.02	.1	4
Apricots	85.3	51	1.0	.2	12.8	.6	.7	17	.5	23	1	281	1,620	.03	.04	.6	10
Bananas	75.7	85	1.1	.2	22.2	.5	.8	8	.7	26	1	370	114	.05	.06	.7	10
Blackberries	84.5	58	1.2	.9	12.9	4.1	.5	32	.9	19	1	170	120	.03	.04	.4	21
Blueberries	83.2	62	.7	.5	15.3	1.5	.3	15	1.0	13	1	81	60	.03	.06	.5	14
Cherries	83.7	58	1.2	.3	14.3	.2	.5	22	.4	19	2	191	600	.05	.06	.4	10
Cranberries	87.9	46	.4	.7	10.8	1.4	.2	14	.5	10	2	82	24	.03	.02	.1	11
Grapefruit	88.4	41	.5	.1	10.6	.2	.4	16	.4	16	1	135	48	.04	.02	.2	38
Orange	86.0	49	1.0	.2	12.2	.5	.6	41	.4	20	1	200	120	.10	.04	.4	50
Strawberries	89.9	37	.7	.5	8.4	1.3	.5	21	1.0	21	1	164	36	.03	.07	.6	59
Av. (APB)	84.9	54.5	.8	.4	13.4	1.1	.5	19.3	.61	17.7	1.2	178.4	280	.042	.043	.4	22.7
Av. (ZMB)	0	361	5.2	2.6	88.7	7.3	3.3	128	4.0	117	7.9	1,181	1,854	.28	.28	2.65	150

Vegetable (underground)

Beets	87.3	43	1.6	.1	9.9	.8	1.1	16	.7	33	60	335	12	.03	.05	.4	10
Carrots	88.2	42	1.1	.2	9.7	1.0	.8	37	.7	36	47	341	6,600	.06	.05	.6	8

Table 5 (continued)
PROXIMATE ANALYSES OF CONVENTIONAL PLANT FOODS
(per 100g)

Food item	H$_2$O (g)	Cal	Prot (g)	Fat (g)	Total Carb (g)	Fiber (g)	Ash (g)	Ca (mg)	P (mg)	Fe (mg)	Na (mg)	K (mg)	β-Car (μg)	Thia (mg)	Rib (mg)	Nia (mg)	Vit C (mg)
Garlic	61.3	137	6.2	.2	30.8	1.5	1.5	29	202	1.5	19	529	Tr	.25	.08	.5	15
Onions	89.1	38	1.5	.1	8.7	.6	.6	27	36	.5	10	157	24	.03	.04	.2	10
Radishes	94.5	17	1.0	.1	3.6	.7	.8	30	31	1.0	18	322	6	.03	.03	.3	26
Av. (APB)	84.1	55	2.3	.16	12.5	.9	.96	28	68	.88	31	337	1,328	.08	.05	.4	14
Av. (ZMB)	0	346	14.5	1.0	78.6	5.7	6.0	176	428	5.5	194	2,119	8,352	.50	.31	2.5	88
									Roots								
Jerusalem artichoke	79.8	—	2.3	.1	16.7	.8	1.1	14	78	3.4	—	—	12	.20	.06	1.3	4
Parsnip	79.1	76	1.7	.5	17.5	2.0	1.2	50	77	.7	12	541	18	.08	.09	.2	16
Potato	79.8	76	2.1	.1	17.1	.5	.9	7	53	.6	3	407	Tr	.10	.04	1.5	20
Rutabaga	87.0	46	1.1	.1	11.0	1.1	.8	66	39	.4	5	239	348	.07	.07	1.1	43
Salsify	77.6	—	2.9	.6	18.0	1.8	.9	47	66	1.5	—	380	6	.04	.04	.3	11
Sweetpotato	70.6	114	1.7	.4	26.3	.7	1.0	32	47	.7	10	243	5,280	.10	.06	.6	21
Taro	73.0	98	1.9	.2	23.7	.8	1.2	28	61	1.0	7	514	12	.13	.04	1.1	4
Turnip	91.5	30	1.0	.2	6.6	.9	.7	39	30	.5	49	268	Tr	.04	.07	.6	36
Yam	73.5	101	2.1	.2	23.2	.9	1.0	20	69	.6	—	600	Tr	.10	.04	.5	9
Yambean	85.1	55	1.4	.2	12.8	.7	.5	15	18	.6	—	—	Tr	.04	.03	.3	20
Av. (APB)	79.7	74	1.8	.3	17.3	1.0	.9	32	54	1.0	14	399	568	.09	.05	.8	18
Av. (ZMB)	0	365	8.9	1.5	85.2	4.9	4.4	158	266	4.9	69	1,966	2,798	.44	.25	3.9	89
									Cereals								
Barley	11.1	349	8.2	1.0	78.8	.5	.9	16	189	2.0	3	160	0	.12	.05	3.1	0
Buckwheat	11.0	335	11.7	2.4	72.9	9.9	2.0	114	282	3.1	—	448	0	.60	—	4.4	0

Millet	11.8	327	9.9	2.9	72.9	2.5	20	311	6.8	—	430	0	.73	.38	2.3	0
Popcorn	9.8	362	11.9	4.7	72.1	1.5	10	264	2.5	3	—	—	.39	.11	2.1	0
Rice	12.0	360	7.5	1.9	77.4	1.2	32	221	1.6	9	214	0	.34	.05	4.7	0
Rye	11.	334	12.1	1.7	73.4	1.8	38	376	3.7	1	467	0	.43	.22	1.6	0
Sorghum	11.	332	11.0	3.3	73.0	1.7	28	287	4.4	—	350	0	.38	.15	3.9	0
Wheat. red	13.0	330	14.0	2.2	69.1	1.7	36	383	3.1	3	370	0	.57	.12	4.3	0
Wheat, white	11.5	335	9.4	2.0	75.4	1.7	36	394	3.0	3	390	0	.53	.12	5.3	0
Wildrice	8.5	353	14.1	.7	75.3	1.4	19	339	4.2	7	220	0	.45	.63	6.2	0
Av. (APB)	11.1	342	11.0	2.3	74.0	1.6	35	305	3.4	4	339	0	.45	.20	3.8	0
Av. (ZMB)	0	385	12.4	2.6	83.2	1.8	39	343	3.8	4	381	0	.51	.22	4.3	0
Pulses (dry)																
Bean (lima)	10.3	345	20.4	1.6	64.0	3.7	72	385	7.8	4	1,529	Tr	.48	.17	1.9	—
Bean (mung)	10.7	340	24.2	1.3	60.3	3.5	118	340	7.7	6	1,028	48	.38	.21	2.6	—
Bean (red)	10.4	343	22.5	1.5	61.9	3.7	110	406	6.9	10	984	12	.51	.20	2.3	—
Broadbean	11.9	338	25.1	1.7	58.2	3.1	102	391	7.1	—	—	42	.50	.30	2.5	—
Chickpea	10.7	360	20.5	4.8	61.0	3.0	150	331	6.9	26	797	30	.31	.15	2.0	—
Cowpea	10.5	343	22.8	1.5	61.7	3.5	74	426	5.8	35	1,024	18	1.05	.21	2.2	—
Lentil	11.1	340	24.7	1.1	60.1	3.0	79	377	6.8	30	790	36	.37	.22	2.0	—
Peanut	5.6	564	26.0	47.5	18.6	2.3	69	401	2.1	5	674	—	1.14	.13	17.2	0
Pea	11.7	340	24.1	1.3	60.3	2.6	64	340	5.1	35	1,005	72	.74	.29	3.0	—
Soybean	10.0	403	34.1	17.7	33.5	4.7	226	554	8.4	5	1,677	48	1.10	.31	2.2	—
Av. (APB)	10.3	372	24.4	8.0	54.0	3.3	106	395	6.5	17	1,056	34	.66	.22	3.8	0
Av. (ZMB)	0	415	27.2	8.9	60.2	3.7	118	440	7.2	19	1,177	37	.73	.25	4.2	0
Nuts																
Almond	4.7	598	18.6	54.2	19.5	3.0	234	504	4.7	4	773	0	.24	.92	3.5	Tr
Brazilnut	4.6	654	14.3	66.9	10.9	3.3	186	693	3.4	1	715	Tr	.96	.12	1.6	—
Cashew	5.2	561	17.2	45.7	29.3	2.6	38	373	3.8	15	464	60	.43	.25	1.8	—
Hickory	3.3	673	13.2	68.7	12.8	2.0	Tr	360	2.4	—	—	—	—	—	—	—
Macadamia	3.0	691	7.8	71.6	15.9	1.7	48	161	2.0	—	264	0	.34	.11	1.3	0
Pecan	3.4	687	9.2	71.2	14.6	1.6	73	289	2.4	Tr	603	78	.86	.13	.9	2

Table 5 (continued)
PROXIMATE ANALYSES OF CONVENTIONAL PLANT FOODS
(per 100g)

Food item	H$_2$O (g)	Cal	Prot (g)	Fat (g)	Total Carb (g)	Fiber (g)	Ash (g)	Ca (mg)	P (mg)	Fe (mg)	Na (mg)	K (mg)	β-Car (μg)	Thia (mg)	Rib (mg)	Nia (mg)	Vit C (mg)
Pilinut	6.3	669	11.4	71.1	8.4	2.7	2.8	140	554	3.4	3	489	24	.88	.09	.5	Tr
Pinon	3.1	635	13.0	60.5	20.5	1.1	2.9	12	604	5.2	—	—	18	1.28	.23	4.5	Tr
Pistachio	5.3	594	19.3	53.7	19.0	1.9	2.7	131	500	7.3	—	972	138	.67	—	1.4	0
Walnut	3.1	628	20.5	59.3	14.8	1.7	2.3	Tr	570	6.0	3	460	180	.22	.11	.7	—
Av. (APB)	4.2	639	14.4	62.3	16.6	1.7	2.5	86	461	4.1	4	592	55	.65	.24	1.8	Tr
Av. (ZMB)	0	667	15.0	65.0	17.3	1.8	2.6	90	481	4.3	4	618	57	.68	.25	1.9	Tr

[a] APB, Average on an "as-purchased basis".

[b] ZMB, Average on a zero-moisture basis.

[c] tr, Trace.

REFERENCES

1. Council of Scientific and Industrial Research, *The Wealth of India*, 11 vols., New Delhi, 1948—1976.
2. **Grieve, M.**, *A Modern Herbal*, reprint 1974, Hafner Press, New York, 1931, 916.
3. **Watt, J. M. and Breyer-Brandwijk, M. G.**, *The Medicinal and Poisonous Plants of Southern and Eastern Africa*, 2nd ed., E. & S. Livingstone, Edinburgh, 1962, 1457.
4. **Hartwell, J. L.**, Plants used against cancer. A survey, *Lloydia*, p. 30, 1967—1971.
5. **Burkill, J. H.**, *A Dictionary of the Economic Products of the Malay Peninsula*, 2 vols., Art Printing Works, Kuala Lumpur, 1966.
6. **Mitchell, J. C. and Rook, A.**, *Botanical Dermatology*, Greenglass, Vancouver, 1979, 787.
7. **Guenther, E.**, *The Essential Oils*, 6 vols., D Van Nostrand, Toronto, 1948—1952.
8. **Uphof, J. C. Th.**, *Dictionary of Economic Plants*, Verlag von J. Cramer, Lehre, 1968, 591.
9. **Altschul, S., von R.**, *Drugs and Foods from Little-Known Plants*, Harvard University Press, Cambridge, 1973, 366.
10. **Perdue, R. E., Jr. and Hartwell, J. L., Eds.**, Plants and cancer, *Proc. 16th Annu. Meet. Soc. Econ. Bot. Cancer Treatment Rep.*, 60(8), 973, 1976.
11. **Lewis, W. H. and Elvin-Lewis, M. P. F.**, *Medical Botany*, John Wiley & Sons, New York, 1977, 515.
12. **Keys, J. D.**, *Chinese Herbs, Their Botany, Chemistry and Pharmacodynamics*, Chas. E. Tuttle, Tokyo, 1976, 388.
13. **Fernald, M. L., Kinsey, A. C., and Rollins, R. C.**, *Edible Wild Plants of Eastern North America*, Rev. ed., Harper and Brothers, New York, 1958, 452.
14. **Kingsbury, J. M.**, *Poisonous Plants of the United States and Canada*, Prentice-Hall, Englewood Cliffs, N.J., 1964, 626.
15. **Duke, J. A.**, Phytotoxin tables, *CRC Crit. Rev. Toxicol.*, 5(3), 189, 1977.
16. **Perry, L. M.**, *Medicinal Plants of East and Southeast Asia*, MIT Press, Cambridge, 1980, 620.
17. **Morton, J.**, *Major Medicinal Plants*, Charles C Thomas, Springfield, Ill., 1977, 431.
18. **Krochmal, A. and Krochmal, C.**, *A Guide to the Medicinal Plants of the United States*, 3rd ed., Quadrangle/The New York Times Book Company, New York, 1975, 259.
19. **Millspaugh, C. F.**, *American Medicinal Plants*, Dover Publications, New York, 1974, 806.
20. *The Merck Index, An Encyclopedia of Chemicals and Drugs*, 8th ed., Merck & Company, Rahway, N.J., 1968.
21. **Duke, J. A.**, *Nutrition File*, Computer index to nutritional data in 21 a to e, 1981.
21a. **Leung, W., Woot-Tseun, Busson, F., and Jardin, C.**, Food Composition Table for Use in Africa, Food and Agriculture Organization and U.S. Department of Health, Education and Welfare, 1968, 306.
21b. **Leung, W., Woot-Tseun, Butrum, R. R., and Chang, F. H.**, I. Proximate composition mineral and vitamin contents of East Asian food, in Food Composition Table for Use in East Asia, Food and Agriculture Organization and U.S. Department of Health, Education and Welfare, 1972, 334.
21c. **Leung, W., Woot-Tseun, and Flores, M.**, Tabla de Composicion de Alimentos Para Uso en America Latina, National Institutes of Health, Bethesda, Md., 1961, 132.
21d. **Watt, B. K. and Merrill, A. L.**, Composition of Foods, Agriculture Handbook No. 8, ARS, U.S. Department of Agriculture, Washington, D.C., 1963, 189.
21e. **Miller, D. F.**, Composition of Cereal Grains and Forages, Publ. 585, National Academy of Sciences-National Research Council, Washington, D.C., 1958, 663.
22. **Williams, L. O.**, The useful plants of Central America, *Ceiba*, 24(1-2), 342, 1981.
23. **Philips, H. J.**, Lebanese Folk Cures, thesis, Columbia University, New York, 1955.
24. **Mors, W. B. and Rizzini, C. T.**, *Useful Plants of Brazil*, Holden-Day, San Francisco, 1966, 166.
25. Dictionary of Chinese Traditional Medicine, Kiangsu Institute Modern Medicine, *Encyclopedia of Chinese Drugs*, 2 vols., Shanghai Scientific and Technical Publications, Shanghai, 1977, 2754.
26. **Kirtikar, K. R., Basu, B. D., and I. C. S.**, *Indian Medicinal Plants*, 4 vols. text, 4 vols. plates, 2nd ed., reprint, Jayyed Press, New Delhi, 1975, 6.
27. **Steinmetz, E. F.**, *Codex Vegetabilis. No Pagination*, E. F. Steinmetz, Amsterdam, 1957.
28. **Tierra, M.**, *The Way of Herbs*, Unity Press, Santa Cruz, Calif., 1980.
29. **Leung, A. Y.**, *Encyclopedia of Common Natural Ingredients Used in Food, Drugs, and Cosmetics*, John Wiley & Sons, New York, 1980, 409.
30. **Hutchens, A. R.**, *Indian Herbalogy of North America*, Merco, Ontario, Can., 1973, 382.
31. **Lawrence, B. M.**, *Essential Oils 1976—1977*, Allured Publishing, Wheaton, Ill., 1978, 175.
32. **Duke, J. A. and Wain, K. K.**, *Medicinal Plants of the World*, Computer index with more than 85,000 entries, 3 vols., 1981, 1654.
33. **List, P. H. and Horhammer, L.**, *Hager's Handbuch der Pharmazeutischen Praxis*, Vols. 2 to 6, Springer-Verlag, Berlin, 1969—1979.
34. **Hardin, J. W. and Arena, J. M.**, *Human Poisoning from Native and Cultivated Plants*, 2nd ed., Duke University Press, Durham, 1974, 194.

35. **Wyman, D.,** *Wyman's Gardening Encyclopedia,* Macmillan, New York, 1974, 1222.
36. Liberty Hyde Bailey Hortorium, *Hortus III,* Macmillan, New York, 1976, 1290.
37. **Tyler, V. E.,** *The Honest Herbal — A Sensible Guide to the Use of Herbs and Related Remedies,* George F. Stickley, Philadelphia, 1982, 263.
38. **Duke, J. A.,** *Medicinal Plants of the Bible,* 1st ed., Conch Publications, New York, 1983.
39. **Gibbons, E.,** *Stalking the Healthful Herbs,* Field Guide Ed., David McKay, New York, 1970.
40. **Duke, J. A.,** *Handbook of Legumes of World Economic Importance,* Plenum Press, New York, 1981, 345.
41. **Duke, J. A. and Ayensu, E. S.,** *Medicinal Plants of China,* 1st ed., Reference Publications, Algonac, Mich., 1985.
42. **Morton, J. F.,** *Atlas of Medicinal Plants of Middle America,* Charles C Thomas, Springfield, Ill., 1981, 1420.
43. **Duke, J. A.,** Amerindian Medicinal Plants, Typescript, Fulton, Md., 1983.
44. **Duke, J. A.,** Ayurvedic Medicinal Plants, Typescript, Fulton, Md., 1983.
45. **Erichsen-Brown, C.,** *Use of Plants for the Past 500 Years,* Breezy Creeks Press, Aurora, Can., 1979, 512.
46. **Morton, J. F.,** *Folk Remedies of the Low Country,* E. A. Seemann Publishing, Miami, 1974, 176.
47. **Rose, J.,** *Herbs and Things,* Grosset & Dunlap, New York, 1972, 323.
48. **Parvati, J.,** *Hygieia A Woman's Herbal, A Freestone Collective Book,* Bookpeople, Berkeley, 1978, 248.
49. **Rose, J.,** *Herbal Guide to Inner Health,* Grosett & Dunlap, New York, 1979, 239.
50. **Morton, J. F.,** Is there a safer tea, *Morris Arboretum Bull.,* 26(2), 24, 1975.
51. **Anon.,** *Herbal Highs,* p. 16.
52. **Schleiffer, H.,** *Sacred Narcotic Plants of the New World Indians,* Hafner Press, New York, 1973, 156.
53. **Schleiffer, H.,** *Narcotic Plants of the Old World,* Lubrecht & Cramer, Monticello, N.Y., 1979, 193; reviewed by Duke, J. A., *Taxon,* 29(4), 553, 1980.
54. **Emboden, W. A., Jr.,** *Narcotic Plants,* Macmillan, New York, 1972, 168.
55. **Siegel, R. K.,** Herbal intoxication, *J.A.M.A.,* 236(5), 473, 1976.
56. **Perkins, K. D. and Payne, W. W.,** *Guide to the Poisonous and Irritant Plants of Florida,* Fla. Coop. Ext. Serv., University of Florida, Gainesville, 1978.
57. **Garcia Barriga, H.,** *Flora Medicinal de Colombia,* 2 vols., Institututo Ciencias Naturales, Universidad Nacional, Bogota, D.E., Columbia, 1974 and 1975.
58. **Schultes, R. L. and Hofmann, A.,** *The Botany and Chemistry of Hallucinogens,* Charles C Thomas, Springfield, Ill., 1973, 267.
59. **Atal, C. K. and Kapur, B. M., Eds.,** *Cultivation and Utilization of Medicinal Plants,* Regional Research Laboratory, Jammu-Tawi, India, 1982, 877.
60. **Duke, J. A.,** *Isthmian Ethnobotanical Dictionary,* 3rd ed., Harrod and Company, Baltimore, Md., 1972.
61. **Duke, J. A.,** Herbs of Dubious Salubrity. Information Summary on ca 100 Herbs, assisted with the use section by C. F. Reed, unpublished draft.
62. Food and Drug Administration, *Health Foods Bus.,* June 1978.
63. **Duke, J. A.,** *Handbook of Energy Species, Public Domain,* Plenum Press, submitted.
64. **Purseglove, J. W., Brown, E. G., Green, C. L., and Robbins, S. R. J.,** *Spices,* 2 vols., Longman, London, 1981.
65. **Duke, J. A.,** Herbs as a small farm enterprise and the value of aromatic plants as economic intercrops, in *Research for Small Farms,* Proc. Special Symp., Kerr, H. W., Jr. and Knutson, L., Eds., USDA Misc. Publ. No. 1422, 1982, 76.
66. **Duke, J. A.,** Vegetarian vitachart, *Q. J. Crude Drug Res.,* 15, 45, 1977.
67. American Heart Association, *Nat. Foods Merchandizer,* 5, 20, 1983.
68. **Farnsworth, N. R.,** *Seminar on New Crops, handout,* National Academy of Sciences, Washington, D.C., December 10, 1980.
69. **Sherry, C. J. and Koontz, J. A.,** *Pharmacologic studies of "catnip tea": the hot water extract of Nepeta cataria, Q. J. Crude Drug Res.,* 17(2), 68, 1979.
70. **Duke, J. A.,** Taxonomist and toxicologist, presented at 2nd Annu. Herb Symp., Santa Cruz, Calif., September 15, 1978.
71. *Aphrodisia,* 1982/1983 Catalog.
72. **Ayensu, E. S.,** *Medicinal Plants of the West Indies,* Reference Publication, Algonac, Mich., 282, 1981.
73a. **Plowman, T.,** Chamairo: mussatia hyacinthia — an admixture to coca from Amazonian Peru and Bolivia, *Bot. Mus. Leafl. Harvard U.,* 28(3), 253, 1980.
73b. **Chandler, R. F., Hooper, S. N., and Harvey, M. J.,** Ethnobotany and phytochemistry of yarrow, Achillea millefolium, Compositae, *Econ. Bot.,* 36(2), 203, 1982.
74. **Leppik, E. E.,** unpublished typescript, U.S. Department of Agriculture, Washington, D.C.
75. **Hooper, D.,** Useful Plants and Drugs of Iran and Iraq, Bot. Series, Publ. 387, Field Museum, Chicago, 9(3), 71, 1937.

76. **Morgan, G. R.,** The ethnobotany of sweet flag among North American Indians, *Bot. Mus. Leafl. Harvard U.,* 28(3), 235, 1980.

77. **Atal, C. K.,** *Chemistry and Pharmacology of Vasicine — A New Oxytocic Abortifacient,* RRL-Jammu, 1980, 155.

78. Regional Research Lab, *Jammu Newsl.,* 7, 21, 1980.

79. Conservation Foundation Letter, Medicinal Plants Need Extensive Safeguarding, *Newsletter,* November 1982.

80. **Schultes, R. E. and Farnsworth, N. R.,** Ethnomedical, botanical and phytochemical aspects of natural hallucinogens, *Bot. Mus. Leafl. Harvard U.,* 28(2), 123, 1980.

81. **Duke, J, A.,** Aloe: the facts behind the folklore, *Equus,* 67, 14, 1980.

82. **Mahmood, I., Masood, A., Saxena, S. K., and Husain, S. I.,** Effects of some plant extracts on the mortality of *Meloidogyne incognita* and *Rotylenchus reniformis, Acta Bot. Indica,* 7(2), 129, 1979.

83. **Gupta, K.,** Aloes compound (a herbal drug) in functional sterility, in Proc. 16th All India Obstet. and Gynecol. Congr., New Delhi, 1972.

84. **Boulos, L.,** *Medicinal Plants of North Africa,* Reference Publication, Algonac, Mich., 1983, 288.

85. **Embong, M. B., Hadziyev, D., and Molnar, S.,** Essential oils from spices grown in Alberta, dill seed oil, *Anethum graveolens* L. (Umbelliferae), *Can. Inst. Technol. J.,* 10, 208, 1977.

86. **Escher, S., Keller, U., and Willhalm, B.,** New phellandrene derivatives from the root oil of *Angelica archangelica* L., *Helvetica Chim. Acta,* 62(7), 2061, 1979.

87. **Sung, C. P., Baker, A. P., Holden, D. A., Smith, W. J., and Shakrin, L. W.,** Effects of extracts of *Angelica polymorpha* on reaginic antibody production, *J. Nat. Prod.,* 45(4), 398, 1982.

88. **Garg, S. K., Gupa, S. R., and Sharma, N. D.,** Coumarins from *Apium graveolens* seeds, *Phytochemistry,* 18(9), 1580, 1979.

89. **Marsh, A. C., Moss, M. K., and Murphy, E. W.,** Composition of Foods, Spices and Herbs, Agric. Handbook No. 8-2, 1977.

90. **Kupchan, S. M., Hemingway, R. J., and Doskotch, R. W.,** Tumor inhibitors. IV. Apocannoside and cymarin, the cytotoxic principals of *Apocynum cannabinum, J. Med. Chem.,* 7, 803, 1964.

91. **Duke, J. A.,** Papaveraceous polyclave, *CRC Crit. Rev. Toxicol.,* 3, 1, 1974.

92. **Mix, D. B., Guinaudeau, M., and Shamma, M.,** The aristolochic acids and aristolactans, *J. Nat. Prod.,* 45, 657, 1982.

93. **Mosca, A. M. L. and Costanzo, M. P. C.,** Antifungal activity in vitro of ''Arnica montana'' L., *Allionia,* 24, 85, 1980/81.

94. **Hocking, G. M.,** American *Arnica* in medicine, *Chemurgic Dig.,* 4(1), 10, 1945.

95. **Herrman, H. D., Willuhn, G., and Hausen, B. M.,** Helenalinmethacrylate, a new pseudoguzianolide from the flowers of *Arnica montana* L. and the sensitizing capacity of their sesquiterpene lactones, *Planta Medica,* 34, 299, 1978.

96. **Kaul, V. K., Nigam, S. S., and Banerjee, A. K.,** Insecticidal activity of some essential oils, *Indian J. Pharm.,* 40(1), 22, 1978.

97. **Anon.,** Scientist probe secrets of Sauce Bearnaise, *Science,* 204, 388, 1979.

98. **Farnsworth, N. R. and Morris, R. W.,** Higher plants — the sleeping of drug development, *Am. J. Pharm.,* 147, 46, 1976.

99. **Gaertner, E. E.,** The history and use of milkweed (*Asclepias syriaca* L.), *Econ. Bot.,* 33, 119, 1980.

100. **Hamel, P. and Chiltoskey, M.,** *Cherokee Plants and Their Uses,* Herald Publishing, Sylva, N.C., 1975.

101. **Farnsworth, N. R.,** Drugs from higher plants, *Tilc Till,* 55, 33, 1969.

102. **Kilmer, F. B.,** Drug clerks and belladonna, *J. Am. Pharm. Assoc.,* 13(12), 1131, 1924.

103. **de Rios, M. D.,** Ayahuasca — the healing vine, *Int. J. Soc. Psychiatry,* 17(4), 1, 1971.

104. **Reichel-Dolmatoff, G.,** Notes on the cultural extent of the use of yaje (*Banisteriopsis caapi*) among the Indians of Vaupes, Colombia, *Econ. Bot.,* 24(1), 32, 1970.

105. **Cardenas, M.,** *Manual de Plantas Economicas de Bolivia,* Imprenta Icthus, Cochabamba, Bolivia, 1969, 421.

106. **Kawanishi, K., Uhara, Y., and Hashimoto, Y.,** Shihunine and dihydroshihunine from *Banisteriopsis caapi, J. Nat. Prod.,* 45(5), 637, 1982.

107. **Ayensu, E. S.,** *Medicinal Plants of West Africa,* Reference Publication, Algonac, Mich., 1978, 330.

108. **Anon.,** Hippy letter to J. A. Duke, reactions to borage.

109. **Plowman, T.,** Brunfelsia in ethnomedicine, *Bot. Mus. Leafl. Harvard U.,* 25(10), 289, 1977.

110. **Anon.,** *Chem. Mark. Rep.,* December 12, 1977.

111. **Schultes, R. E.,** De *Plantis toxicariis* e mundo novo tropicale commentationes. XXVI. Ethnopharmocological notes on the flora of northwestern South America, *Bot. Mus. Leafl. Harvard U.,* 28, 1, 1980.

112. Weed Science Society of America, Newsletter, October 1980.

113. **Anon.,** *Alberta Agric. News,* 5(10), January 1980.

114. **Anon.,** *Indian For. Leafl.,* 95, 6, 1947.

115. **Anon.,** *Chem. Mark. Rep.,* October 1, 1979.
116. **Buccellato, F.,** Ylang survey, *Perfumer Flavorist,* 7(4), 9, 1982.
117. **Buchanan, R. L.,** Toxicity of spices containing methylenedioxybenzene derivatives: a review, *J. Food Safety,* 1, 275, 1978.
118. **Anon.,** Chymopapain approved, FDA Drug Bull. 12, December 1982.
119. **Flath, R. A. and Forrey, R. R.,** Volatile components of papaya (*Carica papaya* L., Solo variety), *J. Agric. Food Chem.,* 25(1), 103, 1977.
120. **Atzorn, R., Weiler, E. W., and Lenk, M. H.,** Formation and distribution of sennosides in *Cassia angustifolia,* as determined by a sensitive and specific radioassay, *Planta Medica,* 41(1), 1, 1981.
121. **Anton, R. and Haag-Berrurier, M.,** Therapeutic use of natural anthraquinone for other than laxative actions, *Pharmacology,* 20 (Suppl. 1), 104, 1980.
122. **Getahun, A. and Krikorian, A. D.,** Chat: coffee's rival from Harar, Ethiopia. I. Botany, cultivation, and use, *Econ. Bot.,* 27, 353, 1973.
123. **Krikorian, A. D. and Getahun, A.,** Chat: coffee's rival from Harar, Ethiopia. II. Chemical composition, *Econ. Bot.,* 27(4), 378, 1973.
124. **Hammouda, E. S. M.,** Effect of dietary khat extract on the testis of the white leghorn *Gallus domesticus, Fac. Sci. (King Abdul Aziz U.) Bull.,* 2, 17, 1978.
125. **Anon.,** *Bull. Narcotics,* 32(3), 1, 1980. (Special issue on Khat).
126. **Dharmananda, S.,** Herbs for women, *Health Foods Bus.,* August, 59, 1979.
127. **Quisumbing, E.,** *Medicinal Plants of the Philippines,* Tech. Bull. 16, Phil. Dept. Agr., Nat. Res., Manila, 1951, 1234.
128. **Chai, M. A. P.,** Gotu Kola, *Herbalist,* February 6 to 8, 1979.
129. **Mowrey, D.,** Questions and answers, *The Herbalist,* February 23, 1979.
130. **Anon.,** *WHO Chronicle,* 31, 428, 1977.
131. **Takaoka, D., Takaoka, K., Oshita, T., and Hiroi, M.,** Sesquiterpene alcohols in camphor oils, *Phytochemistry,* 15, 425, 1976.
132. **Mallavarapu, G. R. and Row, L. R.,** Chemical constituents of some cucurbitaceae plants, *Indian J. Chem.,* 17(4), 417, 1979.
133. **Nag, T. N. and Harsh, M. L.,** Arid zone plants of Rajasthan — a source of steroidal sapogenins, *Acta Botanica Indica,* 10(1), 8, 1982.
134. **Rizvi, S. J. H., Jaiswal, Y., Muierji, D., and Mathur, S. N.,** 1,3,7-Trimethylxanthine — a new natural herbicide: its mode of action, abstract, *Plant Physiol.,* 65(6) (Suppl.), 99, 1980.
135. **Palotti, G.,** The time for a Coca Cola may not be right, *Ind. Aliment.,* 16(12), 146, 1977.
136. **Tiscornia, E., Centi-Grossi, M., Tassi-Micco, C., and Evangelisti, F.,** The sterol fraction of coffee (*Coffea arabica* L.), *Rev. Hal. Sostanze Grasse,* 56(8), 283, 1979.
137. *FDA By-Lines No. 2,* April 1981.
138. **Jacobson, M.,** (as quoted by Solomon, G. L.), Anti-caffeine crusaders raise questions about its safety, *Wash. Star,* December 20, 1978.
139. **Anon.,** *Sci. News,* February 12, 1983.
140. **Ponte, L.,** All about caffeine, *Readers Dig.,* January, 72, 1983.
141. **Anon.,** FDA panel advises dropping GRAS status for caffeine in colas, *Chem. Mark. Rep.,* June 26, 1976.
142. **Pfander, H. and Schurtenberger, H.,** Biosynthesis of C_{20}-carotenoids in crocus sativus, *Phytochemistry,* 21, 1039, 1982.
143. **Bristol, M. L.,** Tree Datura drugs of the Colombian Sibundoy, *Bot. Mus. Leafl. Harvard U.,* 22(5), 165, 1969.
144. **El-Dirdiri, N. I., Wasfi, I. A., Adams, S. E., and Edds, G. T.,** Toxicity of *Datura stramonium* to sheep and goats, *Vet. Human Toxicity,* 23(4), 241, 1981.
145. **Anon.,** Plant poisonings — New Jersey, *Morbidity Mortality Wkly. Rep.,* 30(6), 1, 1981.
146. **Walter, W. G. and Khanna, P. N.,** Chemistry of the Aroids. I. Dieffenbachia seguine, amoena and picta, *Econ. Bot.,* 26, 364, 1972.
147. **Farnsworth, N. F.,** Folk medicines — fantasy, fiction or fact, *J. Pharm.,* 3, 49, 1962.
148. **Plowman, T.,** Folk uses of new world aroids, *Econ. Bot.,* 23, 97, 1969.
149. **Farnsworth, N. R. and Bingel, A. S.,** Problems and prospects of discovering new drugs from higher plants by pharmacological screening, in *New Natural Products and Plant Drugs with Pharmacological, Biological or Therapeutical Activity,* Wagnar, H. and Wolff, P., Eds., Springer-Verlag, New York, 1977, 1.
150. **Farnsworth, N. R.,** The plant kingdom — supplier of steroids, *Tile Till,* 53, 55, 1967.
151. **Sullivan, G.,** Occurrence of umbelliferone in the seeds of *Dipteryx odorata, J. Agric. Food Chem.,* 30, 609, 1982.
152. **Luanratana, O. and Griffin, W. J.,** Alkaloid biosynthesis in a Duboisia hybrid, *J. Nat. Prod.,* 45, 551, 1982.

153. **Vereshchagin, I. A., Geskina, O. D., and Bukhteeva, R. R.,** Increasing of antibiotic therapy efficiency with adaptogens in children suffering from dysentery and proteus infection, *Antibiotiki (Mosc.)*, 27(1), 65, 1982.

154. **Baranov, A. I.,** Medicinal uses of ginseng and related plants in the Soviet Union: recent trends in the Soviet literature, *J. Ethnopharm.*, 6, 339, 1982.

155. Forest Service, *Range Plant Handbook*, U.S. Department of Agriculture, Washington, D.C., 1937.

156. **Train, P., Henrichs, J. P., and Archer, W. A.,** *Medicinal Uses of Plants by Indian Tribes of Nevada*, reissued by Quarterman Publ., Lawrence, Mass., 1957, 139.

157. **Holm, L. G., Plucknett, D. L., Pancho, J. V., and Herberger, J. P.,** *The World's Worst Weeds*, University Press of Hawaii, Honolulu, 1977, 609.

158. **Wesley, F.,** Freud and cocaine (letter), *Am. Sci.*, 71, 12, 1983.

159. **Calvin, M.,** (as quoted or interpreted by T. H. M.), The petroleum plant: perhaps we can grow gasoline, *Science*, 194, 46, 1976.

160. **Sachs, R. M., Low, C. B., MacDonald, J. D., Awad, A. R., and Sully, M. J.,** Euphorbia lathyris: a potential source of petroleum-like products, *Cal. Agric.*, 35(7/8), 29, 1981.

161. **de Montmorency, A.,** *Spotlight*, July 14, 1980.

162. **Kinghorn, A. D.,** Characterization of an irritant 4-deoxyphorbol diester from *Euphorbia tirucalli*, *J. Nat. Prod.*, 42, 112, 1979.

163. **Seher, A. and Ivanov, S. A.,** Natural antioxidants. II. Antioxidants in the fatty oil of Foeniculum vulgare Miller, *Fette Seifen Anstrichmittel*, 78(6), 224, 1976.

164. **Denee, R. and Huizing, H. F.,** Purification and separation of anthracene derivatives on a polystyrene-divinylbenzene copolymer, *J. Nat. Prod.*, 44, 257, 1981.

165. **Piampiano, C.,** U.S. Patent 4,219,579, 1980.

166. **Scully, V.,** *A Treasury of American Indian Herbs*, Bonanza Books, New York, 1970, 306.

167. **Sokoloff, M. D., Funaoka, K., Toyomizu, M., Saelhof, C. C., and Bird, L.,** The oncostatic factors present in *Gleditsia triacanthos*, *Growth*, 28, 97, 1964.

168. **Mitscher, L. A., Park, Y. H., Clark, D., and Beal, J. L.,** Antimicrobial agents from higher plants. Antimicrobial isoflavonoids and related substances from *Glycyrrhiza glabra* L. var. typica, *J. Nat. Prod.*, 43(2), 259, 1980.

169. **Avery, J. L.,** Nature's medicine chest, *Country Mag.*, 4(3), 46, 1983.

170. **Samy, M. S.,** Chemical and nutritional studies on roselle seeds (*Hibiscus sabdariffa* L.) *Zeitr. Ernah-rungwise*, 19, 47, 1980.

171. **Ahmad, M. V., Husain, S. K., Ahmad, J., and Osman, S. M.,** Hibiscus sabdariffa seed oil: a re-investigation, *J. Sci. Food Agric.*, 4, 424, 1979.

172. **Salama, R. B. and Ibrahim, S. A.,** Ergosterol in *Hibiscus sabdariffa* seed oil, *Planta Medica*, 36, 221, 1979.

173. **Little, E. L., Jr. and Wadsworth, F. H.,** Common Trees of Puerto Rico and the Virgin Islands, USDA Handbook #249, U.S. Department of Agriculture, Washington, D.C., 1964, 548.

174. **Kinghorn, A. D., Ed.,** *Toxic Plants*, Columbia University Press, New York, 1979, 195.

175. **Bradford, I.,** Hops and hop products, *Chem. Ind.*, No. 24, 88, 1979.

176. **Moerman, D. E.,** *Geraniums for the Iroquois*, Reference Publications, Algonac, Mich., 1982, 242.

177. **Cushman, M., DeKow, F. W., and Jacobsen, L. B.,** Conformations, DNA binding parameters and antileukemic activity of certain cytotoxic protoberberine alkaloids, *J. Med. Chem.*, 22(3), 331, 1979.

178. **Nepelkopf, E.,** as quoted in Dateline San Diego (WNS), on file in U.S. Department of Agriculture, Washington, D.C., 1976.

179. **Hocking, G. M.,** Henbane-healing herb of Hercules and of Apollo, *Econ. Bot.*, 1, 306, 1947.

180. **Vickery, A. R.,** Traditional uses and folklore of Hypericum in the British Isles, *Econ. Bot.*, 35(3), 289, 1981.

181. **Porter, R. H.,** Mate-South American or Paraguay tea, *Econ. Bot.*, 4, 37, 1950.

182. **Hofmann, A.,** The active principles of the seeds of *Rivea corymbosa* and *Ipomoea violacea*, *Bot. Mus. Leafl. Harvard U.*, 20, 194, 1963.

183. **Ochse, J. J.,** *Vegetables of the Dutch East Indies*, (reprint), A. Asher & Co., Hacquebard, Amsterdam, 1980.

184. **Gaydou, A. M., Menet, L., Ravelojaona, G., and Geneste, P.,** Energy resources of plant origin in Madagascar: ethyl alcohol and seed oils, *Oleagineux*, 37(3), 135, 1982.

185. **Tewari, J. P. and Shukla, I. K.,** Inhibition of infectivity of 2 strains of watermelon mosaic virus by latex of some angiosperms, *Geobios*, 9, 124, 1982.

186. **Duke, J. A.,** Narcotic plants of the Old World (review), *Taxon*, 29, 553, 1980.

187. **NAS,** *Atlas of Nutritional Data on United States and Canadian Feeds*, National Academy of Science, Washington, D.C.

188. **Plowman, T., Gyllenhaal, L. O., and Lindgren, J. E.,** *Latua pubiflora*, magic plant from southern Chile, *Bot. Mus. Leafl., Harvard U.*, 23, 61, 1971.

189. **Pech, B. and Bruneton, J.,** Alkaloids of noble laurel, *Laurus nobilis, J. Nat. Prod.,* 45(5), 560, 1982.

190. **El-Feraly, F. S. and Benigni, D. A.,** Sesquiterpene lactones of *Laurus nobilis* leaves, *J. Nat. Prod.,* 43, 527, 1980.

191. **Garmon, L.,** Kitchen ecology: cukes, spice, bugs, *Sci. News,* September 12, 1982.

192. **Gorkin, R. A.,** Light on the shroud?, *Science,* 201, 1078, 1978.

193. **Lawrence, B. M.,** *Essential Oils 1978,* Allured Publishing, Wheaton, Ill., 1979, 192.

194. **Shihata, I. M., Hassan, A. G., and Mayah, G. Y.,** Pharmacological effects of *Lawsonia inermis* leaves (el-henna), *Egypt. J. Vet. Sci.,* 15, 31, 1978.

195. **Mahmoud, Z. F., Salam, N. A. A., and Khafagy, S. M.,** Constituents of henna leaves (*Lawsonia inermis* L.) growing in Egypt, *Fitoterapia,* 51, 153, 1980.

196. **Chakrabartty, T., Poddar, G., and St. Pyrek, J.,** Isolation of di hydroxy lupene and di hydroxy lupane from the bark of *Lawsonia inermis, Phytochemistry,* 21(7), 1814, 1982.

197. **Tschesche, R., Diederich, A., and Jha, H. C.,** Caffeic acid 4-rutinoside from *Leonurus cardiaca* leaves and flowers, *Phytochemistry,* 19(12), 2783, 1980.

198. Food and Drug Administration, *Washington Post,* January 11, 1982.

199. **Hayashi, K., Wada, K., Mitsubashi, H., Bando, H., Takase, M., Terada, S., Koide, Y., Alba, T., Narita, T., and Mizuno, D.,** Antitumor active glycosides from condurango cortex, *Chem. Pharm. Bull. (Tokyo),* 28(6), 1954, 1980.

200. **Fujita, S. I. and Fujita, Y.,** Studies on the essential oils of the genus *Mentha.* VII. On the high boiling components of the essential oil of *Mentha pulegium* Linn., *J. Agric. Chem. Soc.,* 46, 303, 1972.

201. **Schultes, R. E.,** A new narcotic genus from the Amazon slope of the Colombian Andes, *Bot. Mus. Leafl., Harvard U.,* 17, 1, 1955.

202. **Bristol, M. L.,** Tree Datura drugs of the Colombian Sibundoy, *Bot. Mus. Leafl., Harvard U.,* 22, 165, 1969.

203. **Schultes, R. E.,** Evolution of the identification of the major South American narcotic plants, *Bot. Mus. Leafl., Harvard U.,* 26, 311, 1978.

204. **Carvalho Filho, O. M. and Salviano, L. M. C.,** Evidence of inhibitory action of jurema-preta on ''in vitro'' fermentation of forage grasses, *Bol. Pesquisa* (Centr. Pesq. Agropec. Tropic. Semi-arido), 11, 1, 1982.

205. **Williams, L. O.,** Bayberry wax and bayberries, *Econ. Bot.,* 12, 103, 1958.

206. **Anon.,** Miscellaneous, *Chem. Mark. Rep.,* December 15, 1975.

207. **Buchanan, R. L., Goldstein, S., and Budroe, J. D.,** Examination of chili pepper and nutmeg oleoresins using the Salmonella/Mammalian microsome mutagenicity assay, *J. Food Sci.,* 47(1), 330, 1982.

208. **Wahlberg, I., Hjelte, M. B., Karlsson, K., and Enzell, C. R.,** Constituents of commercial tolu balsam, *Acta Chem. Scand.,* 25, 3285, 1971.

209. **Piozzi, F., Marino, M. L., Fuganti, C., and Di Martino, A.,** Occurrence of nonbasic metabolites in Amaryllidaceae, *Phytochemistry,* 8, 1745, 1969.

210. **Krochmal, A., Walters, R. S., and Doughty, R. M.,** *A Guide to Medicinal Plants of Appalachia,* Agric. Handbook No. 400, 1971, 291.

211. **Duke, J.,** Catnip an all-purr-puss tonic, *Wireless Flash No. 139,* December 7, 1981.

212. **Harney, J. W., Barofsky, I. M., and Leary, J. D.,** Behavioral and toxicological studies of cyclopentanoid monoterpenes from *Nepeta cataria, J. Nat. Prod.,* 41(4), 367, 1979.

213. **Tittel, G. and Wagner, H.,** Qualitative and quantitative analyses of glycoside heart drugs by the HPLC method. III. Qualitative and quantitative HPLC analysis of the cardenolides and other preparations from *Nerium oleander, Planta Medica,* 43(3), 252, 1981.

214. **Khafagy, S. M. and Metwally, A. M.,** Phytochemical study of *Nicotiana glauca* R. Grah. grown in Egypt, *J. Phar. Sci. (U.A.R.),* 9, 83, 1968.

215. **Rindl, M. and Sapiro, M. L.,** A chemical investigation of the constituents of *Nicotiana glauca* R. Grah (Solanaceae) (wild tobacco), Onderstepoort, *J. Vet. Sci. Anim. Ind.,* 22(2), 301, 1949.

216. **Keeler, R. F., Balls, L. D., and Panter, K.,** Teratogenic effects of *Nicotiana glauca* and concentration of anabasine, the suspect teratogen in plant parts, *Cornell Vet.,* 71(1), 47, 1981.

217. **Agaceta, L. M., Dumag, P. V., and Batolos, J. A.,** Studies on the control of snail vectors of fascioliasis: mollucicidal activity of some indigenous plants, *NSDB Technol. J.: Abstr. Trop. Agric.,* 6(2), 30, 1981.

218. OTA Symp., An Alternate Use for Tobacco Agriculture: Proteins for Food Plus a Safer Smoking Material, Office of Technology Assessment, Washington, D.C., 1982.

219. **Kung, S. D. and Tso, T. C.,** Tobacco as a potential food source and smoke material: soluble protein content, extraction, and amino acid composition, *J. Food Sci.,* 43, 1844, 1978.

220. **Darrah, H. H.,** *The Cultivated Basils,* Buckeye Printing, Independence, Mo., 1980, 40.

221. **Deshpande, R. S. and Ripnis, H. P.,** Insecticidal activity of *Ocimum basilicum* Linn, *Pesticides (PSTDAN),* 11(5), 11, 1977.

222. **Bowers, W. S. and Nishida, R.,** Juvocimenes: potent juvenile hormone mimics from sweet basil, *Science,* 209(4460), 1030, 1980.

223. **Hu, S. Y.,** A contribution to our knowledge of ginseng, *Am. J. Chem. Med.,* 5(1), 1, 1976.

224. **Li, C. P.,** A new medical trend in China, *Am. J. Chem. Med.,* 3(3), 213, 1975.

225. **Tong, L. S. and Chao, C. Y.,** Effects of ginsenoside Rg$_1$ of Panax ginseng on mitosis in human blood lymphocytes in vitro, *Am. J. Chem. Med.,* 8(3), 254, 1980.

226. **Kim, C., Chos, H., Kim, C. C., Kim, J. K., Kim, M. S., Ahn, B. T., and Park, H. J.,** Influence of ginseng on mating behavior of male rats, *Am. J. Clin. Med.,* p. 163, 1976.

227. **Han, B. H., Park, M. H., Woo, L. K., Woo, W. S., and Han, Y. N.,** Studies on antioxidant components of Korean ginseng(l), *Korean Biochem. J.,* 12(1), 33, 1979.

228. **Williams, L. and Duke, J. A.,** Growing ginseng, Farmer's Bull. No. 2201, U.S. Department of Agriculture, 1978, 8.

229. **Siegel, R. K.,** Ginseng abuse syndrome, *JAMA,* 241, 1614, 1979.

230. **Farnsworth, N. R. and Bederka, J. P.,** Ginseng, fantasy, fiction or fact, in A Collection of Articles Concerning the Current and Future Importance of Medicinal Plants, presented to NAS, December 10, 1980.

231. **Anon.,** Ginseng: a cacophony of confusion, *Health Foods Bus.,* February, 39, 1979.

232. **Lui, J. H. C. and Staba, E. J.,** The ginsenosides of various ginseng plants and selected products, *J. Nat. Prod.,* 43(3), 340, 1980.

233. **Gross, S.,** 1st Annual Herb Trade Association National Herb Symp., University of California, Santa Cruz, September 17 and 18, 1977.

234. **Duke, J.,** Sang's Swan Song, composed Santa Cruz, 1977.

235. **Anon.,** Love potions you can make at home, *Globe,* May 11, 1982.

236. **Watson, S. J., Berger, P. A., Akil, H., Mills, M. J., and Barchas, J. D.,** Effects of naloxone on schizophrenia: reduction in hallucinations in a subpopulation of subjects, *Science,* 201, 73, 1978.

237. **Faden, A. J. and Holaday, J. W.,** Opiate antagonists: a role in the treatment of hypovolemic shock, *Science,* 205, 317, 1979.

238. **Gessa, G. L., Paglietti, E., and Pellegrini Quarantotti, B.,** Induction of copulatory behavior in sexually inactive rats by naloxone, *Science,* 204, 203, 1979.

239. **Anon.,** Codeine plant, *Chem. Mark. Rep.,* June 6, 1977.

240. **Seddigh, M., Jolliff, G. D., Calhoun, W., and Crane, J. M.,** Papaver bracteatum, potential commercial source of thebaine, *Econ. Bot.,* 36(4), 433, 1982.

241. **Shamma, M., Morwarid, A. M., and Viroben, G.,** Nutritional constituents of Papaver bracteatum seed-meal, *Ind. J. Anim. Sci.,* 48(11), 830, 1978.

242. **Kettenes Vanden Bosch, J. J., Salemink, C. A., and Khan, J.,** Biological activity of the alkaloids of Papaver bracteatum Lindl., *J. Ethnopharm.,* 3, 21, 1981.

243. **Duke, J. A.,** Utilization of Papaver, *Econ. Bot.,* 27, 390, 1973.

244. **Anon.,** *Bull. Narcotics,* 33(3), 29, 1981.

245. **Anon.,** *Bull. Narcotics,* 32(1), 45, 1980.

246. **Anon.,** Opiate's shifting scene, *Chem. Bus.,* February 7, 7, 1983.

247. **Schultes, R. E.,** Planta Columbianae. II. Yoco: a stimulant of southern Colombia, *Bot. Mus. Leafl. Harvard U.,* 10, 301, 1942.

248. **Anon.,** Capsules, more than a reputation, *Time,* December 14, 1981.

249. **Ross, S. A., Megalla, S. E., Dishay, D. W., and Awad, A. M.,** Studies for determining antibiotic substances in some Egyptian plants. I. Screening for antimicrobial activity, *Fitoterapia,* 51(6), 303, 1981.

250. **Black, W. L. and Parker, K. W.,** *Toxicity Tests on African Rue (Peganum harmala L.)* Bull. 240, Ag. Exp. Station, N.M., 1936.

251. **Ishikura, N.,** Anthocyanins and flavons in leaves and seeds of Perilla plant, *Agric. Biol. Chem.,* 45(8), 1855, 1981.

252. **Wilson, B. J., Garst, J. E., Linnabary, R. D., and Channel, R. B.,** Perilla ketone: a potent lung toxin from the mint plant, *Perilla frutescens* Britton, *Science,* 197, 573, 1977.

253. **Wilson, B. J., Garst, J. E., and Linnabary, R. D.,** Pulmonary toxicity of 3-substituted furans from the mint plant *Perilla frutescens* Britton, *Toxicol. Appl. Pharmacol.,* 45(1), 300, 1978.

254. **Okazaki, N., Matsunaki, M., Kondo, M., and Okamoto, K.,** Contact dermatitis due to beefsteak plant *Perilla frutescens* var. *acuta, Skin Res.,* 24(2), 250, 1982.

255. **Stern, P.,** phone conversation, 1983.

256. **Schultes, R. E.,** *Bot. Mus. Leafl. Havard U.,* 28(1), 81, 1980.

257. **Bruns, K. and Kohler, M.,** Uber die zusammensetzung des boldoblatterols, *Parf. Kosm.,* 55, 225, 1975.

258. **Davis, K. L., Mohs, R. C., Rinklenberg, J. R., Pfefferbaum, A., Hollister, L. E., and Kopell, B. S.,** Physostigmine: improvement of long-term memory processes in normal humans, *Science,* 201, 272, 1978.

259. **Bartus, R. T.,** Physostigmine and recent memory: effects in young and aged nonhuman primates, *Science,* 206, 1087, 1979.

260. **McHale, D., Laurie, W. A., and Woof, M. A.,** Composition of West Indian bay oils, *Food Chem.,* 2(1), 19, 1977.

261. **Duke, J. A.,** Herbs for the liqueur connoisseur, unpublished draft on file at U.S. Department of Agriculture, 1977.
262. **Gibbs, R. S.,** *Chemotaxonomy of Flowering Plants,* 4 vols., McGill-Queen's University Press, Montreal, 1974.
263. **Kunzemann, J. and Herrmann, K.,** Isolation and identification of flavon(ol)-0-glycosides in caraway (*Carum carvi* L.), fennel (*Foeniculum vulgare* Mill.), anise (*Pimpinella anisum* L.) and coriander (*Coriandrum sativum* L.), and of flavon-C-glycosides in anise. I. Phenolics of spices, *Zeits. Lebensmittel-Unterschung und-Forschung,* 164(3), 194, 1977.
264. **El-Moghazi, A. M., Ali, A. A., Ross, S. A., and Mottaleb, M. A.,** Flavonoids luteolin, luteolin 7-0-glucoside, luteolin 7-0-xyloside of *Pimpinella anisum* fruits growing in Egypt, *Herba Polonica,* 27(1), 13, 1981.
265. **Eiseman, F. B., Jr.,** personal communication, December 28, 1982.
266. **Duve, R. N.,** Highlights of the chemistry. and pharmacology of yagona (*Piper methysticum*), *Fiji Agric. J.,* 38(2), 81, 1976.
267. **Chaudhuri, T. and Sen, C.,** Effects of some plant extracts on three sclerotia-forming fungal pathogens, *Zeits. Pflanzenkrank. Pflanzenschutz,* 89(10), 582, 1982.
268. *Dispensatory of the United States of America,* 21st ed., 1962.
269. Smile Herb Shop, *Tea Here Now,* a retail catalog, College Park, Md.
270. **Modell, W.,** *Drugs of Choice — 1976,* C. V. Mosby, St. Louis, 1977, 897.
271. **Earle, F. R. and Jones, Q.,** Analyses of seed samples from 113 plant families, *Econ. Bot.,* 16(4), 221, 1962.
272. **Jones, Q. and Earle, F. R.,** Chemical analyses of seeds. II. Oil and protein content of 759 species, *Econ. Bot.,* 20(2), 127, 1966.
273. **Gohl, B.,** Tropical Feeds, FAO Animal Production and Health Series No. 12, Food and Agriculture Organization, Rome, 1981, 529.
274. **Simpson, B. B.,** *Mesquite: Its Biology in Two Desert Ecosystems,* Dowden, Hutchinson & Ross, Stroudsburg, 1977.
275. **Felker, P. and Bandurski, R. S.,** Uses and potential uses of leguminous trees for minimal energy input agriculture, *Econ. Bot.,* 33(2), 172, 1978.
276. **Milton, J., Fletcher, H., Ed.,** *Paradise Lost,* Poetical Works, 4 vols., 1943—1948.
277. The Conservation Foundation Letter, November 1982.
278. **Gillis, W. T.,** Poison-ivy and its kin, *Arnoldia,* 35(2), 93, 1975.
279. **Farnsworth, N. R.,** Credible or incredible, *Till Till,* 54, 58, 1968.
280. **Inatani, R., Nakatani, N., Fuwa, H., and Seto, H.,** Structure of a new antioxidative phenolic diterpene isolated from rosemary (*Rosmarinus officinalis* L.), *Agric. Biol. Chem.,* 46(6), 1661, 1982.
281. **Camazine, S. and Bye, R. A.,** A study of the medical ethnobotany of the Zuni Indians of New Mexico, *J. Ethnopharm.,* 2, 365, 1980.
282. **Wasson, R. G.,** A new Mexican psychotropic drug from the mint family, *Bot. Mus. Leafl., Harvard U.,* 20, 77, 1962.
283. **Phelan, J. T. and Juardo, J.,** Chemosurgical management of carcinoma of the external ear, *Surg. Gynecol. Obstet.,* 117, 224, 1963.
284. **Demole, E., Demole, C., and Enggist, P.,** A chemical investigation of the volatile constituents of East Indian sandalwood oil (*Santalum album* L.), *Helv. Chim. Acta,* 59, 737, 1976.
285. **Goulart, E. G., da Costa, J. A. R., Brazil, R. P., Gilbert, B., Lopes, J. N. C., Sarti, S. J., Vichnewski, W., and Thames, A. W.,** Inhibitory action of natural and synthetic products on the development of the free living stages of *Strongyloides stercoralis* and hookworms, *Rev. Bras. Med.,* 31(5), 303, 1974.
286. **Benedetti, M. S., Malnoe, A., and Broillet, L.,** *Toxicology,* 7, 69, 1977.
287. **San Martin, R., Granger, R., Adzet, T., Passer, J., and Tevlade-Arbousset, M. G.,** Chemical polymorphism in two Mediterranean labiates, *Satureja montana* L. and *Satureja obovata* Lag., *Plant Med. Phytother.,* 7, 95, 1973.
288. **Terhune, S. J., Hogg, J. W., and Lawrence, B. M.,** B-spathulene: a new sesquiterpene in Schinus molle oil, *Phytochemistry,* 13, 865, 1974.
289. **Hashim, F. M., El-Hossary, G. A., and El-Sakhawy, F. S.,** Preliminary phytochemical study of *Schinus molle* L. growing in Egypt, *Egyptian J. Pharm. Sci.,* 19(1/4), 235, 1978.
290. **Salamon, J.,** Many gourmets get the itch to sample pink peppercorns, *Wall Street J.,* December 21, 1981.
291. **Morton, J. F.,** Brazilian pepper — its impact on people, animals, and the environment, *Econ. Bot.,* 32(4), 353, 1979.
292. **Anon.,** Marula fruit, pachyderms get pickled, *Columbus Dispatch,* April 4, 1972.
293. **Petershofer-Halbmayer, H., Kubelka, O., Jurenitsch, J., and Kubelka, W.,** Isolation of hordenine (''cactine'') from *Selenicereus grandiflorus* (L.) Britt. & Rose and *Selenicereus pteranthus* (Link & Ott.) Britt. & Rose, *Sci. Pharm.,* 50, 29, 1982.

294. **Wagner, H. and Flachsbarth, H.,** A new antiphlogistic principle from *Sabal serrulata* I, *Pl. Med.,* 41, 244, 1981.

295. **Verbiscar, A. J. and Banigan, T. F.,** Composition of jojoba seeds and foliage, *J. Agric. Food Chem.,* 26(6), 1456, 1978.

296. **Anon.,** Stopping the desert profitability, *Economist,* March 31, 85, 1982.

297. National Academy of Sciences, *Products from Jojoba: A Promising New Crop for Arid Lands,* Comm. Jojoba Utilization, National Academy of Sciences, Washington, D.C., 1975, 30.

298. **Utz, W. J., O'Connell, P. L., Storey, R., and Bower, N. W.,** Nutritional evaluation of the jojoba plant: elemental analysis of the seed meal, *J. Agr. Food Chem.,* 30(2), 392, 1982.

299. **Swingle, R. S.,** *Prog. Agric. Az.,* 32(3), 25, 1981.

300. **Willuhn, G. and May, S.,** Triterpenes and steroids in callus cultures of *Solanum dulcamara, Planta Med.,* 46, 153, 1982.

301. **Khanna, P. and Rathore, A. K.,** Diosgenin & solasodine from *Solanum nigrum* L. complex, *Indian J. Exp. Biol.,* 15(9), 808, 1977.

302. **Weller, R. F. and Phipps, R. H.,** A review of black nightshade (*Solanum nigrum*), *Prot. Ecol.,* 1(2), 121, 1979.

303. **Palz, W. and Chartier, P., Eds.,** *Energy from Biomass in Europe,* Applied Science Publishers, London, 1980, 234.

304. **Wallace, J., Wallechinsky, D., and Wallace, A.,** Significa, *Parade Mag.,* January 9, 1983.

305. *New Scientist,* July 15, 14, 1976.

306. **Takechi, M. and Tanaka, Y.,** Purification and characterization of antiviral substance eugeniin from the bud of Syzygium aromatica aromaticum, *Planta Medica,* 42(1), 69, 1981.

307. **Borino, B.,** 1000-Year-old Inca cancer cure works, *Globe,* September 15, 1981.

308. **Perdue, R. E.,** KB cell culture. I. Role in discovery of antitumor agents from higher plants, *J. Nat. Prod.,* 45(4), 418, 1982.

309. **Prakash, L. and Singh, R.,** Chemical constituents of stem bark and root heartwood of *Tabebuia pentaphylla* (Linn.) Hemsl. (Bignoniaceae), *Pharmazie,* 35(12), 813, 1980.

310. **Sieber, S. M., Mead, J. A. R., and Adamson, R. H.,** Pharmacology of antitumor agents from higher plants, *Cancer Treatment Rep.,* 60(8), 1127, 1976.

311. **Brewer, G. J. and Hall, H. J.,** A feeding deterrent effect of a water extract of tansy (*Tanacetum vulgare* L., Compositae) on three lepidopterous larvae, *J. Kan. Entomol. Soc.,* 54(4), 733, 1981.

312. **Tetenyi, P., Kaposi, P., and Hethelyi, E.,** Variations in the essential oils of "*Tanacetum vulgare*", *Phytochemistry,* 14, 1539, 1975.

313. **Appendino, G., Gariboldi, P., and Nano, G. M.,** Crispolide an unusual hydro per oxy sesquiterpene from *Tanacetum vulgare, Phytochemistry (Oxf.),* 21(5), 1099, 1982.

314. **Keindorf, A. and Keindorf, H. J.,** Hyperoestrogenism in cattle (due to *Tanacetum vulgare* in pasture), *Monatshefte Veterinarmed.,* 33(12), 453, 1978.

315. **Anon.,** (Jake's Page), A weed by any other name, *Science 83,* April 82, 1983.

316. **Wat, C. K., Biswas, R. K., Graham, E. A., Bohm, L., and Towers, G. H. N.,** Ultraviolet-mediated cytotoxic activity of phenylheptatriyne from *Bidens pilosa* L., *J. Nat. Prod.,* 42, 103, 1979.

317. **James, C.,** Mysteries: why do warts disappear, *Science 83,* April, 96, 1983.

318. **Sutton, R. H.,** Cocoa poisoning in a dog, *Vet. Rec.,* 109 (25/26), 563, 1981.

319. **Van den Broucke, C. O. and Lemli, J. A.,** Pharmacological and chemical investigation of thyme liquid extracts, *Planta Medica,* 41(2), 129, 1981.

320. **Suffness, M.,** National Cancer Institute, personal communication, 1983.

321. **Karawya, M. S., Wassel, G. M., Baghdad, H. H., and Ammar, N. M.,** Mucilaginous contents of certain Egyptian plants, *Planta Medica,* 38(1), 73, 1980.

322. **Nagarajan, S., Jain, H. C., and Aulakh, G. S.,** Indigenous plants used in fertility control, in *Cultivation and Utilization of Medicinal Plants,* Atal, C. K. and Kapur, B. M., Eds., CSIR, Jammu-Tawi, 1982, 877.

323. **Singhal, P. C., Gupta, R. K., and Joshi, L. D.,** Hypocholesterolemic effect of Trigonella foenum-graceum (Methi), *Curr. Sci.,* 51(3), 136, 1982.

324. **Khurana, S. K., Krishnamoorthy, V., Parmar, V. S., Sanduja, R., and Chawla, H. L.,** 3 4 7 Tri methyl coumarin from Trigonella-foenum-graecum stems, *Phytochemistry (Ox.),* 21(8), 2145, 1982.

325. **Roder, E., Wiedenfeld, H., and Jost, E. J.,** Tussilagine — a new pyrrolizidine alkaloid from *Tussilago farfara, Planta Medica,* 43(3), 99, 1981.

326. **Rosberger, D. F., Resch, J. F., and Meinwald, J.,** The occurrence of senecionine in *Tussilago farfara, Mitt. Geb. Lebensmitt. Hyg.,* 72(4), 1981.

327. **Bricklin, M., Ed.,** *The Practical Encyclopedia of Natural Healing,* Rodale Press, Emmaus, 1976, 582.

328. **Buttery, R. G., Black, D. R., Guadagni, D. G., Ling, L. C., Connolly, G., and Teranishi, R.,** California bay oil. I. Constituents, odor properties, *J. Agric. Food Chem.,* 22(5), 773, 1974.

329. **MacGregor, J. T., Layton, L. L., and Buttery, R. G.,** California bay oil. II. Biological effects of constituents, *J. Agric. Food Chem.,* 22(5), 777, 1974.

330. **Curtis, F. E.,** The pregnant onion, *Org. Gard. Farm,* May, 76, 1977.
331. **Balbaa, S. J., Khafagy, S. M., Khayyal, S. E., and Girgis, A. B.,** TLC-spectophotometric assay of the main glycosides of red squill, *J. Nat. Prod.,* 42(5), 522, 1979.
332. **Nikolaeva, V. G.,** Plants used in folk medicine in the USSR for diseases of kidneys and the urinary system, *Rastitel Nye Resursy,* 12(2), 307, 1976.
333. **Keough, C.,** Mistletoe medicine, *Org. Gardening,* December, 124, 1979.
334. **Bloksma, N., Schmelermann, P., de Reuver, M., van Dijk, H., and Willers, J.,** Stimulation of humoral and cellular immunity by Viscum preparations, *Planta Medica,* 46, 221, 1982.
335. **Duke, J. A.,** as quoted in the Washington Post, 1975.
336. **Brown, R. G. and Brown, M. L.,** *Woody Plants of Maryland,* Port City Press, Baltimore, 1972, 347.
337. **Correll, D. S. and Correll, H. B.,** *Aquatic and Wetland Plants of Southwestern United States,* 2 vols., Stanford University Press, Stanford, 1975, 1777.
338. **Degener, O. and Degener, I.,** *Flora Hawaiiensis or New Illustrated Flora of the Hawaiian Islands,* Books 1 to 6, Published by the Author, looseleaf, 1933—1963.
339. **Dimitri, M. J., Ed.,** *Enciclopedia Argentina de Agricultura y Jardineria,* 2 vols., Editorial Acme, Buenos Aires, 1980, 1161.
340. **Hatton, R. G.,** *Handbook of Plant and Floral Ornament,* Dover Publications, 1960, 539.
341. **Little, E. L., Jr., Woodbury, R. O., and Wadsworth, F. H.,** Trees of Puerto Rico and the Virgin Islands, Vol. 2, Agric. Handbook No. 449, U.S. Department of Agriculture, Washington, D.C., 1974, 1024.
342. **Nasir, E. and Ali, S. I., Eds.,** *Flora of West Pakistan, No. 100. Papilionaceae,* Ferozsons, Karachi, 1977, 389.
343. Agricultural Research Service, USDA, Selected Weeds of the United States, Agric. Handbook No. 366, USGPO, Washington, D.C., 1970, 463.
344. **Elliott, D. B.,** *Roots An Underground Botany and Forager's Guide,* Chatham Press, Old Greenwich, Conn., 1976, 128.
345. **Phillips, E. P.,** *The Weeds of South Africa,* Union S. Africa Dept. Agric. Div. Bot. Series 41, Government Printer, Pretoria, 1938, 229.
346. **Chuang, C. C. and Huang, C.,** *The Leguminosae of Taiwan for Pasture and Soil Improvement,* Dept. Bot. Nat. Taiwan U. Animal Industry Series No. 7, Taipei, Taiwan, 1965, 286.
347. **Vilmorin-Andrieux, M.,** *The Vegetable Garden,* 1st reprint, Jeavons-Leler Press, Palo Alto, 1976 (First published in 1885), 620.
348. **Strausbaugh, P. D. and Core, E. L.,** *Flora of West Virginia,* 4 vols., West Virginia University, Morganton, 1952—1964, 1075.
349. **Wilbur, R. L.,** *The Leguminous Plants of North Carolina,* Tech. Bull. No. 151, N.C. Agricultural Experiment Station, Raleigh, 1963, 294.
350. **Britton, N. L. and Brown, A.,** *An Illustrated Flora of the United States,* 3 vols., 1896, 1897, 1898.
351. **Duke, J. A.,** Growing Ginseng, Farmers Bull. No. 2201, U.S. Department of Agriculture, Washington, D.C., 1981.
352. **Cheney,** *Econ. Bot.,* 1, 243, 1947.
353. NIOSH, Registry of Toxic Effects of Chemical Substances, National Institute of Occupational Safety and Health, Rockville, Md., 1975.

LIST OF FIGURES AND CREDITS*

* Figure numbers correspond to plant number in text. Some plants are not illustrated.

LIST OF FIGURES AND CREDITS (continued)

LIST OF FIGURES AND CREDITS (continued)

LIST OF FIGURES AND CREDITS (continued)

INDEX

A

E

F

L

M

R